JENDY DOAN
12:30 - 1:50

FINITE MATHEMATICS
WITH APPLICATIONS

FOR BUSINESS AND
SOCIAL SCIENCES
SIXTH EDITION

Abe Mizrahi
Indiana University Northwest

Michael Sullivan
Chicago State University

John Wiley & Sons, Inc.
New York   Chichester   Brisbane   Toronto   Singapore

To Our Parents with Gratitude

ACQUISITIONS EDITOR / Robert Macek
DEVELOPMENTAL EDITOR / Joan Carrafiello
PRODUCTION MANAGER / Katharine Rubin
DESIGNER / Kevin Murphy
PRODUCTION SUPERVISOR / Nancy Prinz
MANUFACTURING MANAGER / Lorraine Fumoso
COPY EDITING MANAGER / Deborah Herbert
ILLUSTRATION / John Balbalis
PART OPENING PHOTO / Maurits Cornelis Escher
Rippled Surface, Private Collection
1950, Lithograph / Art Resource.
COVER ART / Roy Wiemann

Library of Congress Cataloging-in-Publication Data:

Mizrahi, Abe.
    Finite mathematics with applications for business and social
sciences / Abe Mizrahi, Michael Sullivan. — 6th ed.
        p.   cm.
    Includes index.
    ISBN 0-471-54744-1
    1. Business mathematics.   2. Social sciences — Mathematics.
3. Mathematics.   I. Sullivan, Michael, 1942 –   .   II. Title.
HF5695.M66   1992
519 — dc20   91-27562
                                                                CIP

Printed in the United States of America

10 9 8 7 6 5 4 3 2 1

PREFACE

The first edition of this book was published in 1973. At that time our purpose was to present a low-level approach to the mathematics required in business and the social sciences, giving emphasis to real-world applications from these fields. In subsequent editions this remained our purpose, as it does in this, the sixth edition. However, in this edition we have enlarged our range of applications to include applications to computers and the field of computer science.

ORGANIZATION

This edition is divided into three independent parts: Linear Algebra, Probability, and Discrete Mathematics. Depending on the interests of the audience, an instructor can start a course at the beginning of any of the three parts. The chart on page v illustrates the interaction of the chapters.

Part One, Linear Algebra, contains five chapters: Chapter 1 lays the foundation and is considered a review chapter. Chapter 2 presents a discussion of matrices, beginning with systems of linear equations. Chapters 3 and 4 concentrate on linear programming and applications. Chapter 5 contains some of the mathematics found in finance.

Part Two, Probability, contains six chapters. Chapter 6 introduces sets and elementary counting techniques. Chapter 7 deals with probability. Chapter 8 examines more topics in probability, conditional probability, independence, Bayes' formula, the binomial probability model, random variables, and expectation. Chapter 9 provides an introduction to statistics; Chapter 10 discusses Markov chains. Chapter 11 provides an application to game theory.

Part Three, Discrete Mathematics, contains three chapters. Chapter 12 discusses logic and its application to circuit design. Chapter 13 introduces relations, functions, sequences, and mathematical induction, emphasizing each one's role in computer science. Chapter 14 presents an introduction to graphs and networks.

PREREQUISITES

We have made every effort to keep the formal prerequisites to a minimum. We assume only a first year of high school algebra. We review in Chapter 1 those topics that are typically weak spots for students.

CHANGES TO THE SIXTH EDITION

Among the changes in this edition, the following are the most significant.

The stock of exercises from the previous editions has been significantly expanded. Problems at the end of most sections are divided into four parts, A, B, C, and Applications.

Chapter 1: The section on straight lines is rewritten. The section on systems of equations is moved to Chapter 2.

Chapter 2: The section on solving of linear systems is completely rewritten. By starting with a careful presentation of linear equations in two and three variables, the idea and use of matrices becomes a more easily assimilable idea.

The concept of a parameter is discussed in Section 2.1 to enable the reader to better understand the idea of infinite solutions.

Chapter 3: Minor changes.

Chapter 4: Minor changes.

Chapter 5: The text has been completely rewritten with an emphasis on the hand calculator to solve problems, even though tables are included in the appendix. Geometric Series are used to derive some of the formulas.

Chapter 6: *Sets: Counting Techniques.* This is Chapter 6 from the fifth edition plus some of the material involving advanced counting techniques from Chapter 8 in the fifth edition.

Chapter 7: *Introduction to Probability.* This chapter is part of Chapter 7 from the fifth edition except that we stop at conditional probability.

Chapter 8: The discussion of this chapter starts with conditional probability and ends with expectation. The only new material is the section on random variables. As a result, we use random variables to introduce expectation.

Chapter 9: Section 9.3, Other Graphical Techniques, is new. Here we teach students to use pie charts and bar graphs in presenting data.

Chapter 10: Section 10.5, Solving Nonstrictly Determined Matrix Games Using Linear Programming, is entirely new. It ties together the chapter on linear programming and games.

Chapter 11: Minor changes.

Chapter 12: Minor changes.

Chapter 13: Part of the material included in the previous edition has been removed to make the chapter more readable and in tune with the preceding chapters.

FEATURES

* More than 400 detailed examples, clearly illustrating concepts and techniques.

* The text features a rich selection of applications from many disciplines. This edition includes several new applications to business.

* Procedures and processes for solving problems are given as a series of steps and are prominently displayed in boxes. See, for example, page 154 and page 384.

* More than 2800 exercises, ranging from drill to challenging. Many of these are applied-type problems.

* Each chapter contains a review featuring a list of Important Terms and Formulas from the chapter, True-False and Fill-in-the-Blank Questions, a collection of Review Exercises, and, when appropriate, Mathematical Questions taken from CPA, CMA, and actuary exams.

* Important definitions are presented in boldface; formulas and theorems are placed in a box, screened in color.

* Answers to the Odd-Numbered Problems appear at the end of the book.

* Flowcharts are utilized whenever possible to outline procedures. See, for example, page 196.

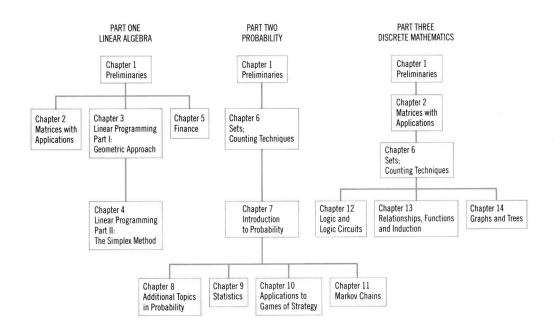

PART ONE
LINEAR ALGEBRA

Chapter 1
Preliminaries

Chapter 2
Matrices with
Applications

Chapter 3
Linear Programming
Part I:
Geometric Approach

Chapter 5
Finance

Chapter 4
Linear Programming
Part II:
The Simplex Method

PART TWO
PROBABILITY

Chapter 1
Preliminaries

Chapter 6
Sets;
Counting Techniques

Chapter 7
Introduction
to Probability

Chapter 8
Additional Topics
in Probability

Chapter 9
Statistics

Chapter 10
Applications to
Games of Strategy

Chapter 11
Markov Chains

PART THREE
DISCRETE MATHEMATICS

Chapter 1
Preliminaries

Chapter 2
Matrices with
Applications

Chapter 6
Sets;
Counting Techniques

Chapter 12
Logic and
Logic Circuits

Chapter 13
Relationships, Functions
and Induction

Chapter 14
Graphs and Trees

SUPPLEMENTS

* An Instructor's Manual presents worked-out solutions to both the even and the odd-numbered problems.
* A Student Solutions Manual contains worked-out solutions to the odd-numbered problems.
* Computer-Generated Test Bank. This microcomputer testing system contains questions prepared by the authors and is available for Macintosh, IBM-PC, and for most compatibles.

ACKNOWLEDGMENTS

We thank the many students at Indiana University Northwest and Chicago State University for their comments and criticisms.

We also thank these reviewers for their suggestions and contributions to this and earlier editions:

Judy Barclay, Cuesto College; Charles Barker, DeAnza College; Carole Bernett, William Rainey Harper College; William Blair, Northern Illinois University; W. J. Bonini, Western Wyoming College; Gary G. Cochell, Calver-Stockton College; Portia Cornell, University of Redlands; Milton D. Cox, Miami University; Neale Fadden, Belleville Area College; Eugene Franks, Dyke College; James C. Fraventhal, SUNY at Stony Brook; Cathy Frey, Norwich University; Dennis Freeman, Montgomery College; Samuel Graff, John Jay College of Criminal Justice; Gerald Hahn, College of St. Thomas; Joseph Hansen, Northeastern University; Herbert Hethcote, University of Iowa; Brian Hickey, First Central College; Wayne Hiles, Kankakee Community College; Roy Honde, West Valley College; Martin Kotler, Pace University; James E. Lightner, Western Maryland College; Rick Lottmann, Sonoma State College; Aima McKinney, North Florida Junior College; Jeffrey McLean, College of St. Thomas; Kishore Marathe, Brooklyn College; Charles J. Miller, Foothill College; Robert Muksian, Bryant College; Henry J. Osner, Modesto Junior College; Willard A. Parker, Kansas State University; Elaine Pavelka, Morton College; Carlos A. Pereira, The King's College; Dana Piens, Rochester Community College; George H. Potter, Schenectady Community College; Thomas Radin, San Juaquin Delta Community College; Eric E. Robinson, Ithaca College; Gerald Root, American International College; Arthur Schlissel, John Jay College of Criminal Justice; Charles Sheumaker, University of S.W. Louisiana; Carol Slinger, Marian College; Richard Stansfield, Jefferson College; Olive D. Taylor, Meredith College; Elizabeth E. Taylor, University of Richmond; Luther Truell, Victoria College; Ernest True, Norwich College; Jack Wadhaus, Golden West College; Hubert Walezak, College of St. Thomas; Robert Wheeler, Northern Illinois University; John L. Whitcomb, University of North Dakota; John A. Winn, Hofstra University; M. Yablon, John Jay College/CUNY

We especially thank William Hosch and Henry Wyzinski of Indiana University Northwest and Richard Staum of Kings Borough Community College, who read the manuscript in its entirety and made valuable suggestions, and Mitzi Chaffer from Central Michigan University for allowing us to use some of her problems.

We also appreciate the skill and patience of Leona M. Lashenik who typed the revision.

We are grateful to the following organizations who permitted the reproduction of actual questions from previous professional examinations.

American Institute of Certified Public Accountants (CPA Exams)

Educational Testing Service (Actuary Exams)

Institute of Management Accounting of the National Association of Accountants (CMA Exams)

Again, we are indebted to the people at John Wiley whose talent played a significant part in the publication of this book. We particularly want to single out Joan Carrafiello, Deborah Herbert, Katharine Rubin, Nancy Prinz and Bob Macek for their assistance.

We assume equal responsibility for the book's strengths and weaknesses, and welcome comments and suggestions for its improvement.

Abe Mizrahi
Michael Sullivan

ABOUT THE AUTHORS

Abe Mizrahi, received his doctorate in mathematics from the Illinois Institute of Technology in 1965. He is currently Professor of Mathematics at Indiana University Northwest. Professor Mizrahi and Professor Sullivan are coauthors of *Mathematics for Business and Social Sciences,* fourth edition, John Wiley & Sons, Inc., 1988, as well as *Calculus with Analytic Geometry,* third edition, Wadsworth Publishing Company, 1990. Dr. Mizrahi is a member of the Mathematics Association of America. Articles he has written explore topics in math education and the applications of mathematics to economics. Professor Mizrahi has served on many CUPM committees and was a panel member on the CUPM Committee on Applied Mathematics in the Undergraduate Curriculum. Professor Mizrahi is a recipient of many NSF grants and has been a consultant to a number of businesses and federal agencies.

Michael Sullivan, received his doctorate in mathematics from the Illinois Institute of Technology in 1967. Since 1965 he has been teaching at Chicago State University in the Department of Mathematics and Computer Science, holding the rank of professor. Dr. Sullivan is a member of the American Mathematical Society, the Mathematics Association of America, and Sigma Xi. He has served on CUPM curriculum committees and is a member of the Illinois Section MAA High School Lecture Committee.

CONTENTS

* This section may be omitted without loss of continuity.

* This section may be omitted without loss of continuity.

PART ONE

LINEAR ALGEBRA

1

PRELIMINARIES

* This section may be omitted without loss of continuity.

1.1

REAL NUMBERS

SETS

We begin with the idea of a *set*. A *set* is a collection of objects considered as a whole. The objects of a set S are called *elements* of S, or *members* of S. The set that has no elements, called the *empty set* or *null set,* is denoted by the symbol $\varnothing$.

If a is an element of the set S, we write $a \in S$, which is read "*a* is an element of *S*" or "*a* is in *S*." To indicate that a is not an element of S, we write $a \notin S$, which is read "*a* is not an element of *S*" or "*a* is not in *S*."

Ordinarily, a set S can be written in either of two ways. These two methods are illustrated by the following example: Consider the set D that has the elements

$$0, 1, 2, 3, 4, 5, 6, 7, 8, 9$$

In this case, we write

$$D = \{0, 1, 2, 3, 4, 5, 6, 7, 8, 9\}$$

This expression is read "D is the set consisting of the elements 0, 1, 2, 3, 4, 5, 6, 7, 8, 9." Here, we list or display the elements of the set D.

Another way of writing this same set D is to write

$$D = \{x | x \text{ is a nonnegative integer less than } 10\}$$

This is read "D is the set of all x such that x is a nonnegative integer less than 10." Here, we have described the set D by giving a property that every element of D has and that no element not in D can have.

SETS OF NUMBERS

One of the most frequently used set of numbers in finite mathematics is the set of *counting numbers,* namely,

$$N = \{1, 2, 3, 4, \ . \ . \ .\}$$

These are sometimes referred to as *positive integers.* As the name *counting numbers* implies, they are used to count things. For example, there are 26 letters in our alphabet, and there are 100 cents in a dollar.

Another important collection of numbers is the set of *integers,*

$$Z = \{0, 1, -1, 2, -2, \ldots\}$$

Integers enable us to handle certain types of situations. For example, if a company shows a loss of $3 per share, we might decide to denote this loss as a gain of $-$3. Then, if this same company shows a profit of $4 per share next year, the profit over the 2 year period can be obtained by adding -3 to 4, getting a profit of $1.

However, integers do not enable us to solve *all* problems. For example, can we use an integer to answer the question, "What part of a dollar is 49¢?" Or, can we use an integer to represent the length of a city lot (in feet) if we end up with a length more than 125 feet and less than 126 feet?

The answer to both these questions is "No"! We need new or different numbers to handle such situations. These new numbers are called *rational numbers.*

Rational numbers are ratios of integers, $Q = \left\{ \dfrac{a}{b} \ \middle| \ a \in Z, b \in Z, \text{ and } b \neq 0 \right\}$,

for a rational number $\dfrac{a}{b}$, the integer a is called the *numerator,* and the integer b, which cannot be zero, is called the *denominator.*

Thus to answer the first question, "What part of a dollar is 49¢?" we can say $\frac{49}{100}$.

Examples of rational numbers are $\frac{3}{4}, \frac{5}{3}, -\frac{2}{7}, \frac{100}{3}, -\frac{8}{3}$. Since the ratio of any integer to 1 is that integer, the integers are also rational numbers. Thus, $\frac{3}{1} = 3, -\frac{2}{1} = -2, \frac{0}{1} = 0$.

In this book, we will always write rational numbers in *lowest terms;* that is, the numerator and the denominator will contain no common factors. Thus, $\frac{4}{6}$, which is not in lowest terms, will be written as $\frac{2}{3}$, which is in lowest terms.

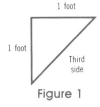

1 foot

1 foot

Third side

Figure 1

In some situations, even a rational number will not accurately describe the situation. For example, if you have an isosceles right triangle in which the two equal sides are 1 foot long, can the length of the third side be expressed as a rational number? See Figure 1.

And how about the value of π (Greek letter "pi")? The Greeks first learned that for all circles the ratio of their circumferences to their diameters has a common value. Can this common value, which we call π, be represented by a rational number? See Figure 2.

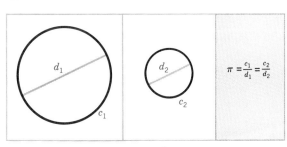

Figure 2

As it happens, the answer to both these questions is "No." For the first question, by the Pythagorean theorem* the length of the third side is $\sqrt{2}$. However, $\sqrt{2}$ is not a rational number—it is not the ratio of two integers. The well-known symbol assigned to the ratio of the circumference to the diameter of a circle is π, which also cannot be expressed as the ratio of two integers. Such numbers as $\sqrt{2}$, π, and $\sqrt[3]{5}$ are called *irrational numbers*—numbers that are not rational.

The irrational numbers and the rational numbers together form the *real numbers.*

DECIMALS AND PERCENTS

To represent each real number, we use what is commonly referred to as a *decimal representation,* or simply a *decimal.* The table that follows lists the decimal equivalents of some frequently encountered rational numbers.

Rational number	$\frac{1}{2}$	$\frac{1}{3}$	$\frac{2}{3}$	$\frac{1}{4}$	$\frac{3}{4}$	$\frac{1}{5}$	$\frac{2}{5}$	$\frac{3}{5}$	$\frac{4}{5}$	$\frac{1}{8}$	$\frac{3}{8}$	$\frac{5}{8}$	$\frac{7}{8}$
Decimal equivalent	0.5	0.333 . . .	0.666 . . .	0.25	0.75	0.2	0.4	0.6	0.8	0.125	0.375	0.625	0.875

The decimal equivalent of any rational number is obtained by long division: Divide the denominator into the numerator. For example, we find that the decimal equivalent of $\frac{7}{66}$ is 0.10606 . . . as follows:

$$
\begin{array}{r}
0.10606 \\
66\overline{)7.00000} \\
6\,6 \\
\hline
400 \\
396 \\
\hline
400 \\
396 \\
\hline
\text{etc.}
\end{array}
$$

Observe the following fact: The decimal equivalent of a rational number is always one of two types: *(1)* terminating or ending ($\frac{3}{4}$, $\frac{4}{5}$, etc.) or *(2)* eventually repeating ($\frac{2}{3}$—the 6's repeat; $\frac{7}{66}$—eventually the block 06 repeats).

At first, it may appear that these two types account for all possible decimals. However, it is relatively easy to construct a decimal that neither terminates nor eventually repeats. For example, the decimal

$$0.123456789101112 \text{ . . .}$$

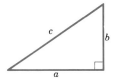

* The Pythagorean theorem states that in a right triangle the square of the length of the side opposite the right angle equals the sum of the squares of the lengths of the other two sides. Thus, in the illustration $c^2 = a^2 + b^2$.

where we write down the positive integers one after the other, will neither terminate nor eventually repeat.

In fact, there are an infinite number of such decimals, and they represent the irrational numbers. For example, the real numbers

$$\sqrt{2} = 1.414213 \ldots \qquad \text{and} \qquad \pi = 3.14159 \ldots$$

are irrational, since they have decimal representations that neither terminate nor eventually repeat.

Thus, *real numbers* and *all possible decimals* are equivalent concepts. It is this feature of real numbers that provides the "practicality" of real numbers.* In the physical world a changing magnitude, such as the length of a heated rod or the velocity of a particle, is assumed to pass through every possible magnitude from the initial one to the final one. Since the precise measurement of a magnitude is naturally given by a decimal, the logical equivalent of all possible magnitudes is all possible decimals (real numbers).

In practice, it is usually necessary to represent irrational real numbers by approximations. For example, using the symbol $\approx$, read "approximately equal to," we can write

$$\sqrt{2} \approx 1.4142 \qquad \text{and} \qquad \pi \approx 3.1416$$

In general, any real number can be approximated to within any closeness by a rational number. For example, $\pi = 3.14159265 \ldots$ can be approximated within $\frac{1}{100}$ by 3.14, or within $\frac{1}{10,000}$ by 3.1416.

One of the main applications of decimals comes from problems involving *percents*. Percents are widely used; in consumer mathematics, interest rates, tax rates, and discounts are often given as percents, as will be shown in a later chapter. The word *percent* means "per hundred." This idea is used in the basic definition of percents; if the symbol % represents "percent," then

$$1\% = \tfrac{1}{100} = 0.01$$

For example,

$$15\% = 15(1\%) = 15(0.01) = 0.15$$

Example 1
Convert each percent to decimal

(a) $58\% = 58(1\%) = 58(0.01) = 0.58$

(b) $7.2\% = 0.072$

(c) $0.3\% = 0.003$ ■

A decimal can be converted to a percent in much the same way.

---

* Other number systems exist that have many applications (such as the complex numbers). In this book, however, we limit our discussion to problems in which only real numbers are used.

Example 2
Convert each decimal to a percent

(a) $0.24 = 24(0.01) = 24(1\%) = 24\%$
(b) $0.742 = 74.2(0.01) = 74.2(1\%) = 74.2\%$
(c) $3.4 = 340\%$ ■

The idea should be clear — to change from a decimal to a percent — move the decimal point two places to the right and add a percent symbol. Reverse this procedure to get from a percent to its decimal equivalent.

The following examples show how to work the various types of problems involving percent.

Example 3
Find 24% of 460.

The word "of" translates as "times," with 24% of 460 given by

$$(24\%)(460) = (0.24)(460) = 110.4$$ ■

Example 4
What percent of 800 is 65?

Here we are asking, what fraction of 800 is 65.

The answer is $\frac{65}{800} = 0.08125$.

We convert this decimal to percent

$$0.08125 = 8.125\%$$ ■

Example 5
42 is 6% of what number?

Denote by N the unknown number. Then 6% of N is $6\% \cdot N$ or $0.06N$. Since 6% of the number is 42, we have

$$0.06N = 42$$

Divide both sides by 0.06

$$N = \frac{42}{0.06}$$
$$N = 700$$ ■

Example 6
A resident of Illinois has base income, after adjustment for deductions, of $18,000. The state income tax on this base income is $2\frac{1}{2}\%$. What tax is due?

Solution

We must find $2\frac{1}{2}\%$ of $18,000. We convert $2\frac{1}{2}\%$ to its decimal equivalent and then multiply by $18,000.

$$2\frac{1}{2}\% \text{ of } \$18,000 = (0.025)(\$18,000) = \$450$$

The state income tax is $450.00.   ∎

PROPERTIES OF REAL NUMBERS

As an aid to your review of real numbers and their properties, we list several important rules and notations. The letters a, b, c, d represent real numbers; any exceptions will be noted as they occur.

1. *Commutative Laws*
   (a) $a + b = b + a$ (b) $a \cdot b = b \cdot a$

2. *Associative Laws*
   (a) $a + b + c = (a + b) + c = a + (b + c)$
   (b) $a \cdot b \cdot c = (a \cdot b) \cdot c = a \cdot (b \cdot c)$

3. *Distributive Law*
   $a \cdot (b + c) = (a \cdot b) + (a \cdot c)$

4. *Arithmetic of Ratios*

   (a) $\dfrac{a}{b} = a \cdot \dfrac{1}{b}$ $b \neq 0$

   (b) $\dfrac{a}{b} + \dfrac{c}{d} = \dfrac{(a \cdot d) + (b \cdot c)}{b \cdot d}$ $b \neq 0, \quad d \neq 0$

   (c) $\dfrac{a}{b} \cdot \dfrac{c}{d} = \dfrac{a \cdot c}{b \cdot d}$ $b \neq 0, \quad d \neq 0$

   (d) $\dfrac{a}{b} \div \dfrac{c}{d} = \dfrac{a}{b} \cdot \dfrac{d}{c} = \dfrac{a \cdot d}{b \cdot c}$ $b \neq 0, \quad c \neq 0, \quad d \neq 0$

5. *Rules for Division*

   $0 \div a = 0$ or $\dfrac{0}{a} = 0$ for any real number a different from 0

   $a \div 0$ or $\dfrac{a}{0}$ is undefined for any real number a, including 0; *never divide by zero!*

   $a \div a = 1$ for any real number a different from 0

6. *Cancellation Laws*
   (a) If $a \cdot c = b \cdot c$ and c is not 0, then $a = b$.

   (b) If b and c are not 0, then $\dfrac{a \cdot c}{b \cdot c} = \dfrac{a}{b}$.

7. *Product Law*
   If $a \cdot b = 0$, then either $a = 0$ or $b = 0$.

8. *Rules of Signs*
   (a) $a \cdot (-b) = -(a \cdot b)$ (b) $(-a) \cdot b = -(a \cdot b)$
   (c) $(-a) \cdot (-b) = a \cdot b$ (d) $-(-a) = a$

9. *Agreements: Notations*
   (a) Given $a \cdot b + c$ or $c + a \cdot b$, we agree to multiply $a \cdot b$ first, and then add c.
   (b) A mixed number $3\frac{5}{8}$ means $3 + \frac{5}{8}$; 3 *times* $\frac{5}{8}$ is written as $3(\frac{5}{8})$ or $(3)(\frac{5}{8})$ or $3 \cdot \frac{5}{8}$.

10. *Exponents*
    For any positive integer n and any real number x, we define
    $$x^1 = x, \quad x^2 = x \cdot x, \quad \ldots, \quad x^n = \underbrace{x \cdot x \cdot \ldots \cdot x}_{n \text{ factors}}$$
    For any real number $x \neq 0$, we define
    $$x^0 = 1, \quad x^{-1} = \frac{1}{x}, \quad x^{-2} = \frac{1}{x^2}, \quad \ldots, \quad x^{-n} = \frac{1}{x^n}$$

11. *Laws of Exponents* (a, x are real numbers; n, m are integers)
    (a) $x^n \cdot x^m = x^{n+m}$ (b) $(x^n)^m = x^{nm}$
    (c) $(ax)^n = a^n \cdot x^n$ (d) $\left(\frac{x}{a}\right)^n = \frac{x^n}{a^n}$ $a \neq 0$
    (e) $\frac{x^n}{x^m} = x^{n-m}$ $x \neq 0$

12. *Roots*
    For a positive integer q, the qth root of x, $\sqrt[q]{x}$, is a symbol for the real number that, when raised to the power q, equals x. If q is even and x is positive, then $\sqrt[q]{x}$ is declared to be positive. For example,
    $$\sqrt[3]{8} = 2 \quad \text{since } 2^3 = 8 \qquad \sqrt[2]{64} = 8 \quad \text{since } 8^2 = 64$$
    Here $\sqrt[3]{x}$ is called the *cube root* of x and $\sqrt[2]{x}$ is called the *square root* of x. Usually, we abbreviate square roots by $\sqrt{\ }$, dropping the ^{2}. *Be careful! No meaning is assigned to even roots of negative numbers*, since any real number raised to an even power is nonnegative. For example, $\sqrt{4} = 2$, whereas $\sqrt{-4}$ has no meaning in the set of real numbers. On the other hand, $\sqrt[3]{27} = 3$, while $\sqrt[3]{-64} = -4$ since $(-4)^3 = -64$. Thus, meaning is given to odd roots of negative numbers. Finally, following the usual convention, even roots of positive numbers are always positive. Thus, even though $(2)^2 = 4$ and $(-2)^2 = 4$, only the positive root 2 equals $\sqrt{4}$. That is, $\sqrt{4} = 2$.

Example 7

(a) $\frac{2}{3} \cdot \frac{9}{4} = \frac{18}{12} = \frac{6 \cdot 3}{6 \cdot 2} = \frac{3}{2}$ (b) $\frac{1}{2} + \frac{3}{4} = \frac{2}{4} + \frac{3}{4} = \frac{5}{4}$
(c) $(-3)^2 = 9$ (d) $-3^2 = -9$
(e) $3 + 4 \cdot 2 = 3 + 8 = 11$ (f) $(3+4)^2 = 7^2 = 49$

POSITIVE AND NEGATIVE NUMBERS

The real numbers can be divided into three nonempty sets that have no elements in common: *(1)* the set of positive real numbers; *(2)* the set with just 0 as a member; and *(3)* the set of negative real numbers. We list some familiar properties:

1. Any real number is either positive or negative or equal to zero.
2. The sum and product of two positive numbers is positive.
3. The product of two negative numbers is positive.
4. The product of a positive number and a negative number is negative.

For real numbers a and b, a is less than b ($a < b$) or b is greater than a ($b > a$) if and only if the difference $b - a$ is a positive real number. For example, $2 < 7$ and $-3 > -6$. It is easy to conclude that

$$a \text{ is positive if and only if } a > 0$$
$$a \text{ is negative if and only if } a < 0$$

As a result, if $a < b$, there is a positive number p so that $a + p = b$. If a is less than or equal to b, we write $a \le b$. If a is greater than or equal to b, we write $a \ge b$. If $a < c$ and $c < b$, we write $a < c < b$. This says that c is between a and b. Similarly, if $a \le c$ and $c \le b$, we write $a \le c \le b$. The notations $a \le c < b$ and $a < c \le b$ are given similar interpretations.

Inequalities obey the following laws:

1. *Addition Law*
   If $a \le b$, then $a + c \le b + c$ for any choice of c. That is, the addition of a number to each side of an inequality will not affect the sense or direction of the inequality.

2. *Multiplication Laws*
   (a) If $a \le b$ and $c > 0$, then $a \cdot c \le b \cdot c$.
   (b) If $a \le b$ and $c < 0$, then $a \cdot c \ge b \cdot c$.

   When multiplying each side of an inequality by a number, the sense or direction of the inequality remains the same if we multiply by a positive number; it is reversed if we multiply by a negative number.

3. *Division Laws*
   (a) If $a > 0$, then $\dfrac{1}{a} > 0$. That is, the reciprocal of a positive number is positive.

   (b) If a, b are positive and if $a < b$, then $\dfrac{1}{a} > \dfrac{1}{b}$.

4. *Trichotomy Law*
   For any two real numbers a, b, one and only one of the following is true: $a < b$, $a = b$, or $a > b$.

Example 8

(a) Since $2 < 3$, then $2 + 5 < 3 + 5$, or $7 < 8$.

(b) Since $2 < 3$ and $6 > 0$, then $2 \cdot 6 < 3 \cdot 6$, or $12 < 18$.

(c) Since $2 < 3$ and $-4 < 0$, then $2 \cdot (-4) > 3 \cdot (-4)$, or $-8 > -12$.

(d) Since $3 > 0$, then $\dfrac{1}{3} > 0$.

(e) Since $2 < 3$, then $\dfrac{1}{2} > \dfrac{1}{3}$. ■

COORDINATES

Real numbers can be represented geometrically on a horizontal line. We begin by selecting an arbitrary point O, called the *origin,* and associate it with the real number 0. We then establish a scale by marking off line segments of equal length (units) on each side of 0. By agreeing that the positive direction is to the right of 0 and the negative direction is to the left of 0, we can successively associate the integers $1, 2, 3, \ldots$ with each mark to the right of 0 and the integers $-1, -2, -3, \ldots$ with each mark to the left of 0. See Figure 3.

Figure 3

By subdividing these segments, we can locate rational numbers such as $\tfrac{1}{2}$ and $-\tfrac{3}{2}$. The irrational numbers are located by geometric construction (as in the case of $\sqrt{2}$) or by other means. In this way, every point P on the line is associated with a unique real number x, called the *coordinate of P* (see Figure 4). Coordinates establish an ordering for the real numbers; that is, if a and b are coordinates of two points P and Q, respectively, then $a < b$ means that P lies to the left of Q on the line.

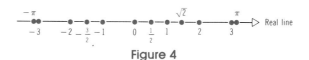

Figure 4

VARIABLES

A *variable* is a symbol (usually a letter x, y, etc.) used to represent any real number. An *equation* is a statement involving one or more variables and an "equals" sign ($=$). To *solve* an equation means to find all possible numbers that the variables can assume to make the statement true. The set of all such numbers is called the *solution.* Two equations with the same solution are called *equivalent equations.*

Example 9
Solve the equation: $3x - 8 = x + 4$

Solution
We use the properties of real numbers and proceed as follows:

$$
\begin{aligned}
3x - 8 &= x + 4 \\
2x - 8 &= 4 && \text{Subtract } x \text{ from both sides} \\
2x &= 12 && \text{Add 8 to both sides} \\
x &= 6 && \text{Divide each side by 2}
\end{aligned}
$$

The solution is $x = 6$. ■

Example 10

Solve the equation: $x^2 + x - 12 = 0$

Solution

We factor the left side, obtaining

$$(x - 3)(x + 4) = 0$$

The Product Law states that either the first factor or the second factor must equal zero. Thus,

$$
\begin{aligned}
x - 3 &= 0 \quad \text{or} \quad x + 4 = 0 \\
x &= 3 \quad \text{or} \quad x = -4
\end{aligned}
$$

The solutions to the equation are $x = 3$ or $x = -4$. ■

INEQUALITIES

An *inequality* is a statement involving one or more variables and one of the inequality symbols ($<, \leq, >, \geq$). To *solve* an inequality means to find all possible numbers that the variables can assume to make the statement true. The set of all such numbers is called the *solution*. Two inequalities with the same solution are called *equivalent inequalities*.

To find the solution of an inequality, we apply the laws for inequalities.

Example 11

Solve the inequality: $x + 2 \leq 3x - 5$

Solution

$$
\begin{aligned}
x + 2 &\leq 3x - 5 \\
-2x &\leq -7 && \text{Subtract 2 and then } 3x \text{ from both sides} \\
x &\geq \frac{7}{2} && \text{Multiply by } -\tfrac{1}{2} \text{ and remember to reverse the inequality because we multiplied by a negative number}
\end{aligned}
$$

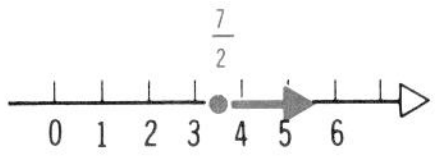

Figure 5

The solution is the set of all real numbers to the right of $\frac{7}{2}$, including $\frac{7}{2}$ (see Figure 5). ■

In graphing the solution to an inequality, our practice will be to use a filled-in circle (●) if the number is included (such as $\frac{7}{2}$ in Figure 5) and an open circle (○) if the number is to be excluded.

An inequality, such as $a < x < b$, is the standard way of stating that x lies between a and b. For example, the inequality

$$3 < x < 5$$

states that x lies between 3 and 5.

Exercise 1.1

Answers to Odd-Numbered Problems begin on page A-1.

A   In Problems 1–8 represent each rational number as a decimal.

1. $\frac{1}{2}$	**2.** $\frac{3}{4}$	**3.** $\frac{13}{8}$	**4.** $\frac{15}{8}$
5. $\frac{4}{3}$	**6.** $\frac{5}{3}$	**7.** $\frac{1}{6}$	**8.** $\frac{5}{6}$

In Problems 9–16 write each decimal as a percent.

9. 0.45	**10.** 0.85	**11.** 1.12	**12.** 1.25
13. 0.06	**14.** 0.07	**15.** 0.0025	**16.** 0.0015

In Problems 17–24 write each percent as a decimal.

17. 42%	**18.** 7.25%	**19.** 0.2%	**20.** 300%
21. 0.001%	**22.** 4.3%	**23.** 73.4%	**24.** 92%

In Problems 25–32 write each rational number in lowest terms.

25. $\frac{2}{40}$	**26.** $\frac{25}{45}$	**27.** $\frac{6}{8}$	**28.** $\frac{8}{24}$
29. $\frac{52}{54}$	**30.** $\frac{24}{64}$	**31.** $\frac{162}{243}$	**32.** $\frac{81}{54}$

In Problems 33–40 calculate the indicated quantity.

33. 15% of 1000	**34.** 20% of 500	**35.** 18% of 100
36. 10% of 50	**37.** 210% of 50	**38.** 135% of 1000
39. 10.5% of $115.00	**40.** 7.5% of $142.15	

B   In Problems 41–58 find the solution x of each equation.

41. $2x + 5 = 7$	**42.** $x + 6 = 2$	**43.** $6 - x = 0$
44. $6 + x = 0$	**45.** $3(2 - x) = 9$	**46.** $5(x + 1) = 10$
47. $4x + 3 = 2x - 5$	**48.** $5x - 8 = 2x + 1$	**49.** $\dfrac{4x}{3} + \dfrac{x}{3} = 5$
50. $\dfrac{2x}{5} + \dfrac{x}{5} = 9$	**51.** $\dfrac{3x - 5}{x - 3} = 1$	**52.** $\dfrac{2x + 1}{x - 1} = 3$
53. $x^2 - x - 12 = 0$	**54.** $x^2 + 7x = 0$	**55.** $x^2 - 5x + 6 = 0$
56. $x^2 - x - 6 = 0$	**57.** $x^2 - 16 = 0$	**58.** $x^2 - 6x + 9 = 0$

In Problems 59–70 find x.

59. $x = 3^2$ **60.** $x = 2^3$ **61.** $x = 2^{-3}$ **62.** $x = 3^{-2}$

63. $x = -3^2$ **64.** $x = (-3)^2$ **65.** $x^3 = 8$ **66.** $x^5 = -32$

67. $2^x = 4$ **68.** $3^x = 81$ **69.** $4^x = 4^5$ **70.** $2^x = 2^{10}$

In Problems 71–74 replace the * by $<$, $>$, or $=$.

71. $\frac{1}{3} * 0.33$ **72.** $\frac{1}{4} * 0.25$ **73.** $3 * \sqrt{9}$ **74.** $\pi * \frac{22}{7}$

In Problems 75–82 find the solution.

75. $3x + 5 \le 2$ **76.** $14x - 21x + 16 \le 3x - 2$

77. $3x + 5 \ge 2$ **78.** $4 - 5x \ge 3$

79. $-3x + 5 \le 2$ **80.** $8 - 2x \le 5x - 6$

81. $6x - 3 \ge 8x + 5$ **82.** $-3x \le 2x + 5$

C   In Problems 83–90 graph each inequality.

83. $x + 3 > 4x - 6$ **84.** $2x + 1 < 3x - 2$

85. $2x - 2 \le x + 3$ **86.** $3x - 1 \ge 2x + 5$

87. $3x - \frac{1}{2} < 7$ **88.** $\frac{3}{2x} - \frac{2}{3} < \frac{5}{6}$

89. $\frac{9}{4} < \frac{5}{2} + \frac{2}{3x}$ **90.** $3(x + 2) + 6 < 6 [2 + 3(x - 1)]$

91. 30% of a number is 210. Find the number.

92. x% of 350 is 84. Find x.

93. In 1974 the average cost of one-half gallon of milk was 59 cents; by 1990 the average had risen to 99 cents. Find the percentage increase.

94. In 1974 the average cost of a 1 lb loaf of bread was 33 cents; by 1990 the average had risen to 99 cents. Find the percentage increase.

1.2

RECTANGULAR COORDINATES AND STRAIGHT LINES

RECTANGULAR COORDINATES
GRAPHS OF LINEAR EQUATIONS IN TWO VARIABLES
SLOPE OF A LINE
OTHER EQUATIONS OF LINES
APPLICATION

RECTANGULAR COORDINATES

Consider two lines, one horizontal and the other vertical. Call the horizontal line the *x-axis* and the vertical line the *y-axis*. Assign coordinates to points on these lines, as described previously, by using their point of intersection as the origin O and using a convenient scale on each. We follow the usual convention that points on the x-axis to the right of O are associated with positive real numbers, those to the left of O with negative numbers, those on the y-axis above O are associated with positive real numbers, and those below O with negative real numbers. This gives the origin a value of zero on both the x-axis and the y-axis. See Figure 6.

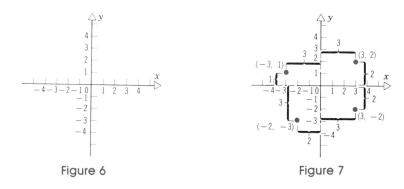

| Figure 6 | Figure 7 |

Any point P in the plane formed by the x-axis and y-axis can then be located by using an *ordered pair* of real numbers. Let x denote the signed distance of P from the y-axis (signed in the sense that if P is to the right of the y-axis, then $x > 0$ and if P is to the left of the y-axis, then $x < 0$); and let y denote the signed distance of P from the x-axis. The ordered pair (x, y), the *coordinates of P*, then gives us enough information to locate the point P. We can assign ordered pairs of real numbers to every point P, as shown in Figure 7.

If (x, y) are the coordinates of a point P, then x is called the *x-coordinate* of P and y is the *y-coordinate* of P. For example, the coordinates of the origin O are $(0, 0)$. The x-coordinate of any point on the y-axis is 0; the y-coordinate of any point on the x-axis is 0.

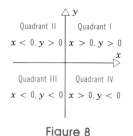

Figure 8

The coordinate system described here is a *rectangular* or *Cartesian coordinate system** and divides the plane into four sections called *quadrants* (see Figure 8).

In quadrant I, both the x-coordinate and the y-coordinate of all points are positive; in quadrant II, $x < 0$ and $y > 0$; in quadrant III, both x and y are negative; and in quadrant IV, $x > 0$ and $y < 0$.

GRAPHS OF LINEAR EQUATIONS IN TWO VARIABLES
A *linear equation in two variables* can be written in *standard form* as

$$Ax + By = C \qquad (1)$$

---

* RENÉ DESCARTES (1596–1650), after whom the Cartesian coordinate system was named, was born in La Haye, France. After spending several years participating in various wars, he published a treatise introducing analytic geometry. In addition to being a mathematician, he was a respected philosopher and theologian.

with A and B not both zero. Examples of linear equations are

$3x - 5y - 6 = 0$ This equation can be written as
                      $3x - 5y = 6$ $A = 3, B = -5, C = 6$

$-3x = 2y - 1$ This equation can be written as
                      $-3x - 2y = -1$ $A = -3, B = -2, C = -1$

$y = \frac{3}{4}x - 5$ Here we can write
                      $-\frac{3}{4}x + y = -5$ or $3x - 4y = 20,$
                      $A = 3, B = -4, C = 20$

$y = -5$ Here we can write
                      $0 \cdot x + y = -5,$ $A = 0, B = 1, C = -5$

$x = 4$ Here we can write
                      $x + 0 \cdot y = 4,$ $A = 1, B = 0, C = 4$

The *graph* of an equation is the set of all points (x, y) whose coordinates satisfy the equation. For example $(0, 4)$ is in the graph of the equation $3x + 4y = 16$, because when x is replaced by zero and y is replaced by 4 in the equation we get

$$3 \cdot 0 + 4 \cdot 4 = 16$$

which is an equality.

To *graph an equation* means to locate on a rectangular coordinate system all the points (x, y) whose coordinates satisfy the equation. It can be shown that if A, B, and C are real numbers, with A and B not both zero, then the graph of the equation

$$Ax + By = C$$

is a *straight line.* Conversely, any straight line is the graph of an equation of this form. Since two points uniquely determine a straight line, it is sufficient to obtain any two points satisfying equation (1) to establish the full graph of the solution. The easiest two points to plot are the *intercepts* of the line.

Intercepts The points at which the graph of a line L crosses the axes are called *intercepts*. The *x-intercept* is the point at which the line crosses the *x*-axis, and the *y-intercept* is the point at which the line crosses the *y*-axis.

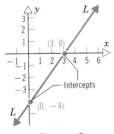

Figure 9

To find the intercepts, perform the following:

Step 1: Set $y = 0$ and solve for x. This determines the *x*-intercept of the line.
Step 2: Set $x = 0$ and solve for y. This determines the *y*-intercept of the line.

For example, the line L in Figure 9 has *x*-intercept $(3, 0)$ and *y*-intercept $(0, -4)$.

Example 1

Find the intercepts of the line $2x + 3y = 6$. Graph this line.

Solution

To find the point at which the graph crosses the x-axis—that is, to find the x-intercept—we need to find the number x for which $y = 0$. Thus, we set $y = 0$ to get

$$2x + 3(0) = 6$$
$$2x = 6$$
$$x = 3$$

The x-intercept is $(3, 0)$. To find the y-intercept, we set $x = 0$ and solve for y:

$$2(0) + 3y = 6$$
$$3y = 6$$
$$y = 2$$

The y-intercept is $(0, 2)$.

We now know two points on the line: $(3, 0)$ and $(0, 2)$. See Figure 10 for the graph.

Note that all points on the line satisfy the equation, for example $(\frac{3}{2}, 1)$, $(-3, 4)$, etc., and all points off the line fail to satisfy the equation. ∎

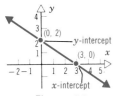

Figure 10

Example 2

Graph the equation $y = 2x - 5$.

From the intercepts listed in the table we obtain the graph shown in Figure 11. It is sometimes wise to find a third point as a check.

x	y
0	5
$-\frac{5}{2}$	0
10	25

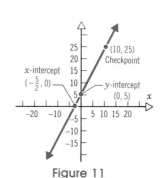

Figure 11 ∎

Example 3

Graph the equation $y - x = 0$.

This is a linear equation equivalent to $-x + y = 0$ with $A = -1, B = 1, C = 0$. If we attempt to find the intercepts, we encounter a slight difficulty since when $y = 0, x = 0$ and vice versa. This is so since the line contains $(0, 0)$ and this is the only intercept. Since we need another point, let $x = 1$ (chosen arbitrarily for ease of calculation) then $y = 1$. See Figure 12.

x	y
0	0
1	1
3	3

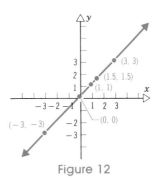

Figure 12

SLOPE OF A LINE

To enable us to understand the graph of a straight line better, we now introduce the idea of the *slope of a line.*

Slope of a Line **Let P and Q be two distinct points with coordinates (x_1, y_1) and (x_2, y_2), respectively. The *slope m* of the line L containing P and Q is defined by the formula***

$$m = \frac{y_2 - y_1}{x_2 - x_1} \qquad \text{if} \quad x_1 \neq x_2$$

If $x_1 = x_2$, the slope m of L is *undefined* (since this results in division by zero) and L is a *vertical line.*

We can also write the slope m of a nonvertical line as

$$m = \frac{\text{change in } y}{\text{change in } x} = \frac{\Delta y}{\Delta x}$$

That is, the slope m of a nonvertical line L is the ratio of the change in the

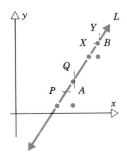

* The following argument, involving similar triangles, shows that the slope of a line L is independent of what two distinct points are used: Let L be a nonvertical line joining P and Q and let X and Y be any other two distinct points on L. Construct the triangles depicted in the figure. Since triangle PQA is similar to triangle XYB (why?), it follows that the lengths of the corresponding sides are in proportion. That is, $|AQ|/|BY| = |AP|/|BX|$ or $|AQ|/|AP| = |BY|/|BX|$. But the slope m of L is $|AQ|/|AP|$, and by the foregoing equality, we see that $m = |BY|/|BX|$. In other words, since X and Y are *any* two points, the slope m of a line L is independent of what points on L are used to compute m.

y-coordinates from P to Q to the change in the x-coordinates from P to Q. In other words, the slope m of a line L equals the "rise over run" of the line. See Figure 13.

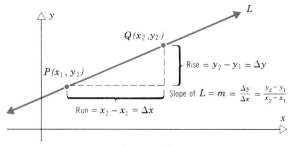

Figure 13

Since

$$\frac{y_2 - y_1}{x_2 - x_1} = \frac{y_1 - y_2}{x_1 - x_2}$$

the result is the same whether the changes are computed from P to Q or from Q to P.

Example 4
The slope m of the line containing the points $(1, 2)$ and $(5, -3)$ may be computed as

$$m = \frac{(-3) - (2)}{(5) - (1)} = \frac{-5}{4} \quad \text{or as} \quad m = \frac{(2) - (-3)}{(1) - (5)} = \frac{5}{-4} = \frac{-5}{4} \quad \blacksquare$$

To get a better idea of the meaning of the slope of a line L, consider the following example.

Example 5
Compute the slopes of the lines L_1, L_2, L_3, and L_4 containing the following pairs of points. Graph each line.

$$
\begin{array}{lll}
L_1: & P = (2, 3) & Q_1 = (-1, -2) \\
L_2: & P = (2, 3) & Q_2 = (3, -1) \\
L_3: & P = (2, 3) & Q_3 = (5, 3) \\
L_4: & P = (2, 3) & Q_4 = (2, 5)
\end{array}
$$

Solution
Let m_1, m_2, m_3, and m_4 denote the slopes of the lines L_1, L_2, L_3, and L_4, respectively. Then

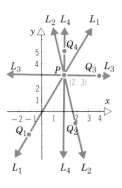

$$m_1 = \frac{-2 - 3}{-1 - 2} = \frac{-5}{-3} = \frac{5}{3} \qquad \text{A rise of 5 over a run of 3}$$

$$m_2 = \frac{-1 - 3}{3 - 2} = \frac{-4}{1} = -4$$

$$m_3 = \frac{3 - 3}{5 - 2} = \frac{0}{3} = 0$$

m_4 is undefined since $x_1 = x_2 = 2$

The graphs of these lines are given in Figure 14.

As Figure 14 indicates, when the slope m of a line is positive, the line *slants upward* from left to right (L_1); when the slope m is negative, the line *slants downward* from left to right (L_2); when the slope $m = 0$, the line is horizontal (L_3); and when the slope m is undefined, the line is vertical (L_4). Figure 15 illustrates the slopes of several lines. Note the pattern.

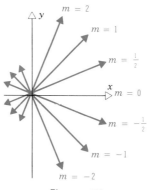

Figure 15

Thus, the slope of a line is an indicator of its direction. We will see later that two nonvertical lines are *parallel*—that is, they have the same direction—if and only if they have the same slope.

OTHER EQUATIONS OF LINES

With the aid of Figure 14 we are able to obtain the two simplest equations of a line. We observe that for a vertical line, L_4, the slope is undefined and that all x values of points on the line are equal, whereas y can take on any value. Therefore, in general:

A vertical line is given by the equation

$$x = a$$

where a is a given real number.

Similarly, the line L_3 is a horizontal line with slope 0. All y values of the points on the line are equal, whereas x can take on any value. Therefore, in general:

A horizontal line is given by the equation

$$y = b$$

where b is a given real number.

Example 6

The graph of the equation $x = 3$ is a vertical line (see Figure 16a). The graph of the equation $y = 3$ is a horizontal line (see Figure 16b).

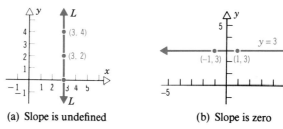

(a) Slope is undefined   (b) Slope is zero

Figure 16

It is important to note the distinction between the phrases "the slope is zero" (true for horizontal lines) and "the slope is not defined" (true for vertical lines).

Now let L be a nonvertical line with slope m and containing (x_1, y_1). For (x, y) any other point on L, we have

$$m = \frac{y - y_1}{x - x_1} \quad \text{or} \quad y - y_1 = m(x - x_1)$$

Point–Slope Form An equation of a nonvertical line of slope m that passes through the point (x_1, y_1) is

$$y - y_1 = m(x - x_1)$$

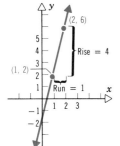

Figure 17

Example 7

An equation of the line with slope 4 and passing through the point $(1, 2)$ is

$$y - 2 = 4(x - 1)$$
$$y = 4x - 2$$

See Figure 17.

Example 8

Find an equation of the line L passing through the points $(2, 3)$ and $(-4, 5)$. Write it in standard form. Graph the line.

Solution

Since two points are given, we first compute the slope of the line:

$$m = \frac{5-3}{-4-2} = \frac{2}{-6} = \frac{-1}{3}$$

Using the point $(2, 3)$ (we could use the other point instead, if we wished), and the fact that the slope $m = \frac{-1}{3}$, the point–slope equation of the line is

$$y - 3 = \frac{-1}{3}(x - 2)$$

We simplify this equation by multiplying both sides by 3 and collecting terms.

$$3(y - 3) = 3(-1/3)(x - 2) \qquad \text{Multiply by 3}$$
$$3y - 9 = -1(x - 2)$$
$$3y - 9 = -x + 2$$
$$x + 3y = 11$$

See Figure 18 for the graph.

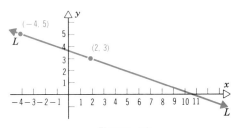

Figure 18

Another useful equation of a line is obtained when the slope m and y-intercept $(0, b)$ are known. Since in this event we know both the slope m of the line and a point $(0, b)$ on the line, we may use the point–slope form to obtain the following equation:

$$y - b = m(x - 0) \qquad \text{or} \qquad y = mx + b$$

Slope–Intercept Form An equation of a line L with slope m and y-intercept $(0, b)$ is

$$y = mx + b$$

When the equation of a line is written in slope–intercept form, it is easy to find the slope m and y-intercept $(0, b)$ of the line. For example, suppose the equation of a line is

$$y = -2x + 3$$

Compare it to $y = mx + b$:

$$y = -2x + 3$$
$$\uparrow \quad \uparrow$$
$$y = \ \ mx + b$$

The slope of this line is -2 and its y-intercept is $(0, 3)$. Let's look at another example.

Example 9

Find the slope m and y-intercept $(0, b)$ of the line L given by $2x + 4y = 8$. Graph the line.

Solution

To obtain the slope and y-intercept, we transform the equation to its slope–intercept form. Thus, we need to solve for y:

$$2x + 4y = 8$$
$$4y = -2x + 8$$
$$y = \left(\frac{-1}{2}\right)x + 2$$

The coefficient of x, $-\frac{1}{2}$, is the slope, and the y-intercept is $(0, 2)$. To graph this line, we need two points. Normally, the easiest points to locate are the intercepts. The y-intercept is $(0, 2)$. To obtain the x-intercept, set $y = 0$ and solve for x. When $y = 0$, we have

$$2x = 8$$
$$x = 4$$

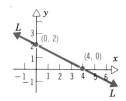

Figure 19

Thus, the intercepts are $(4, 0)$ and $(0, 2)$, as shown in Figure 19. ■

SUMMARY

1. Given the general equation of a line, information can be found about the line:
   (a) Place the equation in slope–intercept form $y = mx + b$ to find the slope m and y-intercept $(0, b)$.
   (b) Let $y = 0$ and solve for x to find the x-intercept.

2. Given information about a straight line, the equation of the line can be found by using the appropriate form of the equation.

Given	Use	Equation
Point (x_1, y_1) Slope m	Point–slope form	$y - y_1 = m(x - x_1)$
Two points $(x_1, y_1), (x_2, y_2)$	If $x_1 = x_2$, use vertical equation	$x = x_1$
	If $x_1 \neq x_2$, find the slope	$y - y_1 = \left(\dfrac{y_2 - y_1}{x_2 - x_1}\right)(x - x_1)$
	$m = \dfrac{y_2 - y_1}{x_2 - x_1}$ and use the point–slope form	$= m(x - x_1)$
Slope m, y-intercept $(0, b)$	Slope–intercept form	$y = mx + b$

APPLICATION
Many situations in business use linear equations. The next example illustrates how to construct such equations.

Example 10
When a company sold watches at \$50 each, the demand was 1500 watches per month. After the price rose to \$60 each, the demand dropped to 1200 watches per month. Assuming that the price p is linearly related to the demand q (quantity of watches); (we use the phrase "linearly related" to mean that the two variables are related by a linear equation):

(a) Find the slope of the line using the two data points.
(b) Find an equation of the line relating price to quantity. (Write the final answer in the form $p = mq + b$; use q to represent x and p to represent y).
(c) Graph the demand equation.

Solution
(a) From the information given we obtain the two data points (1500, 50) and (1200, 60). Therefore the slope is
$$m = \frac{60 - 50}{1200 - 1500} = -\frac{1}{30}$$

(b) The equation becomes
$$p - p_1 = m(q - q_1) \qquad \text{Point–slope equation}$$
$$p - 50 = -\tfrac{1}{30}(q - 1500) \qquad p_1 = 50, \ q_1 = 1500$$
$$p = -\tfrac{1}{30}q + 100$$

(c)   The graph of the linear equation $p = -\frac{1}{30}q + 100$ is given in Figure 20a. However, since p and q represent nonnegative quantities, we restrict the graph to the first quadrant as shown in Figure 20b.

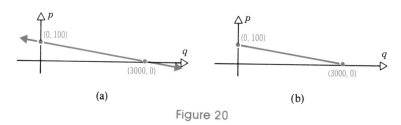

(a)   (b)

Figure 20   ▪

Economists refer to the equation $p = -\frac{1}{30}q + 100$ as a *linear demand equation*.

Exercise 1.2   *Answers to Odd-Numbered Problems begin on page A-1.*

A   **1.** Find the coordinates of each point in the figure below.

   2. Locate the points $(3, -2)$, $(-2, -3)$, $(5, 0)$, $(6, -2)$, and $(-2, 1)$ using the figure below.

In Problems 3–6 use the above figure, where $A = (a_1, a_2)$, $B = (b_1, b_2)$, $C = (c_1, c_2)$, and $F = (f_1, f_2)$, to compute each quantity.

3. $\dfrac{f_2 - a_2}{f_1 - a_1}$ **4.** $\dfrac{f_2 - c_2}{f_1 - c_1}$ **5.** $\dfrac{a_2 - c_2}{a_1 - c_1}$ **6.** $\dfrac{a_2 - b_2}{a_1 - b_1}$

In Problems 7–10 copy the tables at the right and fill in the missing values of the given equations. Use these points to graph each equation.

7. $y = x - 3$

x	0		2	-2	4	-4
y		0				

8. $y = -3x + 3$

x	0		2	-2	4	-4
y		0				

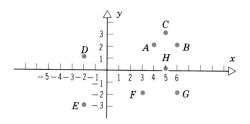

9. $2x - y = 6$

x	0		2	-2	4	-4
y		0				

10. $x + 3y = 9$

x	0		2	-2	4	-4
y		0				

In Problems 11–20 express the following linear equations in standard form: $Ax + By = C$

11. $6x + 2 = 4y + 3$

12. $y = 3x - 1$

13. $2 + 4y + 6x - 3$

14. $2y + 3x = 4 + 2x$

15. $2(x + 3) = y - 1$

16. $1 + 6(y - 2) = 3(x + 4)$

17. $x = -2$

18. $y = 3$

19. $3[2 - 5(x + 2y)] - 8 = 0$

20. $2[1 - 3(x + y)] - 4 = 0$

B   In Problems 21–30 find the intercepts of the following lines.

21. $2x + 2y = 6$

22. $^\hat{}3x - 4y = -12$

23. $-x + 5y = 30$

24. $-2x - 3y = 10$

25. $2 + 3y = 6x - 8$

26. $x + y = 3x + 4$

27. $12x - 3y - 24 = 0$

28. $6(x - 2) + y = 3$

29. $x = 4$

30. $y = 5$

In Problems 31–44 find the intercepts of the following lines. Graph each line.

31. $x + y = 5$

32. $x - y = 2$

33. $x + 3y = 12$

34. $x - 5y = 20$

35. $-4x - 3y = 24$

36. $3x - 5y = 30$

37. $y = 2x + 5$

38. $y = 5x + 7$

39. $y = \frac{1}{3}x + 8$

40. $y = -\frac{1}{3}x + 9$

41. $6x - 2 + y = 7$

42. $x - 3y + 4 = x + 2$

43. $x = 4$

44. $y = -3$

In Problems 45–54 graph the set of points (x, y) that obey the given equation.

45. $y = 3x$

46. $y = 4x$

47. $y = 2x - 3$

48. $y = 2x + 1$

49. $y = 0$

50. $x = 0$

51. $y = -2x - 3$

52. $y = -2x - 4$

53. $3x + 2y + 6 = 0$

54. $2x - 3y + 12 = 0$

55. Graph the equations in Problems 47 and 48 on the same coordinate system. Do you notice anything?

56. Graph the equations in Problems 51 and 52 on the same coordinate system. Do you notice anything?

In Problems 57–62 find the slope m of the line joining the given pair of points.

57. $P = (2, 3)$ $Q = (0, 1)$ **58.** $P = (1, 1)$ $Q = (5, -6)$
59. $P = (-3, 0)$ $Q = (-5, -4)$ **60.** $P = (4, -3)$ $Q = (0, 0)$
61. $P = (0.1, 0.3)$ $Q = (1.5, 4.0)$ **62.** $P = (-3, -2)$ $Q = (6, -2)$

In Problems 63–74 find the equation for the line having the given properties. Write the equation in the standard form: $Ax + By = C$.

63. Slope $= 2$; passing through $(-2, 3)$
64. Slope $= 3$; passing through $(4, -3)$
65. Slope $= -\frac{2}{3}$; passing through $(1, -1)$
66. Slope $= \frac{1}{2}$; passing through $(3, 1)$
67. Passing through $(1, 3)$ and $(-1, 2)$
68. Passing through $(-3, 4)$ and $(2, 5)$
69. Slope $= -3$; y-intercept $= (0, 3)$
70. Slope $= -2$; y-intercept $= (0, -2)$
71. x-intercept $= (2, 0)$; y-intercept $= (0, -1)$
72. x-intercept $= (-4, 0)$; y-intercept $= (0, 4)$
73. Slope undefined; passing through $(1, 4)$
74. Slope undefined; passing through $(2, 1)$

In Problems 75–80 find the slope and y-intercept of the given line. Graph each line.

75. $3x - 2y = 6$ **76.** $4x + y = 2$ **77.** $x + 2y = 4$
78. $-x - y = 4$ **79.** $x = 4$ **80.** $y = 3$

C   In Problems 81–84 find an equation of the line shown in the graphs below.

81.

82.

83.

84.

APPLICATIONS

85. Temperature Conversion The relationship between Celsius (°C) and Fahrenheit (°F) for measuring temperature is linear. Find an equation relating °C and °F if 0°C corresponds to 32°F and 100°C corresponds to 212°F. Use the equation to find the Celsius measure of 70°F.

86. Demand Equations When a company sold a hand calculator at $20 each, the demand was 2500 calculators per month. After the price rose to $25 each, the demand dropped off to 2000 calculators per month. Assuming that the demand q (number of calculators) is linearly related to the price p,

(a) Find the slope of the line using the two data points.

(b) Find an equation of the line relating price to quantity. Write the final answer in the form $q = mp + b$.

87. Cost Equations Suppose the cost to produce 5 units of a product is $20 and the cost of 10 units is $36. If the cost C is linearly related to the quantity q, find a linear equation relating C to q. Find the cost to produce 4 units.

88. Production Cost Given the data in the table.

Level of production	0	20	40
Cost	120	140	160

Assuming that the cost increases linearly with the production level, write an equation that will give the cost (y) in terms of the production level (x).

89. Weight–Height Relation Suppose the weights (w) of male college students are linearly related to their heights (h). If a student 62 inches tall weighs 120 pounds and a student 72 inches tall weighs 170 pounds, write an equation to express weight in terms of height.

90. Disease Propagation Research indicates that in a controlled environment, the number of diseased mice will increase linearly each day after one of the mice in the cage is infected with a particular type of disease causing germ. There were 8 diseased mice 4 days after the first exposure and 14 diseased mice after 6 days. Write an equation that will give the number of diseased mice after any given number of days. If there were 40 mice in the cage, how long will it take until they are all infected?

91. Oil Depletion Oil is pumped from an oil field at a constant rate each year, so that its oil reserves have been decreasing linearly with time. Geologists estimate that the field reserves were 400,000 barrels in 1980 and 320,000 barrels in 1990.

(a) Write an equation describing the amount of oil left in the field at any time.

(b) If the trend continues, when will the oil well dry out?

92. Product Promotion A cereal company finds that the number of people who will buy one of its products the first month it is introduced is linearly related to the amount of money it spends on advertisement. If it spends $400,000 on advertising, 100,000 boxes of cereal will be sold, and if it spends $600,000, 140,000 boxes will be sold.

(a) Write an equation describing the relation between the amount spent on advertising and the number of boxes sold.

(b) How much advertising is needed to sell 200,000 boxes of cereal?

93. College Degrees The percent of people over 55 who have college degrees is summarized in the following table:

Year (t)	1970	1975	1980	1985	1990
Percent with college degree (y)	30	34	38	42	46

(a) Find an equation of the line through the points (1970, 30) (1990, 46).

(b) If the trend continues, estimate the percentage of people over 55 who will have a college degree by the year 2000.

94. Water Preservation Since the beginning of the month a local reservoir has been losing water at a constant rate. On the 10th of the month, the reservoir held 300 million gallons of water and on the 18th it held only 262 million gallons.

(a)   Write an equation that will give the amount of water in the reservoir at any time.

(b)   How much water was in the reservoir on the 14th of the month?

95. SAT Scores The average SAT scores of incoming freshmen at a midwestern college have been declining at a constant rate in recent years. In 1985, the average SAT score was 592, while in 1989 the average SAT score was only 564.

(a)   Write an equation that will give the average SAT score at any time.

(b)   If the trends continue, what will the average SAT score of incoming freshmen be in 1994?

1.3

PARALLEL AND INTERSECTING LINES

PARALLEL LINES
INTERSECTING LINES
PERPENDICULAR LINES

PARALLEL LINES

Let L and M be two lines. Exactly one of the following three relationships must hold for the two lines L and M:

1. All the points on L are the same as the points on M.
2. L and M have no points in common.
3. L and M have exactly one point in common.

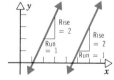

Figure 21

If the first relationship holds, the lines L and M are called *identical lines.* In this case, their slopes and their intercepts will be the same.

When two lines (in a plane) have no points in common, they are said to be *parallel.* Look at Figure 21.

Vertical lines that are distinct are parallel.

We can use the slope of a straight line to determine if a line is parallel to another line.

> Two distinct nonvertical lines L and M with slopes m_1 and m_2, respectively, are
>
> Parallel if and only if $m_1 = m_2$ (written $L \| M$)

Example 1

Show that the lines given by the equations below are parallel:

$$L: \quad 2x + 3y = 6 \qquad M: \quad 4x + 6y = 0$$

Solution

To see if these lines have equal slopes, we put each equation into slope–intercept form:

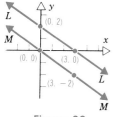

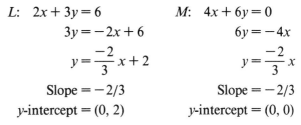

$$
\begin{array}{ll}
L: \quad 2x + 3y = 6 & M: \quad 4x + 6y = 0 \\
\quad\quad 3y = -2x + 6 & \quad\quad 6y = -4x \\
\quad\quad y = \dfrac{-2}{3}x + 2 & \quad\quad y = \dfrac{-2}{3}x
\end{array}
$$

Slope $= -2/3$ Slope $= -2/3$

y-intercept $= (0, 2)$ y-intercept $= (0, 0)$

Figure 22

Since each has slope $-\frac{2}{3}$ and different y-intercepts, the lines are parallel. See Figure 22. ∎

INTERSECTING LINES

If two lines L and M have exactly one point in common, then L and M are said to *intersect,* and the common point is called the *point of intersection.*

Example 2

Find the point of intersection of the two lines:

$$L: \quad x + y = 5 \qquad M: \quad 2x + y = 6$$

Let the coordinates of the point of intersection of L and M be (x_0, y_0). Since (x_0, y_0) is on both L and M, we must have

$$x_0 + y_0 = 5 \quad \text{and} \quad 2x_0 + y_0 = 6$$

Solving for y_0 in each equation, we get

$$y_0 = 5 - x_0 \qquad y_0 = 6 - 2x_0$$

Setting these equal, we obtain

$$5 - x_0 = 6 - 2x_0$$
$$x_0 = 1$$

Since $x_0 = 1$, then $y_0 = 5 - x_0 = 4$. Thus, the point of intersection of L and M is $(1, 4)$. To check this result, we verify that $(1, 4)$ is on both L and M:

$$
\begin{array}{ll}
x + y = 5 & 2x + y = 6 \\
1 + 4 = 5 & 2 \cdot 1 + 4 = 6
\end{array}
$$

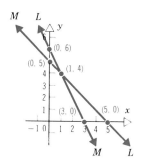

Figure 23

This verifies that $(1, 4)$ is the point of intersection. The graphs of the lines are given in Figure 23. ∎

Example 3

Without solving, determine if the pairs of equations are identical, parallel, or intersect.

(a) $x + y = -2$ (b) $x + y = -2$ (c) $x + y = -2$

  $4x + 4y = -8$ $2x + 2y = -14$ $2x - y = -4$

Solution

Put each equation in slope–intercept form:

(a) $x + y = -2$ $\qquad\qquad$ $4x + 4y = -8$

  $y = -x - 2$ $\qquad\qquad$ $4y = -4x - 8$

  $m = -1, b = -2$ $\qquad$ $y = -x - 2$

  $\qquad\qquad\qquad\qquad$ $m = -1, b = -2$

The lines are identical since they have equal slopes, -1, and the same y-intercept $(0, -2)$.

(b) $x + y = -2$ $\qquad\qquad$ $2x + 2y = -14$

  $y = -x - 2$ $\qquad\qquad$ $2y = -2x - 14$

  $m = -1, b = -2$ $\qquad$ $y = -x - 7$

  $\qquad\qquad\qquad\qquad$ $m = -1, b = -7$

The lines are parallel since they have equal slopes (-1) and different y-intercepts $(0, -2)$ and $(0, -7)$.

(c) $x + y = -2$ $\qquad\qquad$ $2x - y = -4$

  $y = -x - 2$ $\qquad\qquad$ $y = 2x + 4$

  $m = -1, b = -2$ $\qquad$ $m = 2, b = 4$

The lines intersect since they have unequal slopes $(-1$ and $2)$. ∎

PERPENDICULAR LINES

When two lines intersect and form a right angle they are said to be perpendicular.

> Two distinct nonvertical lines L and M with slopes m_1 and m_2, respectively, are
>
> Perpendicular if and only if $m_1 = -\dfrac{1}{m_2}$ (written $L \perp M$)
>
> That is, the slopes are negative reciprocals.

Example 4
Show that the lines given by the equations below are perpendicular.
$$L: x - 2y = 6 \qquad M: 2x + y = 1$$

Solution
To see if these are perpendicular, check the slope of each:

$$
\begin{array}{ll}
x - 2y = 6 & 2x + y = 1 \\
-2y = -x + 6 & y = -2x + 1 \\
y = \tfrac{1}{2}x - 3 & \text{slope} = m_2 = -2 \\
\text{slope} = m_1 = \tfrac{1}{2} &
\end{array}
$$

Since $m_1 = -\dfrac{1}{m_2}$, the lines are perpendicular.

Example 5
Given the line $x - 4y = 8$, find the equation of the line that passes through (2.1) and is
(a)   parallel to the given line.
(b)   perpendicular to the given line.

Solution
First find the slope of the line $x - 4y = 8$ by putting in the form $y = mx + b$

$$
\begin{aligned}
x - 4y &= 8 \\
-4y &= -x + 8 \\
y &= \tfrac{1}{4}x - 2
\end{aligned}
$$

so the slope of the line is $\tfrac{1}{4}$.

(a)   Since the line we are seeking is parallel to the given line, it should have the same slope. Thus we have the slope of the line and a point on the line. Using the point–slope equation of a straight line gives

$$
\begin{aligned}
y - y_1 &= m(x - x_1) \qquad m = \tfrac{1}{4} \qquad x_1 = 2 \qquad y_1 = 1 \\
y - 1 &= \tfrac{1}{4}(x - 2) \\
y &= \tfrac{1}{4}x + \tfrac{1}{2}
\end{aligned}
$$

(b)   The slope of the line perpendicular to the given line is the negative reciprocal of $\tfrac{1}{4}$, that is, -4. We now know the slope of the line and a point on the line. Using the point–slope equation of a line we have

$$
\begin{aligned}
y - y_1 &= m(x - x_1) \qquad m = -4 \qquad x_1 = 2 \qquad y_1 = 1 \\
y - 1 &= -4(x - 2) \\
y &= -4x + 9
\end{aligned}
$$

Exercise 1.3

Answers to Odd-Numbered Problems begin on page A-3.

A   In Problems 1–4 show that the lines are parallel by verifying that the slope of each line is equal and that the y-intercepts are unequal.

1. $x + y = 10$
    $3x + 3y = 1$

2. $x - y = 5$
    $-2x + 2y = 7$

3. $2x - 3y + 8 = 0$
    $6x - 9y + 2 = 0$

4. $4x - 2y + 7 = 0$
    $-2x + y + 2 = 0$

In Problems 5–18 find the point of intersection of each pair of lines.

5. $x + y = 5$
    $3x - y = 7$

6. $2x + y = 7$
    $x - y = -4$

7. $3x - 4y = 1$
    $x - 2y = -4$

8. $4x + 3y = 2$
    $2x - y = -1$

9. $3x - 2y + 5 = 0$
    $3x + y - 2 = 0$

10. $4x + y - 6 = 0$
    $4x - 2y = 0$

11. $2x - 3y + 4 = 0$
    $3x + 2y - 7 = 0$

12. $3x - 4y - 2 = 0$
    $2x + 5y - 9 = 0$

13. $3x - 4y + 8 = 0$
    $2x + y - 2 = 0$

14. $5x + 2y - 15 = 0$
    $2x - 3y - 6 = 0$

15. $-2x + 3y - 7 = 0$
    $3x + 2y - 9 = 0$

16. $-3x + 4y - 10 = 0$
    $2x - 3y + 7 = 0$

17. $3x + 2y + 6 = 0$
    $5x - 2y + 10 = 0$

18. $4x + 3y + 1 = 0$
    $2x + 5y - 3 = 0$

In Problems 19–26 without solving, determine whether the pair of equations are identical, parallel, or intersecting.

19. $L:$ $2x - 3y + 6 = 0$
    $M:$ $4x - 6y + 7 = 0$

20. $L:$ $4x - y + 2 = 0$
    $M:$ $3x + 2y = 0$

21. $L:$ $-2x + 3y + 6 = 0$
    $M:$ $4x - 6y - 12 = 0$

22. $L:$ $2x + 3y - 5 = 0$
    $M:$ $5x - 6y + 1 = 0$

23. $L:$ $3x - 3y + 10 = 0$
    $M:$ $x + y - 2 = 0$

24. $L:$ $2x - 5y - 1 = 0$
    $M:$ $x - 2y - 1 = 0$

25. $L:$ $3x - 2y - 4 = 0$
    $M:$ $-6x + 4y - 7 = 0$

26. $L:$ $8x - 2y + 4 = 0$
    $M:$ $-4x + y - 2 = 0$

In Problems 27–32 show that the lines are perpendicular by verifying that the slopes are negative reciprocals of each other.

27. $x - 3y = 2$
    $6x + 2y = 5$

28. $2x + 3y = 4$
    $9x - 6y = 1$

29. $x + 2y = 7$
    $2x - y = 15$

30. $4x + 12y = 3$
    $15x - 5y = -1$

31. $3x + 12y = 2$
    $4x - y = -2$

32. $20x - 2y = -7$
    $x + 10y = 8$

B   33. Find an equation of the line passing through $(1, 2)$ and parallel to $2x - y = 6$.

34. Find an equation of the line passing through $(-1, 3)$ and parallel to $x + y = 4$.

35. Find the equation of the line passing through $(0, 0)$ and parallel to $y = 3x - 2$.

36. Find the equation of the line passing through $(-5, -7)$ and parallel to $x = y$.

37. Find the equation of the line passing through $(-1, -2)$ and perpendicular to $y = 2x - 5$.

38. Find the equation of the line passing through $(-1, -2)$ and perpendicular to $6x - 2y + 5 = 0$.

39. Find the equation of the line passing through $(\frac{-1}{3}, \frac{4}{3})$ and perpendicular to $y = 2x - 5$.

40. Find the equation of the line passing through $(\frac{-2}{3}, \frac{3}{3})$ and perpendicular to $y = 3x - 15$.

41. Find t so that $tx - 4y + 3 = 0$ is perpendicular to $2x + 2y - 5 = 0$.

42. Find t so that $(1, 2)$ is a point on the line $tx - 3y + 4 = 0$.

43. Find the equation of the line passing through $(-2, -5)$ and perpendicular to the line through $(-2, 9)$ and $(3, -10)$.

44. Find the equation of the line passing through $(-2, -5)$ and perpendicular to the line through $(-4, 5)$ and $(2, -1)$.

45. Find the equation of the horizontal line passing through $(-1, -3)$.

46. Find the equation of the vertical line passing through $(-2, 5)$.

C 47. Show that $b(x - x_0) - a(y - y_0) = 0$ is the equation of a line through (x_0, y_0) perpendicular to $ax + by = c$.

48. Verify that the points $A = (1, 3)$, $B = (2, 1)$, $C = (8, 4)$ form a right triangle.

1.4

APPLICATIONS*

SIMPLE INTEREST
BREAK-EVEN POINT
PREDICTION
ECONOMICS

SIMPLE INTEREST

A knowledge of interest — whether on money borrowed or on money saved — is of ultimate importance today. The old adage "Neither a lender nor a borrower be" is not true in this age of charge accounts and golden passbook savings plans.

Interest is money paid for the use of money. The total amount of money borrowed (whether by an individual from a bank in the form of a loan or by a bank from an individual in the form of a savings account) is called the *principal*. The *rate of interest* is the amount charged for the use of the principal for a given period of time (usually on a yearly or *per annum* basis). The rate of interest is generally expressed as a *percent*.

* This section may be omitted without loss of continuity.

Simple Interest *Simple interest* **is interest computed on the principal for the entire period it is borrowed.**

In general, if a principal of P dollars is borrowed at a simple annual rate of interest r, expressed as a decimal, for a period of t years, the interest I charged is

$$I = Prt$$

The amount A owed at the end of a period of time is the sum of the principal and the interest. That is,

$$A = P + I = P + Prt$$

Example 1
If $250 is borrowed for 9 months at a simple interest rate of 8% per annum, what will be the interest charged?

Solution
The actual period the money is borrowed for is 9 months, or $\frac{3}{4}$ of a year. Thus, the interest charged will be the product of the principal ($250) times the annual rate of interest (0.08) times the period of time held expressed in years ($\frac{3}{4}$):

$$\text{Interest charged} = \$(250)(0.08)\left(\frac{3}{4}\right) = \$15$$

Example 2
If $500 is borrowed at a simple interest rate of 10% per annum, the amount A due after t years is

$$A = \$500 + \$500(.10)t = \$500 + \$50t$$

Thus, the amount due after 2 years is

$$A = \$500 + \$50(2) = \$600$$

The amount due after 6 months ($\frac{1}{2}$ year) is

$$A = \$500 + \$50(\tfrac{1}{2}) = \$525$$

The equation

$$A = 500 + 50t$$

is a linear equation in which A and t are the variables. If we graph this equation using A for the vertical axis and t for the horizontal axis, we can see how the amount A changes over time (see Figure 24). The slope of the line (50) equals the constant annual interest due on the loan.

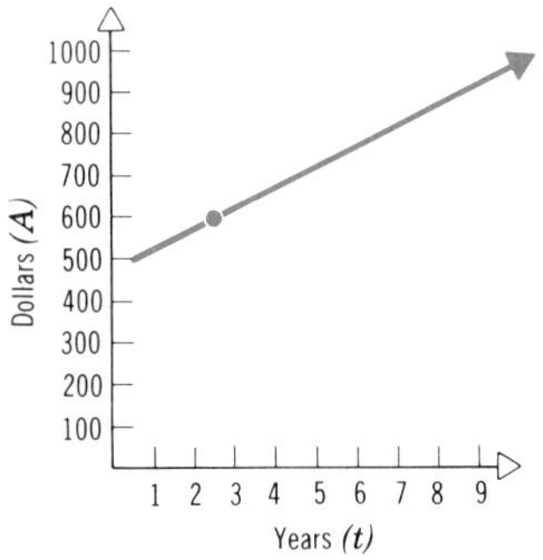

Figure 24

BREAK-EVEN POINT

In many businesses, the relation between the cost C of production and the number x of items produced can be expressed as a linear equation. Similarly, sometimes the relation between the revenue R obtained from sales and the number x of items produced can be expressed as a linear equation. When the cost C of production exceeds the revenue R from sales, the business is operating at a loss; when the revenue R exceeds the cost C, there is a profit; and when the revenue R and the cost C are equal, there is no profit or loss—the point at which

$$R = C$$

is usually referred to as the *break-even point.*

Example 3

Sweet Delight Candies, Inc., has daily fixed costs from salaries and building operations of $300. Each pound of candy produced costs $1 and is sold for $2 per pound. What is the break-even point—that is, how many pounds of candy must be sold daily to guarantee no loss and no profit?

Solution

The cost C of production is the fixed cost plus the variable cost of producing x pounds of candy at $1 per pound. Thus,

$$\text{Cost} = \text{variable cost} + \text{fixed cost}$$
$$C = \$1 \cdot x + \$300 = x + 300$$

The revenue R realized from the sale of x pounds of candy at $2 per pound is

$$R = \$2 \cdot x = 2x$$

The break-even point is the point where these two lines intersect. Setting $R = C$, we find

$$2x = x + 300$$
$$x = 300$$

That is, 300 pounds of candy must be sold in order to break even.

In Figure 25, we see a graphic interpretation of the break-even point for Example 3. Note that for $x > 300$, the revenue R always exceeds the cost C so that a profit results. Similarly, for $x < 300$, the cost exceeds the revenue, resulting in a loss.

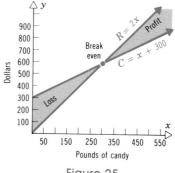

Figure 25

Example 4

Pricing Candy   After negotiations with employees of Sweet Delight Candies and an increase in the price of chocolate, the daily cost C of production for x pounds of candy is

$$C = \$1.05x + \$330$$

(a)   If each pound of candy is sold for $2, how many pounds must be sold daily to break even?

(b)   If the selling price is increased to $2.25 per pound, what is the break-even point?

(c)   If it is known that at least 325 pounds of candy can be sold daily, what price should be charged per pound to guarantee no loss?

Solution

(a)   If each pound is sold for $2, the revenue R from sales is

$$R = \$2x$$

where x represents the number of pounds sold. When we set $R = C$, we find that the break-even point is the solution of

$$2x = 1.05x + 330$$
$$0.95x = 330$$
$$x = \frac{33{,}000}{95} = 347.37$$

Thus, if 347 pounds of candy are sold, a loss is incurred; if 348 pounds are sold, a profit results.

(b)   If the selling price is increased to $2.25 per pound, the revenue R from sales is

$$R = \$2.25x$$

The break-even point is the solution of

$$2.25x = 1.05x + 330$$
$$1.2x = 330$$
$$x = \frac{3300}{12} = 275$$

With the new selling price, the break-even point is 275 pounds.

(c)   If we know that at least 325 pounds of candy will be sold daily, the price per pound p needed to guarantee no loss (i.e., to guarantee at worst a break-even point) is the solution of

$$325p = (1.05)(325) + 330$$
$$325p = 671.25$$
$$p = \$2.07$$

We should charge at least \$2.07 per pound to guarantee no loss, provided at least 325 pounds will be sold. ∎

Example 5

A producer sells items for \$0.30 each. If the cost for production is

$$C = \$0.15x + \$105$$

where x is the number of items sold, find the break-even point. If the cost can be changed to

$$C = \$0.12x + \$110$$

would it be advantageous?

Solution

The revenue R received is

$$R = \$0.3x$$

The break-even point is the solution of

$$0.3x = 0.15x + 105$$
$$0.15x = 105$$
$$x = 700$$

Thus, for the first cost, the break-even point is 700 items.

To determine the answer to the second part, we find that the break-even point at the new cost is $x = 611.11$. The old break-even point was $x = 700$. Thus, the new cost will require fewer items to be sold in order to break even. Management should probably change over to the new cost. See Figure 26.

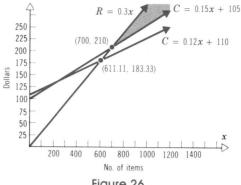

Figure 26

PREDICTION

Linear equations are sometimes used as predictors of future results. Let's look at an example.

Example 6

In 1987 the cost of an average home was $60,000. One year later the average home sold for $66,000. Assuming this pattern continues, that is, assuming that the increase will remain at $6000 per year, develop a formula for predicting the cost of an average home in 1991. What will it cost in 1996?

Solution

We agree to let x represent the year and y represent the cost. We seek a relationship between x and y. Two points on the graph of the equation relating x and y are

$$(1987, 60,000) \quad \text{and} \quad (1988, 66,000)$$

The assumption that the rate of increase remains constant tells us that the equation relating x and y is linear. The slope of this line is

$$\frac{66,000 - 60,000}{1988 - 1987} = 6000$$

Using this fact and the point $(1987, 60,000)$, the point–slope form of the equation of the line is

$$y - 60,000 = 6000(x - 1987)$$
$$y = 60,000 + 6000(x - 1987)$$

For $x = 1991$, we find the cost of an average home to be

$$y = 60,000 + 6000(1991 - 1987)$$
$$= 60,000 + 6000(4)$$
$$= \$84,000$$

For $x = 1996$, we find

$$y = 60{,}000 + 6000(9) = \$114{,}000$$

Figure 27 illustrates the situation.

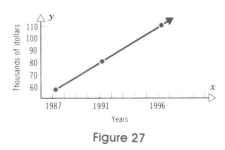

Figure 27

These predictions of future cost are based on the assumption that annual increases remain constant. If this assumption is not accurate, our predictions will be incorrect.

ECONOMICS

The *supply equation* in economics is used to specify the amount of a particular commodity that sellers have available to offer in the market at various prices. The *demand equation* specifies the amount of a particular commodity that buyers are willing to purchase at various prices.

An increase in price p usually causes an increase in the supply S and a decrease in demand D. On the other hand, a decrease in price brings about a decrease in supply and an increase in demand. The *market price* is defined as the price at which supply and demand are equal.

Example 7

Market Price of Flour The supply and demand for flour during the period 1920–1935 were estimated as being given by the equations

$$S = 0.8p + 0.5 \qquad D = -0.4p + 1.5$$

where p is measured in dollars and S and D are measured in 50 pound units of flour. Find the market price and graph the supply and demand equations.

Solution

The market price is the point of intersection of the two lines. Thus, the market price p is the point of intersection of

$$0.8p + 0.5 = S \qquad \text{and} \qquad -0.4p + 1.5 = D$$

So,

$$0.8p + 0.5 = -0.4p + 1.5$$
$$1.2p = 1$$
$$p = 0.83$$

The graphs are shown in Figure 28.

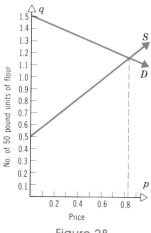

Figure 28

Exercise 1.4   *Answers to Odd-Numbered Problems begin on page A-3.*

A   *Simple Interest Problems*

1. Suppose you borrow $1000 at a simple interest rate of 18% per annum.
    (a)   What is the amount A due after t years?
    (b)   How much is due after 6 months?
    (c)   How much is due after 1 year?
    (d)   How much is due after 2 years?

2. Rework Problem 1 if you borrow $4000 at a simple interest rate of 14%.

Break-Even Problems.   In Problems 3–6 find the break-even point for the cost C of production and the revenue R. Graph each result.

3. $C = \$10x + \$600$ $R = \$30x$
4. $C = \$5x + \$200$ $R = \$8x$
5. $C = \$0.2x + \$50$ $R = \$0.3x$
6. $C = \$1800x + \$3000$ $R = \$2500x$

7. A manufacturer produces items at a daily cost of $0.75 per item and sells them for $1 per item. The daily operational overhead is $300. What is the break-even point? Graph your result.

8. If the manufacturer of Problem 7 is able to reduce the cost per item to $0.65, but with a resultant increase to $350 in operational overhead, is it advantageous to do so? Graph your result.

Prediction Problems

9. Suppose the sales of a company are given by

$$S = \$5000x + \$80,000$$

where x is measured in years and $x = 0$ corresponds to the year 1987.
    (a)   Find S when $x = 0$.
    (b)   Find S when $x = 3$.

(c)   Find the predicted sales in 1992, assuming this trend continues.
(d)   Find the predicted sales in 1995, assuming this trend continues.

10.   Rework Problem 9 if the sales of the company are given by

$$S = \$3000x + \$60,000$$

Economics Problems.   In Problems 11 – 14 find the market price for each pair of supply and demand equations.

11. $S = p + 1$ $D = 3 - p$
12. $S = 2p + 3$ $D = 6 - p$
13. $S = 20p + 500$ $D = 1000 - 30p$
14. $S = 40p + 300$ $D = 1000 - 30p$

APPLICATIONS   15.   Market Price of Sugar   The supply and demand equations for sugar from 1890 to 1915 were estimated by H. Schulz to be given by

$$S = 0.7p + 0.4 \qquad D = -0.5p + 1.6$$

Find the market price. What quantity of supply is demanded at this market price? Graph both the supply and demand equations. Interpret the point of intersection of the two lines.

16.   Supply and Demand Equations   The market price for a certain product is $5.00 per unit and occurs when 14,000 units are produced. At a price of $1, no units are manufactured and, at a price of $19.00, no units will be purchased. Find the supply and demand equations, assuming they are linear.

17.   Subcontracting   Assume that a manufacturer can purchase a needed component from a supplier at a cost of $8 per unit, or it can invest $20,000 in equipment and produce the item at a cost of $5.50 per unit.
      Determine the quantity of components for which total costs are equal for the two options.
      Which option should the manufacturer choose if 6000 units are required? What is their minimum cost for 6000 units?

18.   Equipment Investment   A firm has two machines from which it can choose in producing a new product. One automated machine costs $200,000 and produces items at a cost of $4 per unit. Another semi-automated machine costs $75,000 and produces items at a cost of $5.25 per unit.
      What volume of output makes the two machines equally costly?
      If 80,000 units are to be produced, which machine is the least costly?
      What is the minimum cost for 80,000 units?

19.   Subcontracting   An electronics firm needs a special microprocessor for a microcomputer it manufactures. Three alternatives have been identified for satisfying its needs: it can purchase the microprocessors from a supplier at a cost of $10 each or the firm could purchase either of two pieces of automated equipment and manufacture the microprocessors. One piece of equipment costs $80,000 and has variable costs of $8 per microprocessor. The second piece of equipment costs $120,000 and would result in variable costs of $5 per unit. Determine the minimum cost alternatives for different ranges of output by graphing the cost functions for the three alternatives.

CHAPTER REVIEW

Important Terms
and Formulas

set	inequality	straight line
empty set	x-axis	slope of a line
counting numbers	y-axis	vertical line
integers	rectangular coordinates	parallel lines
rational numbers	ordered pair	intersecting lines
numerator	x-coordinate	perpendicular lines
denominator	y-coordinate	*simple interest
irrational numbers	quadrants	*principal
real numbers	x-intercept	*rate of interest
decimals	y-intercept	*amount
percents	linear equation in two	*break-even point
positive number	variables	
negative number	graph	

$$a + b = b + a$$
$$a \cdot b = b \cdot a$$
$$a + b + c = (a + b) + c = a + (b + c)$$
$$a \cdot b \cdot c = (a \cdot b) \cdot c = a \cdot (b \cdot c)$$
$$a \cdot (b + c) = (a \cdot b) + (a \cdot c)$$
$$x^n \cdot x^m = x^{n+m}$$
$$(x^n)^m = x^{nm}$$
$$(ax)^n = a^n \cdot x^n$$

$$\left(\frac{x}{a}\right)^n = \frac{x^n}{a^n} \qquad a \neq 0$$
$$\frac{x^n}{x^m} = x^{n-m} \qquad x \neq 0$$
$$m = \frac{y_2 - y_1}{x_2 - x_1}$$
$$y = mx + b$$
$$y - y_1 = m(x - x_1)$$

True – False
Questions

(Answers on page A-4)

T F   1. In the slope–intercept equation of a line, $y = mx + b$, m is the slope and b is the x-intercept.

T F   2. The graph of the equation $Ax + By = C$, where A, B, C are real numbers and A, B are not both zero, is a straight line.

T F   3. The y-intercept of the line $2x - 3y + 6 = 0$ is $(0, 2)$.

T F   4. The slope of the line $2x - 4y + 7 = 0$ is $-\frac{1}{2}$.

T F   5. Parallel lines always have the same intercepts.

T F   6. Intersecting lines have different slopes.

T F   7. Perpendicular lines have slopes that are reciprocals of each other.

T F   8. A linear relation between two variables can always be graphed as a line.

T F   9. All lines with equal slopes are distinct.

T F   10. All vertical lines have positive slope.

* From optional section.

Fill in the Blanks   *(Answers on page A-4)*

1. If (x, y) are rectangular coordinates of a point, the number x is called the _____and y is called the _____.

2. The decimal equivalent of a rational number is either _____ or _____.

3. The slope of a vertical line is _____; the slope of a horizontal line is _____.

4. If a line slants downward as it moves from left to right, its slope will be a _____number.

5. If two lines have the same slope but different y-intercepts, they are _____.

6. If two lines have the same slope and the same y-intercept, they are said to be _____.

7. Lines that intersect at right angles are said to be _____ to each other.

8. Distinct lines that have different slopes are _____ lines.

9. The graph of a linear equation is always _____.

10. If the correspondence between two variables t and w is linear, then their relationship can be described by an equation _____.

Review Exercises   *Answers to Odd-Numbered Problems begin on page A-4.*

In Problems 1–6 find the solution x of each equation.

A

1. $3x + 6 = 2x - 1$

2. $-3x - 2 = 2x + 8$

3. $-2(x + 3) = x + 5$

4. $2x - 3 = -2(x + 2)$

5. $\dfrac{4x - 1}{x + 2} = 5$

6. $\dfrac{3x + 2}{2x - 1} = 1$

In Problems 7–10 find the solution of each inequality.

7. $2x - 1 \le 5$

8. $8x + 1 \ge 9$

9. $3x + 7 \ge -2x + 2$

10. $-3x + 4 \le 2x - 1$

In Problems 11–14 graph each linear equation.

11. $y = -2x + 3$

12. $y = 6x - 2$

13. $2y = 3x + 6$

14. $3y = 2x + 6$

In Problems 15–18 find an equation for the line containing each pair of points. Give your answer in the standard form: $Ax + By = C$.

15. $P = (1, 2)$ $Q = (-3, 4)$

16. $P = (-1, 3)$ $Q = (1, 1)$

17. $P = (0, 0)$ $Q = (-2, 3)$

18. $P = (-2, 3)$ $Q = (0, 0)$

In Problems 19–22 find an equation for the line. Give your answer in the standard form: $Ax + By = C$.

19. Slope is 2;
x-intercept is $(-1, 0)$

20. Slope is -1;
y-intercept is $(0, 1)$

21. Passing through $(1, 3)$ with slope 1

22. Passing through $(2, -1)$ with slope -2

In Problems 23–26 find the slope and y-intercept of each line. Graph each line.

23. $-9x - 2y + 18 = 0$

24. $-4x - 5y + 20 = 0$

25. $4x + 2y - 9 = 0$

26. $3x + 2y - 8 = 0$

In Problems 27–32, without solving, determine whether each pair of equations are identical, parallel, or intersect.

27. $3x - 4y + 12 = 0$
$6x - 8y + 9 = 0$

28. $2x + 3y + 5 = 0$
$4x + 6y + 10 = 0$

29. $x - y + 2 = 0$
$3x - 4y + 12 = 0$

30. $2x + 3y - 5 = 0$
$x + y - 2 = 0$

31. $4x + 6y + 12 = 0$
$2x + 3y + 6 = 0$

32. $3x - y = 0$
$6x - 2y + 5 = 0$

APPLICATIONS

33. Investment Problem Mr. and Mrs. Byrd have just retired and find that they need $10,000 per year to live on. Fortunately, they have a nest egg of $90,000 which they can invest in somewhat risky B-rated bonds at 16% interest per year or in a well-known bank at 6% per year. How much money should they invest in each so that they realize exactly $10,000 in income each year?

34. Mixture Problem One solution is 20% acid and another is 12% acid. How many cubic centimeters of each solution should be mixed to obtain 100 cc of a solution that is 15% acid?

35. The annual sales of Motors, Inc., for the past 5 years are listed in the table.

Year	Units Sold (in thousands)
1987	3400
1988	3200
1989	3000
1990	2800
1991	2600

(a) Graph this information using the x-axis for years and the y-axis for units sold. (For convenience, use different scales on the axes.)

(b) Draw a line L that passes through two of the points and comes close to passing through the remaining points.

(c) Find the equation of this line L.

(d) Using this equation of the line, what is your estimate for units sold in 1992?

36. Attendance at a Dance A church group is planning a dance in the school auditorium to raise money for its school. The band they will hire charges $500; the advertising costs are estimated at $100; and food is supplied at the rate of $2.00 per person. The church group would like to clear at least $900 after expenses.

(a) Determine how many people need to attend the dance for the group to break even if tickets are sold at $5 each.

(b) Determine how many people need to attend in order to achieve the desired profit if tickets are sold for $5 each.

(c) Answer the above two questions if the tickets are sold for $6 each.

Mathematical Questions

From CPA and CMA Exams (Answers on page A-5)

1. *CPA Exam*

   The Oliver Company plans to market a new product. Based on its market studies, Oliver estimates that it can sell 5500 units in 1976. The selling price will be $2.00 per unit. Variable costs are estimated to be 40% of the selling price. Fixed costs are estimated to be $6000. What is the break-even point?

   (a) 3750 units   (b) 5000 units
   (c) 5500 units   (d) 7500 units

2. *CPA Exam*

   The Breiden Company sells rodaks for $6.00 per unit. Variable costs are $2.00 per unit. Fixed costs are $37,500. How many rodaks must be sold to realize a profit before income taxes of 15% of sales?

   (a) 9375 units   (b) 9740 units
   (c) 11,029 units   (d) 12,097 units

3. *CPA Exam*

   Given the following notations, what is the break-even sales level in units?

   $$SP = \text{Selling price per unit}$$
   $$FC = \text{Total fixed cost}$$
   $$VC = \text{Variable cost per unit}$$

   (a) $\dfrac{SP}{FC \div VC}$ (b) $\dfrac{FC}{VC \div SP}$ (c) $\dfrac{VC}{SP - FC}$ (d) $\dfrac{FC}{SP - VC}$

4. *CPA Exam*

   At a break-even point of 400 units sold, the variable costs were $400 and the fixed costs were $200. What will the 401st unit sold contribute to profit before income taxes?

   (a) $0   (b) $0.50   (c) $1.00   (d) $1.50

Use the following information to answer Problems 5–8:

Akron, Inc. owns 80% of the capital stock of Benson Company and 70% of the capital stock of Cashin, Inc. Benson Company owns 15% of the capital stock of Cashin, Inc. Cashin, Inc., in turn, owns 25% of the capital stock of Akron, Inc. These ownership interrelationships are illustrated in the following diagram:

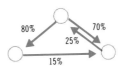

Net income before adjusting for interests in intercompany net income for each corporation follows:

Akron, Inc.	$190,000
Benson Co.	$170,000
Cashin, Inc.	$230,000

The following notations relate to items 5 through 8. Ignore all income tax considerations.

A_e = Akron's consolidated net income; that is, its net income plus its share of the consolidated net income of Benson and Cashin

B_e = Benson's consolidated net income; that is, its net income plus its share of the consolidated net income of Cashin

C_e = Cashin's consolidated net income; that is, its net income plus its share of the consolidated income of Akron

5. *CPA Exam*
    The equation, in a set of simultaneous equations, which computes A_e is:
    (a) $A_e = .75(190,000 + .8B_e + .7C_e)$
    (b) $A_e = 190,000 + .8B_e + .7C_e$
    (c) $A_e = .75(190,000) + .8(170,000) + .7(230,000)$
    (d) $A_e = .75(190,000) + .8B_e + .7C_e$

6. *CPA Exam*
    The equation, in a set of simultaneous equations, which computes B_e is:
    (a) $B_e = 170,000 + .15C_e - .75A_e$
    (b) $B_e = 170,000 + .15C_e$
    (c) $B_e = .2(170,000) + .15(230,000)$
    (d) $B_e = .2(170,000) + .15C_e$

7. *CPA Exam*
    Cashin's minority interest in consolidated net income is:
    (a) $.15(230,000)$ (b) $230,000 + .25A_e$
    (c) $.15(230,000) + .25A_e$ (d) $.15C_e$

8. *CPA Exam*
    Benson's minority interest in consolidated net income is:
    (a) $34,316   (b) $25,500
    (c) $45,755   (d) $30,675

9. *CPA Exam*
    A graph is set up with "depreciation expense" on the vertical axis and "time" on the horizontal axis. Assuming linear relationships, how would the graphs for straight-line and sum-of-the-years'-digits depreciation, respectively, be drawn?
    (a) Vertically and sloping down to the right
    (b) Vertically and sloping up to the right
    (c) Horizontally and sloping down to the right
    (d) Horizontally and sloping up to the right

The following statement applies to items 10–12:

In analyzing the relationship of total factory overhead with changes in direct labor hours, the following relationship was found to exist: $Y = $1000 + $2X$.

10. *CMA Exam*

The relationship as shown above is:

    (a) Parabolic   (b) Curvilinear
    (c) Linear   (d) Probabilistic
    (e) None of the above

11. *CMA Exam*

Y in the above equation is an estimate of:

    (a) Total variable costs   (b) Total factory overhead
    (c) Total fixed costs   (d) Total direct labor hours
    (e) None of the above

12. *CMA Exam*

The $2 in the equation is an estimate of:

    (a) Total fixed costs   (d) Fixed costs per direct labor hour
    (b) Variable costs per direct labor hour   (e) None of the above
    (c) Total variable costs

2

MATRICES WITH APPLICATIONS

2.1

REVIEW: SYSTEMS OF LINEAR EQUATIONS

SYSTEMS IN TWO VARIABLES
SUBSTITUTION METHOD
ELIMINATION BY ADDITION METHOD
SYSTEMS IN THREE VARIABLES

In Chapter 1 we found the break-even point and the market price by finding the point of intersection of two linear equations. The geometric problem of finding the point where two given lines intersect corresponds to the algebraic problem of solving two linear equations in two unknowns. In this section we review the methods of substitution and elimination, the generalization of which will lead naturally into the study of mathematical objects called *matrices,* the topic of this chapter.

SYSTEMS IN TWO VARIABLES

We start with an example. A company makes two kinds of machine parts. The first requires 4 hours of labor and 12 pounds of metal; the second uses 3 hours of labor and only 4 pounds of metal. On a given product run, there are 120 labor hours and 240 pounds of metal available. How many of each kind of part should be made to utilize all their resources?

Let

$$x = \text{the number of the first kind of part}$$
$$y = \text{the number of the second kind of part}$$

Then

$$4x + 3y = 120 \text{ labor hours}$$
$$12x + 4y = 240 \text{ pounds of metal}$$

What we now have is a system of two linear equations with two unknowns. To solve this system we find all ordered pairs that satisfy both equations simultaneously. In general, we will be interested in solving a *system of two linear equations in two unknowns* x and y in the form of

$$A_1 x + B_1 y = C_1$$
$$A_2 x + B_2 y = C_2$$

where $A_1, B_1, C_1, A_2, B_2, C_2$ are real numbers. A *solution* (x_0, y_0) of such a system is an ordered pair of real numbers that satisfy both equations when x is replaced by x_0 and y is replaced by y_0. To solve a system is to find all such ordered pairs (x_0, y_0).

We can solve systems of equations by using two basic methods:

Method 1: substitution.
Method 2: elimination by addition.

SUBSTITUTION METHOD

We start by outlining the steps for the *substitution method.*

Step 1: Pick one of the equations and solve for one of the unknowns in terms of the other.

Step 2: Substitute this expression for the same unknown in the other equation.

Step 3: Solve this equation.

Step 4: Use the solution found in Step 3 in either of the original equations to get the value of the other unknown.

Let's look at an example.

Example 1

Use the substitution method to find the solution of the system:

$$2x + y = -6$$
$$4x - 2y = -4$$

Solution

Step 1: We choose to solve for y in the first equation since this results in the least amount of work

$$2x + y = -6$$
$$y = -2x - 6$$

Step 2: We replace y in the second equation by this expression:

$$4x - 2(-2x - 6) = -4$$

Step 3: Simplify and solve this equation:

$$4x + 4x + 12 = -4$$
$$8x = -16$$
$$x = -2$$

Step 4: Replace x by -2 in the first equation:

$$2(-2) + y = -6$$
$$-4 + y = -6$$
$$y = -2$$

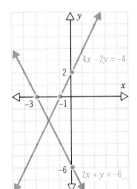

Figure 1

The solution of the system is $x = -2$, $y = -2$ or $(-2, -2)$, as shown in Figure 1.

Check:

$$2x + y = -6 \qquad\qquad 4x - 2y = -4$$
$$2(-2) + (-2) \stackrel{?}{=} -6 \qquad 4(-2) - 2(-2) \stackrel{?}{=} -4$$
$$-6 = -6 \checkmark \qquad\qquad -4 = -4 \checkmark$$

ELIMINATION BY ADDITION METHOD

We next introduce the *elimination by addition method.* The object is to eliminate all but one of the variables from each of the equations in order to obtain an *equivalent system* (i.e., a system with the same solution as the original system) in which the solution is obvious.

There are three legitimate operations that we can perform to transform an original system of linear equations into an equivalent system:

1. The order in which the equations are written can be changed.
2. Both sides of any equation can be multiplied by a nonzero constant.
3. Both sides of any equation can be multiplied by a constant and the result added to one of the other equations.

Operations 2 and 3 are used extensively in the next set of examples. Operation 1 will be of use when we deal with larger systems.

Example 2

Solve the system of equations using elimination by addition.

$$6x + 5y = 21$$
$$3x - 4y = 30$$

Solution

Our objective is to use the operations just outlined to eliminate by addition one of the variables, thus obtaining a system with an obvious solution. We choose to eliminate the variable y. If we multiply the first equation by 4 and the second by 5 (operation 2),

$$4(6x + 5y) = 4(21)$$
$$5(3x - 4y) = 5(30)$$

We obtain an equivalent system

$$24x + 20y = 84$$
$$15x - 20y = 150$$

When we add the first equation to the second equation (operation 3) the terms involving y drop out and we obtain

$$39x = 234$$

Dividing both sides by 39 we obtain

$$x = 6$$

We can now find y by substituting $x = 6$ into the first equation.

$$24(6) + 20y = 84$$
$$144 + 20y = 84$$
$$20y = -60$$
$$y = -3$$

The solution is $(6, -3)$.

Check:

$$6x + 5y = 21 \qquad\qquad 3x - 4y = 30$$
$$6(6) + 5(-3) \overset{?}{=} 21 \qquad 3(6) - 4(-3) \overset{?}{=} 30$$
$$21 = 21 \checkmark \qquad\qquad 30 = 30 \checkmark$$

The lines intersect at the point $(6, -3)$, as shown in Figure 2. ■

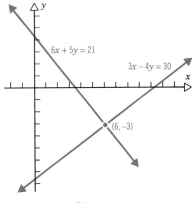

Figure 2

The next two examples illustrate systems of equations that have either no solution or infinitely many solutions.

Example 3
Solve

$$4x - 2y = 9$$
$$-6x + 3y = 5$$

Solution
We apply the elimination method by multiplying the first equation by 3 and the second by 2, we obtain an equivalent system.

$$12x - 6y = 27$$
$$-12x + 6y = 10$$

When we add these equations the terms involving both x and y drop out, leaving us with

$$0 = 37$$

which is clearly impossible. ■

We say that the given equations form an *inconsistent system,* that is, they have no common solution (x, y). This means that the lines they represent are parallel, as Figure 3 illustrates.

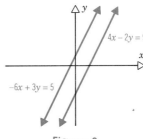

Figure 3

Example 4
Solve

$$10x + 5y = 15$$
$$6x + 3y = 9$$

Solution
Multiply the first equation by 3 and the second by -5.

$$30x + 15y = 45$$
$$-30x - 15y = -45$$

When we add the two equations we get

$$0 = 0$$

which is true for all x and y.   ■

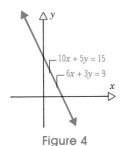

Figure 4

Such a system is a *dependent system:* dependent because the first equation can be obtained from the second by multiplying by $\frac{5}{3}$. Both equations have the same straight-line graph shown in Figure 4, and every point on this line is a common solution to the given equations, and the system has infinitely many solutions.

To obtain the infinitely many solutions we proceed as follows. If we set $x = t$, then using either equation, we obtain $y = -2t + 3$; that is, $(t, -2t + 3)$ is a solution for any real number t. The variable t is called a *parameter;* replacing it with any real number produces a particular solution to the system. For example, if we let $t = 0$, then $x = 0$ and $y = 3$ and the solution is $(0, 3)$. If we set $t = 1$, then $x = 1$ and $y = 1$ and another solution is $(1, 1)$ and so on for any choice of t.

These examples illustrate all the possible cases that can arise while solving a system of two linear equations in two unknowns. The results may be summarized as follows:

Systems with a unique solution:
The system has only one solution, and the corresponding lines intersect in one point.

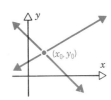

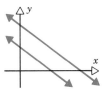

Inconsistent systems: The system has no solution, and the corresponding lines are parallel.

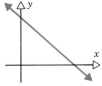

Dependent systems: The system has infinitely many solutions, and the corresponding lines coincide.

Example 5

Application   Nutt's Nuts, a store that specializes in selling nuts, sells cashews for $1.50 per pound and peanuts for $0.80 per pound. At the end of the month it is found that the peanuts are not selling well. In order to sell 30 pounds of peanuts more quickly, the store manager decides to mix the 30 pounds of peanuts with some cashews and sell the mixture of peanuts and cashews for $1.00 a pound. How many pounds of cashews should be mixed with the peanuts so that the profit remains the same?

Solution

There are two unknowns: the number of pounds of cashews (call this x) and the number of pounds of the mixture (call this y). Since we know that the number of pounds of cashews plus 30 pounds of peanuts equals the number of pounds of the mixture, we can write

$$y = x + 30$$

Also, in order to keep profits the same, we must have

$$\begin{pmatrix} \text{Pounds of} \\ \text{cashews} \end{pmatrix} \cdot \begin{pmatrix} \text{Price per} \\ \text{pound} \end{pmatrix} + \begin{pmatrix} \text{Pounds of} \\ \text{peanuts} \end{pmatrix} \cdot \begin{pmatrix} \text{Price per} \\ \text{pound} \end{pmatrix}$$
$$= \begin{pmatrix} \text{Pounds of} \\ \text{mixture} \end{pmatrix} \cdot \begin{pmatrix} \text{Price per} \\ \text{pound} \end{pmatrix}$$

That is,

$$(1.50)x + (0.80)(30) = (1.00)y$$
$$\tfrac{3}{2}x + 24 = y$$

Thus, we have a system of two linear equations in two unknowns to solve, namely,

$$y = x + 30$$
$$y = \tfrac{3}{2}x + 24$$

Using the substitution method, we find

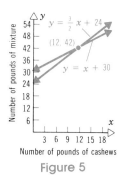

Figure 5

$$\tfrac{3}{2}x + 24 = x + 30$$
$$\tfrac{1}{2}x = 6$$
$$x = 12$$

The store manager should mix 12 pounds of cashews with 30 pounds of peanuts. See Figure 5. ∎

SYSTEMS IN THREE VARIABLES

Any equation that can be written in the form

$$ax + by = c$$

where a, b, and c are constants (a, b, not both zero) is called a *linear equation in two variables*. Similarly, any equation that takes the form

$$ax + by + cz = d$$

where a, b, c, and d are constants (a, b, c not all zero) is called a *linear equation in three variables*.

A linear system consisting of three linear equations in three variables x, y, and z has the general form:

$$\begin{aligned} a_1x + b_1y + c_1z &= d_1 \\ a_2x + b_2y + c_2z &= d_2 \\ a_3x + b_3y + c_3z &= d_3 \end{aligned} \qquad (1)$$

Higher-order systems are introduced in a similar way. For now we will restrict our discussion to systems of equations having the same number of equations as variables. Later on in this chapter, we will relax this restriction.

A solution (x_0, y_0, z_0) of equations (1) is an *ordered triplet* of real numbers $x = x_0$, $y = y_0$, $z = z_0$ that satisfies all equations. To *solve* a system is to find all such ordered triplets. If operations are performed on a system and the new system has the same solution as the original ones, we say that the systems are *equivalent*.

Just as a linear equation in two variables represents a straight line in the plane, it

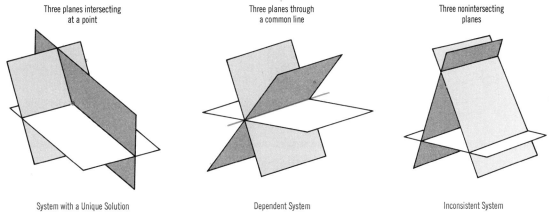

Figure 6

can be shown that a linear equation $ax + by + cz = d$, (a, b, and c not simultaneously equal to zero) in three variables represents a plane in a three-dimensional space. Each equation in (1) represents a plane in a three-dimensional space.

The *solution(s) of the system* is the point(s) of intersection of the three planes defined by the three linear equations of the system. Figure 6 shows several of the ways in which three planes can intersect. As in the two-variable case, we see that a system of three linear equations in three variables has one solution, infinitely many solutions, or no solutions.

The procedure we consider first for solving a system such as (1) is an extension of the elimination by addition method considered for systems with two variables. Using the three operations outlined in the last section, we attempt to eliminate all but one of the variables from each equation.

Example 6

Solve the following system using elimination by addition:

$$3x + 2y - 5z = 19 \tag{2}$$
$$2x - 3y + 3z = -15 \tag{3}$$
$$5x - 4y - 2z = -2 \tag{4}$$

Solution

Because of the convenient coefficients of y in equations (2) and (4) we choose to eliminate y by multiplying equation (2) by 2 and add to equation (4)

$$
\begin{array}{ll}
6x + 4y - 10z = 38 & \text{2(equation 2)} \\
\underline{5x - 4y - 2z = -2} & \\
11x - 12z = 36 & \tag{5}
\end{array}
$$

To eliminate y (the same variable) from equation (2) and (3) we multiply equation (2) by 3 and equation (3) by 2 and add

$$
\begin{array}{ll}
9x + 6y - 15z = 57 & \text{3(equation 2)} \\
\underline{4x - 6y + 6z = -30} & \text{2(equation 3)} \\
13x - 9z = 27 & \tag{6}
\end{array}
$$

Equations (5) and (6) form a system of equations in two variables.

$$11x - 12z = 36 \tag{5}$$
$$13x - 9z = 27 \tag{6}$$

Next we eliminate z,

$$
\begin{array}{ll}
33x - 36z = 108 & \text{3(equation 5)} \\
\underline{-52x + 36z = -108} & \text{$-$4(equation 6)} \\
-19x = 0 & \\
x = 0 &
\end{array}
$$

Substitute $x = 0$ back into either equation (5) or (6).

$$11x - 12z = 36 \tag{5}$$
$$11(0) - 12z = 36$$
$$-12z = 36$$
$$z = -3$$

Substitute $x = 0$, $z = -3$ back into any of the original equations (we choose 2).

$$3x + 2y - 5z = 19$$
$$3(0) + 2y - 5(-3) = 19$$
$$2y + 15 = 19$$
$$2y = 4$$
$$y = 2$$

The solution to the original system is $x = 0$, $y = 2$, $z = -3$, or $(0, 2, -3)$.

To check the solution, we must check each equation in the original system.

$$3x + 2y - 5z = 19 \qquad\qquad 2x - 3y + 3z = -15$$
$$3(0) + 2(2) - 5(-3) \stackrel{?}{=} 19 \qquad 2(0) - 3(2) + 3(-3) \stackrel{?}{=} -15$$
$$0 + 4 + 15 = 19 \checkmark \qquad\qquad 0 - 6 + (-9) = -15 \checkmark$$
$$5x - 4y - 2z = -2$$
$$5(0) - 4(2) - 2(-3) \stackrel{?}{=} -2$$
$$0 - 8 + 6 = -2 \checkmark \qquad\qquad\qquad \blacksquare$$

Solving a System of Three Equations

To solve a system of three linear equations in three variables, proceed as follows:

1. Eliminate one of the variables from two of the equations. Eliminate the *same* variable from a *different* pair of equations. The result is a system of two equations in the remaining two variables.
2. Solve the system of two equations. We now have a partial solution consisting of values for two of the variables.
3. Substitute the partial solution found in Step 2 into any of the original equations, and solve for the variable that was eliminated in Step 1.

APPLICATION

A system of three linear equations in three variables can often be used to solve a practical problem with three unknown quantities.

Example 9

FoodPerfect Corporation manufactures three models of the Perfect Foodprocessor. Each Model A processor requires 30 minutes of electrical assembly, 40 minutes of mechanical assembly, and 30 minutes of testing; each Model B requires 20 minutes of electrical assembly, 50 minutes of mechanical assembly, and 30 minutes of testing; and each Model C requires 30 minutes of electrical assembly, 30 minutes of mechanical assembly, and 20 minutes of testing. If 2500 minutes of electrical assembly, 3500 minutes of mechanical assembly, and 2400 minutes of testing are used in one day, how many of each model will be produced?

Let

$$x = \text{the number of Model As produced}$$
$$y = \text{the number of Model Bs}$$
$$z = \text{the number of Model Cs}$$

Using a table,

	Model			
	A	**B**	**C**	**Available**
Electrical assembly	30	20	30	2500
Mechanical assembly	40	50	30	3500
Testing	30	30	20	2400

The resulting equations are

$$30x + 20y + 30z = 2500 \quad \text{or} \quad 3x + 2y + 3z = 250 \qquad (1)$$
$$40x + 50y + 30z = 3500 \quad \text{or} \quad 4x + 5y + 3z = 350 \qquad (2)$$
$$30x + 30y + 20z = 2400 \quad \text{or} \quad 3x + 3y + 2z = 240 \qquad (3)$$

Eliminating x,

$$
\begin{array}{llr}
12x + 8y + 12z = 1000 & 4(\text{equation 1}) & (4) \\
\underline{-12x - 15y - 9z = -1050} & -3(\text{equation 2}) & (5) \\
-7y + 3z = -50 & & (6)
\end{array}
$$

$$
\begin{array}{llr}
3x + 2y + 3z = 250 & & (1) \\
\underline{-3x - 3y - 2z = -240} & -(\text{equation 3}) & (7) \\
-y + z = 10 & & (8)
\end{array}
$$

Solve (6) and (8) to find y and z.

$$
\begin{array}{llr}
-7y + 3z = -50 & & (6) \\
\underline{7y - 7z = -70} & -7(\text{equation 8}) & (9) \\
-4z = -120 & & \\
z = 30 & &
\end{array}
$$

$$
\begin{array}{lr}
-y + 30 = 10 & (8) \\
y = 20 &
\end{array}
$$

Substituting $y = 20$, $z = 30$ in (1),

$$3x + 2(20) + 3(30) = 250$$
$$3x + 40 + 90 = 250$$
$$x = 40$$

Thus, in one day,

	40 Model A
	20 Model B
and	30 Model C

Perfect Foodprocessors are produced. ■

Exercise 2.1

Answers to Odd-Numbered Exercises begin on page A-5.

A In Problems 1–20 solve the given systems by using the substitution method.

1. $2x - 3y = 14$
$x = y + 10$

2. $3x - y = 7$
$4x - 5y = 2$

3. $2x + 3y = 5$
$4x + 7y = 11$

4. $2x - 4y = 16$
$3x - 5y = 19$

5. $x + 2y = 1$
$4x - y = 13$

6. $4x - 3y = 1$
$x - y = -1$

7. $x - 3y = 14$
$x - 2 = 0$

8. $x - 3y = -2$
$2x - 8y = -10$

9. $6x + 6y = 6$
$-2x - y = 3$

10. $2x + 3y = 4$
$x - y = 5$

11. $x - 3y = -1$
$2x - 3y = 4$

12. $2x + y = -3$
$4x - y = 15$

13. $x - 2y = 4$
$2x - 10y = -1$

14. $x + y = 1$
$3x + 2y = 0$

15. $5x - 4y = 0$
$x - 2y = 0$

16. $3x + 4y = -4$
$x + 8y = 0$

17. $2x + 3y = -3$
$5x - 2y = 21$

18. $3x + 4y = 1$
$4x - y = 14$

19. $2x - 7y = -28$
$7x + 2y = -53$

20. $2x - 6y = -17$
$8x + 3y = 13$

In Problems 21–40 solve the given systems by using the elimination method.

21. $x - y = 16$
$x - 3y = 2$

22. $x - y = 8$
$x + y = 6$

23. $3x + 7y = -1$
$9x - y = 19$

24. $2x - 5y = 13$
$x + 2y = 11$

25. $5x + 3y = 34$
$3x + 5y = 30$

26. $2x + 3y = -4$
$5x + 7y = -10$

27. $-6x - 3y = 6$
$2x + y = 2$

28. $3x = 9$
$x - y = 7$

29. $3x - 6y = -4$
$-2y = 2 - x$

30. $7x + 9y = 12$
$3x + 6y = -5$

31. $3x - 8y = -2$
$5x + 3y = 13$

32. $6x - 8y = 5$
$8x + 12y = 7$

33. $7x + 3y = 10$
$3x + 5y = -7$

34. $3x + 5y = 13$
$12x - 5y = 2$

35. $3x - 2y = 8$
$3x + 5y = -41$

36. $5x + 7y = -50$
$5x - 8y = 25$

37. $18x + 4y = 2$
    $-27x - 6y = -3$

38. $-3x + 4y = 12$
    $6x - 7y = -18$

39. $\frac{1}{2}x - \frac{1}{3}y = \frac{3}{2}$
    $\frac{4}{3}x + y = 5$

40. $-\frac{1}{6}x + y = \frac{1}{6}$
    $-\frac{4}{3}x + 3y = \frac{1}{3}$

B In Problems 41–46 use either the substitution method or the elimination method to solve each system. If the system is dependent, then write the solution set as an ordered pair with the parameter $t = x$ (see Example 4).

41. $4x + y = -3$
    $-8x - 2y = 6$

42. $2x + 3y = -4$
    $3x + 2y = -1$

43. $\frac{2}{3}x + \frac{1}{6}y = 5$
    $\frac{1}{3}x - \frac{1}{3}y = 2$

44. $0.8x + 0.2y = 1.2$
    $2.8x + 0.7y = 4.2$

45. $13x - 4y = -66$
    $5x + 2y = 10$

46. $17.05x - 3.24y = 22.63$
    $3.21x - 4.56y = 3.96$

C In Problems 47–60 use the elimination method to solve each system.

47. $2x + y + 6z = 3$
    $x - y + 4z = 1$
    $3x + 2y - 2z = 2$

48. $5x - 7y + 4z = 2$
    $3x + 2y - 2z = 3$
    $2x - y + 3z = 4$

49. $x + 2y + z = 4$
    $-x + y + z = -3$
    $x + 5y + 3z = 1$

50. $r - s = 2$
    $s - 2t = 0$
    $r + t = 1$

51. $b - 4c = 5$
    $-a - b + c = \frac{3}{2}$
    $a + b + 2c = 0$

52. $2x - y + z = 3$
    $-x + 2y - z = 4$
    $3x + y - 2z = -1$

53. $3x + 2y + z = 4$
    $4x - 3y + 2z = 11$
    $x - 5y - 4z = 7$

54. $2x - 5y + 3z = 2$
    $3x + 2y + z = 10$
    $x + y - z = 6$

55. $x + y = 5$
    $x + y + z = 3$
    $y + 2z = 2$

56. $x + y = 2$
    $x + z = 5$
    $y + z = 5$

57. $5x - 3y = -1$
    $3x + z = 1$
    $2y + z = 2$

58. $4x - 3y + z = 4$
    $3y + 4z = 2$
    $2x + 4y = -10$

59. $x + y + 2z = 4$
    $x - y + z = 1$
    $-x - 2y + z = -2$

60. $x + y + 3z - 9 = 0$
    $3x + 2y - z - 6 = 0$
    $2x - y + z - 5 = 0$

APPLICATIONS 61. **Mixture** Sweet Delight Candies, Inc., sells boxes of candy consisting of creams and caramels. Each box sells for $4.00 and holds 50 pieces of candy (all pieces are the same size). If the caramels cost $0.05 to produce and the creams cost $0.10 to produce, how many caramels and creams should be in each box for no profit or loss? Would you increase or decrease the number of caramels in order to obtain a profit?

62. Mixture The manager of Nutt's Nuts regularly sells cashews for $1.50 per pound, pecans for $1.80 per pound, and peanuts for $0.80 per pound. How many pounds of cashews and pecans should be mixed with 40 pounds of peanuts to obtain a mixture of 100 pounds that will sell at $1.25 a pound so that the profit or loss is unchanged?

63. Investment Mr. Nicholson has just retired and needs $6000 per year in income to live on. He has $50,000 to invest and can invest in AA bonds at 15% annual interest or in Savings and Loan Certificates at 7% interest per year. How much money should be invested in each so that he realizes exactly $6000 in income per year?

64. Investment Mr. Nicholson finds after 2 years that because of inflation he now needs $7000 per year to live on. How should he transfer his funds to achieve this amount? (Use the data from Problem 63.)

65. Joan has $1.65 in her piggy bank. She knows she only placed nickels and quarters in the bank and she knows that, in all, she put 13 coins in the bank. Can she find out how many nickels she has without breaking her bank?

66. Mixture A coffee manufacturer wants to market a new blend of coffee that will cost $2.90 per pound by mixing $2.75 per pound coffee and $3 per pound coffee. What amounts of the $2.75 per pound coffee and $3 per pound coffee should be blended to obtain the desired mixture? [*Hint:* Assume the total weight of the desired blend is 100 pounds.]

67. Mixture One solution is 15% acid and another is 5% acid. How many cubic centimeters of each should be mixed to obtain 100 cc of a solution that is 8% acid?

68. Investment A bank loaned $10,000, some at an annual rate of 8% and some at an annual rate of 18%. If the income from these loans was $1000, how much was loaned at 8%? How much at 18%?

69. Receipts The Star Theater wants to know whether the majority of its patrons are adults or children. During a week in July, 5200 tickets were sold and the receipts totaled $11,875. The adult admission is $2.75 and the children's admission is $1.50. How many adult patrons were there?

70. Mixture A candy manufacturer sells a candy mix for $1.10 a pound. It is composed of two candies, one worth $0.90 a pound, the other worth $1.50 a pound. How many pounds of each candy is in 60 pounds of mixture?

71. Finance Two investments are made totaling $50,000. In 1 year the first investment yields a profit of 10%, whereas the second yields a profit of 12%. Total profit for this year is $5250. Find the amount initially put into each investment.

72. Finance Laura invested part of her money at 8% and the rest at 10%. The income from both investments totaled $3200. If she interchanged her investments, her income would have totaled $2800. How much did she have in each investment?

73. Agriculture The Smith farm has 1000 acres of land to be used to raise corn and soybeans. The cost of cultivating the corn and the soybeans is $62 and $44 per acre, respectively. Mr. Smith has a budget of $45,800 to use for cultivating these crops. If Mr. Smith wishes to use all the land and all the money budgeted, how many acres of each crop should he plant?

74. Diet Preparation A farmer prepares feed for livestock by combining two grains. Each unit of the first grain contains 2 units of protein and 5 units of iron, while each unit of the second grain contains 4 units of protein and 1 unit of iron. Determine the number of units of each kind of grain the farmer needs to feed each animal daily if each animal must have 10 units of protein and 16 units of iron each day?

75. **Purchasing** David owns a restaurant and wants to order 200 dinner sets. One design costs $20 per set and the other costs $15 per set. He has $3200 with which to buy the sets. How many of each type of set should he buy if he is to use all the money and acquire 200 sets?

76. **Air Pollution** A chemical company produces two products, A and B. The manufacturing process emits 2 ft³ of carbon monoxide and 4 ft³ of sulfur dioxide per unit of product A, and 4 ft³ of carbon monoxide and 6 ft³ of sulfur dioxide per unit of product B. To comply with federal pollution regulations they may emit a maximum of 140 ft³ of carbon monoxide and 240 ft³ of sulfur dioxide per day. Find production levels of A and B that maximize production while complying with emission regulations.

77. **Diet Preparation** A hospital dietician is planning a meal consisting of three foods whose ingredients are summarized as follows:

	Units of		
	Food I	**Food II**	**Food III**
Units of protein	10	5	15
Units of carbohydrates	3	6	3
Units of iron	4	4	6

Determine the number of units of each food needed to create a meal containing 100 units of protein, 50 units of carbohydrates, and 50 units of iron.

78. **Production** A luggage manufacturer produces three types of luggage; economy, standard, and deluxe. The company produces 1000 pieces of luggage at a cost of $20, $25, and $30 for the economy, standard, and deluxe luggage, respectively. The manufacturer has a budget of $20,700. Each economy luggage requires 6 hours of labor, each standard luggage requires 10 hours of labor, and each deluxe model requires 20 hours of labor. The manufacturer has a maximum of 6800 hours of labor available. If the manufacturer sells all the luggage, uses the entire budget and all the available labor, how many of each luggage should be produced?

79. **Packaging** A recreation center wants to purchase albums to be used in the center. There is no requirement as to the artists. The only requirement is that they purchase 40 rock albums, 32 western albums, and 14 blues albums. There are three different shipping packages offered by the record company. They are an *assorted* carton, containing 2 rock albums, 4 western albums, and 1 blues album; a *mixed* carton containing 4 rock and 2 western albums; and a *single* carton containing 2 blues albums. What combination of these packages is needed to fill the center's order?

80. **Mixture** Suppose that a store has three sizes of cans of nuts. The *large* size contains 2 pounds of peanuts and 1 pound of cashews. The *mammoth* size contains 1 pound of walnuts, 6 pounds of peanuts, and 2 pounds of cashews. The *giant* size contains 1 pound of walnuts, 4 pounds of peanuts, and 2 pounds of cashews. Suppose that the store receives an order for 5 pounds of walnuts, 26 pounds of peanuts, and 12 pounds of cashews. How can it fill this order with the given sizes of cans?

81. **Mixture** Suppose that the store in Problem 80 receives a new order for 6 pounds of walnuts, 34 pounds of peanuts, and 15 pounds of cashews. How can this order be filled with the given cans?

82. Mixture Sally's Girl Scout troop is selling cookies for the Christmas season. There are three different kinds of cookies in three different containers: *bags* that hold 1 dozen chocolate chip and 1 dozen oatmeal; *gift boxes* that hold 2 dozen chocolate chip, 1 dozen mints, and 1 dozen oatmeal; and *cookie tins* that hold 3 dozen mints and 2 dozen chocolate chip. Sally's mother is having a Christmas party and wants 6 dozen oatmeal cookies, 10 dozen mints, and 14 dozen chocolate chip cookies. How can Sally fill her mother's order?

2.2

SYSTEMS OF LINEAR EQUATIONS AND AUGMENTED MATRICES

MATRICES
MATRIX REPRESENTATION OF A SYSTEM OF LINEAR EQUATIONS
ROW OPERATIONS

If we try to use the elimination by addition method to solve systems of equations with a large number of equations and unknowns, we will find that it is very tedious and prone to errors. (Try to solve a system with seven equations and seven unknowns and you will understand why.) We need a procedure that reduces our work and at the same time gives a more efficient approach to solving systems of equations.

We can generalize this method, and make it more efficient, by the introduction of a mathematical form called a *matrix*. The advantage of using matrices to solve a system of equations will become clear later.

MATRICES
Let us begin with a definition of a matrix.

A *matrix* is a rectangular array of numbers enclosed by a pair of brackets

Examples of matrices are

$$(a) \begin{bmatrix} 1 & 3 \\ -2 & 4 \end{bmatrix} \quad (b) \begin{bmatrix} 4 & 1 & -3 \\ 2 & 1 & 2 \end{bmatrix} \quad (c) \begin{bmatrix} 4 & 3 \end{bmatrix}$$

The numbers in a matrix are called *elements* of the matrix. The matrix (a) above has two *rows* and two *columns;* the matrix (b) has two rows and three columns; the matrix (c) has one row and two columns. For example, in matrix (a), the element 4 is in row 2 column 2; in matrix (b), the element -3 is in row 1, column 3. The *diagonal entries* of a matrix are those elements for which the row number and column number are equal. Thus, in matrix (a), 1 and 4 are the diagonal entries.

MATRIX REPRESENTATION OF A SYSTEM OF LINEAR EQUATIONS

Consider the following two systems of two linear equations in two unknowns.

$$\begin{array}{c} x + 4y = 14 \\ 3x - 2y = 0 \end{array} \quad \text{and} \quad \begin{array}{c} u + 4v = 14 \\ 3u - 2v = 0 \end{array}$$

We make the observation that, except for the symbols used to represent the unknowns, these two systems are identical. That is, it is the coefficients of the unknowns and the numbers that appear to the right of the equal sign that distinguish one system from another. As a result, we can dispense altogether with the letters used to symbolize the unknowns—provided we have some means of keeping track of them. A matrix serves us well in this regard.

The system

$$x + 4y = 14$$
$$3x - 2y = 0$$

can be represented by the matrix

	Column One	Column Two	Column Three
Row One	1	4	14
Row Two	3	−2	0

Here it is understood that column one contains the coefficients of the unknown x, column two contains the coefficients of the unknown y, and column three contains the numbers to the right of the equal sign. Each row of the matrix represents an equation of the system. Although not required, it has become customary to place a vertical bar in the matrix as a reminder of the equal sign.

The matrix used to represent a system of linear equations is called the *augmented matrix* of the system.

Example 1

System of Linear Equations	Augmented Matrix
(a) $\begin{aligned} 5x - 2y &= 5 \\ 2x - y &= -4 \end{aligned}$	$\begin{bmatrix} 5 & -2 & 5 \\ 2 & -1 & -4 \end{bmatrix}$
(b) $\begin{aligned} 3x + 4y + 3z &= 10 \\ x + y - z &= 1 \\ x + 6y &= 4 \end{aligned}$	$\begin{bmatrix} 3 & 4 & 3 & 10 \\ 1 & 1 & -1 & 1 \\ 1 & 6 & 0 & 4 \end{bmatrix}$
(c) $\begin{aligned} x + 3y + 4z &= 2 \\ -y - 3z &= 0 \\ 2y + 5z &= -6 \end{aligned}$	$\begin{bmatrix} 1 & 3 & 4 & 2 \\ 0 & -1 & -3 & 0 \\ 0 & 2 & 5 & -6 \end{bmatrix}$
(d) $\begin{aligned} x &= 2 \\ -y - 3z &= 0 \\ 4y + 2z &= -6 \end{aligned}$	$\begin{bmatrix} 1 & 0 & 0 & 2 \\ 0 & -1 & -3 & 0 \\ 0 & 4 & 2 & -6 \end{bmatrix}$

(e)
$$\begin{aligned} x &= 1 \\ y - 3z &= 0 \\ 2z &= -6 \end{aligned}$$
$$\left[\begin{array}{ccc|c} 1 & 0 & 0 & 1 \\ 0 & 1 & -3 & 0 \\ 0 & 0 & 2 & -6 \end{array}\right]$$

(f)
$$\begin{aligned} x &= 2 \\ y &= 0 \\ z &= -4 \end{aligned}$$
$$\left[\begin{array}{ccc|c} 1 & 0 & 0 & 2 \\ 0 & 1 & 0 & 0 \\ 0 & 0 & 1 & -4 \end{array}\right]$$ ∎

Given an augmented matrix, we can write the corresponding system of equations.

Example 2

Write a system of linear equations for each augmented matrix.

(a) $\left[\begin{array}{cc|c} 4 & -1 & 6 \\ 3 & 1 & 2 \end{array}\right]$ (b) $\left[\begin{array}{ccc|c} 3 & 2 & 5 & 3 \\ -1 & 4 & 0 & 2 \\ 5 & 1 & 6 & 1 \end{array}\right]$

Solution

(a) If we use x and y as unknowns, the system of equations is:

$$\begin{aligned} 4x - y &= 6 \\ 3x + y &= 2 \end{aligned}$$

(b) If we use x_1, x_2, and x_3 as unknowns, the systems of equations is:

$$\begin{aligned} 3x_1 + 2x_2 + 5x_3 &= 3 \\ -x_1 + 4x_2 &= 2 \\ 5x_1 + x_2 + 6x_3 &= 1 \end{aligned}$$ ∎

ROW OPERATIONS

We begin by going through the solution of a system of two linear equations in two unknowns:

$$\begin{aligned} x + 4y &= 14 \\ 3x - 2y &= 0 \end{aligned} \tag{1}$$

To find a solution, we multiply the first equation by 3, obtaining $3x + 12y = 42$, and then subtract it from the second:

$$\begin{array}{r} 3x - 2y = 0 \\ 3x + 12y = 42 \\ \hline -14y = -42 \end{array}$$

The original system of equations (1) may now be written as the equivalent system

$$\begin{aligned} x + 4y &= 14 \\ -14y &= -42 \end{aligned}$$

Dividing the second equation by -14, we get

$$x + 4y = 14$$
$$y = 3$$

To find x, we multiply the second equation by -4 and add it to the first. The result is

$$x = 2$$
$$y = 3 \qquad (2)$$

This system of equations has the obvious solution $x = 2$, $y = 3$ and is equivalent to the original system (1), so that the solution of the original system (1) is also $x = 2$, $y = 3$.

We obtained the final system (2) from the original system (1) by a series of operations. Because of its simplicity, the original system could have been solved more quickly by either the substitution method or the elimination by addition method (refer to Section 2.1), but we chose the above operations because the *pattern* of the solution shown provides another method for solving a system of equations. The advantages of this third method are:

1. It is algorithmic in character; that is, it consists of repetitive steps so that it can be programmed on a computer.
2. It works on any system of linear equations.

Let's repeat the steps we took to solve this system of equations, but now we will manipulate the augmented matrix instead of the equations. For convenience, the equations are listed next to the augmented matrix.

$$\begin{bmatrix} 1 & 4 & | & 14 \\ 3 & -2 & | & 0 \end{bmatrix} \qquad \begin{matrix} x + 4y = 14 \\ 3x - 2y = 0 \end{matrix}$$

Multiplying the first *equation* by 3 and subtracting it from the second *equation* corresponds to multiplying the first *row* of the matrix by -3 and adding the result to the second *row*. We replace the second row by the result of this. The result is the matrix

$$\begin{bmatrix} 1 & 4 & | & 14 \\ 0 & -14 & | & -42 \end{bmatrix} \qquad \begin{matrix} x + 4y = 14 \\ -14y = -42 \end{matrix}$$

Next, divide the second row of this last matrix by -14 to obtain

$$\begin{bmatrix} 1 & 4 & | & 14 \\ 0 & 1 & | & 3 \end{bmatrix} \qquad \begin{matrix} x + 4y = 14 \\ y = 3 \end{matrix}$$

Finally, multiply the second row of this matrix by -4 and add it to the first row to obtain

$$\begin{bmatrix} 1 & 0 & | & 2 \\ 0 & 1 & | & 3 \end{bmatrix} \qquad \begin{matrix} x = 2 \\ y = 3 \end{matrix}$$

When manipulations such as the foregoing are performed on a matrix, they are called *elementary row operations.* We now list the three basic types of row operations.

Elementary Row Operations

1. The interchange of any 2 rows of a matrix.
2. The replacement of any row of a matrix by a nonzero multiple of that same row.
3. The replacement of any row of a matrix by the sum of that row and a multiple of some other row.

An example of each elementary row operation is given below.

Example 3

$$A = \begin{bmatrix} 3 & 4 & -3 \\ 7 & -\frac{1}{2} & 0 \end{bmatrix}$$

1. The matrix obtained by interchanging the first and second rows of A is

$$\begin{bmatrix} 7 & -\frac{1}{2} & 0 \\ 3 & 4 & -3 \end{bmatrix}$$

We denote this operation by writing $R_1 = r_2, R_2 = r_1$, where r_1 and r_2 denotes the "old" rows 1 and 2, and R_1 and R_2 denotes the "new" rows 1 and 2.

2. The matrix obtained by multiplying row 2 of A by 5 is

$$\begin{bmatrix} 3 & 4 & -3 \\ 35 & -\frac{5}{2} & 0 \end{bmatrix}$$

We denote this operation by writing $R_2 = 5r_2$, where r_2 denotes the "old" row 2 and R_2 denotes the "new" row 2.

3. The matrix obtained from A by adding 3 times row 1 to row 2 is

$$\begin{bmatrix} 3 & 4 & -3 \\ 3 \cdot 3 + 7 & 3 \cdot 4 + -\frac{1}{2} & 3 \cdot (-3) + 0 \end{bmatrix} = \begin{bmatrix} 3 & 4 & -3 \\ 16 & \frac{23}{2} & -9 \end{bmatrix}$$

We denote this operation by writing $R_2 = 3r_1 + r_2$. ∎

Let's see how row operations are used to solve a system of equations.

Example 4
Solve the system of equations:

$$\begin{aligned} x - y &= 2 \\ 2x - 3y &= 2 \end{aligned} \tag{1}$$

Solution
First, we write the augmented matrix:

$$\begin{bmatrix} 1 & -1 & | & 2 \\ 2 & -3 & | & 2 \end{bmatrix} \qquad \begin{aligned} x - y &= 2 \\ 2x - 3y &= 2 \end{aligned} \tag{2}$$

Our objective is to use the elementary row operations to try to transform (2) into the form.

$$\left[\begin{array}{cc|c} 1 & 0 & h \\ 0 & 1 & k \end{array}\right] \qquad (3)$$

Where h and k are real numbers. Once we are able to obtain (3) the solution is obvious since the system of equations corresponding to (3) is

$$\begin{array}{cc} 1x + 0y = h & x = h \\ 0x + 1y = k & \text{or} & y = k \end{array}$$

With this objective in mind, let's see if we can transform (2) into (3). Perform the row operation

$$R_2 = -2r_1 + r_2$$

(This has the effect of leaving row 1 fixed and getting a 0 in row 2, column 1.)

$$\left[\begin{array}{cc|c} 1 & -1 & 2 \\ 0 & -1 & -2 \end{array}\right] \qquad \begin{array}{c} x - y = 2 \\ -y = -2 \end{array}$$

Perform the row operation

$$R_2 = -r_2$$

(This has the effect of getting a 1 in row 2, column 2.)

$$\left[\begin{array}{cc|c} 1 & -1 & 2 \\ 0 & 1 & 2 \end{array}\right] \qquad \begin{array}{c} x - y = 2 \\ y = 2 \end{array}$$

Perform the row operation

$$R_1 = r_2 + r_1$$

(This has the effect of leaving row 2 fixed and getting a 0 in row 1, column 2.)

$$\left[\begin{array}{cc|c} 1 & 0 & 4 \\ 0 & 1 & 2 \end{array}\right] \qquad \begin{array}{c} x = 4 \\ y = 2 \end{array} \qquad (4)$$

We stop here since, from the augmented matrix (4), we obtain the system

$$\begin{array}{cc} x + 0y = 4 & x = 4 \\ 0x + y = 2 & \text{or} & y = 2 \end{array}$$

which gives the solution $x = 4$, $y = 2$. ■

The pattern of steps used in the previous examples can be followed to solve any system of linear equations. Let's list those steps.

Solving a System of Equations Using Matrices

Step 1. Write the augmented matrix corresponding to the system. Remember that the constants must be to the right of the equal sign.

Step 2. Perform row operations to get the entry 1 in row 1, column 1.

Step 3. Perform row operations that leave the entry 1 obtained in Step 2 undisturbed while getting 0's in the rest of column 1.

Step 4. Perform row operations to get a 1 in row 2, column 2 (if possible) without disturbing column 1. If this is not possible, obtain a 1 in the row 2, column 3 position.

Step 5. Perform row operations to get 0's in the rest of column 2 without disturbing column 1 or the entry 1 obtained in Step 4.

Step 6. Continue in this way up to and including the last row of the matrix if possible.

Here is an example of three linear equations in three unknowns.

Example 5

Solve the system of equations:

$$x + y + z = 6$$
$$3x + 2y - z = 4$$
$$3x + y + 2z = 11$$

Solution

Step 1. The augmented matrix corresponding to the system is

$$\left[\begin{array}{ccc|c} 1 & 1 & 1 & 6 \\ 3 & 2 & -1 & 4 \\ 3 & 1 & 2 & 11 \end{array}\right]$$

Step 2. Since a 1 appears in row 1, column 1, we can skip to Step 3.

Step 3. Perform the row operations*

$$R_2 = -3r_1 + r_2 \quad \text{and} \quad R_3 = -3r_1 + r_3$$

Notice that these operations leave row 1 undisturbed, while getting 0's in the rest of column 1:

$$\left[\begin{array}{ccc|c} 1 & 1 & 1 & 6 \\ 0 & -1 & -4 & -14 \\ 0 & -2 & -1 & -7 \end{array}\right]$$

---

* You should convince yourself that doing these simultaneously is the same as doing the first followed by the second.

Step 4. To get a 1 in row 2, column 2, we use

$$R_2 = (-1)r_2$$

Notice that column 1 remains undisturbed:

$$\left[\begin{array}{ccc|c} 1 & 1 & 1 & 6 \\ 0 & 1 & 4 & 14 \\ 0 & -2 & -1 & -7 \end{array}\right]$$

Step 5. To get 0's in the rest of column 2, we use

$$R_1 = -r_2 + r_1$$
$$R_3 = 2r_2 + r_3$$

Notice that row 2 is left undisturbed:

$$\left[\begin{array}{ccc|c} 1 & 0 & -3 & -8 \\ 0 & 1 & 4 & 14 \\ 0 & 0 & 7 & 21 \end{array}\right]$$

Step 6. Continuing, we seek a 1 in row 3, column 3, so we use

$$R_3 = \tfrac{1}{7}r_3$$

Notice that this is the only choice available to us, since any other choice would disturb the 0's and 1's already obtained:

$$\left[\begin{array}{ccc|c} 1 & 0 & -3 & -8 \\ 0 & 1 & 4 & 14 \\ 0 & 0 & 1 & 3 \end{array}\right]$$

To get 0's in the rest of column 3, we use

$$R_1 = 3r_3 + r_1$$
$$R_2 = -4r_3 + r_2$$

The result is

$$\left[\begin{array}{ccc|c} 1 & 0 & 0 & 1 \\ 0 & 1 & 0 & 2 \\ 0 & 0 & 1 & 3 \end{array}\right]$$

The solution of the system is $x = 1$, $y = 2$, $z = 3$.　■

The method outlined here also reveals systems that have no solution or an infinite number of solutions.

Example 6
Solve the system

$$3x - 6y = 4$$
$$6x - 12y = 5$$

The augmented matrix representing this system is

$$\begin{bmatrix} 3 & -6 & | & 4 \\ 6 & -12 & | & 5 \end{bmatrix}$$

To get a 1 in row 1, column 1 we use $R_1 = \frac{1}{3}r_1$:

$$\begin{bmatrix} 1 & -2 & | & \frac{4}{3} \\ 6 & -12 & | & 5 \end{bmatrix}$$

To get 0 in row 2, column 1, use $R_2 = -6r_1 + r_2$:

$$\begin{bmatrix} 1 & -2 & | & \frac{4}{3} \\ 0 & 0 & | & -3 \end{bmatrix}$$

Let's stop here to look at the actual system of equations:

$$x - 2y = \frac{4}{3}$$
$$0x + 0y = -3$$

The second equation can never be true—no matter what the choice of x and y. Hence, there are no numbers x and y that can obey both equations. That is, the system has no solution. ∎

Example 7
Solve the system

$$2x - 3y = 5$$
$$4x - 6y = 10$$

Solution
The augmented matrix representing this system is

$$\begin{bmatrix} 2 & -3 & | & 5 \\ 4 & -6 & | & 10 \end{bmatrix}$$

To get a 1 in row 1, column 1, we use $R_1 = \frac{1}{2}r_1$:

$$\begin{bmatrix} 1 & -\frac{3}{2} & | & \frac{5}{2} \\ 4 & -6 & | & 10 \end{bmatrix}$$

To get a 0 in column 1, row 2, we use $R_2 = -4r_1 + r_2$:

$$\begin{bmatrix} 1 & -\frac{3}{2} & | & \frac{5}{2} \\ 0 & 0 & | & 0 \end{bmatrix}$$

The system of equations looks like

$$x - \frac{3}{2}y = \frac{5}{2}$$
$$0x + 0y = 0$$

The second equation is true for any choice of x and y. Hence, all numbers x and y that obey the first equation are solutions of the system. Since any point on the line $x - \frac{3}{2}y = \frac{5}{2}$ is a solution, there are an infinite number of solutions.

We can obtain the infinite number of solutions by applying the method used in section 2.1; that is, by introducing a parameter. Look at the equation $x - \frac{3}{2}y = \frac{5}{2}$; we can solve for either variable in terms of the other. We choose to solve for x in terms of y to get $x = \frac{3}{2}y + \frac{5}{2}$.

We introduce the parameter t, a real number and we let $y = t$, obtaining,

$$x = \frac{3}{2}t + \frac{5}{2}$$
$$y = t$$

which is the solution to the original system. Some solutions are

If $t = 0$, then $x = \frac{5}{2}$ and $y = 0$. Thus $x = \frac{5}{2}$ and $y = 0$ is a solution.

If $t = 1$, then $x = 4$ and $y = 1$. Thus $x = 4$ and $y = 1$ is a solution.

If $t = 5$, then $x = 10$ and $y = 5$. Thus $x = 10$ and $y = 5$ is a solution.

If $t = -3$, then $x = -2$ and $y = -3$. Thus $x = -2$ and $y = -3$ is a solution.

And so on for any choice of t. ∎

We close with a capsule summary of what to expect from any system of two equations with two unknowns.

SUMMARY

After completing the steps outlined earlier, one of the following matrices will result for a system of two linear equations with two unknowns:

$$\begin{bmatrix} 1 & 0 & | & c \\ 0 & 1 & | & d \end{bmatrix}$$ **Unique solution:** $x = c,\ y = d$

$$\begin{bmatrix} a & b & | & c \\ 0 & 0 & | & 0 \end{bmatrix}$$ **Infinite number of solutions:** $ax + by = c$

$$\begin{bmatrix} a & b & | & c \\ 0 & 0 & | & \text{nonzero number} \end{bmatrix}$$ **No solution**

Exercise 2.2   *Answers to Odd-Numbered Problems begin on page A-6.*

A   In Problems 1–10 write the augmented matrix of each system.

1. $2x - 3y = 5$ **2.** $4x + y = 5$
$\quad\ x - y = 3$ $\quad\ 2x + y = 5$

3. $2x + y + 6 = 0$ **4.** $\quad\quad x - y = -3$
$\quad\quad x + y = -1$ $\quad\ 4x - y + 2 = 0$

5. $2x - y - z = 0$ **6.** $\quad x + y + z = 3$
$\quad\ x - y - z = 1$ $\quad\ 2x + z = 0$
$\quad\ 3x - y = 2$ $\quad\ 3x - y - z = 1$

7. $\quad 2u - 3v + w - 7 = 0$ **8.** $5r - 3s + 6t + 1 = 0$
$\quad\quad u + v - w = 1$ $\quad\quad -r - s + t = 1$
$\quad 2u + 2v - 3w + 4 = 0$ $\quad\ 2r + 3s + 5 = 0$

9. $4x_1 - x_2 + 2x_3 - x_4 = 4$
    $x_1 + x_2 + 6 = 0$
    $2x_2 - x_3 + x_4 = 5$

10. $3x_1 - 5 = 0$
    $x_1 - x_2 + x_3 = 6$
    $2x_2 + x_3 + 4 = 0$

In Problems 11–14 write a system of linear equations for each augmented matrix.

11. $\begin{bmatrix} 3 & -1 & | & 4 \\ 2 & 5 & | & 2 \end{bmatrix}$

12. $\begin{bmatrix} 4 & 1 & | & 7 \\ -2 & 5 & | & 5 \end{bmatrix}$

13. $\begin{bmatrix} 0 & 5 & 2 & | & 3 \\ 0 & 1 & 3 & | & 1 \\ 1 & 2 & 4 & | & 6 \end{bmatrix}$

14. $\begin{bmatrix} 0 & 2 & 4 & | & 9 \\ 1 & 3 & 6 & | & 3 \\ 0 & 1 & 5 & | & 1 \end{bmatrix}$

In Problems 15–18 state the row operation used to transform the matrix on the left to the one on the right. Describe each row operation using r_1 and r_2 for the old rows and R_1 and R_2 for the new rows.

15. $\begin{bmatrix} 3 & 2 & 1 \\ 2 & 1 & 0 \end{bmatrix} \quad \begin{bmatrix} 2 & 1 & 0 \\ 3 & 2 & 1 \end{bmatrix}$

16. $\begin{bmatrix} 3 & 2 & 1 \\ 2 & 1 & 0 \end{bmatrix} \quad \begin{bmatrix} 1 & 1 & 1 \\ 2 & 1 & 0 \end{bmatrix}$

17. $\begin{bmatrix} 3 & 2 & 1 \\ 2 & 1 & 0 \end{bmatrix} \quad \begin{bmatrix} 12 & 8 & 4 \\ 2 & 1 & 0 \end{bmatrix}$

18. $\begin{bmatrix} 3 & 2 & 1 \\ 2 & 1 & 0 \end{bmatrix} \quad \begin{bmatrix} 3 & 2 & 1 \\ 4 & 2 & 0 \end{bmatrix}$

B   In Problems 19–24 perform the indicated row operation on the matrix

$$\begin{bmatrix} 3 & 6 & 9 \\ 0 & 1 & 4 \\ 1 & 0 & 2 \end{bmatrix}$$

19. $R_1 = r_3, \quad R_3 = r_1$

20. $R_1 = \tfrac{1}{3}r_1$

21. $R_1 = (-3)r_3 + r_1$

22. $R_1 = r_2 + r_1$

23. $R_2 = \tfrac{1}{3}r_1 + r_2$

24. $R_3 = -\tfrac{1}{3}r_1 + r_3$

In Problems 25–52 solve each system of equations using the method of row operations.

25. $x + y = 6$
    $2x - y = 0$

26. $x - y = 2$
    $2x + y = 1$

27. $2x + y = 5$
    $x - y = 1$

28. $3x + 2y = 7$
    $x + y = 3$

29. $2x + 3y = 7$
    $3x - y = 5$

30. $2x - 3y = 5$
    $3x + y = 2$

31. $5x - 7y = 31$
    $3x + 2y = 0$

32. $2x + 8y = 17$
    $3x - y = 1$

33. $2x - 3y = 0$
    $4x + 9y = 5$

34. $3x - 4y = 3$
    $6x + 2y = 1$

35. $4x - 3y = 4$
    $2x + 6y = 7$

36. $3x - 5y = 3$
    $6x + 10y = 10$

37. $\tfrac{1}{2}x + \tfrac{1}{3}y = 2$
    $x + y = 5$

38. $x - \tfrac{1}{4}y = 0$
    $\tfrac{1}{2}x + \tfrac{1}{2}y = \tfrac{5}{2}$

39. $x + y = 1$
    $3x - 2y = \tfrac{4}{3}$

40. $4x - y = \tfrac{11}{4}$
    $3x + y = \tfrac{5}{2}$

41. $2x + y + z = 6$
    $x - y - z = -3$
    $3x + y + 2z = 7$

42. $x + y + z = 5$
    $2x - y + z = 2$
    $x + 2y - z = 3$

43.
$$x + y - z = -2$$
$$3x + y + z = 0$$
$$2x - y + 2z = 1$$

44.
$$2x - y - z = -5$$
$$x + y + z = 2$$
$$x + 2y + 2z = 5$$

45.
$$2x + y - z = 2$$
$$x + 3y + 2z = 1$$
$$x + y + z = 2$$

46.
$$2x + 2y + z = 6$$
$$x - y - z = -2$$
$$x - 2y - 2z = -5$$

47.
$$x + y - z = 0$$
$$2x + 4y - 4z = -1$$
$$2x + y + z = 2$$

48.
$$x + y - z = 0$$
$$4x + 2y - 4z = 0$$
$$x + 2y + z = 0$$

49.
$$3x + y - z = \tfrac{2}{3}$$
$$2x - y + z = 1$$
$$4x + 2y = \tfrac{8}{3}$$

50.
$$x + y = 1$$
$$2x - y + z = 1$$
$$x + 2y + z = \tfrac{8}{3}$$

51.
$$x_1 + x_2 + x_3 + x_4 = 4$$
$$2x_1 - x_2 + x_3 = 0$$
$$3x_1 + 2x_2 + x_3 - x_4 = 6$$
$$x_1 - 2x_2 - 2x_3 + 2x_4 = -1$$

52.
$$x_1 + x_2 + x_3 + x_4 = 4$$
$$-x_1 + 2x_2 + x_3 = 0$$
$$2x_1 + 3x_2 + x_3 - x_4 = 6$$
$$-2x_1 + x_2 - 2x_3 + 2x_4 = -1$$

C In Problems 53–64 discuss each system of equations. Determine whether the system has a unique solution, no solution, or infinitely many solutions. Use matrix techniques.

53.
$$x - y = 5$$
$$2x - 2y = 6$$

54.
$$4x + y = 5$$
$$8x + 2y = 10$$

55.
$$2x - 3y = 6$$
$$4x - 6y = 12$$

56.
$$2x - 3y = 6$$
$$4x - 6y = 8$$

57.
$$5x - 6y = 1$$
$$-10x + 12y = 0$$

58.
$$3x + 4y = 7$$
$$x - y = 2$$

59.
$$2x + 3y = 5$$
$$4x + 4y = 8$$

60.
$$2x - y = 0$$
$$4x - 2y = 0$$

61.
$$2x - y - z = 0$$
$$x - y - z = 1$$
$$3x - y - z = 2$$

62.
$$x + y + z = 3$$
$$2x + y + z = 0$$
$$3x + y + z = 1$$

63.
$$2x_1 - x_2 + x_3 = 6$$
$$3x_1 - x_2 + x_3 = 6$$
$$4x_1 - 2x_2 + 2x_3 = 12$$

64.
$$x_1 - x_2 + x_3 = 2$$
$$2x_1 - 3x_2 + x_3 = 0$$
$$3x_1 - 3x_2 + 3x_3 = 6$$

APPLICATIONS **65. Wage Scheduling** MultiCraft Inc. has 2000 dockworkers and 3000 office workers with a total pay of $46,000 per hour. MultiCraft decided to add an additional amount to the hourly payroll in order to give the dockworkers a 7% per hour raise and office workers a 6% per hour raise. If the amount added to the hourly payroll was $2980, what was the average hourly wage for the dockworkers and office workers?

66. Pollution Three industries together produce a pollution count of 600. Industry I produces twice as much pollution as Industry II. If industry I would cut back its pollution by 50%, then the total count would be 500. Find the pollution count for each industry.

67. Investment An amount of $5000 is put into three investments at rates of 6%, 7%, and 8% per annum, respectively. The total annual income is $358. The income from the first two investments is $70 more than the income from the third investment. Find the amount of each investment.

68. Investment An amount of $6500 is placed in three investments at rates of 6%, 8%, and 9% per annum, respectively. The total annual income is $480. If the income from the third investment is $60 more than the income from the second investment, find the amount of each investment.

69. Production A citrus company completes the preparation of its products by cleaning, filling, and labeling bottles. Each case of orange juice requires 10 minutes in the cleaning machine, 4 minutes in the filling machine, and 2 minutes in the labeling machine. For each case of tomato juice, the times are 12 minutes of cleaning, 4 minutes for filling, and 1 minute to label. Pineapple juice requires 9 minutes of cleaning, 6 minutes of filling, and 1 minute of labeling per case. If the company runs the cleaning machine for 398 minutes, the filling machine for 164 minutes, and the labeling machine for 58 minutes, how many cases of each type of juice are prepared?

70. Production The manufacture of an automobile requires painting, drying, and polishing. The Rome Motor Company produces three types of cars: the Centurion, the Tribune, and the Senator. Each Centurion requires 8 hours for painting, 2 hours for drying, and 1 hour for polishing. A Tribune needs 10 hours for painting, 3 hours for drying, and 2 hours for polishing. It takes 16 hours of painting, 5 hours of drying, and 3 hours of polishing to prepare a Senator. If the company uses 240 hours for painting, 69 hours for drying, and 41 hours for polishing in a given month, how many of each type of car are produced?

2.3

SOLVING SYSTEMS OF *m* LINEAR EQUATIONS IN *n* UNKNOWNS

REDUCED ROW-ECHELON FORM
INFINITE NUMBER OF SOLUTIONS
SUMMARY
APPLICATION

We learned in the previous section that a system of two linear equations in two unknowns will have either one solution, no solution, or infinitely many solutions. As it turns out, no matter how many equations are in a system of linear equations and no matter how many unknowns a system has, these are still the only three possibilities that arise.

Thus, for example, the system of three linear equations in four unknowns

$$x_1 + 3x_2 + 5x_3 + x_4 = 2$$
$$2x_1 + 3x_2 + 4x_3 + 2x_4 = 1$$
$$x_1 + 2x_2 + 3x_3 + x_4 = 1$$

will have either one solution, no solution, or infinitely many solutions.

In general, a system of m linear equations in the n unknowns $x_1, x_2, \ldots, x_n$ is of the form

$$a_{11}x_1 + a_{12}x_2 + \cdots + a_{1n}x_n = b_1$$
$$a_{21}x_1 + a_{22}x_2 + \cdots + a_{2n}x_n = b_2$$
$$a_{31}x_1 + a_{32}x_2 + \cdots + a_{3n}x_n = b_3$$

$$a_{i1}x_1 + a_{i2}x_2 + \cdots + a_{in}x_n = b_i$$

$$a_{m1}x_1 + a_{m2}x_2 + \cdots + a_{mn}x_n = b_m$$

where a_{ij} and b_i are real numbers, $i = 1, 2, \ldots, m, j = 1, 2, \ldots, n$.

This is a system of *linear* equations because the unknowns $x_1, x_2, \ldots, x_n$ all appear to the first power and there are no products of unknowns. Finally, if we count the number of equations, we conclude that there are m of them, each containing n unknowns $x_1, x_2, \ldots, x_n$.

Solution A *solution* of a system of m equations in n unknowns $x_1, x_2, \ldots, x_n$ is any ordered set $(x_1, x_2, \ldots, x_n)$ of real numbers for which *each* of the m equations of the system is satisfied.

REDUCED ROW-ECHELON FORM

A system of m linear equations in n unknowns will have either one solution, no solution, or infinitely many solutions. We can determine which of these possibilities occurs and, if solutions exist, find them by performing row operations on the augmented matrix of the system until we arrive at a matrix that has the following configuration:

1. Any rows that consist entirely of zeros are located at the bottom of the matrix.

2. The first nonzero element in each row is 1 and it has 0's above it and below it.

3. The leftmost 1 in any row is to the right of the leftmost 1 in the row above.

The matrix having this configuration (there is only one) is called the *reduced row-echelon form* of the augmented matrix.

Example 1
The matrix below

$$\left[\begin{array}{ccc|c} 1 & 0 & 0 & 3 \\ 0 & 1 & 0 & 8 \\ 0 & 0 & 1 & -4 \end{array} \right]$$

is the reduced row-echelon form of a system of three linear equations in three unknowns. If the unknowns are x_1, x_2, x_3, this matrix represents the system of equations

$$x_1 = 3$$
$$x_2 = 8$$
$$x_3 = -4$$

Thus, the system has one solution $(3, 8, -4)$.   ■

Example 2

The matrix below

$$\begin{bmatrix} 1 & 0 & 3 & | & 0 \\ 0 & 1 & 2 & | & 0 \\ 0 & 0 & 0 & | & 1 \\ 0 & 0 & 0 & | & 0 \end{bmatrix}$$

is the reduced row-echelon form of a system of four equations in three unknowns. If the unknowns are x_1, x_2, x_3, the equation represented by the third row is

$$0 \cdot x_1 + 0 \cdot x_2 + 0 \cdot x_3 = 1 \qquad \text{or} \qquad 0 = 1$$

Since $0 = 1$ is meaningless, we conclude the system has no solution.   ■

Example 3

The matrix below

$$\begin{bmatrix} 1 & 0 & 0 & 2 & | & 5 \\ 0 & 1 & 0 & 1 & | & 2 \\ 0 & 0 & 1 & 3 & | & 4 \end{bmatrix}$$

is the reduced row-echelon form of a system of three equations in four unknowns. If x_1, x_2, x_3, x_4 are the unknowns, the system of equations is

$$\begin{aligned} x_1 + 2x_4 &= 5 \\ x_2 + x_4 &= 2 \\ x_3 + 3x_4 &= 4 \end{aligned} \qquad \text{or} \qquad \begin{aligned} x_1 &= 5 - 2x_4 \\ x_2 &= 2 - x_4 \\ x_3 &= 4 - 3x_4 \end{aligned}$$

Thus, the system has infinitely many solutions. One way to represent the infinitely many solutions is to use a parameter, say t, and set the variable x_4 equal to it. Thus for any real number t,

$$x_4 = t$$
$$x_1 = 5 - 2t$$
$$x_2 = 2 - t$$
$$x_3 = 4 - 3t$$

represents a solution. For example,

$$\text{If } t = 0, \text{ then } x_4 = 0, x_1 = 5, x_2 = 2, x_3 = 4$$
$$\text{If } t = 1, \text{ then } x_4 = 1, x_1 = 3, x_2 = 1, x_3 = 1$$
$$\text{If } t = 2, \text{ then } x_4 = 2, x_1 = 1, x_2 = 0, x_3 = -2$$

and so on.   ■

Example 4

The following matrices are not in reduced row-echelon form:

$$\begin{bmatrix} 1 & 0 & 0 \\ 0 & 0 & 0 \\ 0 & 1 & 0 \end{bmatrix}$$ The second row contains all 0's and the third does not — this violates the rule that states any rows containing all zeros are at the bottom.

$$\begin{bmatrix} 1 & 0 & 2 & 4 \\ 0 & 0 & 2 & 3 \\ 0 & 0 & 0 & 1 \end{bmatrix}$$ The first nonzero element in row 2 is not a 1.

$$\begin{bmatrix} 1 & 0 & 0 \\ 0 & 0 & 1 \\ 0 & 1 & 0 \end{bmatrix}$$ The leftmost 1 in the third row is not to the right of the leftmost 1 in the row above it.   ■

Let's review the procedure for getting the reduced row-echelon form of a matrix.

Example 5

Find the reduced row-echelon form of

$$A = \begin{bmatrix} 1 & -1 & \bigm| & 2 \\ 2 & -3 & \bigm| & 2 \\ 3 & -5 & \bigm| & 2 \end{bmatrix}$$

Solution

The entry in row 1, column 1, namely a_{11}, is 1. Thus, we proceed to obtain a matrix in which all the entries in column 1 except a_{11} are 0. We can obtain such a matrix by performing the row operations

$$R_2 = -2r_1 + r_2$$
$$R_3 = -3r_1 + r_3$$

The new matrix is

$$\begin{bmatrix} 1 & -1 & \bigm| & 2 \\ 0 & -1 & \bigm| & -2 \\ 0 & -2 & \bigm| & -4 \end{bmatrix}$$

We want the entry in row 2, column 2, namely a_{22}, to be equal to 1. By the operations $R_2 = (-1)r_2$, we obtain

$$\left[\begin{array}{rr|r} 1 & -1 & 2 \\ 0 & 1 & 2 \\ 0 & -2 & -4 \end{array}\right]$$

Now column 2 should have 0's except for $a_{22} = 1$. This can be accomplished by applying the row operations

$$R_1 = r_2 + r_1 \qquad R_3 = 2r_2 + r_3$$

The new matrix is

$$\left[\begin{array}{rr|r} 1 & 0 & 4 \\ 0 & 1 & 2 \\ 0 & 0 & 0 \end{array}\right]$$

This is the reduced row-echelon form of A.   ■

Example 6
Solve the system

$$x - y = 2$$
$$2x - 3y = 2$$
$$3x - 5y = 2$$

Solution
The augmented matrix of this system is

$$\left[\begin{array}{rr|r} 1 & -1 & 2 \\ 2 & -3 & 2 \\ 3 & -5 & 2 \end{array}\right]$$

Following the solution to Example 5, we place this matrix in reduced row-echelon form.

$$\left[\begin{array}{rr|r} 1 & 0 & 4 \\ 0 & 1 & 2 \\ 0 & 0 & 0 \end{array}\right]$$

We conclude that the system has the solution $x = 4$, $y = 2$.   ■

Example 7
Solve the system

$$2x_1 - 3x_2 + 2x_3 = 1$$
$$x_1 - x_2 + 2x_3 = 2$$
$$3x_1 - 5x_2 + 2x_3 = -3$$
$$-4x_1 + 12x_2 + 8x_3 = 10$$

Solution

We need to find the reduced row-echelon form of the augmented matrix of this system, namely,

$$\left[\begin{array}{ccc|c} 2 & -3 & 2 & 1 \\ 1 & -1 & 2 & 2 \\ 3 & -5 & 2 & -3 \\ -4 & 12 & 8 & 10 \end{array}\right]$$

To obtain a 1 in row 1, column 1, we interchange rows 1 and 2: $R_1 = r_2$, $R_2 = r_1$. The new matrix is

$$\left[\begin{array}{ccc|c} 1 & -1 & 2 & 2 \\ 2 & -3 & 2 & 1 \\ 3 & -5 & 2 & -3 \\ -4 & 12 & 8 & 10 \end{array}\right]$$

To obtain 0's in column 1 under $a_{11} = 1$, we use the row operations

$$R_2 = -2r_1 + r_2 \qquad R_3 = -3r_1 + r_3 \qquad R_4 = 4r_1 + r_4$$

The new matrix is

$$\left[\begin{array}{ccc|c} 1 & -1 & 2 & 2 \\ 0 & -1 & -2 & -3 \\ 0 & -2 & -4 & -9 \\ 0 & 8 & 16 & 18 \end{array}\right]$$

To get $a_{22} = 1$, we use $R_2 = -r_2$, obtaining

$$\left[\begin{array}{ccc|c} 1 & -1 & 2 & 2 \\ 0 & 1 & 2 & 3 \\ 0 & -2 & -4 & -9 \\ 0 & 8 & 16 & 18 \end{array}\right]$$

To obtain 0's in column 2 (except for $a_{22} = 1$), we use

$$R_1 = r_2 + r_1 \qquad R_3 = 2r_2 + r_3 \qquad R_4 = -8r_2 + r_4$$

The new matrix is

$$\left[\begin{array}{ccc|c} 1 & 0 & 4 & 5 \\ 0 & 1 & 2 & 3 \\ 0 & 0 & 0 & -3 \\ 0 & 0 & 0 & -6 \end{array}\right] \tag{1}$$

There is no need to reduce the above matrix any further, since the third row yields the equation

$$0\,x_1 + 0\,x_2 + 0\,x_3 = -3$$

or

$$0 = -3$$

We conclude the system has no solution.   ∎

INFINITE NUMBER OF SOLUTIONS

We have seen several examples of systems of linear equations that have an infinite number of solutions. Let's look at a few more examples.

Example 8
Solve the system

$$x_1 + x_2 + x_3 = 7$$
$$x_1 - x_2 - 3x_3 = 1$$

Solution
The augmented matrix of the system is

$$\begin{bmatrix} 1 & 1 & 1 & | & 7 \\ 1 & -1 & -3 & | & 1 \end{bmatrix}$$

The reduced row-echelon form (as you should verify) is

$$\begin{bmatrix} 1 & 0 & -1 & | & 4 \\ 0 & 1 & 2 & | & 3 \end{bmatrix}$$

The system of equations represented by this matrix are

$$x_1 - x_3 = 4$$
$$x_2 + 2x_3 = 3 \tag{2}$$

We can assign any value to x_3 and use it to compute values of x_1 and x_2. Thus, the system has infinitely many solutions. ∎

If we rearrange the equations in (2) in the form

$$x_1 = 4 + x_3$$
$$x_2 = 3 - 2x_3$$

it is easier to see the role the unknown x_3 plays. In fact, because the remaining unknowns are expressed in terms of x_3, we treat x_3 as a parameter.

The next example illustrates a system having an infinite number of solutions with two parameters.

Example 9
Solve the system

$$x_1 + x_2 + 2x_3 + 2x_4 = 2$$
$$x_1 + x_3 + x_4 = 0$$
$$x_2 + x_3 + x_4 = 2$$

Solution
The augmented matrix of the system is

$$\begin{bmatrix} 1 & 1 & 2 & 2 & 2 \\ 1 & 0 & 1 & 1 & 0 \\ 0 & 1 & 1 & 1 & 2 \end{bmatrix}$$

The reduced row-echelon form (as you should verify) is

$$\begin{bmatrix} 1 & 0 & 1 & 1 & 0 \\ 0 & 1 & 1 & 1 & 2 \\ 0 & 0 & 0 & 0 & 0 \end{bmatrix}$$

The equations represented by this system are

$$x_1 + x_3 + x_4 = 0$$
$$x_2 + x_3 + x_4 = 2$$

We can rewrite this system in the form

$$x_1 = -x_3 - x_4$$
$$x_2 = 2 - x_3 - x_4 \tag{3}$$

The system has infinitely many solutions, obtained by assigning the two parameters $x_3 = s$ and $x_4 = t$ arbitrary values. Some choices are shown in the table.

s	t	x_3	x_4	x_1	x_2	Solution (x_1, x_2, x_3, x_4)
0	0	0	0	0	2	$(0, 2, 0, 0)$
1	0	1	0	-1	1	$(-1, 1, 1, 0)$
0	2	0	2	-2	0	$(-2, 0, 0, 2)$

Parameters are not unique. We could have chosen x_1 and x_4 as parameters by rewriting the equations (2) in the following manner.
From the first equation

$$x_3 = -x_1 - x_4$$

We can replace x_3 in the second equation by the above to produce

$$x_2 = 2 - x_3 - x_4 = 2 + x_1 + x_4 - x_4$$

We then get the system

$$x_2 = 2 + x_1$$
$$x_3 = -x_1 - x_4$$

showing the role of $x_1 = s$ and $x_4 = t$ as parameters.

SUMMARY
To solve a system of m linear equations in n unknowns

1. Write its augmented matrix.
2. Compute the reduced row-echelon form of the augmented matrix.
3. If this matrix has a row of zeros, except for a nonzero entry in the last column, the system has no solution.
4. Otherwise determine from this matrix whether the system has a unique solution or infinitely many solutions.

APPLICATION

Example 10

Mixing Acids In a chemistry laboratory one solution is 10% hydrochloric acid (HCl), a second solution contains 20% HCl, and a third contains 40% HCl. How many liters of each should be mixed to obtain 100 liters of 25% HCl?

Solution
Let x_1, x_2, and x_3 represent the number of liters of 10%, 20%, and 40% solutions of HCl, respectively. Since we want 100 liters in all and the amount of HCl obtained from each solution must sum to 25% of 100, or 25 liters, we must have

$$x_1 + x_2 + x_3 = 100$$
$$0.1x_1 + 0.2x_2 + 0.4x_3 = 25$$

No. of Liters 10% Solution	No. of Liters 20% Solution	No. of Liters 40% Solution
0	75	25
10	60	30
12	57	31
16	51	33
20	45	35
25	37.5	37.5
26	36	38
30	30	40
36	21	43
38	18	44
46	6	48
50	0	50

Thus, our problem is to solve a system of two equations in three unknowns.
By matrix techniques, we obtain the solution

$$x_1 = 2x_3 - 50$$
$$x_2 = -3x_3 + 150$$

(4)

where x_3 can represent any real number. Now the practical considerations of this problem lead us to the conditions that $x_1 \geq 0, x_2 \geq 0, x_3 \geq 0$. From (4) we see that we must have $x_3 \geq 25$ and $x_3 \leq 50$, since otherwise $x_1 < 0$ or $x_2 < 0$. Some possible solutions are listed in the table on page 85. The final determination by the chemistry laboratory will more than likely be based on the amount and availability of one acid solution versus others. ■

Exercise 2.3

Answers to Odd-Numbered Problems begin on page A-6.

A   In Problems 1–10, tell whether the given matrix is in reduced row-echelon form.

1. $\begin{bmatrix} 1 & 2 & 3 \\ 0 & 0 & 0 \\ 0 & 0 & 1 \end{bmatrix}$

2. $\begin{bmatrix} 1 & 2 & 3 \\ 0 & 0 & 0 \\ 0 & 0 & 0 \end{bmatrix}$

3. $\begin{bmatrix} 1 & 1 & 0 \\ 0 & 1 & 0 \end{bmatrix}$

4. $\begin{bmatrix} 1 & 0 & 3 \\ 0 & 1 & 0 \end{bmatrix}$

5. $\begin{bmatrix} 0 & 1 \\ 1 & 0 \end{bmatrix}$

6. $\begin{bmatrix} 0 & 1 & 0 \\ 0 & 0 & 1 \\ 0 & 0 & 0 \end{bmatrix}$

7. $\begin{bmatrix} 0 & 0 & 1 \\ 0 & 0 & 0 \end{bmatrix}$

8. $\begin{bmatrix} 0 & 0 \\ 0 & 0 \end{bmatrix}$

9. $\begin{bmatrix} 1 & 0 & 0 & 0 & 0 \\ 0 & 0 & 1 & 2 & 0 \\ 0 & 0 & 0 & 0 & 1 \\ 0 & 0 & 0 & 0 & 0 \end{bmatrix}$

10. $\begin{bmatrix} 1 & 1 & 0 \\ 0 & 0 & 2 \end{bmatrix}$

In Problems 11–20 the reduced row-echelon form of the augmented matrix of a system of equations is given. Tell whether the system has one solution, no solution, or infinitely many solutions.

11. $\left[\begin{array}{cc|c} 1 & 1 & 1 \\ 0 & 0 & 0 \end{array}\right]$

12. $\left[\begin{array}{cc|c} 1 & 0 & 0 \\ 0 & 0 & 1 \end{array}\right]$

13. $\left[\begin{array}{ccc|c} 1 & 0 & 0 & 0 \\ 0 & 1 & 0 & 0 \\ 0 & 0 & 1 & 6 \end{array}\right]$

14. $\left[\begin{array}{ccc|c} 1 & 0 & -2 & 6 \\ 0 & 1 & 3 & 1 \end{array}\right]$

15. $\left[\begin{array}{ccc|c} 1 & 2 & 0 & 1 \\ 0 & 0 & 1 & 2 \\ 0 & 0 & 0 & 0 \end{array}\right]$

16. $\left[\begin{array}{ccc|c} 1 & 2 & 0 & 0 \\ 0 & 0 & 1 & 0 \\ 0 & 0 & 0 & 1 \end{array}\right]$

17. $\left[\begin{array}{cccc|c} 1 & 0 & 1 & -1 & 0 \\ 0 & 1 & 2 & 1 & 1 \\ 0 & 0 & 0 & 0 & 0 \end{array}\right]$

18. $\left[\begin{array}{ccc|c} 1 & 0 & 0 & -1 \\ 0 & 1 & 0 & 3 \\ 0 & 0 & 1 & 4 \\ 0 & 0 & 0 & 0 \end{array}\right]$

19. $\left[\begin{array}{ccc|c} 1 & 0 & -1 & 1 \\ 0 & 1 & 2 & 1 \end{array}\right]$

20. $\left[\begin{array}{cc|c} 1 & 0 & 1 \\ 0 & 1 & 1 \\ 0 & 0 & 0 \end{array}\right]$

B   In Problems 21–30 solve each system of equations by finding the reduced row-echelon form of the augmented matrix.

21. $\quad x + y = 3$
$\quad\quad 2x - y = 3$

22. $\quad x - y = 5$
$\quad\quad 2x + 3y = 15$

23. $\quad 3x - 3y = 12$
$\quad\quad 3x + 2y = -3$

24. $\quad 6x + y = 8$
$\quad\quad x - 3y = -5$

25. $\quad 3x_1 - 4x_2 = 1$
$\quad\quad 5x_1 + 2x_2 = 19$

26. $\quad 2x_1 + 3x_2 = 8$
$\quad\quad 2x_1 - x_2 = 12$

27. $\quad 2x_1 + 3x_2 = 5$
$\quad\quad 2x_1 - x_2 = 7$

28. $\quad 3x_1 - 4x_2 = 7$
$\quad\quad 5x_1 + 2x_2 = 3$

29. $\quad x_1 - x_2 = 1$
$\quad\quad x_2 - x_3 = 6$
$\quad\quad x_1 + x_3 = -1$

30. $\quad 2x_2 + x_3 = 1$
$\quad\quad 2x_1 - x_2 + 3x_3 = 0$
$\quad\quad x_1 + 2x_2 - x_3 = 5$

C   In Problems 31–38 solve each system of equations by finding the reduced row-echelon form of the augmented matrix.

31. $\quad x_1 + x_2 = 7$
$\quad\quad x_2 - x_3 + x_4 = 5$
$\quad\quad x_1 - x_2 + x_3 + x_4 = 6$
$\quad\quad x_2 - x_4 = 10$

32. $\quad x_1 + x_2 + x_3 + x_4 = 0$
$\quad\quad 2x_1 - x_2 - x_3 + x_4 = 0$
$\quad\quad x_1 - x_2 - x_3 + x_4 = 0$
$\quad\quad x_1 + x_2 - x_3 - x_4 = 0$

33. $\quad x_1 + 2x_2 + 3x_3 - x_4 = 0$
$\quad\quad 3x_1 - x_4 = 4$
$\quad\quad x_2 - x_3 - x_4 = 2$

34. $\quad 2x_1 - 3x_2 + 4x_3 = 7$
$\quad\quad x_1 - 2x_2 + 3x_3 = 2$

35. $\quad x_1 - x_2 + x_3 = 5$
$\quad\quad 2x_1 - 2x_2 + 2x_3 = 8$

36. $\quad x_1 + x_2 + x_3 = 3$
$\quad\quad x_1 - x_2 + x_3 = 7$
$\quad\quad x_1 - x_2 - x_3 = 1$

37. $\quad 3x_1 - x_2 + 2x_3 = 3$
$\quad\quad 3x_1 + 3x_2 + x_3 = 3$
$\quad\quad 3x_1 - 5x_2 + 3x_3 = 12$

38. $\quad x_1 + x_2 - x_3 = 12$
$\quad\quad 3x_1 - x_2 = 1$
$\quad\quad 2x_1 - 3x_2 + 4x_3 = 3$

APPLICATIONS   **39.** Mixtures A chemistry laboratory has available three kinds of hydrochloric acid (HCl): 10%, 30%, and 50% solutions. How many liters of each should be mixed to obtain 100 liters of 25% HCl? Provide a table showing at least 6 of the possible solutions.

40. Mixtures Repeat Problem 39 if the mixture is to be 100 liters of 40% HCl.

41. Inventory Control An art teacher finds that colored paper can be bought in three different packages. The first package has 20 sheets of white paper, 15 sheets of blue paper, and 1 sheet of red paper. The second package has 3 sheets of blue paper and 1 sheet of red paper. The last package has 40 sheets of white paper and 30 sheets of blue paper. Suppose he needs 200 sheets of white paper, 180 sheets of blue paper, and 12 sheets of red paper. How many of each type of package should he order?

42. Inventory Control The same art teacher in Problem 41 decides to do a project requiring 88 red sheets of paper, 51 green sheets, and 69 blue sheets. A jumbo pack contains 24 red sheets, 12 blue sheets, and 12 green sheets. The assorted pack contains

15 blue sheets and 9 green sheets. The specialized pack has 10 red sheets. How many of each type of pack will he have to order?

43. Inventory Control An interior decorator has ordered 12 cans of sunset paint, 35 cans of brown, and 18 cans of fuchsia. The paint store has special pair packs, containing 1 can each of sunset and fuchsia; darkening packs containing 2 cans of sunset, 5 cans of brown, and 2 cans of fuchsia; and economy packs, containing 3 cans of sunset, 15 cans of brown, and 6 cans of fuchsia. How many of each type of pack should the paint store send to the interior decorator?

44. Advertising A soap manufacturer decides to spend a total of $6 million on radio, magazine, and TV advertising. If it spends as much on TV advertising as on magazine and radio together, and the amount spent on magazines and TV combined equals five times that spent on radio, what is the amount to be spent on each type of advertising?

45. Inventory Control Three species of bacteria will be kept in one test tube and will feed on three resources. Each member of the first species consumes three units of the first resource and one unit of the third. Each bacterium of the second type consumes one unit of the first resource and two units each of the second and third. Each bacterium of the third type consumes two units of the first resource and four each of the second and third. If the test tube is supplied daily with 12,000 units of the first resource, 12,000 units of the second resource, and 14,000 units of the third, how many of each species can coexist in equilibrium in the test tube so that all of the supplied resources are consumed?

2.4

MATRIX ALGEBRA

EQUALITY OF MATRICES
ADDITION OF MATRICES
SUBTRACTION OF MATRICES
MULTIPLYING A MATRIX BY A NUMBER

Matrices can be added, subtracted, and multiplied. They also possess many of the algebraic properties of numbers. Matrix algebra is the study of these properties. Its importance lies in the fact that many situations in both pure and applied mathematics deal with rectangular arrays of numbers. In fact, in many branches of business and the biological and social sciences it is useful to express and use a set of numbers in a rectangular array. Let's look at an example.

Example 1

Motors, Inc., produces three models of cars: a sedan, a hardtop, and a station wagon. If the company wishes to compare the units of raw material and the units of labor involved in 1 month's production of each of these models, the rectangular array displayed below may be used to present the data:

	Sedan Model	HardTop Model	Station Wagon Model
Units of Material	23	16	10
Units of Labor	7	9	11

The same information may be written concisely as the matrix

$$\begin{bmatrix} 23 & 16 & 10 \\ 7 & 9 & 11 \end{bmatrix}$$

This matrix, has 2 rows (the units) and 3 columns (the models). The first row represents units of material and the second row represents units of labor. The first, second, and third columns represent the sedan, hardtop, and station wagon models, respectively.   ■

A formal definition of a matrix as a rectangular array of numbers is given below.

Matrix A *matrix A* **is a rectangular array of numbers** a_{ij} **of the form**

$$A = \begin{bmatrix} a_{11} & a_{12} & \cdots & a_{1j} & \cdots & a_{1n} \\ a_{21} & a_{22} & \cdots & a_{2j} & \cdots & a_{2n} \\ \vdots & \vdots & & \vdots & & \vdots \\ a_{i1} & a_{i2} & \cdots & a_{ij} & \cdots & a_{in} \\ \vdots & \vdots & & \vdots & & \vdots \\ a_{m1} & a_{m2} & \cdots & a_{mj} & \cdots & a_{mn} \end{bmatrix}$$

*j*th column ↓   ← *i*th row

This matrix contains $m \cdot n$ **numbers.**

Each number a_{ij} of the matrix A has two indices: the *row index, i,* and the *column index, j*. The symbols $a_{i1}, a_{i2}, \ldots, a_{in}$ represent the numbers in the *i*th row, and the symbols $a_{1j}, a_{2j}, \ldots, a_{mj}$ represent the numbers in the *j*th column. The numbers a_{ij} of a matrix are sometimes referred to as the *entries* or *components* or *elements* of the matrix. The matrix A above, which has m rows and n columns, can be abbreviated by

$$A = [a_{ij}] \qquad i = 1, 2, \ldots, m; \quad j = 1, 2, \ldots, n$$

Dimension The *dimension of a matrix A* **is determined by the number of rows and the number of columns of the matrix. If a matrix** A **has** m **rows and** n **columns, we denote the dimension of** A **by** $m \times n$**, read as "m by n."**

For a 2×3 matrix, remember that the first number 2 denotes the number of rows and the second number 3 is the number of columns. A matrix with 3 rows and 2 columns is of dimension 3×2.

Square Matrix **If a matrix A has the same number of rows as it has columns, it is called a *square matrix*.**

In a matrix $A = [a_{ij}]$ the entries for which $i = j$, namely $a_{11}, a_{22}, a_{33}, a_{44}$, and so on, form the *diagonal of A*.

Example 2

In a recent United States census, the following figures were obtained with regard to the city of Glenwood. Each year 7% of city residents move to the suburbs and 1% of the people in the suburbs move to the city. This situation can be represented by the matrix

$$P = \begin{matrix} \\ \text{City} \\ \text{Suburbs} \end{matrix} \begin{matrix} \text{City} & \text{Suburbs} \\ \begin{bmatrix} 0.93 & 0.07 \\ 0.01 & 0.99 \end{bmatrix} \end{matrix}$$

Here, the entry in row 1, column 2, 0.07, indicates that 7% of city residents move to the suburbs. The matrix P is a square matrix, and its dimension is 2×2. The diagonal entries are 0.93 and 0.99. ∎

Row and Column Matrix **A *row matrix* is a matrix with 1 row of elements. A *column matrix* is a matrix with 1 column of elements. Row matrices and column matrices are sometimes referred to as *row vectors* and *column vectors*, respectively.**

Example 3

The matrices

$$A = [23 \quad 16 \quad 10] \qquad B = [7 \quad 9]$$

$$C = \begin{bmatrix} 23 \\ -1 \\ 7 \end{bmatrix} \qquad D = \begin{bmatrix} 16 \\ 9 \end{bmatrix} \qquad E = [9]$$

have the following dimensions: A, 1×3; B, 1×2; C, 3×1; D, 2×1; E, 1×1. Here A, B, and E are row vectors and C, D, and E are column vectors. ∎

EQUALITY OF MATRICES

As with most mathematical quantities, we now want to determine various relationships between two matrices. We might ask, "When, if at all, are two matrices equal?"

Let's try to arrive at a sound definition for equality of matrices by requiring equal matrices to have certain desirable properties. First, it would seem necessary that two equal matrices have the same dimension — that is, that they both be $m \times n$ matrices. Next, it would seem necessary that their entries be identical numbers. With these two restrictions, we define equality of matrices.

Equality of Matrices **Two matrices A and B are *equal* if they are of the same dimension and if corresponding entries are equal. In this case, we write $A = B$, read as "matrix A is equal to matrix B."**

Example 4

In order for the two matrices

$$\begin{bmatrix} p & q \\ 1 & 0 \end{bmatrix} \quad \text{and} \quad \begin{bmatrix} 2 & 4 \\ n & 0 \end{bmatrix}$$

to be equal, we must have $p = 2$, $q = 4$, and $n = 1$.   ■

Example 5

Let A and B be two matrices given by

$$A = \begin{bmatrix} x+y & 6 \\ 2x-3 & 2-y \end{bmatrix} \qquad B = \begin{bmatrix} 5 & 5x+2 \\ y & x-y \end{bmatrix}$$

Find x and y so that A and B are equal (if possible).

Solution

Both A and B are 2×2 matrices. Thus, $A = B$ if

(a) $x + y = 5$ (b) $6 = 5x + 2$
(c) $2x - 3 = y$ (d) $2 - y = x - y$

Here we have four equations in the two unknowns x and y. From equation (d), we see that $x = 2$. Using this value in equation (a), we obtain $y = 3$. But $x = 2$, $y = 3$ do not satisfy either (b) or (c). Hence, there are *no* values for x and y satisfying all four equations. This means A and B can never be equal.   ■

ADDITION OF MATRICES

Can two matrices be added? And, if so, what is the rule or law for addition of matrices?

Let's return to Example 1. In that example, we recorded 1 month's production of Motors, Inc., by the matrix

$$A = \begin{bmatrix} 23 & 16 & 10 \\ 7 & 9 & 11 \end{bmatrix}$$

Suppose the next month's production is

$$B = \begin{bmatrix} 18 & 12 & 9 \\ 14 & 6 & 8 \end{bmatrix}$$

in which the pattern of recording units and models remains the same.

The total production for the 2 months can be displayed by the matrix

$$C = \begin{bmatrix} 41 & 28 & 19 \\ 21 & 15 & 19 \end{bmatrix}$$

since the number of units of material for sedan models is $41 = 23 + 18$; the number of units of material for hardtop models is $28 = 16 + 12$; and so on.

This leads us to define the sum $A + B$ of two matrices A and B as the matrix consisting of the sum of corresponding entries from A and B.

Addition of Matrices Let $A = [a_{ij}]$ and $B = [b_{ij}]$ be two $m \times n$ matrices. **The** *sum* $A + B$ **is defined as the** $m \times n$ **matrix** $[a_{ij} + b_{ij}]$.

Example 6

(a) $\begin{bmatrix} 23 & 16 & 10 \\ 7 & 9 & 11 \end{bmatrix} + \begin{bmatrix} 18 & 12 & 9 \\ 14 & 6 & 8 \end{bmatrix} = \begin{bmatrix} 23 + 18 & 16 + 12 & 10 + 9 \\ 7 + 14 & 9 + 6 & 11 + 8 \end{bmatrix}$

$= \begin{bmatrix} 41 & 28 & 19 \\ 21 & 15 & 19 \end{bmatrix}$

(b) $\begin{bmatrix} 0.6 & 0.4 \\ 0.1 & 0.9 \end{bmatrix} + \begin{bmatrix} 2.3 & 0.6 \\ 1.8 & 5.2 \end{bmatrix} = \begin{bmatrix} 0.6 + 2.3 & 0.4 + 0.6 \\ 0.1 + 1.8 & 0.9 + 5.2 \end{bmatrix}$

$= \begin{bmatrix} 2.9 & 1.0 \\ 1.9 & 6.1 \end{bmatrix}$ ∎

Notice that it is possible to add two matrices only if their dimensions are the same. Also, the dimension of the sum of two matrices is the same as that of the two original matrices.

The following pairs of matrices cannot be added since they are of different dimensions:

$$A = \begin{bmatrix} 1 & 2 \\ 7 & 2 \end{bmatrix} \qquad \text{and} \qquad B = \begin{bmatrix} 1 \\ -3 \end{bmatrix}$$

$$A = [2 \quad 3] \qquad \text{and} \qquad B = [1 \quad 1 \quad 1]$$

$$A = \begin{bmatrix} -1 & 7 & 0 \\ 2 & \frac{1}{2} & 0 \end{bmatrix} \qquad \text{and} \qquad B = \begin{bmatrix} -1 & 2 \\ 3 & 0 \\ 1 & 5 \end{bmatrix}$$

It turns out that the usual rules for the addition of real numbers (such as the commutative laws and associative laws) are also valid for matrix addition.

Example 7

Let

$$A = \begin{bmatrix} 1 & 5 \\ 7 & -3 \end{bmatrix} \qquad \text{and} \qquad B = \begin{bmatrix} 3 & -2 \\ 4 & 1 \end{bmatrix}$$

Then

$$A + B = \begin{bmatrix} 1 & 5 \\ 7 & -3 \end{bmatrix} + \begin{bmatrix} 3 & -2 \\ 4 & 1 \end{bmatrix} = \begin{bmatrix} 1+3 & 5+(-2) \\ 7+4 & -3+1 \end{bmatrix} = \begin{bmatrix} 4 & 3 \\ 11 & -2 \end{bmatrix}$$

$$B + A = \begin{bmatrix} 3 & -2 \\ 4 & 1 \end{bmatrix} + \begin{bmatrix} 1 & 5 \\ 7 & -3 \end{bmatrix} = \begin{bmatrix} 4 & 3 \\ 11 & -2 \end{bmatrix}$$ ∎

This leads us to formulate the following property:

If A and B are two matrices of the same dimension, then

$$A + B = B + A$$

That is, matrix addition is *commutative.*

The associative law for addition of matrices is also true. Thus:

If A, B, and C, are three matrices of the same dimension, then

$$A + (B + C) = (A + B) + C$$

For a proof of this result, we let

$$A = [a_{ij}] \qquad B = [b_{ij}] \qquad C = [c_{ij}]$$

Then

$$\begin{aligned}
A + (B + C) &= [a_{ij}] + ([b_{ij}] + [c_{ij}]) = [a_{ij}] + ([b_{ij} + c_{ij}]) \\
&= [a_{ij}] + [b_{ij} + c_{ij}] = [a_{ij} + (b_{ij} + c_{ij})] \\
&= [(a_{ij} + b_{ij}) + c_{ij}] = [a_{ij} + b_{ij}] + [c_{ij}] \\
&= ([a_{ij} + b_{ij}]) + [c_{ij}] = ([a_{ij}] + [b_{ij}]) + [c_{ij}] \\
&= (A + B) + C
\end{aligned}$$

In the proof, we used the fact that addition of real numbers is associative. Where?
The fact that addition of matrices is associative means that the notation $A + B + C$ is *not* ambiguous, since $(A + B) + C = A + (B + C)$.

Zero Matrix **A matrix in which all entries are zero is called a *zero matrix.* We use the symbol 0 to represent a zero matrix of any dimension.**

For real numbers, zero has the property that $0 + x = x$ for any x. An important property of a zero matrix is that $A + 0 = A$, provided the dimension of 0 is the same as that of A.

Example 8
Let

$$A = \begin{bmatrix} 3 & 4 & -\frac{1}{2} \\ \sqrt{2} & 0 & 3 \end{bmatrix}$$

Then

$$A + 0 = \begin{bmatrix} 3 & 4 & -\frac{1}{2} \\ \sqrt{2} & 0 & 3 \end{bmatrix} + \begin{bmatrix} 0 & 0 & 0 \\ 0 & 0 & 0 \end{bmatrix}$$

$$= \begin{bmatrix} 3+0 & 4+0 & -\frac{1}{2}+0 \\ \sqrt{2}+0 & 0+0 & 3+0 \end{bmatrix} = \begin{bmatrix} 3 & 4 & -\frac{1}{2} \\ \sqrt{2} & 0 & 3 \end{bmatrix} = A \qquad \blacksquare$$

If A is any matrix, the *negative of A*, denoted by $-A$, is the matrix obtained by replacing each entry in A by its negative.

Example 9

If

$$A = \begin{bmatrix} -3 & 0 \\ 5 & -2 \\ 1 & 3 \end{bmatrix}$$

then

$$-A = \begin{bmatrix} 3 & 0 \\ -5 & 2 \\ -1 & -3 \end{bmatrix} \qquad \blacksquare$$

> For any matrix A, we have the property that
> $$A + (-A) = 0$$

SUBTRACTION OF MATRICES

Now that we have defined the sum of two matrices and the negative of a matrix, it is natural to ask about the *difference* of two matrices. As you will see, subtracting matrices and subtracting numbers are much the same kind of process.

Difference of Two Matrices **Let $A = [a_{ij}]$ and $B = [b_{ij}]$ be two $m \times n$ matrices. The *difference $A - B$ is defined as the $m \times n$ matrix $[a_{ij} - b_{ij}]$.***

Example 10

$$A = \begin{bmatrix} 2 & 3 & 4 \\ 1 & 0 & 2 \end{bmatrix} \quad \text{and} \quad B = \begin{bmatrix} -2 & 1 & -1 \\ 3 & 0 & 3 \end{bmatrix}$$

Then

$$A - B = \begin{bmatrix} 2 & 3 & 4 \\ 1 & 0 & 2 \end{bmatrix} - \begin{bmatrix} -2 & 1 & -1 \\ 3 & 0 & 3 \end{bmatrix}$$

$$= \begin{bmatrix} 2-(-2) & 3-1 & 4-(-1) \\ 1-3 & 0-0 & 2-3 \end{bmatrix} = \begin{bmatrix} 4 & 2 & 5 \\ -2 & 0 & -1 \end{bmatrix} \qquad \blacksquare$$

Notice that the difference $A - B$ is nothing more than the matrix formed by subtracting the entries in B from the corresponding entries in A.

Using the matrices A and B from Example 10, we find that

$$B - A = \begin{bmatrix} -2 & 1 & -1 \\ 3 & 0 & 3 \end{bmatrix} - \begin{bmatrix} 2 & 3 & 4 \\ 1 & 0 & 2 \end{bmatrix}$$

$$= \begin{bmatrix} -2-2 & 1-3 & -1-4 \\ 3-1 & 0-0 & 3-2 \end{bmatrix} = \begin{bmatrix} -4 & -2 & -5 \\ 2 & 0 & 1 \end{bmatrix}$$

Observe that $A - B \neq B - A$, illustrating that matrix subtraction, like subtraction of real numbers, is not commutative.

MULTIPLYING A MATRIX BY A NUMBER

Let's return to the production of Motors, Inc., during the month specified in Example 1. The matrix A describing this production is

$$A = \begin{bmatrix} 23 & 16 & 10 \\ 7 & 9 & 11 \end{bmatrix}$$

Let's assume that for 3 consecutive months, the monthly production remained the same. Then the total production for the 3 months is simply the sum of the matrix A taken 3 times. If we represent the total production by the matrix T, then

$$T = \begin{bmatrix} 23 & 16 & 10 \\ 7 & 9 & 11 \end{bmatrix} + \begin{bmatrix} 23 & 16 & 10 \\ 7 & 9 & 11 \end{bmatrix} + \begin{bmatrix} 23 & 16 & 10 \\ 7 & 9 & 11 \end{bmatrix}$$

$$= \begin{bmatrix} 23+23+23 & 16+16+16 & 10+10+10 \\ 7+7+7 & 9+9+9 & 11+11+11 \end{bmatrix}$$

$$= \begin{bmatrix} 3 \cdot 23 & 3 \cdot 16 & 3 \cdot 10 \\ 3 \cdot 7 & 3 \cdot 9 & 3 \cdot 11 \end{bmatrix} = \begin{bmatrix} 69 & 48 & 30 \\ 21 & 27 & 33 \end{bmatrix}$$

In other words, when we add the matrix A 3 times, we multiply each entry of A by 3. This leads to the following definition.

Scalar Multiplication   **Let $A = [a_{ij}]$ be an $m \times n$ matrix and let c be a real number, called a *scalar*. The product of the matrix A by the scalar c, called *scalar multiplication*, is the $m \times n$ matrix $cA = [ca_{ij}]$.**

When multiplying a matrix by a number, each entry of the matrix is multiplied by the number. Notice that the dimension of A and the dimension of the product cA are the same.

Example 11
For

$$A = \begin{bmatrix} 2 \\ 5 \\ -7 \end{bmatrix} \quad \text{and} \quad B = \begin{bmatrix} 20 & 0 \\ 18 & 8 \end{bmatrix}$$

compute: (a) $3A$ (b) $\tfrac{1}{2}B$

Solution

(a) $3A = 3 \begin{bmatrix} 2 \\ 5 \\ -7 \end{bmatrix} = \begin{bmatrix} 3 \cdot 2 \\ 3 \cdot 5 \\ 3 \cdot (-7) \end{bmatrix} = \begin{bmatrix} 6 \\ 15 \\ -21 \end{bmatrix}$

(b) $\frac{1}{2}B = \frac{1}{2} \begin{bmatrix} 20 & 0 \\ 18 & 8 \end{bmatrix} = \begin{bmatrix} \frac{1}{2} \cdot 20 & \frac{1}{2} \cdot 0 \\ \frac{1}{2} \cdot 18 & \frac{1}{2} \cdot 8 \end{bmatrix} = \begin{bmatrix} 10 & 0 \\ 9 & 4 \end{bmatrix}$ ■

Example 12
For
$$A = \begin{bmatrix} 3 & 1 \\ 4 & 0 \\ 2 & -3 \end{bmatrix} \quad \text{and} \quad B = \begin{bmatrix} 2 & -3 \\ -1 & 1 \\ 1 & 0 \end{bmatrix}$$
compute: (a) $A - B$ (b) $A + (-1) \cdot B$

Solution

(a) $A - B = \begin{bmatrix} 1 & 4 \\ 5 & -1 \\ 1 & -3 \end{bmatrix}$

(b) $A + (-1) \cdot B = \begin{bmatrix} 3 & 1 \\ 4 & 0 \\ 2 & -3 \end{bmatrix} + \begin{bmatrix} -2 & 3 \\ 1 & -1 \\ -1 & 0 \end{bmatrix} = \begin{bmatrix} 1 & 4 \\ 5 & -1 \\ 1 & -3 \end{bmatrix}$ ■

The above example illustrates the result that

$$A - B = A + (-1) \cdot B = A + (-B)$$

We continue our study of scalar multiplication by listing some of its properties.

Let k and h be two real numbers; let $A = [a_{ij}]$, $B = [b_{ij}]$, $i = 1, \ldots, m$, $j = 1, \ldots, n$ matrices of dimension $m \times n$. Then

(I) $k(hA) = (kh)A$

(II) $(k + h)A = kA + hA$

(III) $k(A + B) = kA + kB$

We prove (I) and (II) here; the proof of (III) is left for the reader.

(I)   Here
$$k(hA) = k(h[a_{ij}]) = k([ha_{ij}]) = k[ha_{ij}]$$
$$= [kha_{ij}] = (kh)[a_{ij}] = (kh)A$$

(II)   For this property, we have

$$(k + h)A = (k + h)[a_{ij}] = [(k + h)a_{ij}]$$
$$= [ka_{ij} + ha_{ij}] = [ka_{ij}] + [ha_{ij}]$$
$$= k[a_{ij}] + h[a_{ij}] = kA + hA$$

Properties (I)–(III) are illustrated in the following example.

Example 13
For

$$A = \begin{bmatrix} 2 & -3 & -1 \\ 5 & 6 & 4 \end{bmatrix} \quad \text{and} \quad B = \begin{bmatrix} -3 & 0 & 4 \\ 2 & -1 & 5 \end{bmatrix}$$

show that:

(a) $5[2A] = 10A$ (b) $(4 + 3)A = 4A + 3A$ (c) $3[A + B] = 3A + 3B$

Solution

(a) $5[2A] = 5\begin{bmatrix} 4 & -6 & -2 \\ 10 & 12 & 8 \end{bmatrix} = \begin{bmatrix} 20 & -30 & -10 \\ 50 & 60 & 40 \end{bmatrix}$

$10A = \begin{bmatrix} 20 & -30 & -10 \\ 50 & 60 & 40 \end{bmatrix}$

(b) $(4 + 3)A = 7A = \begin{bmatrix} 14 & -21 & -7 \\ 35 & 42 & 28 \end{bmatrix}$

$4A + 3A = \begin{bmatrix} 8 & -12 & -4 \\ 20 & 24 & 16 \end{bmatrix} + \begin{bmatrix} 6 & -9 & -3 \\ 15 & 18 & 12 \end{bmatrix} = \begin{bmatrix} 14 & -21 & -7 \\ 35 & 42 & 28 \end{bmatrix}$

(c) $3[A + B] = 3\begin{bmatrix} -1 & -3 & 3 \\ 7 & 5 & 9 \end{bmatrix} = \begin{bmatrix} -3 & -9 & 9 \\ 21 & 15 & 27 \end{bmatrix}$

$3A + 3B = \begin{bmatrix} 6 & -9 & -3 \\ 15 & 18 & 12 \end{bmatrix} + \begin{bmatrix} -9 & 0 & 12 \\ 6 & -3 & 15 \end{bmatrix} = \begin{bmatrix} -3 & -9 & 9 \\ 21 & 15 & 27 \end{bmatrix}$ ∎

Exercise 2.4

Answers to Odd-Numbered Problems begin on page A-7.

In Problems 1–8 write the dimension of each matrix.

A

1. $\begin{bmatrix} 3 & 2 \\ -1 & 3 \end{bmatrix}$ 2. $\begin{bmatrix} -1 & 0 \\ 0 & 5 \end{bmatrix}$ 3. $\begin{bmatrix} 2 & 1 & -3 \\ 1 & 0 & -1 \end{bmatrix}$ 4. $\begin{bmatrix} 1 & 2 \\ 2 & 1 \\ 0 & -3 \end{bmatrix}$

5. $\begin{bmatrix} 4 \\ 1 \end{bmatrix}$ 6. $[2 \quad 1 \quad -3]$ 7. $[2]$ 8. $[0]$

In Problems 9–16 determine whether the given statements are true or false. If false, tell why.

9. $\begin{bmatrix} 0 \\ 1 \end{bmatrix} = [0 \quad 1]$

10. $\begin{bmatrix} 3 & 2 \\ -1 & 0 \end{bmatrix} = \begin{bmatrix} 3 & 2 \\ -1 & 4 \end{bmatrix}$

11. $\begin{bmatrix} 5 & 0 \\ 0 & 1 \end{bmatrix}$ is square

12. $\begin{bmatrix} 3 & 2 & 1 \\ 4 & -1 & 0 \end{bmatrix}$ is 3×2

13. $\begin{bmatrix} x & 2 \\ 4 & 0 \end{bmatrix} = \begin{bmatrix} 3 & 2 \\ 4 & 0 \end{bmatrix}$ if $x = 3$

14. $\begin{bmatrix} x & y \\ 0 & 0 \end{bmatrix} = [x \quad y]$

15. $\begin{bmatrix} 5 & 0 \\ 1 & 1 \end{bmatrix} = \begin{bmatrix} 2+3 & 0 \\ 1 & 1 \end{bmatrix}$

16. $\begin{bmatrix} 1 & 0 \\ 0 & 1 \end{bmatrix} = \begin{bmatrix} 3-2 & 3-3 \\ 3-3 & 3-2 \end{bmatrix}$

17. Find x and z so that

$$\begin{bmatrix} x \\ z \end{bmatrix} = \begin{bmatrix} 4 \\ 3 \end{bmatrix}$$

18. Find x, y, and z so that

$$\begin{bmatrix} x+y & 2 \\ 4 & 0 \end{bmatrix} = \begin{bmatrix} 6 & x-y \\ 4 & z \end{bmatrix}$$

19. Find x and y so that

$$\begin{bmatrix} x-2y & 0 \\ -2 & 6 \end{bmatrix} = \begin{bmatrix} 3 & 0 \\ -2 & x+y \end{bmatrix}$$

20. Find x, y, and z so that

$$\begin{bmatrix} x-2 & 3 & 2z \\ 6y & x & 2y \end{bmatrix} = \begin{bmatrix} y & z & 6 \\ 18z & y+2 & 6z \end{bmatrix}$$

In Problems 21–26 perform the indicated operation.

21. $\begin{bmatrix} 2 & 3 \\ 4 & 6 \end{bmatrix} + \begin{bmatrix} 3 & 1 \\ -2 & 3 \end{bmatrix}$

22. $3\begin{bmatrix} 1 & 1 & 3 \\ 2 & 1 & 6 \end{bmatrix} + \begin{bmatrix} 4 & 5 & 6 \\ 1 & 0 & 1 \end{bmatrix}$

23. $2\begin{bmatrix} 1 & 4 \\ 0 & 2 \\ 5 & 0 \end{bmatrix} + 4\begin{bmatrix} 6 & 1 \\ 1 & 0 \\ 1 & 0 \end{bmatrix}$

24. $3\begin{bmatrix} 1 & 0 & 2 \\ 0 & 1 & 1 \\ 1 & -1 & -1 \end{bmatrix} - 2\begin{bmatrix} 1 & 1 & 0 \\ -1 & 1 & 3 \\ 0 & 2 & 0 \end{bmatrix}$

25. $(-3)\begin{bmatrix} 1 & 3 & 4 \\ 2 & -1 & 6 \\ -1 & -1 & -3 \\ 0 & 8 & 0 \end{bmatrix} + \begin{bmatrix} 6 & 4 & 2 \\ -3 & 3 & 5 \\ 6 & 6 & 2 \\ 7 & 1 & 4 \end{bmatrix}$

26. $(-1)\begin{bmatrix} 1 & 0 & 1 & 1 \\ 3 & 0 & 1 & 1 \\ 0 & 1 & 2 & 5 \\ 2 & 0 & 1 & 2 \end{bmatrix} + (-3)\begin{bmatrix} 1 & 0 & 0 & -1 \\ 1 & 0 & 2 & -1 \\ 0 & 3 & 1 & 0 \\ 0 & 0 & 1 & 1 \end{bmatrix}$

B　In Problems 27–36 use the matrices below to find the indicated expression.

$$A = \begin{bmatrix} 2 & -3 & 4 \\ 0 & 2 & 1 \end{bmatrix} \quad B = \begin{bmatrix} 1 & -2 & 0 \\ 5 & 1 & 2 \end{bmatrix} \quad C = \begin{bmatrix} -3 & 0 & 5 \\ 2 & 1 & 3 \end{bmatrix}$$

27. $A + B$ 28. $B + C$ 29. $2A - 3C$
30. $3C - 4B$ 31. $(A + B) - 2C$ 32. $4C + (A - B)$
33. $3A + 4(B + C)$ 34. $(A + B) + 3C$ 35. $2(A - B) - C$
36. $2A - 5(B + C)$

C   37. Find x, y, and z so that

$$[2 \quad 3 \quad -4] + [x \quad y \quad z] = [6 \quad -8 \quad 2]$$

38. Find x and y so that

$$\begin{bmatrix} 3 & -2 & 2 \\ 1 & 0 & -1 \end{bmatrix} + \begin{bmatrix} x-y & 2 & -2 \\ 4 & x & 6 \end{bmatrix} = \begin{bmatrix} 6 & 0 & 0 \\ 5 & 2x+y & 5 \end{bmatrix}$$

39. Let

$$U = \begin{bmatrix} 2 \\ -1 \\ 3 \end{bmatrix} \quad V = \begin{bmatrix} \frac{1}{2} \\ 0 \\ 1 \end{bmatrix} \quad W = \begin{bmatrix} -3 \\ -7 \\ 0 \end{bmatrix}$$

Compute the following:
(a) $U + V$ (b) $U - V$ (c) $\frac{1}{2}(U + V)$
(d) $U + V - W$ (e) $2U - 7V$ (f) $\frac{1}{4}U - \frac{1}{4}V - \frac{1}{4}W$

40. Find a_1, a_2, a_3 which satisfy the following:

$$\begin{bmatrix} 2 \\ 1 \\ 0 \end{bmatrix} + \begin{bmatrix} a_1 \\ a_2 \\ a_3 \end{bmatrix} = \begin{bmatrix} 2 \\ -1 \\ 3 \end{bmatrix}$$

APPLICATIONS   41. XYZ Company produces steel and aluminum nails. One week, 25 gross $\frac{1}{2}$-inch steel nails and 45 gross 1-inch steel nails were produced. Suppose 13 gross $\frac{1}{2}$-inch aluminum nails, 20 gross 1-inch aluminum nails, 35 gross 2-inch steel nails, and 23 gross 2-inch aluminum nails were also made. Write a 2×3 matrix depicting this. Could you also write a 3×2 matrix for this situation?

42. Tami, Carol, and Laura go to the candy store. Tami buys 5 sticks of gum, 2 ice cream cones, and 10 jelly beans. Carol buys 2 sticks of gum, 15 jelly beans, and 2 candy bars. Laura buys 1 stick of gum, 1 ice cream cone, and 4 candy bars. Write a matrix depicting this situation.

43. Survey Use a matrix to display the information given below, which was obtained in a survey of voters. Label the rows and columns.

351	Democrats earning under $15,000
271	Republicans earning under $15,000
73	Independents earning under $15,000
203	Democrats earning over $15,000
215	Republicans earning over $15,000
55	Independents earning over $15,000

44. Sales The sales figures for two car dealers during June showed that Dealer A sold 100 compacts, 50 intermediates, and 40 full-size cars, while Dealer B sold 120 compacts, 40 intermediates, and 35 full-size cars. During July, Dealer A sold 80 compacts,

30 intermediates, and 10 full-size cars, while Dealer B sold 70 compacts, 40 intermediates, and 20 full-size cars. Total sales over the 3-month period of June–August
revealed that Dealer A sold 300 compacts, 120 intermediates, and 65 full-size cars. In
the same 3-month period, Dealer B sold 250 compacts, 100 intermediates, and 80
full-size cars.

(a) Write 2×3 matrices summarizing sales data for June, July, and the 3-month
     period for each dealer.

(b) Use matrix addition to find the sales over the 2-month period for June and July
     for each dealer.

(c) Use matrix subtraction to find the sales in August for each dealer.

2.5

MULTIPLICATION OF MATRICES

PROPERTIES OF MATRIX MULTIPLICATION
IDENTITY MATRIX

While addition of matrices and scalar multiplication are fairly straightforward,
defining the *product* $A \cdot B$ of the two matrices A and B requires a bit more detail.
    We explain first what we mean by the product of a row vector with a column
vector.

If $R = [r_1 r_2 \ \ldots \ r_n]$ is a row vector and $C = \begin{bmatrix} c_1 \\ c_2 \\ \vdots \\ c_n \end{bmatrix}$ is a column vector, then by

the product of R and C we mean the number

$$r_1 c_1 + r_2 c_2 + r_3 c_3 + \cdots + r_n c_n$$

Example 1
If

$$R = [1 \quad 5 \quad 3] \quad \text{and} \quad C = \begin{bmatrix} 2 \\ -1 \\ 4 \end{bmatrix}$$

then the product of R and C is

$$1 \cdot 2 + 5 \cdot (-1) + 3 \cdot 4 = 9$$

■

Notice that for the above product to make sense, if R is a $1 \times n$ row vector, then
C must have dimension $n \times 1$.

Example 2

Let

$$R = [1 \quad 0 \quad 1] \quad \text{and} \quad C = \begin{bmatrix} 0 \\ -11 \\ 0 \end{bmatrix}$$

Then the product of R and C is

$$1 \cdot 0 + 0 \cdot (-11) + 1 \cdot 0 = 0 \qquad \blacksquare$$

Given two matrices A and B, the rows of A can be thought of as row vectors, while the columns of B can be thought of as column vectors. This observation will be used in the following main definition.

Matrix Multiplication **Let $A = [a_{ij}]$ be a matrix of dimension $m \times r$ and let $B = [b_{ij}]$ be a matrix of dimension $r \times n$. The *product* $A \cdot B$ is the matrix of dimension $m \times n$, whose *ij*th entry is the product of the *i*th row of A and the *j*th column of B. That is, the *ij*th entry of $A \cdot B$ is**

$$a_{i1}b_{1j} + a_{i2}b_{2j} + a_{i3}b_{3j} + \cdots + a_{ir}b_{rj}$$

The rule for multiplication of matrices is best illustrated by an example.

Example 3

Find the product $A \cdot B$ if

$$A = \begin{bmatrix} 1 & 3 & -2 \\ 4 & -1 & 5 \end{bmatrix} \quad \text{and} \quad B = \begin{bmatrix} 2 & -3 & 4 & 1 \\ -1 & 2 & 2 & 0 \\ 4 & 5 & 1 & 1 \end{bmatrix}$$

Solution

Since A is 2×3 and B is 3×4, the product $A \cdot B$ will be 2×4. To get, for example, the entry in row 2, column 3 of $A \cdot B$, we take the product of the second row of A with the third column of B. That is,

$$\begin{bmatrix} 1 & 3 & -2 \\ \boxed{4 \quad -1 \quad 5} \end{bmatrix} \begin{bmatrix} 2 & -3 & \boxed{4} & 1 \\ -1 & 2 & \boxed{2} & 0 \\ 4 & 5 & \boxed{1} & 1 \end{bmatrix}$$

We compute

$$4 \cdot 4 + (-1) \cdot 2 + 5 \cdot 1 = 19$$

Thus far, we have

$$A \cdot B = \begin{bmatrix} - & - & - & - \\ - & - & 19 & - \end{bmatrix}$$

To obtain the entry in row 1, column 2, we compute

$$1 \cdot (-3) + 3 \cdot 2 + (-2) \cdot 5 = -3 + 6 - 10 = -7$$

The other entries of $A \cdot B$ are obtained in a similar fashion. The final result—and you should verify this—is

$$A \cdot B = \begin{bmatrix} -9 & -7 & 8 & -1 \\ 29 & 11 & 19 & 9 \end{bmatrix}$$ ■

Let's look at some consequences of the definition of matrix multiplication.

If A is a matrix of dimension $m \times r$ (which has r columns) and B is a matrix of dimension $p \times n$ (which has p rows) and if $r \neq p$, the product $A \cdot B$ is not defined. That is, **multiplication of matrices is possible only if the number of columns of the first equals the number of rows of the second.**

If A is of dimension $m \times r$ and B is of dimension $r \times n$, then the product $A \cdot B$ is of dimension $m \times n$. See Figure 7.

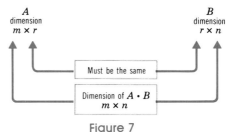

Figure 7

In Example 3, A is of dimension 2×3, B is of dimension 3×4, and we found the product $A \cdot B$ to be of dimension 2×4. Observe that the product $B \cdot A$ is not defined.

Example 4
For

$$A = \begin{bmatrix} 2 & 0 \\ 1 & 5 \end{bmatrix} \quad \text{and} \quad B = \begin{bmatrix} 3 & 2 \\ 1 & 4 \end{bmatrix}$$

compute $A \cdot B$ and $B \cdot A$.

Solution
We observe that the products $A \cdot B$ and $B \cdot A$ are both defined, so

$$A \cdot B = \begin{bmatrix} 6 & 4 \\ 8 & 22 \end{bmatrix} \qquad B \cdot A = \begin{bmatrix} 8 & 10 \\ 6 & 20 \end{bmatrix}$$ ■

Note from Example 4 that even if both $A \cdot B$ and $B \cdot A$ are defined, they may not be equal. We conclude that:

Matrix multiplication is not commutative

A natural question to ask is: Why are matrix products defined this way? As it turns out, such a definition of product is useful in applications. The following example affords one instance.

Example 5
Using the data of 1 month's production of Motors, Inc., from Example 1, Section 2.4, we have

$$A = \begin{bmatrix} \overset{\text{Sedan}}{23} & \overset{\text{Hardtop}}{16} & \overset{\text{Station Wagon}}{10} \\ 7 & 9 & 11 \end{bmatrix} \begin{matrix} \text{Units of material} \\ \text{Units of labor} \end{matrix}$$

Suppose that in this month's production, the cost for each unit of material is $45 and the cost for each unit of labor is $60. What is the total cost to manufacture the sedans, the hardtops, and the station wagons?

Solution
For sedans, the cost is 23 units of material at $45 each, plus 7 units of labor at $60 each, for a total cost of

$$23 \cdot \$45 + 7 \cdot \$60 = 1035 + 420 = \$1455$$

Similarly, for hardtops, the total cost is

$$16 \cdot \$45 + 9 \cdot \$60 = 720 + 540 = \$1260$$

Finally, for station wagons, the total cost is

$$10 \cdot \$45 + 11 \cdot \$60 = 450 + 660 = \$1110$$

We can represent the total cost for sedans, hardtops, and station wagons by the matrix

$$[1455 \quad 1260 \quad 1110]$$

If we represent the cost of units of material and units of labor by the row vector

$$U = [45 \quad 60]$$

the total cost of units for sedans, hardtops, and station wagons will then be given by the product UA.

$$U \cdot A = [45 \quad 60] \begin{bmatrix} 23 & 16 & 10 \\ 7 & 9 & 11 \end{bmatrix}$$
$$= [45 \cdot 23 + 60 \cdot 7 \quad 45 \cdot 16 + 60 \cdot 9 \quad 45 \cdot 10 + 60 \cdot 11]$$
$$= [1455 \quad 1260 \quad 1110]$$

PROPERTIES OF MATRIX MULTIPLICATION
In listing some of the properties of matrix multiplication in this book, we agree to follow the usual convention and write $A \cdot B$ as AB, from now on.

> Let A be a matrix of dimension $m \times r$, let B be a matrix of dimension $r \times p$, and let C be a matrix of dimension $p \times n$. Then matrix multiplication is *associative*. That is,
>
> $$A(BC) = (AB)C$$
>
> The resulting matrix ABC is of dimension $m \times n$.

Notice the limitations that are placed on the dimensions of the matrices in order for multiplication to be possible.

> Let A be a matrix of dimension $m \times r$. Let B and C be matrices of dimension $r \times n$. Then *matrix multiplication distributes over matrix addition.* That is,
>
> $$A(B + C) = AB + AC$$
>
> The resulting matrix $AB + AC$ is of dimension $m \times n$.

IDENTITY MATRIX

A special type of square matrix is the *identity matrix,* which is denoted by I_n. It has the property that all its diagonal entries are 1's and all other entries are 0's. Thus,

$$
I_n = \begin{bmatrix}
1 & 0 & \cdots & 0 & 0 \\
0 & 1 & \cdots & 0 & 0 \\
\vdots & \vdots & & \vdots & \vdots \\
0 & 0 & \cdots & 1 & 0 \\
0 & 0 & \cdots & 0 & 1
\end{bmatrix}
$$

where the subscript n implies that I_n is of dimension $n \times n$. For example,

$$
I_2 = \begin{bmatrix} 1 & 0 \\ 0 & 1 \end{bmatrix}, \qquad I_3 = \begin{bmatrix} 1 & 0 & 0 \\ 0 & 1 & 0 \\ 0 & 0 & 1 \end{bmatrix}, \text{ etc.}
$$

Example 6

For

$$A = \begin{bmatrix} 3 & 2 \\ -4 & 5 \end{bmatrix}$$

compute: (a) AI_2 (b) I_2A

Solution

$$
\text{(a)} \quad AI_2 = \begin{bmatrix} 3 & 2 \\ -4 & 5 \end{bmatrix}\begin{bmatrix} 1 & 0 \\ 0 & 1 \end{bmatrix} = \begin{bmatrix} 3 & 2 \\ -4 & 5 \end{bmatrix} = A
$$

$$
\text{(b)} \quad I_2A = \begin{bmatrix} 1 & 0 \\ 0 & 1 \end{bmatrix}\begin{bmatrix} 3 & 2 \\ -4 & 5 \end{bmatrix} = \begin{bmatrix} 3 & 2 \\ -4 & 5 \end{bmatrix} = A
$$

■

This example can be generalized as follows:

> If A is a matrix of dimension $m \times n$ and if I_n denotes the identity matrix of dimension $n \times n$, and I_m denotes the identity matrix of dimension $m \times m$, then
>
> $$I_m A = A \qquad \text{and} \qquad A I_n = A$$

The reason for stating two formulas is that when the matrix A is not square, care must be taken when forming the products AI and IA. For example, if

$$A = \begin{bmatrix} 1 & 2 \\ 3 & 2 \\ 1 & 1 \end{bmatrix}$$

then A is of dimension 3×2 and

$$AI_2 = A \begin{bmatrix} 1 & 0 \\ 0 & 1 \end{bmatrix} = \begin{bmatrix} 1 & 2 \\ 3 & 2 \\ 1 & 1 \end{bmatrix} \begin{bmatrix} 1 & 0 \\ 0 & 1 \end{bmatrix} = \begin{bmatrix} 1 & 2 \\ 3 & 2 \\ 1 & 1 \end{bmatrix} = A$$

Although the product $I_2 A$ is not defined, we can calculate the product $I_3 A$ as follows:

$$I_3 A = \begin{bmatrix} 1 & 0 & 0 \\ 0 & 1 & 0 \\ 0 & 0 & 1 \end{bmatrix} \begin{bmatrix} 1 & 2 \\ 3 & 2 \\ 1 & 1 \end{bmatrix} = \begin{bmatrix} 1 & 2 \\ 3 & 2 \\ 1 & 1 \end{bmatrix} = A$$

In particular if A is a square matrix of dimension $n \times n$, then $A I_n = I_n A = A$. Thus, for square matrices the identity matrix plays the role that the number 1 plays in the set of real numbers.

Exercise 2.5

Answers to Odd-Numbered Problems begin on page A-7.

A   In Problems 1–10 let A be of dimension 3×4, let B be of dimension 3×3, let C be of dimension 2×3, and let D be of dimension 3×2. Determine which of the following expressions are defined and, for those that are, give the dimension.

1. BA	2. CD	3. AB	4. DC
5. $(BA)C$	6. $A(CD)$	7. $BA + A$	8. $CD + BA$
9. $DC + B$	10. $CB - A$		

In Problems 11–26 use the matrices below to perform the indicated operations.

$$A = \begin{bmatrix} 1 & 2 \\ 0 & 4 \end{bmatrix} \qquad B = \begin{bmatrix} 1 & 2 & 3 \\ -1 & 4 & -2 \end{bmatrix} \qquad C = \begin{bmatrix} 3 & 1 \\ 4 & -1 \\ 0 & 2 \end{bmatrix}$$

$$D = \begin{bmatrix} 1 & 0 & 4 \\ 0 & 1 & 2 \\ 0 & -1 & 1 \end{bmatrix} \qquad E = \begin{bmatrix} 3 & -1 \\ 4 & 2 \end{bmatrix}$$

11. AB	12. DC	13. BC	14. AA
15. $(D + I_3)C$	16. $DC + C$	17. $(DC)B$	18. $D(CB)$

19. EI_2 **20.** I_3D **21.** $(2E)B$ **22.** $E(2B)$

23. $-5E + A$ **24.** $3A + 2E$ **25.** $3CB + 4D$ **26.** $2EA - 3BC$

27. For

$$A = \begin{bmatrix} 1 & -1 \\ 2 & 0 \end{bmatrix} \quad \text{and} \quad B = \begin{bmatrix} 3 & 2 \\ -1 & 4 \end{bmatrix}$$

find AB and BA. Notice that $AB \neq BA$.

B   **28.** Show that, for all values a, b, c, and d, the matrices

$$A = \begin{bmatrix} a & b \\ -b & a \end{bmatrix} \quad \text{and} \quad B = \begin{bmatrix} c & d \\ -d & c \end{bmatrix}$$

are commutative; that is, $AB = BA$.

29. If possible, find a matrix A such that

$$A \begin{bmatrix} 0 & 1 \\ 2 & -1 \end{bmatrix} = \begin{bmatrix} 2 & 1 \\ -1 & 0 \end{bmatrix} \qquad \text{Hint:} \quad \text{Let } A = \begin{bmatrix} a & b \\ c & d \end{bmatrix}.$$

30. For what numbers x will the following be true?

$$[x \quad 4 \quad 1] \begin{bmatrix} 2 & 1 & 0 \\ 1 & 0 & 2 \\ 0 & 2 & 4 \end{bmatrix} \begin{bmatrix} x \\ -7 \\ \frac{5}{4} \end{bmatrix} = 0$$

31. Let

$$A = \begin{bmatrix} 1 & 2 & 5 \\ 2 & 4 & 10 \\ -1 & -2 & -5 \end{bmatrix}$$

Show that $A^2 = 0$. Thus, the rule in the real number system that if $a^2 = 0$, then $a = 0$ does not hold for matrices.

32. What must be true about a, b, c, and d, if we demand that $AB = BA$ for the following matrices?

$$A = \begin{bmatrix} a & b \\ c & d \end{bmatrix} \qquad B = \begin{bmatrix} 1 & 1 \\ -1 & 1 \end{bmatrix}$$

Assume that

$$\begin{bmatrix} a & b \\ c & d \end{bmatrix} \neq \begin{bmatrix} 1 & 0 \\ 0 & 1 \end{bmatrix}$$

33. Let

$$A = \begin{bmatrix} a & b \\ b & a \end{bmatrix}$$

Find a and b such that $A^2 + A = 0$, where $A^2 = AA$.

34. For the matrix

$$A = \begin{bmatrix} a & 1-a \\ 1+a & -a \end{bmatrix}$$

show that $A^2 = AA = I_2$.

C   **35.** Find the vector $[x_1 \ x_2]$ for which

$$[x_1 \ x_2] \begin{bmatrix} \frac{1}{2} & \frac{1}{2} \\ \frac{1}{4} & \frac{3}{4} \end{bmatrix} = [x_1 \ x_2]$$

under the condition that $x_1 + x_2 = 1$. Here, the vector $[x_1 \ x_2]$ is called a *fixed vector* of the matrix

$$\begin{bmatrix} \frac{1}{2} & \frac{1}{2} \\ \frac{1}{4} & \frac{3}{4} \end{bmatrix}$$

36. For a square matrix A, it is possible to find $A \cdot A = A^2$. It is also clear that we can compute

$$A^n = \underbrace{A \cdot \ \cdots \ \cdot A}_{n \ times}$$

Find A^2, A^3, and A^4 for each of the following square matrices:

(a) $A = \begin{bmatrix} 1 & 0 \\ 3 & 2 \end{bmatrix}$ (b) $A = \begin{bmatrix} 3 & 1 \\ -2 & -1 \end{bmatrix}$

(c) $A = \begin{bmatrix} 0 & 1 & 1 \\ 0 & -1 & 2 \\ 6 & 3 & -2 \end{bmatrix}$ (d) $A = \begin{bmatrix} 1 & 0 \\ 0 & 1 \end{bmatrix}$ (e) $A = \begin{bmatrix} \frac{1}{2} & \frac{1}{2} \\ \frac{1}{4} & \frac{3}{4} \end{bmatrix}$

Can you guess what A^n looks like for part (d)? For part (e)?

In Problems 37–46 use the given matrices to perform the indicated operation.

$$A = \begin{bmatrix} 0 & 1 & 0 \\ 0 & 0 & 1 \\ 1 & 0 & 0 \end{bmatrix} \qquad B = \begin{bmatrix} 2 & -1 \\ 3 & 11 \end{bmatrix} \qquad C = \begin{bmatrix} 2 & 0 \\ 4 & 3 \end{bmatrix}$$

37. B^2 **38.** A^2

39. C^3 **40.** $(BC)^2$

41. $B^2 C^2$ **42.** $(CB)^2$

43. If B is a 2×3 matrix and $B \cdot A^2$ is defined, what is the dimension of A?

44. If B is a 4×5 matrix and $A^3 \cdot B$ is defined, what is the dimension of A?

APPLICATIONS   **45.** Suppose a factory is asked to produce three types of products, which we will call P_1, P_2, P_3. Suppose the following purchase order was received: $P_1 = 7$, $P_2 = 12$, $P_3 = 5$. Represent this order by a row vector and call it P:

$$P = [7 \ \ 12 \ \ 5]$$

To produce each of the products, raw materials of four kinds are needed. Call the raw materials M_1, M_2, M_3, and M_4. The matrix below gives the amount of material needed for each product:

$$\begin{array}{c} \\ \\ Q = \begin{matrix} P_1 \\ P_2 \\ P_3 \end{matrix} \end{array} \begin{matrix} M_1 & M_2 & M_3 & M_4 \\ \begin{bmatrix} 2 & 3 & 1 & 12 \\ 7 & 9 & 5 & 20 \\ 8 & 12 & 6 & 15 \end{bmatrix} \end{matrix}$$

Suppose the cost for each of the materials M_1, M_2, M_3, and M_4 is \$10, \$12, \$15, and \$20, respectively. The cost vector is

$$C = \begin{bmatrix} 10 \\ 12 \\ 15 \\ 20 \end{bmatrix}$$

Compute each of the following and interpret each one:

(a) PQ (b) QC (c) PQC

46. Tami went to a department store and purchased 6 pairs of pants, 8 shirts, and 2 jackets. Laura purchased 2 pairs of pants, 5 shirts, and 3 jackets. If the pants are \$5 each, shirts are \$3 each, and jackets are \$9 each, use matrix multiplication to find the amounts spent by Tami and Laura.

2.6

INVERSE OF A MATRIX

REDUCED ROW-ECHELON TECHNIQUE
SOLVING A SYSTEM OF n LINEAR EQUATIONS IN n UNKNOWNS USING INVERSES
MATRICES IN PRACTICE

The inverse of a matrix, if it exists, plays the role that the reciprocal of a number plays in the set of real numbers.

Inverse Let A be a matrix of dimension $n \times n$. A matrix B of dimension $n \times n$ is called an *inverse of A* if $AB = BA = I_n$. We denote the inverse of a matrix A, if it exists, by A^{-1}.

Example 1
Show that $\begin{bmatrix} \frac{1}{2} & -\frac{1}{2} \\ 0 & 1 \end{bmatrix}$ is the inverse of $\begin{bmatrix} 2 & 1 \\ 0 & 1 \end{bmatrix}$.

Solution
Since

$$\begin{bmatrix} 2 & 1 \\ 0 & 1 \end{bmatrix}\begin{bmatrix} \frac{1}{2} & -\frac{1}{2} \\ 0 & 1 \end{bmatrix} = \begin{bmatrix} 1 & 0 \\ 0 & 1 \end{bmatrix}$$

and

$$\begin{bmatrix} \frac{1}{2} & -\frac{1}{2} \\ 0 & 1 \end{bmatrix}\begin{bmatrix} 2 & 1 \\ 0 & 1 \end{bmatrix} = \begin{bmatrix} 1 & 0 \\ 0 & 1 \end{bmatrix}$$

the required condition is met. ■

The next result tells us that a square matrix will not have more than one inverse.

> A square matrix A has at most one inverse. That is, the inverse of a matrix, if it exists, is *unique*.

To verify this, suppose we have two inverses B and C for a matrix A. Then

$$AB = BA = I_n \quad \text{and} \quad AC = CA = I_n$$

Multiplying both sides of $AC = I_n$ on the left by B, we find

$$B(AC) = BI_n = B$$

Similarly, multiplying both sides of $BA = I_n$ on the right by C, we obtain

$$(BA)C = I_n C = C$$

But $B(AC) = (BA)C$. Hence, $B = C$.

Because of the uniqueness of the inverse of a matrix, we are justified in the use of the phrase "the inverse of A" and the notation A^{-1} for that inverse.

What about nonsquare matrices? Can they have inverses? The answer is No. By definition, whenever a matrix has an inverse, it will commute with its inverse under multiplication. So if the nonsquare matrix A had the alleged inverse B, then AB would have to be equal to BA. But the fact that A is not square causes AB and BA to have different dimensions and prevents them from being equal. So such a B could not exist.

> If a matrix is not square, then it does not have an inverse.

The next example provides a technique for finding the inverse of a matrix. Although this technique is not the best, it is illustrative.

Example 2

Find the inverse of the matrix: $A = \begin{bmatrix} 2 & 1 \\ 0 & 1 \end{bmatrix}$.

Solution

We begin by assuming that this matrix has an inverse of the form

$$A^{-1} = \begin{bmatrix} a & b \\ c & d \end{bmatrix}$$

Then the product of A and A^{-1} must be the identity matrix:

$$\begin{bmatrix} 2 & 1 \\ 0 & 1 \end{bmatrix} \begin{bmatrix} a & b \\ c & d \end{bmatrix} = \begin{bmatrix} 1 & 0 \\ 0 & 1 \end{bmatrix}$$

Multiplying the matrices on the left side, we get

$$\begin{bmatrix} 2a + c & 2b + d \\ c & d \end{bmatrix} = \begin{bmatrix} 1 & 0 \\ 0 & 1 \end{bmatrix}$$

The condition for equality requires that

$$2a + c = 1 \qquad 2b + d = 0 \qquad c = 0 \qquad d = 1$$

Thus,

$$a = \tfrac{1}{2} \qquad b = -\tfrac{1}{2} \qquad c = 0 \qquad d = 1$$

Hence, the inverse of

$$A = \begin{bmatrix} 2 & 1 \\ 0 & 1 \end{bmatrix} \qquad \text{is} \qquad A^{-1} = \begin{bmatrix} \tfrac{1}{2} & -\tfrac{1}{2} \\ 0 & 1 \end{bmatrix}$$

We verify that this is the inverse by computing AA^{-1} and $A^{-1}A$

$$AA^{-1} = \begin{bmatrix} 2 & 1 \\ 0 & 1 \end{bmatrix} \begin{bmatrix} \tfrac{1}{2} & -\tfrac{1}{2} \\ 0 & 1 \end{bmatrix} = \begin{bmatrix} 1 & 0 \\ 0 & 1 \end{bmatrix}$$

$$A^{-1}A = \begin{bmatrix} \tfrac{1}{2} & -\tfrac{1}{2} \\ 0 & 1 \end{bmatrix} \begin{bmatrix} 2 & 1 \\ 0 & 1 \end{bmatrix} = \begin{bmatrix} 1 & 0 \\ 0 & 1 \end{bmatrix}$$

Sometimes, a square matrix does not have an inverse.

Example 3
Show that the matrix below does not have an inverse.

$$A = \begin{bmatrix} 0 & 1 \\ 0 & 0 \end{bmatrix}$$

Solution
We proceed as in Example 2 by assuming that A does have an inverse. It will be of the form

$$A^{-1} = \begin{bmatrix} a & b \\ c & d \end{bmatrix}$$

The product of A and A^{-1} must be the identity matrix. Thus,

$$\begin{bmatrix} 0 & 1 \\ 0 & 0 \end{bmatrix} \begin{bmatrix} a & b \\ c & d \end{bmatrix} = \begin{bmatrix} 1 & 0 \\ 0 & 1 \end{bmatrix}$$

Performing the multiplication on the left side, we have

$$\begin{bmatrix} c & d \\ 0 & 0 \end{bmatrix} = \begin{bmatrix} 1 & 0 \\ 0 & 1 \end{bmatrix}$$

But these two matrices can never be equal. We conclude that our assumption that A has an inverse is false. That is, A does not have an inverse.

The procedures used above to find the inverse, if it exists, of a square matrix become quite involved as the dimension of the matrix gets larger. A more efficient method is provided next.

REDUCED ROW-ECHELON TECHNIQUE
We will introduce this technique by looking at a specific example.

Example 4

Find the inverse of the matrix

$$A = \begin{bmatrix} 2 & 1 \\ 0 & 1 \end{bmatrix}$$

Solution

Assuming A has an inverse, we will denote it by

$$X = \begin{bmatrix} x_1 & x_2 \\ x_3 & x_4 \end{bmatrix}$$

Then the product of A and X is the identity matrix of dimension 2×2. That is,

$$AX = I_2$$

$$\begin{bmatrix} 2 & 1 \\ 0 & 1 \end{bmatrix} \begin{bmatrix} x_1 & x_2 \\ x_3 & x_4 \end{bmatrix} = \begin{bmatrix} 1 & 0 \\ 0 & 1 \end{bmatrix}$$

Performing the multiplication on the left yields

$$\begin{bmatrix} 2x_1 + x_3 & 2x_2 + x_4 \\ x_3 & x_4 \end{bmatrix} = \begin{bmatrix} 1 & 0 \\ 0 & 1 \end{bmatrix}$$

This matrix equation can be written as the following system of four equations in four unknowns:

$$2x_1 + x_3 = 1 \qquad 2x_2 + x_4 = 0$$
$$x_3 = 0 \qquad\qquad x_4 = 1$$

By inspection, we find the solution to be

$$x_1 = \tfrac{1}{2} \qquad x_2 = -\tfrac{1}{2} \qquad x_3 = 0 \qquad x_4 = 1$$

Thus, the inverse of A is

$$A^{-1} = \begin{bmatrix} \tfrac{1}{2} & -\tfrac{1}{2} \\ 0 & 1 \end{bmatrix}$$

■

Let's look at what we did more closely. The system of four equations in four unknowns can be written in two blocks as

(a) $2x_1 + x_3 = 1$ (b) $2x_2 + x_4 = 0$
$\qquad x_3 = 0$ $x_4 = 1$

Their augmented matrices are

(a) $\begin{bmatrix} 2 & 1 & | & 1 \\ 0 & 1 & | & 0 \end{bmatrix}$ and   (b) $\begin{bmatrix} 2 & 1 & | & 0 \\ 0 & 1 & | & 1 \end{bmatrix}$

Since the matrix A appears in both (a) and (b), any row operation we perform on (a) and (b) can be performed more easily on the single augmented matrix that combines the two right-hand columns. We denote this matrix by $A|I_2$ and write

$$[A|I_2] = \begin{bmatrix} 2 & 1 & | & 1 & 0 \\ 0 & 1 & | & 0 & 1 \end{bmatrix}$$

If we perform row operations on $[A|I_2]$, just as if we were computing the reduced row-echelon form of A, we get

$$\begin{bmatrix} 1 & 0 & | & \frac{1}{2} & -\frac{1}{2} \\ 0 & 1 & | & 0 & 1 \end{bmatrix}$$

The 2×2 matrix on the right-hand side of the vertical bar is A^{-1}.

This example illustrates the general procedure:

To find the inverse, if it exists, of a square matrix of dimension $n \times n$, follow these steps:

Finding the Inverse of a Matrix

Step 1. Write the matrix $[A|I_n]$.

Step 2. Using row operations, write $[A|I_n]$ in reduced row-echelon form.

Step 3. If the resulting matrix is of the form $[I_n|B]$, that is, if the identity matrix appears on the left side of the bar, then B is the inverse of A. Otherwise, A has no inverse.

Let's work another example.

Example 5

Find the inverse of the matrix

$$A = \begin{bmatrix} 1 & 1 & 2 \\ 2 & 1 & 0 \\ 1 & 2 & 2 \end{bmatrix}$$

Solution

Since A is of dimension 3×3, we use the identity matrix I_3. The matrix $[A|I_3]$ is

$$[A|I_3] = \begin{bmatrix} 1 & 1 & 2 & | & 1 & 0 & 0 \\ 2 & 1 & 0 & | & 0 & 1 & 0 \\ 1 & 2 & 2 & | & 0 & 0 & 1 \end{bmatrix}$$

We proceed to transform this matrix, using row operations:

Use $\begin{aligned} R_2 &= -2r_1 + r_2 \\ R_3 &= -r_1 + r_3 \end{aligned}$ to get $\begin{bmatrix} 1 & 1 & 2 & | & 1 & 0 & 0 \\ 0 & -1 & -4 & | & -2 & 1 & 0 \\ 0 & 1 & 0 & | & -1 & 0 & 1 \end{bmatrix}$

Use $R_2 = (-1)r_2$ to get $\begin{bmatrix} 1 & 1 & 2 & | & 1 & 0 & 0 \\ 0 & 1 & 4 & | & 2 & -1 & 0 \\ 0 & 1 & 0 & | & -1 & 0 & 1 \end{bmatrix}$

Use $\begin{aligned} R_1 &= -r_2 + r_1 \\ R_3 &= -r_2 + r_3 \end{aligned}$ to get $\begin{bmatrix} 1 & 0 & -2 & | & -1 & 1 & 0 \\ 0 & 1 & 4 & | & 2 & -1 & 0 \\ 0 & 0 & -4 & | & -3 & 1 & 1 \end{bmatrix}$

Use $R_3 = (-\frac{1}{4})r_3$ to get
$$\left[\begin{array}{ccc|ccc} 1 & 0 & -2 & -1 & 1 & 0 \\ 0 & 1 & 4 & 2 & -1 & 0 \\ 0 & 0 & 1 & \frac{3}{4} & -\frac{1}{4} & -\frac{1}{4} \end{array}\right]$$

Use $\begin{array}{l} R_1 = 2r_3 + r_1 \\ R_2 = -4r_3 + r_2 \end{array}$ to get
$$\left[\begin{array}{ccc|ccc} 1 & 0 & 0 & \frac{1}{2} & \frac{1}{2} & -\frac{1}{2} \\ 0 & 1 & 0 & -1 & 0 & 1 \\ 0 & 0 & 1 & \frac{3}{4} & -\frac{1}{4} & -\frac{1}{4} \end{array}\right] = [I_3 | A^{-1}]$$

Since the identity matrix I_3 appears on the left side, the matrix appearing on the right is the inverse. That is,

$$A^{-1} = \left[\begin{array}{ccc} \frac{1}{2} & \frac{1}{2} & -\frac{1}{2} \\ -1 & 0 & 1 \\ \frac{3}{4} & -\frac{1}{4} & -\frac{1}{4} \end{array}\right]$$

(You should verify that, in fact, $AA^{-1} = I_3$ and $A^{-1}A = I_3$.)

Example 6

Show that the matrix given below has no inverse.

$$\left[\begin{array}{cc} 3 & 2 \\ 6 & 4 \end{array}\right]$$

Solution

We set up the matrix

$$\left[\begin{array}{cc|cc} 3 & 2 & 1 & 0 \\ 6 & 4 & 0 & 1 \end{array}\right]$$

Use $R_1 = \frac{1}{3}r_1$ to get $\left[\begin{array}{cc|cc} 1 & \frac{2}{3} & \frac{1}{3} & 0 \\ 6 & 4 & 0 & 1 \end{array}\right]$

Use $R_2 = -6r_1 + r_2$ to get $\left[\begin{array}{cc|cc} 1 & \frac{2}{3} & \frac{1}{3} & 0 \\ 0 & 0 & -2 & 1 \end{array}\right]$

The 0's in row 2 tell us we cannot get the identity matrix. This, in turn, tells us the original matrix has no inverse.

SOLVING A SYSTEM OF n LINEAR EQUATIONS IN n UNKNOWNS USING INVERSES

The inverse of a matrix can also be used to solve a system of n linear equations in n unknowns. Let's look at an example.

Example 7

Solve the system of equations

$$\begin{array}{r} x + y + 2z = 1 \\ 2x + y = 2 \\ x + 2y + 2z = 3 \end{array}$$

Solution

If we let

$$A = \begin{bmatrix} 1 & 1 & 2 \\ 2 & 1 & 0 \\ 1 & 2 & 2 \end{bmatrix} \qquad X = \begin{bmatrix} x \\ y \\ z \end{bmatrix} \qquad B = \begin{bmatrix} 1 \\ 2 \\ 3 \end{bmatrix}$$

the above system can be written as

$$AX = B$$

From Example 5, we know A has an inverse, A^{-1}. If we multiply on the left both sides of the equation by A^{-1}, we obtain

$$A^{-1}(AX) = A^{-1}B$$
$$(A^{-1}A)X = A^{-1}B$$
$$I_3 X = A^{-1}B$$
$$X = A^{-1}B$$

$$X = \begin{bmatrix} \frac{1}{2} & \frac{1}{2} & -\frac{1}{2} \\ -1 & 0 & 1 \\ \frac{3}{4} & -\frac{1}{4} & -\frac{1}{4} \end{bmatrix} \begin{bmatrix} 1 \\ 2 \\ 3 \end{bmatrix} = \begin{bmatrix} 0 \\ 2 \\ -\frac{1}{2} \end{bmatrix}$$

Thus, the solution is

$$x = 0 \qquad y = 2 \qquad z = -\tfrac{1}{2}$$

∎

This method of using matrix equations to solve a system of equations is particularly useful for applications in which the constants appearing to the right of the equal sign change while the coefficients of the unknowns on the left side do not. See Problems 33 through 59 for an illustration. See also Section 2.7 on Leontief models for an application.

MATRICES IN PRACTICE

Systems of linear equations arise in business, economics, sociology, chemistry — in fact, in any field that has a quantitative side to it. The systems encountered in practice are often quite large, with 100 equations in 100 unknowns not unusual, making hand calculations with such systems out of the question. Thus computer routines are used to implement work such as row-reducing the augmented matrix. A very popular collection of such routines is the LINPACK package described by Rice.* In more advanced treatments it is shown that solving n equations in n unknowns requires roughly n^3 multiplications and additions. Thus a 100 by 100 system will require 10^6 arithmetic operations. But if we have access to a machine that can perform 100,000 operations per second, the task seems less formidable, since our 100 by 100 system would be solved in 10 seconds. The availability of high-speed computing has greatly enhanced the applicability of matrices, since they can now be used in large-scale problems. See the article by Kolata in *Science*.†

There is an aspect of computer-performed matrix calculations that can at times

---

* John R. Rice, *Matrix Computations and Mathematical Software*, McGraw-Hill, New York, 1981.

† Gina Kolata, "Solving Linear Systems Faster," *Science* (June 14, 1985).

be potentially troublesome in applications. Computers by their nature can perform arithmetic only on decimals that have finitely many nonzero terms. Decimals that are infinite in length are rounded off. For example $\frac{2}{3}$ has the nontermin-
ating decimal expansion .6666 . . . . A machine would round this off and store it
as, say, .6666667 (the actual number of significant places would vary with the
system). This can have consequences for matrix calculations, as we show in the
following example.

Example 8

Suppose we wish to find the row-reduced form of the matrix

$$A = \begin{bmatrix} 1 & \frac{1}{3} \\ 2 & \frac{2}{3} \end{bmatrix}$$

A direct hand calculation shows that the row-reduced form is

$$\begin{bmatrix} 1 & \frac{1}{3} \\ 0 & 0 \end{bmatrix}$$

Now assume we did this on a computer that rounded-off and stored, say, two
significant digits.

Our matrix A would then be represented in the machine as

$$A = \begin{bmatrix} 1 & .33 \\ 2 & .67 \end{bmatrix}$$

If a program were now called upon to row-reduce A, the following steps would
result:

$$R_2 = r_2 - 2r_1 \qquad \begin{bmatrix} 1 & .33 \\ 0 & .01 \end{bmatrix}$$

$$R_2 = \frac{1}{.01} r_2 \qquad \begin{bmatrix} 1 & .33 \\ 0 & 1 \end{bmatrix}$$

$$R_1 = r_1 - .33r_2 \qquad \begin{bmatrix} 1 & 0 \\ 0 & 1 \end{bmatrix}$$

Note that the end result of the computer calculation differs drastically from our
own computed row-reduced form. The problem clearly lies in the rounded representation of the numbers $\frac{1}{3}$ and $\frac{2}{3}$. ∎

Though the above example is simplistic, errors in matrix calculations introduced by round-off or truncation can and do occur and can have disastrous
consequences. Were the matrix in the example above the coefficient matrix of a
system, then the computer program would conclude that the system has a unique
solution, while in reality the system has either infinitely many solutions or no
solution. The subject of error propagation when doing matrix arithmetic on a
computer is of great importance to people who use matrices in practice, since they
want to be assured that their results are meaningful. It is also a subject where much

current work is being done by computer scientists and mathematicians. The books by Rice and by Noble and Daniel* provide further detail.

Exercise 2.6   *Answers to Odd-Numbered Problems begin on page A-8.*

A   In Problems 1–6 show that the given matrices are inverses of each other.

1. $\begin{bmatrix} 1 & 2 \\ 2 & 3 \end{bmatrix} \begin{bmatrix} -3 & 2 \\ 2 & -1 \end{bmatrix}$

2. $\begin{bmatrix} 1 & 5 \\ 2 & 0 \end{bmatrix} \begin{bmatrix} 0 & \frac{1}{2} \\ \frac{1}{5} & -\frac{1}{10} \end{bmatrix}$

3. $\begin{bmatrix} -1 & -2 \\ 3 & 4 \end{bmatrix} \begin{bmatrix} 2 & 1 \\ -\frac{3}{2} & -\frac{1}{2} \end{bmatrix}$

4. $\begin{bmatrix} 1 & 3 \\ 2 & -1 \end{bmatrix} \begin{bmatrix} \frac{1}{7} & \frac{3}{7} \\ \frac{2}{7} & -\frac{1}{7} \end{bmatrix}$

5. $\begin{bmatrix} 1 & 2 & 3 \\ 2 & 3 & 4 \\ 1 & 2 & 1 \end{bmatrix} \begin{bmatrix} -\frac{5}{2} & 2 & -\frac{1}{2} \\ 1 & -1 & 1 \\ \frac{1}{2} & 0 & -\frac{1}{2} \end{bmatrix}$

6. $\begin{bmatrix} 1 & 3 & 3 \\ 1 & 4 & 3 \\ 1 & 3 & 4 \end{bmatrix} \begin{bmatrix} 7 & -3 & -3 \\ -1 & 1 & 0 \\ -1 & 0 & 1 \end{bmatrix}$

In Problems 7–20 find the inverse of each matrix using the reduced row-echelon technique.

7. $\begin{bmatrix} 2 & 5 \\ 1 & 3 \end{bmatrix}$

8. $\begin{bmatrix} 4 & 1 \\ 3 & 1 \end{bmatrix}$

9. $\begin{bmatrix} 1 & -1 \\ 3 & -4 \end{bmatrix}$

10. $\begin{bmatrix} 5 & 3 \\ 3 & 2 \end{bmatrix}$

11. $\begin{bmatrix} 2 & 1 \\ 4 & 3 \end{bmatrix}$

12. $\begin{bmatrix} 2 & 3 \\ 2 & -1 \end{bmatrix}$

13. $\begin{bmatrix} 0 & 0 & 1 \\ 0 & 1 & 0 \\ 1 & 0 & 0 \end{bmatrix}$

14. $\begin{bmatrix} -1 & 1 & 0 \\ 1 & 0 & 2 \\ 3 & 1 & 0 \end{bmatrix}$

15. $\begin{bmatrix} 1 & 1 & -1 \\ 3 & -1 & 0 \\ 2 & -3 & 4 \end{bmatrix}$

16. $\begin{bmatrix} 1 & 1 & 1 \\ 2 & 1 & 1 \\ 1 & 1 & 2 \end{bmatrix}$

17. $\begin{bmatrix} 1 & 1 & -1 \\ 2 & 1 & 1 \\ 1 & 0 & 1 \end{bmatrix}$

18. $\begin{bmatrix} 2 & 3 & -1 \\ 1 & 1 & 1 \\ 0 & 2 & -1 \end{bmatrix}$

19. $\begin{bmatrix} 1 & 1 & 0 & 0 \\ 0 & 1 & -1 & 1 \\ 1 & -1 & 1 & 1 \\ 0 & 1 & 0 & -1 \end{bmatrix}$

20. $\begin{bmatrix} 1 & 2 & -3 & -2 \\ 0 & 1 & 4 & -2 \\ 3 & -1 & 4 & 0 \\ 2 & 1 & 0 & 3 \end{bmatrix}$

In Problems 21–26 show that each matrix has no inverse.

21. $\begin{bmatrix} 4 & 6 \\ 2 & 3 \end{bmatrix}$

22. $\begin{bmatrix} -1 & 2 \\ 3 & -6 \end{bmatrix}$

23. $\begin{bmatrix} -8 & 4 \\ -4 & 2 \end{bmatrix}$

24. $\begin{bmatrix} 2 & 10 \\ 1 & 5 \end{bmatrix}$

25. $\begin{bmatrix} 1 & 1 & 1 \\ 3 & -4 & 2 \\ 0 & 0 & 0 \end{bmatrix}$

26. $\begin{bmatrix} -1 & 2 & 3 \\ 5 & 2 & 0 \\ 2 & -4 & -6 \end{bmatrix}$

In Problems 27–32 find the inverse, if it exists, of each matrix.

27. $\begin{bmatrix} 1 & 1 \\ 1 & 2 \end{bmatrix}$

28. $\begin{bmatrix} 2 & 1 \\ 1 & 1 \end{bmatrix}$

29. $\begin{bmatrix} 3 & -2 \\ 0 & 2 \end{bmatrix}$

30. $\begin{bmatrix} 4 & -1 \\ -1 & 0 \end{bmatrix}$

31. $\begin{bmatrix} 3 & 2 \\ 6 & 4 \end{bmatrix}$

32. $\begin{bmatrix} 4 & 2 \\ 2 & 1 \end{bmatrix}$

---

* B. Noble and J. Daniel, *Applied Linear Algebra,* Prentice-Hall, Englewood Cliffs, N.J., 1977.

In Problems 33–44 solve each system of equations by the method of Example 7.

33. $x + y = 6$
$2x - y = 0$

34. $x - y = 2$
$2x + y = 1$

35. $2x + 3y = 7$
$3x - y = 5$

36. $2x - 3y = 5$
$3x + y = 2$

37. $2x - 3y = 0$
$4x + 9y = 5$

38. $3x - 4y = 3$
$6x + 2y = 1$

39. $\frac{1}{2}x + \frac{1}{3}y = 2$
$x + y = 5$

40. $x - \frac{1}{4}y = 0$
$\frac{1}{2}x + \frac{1}{2}y = \frac{5}{2}$

41. $3x + 7y = 10$
$2x + 5y = 7$

42. $3x + 7y = -4$
$2x + 5y = -3$

43. $3x + 7y = 13$
$2x + 5y = 9$

44. $3x + 7y = 20$
$2x + 5y = 14$

B   In Problems 45–54 solve each system by the method of Example 7, that is, by finding A^{-1}.

45. $2x + y + z = 6$
$x - y - z = -3$
$3x + y + 2z = 7$

46. $x + y + z = 5$
$2x - y + z = 2$
$x + 2y - z = 3$

47. $2x + y - z = 2$
$x + 3y + 2z = 1$
$x + y + z = 2$

48. $2x + 2y + z = 6$
$x - y - z = -2$
$x - 2y - 2z = -5$

49. $3x + y - z = \frac{2}{3}$
$2x - y + z = 1$
$4x + 2y = \frac{8}{3}$

50. $x + y = 1$
$2x - y + z = 1$
$x + 2y + z = \frac{8}{3}$

51. $2x_1 + x_2 - x_3 = 6$
$3x_1 + 2x_2 + 5x_3 = 3$
$x_1 + x_2 + 5x_3 = -2$

52. $x_2 + 3x_3 = -4$
$x_1 + 2x_2 + x_3 = 7$
$x_1 - 2x_2 = 1$

53. $x_1 + 2x_2 + 3x_3 = 9$
$2x_1 + 2x_2 - x_3 = 0$
$3x_1 - 4x_2 - 2x_3 = 4$

54. $2x_1 + 4x_2 = 40$
$4x_1 + 2x_2 = 32$
$x_1 + 2x_3 = 14$

C   In Problems 55–59 solve each system by the method of Example 7, that is, by finding A^{-1}.

55. $x + y + 2z = -2$
$x - 2y + z = 5$
$y - z = 3$

56. $2x_1 + x_3 = 1$
$-x_1 + x_2 + 2x_3 = -1$
$x_1 + x_3 = 5$

57. $x_1 + 2x_2 - 3x_3 = -4$
$2x_1 - x_2 + x_3 - 2x_4 = 7$
$-x_1 - x_2 + x_3 + 4x_4 = 4$
$3x_1 + x_2 - x_3 - 6x_4 = 0$

58. $x_2 + 2x_3 + x_4 = -3$
$-x_1 + 5x_3 + 2x_4 = 5$
$2x_1 + x_2 - x_4 = 3$
$3x_1 + 2x_2 = 1$

59. $x_1 + x_3 + x_4 = 1$
$2x_1 + 3x_2 + 4x_3 + 2x_4 = 3$
$x_2 + 3x_3 + 4x_4 = -2$
$x_1 + 2x_2 + x_3 = 3$

60. Show that the inverse of

$$A = \begin{bmatrix} a & b \\ c & d \end{bmatrix}$$

is given by the formula

$$A^{-1} = \begin{bmatrix} \dfrac{d}{\Delta} & \dfrac{-b}{\Delta} \\ \dfrac{-c}{\Delta} & \dfrac{a}{\Delta} \end{bmatrix}$$

where $\Delta = ad - bc \neq 0$. The number Δ is called the *determinant of A*.

2.7

APPLICATIONS*

LEONTIEF MODELS
CRYPTOGRAPHY
DATA ANALYSIS: METHOD OF LEAST SQUARES

LEONTIEF MODELS

The Leontief models in economics are named after Wassily Leontief, who received the Nobel prize in economics in 1973. These models can be characterized as a description of an economy in which input equals output or, in other words, consumption equals production. That is, the models assume that whatever is produced is always consumed.

Leontief models are of two types: closed, in which the entire production is consumed by those participating in the production; and open, in which some of the production is consumed by those who produce it and the rest of the production is consumed by external bodies.

In the *closed model* we seek the relative income of each participant in the system. In the *open model* we seek the amount of production needed to achieve a forecast demand, when the amount of production needed to achieve current demand is known.

The Closed Model   We begin with an example to illustrate the idea.

Example 1

Three homeowners, Mike, Dan, and Bob, each with certain skills, agreed to pool their talents to make repairs on their houses. As it turned out, Mike spent 20% of

---

* This section may be omitted without loss of continuity.

WASSILY W. LEONTIEF published his first description of the production interdependence of goods and services for an entire economy in 1936. For additional references on Leontief's contributions, see Walter Isard and Phyllis Kaniss, "The 1973 Nobel Prize for Economic Science," *Science* (November 9, 1973) and Wassily W. Leontief, *The Structure of the American Economy, 1919–1935,* Oxford University Press, 1951.

his time on his own house, 40% of his time on Dan's house, and 40% on Bob's house. Dan spent 10% of his time on Mike's house, 50% of his time on his own house, and 40% on Bob's house. Of Bob's time, 60% was spent on Mike's house, 10% on Dan's, and 30% on his own. Now that the projects are finished, they need to figure out how much money each should get for his work, including the work performed on his own house, so that each person comes out even. They agreed in advance that the payment to each one should be approximately $300.00.

Solution

We place the information given in the problem in a 3×3 matrix, as follows:

$$
\begin{array}{r}
\\
\\
\text{Proportion of work done on Mike's house} \\
\text{Proportion of work done on Dan's house} \\
\text{Proportion of work done on Bob's house}
\end{array}
\begin{array}{c}
\text{Work done by} \\
\begin{array}{ccc}
\text{Mike} & \text{Dan} & \text{Bob}
\end{array} \\
\left[\begin{array}{ccc}
0.2 & 0.1 & 0.6 \\
0.4 & 0.5 & 0.1 \\
0.4 & 0.4 & 0.3
\end{array}\right]
\end{array}
$$

Next, we define the unknowns:

$$x_1 = \text{Mike's wages}$$
$$x_2 = \text{Dan's wages}$$
$$x_3 = \text{Bob's wages}$$

For each to come out even will require that the total amount paid out by each one equals the total amount received by each one. Let's analyze this requirement, by looking just at the work done on Mike's house. Mike's wages are x_1. Mike's expenditures for work done on his house are $0.2x_1 + 0.1x_2 + 0.6x_3$. These are required to be equal, so

$$x_1 = 0.2x_1 + 0.1x_2 + 0.6x_3$$

Similarly,

$$x_2 = 0.4x_1 + 0.5x_2 + 0.1x_3$$
$$x_3 = 0.4x_1 + 0.4x_2 + 0.3x_3$$

These three equations can be written compactly as

$$
\begin{bmatrix} x_1 \\ x_2 \\ x_3 \end{bmatrix} =
\begin{bmatrix}
0.2 & 0.1 & 0.6 \\
0.4 & 0.5 & 0.1 \\
0.4 & 0.4 & 0.3
\end{bmatrix}
\begin{bmatrix} x_1 \\ x_2 \\ x_3 \end{bmatrix}
$$

A simple manipulation reduces the system to

$$0.8x_1 - 0.1x_2 - 0.6x_3 = 0$$
$$-0.4x_1 + 0.5x_2 - 0.1x_3 = 0$$
$$-0.4x_1 - 0.4x_2 + 0.7x_3 = 0$$

Solving for x_1, x_2, x_3, we find that

$$x_1 = \tfrac{31}{36}x_3 \qquad x_2 = \tfrac{32}{36}x_3$$

where x_3 is the parameter. To get solutions that fall close to $300, we set $x_3 = 360$.*
The wages to be paid out are therefore

$$x_1 = \$310 \qquad x_2 = \$320 \qquad x_3 = \$360$$ ∎

The matrix in Example 1, namely,

$$\begin{bmatrix} 0.2 & 0.1 & 0.6 \\ 0.4 & 0.5 & 0.1 \\ 0.4 & 0.4 & 0.3 \end{bmatrix}$$

is called an *input – output matrix*.

In the general closed model, we have an economy consisting of n components. Each component produces an *output* of some goods or services, which, in turn, is completely used up by the n components. The proportionate use of each component's output by the economy makes up the input – output matrix of the economy. The problem is to find suitable pricing levels for each component so that total income equals total expenditure.

In general, an input – output matrix for a closed Leontief model is of the form

$$A = [a_{ij}] \qquad i, j = 1, 2, \ldots, n$$

where the a_{ij} represent the fractional amount of goods or services used by i and produced by j. For a closed model, the sum of each column equals 1 (this is the condition that all production is consumed internally) and $0 \leq a_{ij} \leq 1$ for all entries (this is the restriction that each entry is a fraction).

If A is the input – output matrix of a closed system with n components and X is a column vector representing the price of each output of the system, then

$$X = AX$$

represents the requirement that total income equal expenditure.

For example, the first entry of the matrix equality $X = AX$ requires that

$$x_1 = a_{11} x_1 + a_{12} x_2 + \cdots + a_{1n} x_n$$

The right side represents the price paid by component 1 for the goods it uses, while x_1 represents the income of component 1; we are requiring they be equal.

We can rewrite the equation $X = AX$ as

$$X - AX = 0$$
$$I_n X - AX = 0$$
$$(I_n - A)X = 0$$

This matrix equation, which represents a system of equations in which the right-hand side is always 0, is called a *homogeneous system of equations*. It can be shown

---

* Other choices for x_3 are, of course, possible. The choice of which value to use is up to the homeowners. No matter what choice is made, each homeowner comes out even.

that if the entries in the input–output matrix A are positive and if the sum of each column of A equals 1, then this system has a one-parameter solution; that is, we can solve for $n - 1$ of the unknowns in terms of the remaining one, which serves as the parameter. This parameter serves as a "scale factor."

The Open Model For the open model, in addition to internal consumption of goods produced, there is an outside demand for the goods produced. This outside demand may take the form of exportation of goods or the goods needed to support consumer demand. Again, however, we make the assumption that whatever is produced is also consumed.

For example, suppose an economy consists of three industries R, S, and T, and suppose each one produces a single product. We assume that a portion of R's production is used by each of the three industries, while the remainder is used up by consumers. The same is true of the production of S and T. To organize our thoughts, we construct a table that describes the interaction of the use of R, S, and T's production over some fixed period of time. See Table 1.

Table 1

	R	S	T	Consumer	Total
R	50	20	40	70	180
S	20	30	20	90	160
T	30	20	20	50	120

All entries in the table are in appropriate units, say, in dollars. The first row (row R) represents the production in dollars of industry R (input). Out of the total of \$180 worth of goods produced, R, S, and T use \$50, \$20, and \$40, respectively, for the production of their goods, while consumers purchase the remaining \$70 for their consumption (output). Observe that input equals output since everything produced by R is used up by R, S, T, and consumers.

The second and third rows are interpreted in the same way.

An important observation is that the goal of R's production is to produce \$70 worth of goods, since this is the demand of consumers. In order to meet this demand, R must produce a total of \$180, since the difference \$110 is required internally by R, S, and T.

Suppose, however, that consumer demand is expected to change. To effect this change, how much should each industry now produce? For example, in Table 1, current demand for R, S, and T can be represented by a *demand vector:*

$$D_0 = \begin{bmatrix} 70 \\ 90 \\ 50 \end{bmatrix}$$

But suppose marketing forecasts predict that in 3 years the demand vector will be

$$D_3 = \begin{bmatrix} 60 \\ 110 \\ 60 \end{bmatrix}$$

Here the demand for item R has decreased; the demand for item S has significantly increased, and the demand for item T is higher. Given the current total output of R, S, and T at 180, 160, and 120, respectively, what must it be in 3 years to meet this projected demand?

In using input–output analysis to obtain a solution to such a forecasting problem, we take into account the fact that the output of any one of these industries is affected by changes in the other two, since the total demand for say, R, in 3 years depends not only on consumer demand for R, but also on consumer demand for S and T. That is, the industries are interrelated.

The solution of this type of forecasting problem is derived from the *open Leontief model* in input–output analysis.

To obtain the solution, we need to determine how much of each of the three products R, S, and T is required to produce 1 unit of R. For example, to obtain 180 units of R requires the use of 50 units of R, 20 units of S, and 30 units of T (the entries in column 1). Forming the ratios, we find that to produce 1 unit of R requires $\frac{50}{180} = 0.278$ of R, $\frac{20}{180} = 0.111$ of S, and $\frac{30}{180} = 0.167$ of T. If we want, say, x_1 units of R, we will require $0.278x_1$ units of R, $0.111x_1$ units of S, and $0.167x_1$ units of T.

Continuing in this way, we can construct the matrix

$$A = \begin{array}{c} R \\ S \\ T \end{array} \begin{array}{ccc} R & S & T \\ \left[\begin{array}{ccc} 0.278 & 0.125 & 0.333 \\ 0.111 & 0.188 & 0.167 \\ 0.167 & 0.125 & 0.167 \end{array}\right] \end{array}$$

Observe that column 1 represents the amounts of R, S, T required for 1 unit of R; column 2 represents the amounts of R, S, and T required for 1 unit of S; and column 3 represents the amounts of R, S, and T required for 1 unit of T. For example, the entry in row 3, column 2 (0.125), represents the amount of T needed to produce 1 unit of S.

As a result of placing the entries this way, if

$$X = \begin{bmatrix} x_1 \\ x_2 \\ x_3 \end{bmatrix}$$

represents the total output required to obtain a given demand, the product AX represents the amounts of R, S, and T required for internal consumption. The condition that production = consumption requires that

Internal consumption + Consumer demand = Total output

* The entries in A can be checked by using the requirement that $AX + D = X$, for $D =$ initial demand and $X =$ total output. For our example, it must happen that

Internal consumption	+	Consumer demand	=	Total output
AX	+	D	=	X
$\begin{bmatrix} 0.278 & 0.125 & 0.333 \\ 0.111 & 0.188 & 0.167 \\ 0.167 & 0.125 & 0.167 \end{bmatrix}\begin{bmatrix} 180 \\ 160 \\ 120 \end{bmatrix}$	$+$	$\begin{bmatrix} 70 \\ 90 \\ 50 \end{bmatrix}$	$=$	$\begin{bmatrix} 180 \\ 160 \\ 120 \end{bmatrix}$

In terms of the matrix A, the total output X, and the demand vector D, this requirement is equivalent to the equation

$$AX + D = X$$

In this equation, we seek to find X for a prescribed demand D. The matrix A is calculated as above for some initial production process.*

Example 2

For the data given in Table 1, find the total output X required to achieve a future demand of

$$D_3 = \begin{bmatrix} 60 \\ 110 \\ 60 \end{bmatrix}$$

Solution

We need to solve for X in

$$AX + D_3 = X$$
$$X - AX = D_3$$

Total Output $-$ Internal Consumption $=$ Consumer Demand

$$IX - AX = D_3$$
$$(I - A)X = D_3$$

Solving for X, we have

$$X = [I - A]^{-1} \cdot D_3$$

$$= \begin{bmatrix} 0.722 & -0.125 & -0.333 \\ -0.111 & 0.812 & -0.167 \\ -0.167 & -0.125 & 0.833 \end{bmatrix}^{-1} \begin{bmatrix} 60 \\ 110 \\ 60 \end{bmatrix}$$

$$= \begin{bmatrix} 1.6048 & 0.3568 & 0.7131 \\ 0.2946 & 1.3363 & 0.3857 \\ 0.3660 & 0.2721 & 1.4013 \end{bmatrix} \begin{bmatrix} 60 \\ 110 \\ 60 \end{bmatrix}$$

$$= \begin{bmatrix} 178.322 \\ 187.811 \\ 135.969 \end{bmatrix}$$

Thus, the total output of R, S, and T required for the forecast demand D_3 is

$$x_1 = 178.322 \qquad x_2 = 187.811 \qquad x_3 = 135.969 \qquad \blacksquare$$

The general open model can be described as follows:

Suppose there are n industries in the economy. Each industry produces some goods or services, which are partially consumed by the n industries, while the rest are used to meet a prescribed current demand. Given the output required of each industry to meet current demand, what should the output of each industry be to meet some different future demand?

> The matrix $A = [a_{ij}]$, $i, j = 1, \ldots, n$, of the open model is defined to consist of entries a_{ij}, where a_{ij} is the amount of output of industry j required for one unit of output of industry i. If X is a column vector representing the production of each industry in the system and D is a column vector representing future demand for goods produced in the system, then
>
> $$X = AX + D$$

From the equation above, we find

$$[I_n - A]X = D$$

It can be shown that the matrix $I_n - A$ has an inverse, provided each entry in A is positive and the sum of each column in A is less than 1. Under these conditions, we may solve for X to get

$$X = [I_n - A]^{-1} \cdot D$$

This form of the solution is particularly useful since it allows us to find X for a variety of demands D by doing one calculation: $[I_n - A]^{-1}$.

We conclude by noting that the use of an input–output matrix to solve forecasting problems assumes that each industry produces a single commodity and that no technological advances take place in the period of time under investigation (in other words, the proportions found in the matrix A are fixed).

Exercise 2.7A

Answers to Odd-Numbered Problems begin on page A-8.

A In Problems 1–4 find the relative wages of each person for the given closed input–output matrix. In each case, take the wages of C to be the parameter and use $x_3 = C$'s wages = $10,000.

1.
$$\begin{array}{c} \\ A \\ B \\ C \end{array} \begin{array}{ccc} A & B & C \\ \left[\begin{array}{ccc} \frac{1}{2} & \frac{1}{3} & \frac{1}{4} \\ \frac{1}{4} & \frac{1}{3} & \frac{1}{4} \\ \frac{1}{4} & \frac{1}{3} & \frac{1}{2} \end{array}\right] \end{array}$$

2.
$$\begin{array}{c} \\ A \\ B \\ C \end{array} \begin{array}{ccc} A & B & C \\ \left[\begin{array}{ccc} \frac{1}{4} & \frac{2}{3} & \frac{1}{2} \\ \frac{1}{2} & \frac{1}{6} & \frac{1}{4} \\ \frac{1}{4} & \frac{1}{6} & \frac{1}{4} \end{array}\right] \end{array}$$

3.
$$\begin{array}{c} \\ A \\ B \\ C \end{array} \begin{array}{ccc} A & B & C \\ \left[\begin{array}{ccc} 0.2 & 0.3 & 0.1 \\ 0.6 & 0.4 & 0.2 \\ 0.2 & 0.3 & 0.7 \end{array}\right] \end{array}$$

4.
$$\begin{array}{c} \\ A \\ B \\ C \end{array} \begin{array}{ccc} A & B & C \\ \left[\begin{array}{ccc} 0.4 & 0.3 & 0.2 \\ 0.2 & 0.3 & 0.3 \\ 0.4 & 0.4 & 0.5 \end{array}\right] \end{array}$$

5. For the three industries R, S, and T in the open Leontief model of Example 2, compute the total output vector X if the forecast demand vector is

$$D_2 = \begin{bmatrix} 80 \\ 90 \\ 60 \end{bmatrix}$$

6. Rework Problem 5 if the forecast demand vector is

$$D_4 = \begin{bmatrix} 100 \\ 80 \\ 60 \end{bmatrix}$$

APPLICATIONS

7. **Wage Distribution** A society consists of four individuals: a farmer, a builder, a tailor, and a rancher (who produces meat products). Of the food produced by the farmer, $\frac{3}{10}$ is used by the farmer, $\frac{2}{10}$ by the builder, $\frac{2}{10}$ by the tailor, and $\frac{3}{10}$ by the rancher. The builder's production is utilized 30% by the farmer, 30% by the builder, 10% by the tailor, and 30% by the rancher. The tailor's production is used in the ratios $\frac{3}{10}$, $\frac{3}{10}$, $\frac{1}{10}$, and $\frac{3}{10}$ by the farmer, builder, tailor, and rancher. Finally, meat products are used 20% by each of the farmer, builder, and tailor, and 40% by the rancher. What are the relative wages of each if the rancher's wages are scaled at $10,000?

8. **Wage Distribution** If in Problem 7 the meat production utilization changes so that it is used equally by all four individuals, while everyone else's production utilization remains the same, what are the relative wages?

9. **Production Forecasting** Suppose the interrelationships between the production of two industries R and S in a given year are given in the table:

	R	S	Current Consumer Demand	Total Output
R	30	40	60	130
S	20	10	40	70

If the forecast demand in 2 years is

$$D_2 = \begin{bmatrix} 80 \\ 40 \end{bmatrix}$$

what should the total output X be?

CRYPTOGRAPHY

Our second application is to *cryptography,* the art of writing or deciphering secret codes. We begin by giving examples of elementary codes.

Example 1

A message can be encoded by associating each letter of the alphabet with some other letter of the alphabet according to a prescribed pattern. For example, we might have

A B C D E F G H I J K L M N O P Q R S T U V W X Y Z
↓ ↓
C D E F G H I J K L M N O P Q R S T U V W X Y Z A B

With the above code, the word *BOMB* would become DQOD.   ■

Example 2

Another code may associate numbers with the letters of the alphabet. For example, we might have

A B C D E F G H I J K L M N O P Q R S T U V W X Y Z
↓ ↓
26 25 24 23 22 21 20 19 18 17 16 15 14 13 12 11 10 9 8 7 6 5 4 3 2 1

In this code, the word *PEACE* looks like 11 22 26 24 22.   ∎

Both the above codes have one important feature in common. The association of letters with the coding symbols is made using a one-to-one correspondence so that no possible ambiguities can arise.

Suppose we want to encode the following message:

BEWARE   THE   IDES   OF   MARCH

If we decide to divide the message into pairs of letters, the message becomes:

BE   WA   RE   TH   EI   DE   SO   FM   AR   CH

(If there is a letter left over, we arbitrarily assign Z to the last position.) Using the correspondence of letters to numbers given in Example 2, and writing each pair of letters as a column vector, we obtain

$$\begin{bmatrix} B \\ E \end{bmatrix} = \begin{bmatrix} 25 \\ 22 \end{bmatrix} \quad \begin{bmatrix} W \\ A \end{bmatrix} = \begin{bmatrix} 4 \\ 26 \end{bmatrix} \quad \begin{bmatrix} R \\ E \end{bmatrix} = \begin{bmatrix} 9 \\ 22 \end{bmatrix} \quad \begin{bmatrix} T \\ H \end{bmatrix} = \begin{bmatrix} 7 \\ 19 \end{bmatrix} \quad \text{etc.}$$

Next, we arbitrarily choose a 2×2 matrix A, which we know has an inverse A^{-1} (the reason for this is seen later). Suppose we choose

$$A = \begin{bmatrix} 2 & 3 \\ 1 & 2 \end{bmatrix}$$

Its inverse is

$$A^{-1} = \begin{bmatrix} 2 & -3 \\ -1 & 2 \end{bmatrix}$$

Now, we transform the column vectors representing the message by multiplying each of them on the left by the matrix A:

$$A \begin{bmatrix} B \\ E \end{bmatrix} = A \begin{bmatrix} 25 \\ 22 \end{bmatrix} = \begin{bmatrix} 116 \\ 69 \end{bmatrix}$$

$$A \begin{bmatrix} W \\ A \end{bmatrix} = A \begin{bmatrix} 4 \\ 26 \end{bmatrix} = \begin{bmatrix} 86 \\ 56 \end{bmatrix}$$

$$A \begin{bmatrix} R \\ E \end{bmatrix} = A \begin{bmatrix} 9 \\ 22 \end{bmatrix} = \begin{bmatrix} 84 \\ 53 \end{bmatrix} \quad \text{etc.}$$

The coded message is

116   69   86   56   84   53   etc.

To decode or unscramble the above message, pair the numbers in 2×1 column vectors. Multiply each of these column vectors by A^{-1} on the left:

$$A^{-1}\begin{bmatrix}116\\69\end{bmatrix}=\begin{bmatrix}25\\22\end{bmatrix}$$

$$A^{-1}\begin{bmatrix}86\\56\end{bmatrix}=\begin{bmatrix}4\\26\end{bmatrix}\qquad\text{etc.}$$

By reassigning letters to these numbers, we obtain the original message.

Example 3
The message to be encoded is

$$\text{THE}\quad\text{END}\quad\text{IS}\quad\text{NEAR}$$

We agree to associate numbers to letters as follows:

A B C D E F G H I J K L M N O P Q R S T U V W X Y Z
↓ ↓
1 2 3 4 5 6 7 8 9 10 11 12 13 14 15 16 17 18 19 20 21 22 23 24 25 26

The encoded message is to be formed of triplets of numbers.

Solution
This time we must divide the message into triplets of letters, obtaining

$$\text{THE}\quad\text{END}\quad\text{ISN}\quad\text{EAR}$$

in order for the encoded message to have triplets of numbers. (If the message required additional letters to complete the triplet, we would have used Z or YZ.)
Now we choose a 3 × 3 matrix such as

$$A=\begin{bmatrix}1&0&0\\3&1&5\\-2&0&1\end{bmatrix}$$

Its inverse is

$$A^{-1}=\begin{bmatrix}1&0&0\\-13&1&-5\\2&0&1\end{bmatrix}$$

The encoded message is obtained by multiplying the matrix A times each column vector of the original message:

$$A\begin{bmatrix}T\\H\\E\end{bmatrix}=\begin{bmatrix}1&0&0\\3&1&5\\-2&0&1\end{bmatrix}\begin{bmatrix}20\\8\\5\end{bmatrix}=\begin{bmatrix}20\\93\\-35\end{bmatrix}$$

$$A\begin{bmatrix}E\\N\\D\end{bmatrix}=\begin{bmatrix}1&0&0\\3&1&5\\-2&0&1\end{bmatrix}\begin{bmatrix}5\\14\\4\end{bmatrix}=\begin{bmatrix}5\\49\\-6\end{bmatrix}$$

$$A \begin{bmatrix} I \\ S \\ N \end{bmatrix} = \begin{bmatrix} 1 & 0 & 0 \\ 3 & 1 & 5 \\ -2 & 0 & 1 \end{bmatrix} \begin{bmatrix} 9 \\ 19 \\ 14 \end{bmatrix} = \begin{bmatrix} 9 \\ 116 \\ -4 \end{bmatrix}$$

$$A \begin{bmatrix} E \\ A \\ R \end{bmatrix} = \begin{bmatrix} 1 & 0 & 0 \\ 3 & 1 & 5 \\ -2 & 0 & 1 \end{bmatrix} \begin{bmatrix} 5 \\ 1 \\ 18 \end{bmatrix} = \begin{bmatrix} 5 \\ 106 \\ 8 \end{bmatrix}$$

The coded message is

$$20 \quad 93 \quad -35 \quad 5 \quad 49 \quad -6 \quad 9 \quad 116 \quad -4 \quad 5 \quad 106 \quad 8 \qquad \blacksquare$$

To decode the message in Example 3, form 3×1 column vectors of the numbers in the coded message and multiply on the left by A^{-1}.

The above are elementary examples of encoding and decoding. Modern-day cryptography uses sophisticated computer-implemented codes that depend on higher-level mathematics. For an interesting survey of current cryptographic techniques see the article "The Mathematics of Public-Key Cryptography" in the August 1979 issue of the *Scientific American.*

Exercise 2.7B   *Answers to Odd-Numbered Problems begin on page A-8.*

A   **1.** Using the correspondence

A B C D E F G H I J K L M N O P Q R S T U V W X Y Z
↓ ↓
1 2 3 4 5 6 7 8 9 10 11 12 13 14 15 16 17 18 19 20 21 22 23 24 25 26

and the matrices

$$\text{(I)} \quad A = \begin{bmatrix} 2 & 3 \\ 1 & 2 \end{bmatrix} \qquad \text{(II)} \quad A = \begin{bmatrix} 1 & 0 & 0 \\ 3 & 1 & 5 \\ -2 & 0 & 1 \end{bmatrix}$$

Encode the following messages:

(a)   MEET ME AT THE CASBAH
(b)   TOMORROW NEVER COMES
(c)   THE MISSION IS IMPOSSIBLE

2. Using the correspondence given in Problem 1 and the matrix

$$A = \begin{bmatrix} 2 & 3 \\ 1 & 2 \end{bmatrix}$$

decode the following messages:

(a)   51   30   27   16   75   47   19   10   48   26
(b)   70   45   103   62   58   38   102   61   88   57

3. Using the correspondence given in Problem 1 and the matrix

$$A = \begin{bmatrix} 1 & 0 & 0 \\ 3 & 1 & 5 \\ -2 & 0 & 1 \end{bmatrix}$$

decode the message

$$25 \quad 195 \quad -29 \quad 6 \quad 135 \quad 9 \quad 14 \quad 183 \quad -2$$

DATA ANALYSIS: METHOD OF LEAST SQUARES

The method of least squares refers to a technique that is often used in data analysis to find the "best" linear equation that fits a given collection of experimental data. (We will see a little later just what we mean by the word "best.") It is a technique employed by statisticians, economists, business forecasters, and most all who try to interpret and analyze data. Before we begin our discussion, we will need to define the transpose of a matrix.

Transpose **Let A be a matrix of dimension $m \times n$. The *transpose of A*, written A^T, is the $n \times m$ matrix obtained from A by interchanging the rows and columns of A.**

Thus, the first row of A^T is the first column of A; the second row of A^T is the second column of A; and so on.

Example 1
If we let

$$A = \begin{bmatrix} 1 & 2 & 3 \\ 0 & -1 & 2 \end{bmatrix}, \quad B = \begin{bmatrix} 1 & 1 \\ 0 & 1 \\ 2 & 3 \end{bmatrix}, \quad \text{and} \quad C = [1 \quad 0 \quad -1]$$

then

$$A^T = \begin{bmatrix} 1 & 0 \\ 2 & -1 \\ 3 & 2 \end{bmatrix}, \quad B^T = \begin{bmatrix} 1 & 0 & 2 \\ 1 & 1 & 3 \end{bmatrix}, \quad \text{and} \quad C^T = \begin{bmatrix} 1 \\ 0 \\ -1 \end{bmatrix}$$

Note how dimensions are reversed when computing A^T: A has the dimension 2×3, while A^T has the dimension 3×2, and so on.

We begin our discussion of the method of least squares by an example:

Suppose a product has been sold over time at various prices and that we have some data that show the demand for the product (in thousands of units) in terms of its price (in dollars). If we use x to represent price and y to represent demand, the data might look like the information in the table below.

Price	x	4	5	9	12
Demand	y	9	8	6	3

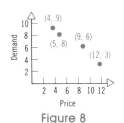

Figure 8

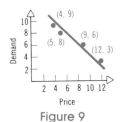

Figure 9

Suppose also that we have reason to assume that y and x are *linearly related.* That is, we assume we can write

$$y = ax + b$$

for some, as yet unknown, a and b. In other words we are assuming that when demand is graphed against price the resulting graph will be a straight line. Our belief that y and x are linearly related might be based on past experience or economic theory. However, if we plot the data from the above table (see Figure 8), it seems pretty clear that no single straight line passes through the plotted points. This may be because:

1. The data may not have been reported accurately, or
2. Our assumption that $y = ax + b$ may not be totally warranted.

Thus, we are looking for an ideal answer to a question about something in the real world where what happens may only roughly (approximately) fit an ideal design. Refer again to Figure 8. Though no straight line will pass through all the above points, we might still ask: "Is there a straight line that best fits the above points"? Intuitively, we seek a line of the sort in Figure 9 that provides a good straight-line approximation to the data. Finding such a line would give us at least an approximate feel for how the demand y is related to the price x.

Our Goal Given a set of noncollinear points, find the straight line that "best" fits these points.

In the process we will need to clarify our use of the word "best." First, we label the data points in Figure 8 from left to right as $(x_1, y_1), (x_2, y_2)$, and so on, so that, for example, (x_2, y_2) is (5, 8). Now suppose the equation of the straight line L in Figure 9 is

$$y = ax + b$$

where we do not know what a and b are. If (x_1, y_1) is actually on the line, it would be true that

$$y_1 = ax_1 + b$$

But, since we can't expect (x_1, y_1) to lie on L, the above relation will not be valid or, expressed another way, there will be a nonzero difference between y_1 and $ax_1 + b$. We designate this difference by r_1. That is,

$$r_1 = y_1 - (ax_1 + b)$$

or

$$y_1 = ax_1 + b + r_1$$

What we have just said about (x_1, y_1), we can repeat for the other data points (x_i, y_i), $i = 2, 3$ and 4. Therefore, since we can't expect y_i to be equal to $ax_i + b$, we will measure the difference by

$$r_i = y_i - (ax_i + b),$$

or equivalently by

$$y_i = ax_i + b + r_i, \qquad i = 1, 2, 3, 4 \tag{1}$$

Geometrically, the r_i's measure the vertical distance between the sought-after line L and the data points (see Figure 10).

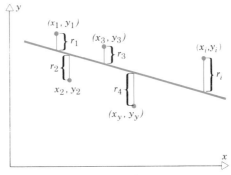

Figure 10

Recall that our problem is to find a line L that "best" fits the data points $(x_i,\ y_i)$. Intuitively, we would seek a line L for which the r_i's are all simultaneously small. Since some r_i's are positive and some may be negative, it will not do to just add them up. To eliminate the signs, we square each r_i. The method of least squares consists of finding a line L for which the sum of the square of the r_i's is as small as possible. Since finding the line $y = ax + b$ is the same as finding a and b, we restate the least-squares problem for this example:

Least Squares Problem   Given the data points $(x_i,\ y_i)$, $i = 1, 2, 3, 4$, find a and b satisfying the equations (1) so that $r_1^2 + r_2^2 + r_3^2 + r_4^2$ is minimized.

A solution to the problem can be neatly expressed using matrix language.

Solution to Least-Squares Problem   The line of best fit

$$y = ax + b \tag{2}$$

to a set of four points $(x_1,\ y_1), (x_2,\ y_2), (x_3,\ y_3), (x_4,\ y_4)$ is obtained by solving the system of two equations in two unknowns

$$A^T A X = A^T Y \tag{3}$$

for a and b, where

$$A = \begin{bmatrix} x_1 & 1 \\ x_2 & 1 \\ x_3 & 1 \\ x_4 & 1 \end{bmatrix}, \qquad X = \begin{bmatrix} a \\ b \end{bmatrix}, \qquad Y = \begin{bmatrix} y_1 \\ y_2 \\ y_3 \\ y_4 \end{bmatrix} \tag{4}$$

Since a derivation that this, in fact, does give the solution would require either calculus or more advanced linear algebra, we omit it.

Example 2

Use least-squares to find the line of best fit for the points

$$(4, 9), (5, 8), (9, 6), (12, 3)$$

Solution

We use equation (4) to set up the matrices A and Y

$$A = \begin{bmatrix} 4 & 1 \\ 5 & 1 \\ 9 & 1 \\ 12 & 1 \end{bmatrix}, \qquad Y = \begin{bmatrix} 9 \\ 8 \\ 6 \\ 3 \end{bmatrix}$$

The equation (3) $A^T A X = A^T Y$ becomes

$$\begin{bmatrix} 4 & 5 & 9 & 12 \\ 1 & 1 & 1 & 1 \end{bmatrix} \begin{bmatrix} 4 & 1 \\ 5 & 1 \\ 9 & 1 \\ 12 & 1 \end{bmatrix} \begin{bmatrix} a \\ b \end{bmatrix} = \begin{bmatrix} 4 & 5 & 9 & 12 \\ 1 & 1 & 1 & 1 \end{bmatrix} \begin{bmatrix} 9 \\ 8 \\ 6 \\ 3 \end{bmatrix}$$

This reduces to a system of two equations in two unknowns:

$$266a + 30b = 166$$
$$30a + 4b = 26$$

The solution, which you can verify, is given by

$$a = \frac{-29}{41}, \qquad b = \frac{484}{41}$$

Hence, the least-squares solution to the problem is given by the straight line

$$y = \frac{-29}{41} x + \frac{484}{41}$$

■

Note, for example, that corresponding to the value $x = 5$, the above line of "best" fit has $y = -29/41 \cdot 5 + 484/41 = 8.27$, while the experimentally observed demand had value 8. Were we to use our computed line to approximate the connection between price and demand, we would predict that a selling price of $x = 6$ would yield a demand of $y = -29/41 \cdot 6 + 484/41 = 7.56$.

The General Least-Squares Problems We presented the method of least squares using a particular example with four data points. We can extend this analysis.

The general least-squares problem consists of finding the best straight line fit to a given set of data points:

$$(x_1, y_1), (x_2, y_2), \ldots, (x_n, y_n) \tag{5}$$

If we let

$$A = \begin{bmatrix} x_1 & 1 \\ x_2 & 1 \\ \cdot & \cdot \\ \cdot & \cdot \\ \cdot & \cdot \\ x_n & 1 \end{bmatrix}, \qquad X = \begin{bmatrix} a \\ b \end{bmatrix}, \qquad Y = \begin{bmatrix} y_1 \\ y_2 \\ \cdot \\ \cdot \\ \cdot \\ y_n \end{bmatrix}$$

then the line of best fit to the data points in (5) is

$$y = ax + b$$

where $X = \begin{bmatrix} a \\ b \end{bmatrix}$ is found by solving the system of two equations in two unknowns

$$A^T A X = A^T Y \qquad (6)$$

It can be shown that the system given by the equation (6) always has a unique solution provided the data points do not all lie on the same vertical line. The least-squares techniques presented here along with more elaborate variations are frequently used today by statisticians and researchers.

Exercise 2.7C   *Answers to Odd-Numbered Problems begin on page A-9.*

A   In Problems 1–6, compute A^T.

1. $A = \begin{bmatrix} 4 & 1 & 2 \\ 3 & 1 & 0 \end{bmatrix}$ 　　　　
2. $A = \begin{bmatrix} 5 & 2 & -1 \\ 1 & 3 & 6 \\ 1 & -1 & 2 \end{bmatrix}$ 　　　　
3. $A = \begin{bmatrix} 1 & 11 \\ 0 & 12 \\ 1 & 4 \end{bmatrix}$

4. $A = \begin{bmatrix} -1 & 6 & 4 \end{bmatrix}$ 　　　　
5. $A = \begin{bmatrix} 8 \\ 6 \\ 3 \end{bmatrix}$ 　　　　
6. $A = \begin{bmatrix} 5 & 3 \\ 3 & 7 \end{bmatrix}$

B   **7.** The following table shows the supply (in thousands of units) of a product at various prices (in dollars).

Price	x	3	5	6	7
Supply	y	10	13	15	16

(a)   Find the least-squares line of best fit to the above data.
(b)   Use the equation of this line to estimate the supply of the product at a price of 8 dollars.

8. Data giving the number of hours a person had studied compared to his or her performance on an exam are given below.

Hours Studied x	Exam Score y
0	50
2	74
4	85
6	90
8	92

(a) Find a least-squares line of best fit to the above data.

(b) What prediction would this line make for a student who studied 9 hours?

9. A business would like to determine the relationship between the amount of money spent on advertising and its total weekly sales. Over a period of 5 weeks it gathers the following data.

Amount Spent on Advertising (in thousands) x	Weekly Sales Volume (in thousands) y
10	50
17	61
11	55
18	60
21	70

Find a least-squares line of best fit to the above data.

10. The following data show the connection between the number of hours a drug has been in a person's body and its concentration in the body.

Number of Hours	Drug Concentration (parts per million)
2	2.1
4	1.6
6	1.4
8	1.0

(a) Fit a least-squares line to the above data.

(b) Use the equation of the line to estimate the drug concentration after 5 hours.

C   11. A matrix is *symmetric* if $A^T = A$. Which of the following matrices are symmetric?

(a) $\begin{bmatrix} 1 & 1 & 2 \\ 1 & 0 & 1 \\ 3 & 2 & 3 \end{bmatrix}$ (b) $\begin{bmatrix} 0 & 1 & 3 \\ 1 & 4 & 7 \\ 3 & 7 & 5 \end{bmatrix}$ (c) $\begin{bmatrix} 1 & 2 & 3 & 0 \\ 2 & 4 & 5 & 0 \\ 3 & 5 & 1 & 0 \end{bmatrix}$

Need a symmetric matrix be square?

12. Show that the matrix $A^T A$ is always symmetric.

CHAPTER REVIEW

Important Terms and Formulas			
	rows	matrix multiplication	*Leontief model (open and closed)
	columns	identity matrix	
	dimension of a matrix	inverse of a matrix	*input–output matrix
	diagonal entries	augmented matrix	*transpose

* From optional section.

square matrix   row operation   *symmetric

vector   reduced row-echelon form

zero matrix   parameter

scalar multiplication

$$A + B = B + A$$
$$A + (B + C) = (A + B) + C$$
$$A + (-A) = 0$$
$$A - B = A + (-1) \cdot B = A + (-B)$$
$$k(hA) = (kh)A$$

$$(k + h)A = kA + hA$$
$$k(A + B) = kA + kB$$
$$A(BC) = (AB)C$$
$$A(B + C) = AB + AC$$

True – False Questions

(Answers on page A-9)

T   F   1. Matrices of the same dimension can always be added.

T   F   2. Matrices of the same dimension can always be multiplied.

T   F   3. A square matrix will always have an inverse.

T   F   4. The reduced row-echelon form of a matrix A is unique.

T   F   5. The identity matrix is a square matrix.

T   F   6. Matrix addition is always defined.

T   F   7. Matrix multiplication is commutative.

T   F   8. A system of equations can be written as a matrix.

T   F   9. There is only one unique matrix representation for a given system of equations.

Fill in the Blanks

(Answers on page A-9)

1. If matrix A is of dimension 3×4 and matrix B is of dimension 4×2, then AB is of dimension _____.

2. A system of three linear equations in three unknowns has either _____ solution, or no solutions, or _____ _____ solutions.

3. If A is a matrix of dimension 3×4, the 3 tells the number of _____ and the 4 tells the number of _____.

4. If $AB = I$, the identity matrix, then B is called the _____ of A.

·5. If we can multiply matrix A and matrix B, to obtain AB, then the number of _____ of A is equal to the number of _____ of B.

6. We use three elementary _____ _____ to reduce a matrix.

7. The identity element of matrix addition is called the _____ _____.

8. The identity element of matrix multiplication is called the _____ _____.

9. If B is a 2×3 matrix and BA^2 is defined, then A is a _____ matrix.

10. If A is a 4×5 matrix and AB^3 is defined, then B is a _____ matrix.

Review
Exercises

Answers to Odd-Numbered Problems begin on page A-9.

In Problems 1–14 compute the given expression for

$$A = \begin{bmatrix} -2 & 0 & 7 \\ 1 & 8 & 3 \\ 2 & 4 & 21 \end{bmatrix} \quad B = \begin{bmatrix} 1 & 3 & 9 \\ 2 & 7 & 5 \\ 3 & 6 & 8 \end{bmatrix} \quad C = \begin{bmatrix} 0 & 1 & 2 \\ 0 & 5 & 1 \\ 8 & 7 & 9 \end{bmatrix}$$

1. $A + B$
2. $B + A$
3. $3(A + B)$
4. $3A + 3B$
5. $3A - 3B$
6. $B - C$
7. $2(5A)$
8. $\frac{3}{2}A$
9. $2A + \frac{1}{2}B - 3C$
10. $A - 2B + 3C$
11. AB
12. BA
13. $(B - A)C$
14. $BC - AC$

In Problems 15–22 find the inverse, if it exists, of each matrix.

15. $\begin{bmatrix} 3 & 0 \\ -2 & 1 \end{bmatrix}$

16. $\begin{bmatrix} 4 & 1 \\ 3 & 1 \end{bmatrix}$

17. $\begin{bmatrix} 1 & 2 & 3 \\ 2 & 4 & 5 \\ 3 & 5 & 6 \end{bmatrix}$

18. $\begin{bmatrix} -1 & 2 & 0 \\ 3 & 2 & -1 \\ 4 & 0 & 3 \end{bmatrix}$

19. $\begin{bmatrix} 4 & 3 & -1 \\ 0 & 2 & 2 \\ 3 & -1 & 0 \end{bmatrix}$

20. $\begin{bmatrix} -6 & 6 & 2 \\ 13 & 3 & 1 \\ 8 & -8 & 8 \end{bmatrix}$

21. $\begin{bmatrix} 1 & 2 & -3 \\ 4 & 6 & 2 \\ -3 & -6 & 9 \end{bmatrix}$

22. $\begin{bmatrix} 9 & 6 & -3 \\ 2 & -6 & 4 \\ -3 & 2 & 1 \end{bmatrix}$

In Problems 23–34 find the solution, if it exists, of each system of linear equations. If the system has infinitely many solutions, list at least three solutions.

23. $2x_1 - x_2 + x_3 = 1$
$x_1 + x_2 - x_3 = 2$
$3x_1 - x_2 + x_3 = 0$

24. $2x_1 + 3x_2 - x_3 = 5$
$x_1 - x_2 + x_3 = 1$
$3x_1 - 3x_2 + 3x_3 = 3$

25. $x_1 - 2x_2 = 6$
$3x_1 + 2x_2 - x_3 = 2$
$4x_1 + 3x_3 = -1$

26. $2x_1 - x_2 + 3x_3 = 5$
$x_1 + 2x_3 = 0$
$3x_1 + 2x_2 + x_3 = -3$

27. $x_1 - 3x_2 = 5$
$3x_2 + x_3 = 0$
$2x_1 - x_2 + 2x_3 = 2$

28. $x_1 - x_3 = 2$
$2x_1 - x_2 = 4$
$x_1 + x_2 + x_3 = 6$

29. $3x_1 + x_2 - 2x_3 = 3$
$x_1 - 2x_2 + x_3 = 4$

30. $2x_1 - x_2 - 3x_3 = 0$
$x_1 - 2x_2 + x_3 = 4$

31. $x_1 + 2x_2 - x_3 = 5$
$2x_1 - x_2 + 2x_3 = 0$

32. $x_1 - x_2 + 2x_3 = 6$
$2x_1 + 2x_2 - x_3 = -1$

33. $2x_1 - x_2 = 6$
$x_1 - 2x_2 = 0$
$3x_1 - x_2 = 6$

34. $x_1 - 2x_2 = 0$
$2x_1 + x_2 = 5$
$x_1 - 3x_2 = -3$

35. What must be true about x, y, z, w, if the matrices

$$A = \begin{bmatrix} x & y \\ z & w \end{bmatrix} \quad \text{and} \quad B = \begin{bmatrix} 1 & 1 \\ -1 & 1 \end{bmatrix}$$

are to commute? That is, $AB = BA$.

36. Let $t = [t_1 \;\; t_2]$, with $t_1 + t_2 = 1$, and let $A = \begin{bmatrix} \frac{1}{4} & \frac{3}{4} \\ \frac{2}{3} & \frac{1}{3} \end{bmatrix}$.
Find t such that $tA = t$.

37. Associate numbers to letters as follows:

1 2 3 4 5 6 7 8 9 10 11 12 13 14 15 16 17 18 19 20 21 22 23 24 25 26
↓ ↓
A B C D E F G H I J K L M N O P Q R S T U V W X Y Z

The matrix A used to encode a message has as its inverse the matrix

$$A^{-1} = \begin{bmatrix} 2 & -3 \\ -1 & 2 \end{bmatrix}$$

Decode the message

11 7 84 51 51 28 66 43 44 29 107 65 64 41

38. Associate numbers to letters as in Problem 37, and use the matrix

$$A = \begin{bmatrix} 1 & 0 & 0 \\ 3 & 1 & 5 \\ -2 & 0 & 1 \end{bmatrix}$$

to encode the message IT'S OVER

3
LINEAR PROGRAMMING PART ONE: GEOMETRIC APPROACH

3.1

INTRODUCTION

Whenever the analysis of a problem leads to minimizing or maximizing a linear expression in which the variable must obey a collection of linear inequalities, a solution may be obtained using linear programming techniques.

Historically, linear programming problems evolved out of the need to solve problems involving resource allocation during World War II by the United States Army. Among those who worked on such problems for the Air Force was George Dantzig*, who later gave a general formulation of the linear programming problem and offered a method for solving it. His technique, called the **simplex method,** is discussed in Chapter 4.

In this chapter, we study ways to solve linear programming problems that involve only two variables. As a result, we can use a geometric approach to solve the problems. But first we need to discuss linear inequalities.

3.2

LINEAR INEQUALITIES

THE GRAPH OF A LINEAR INEQUALITY
SYSTEMS OF LINEAR INEQUALITIES
SOME TERMINOLOGY
APPLICATION

We have already discussed linear equations (linear equalities) in two variables x and y. (Section 1.2). These are equations of the form

$$Ax + By = C \tag{1}$$

where A, B, C are real numbers and A and B are not both zero. If in Equation (1) we replace the equal sign by an inequality symbol, namely, one of the symbols $<, >, \leq, \geq$, we obtain a *linear inequality in two variables x and y.*

For example, the expressions

$$3x + 2y \geq 4, \qquad 2x - 3y < 0, \qquad 3x + 5y > -8$$

are each linear inequalities in two variables. The first of these is called a *nonstrict inequality* since the inequality symbol $\geq$ is nonstrict; the remaining two linear inequalities are *strict.*

---

* GEORGE DANTZIG is one of the pioneering creators of linear programming, which is one of the most important developments in applied mathematics in the last half-century. He developed the simplex method in 1946. A historical perspective worth reading is George Dantzig, "Reminiscences about the Origins of Linear Programming," *Operations Research Letters,* **1,** 2 (April 1982).

THE GRAPH OF A LINEAR INEQUALITY

The *graph of a linear inequality* in two variables x and y is the set of all points (x, y) for which the inequality holds.

Let's look at an example.

Example 1

Graph the inequality: $2x + 3y \geq 6$

Solution

First, we graph the line

$$L: \quad 2x + 3y = 6$$

Any point on the line L obeys the inequality $2x + 3y \geq 6$, since we are seeking all points (x, y) for which $2x + 3y$ is greater than *or equal to* 6. See Figure 1a.

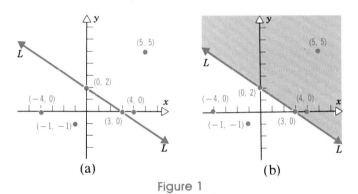

Figure 1

Now, let's test a few points, such as $(-1, -1), (5, 5), (4, 0), (-4, 0)$, to see if they obey the inequality. We do this by substituting the coordinates of each point into the left member of the inequality and determining whether the result is ≥ 6 or < 6.

	$2x$	$+ 3y$		*Conclusion*
$(-1, -1)$:	$2(-1) + 3(-1)$	$= -2 - 3 = -5 < 6$		Not part of graph
$(5, 5)$:	$2(5)$	$+ 3(5)$	$= 25 > 6$	Part of graph
$(4, 0)$:	$2(4)$	$+ 3(0)$	$= 8 > 6$	Part of graph
$(-4, 0)$:	$2(-4) + 3(0)$		$= -8 < 6$	Not part of graph

Notice that the two points $(4, 0)$ and $(5, 5)$ that are part of the graph both lie on one side of L, while the points $(-4, 0)$ and $(-1, -1)$ (not part of the graph) lie on the other side of L. This is not an accident. The graph of the inequality is the shaded region of Figure 1b.   ■

Let's outline the procedure for graphing a linear inequality:

Graphing a Linear Inequality

Step 1: Graph the corresponding linear equation, a line L.

Step 2: Select a point P not on the line L.

Step 3: If the coordinates of this point P satisfy the linear inequality, then all points on the same side of L as the point P satisfy the inequality. If the coordinates of the point P do not obey the linear inequality, then all points on the opposite side of L from P satisfy the inequality.

Points on the line L itself may or may not obey the inequality. Here is an example of a case when the points on L do not satisfy the inequality.

Example 2

Graph the linear inequality: $2x - y < -4$

Solution

The corresponding linear equation is the line

$$L: \quad 2x - y = -4$$

For its graph, see Figure 2*a*.

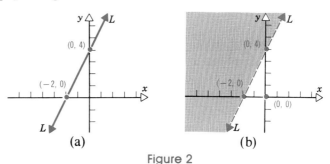

Figure 2

We select a point on either side of L to be tested, for example $(0, 0)$:

$$2(0) - 0 = 0 > -4$$

Since $(0, 0)$ does not obey the inequality, all points on the opposite side of L from $(0, 0)$ are on the graph.

Since no point on L can be on the graph (why?), the graph is the shaded region of Figure 2b and the line L is dashed to indicate that it is not part of the graph. ∎

The set of points belonging to the graph of a linear inequality (for example, the shaded region in Figure 2*b*) is sometimes called a *half-plane.*

Example 3

Graph   (a) $x \le 3$,   (b) $y \ge 4$,   (c) $2x \le y$

Solution

(a) When you are asked to graph an inequality such as $x \leq 3$, you should think of it as an inequality of the form $x + 0 \cdot y \leq 3$ and graph it in the two-dimensional plane. The solution of this inequality is the set of all ordered pairs (x, y) such that $x + 0 \cdot y \leq 3$. Since the coefficient of y is zero, y can assume any real number while x is restricted to real numbers less than or equal to 3. The graph of this inequality is the half-plane in Figure 3a.

(b) In a similar way we graph the inequality $y \geq 4$. Think of it as an inequality of the form $0x + y \geq 4$. Here, x can assume any real number while y is restricted to real numbers greater than or equal to 4. See Figure 3b.

(c) The graph of the inequality $2x \leq y$ is given in Figure 3c.

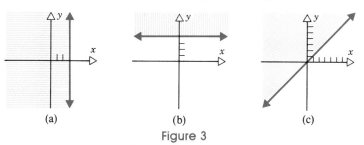

(a)   (b)   (c)

Figure 3

SYSTEMS OF LINEAR INEQUALITIES

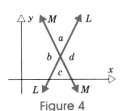

Figure 4

A *system of linear inequalities* is a collection of two or more linear inequalities. To *graph* a system of two inequalities in two variables we locate all points (x, y) that obey each linear inequality of the system. The graph is called the *solution region* of the system.

Let's look at a system of two linear inequalities in two unknowns. Several possible graphs can result. For example, suppose L and M are the lines corresponding to two linear inequalities, and suppose L and M intersect. See Figure 4. Then the two lines L and M divide the plane into four regions a, b, c, and d. One of these regions is the solution of the system.

Example 4
Graph the system:

$$2x - y \leq -4$$
$$x + y \geq -1$$

Solution
The lines corresponding to these linear inequalities are

$$L: \quad 2x - y = -4$$
$$M: \quad x + y = -1$$

The graphs of L and M are shown in Figure 5.

If we graph each linear inequality as a separate problem and then find the region common to the two resulting half-planes, we will have the solution of the system. The heavily shaded region in Figure 6 is the solution.

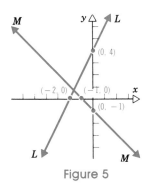

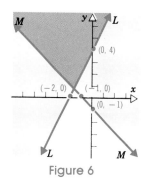

Figure 5

Figure 6   ∎

If the lines L and M are parallel, the system of linear inequalities may or may not have a solution. Examples of such situations are given below.

Example 5
Graph the system:

$$2x - y \le -4$$
$$2x - y \le -2$$

Solution
The lines corresponding to these linear inequalities are

$$L: \quad 2x - y = -4$$
$$M: \quad 2x - y = -2$$

These lines are parallel. Their graphs are shown in Figure 7.

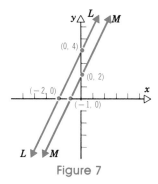

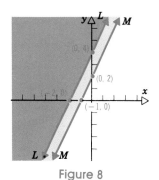

Figure 7

Figure 8

The graphs of the two linear inequalities are shown in Figure 8, and the solution is the heavily shaded region.   ∎

Notice that the solution of this system is the same as that of the single linear inequality $2x - y \le -4$.

Example 6
The solution of the system

$$2x - y \geq -4$$
$$2x - y \leq -2$$

is the heavily shaded region in Figure 9.

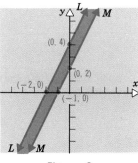

Figure 9

Example 7
The system

$$2x - y \leq -4$$
$$2x - y \geq -2$$

has no solution, as Figure 10 indicates, because the two half-planes have no points in common.

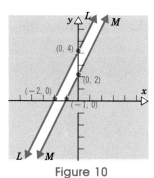

Figure 10

Until now, we have considered systems of only two linear inequalities. The next example is of a system of four linear inequalities. As we will see, the technique for graphing such systems is the same as that used for graphing systems of two linear inequalities in two unknowns.

Example 8

Graph the system:

$$x + y \geq 2$$
$$2x + y \geq 3$$
$$x \geq 0$$
$$y \geq 0$$

Solution

Again, we first graph the four lines:

$$L_1: \quad x + y = 2$$
$$L_2: \quad 2x + y = 3$$
$$L_3: \qquad x = 0$$
$$L_4: \qquad y = 0$$

The graph of the system is the intersection of the four regions determined by each of the four inequalities. See Figure 11. ∎

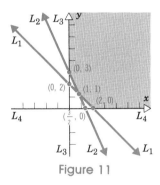

Figure 11

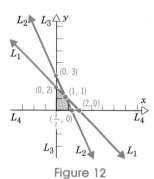

Figure 12

Example 9

Graph the system:

$$x + y \leq 2$$
$$2x + y \leq 3$$
$$x \geq 0$$
$$y \geq 0$$

Solution

Since the lines associated with these linear inequalities are the same as those of the previous example, we proceed directly to the graph. See Figure 12. ∎

SOME TERMINOLOGY

Compare the graphs of the systems of linear inequalities given in Examples 8 and 9. The region in Figure 11 is *unbounded* whereas the region in Figure 12 is *bounded*. A region in the plane is said to be *bounded* if it can be enclosed within some circle with center at (0, 0); if a region cannot be enclosed within such a circle, then it is *unbounded.* See Figure 13.

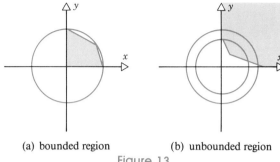

(a) bounded region   (b) unbounded region

Figure 13

The boundary of each of the graphs in Figures 11 and 12 consists of line segments. In fact, the graph of any system of linear inequalities will have line segments as boundaries. The point of intersection of two line segments that form the boundary is called a *vertex of the graph.* For example, the graph of the system given in Example 8 has the vertices (0, 3), (1, 1), and (2, 0). See Figure 11. The graph of the system given in Example 9 has the vertices (0, 2), (0, 0), ($\frac{3}{2}$, 0), (1, 1). See Figure 12.

We will soon see that the vertices of the graph of a system of linear inequalities play a major role in the procedure for solving linear programming problems.

APPLICATION

Example 10

Nutt's Nuts has 75 pounds of cashews and 120 pounds of peanuts. These are to be mixed in 1 pound packages as follows: A low-grade mixture that contains 4 ounces of cashews and 12 ounces of peanuts and a high-grade mixture that contains 8 ounces of cashews and 8 ounces of peanuts.

(a)   Using x to denote the number of the packages of low-grade mixture and using y to denote the number of packages of the high-grade mixture, write down a system of linear inequalities that describes the possible number of each kind of package.

(b)   Graph the system and list its vertices.

Solution

(a)   We begin by naming the variables

$$x = \text{number of packages of low-grade mixture}$$
$$y = \text{number of packages of high-grade mixture}$$

First, we note that the only meaningful values for x and y are nonnegative values. Thus we must restrict x and y so that

$$x \geq 0 \qquad y \geq 0$$

Next, we note that there is a limit to the number of pounds of cashews and peanuts available. First, the total number of pounds of cashews cannot exceed 75 pounds (1200 ounces), and the number of pounds of peanuts cannot exceed 120 pounds (1920 ounces). This means that

$$\begin{pmatrix} \text{Ounces of} \\ \text{cashews} \\ \text{required} \\ \text{for low-grade} \\ \text{mixture} \end{pmatrix} \begin{pmatrix} \text{Number of} \\ \text{packages of} \\ \text{low-grade} \\ \text{mixture} \end{pmatrix} + \begin{pmatrix} \text{Ounces of} \\ \text{cashews} \\ \text{for high-} \\ \text{grade} \\ \text{mixture} \end{pmatrix} \begin{pmatrix} \text{Number of} \\ \text{packages} \\ \text{of high-} \\ \text{grade} \\ \text{mixture} \end{pmatrix} \begin{matrix} \text{cannot} \\ \text{exceed} \end{matrix} \quad 1200$$

$$\begin{pmatrix} \text{Ounces of} \\ \text{peanuts} \\ \text{required} \\ \text{for low-grade} \\ \text{mixture} \end{pmatrix} \begin{pmatrix} \text{Number of} \\ \text{packages of} \\ \text{low-grade} \\ \text{mixture} \end{pmatrix} + \begin{pmatrix} \text{Ounces of} \\ \text{peanuts} \\ \text{for high-} \\ \text{grade} \\ \text{mixture} \end{pmatrix} \begin{pmatrix} \text{Number of} \\ \text{packages of} \\ \text{high-grade} \\ \text{mixture} \end{pmatrix} \begin{matrix} \text{cannot} \\ \text{exceed} \end{matrix} \quad 1920$$

In terms of the data given and the variables introduced, we can write these statements compactly as

$$4x + 8y \leq 1200$$
$$12x + 8y \leq 1920$$

The system of linear inequalities that gives the possible values x and y can take on, is

$$x \geq 0$$
$$y \geq 0$$
$$4x + 8y \leq 1200$$
$$12x + 8y \leq 1920$$

(b) The system of linear inequalities given above can be simplified to the equivalent form

$$x \geq 0 \qquad ①$$
$$y \geq 0 \qquad ②$$
$$x + 2y \leq 300 \qquad ③$$
$$3x + 2y \leq 480 \qquad ④$$

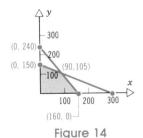

Figure 14

in which, for convenience, we have numbered each linear inequality. The graph of the system is given in Figure 14. The vertices of the graph are the points of intersection of the lines ① and ②, ① and ③, ② and ④, and ③ and ④. The first three are easy to identify by inspection; the last one requires that we solve the system of equations

$$x + 2y = 300$$
$$3x + 2y = 480$$

The vertices are

$$(0, 0), (0, 150), (160, 0), (90, 105)$$

Exercise 3.2   *Answers to Odd-Numbered Problems begin on page A-10.*

A   In Problems 1–10 graph each inequality.

1. $x \geq 0$ 2. $y \geq 0$
3. $x \geq 0, \quad y \geq 0$ 4. $x \leq 0, \quad y \leq 0$
5. $2x - 3y \leq -6$ 6. $3x + 2y \geq 6$
7. $5x + y \leq -10$ 8. $x - 2y > -4$
9. $x \geq 5$ 10. $y \leq -2$

B   In Problems 11–20 graph each system of linear inequalities. Tell whether the graph is bounded or unbounded and list each vertex of the graph.

11. $x \geq 0$ 12. $x \geq 0$ 13. $x \geq 0$
       $y \geq 0$ $y \geq 0$ $y \geq 0$
       $x + y \leq 2$ $2x + 3y \leq 6$ $x + y \geq 2$
                                                            $2x + 3y \leq 6$

14. $x \geq 0$ 15. $x \geq 0$ 16. $x \geq 0$
       $y \geq 0$ $y \geq 0$ $y \geq 0$
       $x + y \geq 2$ $2 \leq x + y$ $2 \leq x + y$
       $2x + 3y \leq 12$ $x + y \leq 8$ $x + y \leq 8$
       $3x + 2y \leq 12$ $2x + y \leq 10$ $1 \leq x + 2y$

17. $x \geq 0$ 18. $x \geq 0$ 19. $x \geq 0$
       $y \geq 0$ $y \geq 0$ $y \geq 0$
       $x + y \geq 2$ $2 \leq x + y$ $1 \leq x + 2y$
       $2x + 3y \leq 12$ $x + y \leq 10$ $x + 2y \leq 10$
       $3x + y \leq 12$ $2x + y \leq 3$

20. $x \geq 0$
       $y \geq 0$
       $1 \leq x + 2y$
       $x + 2y \leq 10$
       $2 \leq x + y$
       $x + y \leq 8$

C   In Problems 21–28 graph each system of linear inequalities. Tell whether the graph is bounded or unbounded and list each vertex of the graph.

21. $x \geq 0$ 22. $x \geq 0$ 23. $x \geq 0$
       $y \geq 0$ $y \geq 0$ $y \geq 0$
       $x + 2y \geq 5$ $4x + y \geq 8$ $8x + y \leq 20$
       $5x + y \geq 16$ $2x + 5y \geq 18$ $6x - 5y \geq -8$
                                  $2x + 3y \geq 14$

24. $x \geq 0$ **25.** $x \geq 0$ **26.** $x \geq 0$
     $y \geq 0$ $y \geq 0$ $y \geq 0$
  $3x - 2y \geq 0$ $3x + 5y \leq 38$ $3x - 8y \geq -17$
  $3x + y \leq 18$ $x - y \leq -6$ $7x - 3y \leq 23$
                               $-4x - 5y \leq 7$

27. $x \geq 0$ **28.** $x \geq 0$
     $y \geq 0$ $y \geq 0$
  $x + 2y \geq 12$ $x + y \leq 4$
   $x + y \geq 8$ $2x + 3y \leq 9$
      $x \geq 2$ $y \leq 2$
      $y \geq 1$ $y \geq \frac{3}{4}x$

APPLICATIONS

29. Mixture   Rework Example 10 if 100 pounds of cashews and 120 pounds of peanuts are available.

30. Mixture   Rework Example 10 if the high-grade mixture contains 10 ounces of cashews and 6 ounces of peanuts.

31. Manufacturing   Mike's Famous Toy Trucks company manufactures two kinds of a toy truck—a dumpster and a tanker. In the manufacturing process, each dumpster requires 3 hours of grinding and 4 hours of finishing, while each tanker requires 2 hours of grinding and 3 hours of finishing. The Company has 2 grinders and 3 finishers, each of whom work 40 hours per week.

    (a) Using x to denote the number of dumpsters and y to denote the number of tankers, write down a system of linear inequalities that describes the possible numbers of each truck that can be manufactured.

    (b) Graph the system and list its vertices.

32. Manufacturing   Repeat Problem 31 if 1 grinder and 2 finishers, each of whom work 40 hours per week, are available.

33. Nutrition   A farmer prepares feed for livestock by combining two types of grain. Each unit of the first grain contains 1 unit of protein and 5 units of iron while each unit of the second grain contains 2 units of protein and 1 unit of iron. Each animal must receive at least 5 units of protein and 16 units of iron each day.

    (a) Write down a system of linear inequalities that describes the possible amounts of each grain the farmer needs to prepare.

    (b) Graph the system and list the vertices.

34. Investment Strategy   Laura wishes to invest up to a total of $40,000 in class AA bonds and stocks. Furthermore, she believes that the amount invested in class AA bonds should be at most one-third of the amount invested in stocks.

    (a) Write down a system of linear inequalities that describes the possible amount of investments in each security.

    (b) Graph the system and list the vertices.

35. Nutrition   To maintain an adequate daily diet, nutritionists recommend the following: at least 85 g of carbohydrate, 70 g of fat, and 50 g of protein. An ounce of food A contains 5 g of carbohydrate, 3 g of fat, and 2 g of protein, while an ounce of food B contains 4 g of carbohydrate, 3 g of fat, and 3 g of protein.

(a) Write down a system of linear inequalities that describes the possible quantities of each food.

(b) Graph the system and list the vertices.

36. Transportation A microwave company has two plants, one on the East Coast and one in the Midwest. It takes 25 hours (packing, transportation, and so on) to transport an order of microwaves from the Eastern plant to its central warehouse and 20 hours from the Midwest plant to its central warehouse. It costs $80 to transport an order from the Eastern plant to the central warehouse and $40 from the Midwestern to its central warehouse. There are 1000 manhours available for packing, transportation, and so on, and $3000 for transportation cost.

(a) Write down a system of linear inequalities that describes the transportation system.

(b) Graph the system and list the vertices.

3.3

A GEOMETRIC APPROACH TO LINEAR PROGRAMMING PROBLEMS

LINEAR PROGRAMMING PROBLEMS
GENERAL RULE FOR LINEAR PROGRAMMING PROBLEMS
PROCEDURE FOR SOLVING LINEAR PROGRAMMING PROBLEMS
MODEL: POLLUTION CONTROL

We begin by restating a portion of Example 10 given in the previous section: Nutt's Nuts has 75 pounds of cashews and 120 pounds of peanuts. These are to be mixed in 1 pound packages as follows: a low-grade mixture that contains 4 ounces of cashews and 12 ounces of peanuts and a high-grade mixture that contains 8 ounces of cashews and 8 ounces of peanuts.

Suppose that in addition to the information given above, we also know what the profit will be on each type of mixture. For example, suppose the profit is $0.25 on each package of the low-grade mixtures and is $0.45 on each package of the high-grade mixture. The question of importance to the manager is "How many packages of each type of mixture should be prepared to maximize the profit?"

If P symbolizes the profit, x is the number of packages of low-grade mixture, and y is the number of high-grade packages, then the question can be restated as "What are the values of x and y so that the expression

$$P = \$0.25x + \$0.45y$$

is a maximum?

LINEAR PROGRAMMING PROBLEMS

The problem above is typical of a *linear programming problem.* It requires that a certain linear expression, the profit, be maximized. This linear expression is called the **objective function.** Furthermore, the problem requires that the maximum profit be achieved under certain restrictions or **constraints,** each of which are linear inequalities involving the variables. The linear programming problem may be restated as

Maximize

$$P = \$0.25x + \$0.45y \qquad \text{Objective function}$$

subject to the conditions that

$$x \geq 0 \qquad \text{Nonnegativity constraint}$$
$$y \geq 0 \qquad \text{Nonnegativity constraint}$$
$$x + 2y \leq 300 \qquad \text{Cashew constraint}$$
$$3x + 2y \leq 480 \qquad \text{Peanut constraint}$$

In general, every linear programming problem has two components:

1. A linear objective function to be maximized or minimized.
2. A collection of linear inequalities that must be satisfied simultaneously.

Linear Programming Problem *A linear programming problem* in two variables, *x* and *y*, consists of *maximizing* or *minimizing* an *objective function*

$$z = Ax + By$$

where *A* and *B* are given real numbers, subject to certain conditions or *constraints* expressible as linear inequalities in *x* and *y*.

Let's look at this more closely. To maximize (or minimize) the quantity $z = Ax + By$ means to locate the point or points (x, y) that make the expression for z the largest (or smallest). But not all points (x, y) are eligible. Only the points that obey *all* the constraints are potential solutions. Hence, we refer to such points as *feasible solutions.*

In a linear programming problem, we want to find the feasible solution that maximizes (or minimizes) the objective function.

Solution of a Linear Programming Problem **By a** *solution to a linear programming problem* **we mean a point (*x*, *y*) in the set of feasible solutions together with the value of the objective function at that point, that maximizes (or minimizes) the objective function.**

If none of the feasible solutions maximize (or minimize) the objective function, or if there are no feasible solutions, then the linear programming problem has no solution.

Example 1
Minimize the quantity

$$z = x + 2y$$

subject to the constraints

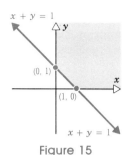

Figure 15

$$x + y \geq 1 \qquad x \geq 0 \qquad y \geq 0$$

Solution

The objective function to be minimized is $z = x + 2y$. The constraints are the linear inequalities

$$x + y \geq 1 \qquad x \geq 0 \qquad y \geq 0$$

The shaded portion of Figure 15 illustrates the set of feasible solutions.

To see if there is a smallest z, we graph $z = x + 2y$ for some choice of z, say, $z = 3$. See Figure 16. By moving the line $x + 2y = 3$ parallel to itself, we can

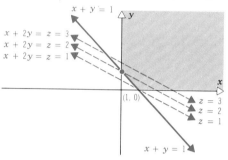

Figure 16

observe what happens for different values of z. Since we want a minimum value for z, we try to move $z = x + 2y$ down as far as possible while keeping some part of the line within the set of feasible solutions. The "best" solution is obtained when the line just touches one corner, or *vertex*, of the set of feasible solutions. If you refer to Figure 16, you will see that the best solution is $x = 1$, $y = 0$, which yields $z = 1$. There is no other feasible solution for which z is smaller.   ■

In Example 1, we can see that the feasible solution that minimizes z occurs at a vertex. This is not an unusual situation. If there is a feasible solution minimizing (or maximizing) the objective function, it is *usually* located at a vertex of the set of feasible solutions.

However, it is possible for a feasible solution that is not a vertex to minimize (or maximize) the objective function. This occurs when the slope of the objective function is the same as the slope of one side of the set of feasible solutions. The following example illustrates this possibility.

Example 2
Minimize the quantity

$$z = x + 2y$$

subject to the constraints

$$x + y \geq 1 \qquad 2x + 4y \geq 3 \qquad x \geq 0 \qquad y \geq 0$$

Solution
Again, we first graph the constraints. The shaded portion of Figure 17 illustrates the set of feasible solutions.

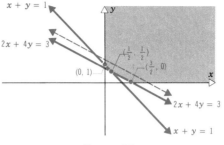

Figure 17

If we graph the objective equation $z = x + 2y$ for some choice of z and move it down, we see that a minimum is reached when $z = \frac{3}{2}$. In fact, any point on the line $2x + 4y = 3$ between $(\frac{1}{2}, \frac{1}{2})$ and $(\frac{3}{2}, 0)$ will minimize the objective function. Of course, the reason any feasible point on $2x + 4y = 3$ minimizes the objective equation $z = x + 2y$ is that these two lines are parallel (both have slope $-\frac{1}{2}$). Thus, this linear programming problem has infinitely many solutions.   ■

The next example illustrates a linear programming problem that has no solution.

Example 3
Maximize the quantity

$$z = x + 2y$$

subject to the constraints

$$x + y \geq 1 \qquad x \geq 0 \qquad y \geq 0$$

Solution
First, we graph the constraints. The shaded portion of Figure 18 illustrates the set of feasible solutions.

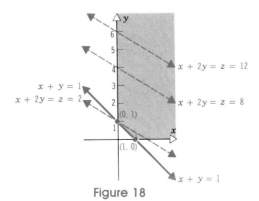

Figure 18

The graphs of the objective function $z = x + 2y$ for $z = 2$, $z = 8$, and $z = 12$ are also shown in Figure 18. Observe that we continue to get larger values for z by moving the graph of the objective function upward. But there is no feasible point that will make z *largest*. No matter how large a value is assigned to z, there is a feasible point that will give a larger value. Since there is no feasible point that makes z largest, we conclude that this linear programming problem has no solution. ■

GENERAL RULE FOR LINEAR PROGRAMMING PROBLEMS

For any linear programming problem that has a solution, the following general result is true:

> If a linear programming problem has a solution, it is located at a vertex of the set of feasible solutions; if a linear programming problem has multiple solutions, at least one of them is located at a vertex of the set of feasible solutions. In either case, the corresponding value of the objective function is unique.

The result stated above requires knowing in advance whether the linear programming problem has a solution. We now state the conditions that will tell when the solution of a linear programming problem does exist.

> **Existence of a Solution** Given a linear programming problem with a feasible solution set R and objective function $z = ax + by$.
>
> **1.** If R is bounded then z has both a maximum and a minimum value on R.
>
> **2.** If R is unbounded and $a \geq 0$, $b \geq 0$, and the constraints include $x \geq 0$ and $y \geq 0$, then z has a minimum value on R but not a maximum. See Example 3.
>
> **3.** If R is the empty set then the linear programming problem has no solution and z has neither a maximum nor a minimum value.

PROCEDURE FOR SOLVING LINEAR PROGRAMMING PROBLEMS

Based on these comments, we can outline a procedure for solving a linear programming problem provided that it has a solution.

Solving a Linear Programming Problem

> **Step 1:** Write an expression for the quantity that is to be maximized or minimized (the objective function).
>
> **Step 2:** Determine all the constraints and graph them.
>
> **Step 3:** List the vertices of the set of feasible solutions.
>
> **Step 4:** Determine the value of the objective function at each vertex.
>
> **Step 5:** Select the optimal solution, that is, the maximum or minimum value of the objective function.

Let's look at some examples.

Example 4

Maximize and minimize the objective function

$$z = x + 5y$$

subject to the constraints

① $x + 4y \le 12$

② $x \le 8$

③ $x + y \ge 2$

④ $x \ge 0$

⑤ $y \ge 0$

Solution

The objective function and the constraints (numbered for convenience) are given (this will not be the case when we do word problems), so we can proceed to graph the constraints. The shaded portion of Figure 19 illustrates the set of feasible solutions. Since this set is bounded, we know a solution exists.

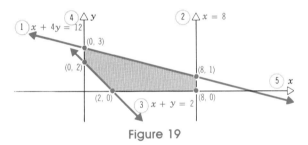

Figure 19

Now we locate the vertices of the set of feasible solutions at the points of intersection of lines ① and ④, ① and ②, ② and ⑤, ③ and ⑤, and ③ and ④. Using methods discussed earlier, we find that the vertices are

$$(0, 3), \quad (8, 1), \quad (8, 0), \quad (2, 0), \quad (0, 2)$$

To find the maximum and minimum value of $z = x + 5y$, we set up a table:

Vertex (x, y)	Value of Objective Function $z = x + 5y$
(0, 3)	$z = 0 + 5(3) = 15$
(8, 1)	$z = 8 + 5(1) = 13$
(8, 0)	$z = 8 + 5(0) = 8$
(2, 0)	$z = 2 + 5(0) = 2$
(0, 2)	$z = 0 + 5(2) = 10$

The maximum value of z is 15, and it occurs at the point $(0, 3)$. The minimum value of z is 2, and it occurs at the point $(2, 0)$.   ∎

Now, let's solve the problem of the cashews and peanuts. (Refer to page 151).

Example 5
Maximize

$$P = 0.25x + 0.45y$$

subject to the constraints

① $x \geq 0$

② $y \geq 0$

③ $x + 2y \leq 300$

④ $3x + 2y \leq 480$

Solution
Before applying the method of this chapter to solve this problem, let's discuss a solution that might be suggested by intuition. Namely, since the profit is higher for the high-grade mixture, you might think that Nutt's Nuts should prepare as many packages of the high-grade mixture as possible. If this were done, then there would be a total of 150 packages (8 ounces divides into 75 pounds of cashews exactly 150 times) and the total profit would be

$$150(0.45) = \$67.50$$

As we will see, this is not the best solution to the problem. This is because there would be several pounds of peanuts left over ($120 - 75 = 45$, to be exact) that would be neither packaged nor sold.

To use more of the peanuts and thus make a higher profit, Nutt's Nuts has to make both high-grade *and* low-grade packages. We are still asking ourselves how many packages of each mixture should be made to obtain the maximum profit, but now we will use linear programming to solve the problem. The graph of the set of feasible solutions is given in Figure 20.

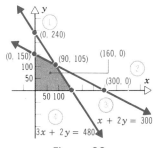

Figure 20

Since this set is bounded, we proceed to locate its vertices. The vertices of the set of feasible solutions are the points of intersection of lines ① and ②, ① and ③, ② and ④, and ③ and ④:

$$(0, 0), \quad (0, 150), \quad (160, 0), \quad (90, 105)$$

(Notice that the points of intersection of lines ① and ④ and lines ② and ③ are not feasible solutions.) It remains only to evaluate the objective equation at each vertex:

Vertex (x, y)	Value of Objective Function $P = (\$0.25)x + (\$0.45)y$
(0, 0)	$P = (0.25)(0) + (0.45)(0) = 0$
(0, 150)	$P = (0.25)(0) + (0.45)(150) = \$67.50$
(160, 0)	$P = (0.25)(160) + (0.45)(0) = \$40.00$
(90, 105)	$P = (0.25)(90) + (0.45)(105) = \$69.75$

Thus, a maximum profit is obtained if 90 packages of low-grade mixture and 105 packages of high-grade mixture are made. The maximum profit obtainable under the conditions described is $69.75.   ■

Example 6

Mike's Famous Toy Trucks manufactures two kinds of toy trucks—a standard model and a deluxe model. In the manufacturing process, each standard model requires 2 hours of grinding and 2 hours of finishing, and each deluxe model needs 2 hours of grinding and 4 hours of finishing. The company has 2 grinders and 3 finishers, each of whom work 40 hours per week. Each standard model toy truck brings a profit of $3 and each deluxe model a profit of $4. Assuming that every truck made will be sold, how many of each should be made to maximize profits?

Solution

First, we name the variables:

$$x = \text{Number of standard models made}$$
$$y = \text{Number of deluxe models made}$$

The quantity to be maximized is the profit, which we denote by P:

$$P = \$3x + \$4y$$

This is the objective function. To manufacture one standard model requires 2 grinding hours and to make one deluxe model requires 2 grinding hours. Thus, the number of grinding hours of x standard and y deluxe models is

$$2x + 2y$$

But the total amount of grinding time available is 80 hours per week. This means we have the constraint

$$2x + 2y \leq 80 \qquad \text{Grinding time constraint}$$

Similarly, for the finishing time we have the constraint

$$2x + 4y \leq 120 \qquad \text{Finishing time constraint}$$

Simplifying each of these constraints and adding the nonnegativity constraints $x \geq 0$ and $y \geq 0$, we may list all the constraints for this problem.

$$x + y \leq 40 \qquad x + 2y \leq 60 \qquad x \geq 0 \qquad y \geq 0$$

Figure 21 illustrates the set of feasible solutions, which is bounded.

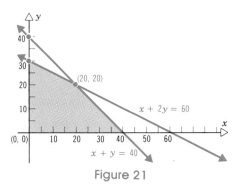

Figure 21

The vertices of the set of feasible solutions are

$$(0, 0), \quad (0, 30), \quad (40, 0), \quad (20, 20)$$

The table lists the corresponding values of the objective equation:

Vertex (x, y)	Value of Objective Function $P = \$3x + \$4y$
$(0, 0)$	$P = 0$
$(0, 30)$	$P = \$120$
$(40, 0)$	$P = \$120$
$(20, 20)$	$P = 3(20) + 4(20) = \$140$

Thus, a maximum profit is obtained if 20 standard trucks and 20 deluxe trucks are manufactured. The maximum profit is $140. ◼

Example 7

Investment Strategy A retired couple have up to $30,000 they wish to invest in fixed-income securities. Their broker recommends investing in two bonds: one a AAA bond yielding 12%; the other a B⁺ bond paying 15%. After some consideration, the couple decide to invest at most $12,000 in the B⁺-rated bond and at least $6000 in the AAA bond. They also want the amount invested in the AAA bond to exceed or equal the amount invested in the B⁺ bond. What should the broker recommend if the couple (quite naturally) want to maximize their return on investment?

Solution
First, we name the variables:

$$x = \text{Amount invested in AAA bond}$$
$$y = \text{Amount invested in B}^+ \text{ bond}$$

The quantity to be maximized — return on investment — which we denote by P, is

$$P = 0.12x + 0.15y$$

This is the objective function. The conditions specified by the problem are:

Up to $30,000 available to invest	$x + y \leq 30,000$
Invest at most $12,000 in B$^+$ bond	$y \leq 12,000$
Invest at least $6000 in AAA bond	$x \geq 6,000$
Amount in AAA bond must exceed or equal amount in B$^+$ bond	$x \geq y$

In addition, we must have the conditions $x \geq 0$ and $y \geq 0$. The total list of constraints is

    ① $x + y \leq 30,000$ ② $y \leq 12,000$ ③ $x \geq 6000$

    ④ $x \geq y$ ⑤ $x \geq 0$ ⑥ $y \geq 0$

Figure 22 illustrates the set of feasible solutions, which is bounded. The vertices of

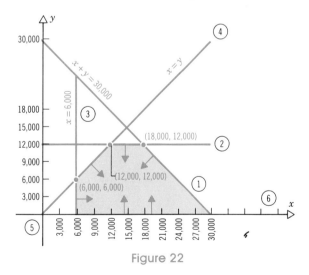

Figure 22

the set of feasible solutions are

  (6000, 0),   (6000, 6000),   (12,000 12,000),   (18,000, 12,000),   (30,000, 0)

The corresponding return on investment at each vertex is:

$$P = 0.12(6000) + 0.15(0) = \$720$$
$$P = 0.12(6000) + 0.15(6000) = 720 + 900 = \$1620$$
$$P = 0.12(12{,}000) + 0.15(12{,}000) = 1440 + 1800 = \$3240$$
$$P = 0.12(18{,}000) + 0.15(12{,}000) = 2160 + 1800 = \$3960$$
$$P = 0.12(30{,}000) + 0.15(0) = \$3600$$

Thus, the maximum return on investment is $3960, obtained by placing $18,000 in the AAA bond and $12,000 in the B$^+$ bond.

Example 8

Urban Economics Model* This example concerns reclaimed land and its allocation into two major uses—agricultural and urban (or nonagricultural). The reclamation of land for urban purposes cost $400 per acre and for agricultural uses, $300. The primal problem is that the reclamation agency wishes to minimize the total cost C of reclaiming the land:

$$C = \$400x + \$300y$$

where $x =$ the number of acres of urban land and $y =$ the number of acres of agricultural land. Although this equation can be minimized by setting both x and y at zero, that is, reclaiming nothing, the problem derives from a number of constraints due to three different groups.

The first is an urban group, which insists that at least 4000 acres of land be reclaimed for urban purposes. The second group is concerned with agriculture and says that at least 5000 acres of land must be reclaimed for agricultural uses. Finally, the third group is concerned only with reclamation and is quite uninterested in the use to which the land will be put. The third group, however, says that at least 10,000 acres of land must be reclaimed. The primal problem and the constraints can, therefore, be written in full as follows:
Minimize

$$C = \$400x + \$300y$$

subject to the constraints

    ① $x \geq 4000$
    ② $y \geq 5000$
    ③ $x + y \geq 10{,}000$
    ④ $x \geq 0$
    ⑤ $y \geq 0$

Figure 23 illustrates the set of feasible solutions, which is not bounded. However, the objective function constants, 400 and 300, are nonnegative; and the

---

* This example is adapted from Maurice Yeates, *An Introduction to Quantitative Analysis in Economic Geography,* McGraw-Hill, New York, 1968.

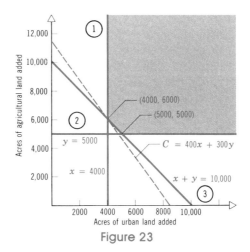

Figure 23

constraints include $x \geq 0$ and $y \geq 0$. We conclude that the objective function C does have a minimum value on this set. That minimum must occur at a vertex. We set up a table:

Vertex (x, y)	Objective Function $C = \$400x + \$300y$
(4000, 6000)	$ 3,400,000
(5000, 5000)	$ 3,500,000

If 4000 acres are devoted to urban purposes, and 6000 acres to agricultural purposes, the cost is a minimum and is

$$C = (\$400)(4000) + (\$300)(6000) = \$3,400,000$$

MODEL: POLLUTION CONTROL

The following model is taken from a paper by Robert E. Kohn.* In this paper, a linear programming model is proposed that can be useful in determining what air pollution controls should be adopted in an airshed. The methodology is based on the premise that air quality goals should be achieved at the least possible cost. Advantages of the model are its simplicity, its emphasis on economic efficiency, and its appropriateness for the kind of data that are already available.

To illustrate the model, consider a hypothetical airshed with a single industry, cement manufacturing. Annual production is 2,500,000 barrels of cement. Although the kilns are equipped with mechanical collectors for air pollution control, they are still emitting 2 pounds of dust for every barrel of cement produced. The industry can be required to replace the mechanical collectors with four-field electrostatic precipitators, which would reduce emissions to 0.5 pound of dust per barrel of cement or with five-field electrostatic precipitators, which would reduce

* R. E. Kohn, "A Mathematical Programming Model for Air Pollution Control," *School Science and Mathematics* (June 1969), pp. 487–499.

emissions to 0.2 pound per barrel. If the capital and operating costs of the four-field precipitator are $0.14 per barrel of cement produced and of the five-field precipitator are $0.18 per barrel, what control methods should be required of this industry? Assume that, for this hypothetical airshed, it has been determined that particulate emissions (which now total 5,000,000 pounds per year) should be reduced by 4,200,000 pounds.

If C represents the cost of control, x is the number of barrels of annual cement production subject to the four-field electrostatic precipitator (cost is $0.14 a barrel of cement produced and pollutant reduction is $2 - 0.5 = 1.5$ pounds of particulates per barrel of cement produced), and y is the number of barrels of annual cement production subject to the five-field electrostatic precipitator (cost is $0.18 a barrel and pollutant reduction is $2 - 0.2 = 1.8$ pounds per barrel of cement produced), then the problem can be stated as follows:

Minimize

$$C = \$0.14x + \$0.18y$$

subject to

$$x + y \leq 2{,}500{,}000$$
$$1.5x + 1.8y \geq 4{,}200{,}000$$
$$x \geq 0$$
$$y \geq 0$$

The first equation states that our objective is to minimize air pollution control costs; the second that barrels of cement production subject to the two control methods cannot exceed the annual production; the third that the particulate reduction from the two methods must be greater than or equal to the particulate reduction target; and the last two expressions mean that we cannot have negative quantities of cement. Figure 24 illustrates a graphic solution to the problem.

The least costly solution would be to install the four-field precipitator on kilns

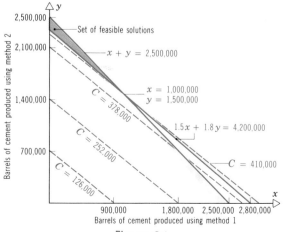

Figure 24

producing 1,000,000 ($x = 1{,}000{,}000$) and the five-field precipitator on kilns producing 1,500,000 ($y = 1{,}500{,}000$) barrels of cement at a cost of $C = \$410{,}000$.
A further analysis of this type of problem is found in Chapter 4.

Exercise 3.3

Answers to Odd-Numbered Problems begin on page A-13.

A   In Problems 1–6 the figure below illustrates the graph of the set of feasible solutions of a
linear programming problem. Find the maximum and minimum values of each objective
function.

1. $z = 2x + 3y$
2. $z = 3x + 27y$
3. $z = x + 8y$
4. $z = 3x + y$
5. $z = x + 6y$
6. $z = x + 5y$

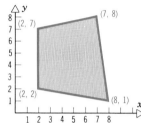

In Problems 7–12 the figure below illustrates the graph of the set of feasible solutions of a
linear programming problem. Find the minimum value of each objective function.

7. $z = 2x + 3y$
8. $z = 3x + 6y$
9. $z = 5x + 2y$
10. $z = 10x + y$
11. $z = 3x + 4y$
12. $z = x + 10y$

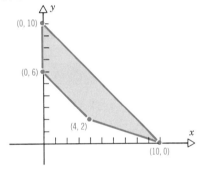

In Problems 13–20 maximize (if possible) the quantity $z = 5x + 7y$ subject to the given
constraints.

13. $x \geq 0$
     $y \geq 0$
     $x + y \leq 2$

14. $x \geq 0$
     $y \geq 0$
     $2x + 3y \leq 6$

15. $x \geq 0$
     $y \geq 0$
     $x + y \geq 2$
     $2x + 3y \leq 6$

16. $x \geq 0$
     $y \geq 0$
     $x + y \geq 2$
     $2x + 3y \leq 12$
     $3x + 2y \leq 12$

17. $x \geq 0$
     $y \geq 0$
     $2 \leq x + y$
     $x + y \leq 8$
     $2x + y \leq 10$

18. $x \geq 0$
     $y \geq 0$
     $2 \leq x + y$
     $x + y \leq 8$
     $1 \leq x + 2y$
     $x + 2y \leq 10$

19. $x \geq 0$
     $y \geq 0$
     $x + 3y \geq 6$

20. $x \geq 0$
     $y \geq 0$
     $2x + 3y \geq 12$

In Problems 21–26 minimize (if possible) the quantity $z = 2x + 3y$ subject to the given constraints.

21. $x \geq 0$
$y \geq 0$
$x + y \geq 2$

22. $x \geq 0$
$y \geq 0$
$2x + y \geq 2$

23. $x \geq 0$
$y \geq 0$
$x + y \geq 2$
$2x + 3y \leq 12$
$3x + y \leq 12$

24. $x \geq 0$
$y \geq 0$
$2 \leq x + y$
$x + y \leq 10$
$2x + 3y \leq 6$

25. $x \geq 0$
$y \geq 0$
$1 \leq x + 2y$
$x + 2y \leq 10$

26. $x \geq 0$
$y \geq 0$
$1 \leq x + 2y$
$x + 2y \leq 10$
$2 \leq x + y$
$x + y \leq 8$

In Problems 27–32 find the maximum and minimum values (if possible) of the given objective function subject to the constraints

$$x \geq 0 \qquad y \geq 0 \qquad x + y \leq 10 \qquad 2x + y \geq 10 \qquad x + 2y \geq 10$$

27. $z = x + y$

28. $z = 2x + 3y$

29. $z = 5x + 2y$

30. $z = x + 2y$

31. $z = 3x + 4y$

32. $z = 3x + 6y$

B **33.** Maximize
$$z = x + 2y$$
subject to
$$x \geq 0$$
$$y \geq 0$$
$$-x + 3y \leq 24$$
$$2x + y \geq 8$$
$$2x + 3y \geq 16$$
$$5x + 3y \leq 60$$

34. Maximize
$$z = 3x + y$$
subject to
$$x \geq 0$$
$$y \geq 0$$
$$-x + y \leq 0$$
$$2x + 3y \leq 12$$
$$3x + y \leq 12$$

35. Minimize
$$z = 5x + 2y$$
subject to
$$x \geq 0$$
$$y \geq 0$$
$$x + y \geq 11$$
$$2x + 3y \geq 24$$
$$x + 3y \leq 18$$

36. Minimize
$$z = 3x + 4y$$
subject to
$$x \geq 0$$
$$y \geq 0$$
$$x + y \leq 12$$
$$5x + 2y \leq 36$$
$$7x + 4y \geq 14$$

C **37.** Maximize
$$z = 5x + 12y$$

38. Minimize
$$z = -10x + 2y$$

subject to $\qquad\qquad\qquad\qquad$ subject to

$$x \geq 0$$
$$y \geq 0$$
$$x + y \leq 10$$
$$-x + y \leq 3$$
$$x - y \leq 3$$
$$x + y \geq 4$$

$$x \geq 0$$
$$y \geq 0$$
$$-2x + y \leq 5$$
$$x + 2y \leq 20$$
$$3x - y \leq 27$$
$$x + y \geq 4$$
$$x \leq 10$$
$$y \leq 10$$

APPLICATIONS

39. Mixture In Example 5 (page 156), if the profit on the low-grade mixture is $0.30 per package and the profit on the high-grade mixture is $0.40 per package, how many packages of each mixture should be made for a maximum profit?

40. Production Scheduling A company produces two types of steel. Type 1 requires 2 hours of melting, 4 hours of cutting, and 10 hours of rolling per ton. Type 2 requires 5 hours of melting, 1 hour of cutting, and 5 hours of rolling per ton. Forty hours are available for melting, 20 for cutting, and 60 for rolling. Each ton of Type 1 produces $240 profit, and each ton of Type 2 yields $80 profit. Find the maximum profit and the production schedule that will produce this profit.

41. Investment Strategy A financial consultant wishes to invest up to a total of $30,000 in two types of securities, one that yields 10% per year and another that yields 8% per year. Furthermore, she believes that the amount invested in the first security should be at most one-third of the amount invested in the second security. What investment program should the consultant pursue in order to maximize income?

42. Investment Strategy An investment broker wants to invest up to $20,000. She can purchase a type A bond yielding a 10% return on the amount invested and she can purchase a type B bond yielding a 15% return on the amount invested. She also wants to invest at least as much in the type A bond as in the type B bond. She will also invest at least $5000 in the type A bond and no more than $8000 in the type B bond. How much should she invest in each type of bond to maximize her return?

43. Production Scheduling A factory manufactures two products, each requiring the use of three machines. The first machine can be used at most 70 hours; the second machine at most 40 hours; and the third machine at most 90 hours. The first product requires 2 hours on Machine 1, 1 hour on Machine 2, and 1 hour on Machine 3; the second product requires 1 hour each on Machines 1 and 2 and 3 hours on Machine 3. If the profit is $40 per unit for the first product and $60 per unit for the second product, how many units of each product should be manufactured to maximize profit?

44. Scheduling Blink appliances has a sale on microwaves and stoves. Each microwave requires 2 hours to unpack and set up, and each stove requires 1 hour. The storeroom space is limited to 50 items. The budget of the store allows only 80 hours of employee time for unpacking and set up. Microwaves sell for $300 each, and stoves sell for $200 each. How many of each should the store order to maximize revenue?

45. Cost Control An appliance repair shop has 5 vacuum cleaners, 12 TV sets, and 18 VCRs to be repaired. The store employs two part-time repairmen. One repairman can repair one vacuum cleaner, three TV sets, and three VCRs in 1 week, while the second repairman can repair one vacuum cleaner, two TV sets and six VCRs in 1 week. The first employee is paid $250 a week and the second employee is paid $220 a week. To minimize the cost, how many weeks should each of the two repairmen be employed?

46. **Transportation** An appliance company has a warehouse and two terminals. To minimize shipping costs the manager must decide how many appliances should be shipped to each terminal. There is a total supply of 1200 units in the warehouse and a demand for 400 units in terminal A and 500 units in terminal B. It costs $12 to ship each unit to terminal A and $16 to ship to terminal B. How many units should be shipped to each terminal in order to minimize cost?

47. **Pollution Control** A chemical plant produces two items A and B. For each item A produced, 2 cubic feet of carbon monoxide and 6 cubic feet of sulfur dioxide are emitted into the atmosphere, whereas to produce item B 4 cubic feet of carbon monoxide and 3 cubic feet of sulfur dioxide are emitted into the atmosphere. Government pollution standards permit the manufacturer to emit a maximum of 3000 cubic feet of carbon monoxide and 5400 cubic feet of sulfur dioxide per week. The manufacturer can sell all of the items that it produces and make a profit of $1.50 per unit for item A and $1.00 per unit for item B. Determine the number of units of each item to be produced each week to maximize profit without exceeding government standards.

48. **Diet Problem** A diet is to contain at least 400 units of vitamins, 500 units of minerals, and 1400 calories. Two foods are available: F_1, which costs $0.05 per unit, and F_2, which costs $0.03 per unit. A unit of food F_1 contains 2 units of vitamins, 1 unit of minerals, and 4 calories; a unit of food F_2 contains 1 unit of vitamins, 2 units of minerals, and 4 calories. Find the minimum cost for a diet that consists of a mixture of these two foods and also meets the minimal nutrition requirements.

49. **Diet Problem** Danny's Chicken Farm is a producer of frying chickens. In order to produce the best fryers possible, the regular chicken feed is supplemented by four vitamins. The minimum amount of each vitamin required per 100 ounces of feed is: Vitamin 1, 50 units; Vitamin 2, 100 units; Vitamin 3, 60 units; Vitamin 4, 180 units. Two supplements are available: Supplement I costs $0.03 per ounce and contains 5 units of Vitamin 1 per ounce, 25 units of Vitamin 2 per ounce, 10 units of Vitamin 3 per ounce, and 35 units of Vitamin 4 per ounce. Supplement II costs $0.04 per ounce and contains 25 units of Vitamin 1 per ounce, 10 units of Vitamin 2 per ounce, 10 units of Vitamin 3 per ounce, and 20 units of Vitamin 4 per ounce. How much of each supplement should Danny buy to add to each 100 ounces of feed in order to minimize his cost, but still have the desired vitamin amounts present?

50. **Optimal Use of Land** A farmer has 70 acres of land available on which to grow some soybeans and some corn. The cost of cultivation per acre, the workdays needed per acre, and the profit per acre are indicated in the table.

	Soybeans	Corn	Total Available
Cultivation cost per acre	$60	$30	$1800
Days of work per acre	3 days	4 days	120 days
Profit per acre	$300	$150	

As indicated in the last column, the acreage to be cultivated is limited by the amount of money available for cultivation costs and by the number of working days that can be put into this part of the business. Find the number of acres of each crop that should be planted in order to maximize the profit.

51. **Mixture** The manager of a supermarket meat department finds that there are 160 pounds of round steak, 600 pounds of chuck steak, and 300 pounds of pork in stock on Saturday morning. From experience, the manager knows that half these quantities can be sold as straight cuts. The remaining meat will have to be ground into hamburger patties and picnic patties for which there is a large weekend demand. Each pound of hamburger patties contains 20% ground round and 60% ground chuck. Each pound of picnic patties contains 30% ground pork and 50% ground chuck. The remainder of each product consists of an inexpensive nonmeat filler that the store has in unlimited quantities. How many pounds of each product should be made if the objective is to maximize the amount of meat used to make the patties?

52. **Production Scheduling** J. B. Rug Manufacturers has available 1200 square yards of wool and 1000 square yards of nylon for the manufacture of two grades of carpeting: high-grade, which sells for $500 per roll, and low-grade, which sells for $300 per roll. Twenty square yards of wool and 40 square yards of nylon are used in a roll of high-grade carpet, and 40 square yards of nylon are used in a roll of low-grade carpet. Forty work-hours are required to manufacture each roll of the high-grade carpet, and 20 work-hours are required for each roll of the low-grade carpet, at an average cost of $6.00 per work-hour. A maximum of 800 work-hours are available. The cost of wool is $5.00 per square yard, and the cost of nylon is $2.00 per square yard. How many rolls of each type of carpet should be manufactured to maximize income? [*Hint:* Income = Revenue from sale − (Production cost for material + labor)]

53. **Production Scheduling** The rug manufacturer in Problem 52 finds that maximum income occurs when no high-grade carpet is produced. If the price of the low-grade carpet is kept at $300 per roll, in what price range should the high-grade carpet be sold so that income is maximized by selling some rolls of each type carpet? Assume all other data remain the same.

CHAPTER REVIEW

Important Terms		
linear inequality in two variables	systems of linear inequalities	linear programming problem
nonstrict linear inequality	unbounded graph	constraints
strict linear inequality	bounded graph	feasible solution
graph of a linear inequality	vertex of the graph	solution of a linear programming problem
half-plane	objective function	

True–False Questions

(Answers on page A-13)

T F 1. The graph of a system of linear inequalities may be bounded or unbounded.

T F 2. The graph of the set of constraints of a linear programming problem, under certain conditions, could have a circle for a boundary.

T F 3. The objective function of a linear programming problem is always a linear equation involving the variables.

T F   4. In a linear programming problem, there may be more than one point that maximizes or minimizes the objective function.

T F   5. Some linear programming problems will have no solution.

T F   6. If a linear programming problem has a solution, it is located at the center of the set of feasible solutions.

Fill in
the Blanks

(Answers on page A-13)

1. The graph of a linear inequality in two variables is called a _____ .

2. In a linear programming problem, the quantity to be maximized or minimized is referred to as the _____ function.

3. The points that obey the collection of constraints of a linear programming problem are called _____ solutions.

4. A linear programming problem will always have a solution if the set of feasible solutions is _____ .

5. If a linear programming problem has a solution, it is located at a _____ of the set of feasible solutions.

Review
Exercises

Answers to Odd-Numbered Problems begin on page A-13.

In Problems 1–4, graph each linear inequality.

A   **1.** $x - 3y < 0$ **2.** $4x + y \geq 8$

    3. $5x + y \geq 10$ **4.** $2x + 3y > 6$

In Problems 5–10, graph each system of linear inequalities. Locate the vertices and tell whether the graph is bounded or unbounded.

5. $x \geq 0, y \geq 0, 3x + 2y \leq 12, x + y \geq 1$.

6. $x \geq 0, y \geq 0, x + y \leq 8, 2x + y \geq 2$.

7. $x \geq 0, y \geq 0, x + 2y \geq 4, 3x + y \geq 6$.

8. $x \geq 0, y \geq 0, 2x + y \geq 4, 3x + 2y \geq 6$.

9. $x \geq 0, y \geq 0, 3x + 2y \geq 6, 3x + 2y \leq 12, x + 2y \leq 8$.

10. $x \geq 0, y \geq 0, x + 2y \geq 2, x + 2y \leq 10, 2x + y \leq 10$.

B   In Problems 11–18 use the constraints below to solve each linear programming problem.

$$x \geq 0$$
$$y \geq 0$$
$$x + 2y \leq 40$$
$$2x + y \leq 40$$
$$x + y \geq 10$$

11. Maximize $z = x + y$ **12.** Maximize $z = 2x + 3y$

13. Minimize $z = 5x + 2y$ **14.** Minimize $z = 3x + 2y$

15. Maximize $z = 2x + y$ **16.** Maximize $z = x + 2y$

17. Minimize $z = 2x + 5y$ **18.** Minimize $z = x + y$

In Problems 19–22 maximize and minimize (if possible) the quantity $z = 15x + 20y$ subject to the given constraints

19.
$$x \geq 0$$
$$y \geq 0$$
$$x \leq 5$$
$$y \leq 8$$
$$3x + 4y \geq 12$$

20.
$$x \geq 0$$
$$y \geq 0$$
$$x \leq 6$$
$$y \leq 6$$
$$3x + 2y \geq 6$$

21.
$$x \geq 0$$
$$y \geq 0$$
$$2x + 3y \leq 22$$
$$x \leq 5$$
$$y \leq 6$$

22.
$$x \geq 0$$
$$y \geq 0$$
$$x + 2y \leq 20$$
$$x + 10y \geq 36$$
$$5x + 2y \geq 36$$

C 23. Maximize
$$z = x + 3y$$
subject to
$$\tfrac{1}{2}x + \tfrac{3}{4}y \leq 550$$
$$x + y \leq 850$$
$$\tfrac{1}{4}x + \tfrac{1}{2}y \leq 350$$
$$x \geq 250$$
$$y \geq 375$$

24. Minimize
$$z = 4x + 3y$$
subject to
$$x \geq 0$$
$$y \geq 0$$
$$x + y \geq 2000$$
$$x \leq 1200$$
$$0.8x - 0.2y \geq 0$$
$$0.9x - 0.1y \geq 0$$

APPLICATIONS

25. Mixture A company makes two kinds of animal food, A and B, which contain two food supplements. It takes 2 pounds of the first supplement and one pound of the second to make a dozen cans of Food A, and 4 pounds of the first supplement and 5 pounds of the second to make a dozen cans of Food B. On a certain day 80 pounds of the first supplement and 70 pounds of the second are available. Maximize company profits if the profit on a dozen cans of Food A is $3.00 and the profit on a dozen cans of Food B is $10.00.

26. Production Scheduling A company sells two types of shoes. The first uses 2 units of leather and 2 units of manmade material and yields a profit of $8 per pair. The second type requires 5 units of leather and 1 unit of manmade material and gives a profit of $10 per pair. If there are 40 units of leather and 16 units of manmade material available, how many pairs of each type of shoe should be sold to maximize profit? What is the maximum profit?

27. Diet Problem Katy needs at least 60 units of carbohydrates, 45 units of protein, and 30 units of fat each month. From each pound of Food A, she receives 5 units of carbohydrates, 3 of protein, and 4 of fat. Food B contains 2 units of carbohydrates, 2 units of protein, and 1 unit of fat per pound. If Food A costs $1.30 per pound and Food B costs $0.80 per pound, how many pounds of each food should Katy buy each month to keep costs at a minimum?

28. Mixture The ACE Meat Market makes up a combination package of ground beef and ground pork for meat loaf. The ground beef is 75% lean (75% beef, 25% fat) and costs the market 70¢ per pound. The ground pork is 60% lean (60% pork, 40% fat) and costs the market 50¢ per pound. If the meat loaf is to be at least 70% lean, how much ground beef and ground pork should be mixed to keep the price per pound at a minimum?

29. Production Scheduling A ski manufacturer makes two types of skis: downhill and cross-country. Using the information given below, how many of each type of ski should be made for a maximum profit to be achieved? What is the maximum profit?

	Downhill	Cross-Country	Maximum Time Available
Manufacturing time per ski	2 hours	1 hour	40 hours
Finishing time per ski	1 hour	1 hour	32 hours
Profit per ski	$70	$50	

30. Production Scheduling Rework Problem 29 if the manufacturing unit has a maximum of 48 hours available.

Mathematical Questions

From CPA and CMA Exams (Answers on page A-14)

Use the following information to answer Problems 1–3:

CPA Exam

The Random Company manufactures two products, Zeta and Beta. Each product must pass through two processing operations. All materials are introduced at the start of Process No. 1. There are no work-in-process inventories. Random may produce either one product exclusively or various combinations of both products subject to the following constraints:

	Process No. 1	Process No. 2	Contribution Margin per Unit
Hours required to produce one unit of:			
Zeta	1 hour	1 hour	$4.00
Beta	2 hours	3 hours	5.25
Total capacity in hours per day	1000 hours	1275 hours	

A shortage of technical labor has limited Beta production to 400 units per day. There are no constraints on the production of Zeta other than the hour constraints in the above schedule. Assume that all relationships between capacity and production are linear, and that all of the above data and relationships are deterministic rather than probabilistic.

1. Given the objective to maximize total contribution margin, what is the production constraint for Process No. 1?
   (a) Zeta + Beta ≤ 1000   (b) Zeta + 2Beta ≤ 1000
   (c) Zeta + Beta ≥ 1000   (d) Zeta + 2Beta ≥ 1000

2. Given the objective to maximize total contribution margin, what is the labor constraint for production of Beta?

(a) Beta $\leq$ 400   (b) Beta $\geq$ 400
(c) Beta $\leq$ 425   (d) Beta $\geq$ 425

3. What is the objective function of the data presented?
    (a) Zeta + 2Beta = $9.25
    (b) ($4.00)Zeta + 3($5.25)Beta = Total contribution margin
    (c) ($4.00)Zeta + ($5.25)Beta = Total contribution margin
    (d) 2($4.00)Zeta + 3($5.25)Beta = Total contribution margin

4. *CPA Exam*
    Williamson Manufacturing intends to produce two products, X and Y. Product X requires 6 hours of time on Machine 1 and 12 hours of time on Machine 2. Product Y requires 4 hours of time on Machine 1 and no time on Machine 2. Both machines are available for 24 hours. Assuming that the objective function of the total contribution margin is $2X + $1Y$, what product mix will produce the maximum profit?
    (a) No units of Product X and 6 units of Product Y
    (b) 1 unit of Product X and 4 units of Product Y
    (c) 2 units of Product X and 3 units of Product Y
    (d) 4 units of Product X and no units of Product Y

5. *CPA Exam*
    Quepea Company manufactures two products, Q and P, in a small building with limited capacity. The selling price, cost data, and production time are given below:

	Product Q	**Product P**
Selling price per unit	$20	$17
Variable costs of producing and selling a unit	$12	$13
Hours to produce a unit	3	1

    Based on this information, the profit maximization objective function for a linear programming solution may be stated as:
    (a) Maximize $20Q + $17P$ (b) Maximize $12Q + $13P$
    (c) Maximize $3Q + $1P$ (d) Maximize $8Q + $4P$

6. *CPA Exam*
    Patsy, Inc., manufactures two products, X and Y. Each product must be processed in each of three departments: machining, assembling, and finishing. The hours needed to produce one unit of product per department and the maximum possible hours per department follow:

Department	Production Hours per Unit		Maximum Capacity in Hours
	X	Y	
Machining	2	1	420
Assembling	2	2	500
Finishing	2	3	600

    Other restrictions follow:

    $$X \geq 50 \qquad Y \geq 50$$

The objective function is to maximize profits where profit = $4X + $2Y. Given the objective and constraints, what is the most profitable number of units of X and Y, respectively, to manufacture?

(a) 150 and 100   (b) 165 and 90
(c) 170 and 80   (d) 200 and 50

7. *CPA Exam*

Milford Company manufactures two models, medium and large. The contribution margin expected is $12 for the medium model and $20 for the large model. The medium model is processed 2 hours in the machining department and 4 hours in the polishing department. The large model is processed 3 hours in the machining department and 6 hours in the polishing department. How would the formula for determining the maximization of total contribution margin be expressed?

(a) $5X + 10Y$ (b) $6X + 9Y$
(c) $12X + 20Y$ (d) $12X(2 + 4) + 20Y(3 + 6)$

8. *CMA Exam*

The Elon Company manufactures two industrial products—X-10, which sells for $90 a unit, and Y-12, which sells for $85 a unit. Each product is processed through both of the company's manufacturing departments. The limited availability of labor, material, and equipment capacity has restricted the ability of the firm to meet the demand for its products. The production department believes that linear programming can be used to routinize the production schedule for the two products.

The following data are available to the production department:

	Amount Required per Unit	
	X-10	Y-12
Direct Material: Weekly supply is limited to 1800 pounds at $12.00 per pound	4 lb	2 lb
Direct Labor:		
Department 1—Weekly supply limited to 10 people at 40 hours each at an hourly cost of $6.00	$\frac{2}{3}$ hour	1 hour
Department 2—Weekly supply limited to 15 people at 40 hours each at an hourly rate of $8.00	$1\frac{1}{4}$ hours	1 hour
Machine Time:		
Department 1—Weekly capacity limited to 250 hours	$\frac{1}{2}$ hour	$\frac{1}{2}$ hour
Department 2—Weekly capacity limited to 300 hours	0 hours	1 hour

The overhead costs for Elon are accumulated on a plantwide basis. The overhead is assigned to products on the basis of the number of direct labor hours required to manufacture the product. This base is appropriate for overhead assignment because most of the variable overhead costs vary as a function of labor time. The estimated overhead cost per direct labor hour is

Variable overhead cost	$ 6.00
Fixed overhead cost	6.00
Total overhead cost per direct labor hour	$12.00

The production department formulated the following equations for the linear programming statement of the problem:

$$A = \text{Number of units of X-10 to be produced}$$
$$B = \text{Number of units of Y-12 to be produced}$$

Objective function to minimize costs:

$$\text{Minimize} \quad Z = 85A + 62B$$

Constraints:

Material	$4A + 2B \le 1800$ lb
Department 1 labor	$\frac{2}{3}A + 1B \le 400$ hours
Department 2 labor	$1\frac{1}{4}A + 1B \le 600$ hours
Nonnegativity	$A \ge 0 \qquad B \ge 0$

(a) The formulation of the linear programming equations as prepared by Elon Company's production department is incorrect. Explain what errors have been made in the formulation prepared by the production department.

(b) Formulate and label the proper equations for the linear programming statement of Elon Company's production problem.

(c) Explain how linear programming could help Elon Company determine how large a change in the price of direct materials would have to be to change the optimum production mix of X-10 and Y-12.

9. *CPA Exam*

Hale Company manufactures products A and B, each of which requires two processes, polishing and grinding. The contribution margin is $3 for Product A and $4 for Product B. The illustration shows the maximum number of units of each product that may be processed in the two departments.

Considering the constraints (restrictions) on processing, which combination of products A and B maximizes the total contribution margin?

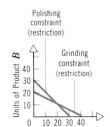

Polishing constraint (restriction)

Grinding constraint (restriction)

Units of Product B

Units of product A

(a) 0 units of A and 20 units of B
(b) 20 units of A and 10 units of B
(c) 30 units of A and 0 units of B
(d) 40 units of A and 0 units of B

10. *CPA Exam*

Johnson, Inc., manufactures product X and product Y, which are processed as follows:

	Type A Machine	Type B Machine
Product X	6 hours	4 hours
Product Y	9 hours	5 hours

The contribution margin is $12 for product X and $7 for product Y. The available time daily for processing the two products is 120 hours for machine Type A and 80 hours for machine Type B. How would the restriction (constraint) for machine Type B be expressed?

(a) $4X + 5Y$
(b) $4X + 5Y \le 80$
(c) $6X + 9Y \le 120$
(d) $12X + 7Y$

11. *CMA Exam*

A small company makes only two products, with the following two production constraints representing two machines and their maximum availability:

$$2X + 3Y \leq 18$$
$$2X + Y \leq 10$$

where

$$X = \text{the units of the first product}$$
$$Y = \text{the units of the second product}$$

If the profit equation is $Z = \$4X + \$2Y$, the maximum possible profit is

(a) $20
(b) $21
(c) $18
(d) $24
(e) Some profit other than those given above

CMA Exam

Questions 12 through 14 are based on the Jarten Company, which manufactures and sells two products. Demand for the two products has grown to such a level that Jarten can no longer meet the demand with its facilities. The company can work a total of 600,000 direct labor hours annually using three shifts. A total of 200,000 hours of machine time is available annually. The company plans to use linear programming to determine a production schedule that will maximize its net return.

The company spends $2,000,000 in advertising and promotion and incurs $1,000,000 for general and administrative costs. The unit sale price for Model A is $27.50; Model B sells for $75.00 each. The unit manufacturing requirement and unit cost data are as shown below. Overhead is assigned on a machine hour (MH) basis.

	Model A		Model B	
Raw material		$ 3		$ 7
Direct labor	1 DLH @ $8	8	1.5 DLH @ $8	12
Variable overhead	.5 MH @ $12	6	2.0 MH @ $12	24
Fixed overhead	.5 MH @ $4	2	2.0 MH @ $4	8
		$19		$51

12. The objective function that would maximize Jarten's net income is

(a) $10.50A + 32.00B$
(b) $8.50A + 24.00B$
(c) $27.50A + 75.00B$
(d) $19.00A + 51.00B$
(e) $17.00A + 43.00B$

13. The constraint function for the direct labor is

(a) $1A + 1.5B \leq 200,000$
(b) $8A + 12B \leq 600,000$
(c) $8A + 12B \leq 200,000$
(d) $1A + 1.5B \leq 4,800,000$
(e) $1A + 1.5B \leq 600,000$

14. The constraint function for the machine capacity is

(a) $6A + 24B \leq 200,000$

(b) $1/.5A + 1.5/2.0B \le 800,000$
(c) $.5A + 2B \le 200,000$
(d) $(.5 + .5)A + (2 + 2)B \le 200,000$
(e) $(0.5 \times 1) + (1.5 \times 2.00) \le (200,000 \times 600,000)$

15. *CPA Exam*

Boaz Company manufactures two models, medium (X) and large (Y). The contribution margin expected is $24 for the medium model and $40 for the large model. The medium model is processed 2 hours in the machining department and 4 hours in the polishing department. The large model is processed 3 hours in the machining department and 6 hours in the polishing department. If total contribution margin is to be maximized, using linear programming, how would the objective function be expressed?

(a) $24X(2 + 4) + 40Y(3 + 6)$
(b) $24X + 40Y$
(c) $6X + 9Y$
(d) $5X + 10Y$

4
LINEAR PROGRAMMING PART TWO: THE SIMPLEX METHOD

4.1

THE SIMPLEX TABLEAU

INTRODUCTION
STANDARD FORM OF A MAXIMUM PROBLEM
SLACK VARIABLES AND THE SIMPLEX TABLEAU

INTRODUCTION

In Chapter 3 we described a geometrical method (using graphs) for solving linear programming problems. Unfortunately, this method is useful only when there are no more than two variables and the number of constraints is small.

If we have a large number of either variables or constraints, it is still true that the optimal solution will be found at a vertex of the set of feasible solutions. In fact, we could find these vertices by writing all the equations corresponding to the inequalities of the problem and then proceeding to solve all possible combinations of these equations. We would of course, have to discard any solutions that are not feasible (because they do not satisfy one or more of the constraints). Then we could evaluate the objective function at the remaining feasible solutions. After all this, we might discover that the problem has no optimal solution after all.

Just how difficult is this procedure? Well, if there were just 4 variables and 7 constraints, we would have to solve all possible combinations of 4 equations chosen from a set of 7 equations—that would be 35 solutions in all. Each of these solutions would then have to be tested for feasibility. So, even for this relatively small number of variables and constraints, the work would be quite tedious. In the real world of applications, it is fairly common to encounter problems with *hundreds,* even *thousands,* of variables and constraints. Of course, such problems must be solved by computer. Even so, choosing a more efficient problem-solving strategy than the geometrical method might reduce the computer's running time from hours to seconds, or, for very large problems, from years to hours.

A more systematic approach would involve choosing a solution at one vertex of the feasible set, then moving from there to another vertex at which the objective function has a better value, and continuing in this way until the best possible value is found. One very efficient and popular way of doing this is the subject of the present chapter: the *simplex method.*

STANDARD FORM OF A MAXIMUM PROBLEM

A linear programming problem in which the objective function is to be maximized is referred to as a *maximum linear programming problem.* Such problems are said to be in *standard form* provided two conditions are met:

Condition 1. All the variables are nonnegative.

Condition 2. All other constraints are written as a linear expression that is less than or equal to a positive constant.

Example 1

Determine which of the following maximum linear programming problems are in standard form.

(a) Maximize

$$z = 5x_1 + 4x_2$$

subject to the constraints

$$3x_1 + 4x_2 \leq 120 \qquad x_1 \geq 0$$
$$4x_1 + 3x_2 \leq 20 \qquad x_2 \geq 0$$

(b) Maximize

$$z = 8x_1 + 2x_2 + 3x_3$$

subject to the constraints

$$4x_1 + 8x_2 \leq 120 \qquad x_1 \geq 0$$
$$3x_2 + 4x_3 \leq 120 \qquad x_2 \geq 0$$

(c) Maximize

$$z = 6x_1 - 8x_2 + x_3$$

subject to the constraints

$$3x_1 + x_2 \leq 10 \qquad x_1 \geq 0$$
$$4x_1 - x_2 \leq 5 \qquad x_2 \geq 0$$
$$x_1 + x_2 + x_3 \geq -3 \qquad x_3 \geq 0$$

(d) Maximize

$$z = 8x_1 + x_2$$

subject to the constraints

$$3x_1 + 4x_2 \geq 2 \qquad x_1 \geq 0$$
$$x_1 + x_2 \leq 6 \qquad x_2 \geq 0$$

Solution

(a) This is a maximum problem containing two variables x_1 and x_2. Since both variables are nonnegative and since the other constraints

$$3x_1 + 4x_2 \leq 120$$
$$4x_1 + x_2 \leq 20$$

Linear expressions   Less than or equal   Positive

are each written as linear expressions less than or equal to a positive constant, we conclude the maximum problem is in standard form.

(b)   This is a maximum problem containing three variables x_1, x_2, and x_3. Since the variable x_3 is not given as nonnegative, the maximum problem is not in standard form.

(c)   This is a maximum problem containing three variables x_1, x_2, and x_3. Each variable is nonnegative. The other constraints

$$3x_1 + x_2 \le 10$$
$$4x_1 - x_2 \le 5$$
$$x_1 + x_2 + x_3 \ge -3$$

contains $x_1 + x_2 + x_3 \ge -3$, which is not a linear expression that is $\le$ a positive constant. Thus, the maximum problem is not in standard form. Notice, however, that by multiplying this constraint by -1, we get

$$-x_1 - x_2 - x_3 \le 3$$

which is in the desired form. Thus, although the maximum problem as stated is not in standard form, it can easily be modified to conform to the requirements of the standard form.

(d)   The maximum problem contains two variables x_1 and x_2 each of which is nonnegative. Of the other constraints, the first one, $3x_1 + 4x_2 \ge 2$ does not conform. Thus, the maximum problem is not in standard form. Notice that we cannot modify this problem to place it in standard form. Even though multiplying by -1 will change the $\ge$ to $\le$, in so doing the 2 will change to -2.   ∎

SLACK VARIABLES AND THE SIMPLEX TABLEAU

To apply the simplex method to a maximum problem, we need to first

1. Introduce *slack variables.*

2. Construct the *initial simplex tableau.*

We will show how these steps are done by working with a specific maximum problem in standard form. (This problem is Example 6 of Chapter 3, page 157.)
   Maximize

$$P = 3x_1 + 4x_2$$

subject to the constraints

$$2x_1 + 4x_2 \le 120 \qquad x_1 \ge 0$$
$$2x_1 + 2x_2 \le 80 \qquad x_2 \ge 0$$

This problem is in standard form for a maximum linear programming problem. When we say that $2x_1 + 4x_2 \le 120$, we mean that there is a number greater than or equal to 0, which we might as well call s_1, such that

$$2x_1 + 4x_2 + s_1 = 120$$

This number s_1 is a variable. It must be nonnegative since it is the difference between 120 and a number that is less than or equal to 120. We call it a *slack variable* since it "takes up the slack" between the left and right sides of the inequality.

Similarly, when we say that $2x_1 + 2x_2 \leq 80$, we are saying that there is a slack variable s_2 such that

$$2x_1 + 2x_2 + s_2 = 80$$

Furthermore, the objective function $P = 3x_1 + 4x_2$ can be rewritten as

$$-3x_1 - 4x_2 + P = 0$$

In effect, we have now replaced our original system of constraints and the objective function by a system of three equations in five unknowns:

$$
\begin{aligned}
2x_1 + 4x_2 + s_1 &= 120 \\
2x_1 + 2x_2 \quad\;\; + s_2 &= 80 \\
-3x_1 - 4x_2 \qquad\quad\; + P &= 0
\end{aligned}
$$

Here, it is understood that each of the five variables is required to be nonnegative. To solve the maximum problem is to find the particular solution (x_1, x_2, s_1, s_2, P) that gives the largest possible value for P. The augmented matrix for this system is

$$
\begin{array}{ccccc}
x_1 & x_2 & s_1 & s_2 & P \\
\end{array}
$$
$$
\left[
\begin{array}{ccccc|c}
2 & 4 & 1 & 0 & 0 & 120 \\
2 & 2 & 0 & 1 & 0 & 80 \\
-3 & -4 & 0 & 0 & 1 & 0
\end{array}
\right]
$$

This augmented matrix is called the *initial simplex tableau* for the problem. Notice that we have written the name of each variable above the column in which its coefficients appear.

So far, we have seen this much of the simplex method:

For a maximum problem in standard form

1. **The constraints are changed from inequalities to equations by the introduction of extra variables — one for each constraint and all nonnegative — called *slack variables.***
2. **These equations, together with one that describes the objective function, are placed in an augmented matrix called the *initial simplex tableau.***

Example 2

The following maximum problems are in standard form. For each one introduce slack variables and set up the initial simplex tableau.

(a) Maximize

$$P = 3x_1 + 2x_2 + x_3$$

subject to the constraints

$$3x_1 + x_2 + x_3 \le 30 \qquad x_1 \ge 0$$
$$5x_1 + 2x_2 + x_3 \le 24 \qquad x_2 \ge 0$$
$$x_1 + x_2 + 4x_3 \le 20 \qquad x_3 \ge 0$$

(b) Maximize

$$P = x_1 + 4x_2 + 3x_3 + x_4$$

subject to the constraints

$$2x_1 + x_2 \le 10 \qquad x_1 \ge 0 \qquad x_2 \ge 0$$
$$3x_1 + x_2 + x_3 + 2x_4 \le 18 \qquad x_3 \ge 0 \qquad x_4 \ge 0$$
$$x_1 + x_2 + x_3 + x_4 \le 14$$

Solution
(a) For each constraint we introduce a nonnegative slack variable to obtain the following equations:

$$3x_1 + x_2 + x_3 + s_1 \qquad\qquad = 30 \qquad x_1 \ge 0 \qquad s_1 \ge 0$$
$$5x_1 + 2x_2 + x_3 \qquad + s_2 \qquad = 24 \qquad x_2 \ge 0 \qquad s_2 \ge 0$$
$$x_1 + x_2 + 4x_3 \qquad\qquad + s_3 = 20 \qquad x_3 \ge 0 \qquad s_3 \ge 0$$

These equations, together with the objective function P, written as $-3x_1 - 2x_2 - x_3 + P = 0$ give the initial simplex tableau:

x_1	x_2	x_3	s_1	s_2	s_3	P	
3	1	1	1	0	0	0	30
5	2	1	0	1	0	0	24
1	1	4	0	0	1	0	20
−3	−2	−1	0	0	0	1	0

(b) For each constraint, we introduce a nonnegative slack variable to obtain the equations

$$2x_1 + x_2 \qquad\qquad + s_1 \qquad\qquad = 10 \qquad x_1 \ge 0, x_2 \ge 0, x_3 \ge 0, x_4 \ge 0$$
$$3x_1 + x_2 + x_3 + 2x_4 \qquad + s_2 \qquad = 18 \qquad s_1 \ge 0, s_2 \ge 0, s_3 \ge 0$$
$$x_1 + x_2 + x_3 + x_4 \qquad\qquad + s_3 = 14$$

These equations, together with the objective function P, written as $-x_1 - 4x_2 - 3x_3 - x_4 + P = 0$ give the initial simplex tableaux:

x_1	x_2	x_3	x_4	s_1	s_2	s_3	P	
2	1	0	0	1	0	0	0	10
3	1	1	2	0	1	0	0	18
1	1	1	1	0	0	1	0	14
−1	−4	−3	−1	0	0	0	1	0

Notice that in each initial simplex tableaux an identity matrix appears under the column headed by the slack variables and the objective function. Notice too that the right-hand column will always contain nonnegative constants.

Exercise 4.1   *Answers to Odd-Numbered Exercises begin on page A-14.*

A   In Problems 1–10 determine which maximum linear programming problems are in standard form. Do not attempt to solve them!

1. Maximize

$$P = 2x_1 + x_2$$

subject to the constraints

$$x_1 + x_2 \le 5 \qquad x_1 \ge 0$$
$$2x_1 + 3x_2 \le 2 \qquad x_2 \ge 0$$

2. Maximize

$$P = 3x_1 + 4x_2$$

subject to the constraints

$$3x_1 + x_2 \le 6 \qquad x_1 \ge 0$$
$$x_1 + 4x_2 \le 4 \qquad x_2 \ge 0$$

3. Maximize

$$P = 3x_1 + x_2 + x_3$$

subject to the constraints

$$x_1 + x_2 + x_3 \le 6 \qquad x_1 \ge 0$$
$$2x_1 + 3x_2 + 4x_3 \le 10$$

4. Maximize

$$P = 2x_1 + x_2 + 4x_3$$

subject to the constraints

$$2x_1 + x_2 + x_3 \le 10 \qquad x_2 \ge 0$$

5. Maximize

$$P = 3x_1 + x_2 + x_3$$

subject to the constraints

$$x_1 + x_2 + x_3 \le 8 \qquad x_1 \ge 0$$
$$2x_1 + x_2 + 4x_3 \ge 6 \qquad x_2 \ge 0$$

6. Maximize

$$P = 2x_1 + x_2 + 4x_3$$

subject to the constraints

$$2x_1 + x_2 + x_3 \le -1 \qquad x_1 \ge 0$$
$$x_2 \ge 0$$

7. Maximize

$$P = 2x_1 + x_2$$

subject to the constraints

$$x_1 + x_2 \ge -6 \qquad x_1 \ge 0$$
$$2x_1 + x_2 \le 4 \qquad x_2 \ge 0$$

8. Maximize

$$P = 3x_1 + x_2$$

subject to the constraints

$$x_1 + 3x_2 \le 4 \qquad x_1 \ge 0$$
$$2x_1 - x_2 \ge 1 \qquad x_2 \ge 0$$

9. Maximize

$$P = 2x_1 + x_2 + 3x_3$$

subject to the constraints

$$x_1 + x_2 - x_3 \le 10 \qquad x_1 \ge 0, \quad x_2 \ge 0$$
$$x_2 + x_3 \le 4 \qquad x_3 \ge 0$$

10. Maximize

$$P = 2x_1 + 2x_2 + 3x_3$$

subject to the constraints

$$x_1 - x_2 + x_3 \le 6 \qquad x_1 \ge 0, \quad x_2 \ge 0$$
$$x_1 \le 4 \qquad x_3 \ge 0$$

In Problems 11–16, each maximum problem is not in standard form. Determine if the problem can be modified so as to be in standard form. If it can, write the modified version.

11. Maximize

$$P = x_1 + x_2$$

subject to the constraints

$$3x_1 - 4x_2 \le -6 \qquad x_1 \ge 0$$
$$x_1 + x_2 \le 4 \qquad x_2 \ge 0$$

12. Maximize

$$P = 2x_1 + 3x_2$$

subject to the constraints

$$-4x_1 + 2x_2 \ge -8 \qquad x_1 \ge 0$$
$$x_1 - x_2 \le 6 \qquad x_2 \ge 0$$

13. Maximize

$$P = x_1 + x_2 + x_3$$

subject to the constraints

$$x_1 + x_2 + x_3 \le 6 \qquad x_1 \ge 0, \quad x_2 \ge 0$$
$$4x_1 + 3x_2 \ge 12 \qquad x_3 \ge 0$$

14. Maximize

$$P = 2x_1 + x_2 + 3x_3$$

subject to the constraints

$$x_1 + x_2 + x_3 \ge -8 \qquad x_1 \ge 0, \quad x_2 \ge 0$$
$$x_1 - x_2 \le -6 \qquad x_3 \ge 0$$

15. Maximize

$$P = 2x_1 + x_2 + 3x_3$$

subject to the constraints

$$-x_1 + x_2 + x_3 \ge -6 \qquad x_1 \ge 0, \quad x_2 \ge 0$$
$$2x_1 - 3x_2 \ge -12 \qquad x_3 \ge 0$$

16. Maximize

$$P = x_1 + x_2 + x_3$$

subject to the constraints

$$2x_1 - x_2 + 3x_3 \le 8 \qquad x_1 \ge 0, \quad x_2 \ge 0$$
$$x_1 - x_2 \ge 6 \qquad x_3 \ge 0$$

B   In Problems 17–24 each maximum problem is in standard form. For each one introduce slack variables and set up the initial simplex tableaux.

17. Maximize

$$P = 2x_1 + x_2 + 3x_3$$

subject to the constraints

$$5x_1 + 2x_2 + x_3 \le 20$$
$$6x_1 + x_2 + 4x_3 \le 24$$
$$x_1 + x_2 + 4x_3 \le 16$$
$$x_1 \ge 0, \quad x_2 \ge 0, \quad x_3 \ge 0$$

18. Maximize

$$P = 3x_1 + 2x_2 + x_3$$

subject to the constraints

$$3x_1 + 2x_2 - x_3 \le 10$$
$$x_1 - x_2 + 3x_3 \le 12$$
$$2x_1 + x_2 + x_3 \le 6$$
$$x_1 \ge 0, \quad x_2 \ge 0, \quad x_3 \ge 0$$

19. Maximize

$$P = 3x_1 + 5x_2$$

20. Maximize

$$P = 2x_1 + 3x_2$$

subject to the constraints

$$2.2x_1 - 1.8x_2 \leq 5$$
$$0.8x_1 + 1.2x_2 \leq 2.5$$
$$x_1 + x_2 \leq 0.1$$
$$x_1 \geq 0, \quad x_2 \geq 0$$

subject to the constraints

$$1.2x_1 - 2.1x_2 \leq 0.5$$
$$0.3x_1 + 0.4x_2 \leq 1.5$$
$$x_1 + x_2 \leq 0.7$$
$$x_1 \geq 0, \quad x_2 \geq 0$$

21. Maximize

$$P = 2x_1 + 3x_2 + x_3$$

subject to the constraints

$$x_1 + x_2 + x_3 \leq 50$$
$$3x_1 + 2x_2 + x_3 \leq 10$$
$$x_1 \geq 0, \quad x_2 \geq 0, \quad x_3 \geq 0$$

22. Maximize

$$P = x_1 + 4x_2 + 2x_3$$

subject to the constraints

$$3x_1 + x_2 + x_3 \leq 10$$
$$x_1 + x_2 + 3x_3 \leq 5$$
$$x_1 \geq 0, \quad x_2 \geq 0, \quad x_3 \geq 0$$

23. Maximize

$$P = 3x_1 + 4x_2 + 2x_3$$

subject to the constraints

$$3x_1 + x_2 + 4x_3 \leq 5$$
$$x_1 + x_2 \leq 5$$
$$2x_1 - x_2 + x_3 \leq 6$$
$$x_1 \geq 0, \quad x_2 \geq 0, \quad x_3 \geq 0$$

24. Maximize

$$P = 2x_1 + x_2 + 3x_3$$

subject to the constraints

$$2x_1 + x_2 + x_3 \leq 2$$
$$x_1 - x_2 \leq 4$$
$$2x_1 + x_2 - x_3 \leq 5$$
$$x_1 \geq 0, \quad x_2 \geq 0, \quad x_3 \geq 0$$

4.2

THE SIMPLEX METHOD: THE MAXIMUM PROBLEM

THE PIVOT OPERATION
THE PIVOTING PROCESS
GEOMETRY OF THE SIMPLEX METHOD
THE UNBOUNDED CASE
SUMMARY OF THE SIMPLEX METHOD

THE PIVOT OPERATION

Before going any further in our discussion of the simplex method, we need to discuss the matrix operation known as *pivoting*. The first thing one does in a pivot operation is to choose a pivot element. However, for now the pivot element will be specified in advance; the method of selecting pivot elements in the simplex tableau will be shown later.

Pivoting   To *pivot* a matrix about a given element—called the *pivot element*—is to apply row operations so that the pivot element is replaced by a 1 and all other entries in the same column—called the *pivot column*—become 0's.

Steps for Pivoting

The correct sequence of steps is:

Step 1. In the *pivot row* (where the pivot element appears), divide each entry by the *pivot element* (we assume it is not 0).

Step 2. Obtain 0's elsewhere in the *pivot column* by performing row operations.

The following example illustrates a pivot operation.

Example 1

Perform a pivot operation on the matrix given below, where the pivot element is circled, and the pivot row and pivot column are marked by arrows:

$$\rightarrow \begin{bmatrix} x_1 & x_2 & s_1 & s_2 & P & \\ 2 & ④ & 1 & 0 & 0 & 120 \\ 2 & 2 & 0 & 1 & 0 & 80 \\ -3 & -4 & 0 & 0 & 1 & 0 \end{bmatrix} \tag{1}$$

Solution

In this matrix, the pivot column is column 2 and the pivot row is row 1. Step 1 of the pivoting procedure tells us to divide row 1 by 4. The row operation specified is $R_1 = \frac{1}{4}r_1$.

$$\begin{bmatrix} x_1 & x_2 & s_1 & s_2 & P & \\ \frac{1}{2} & ① & \frac{1}{4} & 0 & 0 & 30 \\ 2 & 2 & 0 & 1 & 0 & 80 \\ -3 & -4 & 0 & 0 & 1 & 0 \end{bmatrix}$$

Now, to accomplish Step 2, we multiply row 1 by -2 and add it to row 2; in addition, we multiply row 1 by 4 and add it to row 3. The row operations specified are

$$R_2 = -2r_1 + r_2 \qquad R_3 = 4r_1 + r_3$$

The new matrix looks like this:

$$\begin{bmatrix} x_1 & x_2 & s_1 & s_2 & P & & \\ \frac{1}{2} & ① & \frac{1}{4} & 0 & 0 & 30 & x_2 \\ 1 & 0 & -\frac{1}{2} & 1 & 0 & 20 & s_2 \\ -1 & 0 & 1 & 0 & 1 & 120 & P \end{bmatrix} \tag{2}$$

This completes the pivot operation, since the pivot column has been replaced by

$$\begin{matrix} 1 \\ 0 \\ 0 \end{matrix}$$

∎

Just what has the pivot operation done? To see, we look again at the original matrix (1). Observe that the entries in columns s_1, s_2, and P form an identity matrix (I_3, to be exact). This makes it easy to solve for s_1, s_2, and P, using the other variables as parameters:

$$2x_1 + 4x_2 + s_1 = 120 \quad \text{or} \quad s_1 = 120 - 2x_1 - 4x_2$$
$$2x_1 + 2x_2 + s_2 = 80 \quad \text{or} \quad s_2 = 80 - 2x_1 - 2x_2$$
$$-3x_1 - 4x_2 + P = 0 \quad \text{or} \quad P = 3x_1 + 4x_2$$

For additional emphasis, we write s_1, s_2, and P to the right of the corresponding rows of the tableau as follows:

$$
\begin{array}{ccccc}
x_1 & x_2 & s_1 & s_2 & P \\
\end{array}
$$

$$
\rightarrow
\left[
\begin{array}{ccccc|c}
2 & \textcircled{4} & 1 & 0 & 0 & 120 \\
2 & 2 & 0 & 1 & 0 & 80 \\
-3 & -4 & 0 & 0 & 1 & 0 \\
\end{array}
\right]
\begin{array}{c}
s_1 \\
s_2 \\
P
\end{array}
$$

After pivoting, we obtain the matrix (2).

Notice that the identity matrix I_3 now appears in columns 2, 4, and 5, which helps us solve for x_2, s_2, and P in terms of x_1 and s_1:

$$x_2 = 30 - \tfrac{1}{2}x_1 - \tfrac{1}{4}s_1$$
$$s_2 = 20 - x_1 + \tfrac{1}{2}s_1$$
$$P = 120 + x_1 - s_1$$

This is why we wrote x_2, s_2, and P to the right of the matrix (2).

Example 2

Perform another pivot operation on

$$
\begin{array}{ccccc}
x_1 & x_2 & s_1 & s_2 & P \\
\end{array}
$$

$$
\rightarrow
\left[
\begin{array}{ccccc|c}
\tfrac{1}{2} & 1 & \tfrac{1}{4} & 0 & 0 & 30 \\
\textcircled{1} & 0 & -\tfrac{1}{2} & 1 & 0 & 20 \\
-1 & 0 & 1 & 0 & 1 & 120 \\
\end{array}
\right]
\begin{array}{c}
x_2 \\
s_2 \\
P
\end{array}
$$

where the new pivot element has been circled.

Solution

Since the pivot element happens to be a 1 in this case, we skip Step 1. For Step 2, we perform the row operations

$$R_1 = -\frac{1}{2}r_2 + r_1 \qquad R_3 = r_2 + r_3$$

The result is

$$
\begin{array}{ccccc}
x_1 & x_2 & s_1 & s_2 & P \\
\end{array}
$$

$$
\left[
\begin{array}{ccccc|c}
0 & 1 & \tfrac{1}{2} & -\tfrac{1}{2} & 0 & 20 \\
1 & 0 & -\tfrac{1}{2} & 1 & 0 & 20 \\
0 & 0 & \tfrac{1}{2} & 1 & 1 & 140 \\
\end{array}
\right]
\begin{array}{c}
x_2 \\
x_1 \\
P
\end{array}
$$

■

Notice that in Example 2 the identity matrix I_3 now appears in columns 1, 2, and 5 — with shifted columns, as

$$\begin{matrix} 0 & 1 & 0 \\ 1 & 0 & 0 \\ 0 & 0 & 1 \end{matrix}$$

We may now solve for x_2, x_1, and P in terms of s_1 and s_2, as indicated on the right of the matrix.

THE PIVOTING PROCESS

We are finally ready to state the details of the simplex method for solving a maximum linear programming problem. This method requires that the problem be in standard form and that the problem be placed in an initial simplex tableau with slack variables. We begin by looking at a maximum problem in standard form.

Maximize

$$P = 3x_1 + 4x_2$$

subject to the constraints

$$2x_1 + 4x_2 \le 120 \qquad x_1 \ge 0$$
$$2x_1 + 2x_2 \le 80 \qquad x_2 \ge 0$$

After introducing slack variables s_1 and s_2, we set up the initial simplex tableau.

$$2x_1 + 4x_2 + s_1 = 120 \qquad x_1 \ge 0 \qquad s_1 \ge 0$$
$$2x_1 + 2x_2 + s_2 = 80 \qquad x_2 \ge 0 \qquad s_2 \ge 0$$
$$-3x_1 - 4x_2 + P = 0$$

The initial simplex tableau is

x_1	x_2	s_1	s_2	P		
2	4	1	0	0	120	s_1
2	2	0	1	0	80	s_2
−3	−4	0	0	1	0	P

Notice that the bottom row contains the negatives of the coefficients in the objective function and that we have set this row off from the rest of the matrix by a dashed line.

We refer to this row as the *objective row*. From this point on, the simplex method consists of pivoting from one tableau to another until the optimal solution is found. Three questions remain to be answered:

1. How is the pivot element selected?
2. When does the process end?
3. How is the optional solution read from the final tableau?

Pivot Element The *pivot element* for the simplex method is found using two rules:

> **Rule 1:** The *pivot column* is selected by locating the most negative entry in the objective row. If all the entries in this column are negative or zero, the problem is unbounded and there is no solution.
>
> **Rule 2:** Divide each entry in the last column by the corresponding entry (from the same row) in the pivot column. (Ignore any rows in which the pivot column entry is less than or equal to 0.) The row in which the smallest positive ratio is obtained is the *pivot row*.

The *pivot element* is the entry at the intersection of the pivot row and the pivot column.

Note that the pivot element is never in the objective row.

In the example we have been using, we select 4 as the pivot element because -4 is the most negative entry in the last row and $120 \div 4 = 30$ is the smallest positive ratio obtainable by dividing an entry in the last column by the corresponding entry in column 2.

$$
\begin{array}{c}
 \\
\rightarrow \\
 \\

\end{array}
\begin{array}{ccccc|c}
x_1 & x_2 & s_1 & s_2 & P & \\
2 & ④ & 1 & 0 & 0 & 120 \\
2 & 2 & 0 & 1 & 0 & 80 \\
\hline
-3 & -4 & 0 & 0 & 1 & 0
\end{array}
\begin{array}{l}
120 \div 4 = 30 \\
80 \div 2 = 40 \\

\end{array}
$$

After pivoting, we see that the smallest (in fact, only) negative entry in the objective row is the -1 in column 1. We check the row ratios in the last column:

$$
\begin{array}{c}
 \\
\rightarrow \\

\end{array}
\begin{array}{ccccc|c}
x_1 & x_2 & s_1 & s_2 & P & \\
\frac{1}{2} & 1 & \frac{1}{4} & 0 & 0 & 30 \\
① & 0 & -\frac{1}{2} & 1 & 0 & 20 \\
\hline
-1 & 0 & 1 & 0 & 1 & 120
\end{array}
\begin{array}{l}
30 \div \frac{1}{2} = 60 \\
20 \div 1 = 20 \\

\end{array}
$$

Rule 2 tells us to select the pivot element in row 2. After pivoting (as in Example 2), we have

$$
\begin{array}{ccccc|c}
x_1 & x_2 & s_1 & s_2 & P & \\
0 & 1 & \frac{1}{2} & -\frac{1}{2} & 0 & 20 \\
1 & 0 & -\frac{1}{2} & 1 & 0 & 20 \\
\hline
0 & 0 & \frac{1}{2} & 1 & 1 & 140
\end{array}
\begin{array}{l}
x_2 \\
x_1 \\
P
\end{array}
$$

Now there are no negative entries in the objective row. Thus, the rules imply that no further pivots can be performed. But this is not surprising because the problem is now solved. To see why, we write the equation from the objective row, namely,

$$ P = 140 - \frac{1}{2} s_1 - s_2 $$

Since $s_1 \geq 0$ and $s_2 \geq 0$, any positive value of s_1 or s_2 would make the value of P less than 140. By choosing $s_1 = 0$ and $s_2 = 0$, we obtain the largest possible value for P, namely, 140. If we write the equations from the first and second rows, substituting 0 for s_1 and s_2, we have

$$x_2 = 20 - \frac{1}{2} s_1 + \frac{1}{2} s_2 = 20$$

$$x_1 = 20 + \frac{1}{2} s_1 - s_2 = 20$$

In other words, we have found the optimal solution,

$$P = 140 \qquad \text{Maximum}$$

which occurs at

$$x_1 = 20 \qquad x_2 = 20$$

This same solution was found earlier in Chapter 3 (page 158) by the geometrical method.

Now, the reason we choose the most negative entry in the objective row is that it is the negative of the *largest* coefficient in the objective function:

$$P = 3x_1 + 4x_2$$

If we were to set $x_1 = x_2 = 0$, we would obtain $P = 0$ as a first approximation for the profit P. Of course, this is not a very good approximation; it can easily be improved by increasing either x_1 or x_2. But the profit per unit of x_2 is \$4, while the profit per unit of x_1 is only \$3. Hence, we choose to increase x_2 rather than x_1. But what is the largest amount by which x_2 can be increased?

We can answer this question by performing the pivot operation, as we already have in Example 1. The matrix becomes

$$
\begin{array}{cccccc}
x_1 & x_2 & s_1 & s_2 & P & \\
\left[\begin{array}{ccccc|c}
\frac{1}{2} & \boxed{1} & \frac{1}{4} & 0 & 0 & 30 \\
1 & 0 & -\frac{1}{2} & 1 & 0 & 20 \\
\hline
-1 & 0 & 1 & 0 & 1 & 120
\end{array}\right] &
\begin{array}{c}
x_2 \\
s_2 \\
\\
P
\end{array}
\end{array}
$$

and the corresponding equations are

$$x_2 = 30 - \frac{1}{2} x_1 - \frac{1}{4} s_1$$

$$s_2 = 20 - x_1 + \frac{1}{2} s_1$$

$$P = 120 + x_1 - s_1$$

This suggests that x_2 can be as large as 30, if we take both x_1 and s_1 to be 0, in which case P will be 120.

So we chose column 2 because we wanted to increase x_2. But why did we choose row 1, rather than row 2, as the pivot row? Let's see what would have happened if we had chosen row 2 as the pivot row. The result would have been

$$\begin{array}{ccccc}
x_1 & x_2 & s_1 & s_2 & P \\
\end{array}$$

$$\left[\begin{array}{ccccc|c}
-2 & 0 & 1 & -2 & 0 & -40 \\
1 & 1 & 0 & \frac{1}{2} & 0 & 40 \\
\hline
1 & 0 & 0 & 2 & 1 & 160
\end{array}\right]\begin{array}{c} s_1 \\ x_2 \\ P \end{array}$$

But this is not acceptable because of the negative number in the last column. (In effect, this matrix tells us that we could get P to be as large as 160, by setting $x_1 = 0$, $x_2 = 40$, and $s_1 = -40$; but this is not a feasible solution because s_1 is supposed to be greater than or equal to 0.)

The reason we get a feasible solution if we choose row 1 to pivot, and an apparent but false solution if we choose row 2, is because the row ratio $(120 \div 4)$ for row 1 is smaller (hence, better) than the row ratio $(80 \div 2)$ for row 2.

So, the reasoning behind the simplex method is fairly complicated, but the process of "moving to a better solution" is made quite easy simply by following the rules. Briefly, the pivoting strategy works like this:

> **Rule 1** forces us to pivot the variable that, for each unit of increase, will most effectively improve the value of the objective function.
>
> **Rule 2** prevents us from making this variable *too large* to be feasible.

In general, the reasoning used in the above problem shows that

> If there are no negative entries in the objective row, the optimal solution has been found.

We thus have a stopping or termination criterion for the simplex method. The tableau reached when the optimal solution is found is called a *final tableau*.

Let's work another one.

Example 3
Maximize

$$P = 6x_1 + 8x_2 + x_3$$

subject to

$$\begin{aligned}
3x_1 + 5x_2 + 3x_3 &\le 20 & x_1 &\ge 0 \\
x_1 + 3x_2 + 2x_3 &\le 9 & x_2 &\ge 0 \\
6x_1 + 2x_2 + 5x_3 &\le 30 & x_3 &\ge 0
\end{aligned}$$

Solution
Note that the problem is in standard form. Introducing slack variables s_1, s_2, and s_3, the system becomes

$$\begin{aligned}
3x_1 + 5x_2 + 3x_3 + s_1 &= 20 & x_1 &\ge 0 & s_1 &\ge 0 \\
x_1 + 3x_2 + 2x_3 + s_2 &= 9 & x_2 &\ge 0 & s_2 &\ge 0 \\
6x_1 + 2x_2 + 5x_3 + s_3 &= 30 & x_3 &\ge 0 & s_3 &\ge 0 \\
-6x_1 - 8x_2 - x_3 + P &= 0
\end{aligned} \qquad (1)$$

The initial simplex tableau is

$$
\begin{array}{cccccccc}
x_1 & x_2 & x_3 & s_1 & s_2 & s_3 & P & \\
\end{array}
$$

$$
\rightarrow
\left[
\begin{array}{ccccccc|c}
3 & 5 & 3 & 1 & 0 & 0 & 0 & 20 \\
1 & ③ & 2 & 0 & 1 & 0 & 0 & 9 \\
6 & 2 & 5 & 0 & 0 & 1 & 0 & 30 \\
\hline
-6 & -8 & -1 & 0 & 0 & 0 & 1 & 0
\end{array}
\right]
\begin{array}{l}
20 \div 5 = 4 \\
9 \div 3 = 3 \\
30 \div 2 = 15 \\
\\
\end{array}
$$

The pivot column is found by locating the column containing the smallest entry in the objective row (-8 in column 2).

The pivot row is obtained by dividing each entry in the last column by the corresponding entry in the pivot column and selecting the smallest nonnegative ratio. Thus, the second row is the pivot row and the pivot element in that row is the circled element 3. Notice that the column under P in any tableau is the same as in the initial tableau. This is why from now on, we will omit the column for P. After pivoting, the new tableau is

$$
\begin{array}{cccccc}
x_1 & x_2 & x_3 & s_1 & s_2 & s_3 \\
\end{array}
$$

$$
\rightarrow
\left[
\begin{array}{cccccc|c}
④⧸③ & 0 & -\frac{1}{3} & 1 & -\frac{5}{3} & 0 & 5 \\
\frac{1}{3} & 1 & \frac{2}{3} & 0 & \frac{1}{3} & 0 & 3 \\
\frac{16}{3} & 0 & \frac{11}{3} & 0 & -\frac{2}{3} & 1 & 24 \\
\hline
-\frac{10}{3} & 0 & \frac{13}{3} & 0 & \frac{8}{3} & 0 & 24
\end{array}
\right]
\begin{array}{l}
5 \div \frac{4}{3} = 3.75 \\
3 \div \frac{1}{3} = 9 \\
24 \div \frac{16}{3} = 4.5 \\
\\
\end{array}
$$

By the same procedure as before, we determine the next pivot element to be $\frac{4}{3}$, in the upper left corner. After pivoting, we get

$$
\begin{array}{cccccc}
x_1 & x_2 & x_3 & s_1 & s_2 & s_3 \\
\end{array}
$$

$$
\rightarrow
\left[
\begin{array}{cccccc|c}
1 & 0 & -\frac{1}{4} & \frac{3}{4} & -\frac{5}{4} & 0 & \frac{15}{4} \\
0 & 1 & \frac{3}{4} & -\frac{1}{4} & \frac{3}{4} & 0 & \frac{7}{4} \\
0 & 0 & 5 & -4 & ⑥ & 1 & 4 \\
\hline
0 & 0 & \frac{7}{2} & \frac{5}{2} & -\frac{3}{2} & 0 & \frac{73}{2}
\end{array}
\right]
$$

Since we still observe a negative entry in the objective row, we pivot again. (Remember, in finding the pivot element, we ignore rows in which the pivot column contains a negative number—in this case, $-\frac{5}{4}$.) The new tableau is

$$
\begin{array}{cccccc}
x_1 & x_2 & x_3 & s_1 & s_2 & s_3 \\
\end{array}
$$

$$
\left[
\begin{array}{cccccc|c}
1 & 0 & \frac{19}{24} & -\frac{1}{12} & 0 & \frac{5}{24} & \frac{55}{12} \\
0 & 1 & \frac{1}{8} & \frac{1}{4} & 0 & -\frac{1}{8} & \frac{5}{4} \\
0 & 0 & \frac{5}{6} & -\frac{2}{3} & 1 & \frac{1}{6} & \frac{2}{3} \\
\hline
0 & 0 & \frac{19}{4} & \frac{3}{2} & 0 & \frac{1}{4} & \frac{75}{2}
\end{array}
\right]
\begin{array}{l}
x_1 \\
x_2 \\
s_2 \\
\\
P
\end{array}
$$

This is a final tableau; the last column says that

$$
P = \tfrac{75}{2} \qquad x_1 = \tfrac{55}{12} \qquad x_2 = \tfrac{5}{4} \qquad x_3 = 0
$$

We see that $x_3 = 0$, because its column contains a number greater than 0 in the last row. We also see that $s_1 = 0$, $s_2 = \frac{2}{3}$, and $s_3 = 0$. These values can be used to check that these are, in fact, solutions of the system of equations (1).   ■

Example 4

Mike's Famous Toy Trucks specializes in making four kinds of toy trucks: a delivery truck, a dump truck, a garbage truck, and a gasoline truck. Three machines—a metal casting machine, a paint spray machine, and a packaging machine—are used in the production of these trucks. The time, in hours, each machine works to make each type of truck and the profit for each truck are given in Table 1. The maximum time available per week for each machine is: metal casting 4000 hours, paint spray 1800 hours, and packaging 1000 hours. How many of each type truck should be produced to maximize profit? Assume that every truck made is sold.

Table 1

	Metal Casting	Paint Spray	Packaging	Profit
Delivery Truck	2 hours	1 hour	0.5 hour	$0.50
Dump Truck	2.5 hours	1.5 hours	0.5 hour	$1.00
Garbage Truck	2 hours	1 hour	1 hour	$1.50
Gasoline Truck	2 hours	2 hours	1 hour	$2.00

Solution

Let x_1, x_2, x_3, and x_4 denote the number of delivery trucks, dump trucks, garbage trucks, and gasoline trucks, respectively, to be made. If P denotes the profit to be maximized, we have the problem:

Maximize

$$P = 0.5x_1 + x_2 + 1.5x_3 + 2x_4$$

subject to the conditions

$$2x_1 + 2.5x_2 + 2x_3 + 2x_4 \le 4000 \qquad x_1 \ge 0 \qquad x_3 \ge 0$$
$$x_1 + 1.5x_2 + x_3 + 2x_4 \le 1800 \qquad x_2 \ge 0 \qquad x_4 \ge 0$$
$$0.5x_1 + 0.5x_2 + x_3 + x_4 \le 1000$$

Since this problem is in standard form, we introduce slack variables s_1, s_2, and s_3, write the initial simplex tableau, and solve:

x_1	x_2	x_3	x_4	s_1	s_2	s_3		
2	2.5	2	2	1	0	0	4000	s_1
1	1.5	1	②	0	1	0	1800	s_2
0.5	0.5	1	1	0	0	1	1000	s_3
−0.5	−1	−1.5	−2	0	0	0	0	

$$\begin{array}{ccccccc}
x_1 & x_2 & x_3 & x_4 & s_1 & s_2 & s_3 \\
\end{array}$$

$$\left[\begin{array}{ccccccc|c}
1 & 1 & 1 & 0 & 1 & -1 & 0 & 2200 \\
0.5 & 0.75 & 0.5 & 1 & 0 & 0.5 & 0 & 900 \\
0 & -0.25 & \boxed{0.5} & 0 & 0 & -0.5 & 1 & 100 \\
\hline
0.5 & 0.5 & -0.5 & 0 & 0 & 1 & 0 & 1800
\end{array}\right]\begin{array}{c} s_1 \\ x_4 \\ s_3 \\ \\ \end{array}$$

$$\begin{array}{ccccccc}
x_1 & x_2 & x_3 & x_4 & s_1 & s_2 & s_3 \\
\end{array}$$

$$\left[\begin{array}{ccccccc|c}
1 & 1.5 & 0 & 0 & 1 & 0 & -2 & 2000 \\
0.5 & 1 & 0 & 1 & 0 & 1 & -1 & 800 \\
0 & -0.5 & 1 & 0 & 0 & -1 & 2 & 200 \\
\hline
0.5 & 0.25 & 0 & 0 & 0 & 0.5 & 1 & 1900
\end{array}\right]\begin{array}{c} s_1 \\ x_4 \\ x_3 \\ \\ \end{array}$$

This is a final tableau. The maximum profit is $P = \$1900$, and it is attained for

$$x_1 = 0 \qquad x_2 = 0 \qquad x_3 = 200 \qquad x_4 = 800 \qquad \blacksquare$$

The practical considerations of the situation described in Example 4 are that delivery trucks and dump trucks are too costly to produce or too little profit is being gained from their sale. Since the slack variable s_1 has a value of 2000 for maximum P and since s_1 represents the number of hours the metal casting machine is printing no truck (that is, the time the machine is idle), it may be possible to release this machine for other duties.

GEOMETRY OF THE SIMPLEX METHOD

The maximum value (provided it exists) of the objective function will occur at one of the vertices of the feasible region. The simplex method is designed to move from vertex to vertex of the feasible region, at each stage improving the value of the objective function until an optimal solution is found. More precisely, the geometry behind the simplex method is this:

1. A given tableau corresponds to a vertex of the feasible region.
2. The operation of pivoting moves us to an adjacent vertex, where the objective function has a value at least as large as it did at the previous vertex.
3. The process continues until the final tableau is reached—which produces a vertex that maximizes the objective function.

Though drawings of this can be rendered in only two or three dimensions (that is, when the objective function has two or three variables in it), the same interpretation can be shown to hold regardless of the number of variables involved. Let's look at an example.

Example 5
Maximize

$$P = 3x_1 + 5x_2$$

subject to the constraints

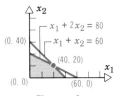

Figure 1

$$x_1 + x_2 \leq 60 \qquad x_1 \geq 0$$
$$x_1 + 2x_2 \leq 80 \qquad x_2 \geq 0$$

The feasible region is shown in Figure 1.
Below, on the left we apply the simplex method, indicating on the right the vertex corresponding to each tableau and the value of the objective function there. The reader is urged to supply the details.

		Tableau				Vertex	Value of $P = 3x_1 + 5x_2$ at the Vertex

x_1	x_2	s_1	s_2	P			
1	1	1	0	0	60	s_1	
1	2	0	1	0	80	s_2	(0, 0)	0
-3	-5	0	0	1	0		

$\downarrow$

$\frac{1}{2}$	0	1	$-\frac{1}{2}$	0	20	s_1	
$\frac{1}{2}$	1	0	$\frac{1}{2}$	0	40	x_2	(0, 40)	200
$-\frac{1}{2}$	0	0	$\frac{5}{2}$	1	200		

$\downarrow$

1	0	2	-1	0	40	x_1	
0	1	-1	1	0	20	x_2	(40, 20)	220
0	0	1	2	1	220			(maximum value)

(final tableau)

THE UNBOUNDED CASE

Thus far in our discussion, it has always been possible to continue to choose pivot elements until the problem has been solved. But it may turn out that all the entries in a column of a tableau are 0 or negative at some stage. If this happens, it means that the problem is *unbounded* and a maximum solution does not exist.

Example 6
Maximize

$$P = x_1 + x_2$$

subject to the constraints

$$-x_1 + x_2 \leq 2$$
$$x_1 - x_2 \leq 2$$
$$x_1 \geq 0$$
$$x_2 \geq 0$$

Solution

The initial simplex tableau is

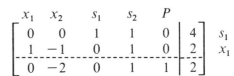

$$\begin{array}{c}\begin{array}{ccccc}x_1 & x_2 & s_1 & s_2 & P\end{array}\\ \left[\begin{array}{ccccc|c}-1 & 1 & 1 & 0 & 0 & 2\\ 1 & -1 & 0 & 1 & 0 & 2\\ \hline -1 & -1 & 0 & 0 & 1 & 0\end{array}\right]\begin{array}{c}s_1\\ s_2\\ {}\end{array}\end{array}$$

When there are two equal smallest negative entries in the last row, you may choose either column as the pivot column. Suppose we arbitrarily choose column 1 to pivot, and the tableau becomes

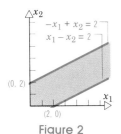

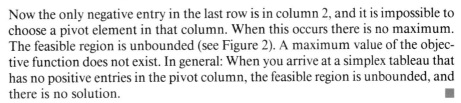

$$\begin{array}{c}\begin{array}{ccccc}x_1 & x_2 & s_1 & s_2 & P\end{array}\\ \left[\begin{array}{ccccc|c}0 & 0 & 1 & 1 & 0 & 4\\ 1 & -1 & 0 & 1 & 0 & 2\\ \hline 0 & -2 & 0 & 1 & 1 & 2\end{array}\right]\begin{array}{c}s_1\\ x_1\\ {}\end{array}\end{array}$$

Now the only negative entry in the last row is in column 2, and it is impossible to choose a pivot element in that column. When this occurs there is no maximum. The feasible region is unbounded (see Figure 2). A maximum value of the objective function does not exist. In general: When you arrive at a simplex tableau that has no positive entries in the pivot column, the feasible region is unbounded, and there is no solution. ■

Figure 2

SUMMARY OF THE SIMPLEX METHOD

The general procedure for solving a maximum linear programming problem using the simplex method can be outlined as follows.

1. The maximum problem is stated in standard form as:
Maximize

$$P = c_1 x_1 + c_2 x_2 + \cdots + c_n x_n$$

subject to the constraints

$$\begin{array}{l}a_{11} x_1 + a_{12} x_2 + \cdots + a_{1n} x_n \le b_1\\ a_{21} x_1 + a_{22} x_2 + \cdots + a_{2n} x_n \le b_2\end{array} \qquad x_1 \ge 0, x_2 \ge 0, \ldots, x_n \ge 0$$

.
.
.

$$a_{m1} x_1 + a_{m2} x_2 + \cdots + a_{mn} x_n \le b_m$$

in which $b_1 > 0$, $b_2 > 0$, $\cdots$ $b_m > 0$.

2. Introduce slack variables $s_1, s_2, \ldots s_m$ so that the constraints take the form of equations

$$\begin{array}{ll}a_{11} x_1 + a_{12} x_2 + \cdots + a_{1n} x_n + s_1 = b_1 & x_1 \ge 0, x_2 \ge 0, \ldots, x_n \ge 0\\ a_{21} x_1 + a_{22} x_2 + \cdots + a_{2n} x_n + s_2 = b_2 & s_1 \ge 0, s_2 \ge 0, \ldots, s_m \ge 0\end{array}$$

.
.
.

$$a_{m1} x_1 + a_{m2} x_2 + \cdots + a_{mn} x_n + s_m = b_m$$

3. Set up the initial simplex tableau

$$
\begin{array}{c}
\begin{array}{cccccccc}
x_1 & x_2 & \cdots & x_n & s_1 & s_2 & \cdots & s_m & P
\end{array} \\
\left[
\begin{array}{ccccccccc|c}
a_{11} & a_{12} & \cdots & a_{1n} & 1 & 0 & \cdots & 0 & 0 & b_1 \\
a_{21} & a_{22} & & a_{2n} & 0 & 1 & \cdots & 0 & 0 & b_2 \\
\cdot & & & & & & & & & \cdot \\
\cdot & & & & & & & & & \cdot \\
\cdot & & & & & & & & & \cdot \\
a_{m1} & a_{m2} & & a_{mn} & 0 & 0 & \cdots & 1 & 0 & b_m \\
\hdashline
-c_1 & -c_2 & & -c_n & 0 & 0 & & 0 & 1 & 0
\end{array}
\right]
\begin{array}{c}
s_1 \\ s_2 \\ \\ \\ \\ s_m \\ P
\end{array}
\end{array}
$$

4. Pivot until
   (a)   All the entries in the objective row are nonnegative. This is a final
         tableau from which a solution can be read.

   or
   (b)   The selection of the pivot column is a column whose entries are all
         negative or zero. In this case the problem is unbounded and there is no
         solution.

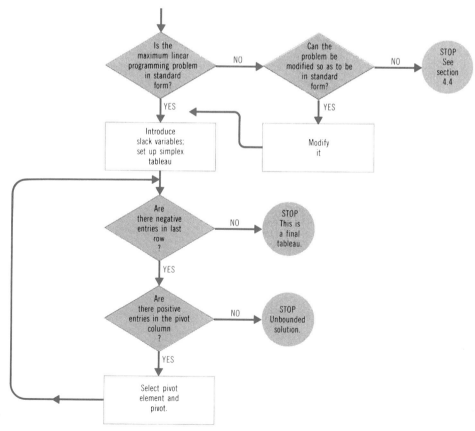

Figure 3

The flowchart in Figure 3 illustrates the steps to be used in solving maximum linear programming problems.

Exercise 4.2

Answers to Odd-Numbered Problems begin on page A-15.

A   In Problems 1–6 perform a pivot operation on each augmented matrix. Write the original system of equations and the final system of equations. The pivot element is circled.

1.

	x_1	x_2	s_1	s_2	P		
	1	②	1	0	0	300	s_1
	3	2	0	1	0	480	s_2
	-1	-2	0	0	1	0	P

2.

	x_1	x_2	s_1	s_2	P		
	1	4	1	0	0	100	s_1
	2	⑤	0	1	0	50	s_2
	-2	-1	0	0	1	0	P

3.

	x_1	x_2	x_3	s_1	s_2	s_3	P		
	1	2	4	1	0	0	0	24	s_1
	2	-1	1	0	1	0	0	32	s_2
	3	②	4	0	0	1	0	18	s_3
	-1	-2	-3	0	0	0	1	0	P

4.

	x_1	x_2	x_3	s_1	s_2	s_3	P		
	1	②	1	1	0	0	0	6	s_1
	2	3	1	0	1	0	0	12	s_2
	1	-2	3	0	0	1	0	0	s_3
	-1	-2	-3	0	0	0	1	0	P

5.

	x_1	x_2	x_3	x_4	s_1	s_2	s_3	s_4	P		
	-3	0	1	0	1	0	0	0	0	20	s_1
	②	0	0	1	0	1	0	0	0	24	s_2
	0	-3	1	0	0	0	1	0	0	28	s_3
	0	-3	0	1	0	0	0	1	0	24	s_4
	-1	-2	-3	-4	0	0	0	0	1	0	P

6.

	x_1	x_2	x_3	x_4	s_1	s_2	s_3	s_4		
	4	7	5	2	1	0	0	0	40	s_1
	②	0	1	4	0	1	0	0	48	s_2
	0	2	-2	8	0	0	1	0	32	s_3
	-2	5	-6	9	0	0	0	1	36	s_4
	-6	-3	-2	-12	0	0	0	0	0	P

In Problems 7–12 determine which of the following statements is true about each tableau:

(a)   It is the final tableau.
(b)   It requires additional pivoting.
(c)   It indicates no solution to the problem.

   If the answer is (a), write down the solution; if the answer is (b), indicate the pivot element.

7.

$$\begin{array}{cccc} x_1 & x_2 & s_1 & s_2 \end{array}$$

$$\left[\begin{array}{cccc|c} 1 & 0 & 1 & -\frac{1}{2} & 20 \\ \frac{1}{2} & 1 & 0 & \frac{1}{4} & 30 \\ \hline -1 & 0 & 0 & 1 & 120 \end{array}\right] \begin{array}{c} s_1 \\ x_2 \\ \end{array}$$

8.

$$\begin{array}{cccc} x_1 & x_2 & s_1 & s_2 \end{array}$$

$$\left[\begin{array}{cccc|c} 1 & 0 & 1 & -\frac{1}{2} & 20 \\ 0 & 1 & -\frac{1}{2} & \frac{1}{2} & 20 \\ \hline 0 & 0 & 1 & \frac{1}{4} & 140 \end{array}\right] \begin{array}{c} x_1 \\ x_2 \\ \end{array}$$

9.

$$\begin{array}{cccc} x_1 & x_2 & s_1 & s_2 \end{array}$$

$$\left[\begin{array}{cccc|c} 0 & \frac{1}{14} & 1 & -\frac{1}{7} & \frac{186}{21} \\ 1 & \frac{12}{7} & 0 & \frac{4}{7} & \frac{32}{7} \\ \hline 0 & \frac{12}{7} & 0 & \frac{32}{7} & \frac{256}{7} \end{array}\right] \begin{array}{c} s_1 \\ x_1 \\ \end{array}$$

10.

$$\begin{array}{cccc} x_1 & x_2 & s_1 & s_2 \end{array}$$

$$\left[\begin{array}{cccc|c} \frac{1}{4} & \frac{1}{2} & 1 & 0 & 10 \\ \frac{7}{4} & 3 & 0 & 1 & 8 \\ \hline -8 & -12 & 0 & 0 & 0 \end{array}\right] \begin{array}{c} s_1 \\ s_2 \\ \end{array}$$

11.

$$\begin{array}{cccc} x_1 & x_2 & s_1 & s_2 \end{array}$$

$$\left[\begin{array}{cccc|c} 1 & -2 & 0 & 4 & 24 \\ 0 & -2 & 1 & 4 & 36 \\ \hline 5 & -10 & 12 & 4 & 20 \end{array}\right] \begin{array}{c} x_1 \\ s_1 \\ \end{array}$$

12.

$$\begin{array}{cccc} x_1 & x_2 & s_1 & s_2 \end{array}$$

$$\left[\begin{array}{cccc|c} 1 & 3 & 1 & 0 & 30 \\ 2 & 1 & 0 & 1 & 12 \\ \hline -2 & -5 & 0 & 0 & 0 \end{array}\right] \begin{array}{c} s_1 \\ s_2 \\ \end{array}$$

In Problems 13–18 use the simplex method to solve each maximum linear programming problem.

13. Maximize

$$P = 5x_1 + 7x_2$$

subject to

$$2x_1 + 3x_2 \le 12 \qquad x_1 \ge 0$$
$$3x_1 + x_2 \le 12 \qquad x_2 \ge 0$$

14. Maximize

$$P = x_1 + 5x_2$$

subject to

$$2x_1 + x_2 \le 10 \qquad x_1 \ge 0$$
$$x_1 + 2x_2 \le 10 \qquad x_2 \ge 0$$

15. Maximize

$$P = 5x_1 + 7x_2$$

subject to

$$x_1 + 2x_2 \le 2 \qquad x_1 \ge 0$$
$$2x_1 + x_2 \le 2 \qquad x_2 \ge 0$$

16. Maximize

$$P = 5x_1 + 4x_2$$

subject to

$$x_1 + x_2 \le 2 \qquad x_1 \ge 0$$
$$2x_1 + 3x_2 \le 6 \qquad x_2 \ge 0$$

17. Maximize

$$P = 3x_1 + x_2$$

subject to

$$x_1 + x_2 \le 2 \qquad x_1 \ge 0$$
$$2x_1 + 3x_2 \le 12 \qquad x_2 \ge 0$$
$$3x_1 + x_2 \le 12$$

18. Maximize

$$P = 3x_1 + 5x_2$$

subject to

$$2x_1 + x_2 \le 4 \qquad x_1 \ge 0$$
$$x_1 + 2x_2 \le 6 \qquad x_2 \ge 0$$

B   In Problems 19–28 use the simplex method to solve each maximum linear programming problem.

19. Maximize

$$P = 2x_1 + x_2 + x_3$$

subject to

$$-2x_1 + x_2 - 2x_3 \le 4 \qquad x_1 \ge 0$$
$$x_1 - 2x_2 + x_3 \le 2 \qquad x_2 \ge 0$$
$$x_3 \ge 0$$

20. Maximize

$$P = 4x_1 + 2x_2 + 5x_3$$

subject to

$$x_1 + 3x_2 + 2x_3 \le 30 \qquad x_1 \ge 0$$
$$2x_1 + x_2 + 3x_3 \le 12 \qquad x_2 \ge 0$$
$$x_3 \ge 0$$

21. Maximize

$$P = 2x_1 + x_2 + 3x_3$$

subject to

$$x_1 + 2x_2 + x_3 \le 25 \qquad x_1 \ge 0$$
$$3x_1 + 2x_2 + 3x_3 \le 30 \qquad x_2 \ge 0$$
$$x_3 \ge 0$$

22. Maximize

$$P = 6x_1 + 3x_2 + 2x_3$$

subject to

$$2x_1 + 2x_2 + 3x_3 \le 30 \qquad x_1 \ge 0$$
$$2x_1 + 2x_2 + x_3 \le 12 \qquad x_2 \ge 0$$
$$x_3 \ge 0$$

23. Maximize

$$P = 2x_1 + 4x_2 + x_3 + x_4$$

subject to

$$2x_1 + x_2 + 2x_3 + 3x_4 \le 12 \qquad x_1 \ge 0$$
$$2x_2 + x_3 + 2x_4 \le 20 \qquad x_2 \ge 0$$
$$2x_1 + x_2 + 4x_3 \le 16 \qquad x_3 \ge 0$$
$$x_4 \ge 0$$

24. Maximize

$$P = 2x_1 + 4x_2 + x_3$$

subject to

$$-x_1 + 2x_2 + 3x_3 \le 6 \qquad x_1 \ge 0$$
$$-x_1 + 4x_2 + 5x_3 \le 5 \qquad x_2 \ge 0$$
$$-x_1 + 5x_2 + 7x_3 \le 7 \qquad x_3 \ge 0$$

25. Maximize

$$P = 2x_1 + x_2 + x_3$$

subject to

$$x_1 + 2x_2 + 4x_3 \le 20 \qquad x_1 \ge 0$$
$$2x_1 + 4x_2 + 4x_3 \le 60 \qquad x_2 \ge 0$$
$$3x_1 + 4x_2 + x_3 \le 90 \qquad x_3 \ge 0$$

26. Maximize

$$P = x_1 + 2x_2 + 4x_3$$

subject to

$$8x_1 + 5x_2 - 4x_3 \le 30 \qquad x_1 \ge 0$$
$$-2x_1 + 6x_2 + x_3 \le 5 \qquad x_2 \ge 0$$
$$-2x_1 + 2x_2 + x_3 \le 15 \qquad x_3 \ge 0$$

27. Maximize

$$P = 10x_1 + 8x_2 + 4x_3$$

subject to

$$15x_1 + 12x_2 + 8x_3 \le 1200 \quad x_1 \ge 0$$
$$5x_1 + 4x_2 + 3x_3 \le 600 \quad x_2 \ge 0$$
$$4x_1 + 4x_2 + 3x_3 \le 480 \quad x_3 \ge 0$$

28. Maximize

$$P = 30x_1 + 40x_2$$

subject to

$$0.4x_1 + 1.2x_2 \le 840$$
$$0.02x_1 + 0.05x_2 \le 46 \quad x_1 \ge 0$$
$$0.2x_1 + 0.3x_2 \le 280 \quad x_2 \ge 0$$

C **29.** Maximize

$$P = x_1 + 2x_2 + 4x_3 + x_4$$

subject to

$$5x_1 + 4x_3 + 6x_4 \le 20 \qquad x_1 \ge 0 \qquad x_3 \ge 0$$
$$4x_1 + 2x_2 + 2x_3 + 8x_4 \le 40 \qquad x_2 \ge 0 \qquad x_4 \ge 0$$

30. Maximize

$$P = x_1 + 2x_2 - x_3 + 3x_4$$

subject to

$$2x_1 + 4x_2 + 5x_3 + 6x_4 \leq 24 \qquad x_1 \geq 0 \qquad x_3 \geq 0$$
$$4x_1 + 4x_2 + 2x_3 + 2x_4 \leq 4 \qquad x_2 \geq 0 \qquad x_4 \geq 0$$

APPLICATIONS

31. Scheduling Products A, B, and C are sold door-to-door. Product A costs $3 per unit, takes 10 minutes to sell (on the average), and costs $0.50 to deliver to the customer. Product B costs $5, takes 15 minutes to sell, and is left with the customer at the time of sale. Product C costs $4, takes 12 minutes to sell, and costs $1.00 to deliver. During any week, a salesperson is allowed to draw up to $500 worth of A, B, and C (at cost) and is allowed delivery expenses not to exceed $75. If a salesperson's selling time is not expected to exceed 30 hours (1800 minutes) in a week, and if the salesperson's profit (net after all expenses) is $1 each on a unit of A or B and $2 on a unit of C, what combination of sales of A, B, and C will lead to maximum profit and what is this maximum profit?

32. Resource Allocations Suppose that a large hospital classifies its surgical operations into three categories according to their length and charges a fee of $600, $900, and $1200, respectively, for each of the categories. The average time of the operations in the three categories is 30 minutes, 1 hour, and 2 hours, respectively; and the hospital has four operating rooms, each of which can be used for 10 hours per day. If the total number of operations cannot exceed 60, how many of each type should the hospital schedule to maximize its revenues?

33. Mixture The Lee refinery blends high and low octane gasoline into three intermediate grades: regular, premium, and super premium. The regular grade consists of 60% high octane and 40% low octane, the premium consists of 70% high octane and 30% low octane, and the super premium consists of 80% high octane and 20% low octane. The company has available 140,000 gallons of high octane and 120,000 gallons of low octane, but can only mix 225,000 gallons. Regular gas sells for $1.20 per gallon, premium sells for $1.30 per gallon, and super premium sells for $1.40 per gallon. How many gallons of each grade should the company mix in order to maximize revenues?

34. Mixture Repeat Problem 33 under the assumption that the combined total number of gallons produced by the refinery cannot exceed 200,000 gallons.

35. Investment A financial consultant has at most $90,000 to invest in stocks, corporate bonds, and municipal bonds. The average yields for stocks, corporate bonds, and municipal bonds is 10%, 8%, and 6%, respectively. Determine how much she should invest in each security to maximize the return on her investments within a year, if she has decided that her investment in stocks should not exceed one half her funds, and that twice her investment in corporate bonds not exceed her investment in municipal bonds.

36. Investment Repeat Problem 35 under the assumption that no more than $25,000 can be invested in stocks.

37. Crop Planning A farmer has at most 200 acres of farm-land suitable for cultivating crops A, B, and C. The cost for cultivating crops A, B, and C are $40, $50, and $30 per acre. The farmer has a maximum of $18,000 available for land cultivation. Crops A, B, and C require 20, 30, and 15 hours per acre of labor, respectively, and there is a maximum of 4200 hours of labor available. If he expects to make a profit of $70, $90, and $50 per acre on crops A, B, and C, respectively, how many acres of each crop should he plant in order to maximize his profit?

38. Crop Planning Repeat Problem 37 if the farmer modified his allocations as follows:

| | Cost of Cultivating | | | Maximum |
	Crop A	Crop B	Crop C	Available
Cost	$30	$40	$20	$12,000
Hours	10	20	18	3,600
Profit	$50	$60	$40	

39. Process Utilization A jean manufacturer makes three types of jeans, each of which goes through three manufacturing phases—cutting, sewing, and finishing. The number of minutes each type of product requires in each of the three phases is given in the following table.

	Cutting	Sewing	Finishing
Jean I	8	12	4
Jean II	12	18	8
Jean III	18	24	12

There are 5200 minutes of cutting time, 6000 minutes of sewing time, and 2200 minutes of finishing time each day. The company can sell all the jeans it makes and make a profit of $3 on each Jean I, $4.50 on each Jean II, and $6 on each Jean III. Determine the number of jeans in each category it should make each day to maximize its profits.

40. Process Utilization A company manufactures three types of toys A, B, and C. Each requires rubber, plastic, and aluminum as given in the table.

Toy	Rubber	Plastic	Aluminum
A	2	2	4
B	1	2	2
C	1	2	4

The company has available 600 units of rubber, 800 units of plastic, and 1400 units of aluminum. The company makes a profit of $4, $3, and $2 on toys A, B, and C, respectively. Assuming all toys manufactured can be sold, determine a production order so that profit is maximum.

41. Mixture Problem Nutt's Nut Company has 500 pounds of peanuts, 100 pounds of pecans, and 50 pounds of cashews on hand. They package three types of 5-pound cans of nuts: Can I contains 3 pounds peanuts, 1 pound pecans, and 1 pound cashews; Can II contains 4 pounds peanuts, $\frac{1}{2}$ pound pecans, and $\frac{1}{2}$ pound cashews; and Can III contains 5 pounds peanuts. The selling price for each can is $8 for Can I, $7 for Can II, and $5 for Can III. How many cans of each kind should be made to maximize revenue?

42. Manufacturing One of the methods used by the Alexander Company to separate copper, lead, and zinc from ores is the flotation separation process. This process consists of three steps: oiling, mixing, and separation. These steps must be applied for 2, 2, and 1 hour, respectively, to produce 1 unit of copper; 2, 3, and 1 hour, respectively, to produce 1 unit of lead; and 1, 1, and 3 hours, respectively, to produce 1 unit of zinc. The oiling and separation phases of the process can be in operation for a maximum of 10 hours a day, while the mixing phase can be in operation for a maximum of 11 hours a day. The Alexander Company makes a profit of $45 per unit of copper, $30 per unit lead, and $35 per unit zinc. The demand for these metals is unlimited. How many units of each metal should be produced daily by use of the flotation process to achieve the highest profit?

43. Manufacturing A wood cabinet manufacturer produces cabinets for television consoles, stereo systems, and radios, each of which must be assembled, decorated, and crated. Each television console requires 3 hours to assemble, 5 hours to decorate, and 0.1 hour to crate and returns a profit of $10. Each stereo system requires 10 hours to assemble, 8 hours to decorate, and 0.6 hour to crate and returns a profit of $25. Each radio requires 1 hour to assemble, 1 hour to decorate, and 0.1 hour to crate and returns a profit of $3. The manufacturer has 30,000, 40,000, and 120 hours available weekly for assembling, decorating, and crating, respectively. How many units of each product should be manufactured to maximize profits?

44. Manufacturing The finishing process in the manufacture of cocktail tables and end tables requires sanding, staining, and varnishing. The time in minutes, required for each finishing process is given in the table below:

	Sanding	Staining	Varnishing
End table	8	10	4
Cocktail table	4	4	8

The equipment required for each process is used on one table at a time and is available for 6 hours each day. If the profit on each cocktail table is $20 and on each end table is $15, how many of each should be manufactured each day in order to maximize profit?

45. Transportation A large TV manufacturer has warehouse facilities for storing its 25″ color TVs in Chicago, New York, and Denver. Each month the city of Atlanta is shipped at most 400 25″ TVs. The cost of transporting each TV to Atlanta from Chicago, New York, and Denver averages $20, $20, and $40, respectively, while the cost of labor required for packing averages $6, $8, and $4, respectively. Suppose $10,000 is allocated each month for transportation costs and $3000 is allocated for labor costs. If the profit on each TV made in Chicago is $50, in New York is $80, and Denver is $40, how should monthly shipping arrangements be scheduled to maximize profits?

4.3

THE SIMPLEX METHOD: THE MINIMUM PROBLEM

STANDARD FORM OF A MINIMUM PROBLEM
DUAL OF A MINIMUM PROBLEM

Thus far in this chapter, we have discussed only maximum linear programming problems, in which the optimal solution yields the largest possible value for the objective function. In this section we discuss *minimum* problems, where the smallest possible value is desired. Among a number of techniques for solving such problems is one developed by John Von Neumann and others, in which the solution (if it exists) of a minimum problem is found by solving a related maximum problem called the *dual problem.*

Before constructing the dual problem, the minimum problem must be placed in *standard form.*

STANDARD FORM OF A MINIMUM PROBLEM

A minimum problem is said to be in standard form provided the following conditions are met:

Condition 1. **All the variables must be nonnegative.**

Condition 2. **All the other constraints must be written with $\geq$ signs. (This is just the opposite of the standard form requirement for a maximum problem.)**

Condition 3. **The objective function to be minimized must be written with nonnegative coefficients.**

Example 1

Determine which of the following minimum problems are in standard form.

(a) Minimize

$$C = 2x_1 + 3x_2$$

subject to the constraints

$$x_1 + 3x_2 \geq 24 \qquad x_1 \geq 0, \quad x_2 \geq 0$$
$$2x_1 + x_2 \geq 18$$

(b) Minimize

$$C = 3x_1 - x_2 + 4x_3$$

subject to the constraints

$$3x_1 + x_2 + x_3 \geq 12 \qquad x_1 \geq 0, \quad x_2 \geq 0$$
$$x_1 + x_2 + x_3 \geq 8 \qquad x_3 \geq 0$$

(c) Minimize

$$C = 2x_1 + x_2 + x_3$$

subject to the constraints

$$x_1 - 3x_2 + x_3 \leq 12 \qquad x_1 \geq 0, \quad x_2 \geq 0$$
$$x_1 + x_2 + x_3 \geq 1 \qquad x_3 \geq 0$$

(d)   Minimize

$$C = 2x_1 + x_2 + 3x_3$$

subject to the constraints

$$-x_1 + 2x_2 + x_3 \geq -2 \qquad x_1 \geq 0, \quad x_2 \geq 0$$
$$x_1 + x_2 + x_3 \geq \quad 6 \qquad x_3 \geq 0$$

Solution
(a)   Since all three conditions are met, this minimum problem is in standard form.
(b)   Conditions 1 and 2 are met but Condition 3 is not, since the coefficient of x_2 in the objective function is negative. Thus, this minimum problem is not in standard form.
(c)   Conditions 1 and 3 are met but condition 2 is not, since the first constraint $x_1 - 3x_2 + x_3 \leq 12$ is not written with a $\geq$ sign. Thus, the minimum problem, as stated, is not in standard form. Notice however, that by multiplying by -1, we can write this constraint as $-x_1 + 3x_2 - x_3 \geq -12$. Written in this way, the minimum problem is in standard form.
(d)   Conditions 1, 2, and 3 are each met, so this minimum problem is in standard form.   ∎

THE DUAL OF A MINIMUM PROBLEM
We illustrate by example how to obtain the dual problem.

Example 2
Obtain the dual problem of the minimum problem:
Minimize

$$C = 300x_1 + 480x_2$$

subject to the conditions

$$x_1 + 3x_2 \geq 0.25 \qquad x_1 \geq 0$$
$$2x_1 + 2x_2 \geq 0.45 \qquad x_2 \geq 0$$

Solution
Observe that the minimum problem is in standard form. We begin by constructing a special matrix for the coefficients of the constraints of this problem without introducing slack variables. As in a simplex tableau, we place the objective function in the last row, but without reversing its signs. The result is

$$\begin{array}{cc} x_1 & x_2 \end{array}$$
$$\left[\begin{array}{cc|c} 1 & 3 & 0.25 \\ 2 & 2 & 0.45 \\ 300 & 480 & 0 \end{array} \right]$$

Similarly, the special matrix for the maximum problem in Example 5 of Chapter 3 (page 156) would be

$$\begin{bmatrix} 1 & 2 & | & 300 \\ 3 & 2 & | & 480 \\ 0.25 & 0.45 & | & 0 \end{bmatrix}$$

Observe that this matrix is the transpose of the previous one; that is, the rows of the first matrix (for the minimum problem) are the columns of the second matrix (for the maximum problem). When a maximum and a minimum problem have this relationship, they are called *dual problems* of each other.

Thus, the dual problem of the given minimum problem is:
Maximize

$$P = 0.25y_1 + 0.45y_2$$

subject to the conditions

$$y_1 + 2y_2 \le 300 \qquad y_1 \ge 0$$
$$3y_1 + 2y_2 \le 480 \qquad y_2 \ge 0$$

This duality relationship is significant because of the following principle:

> **Von Neumann Duality Principle** The optimal solution of a minimum linear programming problem, if the solution exists, has the same value as the optimal solution of the maximum problem that is its dual.

In other words, one way to solve a minimum problem in linear programming is to solve the dual problem. To obtain this dual problem, proceed as follows:

Obtaining the Dual Problem

> **Step 1.** Write the minimum problem in standard form.
>
> **Step 2.** Construct the special matrix from the constraints and the objective function.
>
> **Step 3.** Interchange the rows and columns to form the special matrix of the dual problem.
>
> **Step 4.** Translate this matrix into a maximum problem in standard form.

This process is known as *dualization*. It is illustrated in the next example.

JOHN von NEUMANN (1903–1957) was born in Budapest, Hungary, but spent most of his life at Princeton University and the Institute for Advanced Study. He developed the theory of games at the age of 25 and is largely responsible for inventing the digital computer. Von Neumann, probably the greatest mathematical genius of this century, had a fantastic capacity for doing mental calculations and possessed a photographic memory. He contributed to quantum mechanics, economics, and computer science, and developed a technique that accelerated the production of the first atomic bomb.

Example 3
Find the dual of the minimum problem:
Minimize

$$C = 2x_1 + 3x_2$$

subject to

$$2x_1 + x_2 \geq 6 \qquad x_1 \geq 0$$
$$x_1 + 2x_2 \geq 4 \qquad x_2 \geq 0$$
$$x_1 + x_2 \geq 5$$

Solution
Observe that the minimum problem is in standard form. The special matrix is

$$\begin{bmatrix} 2 & 1 & | & 6 \\ 1 & 2 & | & 4 \\ 1 & 1 & | & 5 \\ 2 & 3 & | & 0 \end{bmatrix}$$

Interchanging rows and columns, we obtain the matrix

$$\begin{bmatrix} 2 & 1 & 1 & | & 2 \\ 1 & 2 & 1 & | & 3 \\ 6 & 4 & 5 & | & 0 \end{bmatrix}$$

This is the special matrix form of the following maximum problem:
Maximize

$$P = 6y_1 + 4y_2 + 5y_3$$

subject to

$$2y_1 + y_2 + y_3 \leq 2 \qquad y_1 \geq 0$$
$$y_1 + 2y_2 + y_3 \leq 3 \qquad y_2 \geq 0$$
$$y_3 \geq 0$$

This maximum problem is in standard form and is the dual of the original problem. ■

Some observations about this example:

1. The variables (x_1, x_2) of the minimum problem have different names from the variables of its dual problem (y_1, y_2, y_3).
2. The minimum problem has three constraints and two variables, while the dual problem has two constraints and three variables. (In general, if a problem has m constraints and n variables, its dual will have n constraints and m variables.)
3. The inequalities defining the constraints are $\geq$ for the minimum problem and $\leq$ for the maximum problem.

4. Since the coefficients in the minimal objective function are positive, the dual problem has nonnegative numbers to the right of the $\leq$ signs.
5. We follow the custom of denoting an objective function by C (for *Cost*) if it is to be minimized and P (for *Profit*) if it is to be maximized.

Example 4

Solve the maximum problem of Example 3 by the simplex method and thereby obtain the solution for the minimum problem.

Solution

We introduce slack variables s_1 and s_2 to get

$$2y_1 + y_2 + y_3 + s_1 = 2$$
$$y_1 + 2y_2 + y_3 + s_2 = 3$$

The initial simplex tableau is

$$\begin{bmatrix} y_1 & y_2 & y_3 & s_1 & s_2 & & \\ ② & 1 & 1 & 1 & 0 & 2 & s_1 \\ 1 & 2 & 1 & 0 & 1 & 3 & s_2 \\ \hline -6 & -4 & -5 & 0 & 0 & 0 & \end{bmatrix}$$

omitting the column for P. We have drawn an extra vertical line in the tableau to set off the columns of s_1 and s_2. (As you will see, the slack variable columns play a special role in the solution of a dual problem.) Now, using the 2 circled above as a pivot element, we pivot and get

$$\begin{bmatrix} y_1 & y_2 & y_3 & s_1 & s_2 & & \\ 1 & \frac{1}{2} & ① & \frac{1}{2} & 0 & 1 & y_1 \\ 0 & \frac{3}{2} & \frac{1}{2} & -\frac{1}{2} & 1 & 2 & s_2 \\ \hline 0 & -1 & -2 & 3 & 0 & 6 & \end{bmatrix}$$

This time, the entry in row 1, column 3, is the pivot. When we pivot, we obtain

$$\begin{bmatrix} y_1 & y_2 & y_3 & s_1 & s_2 & & \\ 2 & 1 & 1 & 1 & 0 & 2 & y_3 \\ -1 & 1 & 0 & -1 & 1 & 1 & s_2 \\ \hline 4 & 1 & 0 & 5 & 0 & 10 & \end{bmatrix}$$

Now all the entries in the objective row are greater than or equal to 0, so this is our final tableau and an optimal solution has been reached. We read from it that the solution to the maximum problem is

$$P = 10 \qquad y_1 = 0 \qquad y_2 = 0 \qquad y_3 = 2$$

The duality principle states that the minimum value of the objective function in the original problem is the same as the maximum value in the dual; that is,

$$C = 10$$

But which values of x_1 and x_2 will yield this minimum value? There are some details of the duality principle and its application that we have omitted here; these concern the relationships between the variables of the original problem and the slack variables used in the solution of the dual problem. As a consequence of these relationships, the entire minimal solution can be read from the right end of the objective row of the final tableau:

$$x_1 = 5 \qquad x_2 = 0 \qquad C = 10$$

Notice in the solution to Example 4 that the value of x_1 is found at the bottom of the column corresponding to s_1 and x_2 is similarly found in the column corresponding to s_2. In general, the values of the slack variables in the maximum tableau are not the same as those of the original variables in the minimum problem. Observe that here

$$x_1 = 5 \qquad x_2 = 0$$

while

$$s_1 = 0 \qquad s_2 = 1$$

We summarize how to solve a minimum linear programming problem below:

Solving a Minimum Problem

Step 1. Write the dual (maximum) problem.

Step 2. Solve this maximum problem by the simplex method.

Step 3. Read the optimal solution for the original problem from the objective row of the final simplex tableau. The variables will appear as the last entries in the columns corresponding to the slack variables.

Step 4. The minimum value of the objective function (C) will appear in the lower right corner of the final tableau; it is equal to the maximum value of the dual objective function (P).

Example 5
Minimize

$$C = 6x_1 + 8x_2 + x_3$$

subject to

$$3x_1 + 5x_2 + 3x_3 \geq 20 \qquad x_1 \geq 0$$
$$x_1 + 3x_2 + 2x_3 \geq 9 \qquad x_2 \geq 0$$
$$6x_1 + 2x_2 + 5x_3 \geq 30 \qquad x_3 \geq 0$$

Solution
This minimum problem is in standard form. The special matrix of this problem is

$$\begin{bmatrix} 3 & 5 & 3 & | & 20 \\ 1 & 3 & 2 & | & 9 \\ 6 & 2 & 5 & | & 30 \\ 6 & 8 & 1 & | & 0 \end{bmatrix}$$

We interchange rows and columns to get

$$\begin{bmatrix} 3 & 1 & 6 & | & 6 \\ 5 & 3 & 2 & | & 8 \\ 3 & 2 & 5 & | & 1 \\ 20 & 9 & 30 & | & 0 \end{bmatrix}$$

The dual problem is:
Maximize

$$P = 20y_1 + 9y_2 + 30y_3$$

subject to

$$\begin{aligned} 3y_1 + y_2 + 6y_3 &\leq 6 & y_1 &\geq 0 \\ 5y_1 + 3y_2 + 2y_3 &\leq 8 & y_2 &\geq 0 \\ 3y_1 + 2y_2 + 5y_3 &\leq 1 & y_3 &\geq 0 \end{aligned}$$

We introduce slack variables s_1, s_2, and s_3. The initial tableau for this problem is

$$\begin{array}{cccccc} y_1 & y_2 & y_3 & s_1 & s_2 & s_3 \\ \left[\begin{array}{cccccc|c} 3 & 1 & 6 & 1 & 0 & 0 & 6 \\ 5 & 3 & 2 & 0 & 1 & 0 & 8 \\ 3 & 2 & 5 & 0 & 0 & 1 & 1 \\ \hline -20 & -9 & -30 & 0 & 0 & 0 & 0 \end{array}\right] \end{array}$$

The final tableau (as you may verify) is

$$\begin{array}{cccccc} y_1 & y_2 & y_3 & s_1 & s_2 & s_3 \\ \left[\begin{array}{cccccc|c} 0 & -1 & 1 & 1 & 0 & -1 & 5 \\ 0 & -\frac{1}{3} & -\frac{19}{3} & 0 & 1 & -\frac{5}{3} & \frac{19}{3} \\ 1 & \frac{2}{3} & \frac{5}{3} & 0 & 0 & \frac{1}{3} & \frac{1}{3} \\ \hline 0 & \frac{13}{3} & \frac{10}{3} & 0 & 0 & \frac{20}{3} & \frac{20}{3} \end{array}\right] \begin{array}{c} s_1 \\ s_2 \\ y_1 \\ \\ \end{array} \end{array}$$

The solution to the maximum problem is

$$P = \frac{20}{3} \qquad y_1 = \frac{1}{3} \qquad y_2 = 0 \qquad y_3 = 0$$

For the minimum problem, the values of x_1, x_2, and x_3 are read as the last entries in the columns under s_1, s_2, and s_3, respectively. Hence, the optimal solution to the minimum problem is

$$x_1 = 0 \qquad x_2 = 0 \qquad x_3 = \frac{20}{3}$$

and the minimum value is $C = \frac{20}{3}$.

Example 6

Transportation Problem The Red Tomato Company operates two plants for canning its tomatoes and has two warehouses for storing the finished products until they are purchased by retailers. The company wants to arrange its shipments from the plants to the warehouses so that the requirements of the warehouses are met and shipping costs are kept at a minimum. The schedule shown in the table represents the per case shipping costs from plant to warehouse.

		Warehouse A	Warehouse B
Plant	I	$0.25	$0.18
	II	$0.25	$0.14

Each week, Plant I can produce at most 450 cases and Plant II can produce no more than 350 cases of tomatoes. Also, each week, Warehouse A requires at least 300 cases and Warehouse B requires at least 500 cases.

Solution

If we represent the number of cases shipped from Plant I to Warehouse A by x_1, from Plant I to Warehouse B by x_2, and so on, the above data can be represented by the following table:

		Warehouse A	Warehouse B	Maximum Available
Plant	I	x_1	x_2	450
	II	x_3	x_4	350
Minimum demand		300	500	

The linear programming problem is stated as follows:
Minimize the cost equation

$$C = 0.25x_1 + 0.18x_2 + 0.25x_3 + 0.14x_4$$

subject to

$$x_1 + x_2 \leq 450 \qquad x_1 \geq 0$$
$$x_3 + x_4 \leq 350 \qquad x_2 \geq 0$$
$$x_1 + x_3 \geq 300 \qquad x_3 \geq 0$$
$$x_2 + x_4 \geq 500 \qquad x_4 \geq 0$$

To get the problem in standard form, we multiply both sides of the first two inequalities by -1:

$$-x_1 - x_2 \geq -450$$
$$-x_3 - x_4 \geq -350$$
$$x_1 + x_3 \geq 300$$
$$x_2 + x_4 \geq 500$$

The special matrix for the minimum problem is

$$\begin{bmatrix} -1 & -1 & 0 & 0 & -450 \\ 0 & 0 & -1 & -1 & -350 \\ 1 & 0 & 1 & 0 & 300 \\ 0 & 1 & 0 & 1 & 500 \\ 0.25 & 0.18 & 0.25 & 0.14 & 0 \end{bmatrix}$$

The dual matrix is

$$\begin{bmatrix} -1 & 0 & 1 & 0 & 0.25 \\ -1 & 0 & 0 & 1 & 0.18 \\ 0 & -1 & 1 & 0 & 0.25 \\ 0 & -1 & 0 & 1 & 0.14 \\ -450 & -350 & 300 & 500 & 0 \end{bmatrix}$$

and the dual (maximum) problem is:
Maximize

$$P = -450y_1 - 350y_2 + 300y_3 + 500y_4$$

subject to

$$-y_1 + y_3 \leq 0.25 \qquad y_1 \geq 0$$
$$-y_1 + y_4 \leq 0.18 \qquad y_2 \geq 0$$
$$-y_2 + y_3 \leq 0.25 \qquad y_3 \geq 0$$
$$-y_2 + y_4 \leq 0.14 \qquad y_4 \geq 0$$

Introducing the slack variables s_1, s_2, s_3, and s_4, we construct the initial simplex tableau and proceed to solve the dual problem:

	y_1	y_2	y_3	y_4	s_1	s_2	s_3	s_4		
	-1	0	1	0	1	0	0	0	0.25	s_1
	-1	0	0	1	0	1	0	0	0.18	s_2
	0	-1	1	0	0	0	1	0	0.25	s_3
$\rightarrow$	0	-1	0	①	0	0	0	1	0.14	s_4
	450	350	-300	-500	0	0	0	0	0	

$$\uparrow$$

	y_1	y_2	y_3	y_4	s_1	s_2	s_3	s_4	
$\rightarrow$	-1	0	①	0	1	0	0	0	0.25
	-1	1	0	0	0	1	0	-1	0.04
	0	-1	1	0	0	0	1	0	0.25
	0	-1	0	1	0	0	0	1	0.14
	450	-150	-300	0	0	0	0	500	70

$$\uparrow$$

$$\rightarrow
\begin{array}{cccc|cccc|c}
y_1 & y_2 & y_3 & y_4 & s_1 & s_2 & s_3 & s_4 & \\
-1 & 0 & 1 & 0 & 1 & 0 & 0 & 0 & 0.25 \\
-1 & \textcircled{1} & 0 & 0 & 0 & 1 & 0 & -1 & 0.04 \\
1 & -1 & 0 & 0 & -1 & 0 & 1 & 0 & 0 \\
0 & -1 & 0 & 1 & 0 & 0 & 0 & 1 & 0.14 \\
\hline
150 & -150 & 0 & 0 & 300 & 0 & 0 & 500 & 145
\end{array}$$

$$
\begin{array}{cccc|cccc|cl}
y_1 & y_2 & y_3 & y_4 & s_1 & s_2 & s_3 & s_4 & & \\
-1 & 0 & 1 & 0 & 1 & 0 & 0 & 0 & 0.25 & y_3\\
-1 & 1 & 0 & 0 & 0 & 1 & 0 & 1 & 0.04 & y_2\\
0 & 0 & 0 & 0 & -1 & 1 & 1 & -1 & 0.04 & s_3\\
-1 & 0 & 0 & 1 & 0 & 1 & 0 & 0 & 0.18 & y_4\\
\hline
0 & 0 & 0 & 0 & 300 & 150 & 0 & 350 & 151 &
\end{array}
$$

$$\quad\quad\quad\quad\quad\quad\quad x_1 \quad x_2 \quad x_3 \quad x_4 \quad C$$

The final tableau yields the solution to the original minimum problem:

$$C = \$151 \qquad x_1 = 300 \qquad x_2 = 150 \qquad x_3 = 0 \qquad x_4 = 350$$

This means Plant I should deliver 300 cases to Warehouse A and 150 cases to Warehouse B; and Plant II should deliver 350 cases to Warehouse B to keep costs at the minimum ($151). ∎

A final note for this section: If you find that the dual of a minimum problem has an unbounded solution, then the minimum problem has no feasible solution; the solution set of its constraints is empty.

Exercise 4.3   *Answers to Odd-Numbered Problems begin on page A-16.*

A   In Problems 1–6 determine which of the given minimum problems are in standard form.

1. Minimize

$$C = 2x_1 + 3x_2$$

subject to the constraints

$$4x_1 - x_2 \geq 2 \qquad x_1 \geq 0$$
$$x_1 + x_2 \geq 1 \qquad x_2 \geq 0$$

2. Minimize

$$C = 3x_1 + 5x_2$$

subject to the constraints

$$3x_1 - x_2 \geq 4 \qquad x_1 \geq 0$$
$$x_1 - 2x_2 \geq 3 \qquad x_2 \geq 0$$

3. Minimize

$$C = 2x_1 - x_2$$

subject to the constraints

$$2x_1 - x_2 \geq 1 \qquad x_1 \geq 0$$
$$-2x_1 \geq -3 \qquad x_2 \geq 0$$

4. Minimize

$$C = 2x_1 + 3x_2$$

subject to the constraints

$$x_1 - x_2 \leq 3 \qquad x_1 \geq 0$$
$$2x_1 + 3x_2 \geq 4 \qquad x_2 \geq 0$$

5. Minimize

$$C = 3x_1 + 7x_2 + x_3$$

6. Minimize

$$C = x_1 - x_2 + x_3$$

subject to the constraints

$$x_1 + x_3 \le 6 \qquad x_1 \ge 0, \quad x_2 \ge 0$$
$$2x_1 + x_2 \ge 4 \qquad x_3 \ge 0$$

subject to the constraints

$$x_1 + x_2 \ge 6 \qquad x_1 \ge 0, \quad x_2 \ge 0$$
$$2x_1 - x_3 \ge 4 \qquad x_3 \ge 0$$

B In Problems 7–10 write the dual problem for each minimum linear programming problem.

7. Minimize

$$C = 2x_1 + 3x_2$$

subject to

$$x_1 + x_2 \ge 2 \qquad x_1 \ge 0$$
$$2x_1 + 3x_2 \ge 6 \qquad x_2 \ge 0$$

8. Minimize

$$C = 3x_1 + 4x_2$$

subject to

$$2x_1 + x_2 \ge 2 \qquad x_1 \ge 0$$
$$2x_1 + x_2 \ge 6 \qquad x_2 \ge 0$$

9. Minimize

$$C = 3x_1 + x_2 + x_3$$

subject to

$$x_1 + x_2 + x_3 \ge 5 \qquad x_1 \ge 0$$
$$2x_1 + x_2 \ge 4 \qquad x_2 \ge 0$$
$$x_3 \ge 0$$

10. Minimize

$$C = 2x_1 + x_2 + x_3$$

subject to

$$2x_1 + x_2 + x_3 \ge 4 \qquad x_1 \ge 0$$
$$x_1 + 2x_2 + x_3 \ge 6 \qquad x_2 \ge 0$$
$$x_3 \ge 0$$

In Problems 11–13 solve each minimum linear programming problem by using the simplex method.

11. Minimize

$$C = 6x_1 + 3x_2$$

subject to

$$x_1 + x_2 \ge 2 \qquad x_1 \ge 0$$
$$2x_1 + 6x_2 \ge 6 \qquad x_2 \ge 0$$

12. Minimize

$$C = 3x_1 + 4x_2$$

subject to

$$x_1 + x_2 \ge 3 \qquad x_1 \ge 0$$
$$2x_1 + x_2 \ge 4 \qquad x_2 \ge 0$$

13. Minimize

$$C = 6x_1 + 3x_2$$

subject to

$$x_1 + x_2 \ge 4 \qquad x_1 \ge 0$$
$$3x_1 + 4x_2 \ge 12 \qquad x_2 \ge 0$$

C In Problems 14–18 solve each minimum linear programming problem by using the simplex method.

14. Minimize

$$C = 2x_1 + 3x_2 + 4x_3$$

subject to

$$x_1 - 2x_2 - 3x_3 \ge -2 \qquad x_1 \ge 0$$
$$x_1 + x_2 + x_3 \ge 2 \qquad x_2 \ge 0$$
$$2x_1 + x_3 \ge 3 \qquad x_3 \ge 0$$

15. Minimize

$$C = x_1 + 2x_2 + x_3$$

subject to

$$x_1 - 3x_2 + 4x_3 \ge 12 \qquad x_1 \ge 0$$
$$3x_1 + x_2 + 2x_3 \ge 10 \qquad x_2 \ge 0$$
$$x_1 - x_2 - x_3 \ge -8 \qquad x_3 \ge 0$$

16. Minimize

$$C = x_1 + 2x_2 + 4x_3$$

subject to

$$
\begin{aligned}
x_1 - x_2 + 3x_3 &\geq 4 & x_1 &\geq 0 \\
2x_1 + 2x_2 - 3x_3 &\geq 6 & x_2 &\geq 0 \\
-x_1 + 2x_2 + 3x_3 &\geq 2 & x_3 &\geq 0
\end{aligned}
$$

17. Minimize

$$C = x_1 + 4x_2 + 2x_3 + 4x_4$$

subject to

$$
\begin{aligned}
x_1 &\geq 0, \quad x_2 \geq 0, \quad x_3 \geq 0, \quad x_4 \geq 0 \\
x_1 + x_3 &\geq 1 \\
x_2 + x_4 &\geq 1 \\
-x_1 - x_2 - x_3 - x_4 &\geq -3
\end{aligned}
$$

18. Minimize

$$C = x_1 + 2x_2 + 3x_3 + 4x_4$$

subject to

$$
\begin{aligned}
x_1 &\geq 0, \quad x_2 \geq 0, \quad x_3 \geq 0, \quad x_4 \geq 0 \\
x_1 + x_3 &\geq 1 \\
x_2 + x_4 &\geq 1 \\
-x_1 - x_2 - x_3 - x_4 &\geq -3
\end{aligned}
$$

APPLICATIONS   **19.** **Mixture Problem** Minimize the cost of preparing the following mixture, which is made up of three foods, I, II, III. Food I costs $2 per unit, Food II costs $1 per unit, and Food III costs $3 per unit. Each unit of Food I contains 2 ounces of protein and 4 ounces of carbohydrate; each unit of Food II has 3 ounces of protein and 2 ounces of carbohydrate; and each unit of Food III has 4 ounces of protein and 2 ounces of carbohydrate. The mixture must contain at least 20 ounces of protein and 15 ounces of carbohydrate.

20. **Advertising** A local appliance store has decided on an advertising campaign utilizing newspaper and radio. Each dollar spent on newspaper advertising is expected to reach 50 people in the "Under $25,000" and 40 in the "Over $25,000" bracket. Each dollar spent on radio advertising is expected to reach 70 people in the "Under $25,000" and 20 people in the "Over $25,000." If the store wants to reach at least 100,000 people in the "Under $25,000" and at least 120,000 in the "Over $25,000" bracket, how should it proceed so that the cost of advertising is minimized?

21. **Diet Preparation** Mr. Jones needs to supplement his diet with at least 50 mg calcium and 8 mg iron daily. The minerals are available in two types of vitamin pills, P and Q. Pill P contains 5 mg calcium and 2 mg iron, while Pill Q contains 10 mg calcium and 1 mg iron. If each P pill costs 3 cents and each Q pill costs 4 cents, how could Mr. Jones minimize the cost of adding the minerals to his diet? What would the daily minimum cost be?

22. **Production Schedule** A company owns two mines. Mine A produces 1 ton of high-grade ore, 3 tons of medium-grade ore, and 5 tons of low-grade ore each day. Mine B produces 2 tons of each grade ore per day. The company needs at least 80 tons of high-grade ore, at least 160 tons of medium-grade ore, and at least 200 tons of low-grade ore. How many days should each mine be operated to minimize costs if it costs $2000 per day to operate each mine?

23. **Production Schedule** Argus Company makes three products: A, B, and C. Each unit of A costs $4, each unit of B costs $2, and each unit of C costs $1 to produce. Argus must produce at least 20 A's, 30 B's, and 40 C's, and cannot produce less than 200 total units of A's, B's, and C's combined. Minimize Argus costs.

24. Diet Planning A health clinic dietician is planning a meal consisting of three foods whose ingredients are summarized as follows:

	One Unit of		
	Food I	Food II	Food III
Units of protein	10	15	20
Units of carbohydrates	1	2	1
Units of iron	4	8	1
Calories	80	120	100

The dietician wishes to determine the number of units of each food to use to create a balanced meal containing at least 40 units of protein, 6 units of carbohydrates, and 12 units of iron, with as few calories as possible.

25. Menu Planning Fresh Starts Catering offers the following lunch menu:

	Menu	
Lunch #1	Soup, salad, sandwich	$6.20
Lunch #2	Salad, pasta	$7.40
Lunch #3	Salad, sandwich, pasta	$9.10

Mrs. Mintz and her friends would like to order 4 bowls of soup, 9 salads, 6 sandwiches, and 5 orders of pasta and keep the cost as low as possible. Compose her order.

26. Inventory Control A department store stocks three brands of toys: A, B, and C. Each unit of brand A occupies 1 square foot of shelf space, each unit of brand B occupies 2 square feet, and each unit of brand C occupies 3 square feet. The store has 120 square feet available for storage. Surveys show that the store should have on hand at least 12 units of brand A and at least 30 units of A and B combined. Each brand A costs the store $8, each unit of B $6, and each unit of brand C $10. Minimize the cost to the store.

4.4

THE SIMPLEX METHOD WITH MIXED CONSTRAINTS; PHASE I/PHASE II

MIXED CONSTRAINTS
THE SIMPLEX METHOD WITH MIXED CONSTRAINTS
THE MINIMUM PROBLEM
EQUALITY CONSTRAINTS
MODEL: POLLUTION CONTROL

Thus far we have only developed the simplex method for solving maximum and minimum problems in standard form. In this section, we develop the simplex method for linear programming problems that cannot be written in standard form.

MIXED CONSTRAINTS

Recall that for a maximum problem in standard form each constraint must be of the form

$$a_1x_1 + a_2x_2 + \cdots + a_nx_n \le b, \qquad b > 0$$

That is, each is a linear expression *less than or equal to a positive constant*. When the constraints are of any other form — greater than or equal to or equal to — we have what are called *mixed constraints*. The following example will illustrate the simplex method for solving mixed constraint problems.

THE SIMPLEX METHOD WITH MIXED CONSTRAINTS

Example 1
Maximize

$$P = 20x_1 + 15x_2$$

subject to the constraints

$$
\begin{aligned}
x_1 + x_2 &\ge 7 & x_1 &\ge 0 \\
9x_1 + 5x_2 &\le 45 & x_2 &\ge 0 \\
2x_1 + x_2 &\ge 8
\end{aligned}
$$

Solution
We first observe this is a maximum problem that is not in standard form. Second, it cannot be modified so as to be in standard form.

> **Step 1.** Write each constraint except the nonnegative constraints as an inequality with the variables on the left side of a $\le$ sign.

To do this, we merely multiply the first and third inequality by -1. The result is that the constraints become

$$
\begin{aligned}
-x_1 - x_2 &\le -7 \\
9x_1 + 5x_2 &\le 45 \\
-2x_1 - x_2 &\le -8
\end{aligned}
$$

> **Step 2.** Introduce nonnegative variables on the left side of each inequality to form an equality.

To do this, we will use the variables, s_1, s_2, s_3 to obtain

$$
\begin{aligned}
-x_1 - x_2 + s_1 &= -7 & s_1 &\ge 0 \\
9x_1 + 5x_2 + s_2 &= 45 & s_2 &\ge 0 \\
-2x_1 - x_2 + s_3 &= -8 & s_3 &\ge 0
\end{aligned}
$$

Step 3. Set up the initial simplex tableau.

$$\begin{array}{ccccc} x_1 & x_2 & s_1 & s_2 & s_3 \end{array}$$

$$\left[\begin{array}{ccccc|c} -1 & -1 & 1 & 0 & 0 & -7 \\ 9 & 5 & 0 & 1 & 0 & 45 \\ -2 & -1 & 0 & 0 & 1 & -8 \\ \hline -20 & -15 & 0 & 0 & 0 & 0 \end{array}\right]$$

This initial tableau represents the solution $x_1 = 0$, $x_2 = 0$, $s_1 = -7$, $s_2 = 45$, $s_3 = -8$. This is not a feasible solution. The reason for this lies in the existence of the two negative constraints in the right-hand column. That is, this tableau represents a solution that causes two of the variables to be negative, in violation of the nonnegativity requirement. Whenever this occurs, the simplex algorithm consists of two phases.

Phase I/Phase II

Step 4. Determine whether Phase I or Phase II applies. Phase I is used whenever negative entries appear in the right-hand column; Phase II is used whenever all the entries in the right-hand column are nonnegative. In determining whether Phase I or Phase II applies, the objective row is ignored.

Step 5. Select the pivot element.
Phase I: The pivot row is the row with the most negative value in the right column. The pivot element is the most negative entry in the pivot row. If all entries in the pivot row are nonnegative, there are no feasible solutions and the problem has no solution.
Phase II: Follow the pivoting strategy given on page 195 of Section 4.2.

For our example, we use Phase I. The pivot row is row 3 (due to the -8). To find the pivot column we choose the smallest negative entry, which is -2. Thus, the pivot column is column 1.

Step 6. Go back to Step 4 unless a final tableau has been reached.

For our example, after pivoting on the entry in row 3, column 1, we get the tableau

$$\begin{array}{ccccc} x_1 & x_2 & s_1 & s_2 & s_3 \end{array}$$

$$\left[\begin{array}{ccccc|c} 0 & \boxed{-\frac{1}{2}} & 1 & 0 & -\frac{1}{2} & -3 \\ 0 & \frac{1}{2} & 0 & 1 & \frac{9}{2} & 9 \\ 1 & \frac{1}{2} & 0 & 0 & -\frac{1}{2} & 4 \\ \hline 0 & -5 & 0 & 0 & -10 & 80 \end{array}\right]$$

Step 4. Phase I applies.

Step 5. The pivot row is row 1; the pivot column is column 2 (or column 5). After pivoting, we get the tableau

$$
\begin{array}{ccccc}
x_1 & x_2 & s_1 & s_2 & s_3 \\
\end{array}
$$

$$
\left[
\begin{array}{ccccc|c}
0 & 1 & -2 & 0 & 1 & 6 \\
0 & 0 & 1 & 1 & 4 & 6 \\
1 & 0 & ① & 0 & -1 & 1 \\
\hline
0 & 0 & -10 & 0 & -5 & 110
\end{array}
\right]
$$

Step 4. Phase II applies.

Step 5. The pivot column is column 3; the pivot row is row 3. After pivoting, we get the tableau

$$
\begin{array}{ccccc}
x_1 & x_2 & s_1 & s_2 & s_3 \\
\end{array}
$$

$$
\left[
\begin{array}{ccccc|c}
2 & 1 & 0 & 0 & -1 & 8 \\
-1 & 0 & 0 & 1 & ⑤ & 5 \\
1 & 0 & 1 & 0 & -1 & 1 \\
\hline
10 & 0 & 0 & 0 & -15 & 120
\end{array}
\right]
$$

Step 4. Phase II applies.

Step 5. The pivot column is column 5; the pivot row is row 2. After pivoting, we get the tableau

$$
\begin{array}{ccccc}
x_1 & x_2 & s_1 & s_2 & s_3 \\
\end{array}
$$

$$
\left[
\begin{array}{ccccc|cl}
\frac{9}{5} & 1 & 0 & \frac{1}{5} & 0 & 9 & x_2 \\
-\frac{1}{5} & 0 & 0 & \frac{1}{5} & 1 & 1 & s_3 \\
\frac{4}{5} & 0 & 1 & \frac{1}{5} & 0 & 2 & s_1 \\
\hline
7 & 0 & 0 & 3 & 0 & 135 &
\end{array}
\right]
$$

This is our final tableau. The maximum value of P is 135, and it is achieved when $x_1 = 0$, $x_2 = 9$, $s_1 = 2$, $s_2 = 0$, $s_3 = 1$. ■

THE MINIMUM PROBLEM

Earlier we presented a method of solving the minimum problem if it was in standard form. In general, a minimum problem can be changed to a maximum problem by realizing that in order to minimize C we must maximize $-C$. The following example illustrates this method.

Example 2
Minimize

$$C = 5x_1 + 6x_2$$

subject to the constraints

$$x_1 \geq 0, \qquad x_2 \geq 0$$
$$x_1 + x_2 \leq 10$$
$$x_1 + 2x_2 \geq 12$$
$$2x_1 + x_2 \geq 12$$
$$x_1 \geq 3$$

Solution

We change our problem from minimizing $C = 5x_1 + 6x_2$ to maximizing $z = -C = -5x_1 - 6x_2$.

Step 1. Write each constraint with $\leq$.

$$x_1 + x_2 \leq 10$$
$$-x_1 - 2x_2 \leq -12$$
$$-2x_1 - x_2 \leq -12$$
$$-x_1 \leq -3$$

Step 2. Introduce nonnegative variables to form equalities:

$$s_1 \geq 0, \qquad s_2 \geq 0, \qquad s_3 \geq 0, \qquad s_4 \geq 0$$
$$x_1 + x_2 + s_1 = 10$$
$$-x_1 - 2x_2 + s_2 = -12$$
$$-2x_1 - x_2 + s_3 = -12$$
$$-x_1 + s_4 = -3$$

Step 3. Set up the initial simplex tableau:

$$
\begin{array}{cccccc}
x_1 & x_2 & s_1 & s_2 & s_3 & s_4 \\
\end{array}
$$

$$
\left[
\begin{array}{cccccc|c}
1 & 1 & 1 & 0 & 0 & 0 & 10 \\
-1 & \boxed{-2} & 0 & 1 & 0 & 0 & -12 \\
-2 & -1 & 0 & 0 & 1 & 0 & -12 \\
-1 & 0 & 0 & 0 & 0 & 1 & -3 \\
\hline
5 & 6 & 0 & 0 & 0 & 0 & 0 \\
\end{array}
\right]
$$

Step 4. Phase I applies.

Step 5. The pivot row is row 2 (or 3) and for this row the pivot column is column 2. After pivoting we get the tableau:

$$
\begin{array}{cccccc}
x_1 & x_2 & s_1 & s_2 & s_3 & s_4 \\
\end{array}
$$

$$
\left[
\begin{array}{cccccc|c}
\frac{1}{2} & 0 & 1 & \frac{1}{2} & 0 & 0 & 4 \\
\frac{1}{2} & 1 & 0 & -\frac{1}{2} & 0 & 0 & 6 \\
\boxed{-\frac{3}{2}} & 0 & 0 & -\frac{1}{2} & 1 & 0 & -6 \\
-1 & 0 & 0 & 0 & 0 & 1 & -3 \\
\hline
2 & 0 & 0 & 3 & 0 & 0 & -36 \\
\end{array}
\right]
$$

Step 4. Phase I applies.

Step 5. The pivot row is row 3, the pivot column is column 1. After pivoting we get the tableau:

$$
\begin{array}{cccccc}
x_1 & x_2 & s_1 & s_2 & s_3 & s_4 \\
\end{array}
$$

$$
\left[
\begin{array}{cccccc|cl}
0 & 0 & 1 & \frac{1}{3} & \frac{1}{3} & 0 & 2 & s_1 \\
0 & 1 & 0 & -\frac{2}{3} & \frac{1}{3} & 0 & 4 & x_2 \\
1 & 0 & 0 & \frac{1}{3} & -\frac{2}{3} & 0 & 4 & x_1 \\
0 & 0 & 0 & \frac{1}{3} & -\frac{2}{3} & 1 & 1 & s_4 \\
\hline
0 & 0 & 0 & \frac{7}{3} & \frac{4}{3} & 0 & -44 & \\
\end{array}
\right]
$$

This is a final tableau. Since the maximum value of $z = -44$, the minimum value of $C = 44$. This occurs when $x_1 = 4$, $x_2 = 4$, $s_1 = 2$, $s_2 = 0$, $s_3 = 0$, $s_4 = 1$ ■

EQUALITY CONSTRAINTS

So far all our constraints used $\leq$ or $\geq$. What can be done if one of the constraints is an equality? One way is to replace the $=$ constraint with two constraints $\leq$ and $\geq$. The next example illustrates this.

Example 3

Minimize

$$C = 7x_1 + 5x_2 + 6x_3$$

subject to the constraints

$$x_1 \geq 0, \qquad x_2 \geq 0, \qquad x_3 \geq 0$$
$$x_1 + x_2 + x_3 = 10$$
$$x_1 + 2x_2 + 3x_3 \leq 19$$
$$2x_1 + 3x_2 \geq 21$$

Solution

We wish to maximize $z = -C = -7x_1 - 5x_2 - 6x_3$ subject to the constraints

$$x_1 \geq 0, \qquad x_2 \geq 0, \qquad x_3 \geq 0$$
$$x_1 + x_2 + x_3 \leq 10$$
$$x_1 + x_2 + x_3 \geq 10$$
$$x_1 + 2x_2 + 3x_3 \leq 19$$
$$2x_1 + 3x_2 \geq 21$$

Step 1. Rewrite the constraints with $\leq$:

$$x_1 + x_2 + x_3 \leq 10$$
$$-x_1 - x_2 - x_3 \leq -10$$
$$x_1 + 2x_2 + 3x_3 \leq 19$$
$$-2x_1 - 3x_2 \leq -21$$

Step 2. Introduce nonnegative variables:

$$s_1 \geq 0, \qquad s_2 \geq 0, \qquad s_3 \geq 0, \qquad s_4 \geq 0$$
$$x_1 + x_2 + x_3 + s_1 = 10$$
$$-x_1 - x_2 - x_3 + s_2 = -10$$
$$x_1 + 2x_2 + 3x_3 + s_3 = 19$$
$$-2x_1 - 3x_2 + s_4 = -21$$

Step 3. Set up the initial simplex tableau:

$$
\begin{array}{ccccccc}
x_1 & x_2 & x_3 & s_1 & s_2 & s_3 & s_4 \\
\end{array}
$$

$$
\left[
\begin{array}{ccccccc|c}
1 & 1 & 1 & 1 & 0 & 0 & 0 & 10 \\
-1 & -1 & -1 & 0 & 1 & 0 & 0 & -10 \\
1 & 2 & 3 & 0 & 0 & 1 & 0 & 19 \\
-2 & \boxed{-3} & 0 & 0 & 0 & 0 & 1 & -21 \\
\hline
7 & 5 & 6 & 0 & 0 & 0 & 0 & 0
\end{array}
\right]
$$

Step 4. Phase I applies.

Step 5. The pivot row is row 4, the pivot column is column 2. Pivoting we get the tableau:

$$
\begin{array}{ccccccc}
x_1 & x_2 & x_3 & s_1 & s_2 & s_3 & s_4 \\
\end{array}
$$

$$
\left[
\begin{array}{ccccccc|c}
\frac{1}{3} & 0 & 1 & 1 & 0 & 0 & \frac{1}{3} & 3 \\
-\frac{1}{3} & 0 & \boxed{-1} & 0 & 1 & 0 & -\frac{1}{3} & -3 \\
-\frac{1}{3} & 0 & 3 & 0 & 0 & 1 & \frac{2}{3} & 5 \\
\frac{2}{3} & 1 & 0 & 0 & 0 & 0 & -\frac{1}{3} & 7 \\
\hline
\frac{11}{3} & 0 & 6 & 0 & 0 & 0 & \frac{5}{3} & -35
\end{array}
\right]
$$

Step 4. Phase I applies.

Step 5. The pivot row is row 2; the pivot column is column 3. Pivoting we get the tableau

$$
\begin{array}{ccccccc}
x_1 & x_2 & x_3 & s_1 & s_2 & s_3 & s_4 \\
\end{array}
$$

$$
\left[
\begin{array}{ccccccc|c}
0 & 0 & 0 & 1 & 1 & 0 & 0 & 0 \\
\frac{1}{3} & 0 & 1 & 0 & -1 & 0 & \frac{1}{3} & 3 \\
\boxed{-\frac{4}{3}} & 0 & 0 & 0 & 3 & 1 & -\frac{1}{3} & -4 \\
\frac{2}{3} & 1 & 0 & 0 & 0 & 0 & -\frac{1}{3} & 7 \\
\hline
\frac{5}{3} & 0 & 0 & 0 & 6 & 0 & -\frac{1}{3} & -53
\end{array}
\right]
$$

Step 4. Phase I applies.

Step 5. The pivot row is row 3, the pivot column is column 1. Pivoting we get the tableau

$$
\begin{array}{ccccccc}
x_1 & x_2 & x_3 & s_1 & s_2 & s_3 & s_4 \\
\end{array}
$$

$$
\left[
\begin{array}{ccccccc|c}
0 & 0 & 0 & 1 & 1 & 0 & 0 & 0 \\
0 & 0 & 1 & 0 & -\frac{1}{4} & \frac{1}{4} & \boxed{\frac{1}{4}} & 2 \\
1 & 0 & 0 & 0 & -\frac{9}{4} & -\frac{3}{4} & \frac{1}{4} & 3 \\
0 & 1 & 0 & 0 & \frac{3}{2} & \frac{1}{2} & -\frac{1}{2} & 5 \\
\hline
0 & 0 & 0 & 0 & \frac{39}{4} & \frac{5}{4} & -\frac{3}{4} & -58
\end{array}
\right]
$$

Step 4. Phase II applies.

Step 5. The pivot column is column 7, the pivot row is row 2. Pivoting we get the tableau:

$$
\begin{array}{ccccccc}
x_1 & x_2 & x_3 & s_1 & s_2 & s_3 & s_4 \\
\end{array}
$$

$$
\left[
\begin{array}{ccccccc|c}
0 & 0 & 0 & 1 & 1 & 0 & 0 & 0 \\
0 & 0 & 4 & 0 & -1 & 1 & 1 & 8 \\
1 & 0 & -1 & 0 & -2 & -1 & 0 & 1 \\
0 & 1 & 2 & 0 & 1 & 1 & 0 & 9 \\
\hline
0 & 0 & 3 & 0 & 9 & 2 & 0 & -52 \\
\end{array}
\right]
\begin{array}{c}
s_1 \\
s_4 \\
x_1 \\
x_2 \\
\\
\end{array}
$$

This is a final tableau. Since the maximum value of z is -52, the minimum value of C is 52. This occurs when $x_1 = 1$, $x_2 = 9$, $x_3 = 0$, $s_1 = 0$, $s_2 = 0$, $s_3 = 0$, $s_4 = 8$.

■

MODEL: POLLUTION CONTROL

In the pollution control model we presented in Chapter 3, we considered an extremely simplified application. A more realistic version is given in the following model.* In this example we merely indicate the complicated nature of attempting to solve a real-world problem. As a result, the linear programming problem is set up, but no solution is actually given.

There are many pollution sources and five (not one) major pollutants in this larger model. The required pollutant reductions in the St. Louis airshed for the year 1970 are given as follows:

Sulfur dioxide	485,000,000 pounds
Carbon monoxide	1,300,000,000 pounds
Hydrocarbons	280,000,000 pounds
Nitrogen oxides	75,000,000 pounds
Particulate matter	180,000,000 pounds

The model includes a wide variety of possible control methods. Among them are the installation of exhaust and crankcase devices on used as well as new automobiles; the substitution of natural gas for coal; the installation of catalytic oxidation systems to convert sulfur dioxide in the stacks of power plants to salable sulfuric acid; and even the municipal collection of leaves as an alternative to burning.

The most contested control method in the St. Louis airshed has been a restriction on the sulfur content of coal. Consider a particular category of traveling grate stokers that burns 3.1% sulfur coal. Let control method 3 be the substitution of 1.8% sulfur coal for the high-sulfur coal in these stokers. The variable X_3 represents the number of tons of 3.1% sulfur coal replaced with low-sulfur coal.

Total cost of this control method is

$$C = (\$2.50)X_3$$

where $2.50 is an estimate of the incremental cost of the low-sulfur coal.

Just as the number of barrels of cement controlled by any process was constrained in our simple example (see Chapter 3), so

* Robert E. Kohn, "Application of Linear Programming to a Controversy on Air Pollution Control," *Management Science*, **17**, 10 (June 1971), pp. B609–B621.

$$X_3 \leq 200{,}000$$

where 200,000 tons is the estimate of the quantity of coal that will be burned in this category of traveling grate stokers in 1970.

For every ton of 3.1% sulfur coal replaced by 1.8% sulfur coal, sulfur dioxide emissions are reduced by

$$\left[\left(\begin{array}{c} 0.031 \\ \text{sulfur} \\ \text{content} \end{array} \right) \left(\begin{array}{c} 2000 \text{ lb} \\ \text{per ton} \\ \text{of coal} \end{array} \right) \left(\begin{array}{c} 0.95 \\ \text{complete} \\ \text{burning} \end{array} \right)(2) \right]$$

$$- \left[(0.944) \left(\begin{array}{c} 0.018 \\ \text{sulfur} \\ \text{content} \end{array} \right) (2000 \text{ lb})(0.95)(2) \right] = 53.2 \text{ lb}$$

where the factor (2) doubles the weight of sulfur burned to get the weight of sulfur dioxide; where the factor (0.944) accounts for the higher BTU content of the low-sulfur coal, which permits 0.944 ton of it to replace 1 ton of the high-sulfur coal; and where (0.95) incorporates an assumption of 95% complete burning. The two expressions within brackets represent emission of sulfur dioxide from 3.1% and 1.8% sulfur coal, respectively. Thus, we have

$$(53.2)X_3 = \text{Pounds of sulfur dioxide reduced}$$

The remaining pollutant reductions are

$$(0.2)X_3 = \text{Pounds of carbon monoxide reduced}$$
$$(0.1)X_3 = \text{Pounds of hydrocarbons reduced}$$
$$(1.1)X_3 = \text{Pounds of nitrogen oxides reduced}$$
$$(12.2)X_3 = \text{Pounds of particulates reduced}$$

The relatively high reduction in particulates reflects not only the fact that 0.944 ton of the 1.8% sulfur coal is burned in place of 1 ton, but also the lower ash content of the substituted coal. (Reduction coefficients are not always positive; low-sulfur coal in a pulverized coal boiler that is equipped with a high-efficiency electrostatic precipitator can cause an increase in particulate emissions. The presence of less sulfur dioxide in the flue gas reduces the chargeability of the particles so that the benefits of the lower ash and higher BTU content may be offset by the reduced efficiency of the electrostatic precipitator.)

The mathematical programming model for 1970 is shown below. Notice that control methods X_1 and X_2 for the cement industry are included (see Chapter 3), as well as control method X_3. The dots represent the remaining 200–300 control methods.

Minimize

$$C = \$0.14X_1 + \$0.18X_2 + \$2.50X_3 + \cdots$$

subject to

$$X_1 + X_2 \leq \qquad 2{,}500{,}000$$
$$X_3 + \cdots \leq \qquad 200{,}000$$
$$\vdots \qquad \vdots$$

$$53.2X_3 + \cdots \geq \quad 485{,}000{,}000 \text{ lb of sulfur dioxide}$$
$$0.2X_3 + \cdots \geq 1{,}300{,}000{,}000 \text{ lb of carbon monoxide}$$
$$0.1X_3 + \cdots \geq \quad 280{,}000{,}000 \text{ lb of hydrocarbons}$$
$$1.1X_3 + \cdots \geq \quad 75{,}000{,}000 \text{ lb of nitrogen oxides}$$
$$1.5X_1 + 1.8X_2 + \cdots \geq \quad 180{,}000{,}000 \text{ lb of particulates}$$
$$X_1, X_2, X_3, \cdots \geq 0$$

The pollution reduction requirements mentioned above appear in the model. In summing the pollutant reductions contributed by the various control methods, we are assuming that all pounds of any pollutant are homogeneous, regardless of where or when they are emitted. This is a limitation of the model because it is dependent on a close correspondence between a pollutant reduction and a specific concentration measured in parts per million or micrograms per cubic meter of that pollutant in the ambient air.

However, where necessary, meteorological sophistication can be incorporated in the model by selective weighting of those sources that seem to have a greater or lesser proportional effect on air quality than others.

Exercise 4.4   *Answers to Odd-Numbered Problems begin on page A-16.*

A   In Problems 1 – 14 use the two-phase method.

1. Maximize

$$P = 4x_1 - 10x_2$$

subject to

$$x_1 \geq 0, \qquad x_2 \geq 0$$
$$x_1 - 4x_2 \geq 4$$
$$2x_1 - x_2 \leq 10$$

2. Maximize

$$P = 2x_1 + x_2$$

subject to

$$x_1 \geq 0, \qquad x_2 \geq 0$$
$$x_1 + 2x_2 \leq 10$$
$$-x_1 + x_2 \geq 2$$

3. Maximize

$$P = 5x_1 + 10x_2$$

subject to

$$x_1 \geq 0, \qquad x_2 \geq 0$$
$$x_1 + x_2 \leq 10$$
$$2x_1 + 3x_2 \geq 12$$

4. Maximize

$$P = 2x_1 + 2x_2$$

subject to

$$x_1 \geq 0, \qquad x_2 \geq 0$$
$$5x_1 + 3x_2 \leq 30$$
$$2x_1 + x_2 \geq 10$$

5. Maximize

$$P = 3x_1 + x_2$$

6. Maximize

$$P = 6x_1 + 2x_2$$

subject to

$$x_1 \geq 0, \quad x_2 \geq 0$$
$$x_1 + x_2 \leq 10$$
$$x_1 + x_2 \geq 5$$

subject to

$$x_1 \geq 0, \quad x_2 \geq 0$$
$$3x_1 + 2x_2 \leq 12$$
$$3x_1 + x_2 \geq 3$$

7. Minimize

$$C = 3x_1 + 2x_2$$

subject to

$$x_1 \geq 0, \quad x_2 \geq 0$$
$$4x_1 + x_2 \leq 200$$
$$2x_1 + x_2 \geq 150$$

8. Minimize

$$C = 2x_1 + 3x_2$$

subject to

$$x_1 \geq 0, \quad x_2 \geq 0$$
$$x_1 + 3x_2 \leq 60$$
$$3x_1 + 5x_2 \geq 180$$

9. Minimize

$$C = 6x_1 - 2x_2$$

subject to

$$x_1 \geq 0, \quad x_2 \geq 0$$
$$x_1 + x_2 \leq 10$$
$$3x_1 + 2x_2 \geq 24$$

10. Minimize

$$C = 4x_1 + 3x_2$$

subject to

$$x_1 \geq 0, \quad x_2 \geq 0$$
$$x_1 + 2x_2 \leq 40$$
$$2x_1 + x_2 \geq 50$$

11. Minimize

$$C = x_1 + 3x_2$$

subject to

$$x_1 \geq 0, \quad x_2 \geq 0$$
$$2x_1 + x_2 \leq 15$$
$$x_1 + x_2 \geq 12$$

12. Minimize

$$C = 3x_1 + 2x_2$$

subject to

$$x_1 \geq 0, \quad x_2 \geq 0$$
$$x_1 + 2x_2 \leq 15$$
$$2x_1 + 3x_2 \geq 24$$

13. Maximize

$$P = 3x_1 + 4x_2$$

subject to

$$x_1 \geq 0, \quad x_2 \geq 0$$
$$x_1 + x_2 \leq 12$$
$$5x_1 + 2x_2 \geq 36$$
$$7x_1 + 4x_2 \geq 14$$

14. Maximize

$$P = 5x_1 + 2x_2$$

subject to

$$x_1 \geq 0, \quad x_2 \geq 0$$
$$x_1 + x_2 \geq 11$$
$$2x_1 + 3x_2 \geq 24$$
$$x_1 + 3x_2 \leq 18$$

B   In Problems 15–16 use the two-phase method.

15. Minimize

$$C = 2x_1 + 3x_2$$

subject to

$$x_1 \geq 0, \quad x_2 \geq 0$$
$$2x_1 + x_2 \leq 20$$
$$2x_1 + x_2 \geq 10$$
$$x_1 + 2x_2 \geq 8$$

16. Minimize

$$C = 20x_1 + 30x_2$$

subject to

$$x_1 \geq 0, \quad x_2 \geq 0$$
$$2x_1 + 10x_2 \geq 80$$
$$6x_1 + 2x_2 \leq 72$$
$$3x_1 + 2x_2 \geq 6$$

C   In Problems 17–28 use the two-phase method

17. Maximize

$$P = 3x_1 + 2x_2 - x_3$$

subject to

$$x_1 \geq 0, \quad x_2 \geq 0, \quad x_3 \geq 0$$
$$x_1 + 3x_2 + x_3 \leq 9$$
$$2x_1 + 3x_2 - x_3 \geq 2$$
$$3x_1 - 2x_2 + x_3 \geq 5$$

18. Maximize

$$P = 3x_1 + 2x_2 - x_3$$

subject to

$$x_1 \geq 0, \quad x_2 \geq 0, \quad x_3 \geq 0$$
$$2x_1 - x_2 - x_3 \leq 2$$
$$x_1 + 2x_2 + x_3 \geq 2$$
$$x_1 - 3x_2 - 2x_3 \leq -5$$

19. Minimize

$$C = 6x_1 + 8x_2 + x_3$$

subject to

$$x_1 \geq 0, \quad x_2 \geq 0, \quad x_3 \geq 0$$
$$3x_1 + 5x_2 + 3x_3 \geq 20$$
$$x_1 + 3x_2 + 2x_3 \geq 9$$
$$6x_1 + 2x_2 + 5x_3 \geq 30$$
$$x_1 + x_2 + x_3 \leq 10$$

20. Minimize

$$C = 2x_1 + x_2 + x_3$$

subject to

$$x_1 \geq 0, \quad x_2 \geq 0, \quad x_3 \geq 0$$
$$3x_1 - x_2 - 4x_3 \leq -12$$
$$x_1 + 3x_2 + 2x_3 \geq 10$$
$$x_1 - x_2 + x_3 \leq 8$$

21. Maximize

$$P = 3x_1 + 2x_2$$

subject to

$$x_1 \geq 0, \quad x_2 \geq 0$$
$$2x_1 + x_2 \leq 4$$
$$x_1 + x_2 = 3$$

22. Maximize

$$P = 6x_1 + 4x_2$$

subject to

$$x_1 \geq 0, \quad x_2 \geq 0$$
$$x_1 + 2x_2 \geq 5$$
$$3x_1 + 4x_2 = 12$$

23. Minimize

$$C = 5x_1 + 4x_2$$

subject to

$$x_1 \geq 0, \quad x_2 \geq 0$$
$$x_1 + x_2 \geq 10$$
$$2x_1 - x_2 = 14$$

24. Minimize

$$C = 3x_1 + 4x_2$$

subject to

$$x_1 \geq 0, \quad x_2 \geq 0$$
$$x_1 + x_2 \geq 20$$
$$x_1 - 2x_2 = 12$$

25. Maximize

$$P = 8x_1 + 12x_2 + 20x_3$$

subject to

$$x_1 \geq 0, \quad x_2 \geq 0, \quad x_3 \geq 0$$
$$x_1 + 3x_2 + 5x_3 \leq 3$$
$$x_1 + x_2 + x_3 = 1$$

26. Maximize

$$P = 2x_1 + x_2 - x_3$$

subject to

$$x_1 \geq 0, \quad x_2 \geq 0, \quad x_3 \geq 0$$
$$x_1 + x_2 + x_3 = 12$$
$$-x_1 + x_2 \geq 1$$

27. Minimize

$$C = 6x_1 + 6x_2 + 3x_3 - 8x_4$$

28. Maximize

$$P = 45x_1 + 27x_2 + 18x_3 + 36x_4$$

subject to

$$x_1 \geq 0, \quad x_2 \geq 0, \quad x_3 \geq 0, \quad x_4 \geq 0$$
$$4x_1 - x_2 + 2x_3 + 2x_4 \leq 4$$
$$x_1 + 2x_2 + 2x_3 - 3x_4 \geq 3$$
$$3x_1 - 3x_3 + 3x_4 \geq 2$$
$$6x_1 - 2x_2 + 4x_4 = 3$$

subject to

$$x_1 \geq 0, \quad x_2 \geq 0, \quad x_3 \geq 0, \quad x_4 \geq 0$$
$$5x_1 + x_2 + x_3 + 8x_4 = 30$$
$$2x_1 + 4x_2 + 3x_3 + 2x_4 = 30$$

APPLICATIONS

29. Shipping Schedule Private Motors, Inc., has two plants, M1 and M2, which manufactures engines; the company also has two assembly plants, A1 and A2, which assemble the cars. M1 can produce at most 600 engines per week. M2 can produce at most 400 engines per week. A1 needs at least 500 engines per week and A2 needs at least 300 engines per week. The following is a table of charges to ship engines to assembly plants.

	A1	A2
M1	$400	$100
M2	$200	$300

How many engines should be shipped each week to each assembly plant from M1? M2? [*Hint:* Consider four variables: x_1 = number of units shipped from M1 to A1, x_2 = number of units shipped from M1 to A2, x_3 = number of units shipped from M2 to A1, and x_4 = number of units shipped from M2 to A2.]

30. Production Quality Oak Tables, Inc., has an individual who does all its finishing work, and it wishes to use him in this capacity at least 36 hours each week. The assembly area can be used at most 48 hours each week. The company has three models of oak tables T1, T2, T3. T1 requires 1 hour for assembly, 2 hours for finishing, and 9 board feet of oak. T2 requires 1 hour for assembly, 1 hour for finishing, and 9 board feet of oak. T3 requires 2 hours for assembly, 1 hour for finishing, and 3 board feet of oak. If we wish to minimize the board feet of oak used, how many of each model should be made?

31. Shipping Schedule A television manufacturer must fill orders from two retailers. The first retailer, R_1, has ordered 55 television sets, while the second retailer, R_2, has ordered 75 sets. The manufacturer has the television sets stored in two warehouses, W_1, and W_2. There are 100 sets in W_1 and 120 sets in W_2. The shipping costs per television set are: $8 from W_1 to R_1; $12 from W_1 to R_2; $13 from W_2 to R_1; $7 from W_2 to R_2. Find the number of television sets to be shipped from each warehouse to each retailer if the total shipping cost is to be a minimum. What is this minimum cost?

32. Shipping Schedule A motorcycle manufacturer must fill orders from two dealers. The first dealer, D_1, has ordered 20 motorcycles, while the second dealer, D_2, has ordered 30 motorcycles. The manufacturer has the motorcycles stored in two warehouses, W_1 and W_2. There are 40 motorcycles in W_1 and 15 in W_2. The shipping costs per motorcycle are as follows: $15 from W_1 to D_1; $13 from W_1 to D_2; $14 from W_2 to D_1; $16 from W_2 to D_2. Under these conditions, find the number of motorcycles to be shipped from each warehouse to each dealer if the total shipping cost is to be held to a minimum. What is this minimum cost?

33. Production Control RCA manufacturing received an order for a machine. The machine is to weigh 150 pounds. The two raw materials used to produce the machine are A, with a cost of $4 per unit, and B, with a cost of $8 per unit. At least 14 units of B and no more than 20 units of A must be used. Each unit of A weighs 5 pounds; each unit of B weighs 20 pounds. How much of each type of raw material should be used for each unit of final product if we wish to minimize cost. [*Hint:* The problem has two solutions (0, 15) and (2, 14) with $C = 120$.]

CHAPTER REVIEW

Important Terms

maximum linear programming problem
standard form
slack variables
initial simplex tableau
objective row
pivot operation

pivot element
pivot column
pivot row
final tableau
simplex method
standard form of a minimum problem

dual problem
duality principle
mixed constraints
Phase I/Phase II

True–False Questions

(Answers on page A-16)

T F   1. In a maximum problem written in standard form each of the constraints, with the exception of the nonnegativity constraints, are written with a $\leq$ symbol.

T F   2. In a maximum problem written in standard form the slack variables are sometimes negative.

T F   3. Once the pivot element is identified in a tableau, the pivot operation causes the pivot element to become a 1 and causes the remaining entries in the pivot column to become 0's.

T F   4. The pivot element is sometimes in the objective row.

T F   5. One way to solve a minimum problem is to first solve its dual, which is a maximum problem.

Fill in the Blanks

(Answers on page A-16)

1. The constraints of a maximum problem in standard form are changed from an inequality to an equation by introducing _____ _____ .

2. The pivot _____ in a standard-form linear programming problem is located by selecting the most negative entry in the objective row.

3. For a minimum problem to be in standard form all the constraints must be written with _____ signs.

4. When the rows of the special matrix of a minimum problem are the columns of the special matrix of a maximum problem, we say the problems are _____ .

5. The _____ _____ _____ principle states that the optimal solution of a minimum linear programming problem, if it exists, has the same value as the optimal solution of the maximum problem, which is its dual.

Review Exercises

A

Answers to Odd-Numbered Problems begin on page A-16.

1. For each simplex tableau below state whether it is a final tableau, or if additional privoting is needed, or if the problem has no solution.

$$
\begin{array}{ccccc}
x_1 & x_2 & s_1 & s_2 & p \\
\end{array}
$$

(a) $\left[\begin{array}{ccccc|c}
1 & 1 & 0 & 0 & 0 & 3 \\
0 & 3 & 1 & 1 & 0 & 2 \\
\hline
0 & -4 & 2 & 3 & 1 & 15
\end{array}\right]$

(b) $\left[\begin{array}{ccccc|c}
1 & 0 & -1 & 1 & 0 & 3 \\
0 & 1 & 3 & 2 & 0 & 2 \\
\hline
0 & 0 & 4 & 3 & 1 & 15
\end{array}\right]$

(c) $\left[\begin{array}{ccccc|c}
-1 & 1 & 0 & 2 & 0 & 4 \\
0 & 0 & 1 & 1 & 0 & 1 \\
\hline
-1 & 1 & 4 & 3 & 1 & 12
\end{array}\right]$

Use the simplex method to solve the following linear programming problems.

2. Maximize
$$P = 0.10x_1 + 0.08x_2$$
subject to
$$x_1 \geq 0, \quad x_2 \geq 0$$
$$x_1 + x_2 \leq 20{,}000$$
$$x_1 \leq \tfrac{1}{2}x_2$$

3. Minimize
$$C = 2x_1 + x_2$$
subject to
$$2x_1 + 2x_2 \geq 8$$
$$x_1 - x_2 \geq 2 \quad x_1 \geq 0, \quad x_2 \geq 0$$

4. Minimize
$$C = 4x_1 + 2x_2$$
subject to
$$x_1 + 2x_2 \geq 4$$
$$x + 4x_2 \geq 6$$
$$x_1 \geq 0, \quad x_2 \geq 0$$

B

5. Maximize
$$P = 100x_1 + 200x_2 + 50x_3$$
subject to the constraints
$$5x_1 + 5x_2 + 10x_3 \leq 1000$$
$$10x_1 + 8x_2 + 5x_3 \leq 2000$$
$$10x_1 + 5x_2 \leq 500$$
$$x_1 \geq 0, \quad x_2 \geq 0, \quad x_3 \geq 0$$

6. Maximize
$$P = x_1 + 2x_2 + x_3$$
subject to the constraints
$$3x_1 + x_2 + x_3 \leq 3$$
$$x_1 - 10x_2 - 4x_3 \leq 20$$
$$x_1 \geq 0, \quad x_2 \geq 0, \quad x_3 \geq 0$$

7. Maximize

$$P = 40x_1 + 60x_2 + 50x_3$$

subject to the constraints

$$2x_1 + 2x_2 + x_3 \le 8$$
$$x_1 - 4x_2 + 3x_3 \le 12$$
$$x_1 \ge 0, \quad x_2 \ge 0, \quad x_3 \ge 0$$

8. Maximize

$$P = 2x_1 + 8x_2 + 10x_3 + x_4$$

subject to the constraints

$$x_1 + 2x_2 + x_3 + x_4 \le 50$$
$$3x_1 + x_2 + 2x_3 + x_4 \le 100$$
$$x_1 \ge 0, \quad x_2 \ge 0, \quad x_3 \ge 0, \quad x_4 \ge 0$$

9. Minimize

$$C = 5x_1 + 4x_2 + 3x_3$$

subject to

$$x_1 + x_2 + x_3 \ge 100$$
$$2x_1 + x_2 \ge 50$$
$$x_1 \ge 0, \quad x_2 \ge 0, \quad x_3 \ge 0$$

10. Minimize

$$C = 2x_1 + x_2 + 3x_3 + x_4$$

subject to

$$x_1 + x_2 + x_3 + x_4 \ge 50$$
$$3x_1 + x_2 + 2x_3 + x_4 \ge 100$$
$$x_1 \ge 0, \quad x_2 \ge 0, \quad x_3 \ge 0, \quad x_4 \ge 0$$

11. Maximize

$$P = 2x_1 + 3x_2$$

subject to

$$x_1 \ge 0, \quad x_2 \ge 0$$
$$2x_1 + x_2 \le 75$$
$$x_1 + x_2 \le 50$$

12. Minimize

$$C = 2x_1 + x_2$$

subject to

$$x_1 \ge 0, \quad x_2 \ge 0$$
$$x_1 + 2x_2 \le 40$$
$$2x_1 + x_2 \ge 50$$

13. Maximize

$$P = 300x_1 + 200x_2 + 450x_3$$

subject to

$$x_1 \ge 0, \quad x_2 \ge 0, \quad x_3 \ge 0$$
$$4x_1 + 3x_2 + 5x_3 \le 140$$
$$x_1 + x_2 + x_3 = 30$$

APPLICATIONS

14. Scheduling An automobile manufacturer must fill orders from two dealers. The first dealer, D_1, has ordered 40 cars, while the second dealer, D_2, has ordered 25 cars. The manufacturer has the cars stored in two locations, W_1 and W_2. There are 30 cars in W_1 and 50 cars in W_2. The shipping costs per car are as follows: $180 from W_1 to D_1; $150 from W_1 to D_2; $160 from W_2 to D_1; $170 from W_2 to D_2. Under these conditions, how many cars should be shipped from each storage location to each dealer so as to minimize the total shipping costs? What is this minimum cost?

15. Optimal Land Use A farmer has 1000 acres of land on which corn, wheat, or soybeans can be grown. Each acre of corn costs $100 for preparation, requires 7 days of labor, and yields a profit of $30. An acre of wheat costs $120 to prepare, requires 10 days of labor, and yields $40 profit. An acre of soybeans costs $70 to prepare, requires 8 days of labor, and yields $40 profit. If the farmer has $10,000 for preparation, and can count on enough workers to supply 8000 days of labor, how many acres should be devoted to each crop to maximize profits?

16. Mixture A brewery manufactures three types of beer—lite, regular, and dark. Each vat of lite beer requires 6 bags of barley, 1 bag of sugar, and 1 bag of hops. Each vat of regular beer requires 4 bags of barley, 3 bags of sugar, and 1 bag of hops. Each vat of dark beer requires 2 bags of barley, 2 bags of sugar, and 4 bags of hops. Each day the brewery has 800 bags of barley, 600 bags of sugar, and 300 bags of hops available. The brewery realizes a profit of $10 per vat of lite beer, $20 per vat of regular beer, and $30 per vat of dark beer. How many vats of lite, regular, and dark beer should be brewed in order to maximize profits? What is the maximum profit?

Mathematical Questions

From CPA Exams (Answers on page A-17)

Use the following information to answer Problems 1–4:

The Ball Company manufactures three types of lamps, which are labeled A, B, and C. Each lamp is processed in two departments—I and II. Total available man-hours per day for departments I and II are 400 and 600, respectively. No additional labor is available. Time requirements and profit per unit for each lamp type are as follows:

	A	B	C
Man-hours required in Department I	2	3	1
Man-hours required in Department II	4	2	3
Profit per unit (Sales price less all variable costs)	$5	$4	$3

The company has assigned you, as the accounting member of its profit planning committee, to determine the number of types of A, B, and C lamps that it should produce in order to maximize its total profit from the sale of lamps. The following questions relate to a linear programming model that your group has developed.

1. The coefficients of the objective function would be
    (a) 4, 2, 3   (b) 2, 3, 1
    (c) 5, 4, 3   (d) 400, 600

2. The constraints in the model would be
    (a) 2, 3, 1   (b) 5, 4, 3
    (c) 4, 2, 3   (d) 400, 600

3. The constraint imposed by the available man-hours in Department I could be expressed as
    (a) $4X_1 + 2X_2 + 3X_3 \leq 400$ (b) $4X_1 + 2X_2 + 3X_3 \geq 400$
    (c) $2X_1 + 3X_2 + 1X_3 \leq 400$ (d) $2X_1 + 3X_2 + 1X_2 \geq 400$

4. The most types of lamps that would be included in the optimal solution would be
    (a) 2   (b) 1
    (c) 3   (d) 0

5. In a system of equations for a linear programming model, what can be done to equalize an inequality such as $3X + 2Y \leq 15$?
    (a) Nothing.   (b) Add a slack variable.
    (c) Add a tableau. (d) Multiply each element by -1.

Use the following information to answer Problems 6 and 7:

The Golden Hawk Manufacturing Company wants to maximize the profits on products A, B, and C. The contribution margin for each product follows:

Product	Contribution Margin
A	$2
B	$5
C	$4

The production requirements and departmental capacities, by departments, are as follows:

Department	Production Requirements by Product (Hours)		
	A	B	C
Assembling	2	3	2
Painting	1	2	2
Finishing	2	3	1

Department	Departmental Capacity (Total Hours)
Assembling	30,000
Painting	38,000
Finishing	28,000

6. What is the profit maximization formula for the Golden Hawk Company?
   (a) $2A + $5B + $4C = X$ (where X = Profit)
   (b) $5A + 8B + 5C \leq 96,000$
   (c) $2A + $5B + $4C \leq X$ (where X = Profit)
   (d) $2A + $5B + $4C = 96,000$

7. What is the constraint for the Painting Department of the Golden Hawk Company?
   (a) $1A + 2B + 2C \geq 38,000$ (b) $2A + $5B + $4C \geq 38,000$
   (c) $1A + 2B + 2C \leq 38,000$ (d) $2A + 3B + 2C \leq 30,000$

8. Watch Corporation manufactures products A, B, and C. The daily production requirements are shown below.

Product	Profit per Unit	Hours Required per Unit per Department		
		Machining	Plating	Polishing
A	$10	1	1	1
B	$20	3	1	2
C	$30	2	3	2
Total Hours per Day per Department		16	12	6

What is Watch's objective function in determining daily production of each unit?
   (a) $A + B + C \leq 60
   (b) $3A + $6B + $7C = 60
   (c) $A + B + C \leq$ Profit
   (d) $10A + $20B + $30C =$ Profit

Questions 9 – 11 are based on a company, that uses a linear programming model to schedule the production of three products. The per-unit selling prices, variable costs, and labor time required to produce these products are presented below. Total labor time available is 200 hours.

Product	Selling Price	Variable Cost	Labor Hours
A	$4.00	$1.00	2
B	$2.00	$ .50	2
C	$3.50	$1.50	3

9. The objective function to maximize the company's gross profit (Z) is
   (a) $4A + 2B + 3.5C = Z$
   (b) $2A + 2B + 3C = Z$
   (c) $5A + 2.5B + 5C = Z$
   (d) $3A + 1.5B + 2C = Z$
   (e) $A + B + C = Z$

10. The constraint of labor time available is represented by
    (a) $2A + 2B + 3C \leq 200$
    (b) $2A + 2B + 3C \geq 200$
    (c) $A + B + C \geq 200$
    (d) $4A + 2B + 3.5C = 200$
    (e) $A/2 + B/2 + C/3 = 200$

11. A linear programming model produces an optimal solution by
    (a) Ignoring resource constraints
    (b) Minimizing production costs
    (c) Minimizing both variable production costs and labor costs
    (d) Maximizing the objective function subject to resource constraints
    (e) Finding the point at which various resource constraints intersect

5

FINANCE

5.1

SIMPLE INTEREST AND SIMPLE DISCOUNT

INTRODUCTION

Although Table 1 in Appendix II will be useful in solving many of the examples and problems in this chapter, you will often find examples and problems in the text for which Table 1 is not extensive enough to be of use. For these problems the use of a financial or a scientific calculator is required. We therefore recommend the purchase of such a calculator or, at least, a calculator with the keys $+, -, \times, \div$, and y^x.

In Appendix I we discuss arithmetic and geometric progressions. These progressions will shed light on the manner in which many of the formulas are derived. We will alert you to these formulas whenever possible.

Very simply, *interest* is money paid for the use of money. You may also think of interest as "rented money" because you pay interest on borrowed money. Interest is the fee you pay to use or "rent" money for a certain period of time the same way you would pay the landlord rent for the use of an apartment.

The total amount of money borrowed, whether by an individual from a bank in the form of a loan or by a bank from an individual in the form of a savings account, is called the *principal.*

The *rate of interest* is the amount charged for the use of the principal for a given length of time, usually on a yearly, or *per annum,* basis. Rates of interest are usually expressed as percentages: 10% per annum, 14% per annum, and so on. However, when using rates of interest in calculations, we use the decimal equivalent: 0.10 for 10%; 0.14 for 14%, and so on.

SIMPLE INTEREST

The easiest type of interest to deal with is called *simple interest.* It is most often used for loans of short duration (less than 1 year).

Simple Interest **Simple interest is interest computed on the principal for the entire period it is borrowed.**

> If a *principal P* is borrowed at a simple interest rate of *r* per annum (where *r* is a decimal) for a period of *t* years, the interest charge *I* is
>
> $$I = Prt \tag{1}$$

Example 1

A loan of $250 is made for 9 months at a simple interest rate of 10% per annum. What is the interest charge?

Solution

The actual period the money is borrowed for is 9 months, which is $\frac{3}{4}$ of a year. The interest charge is the product of the amount borrowed, $250, the annual rate of interest, 0.10, and the length of time in years, $\frac{3}{4}$. Thus,

$$I = Prt$$

$$\text{Interest charge} = (\$250)(0.10)\left(\frac{3}{4}\right) = \$18.75 \qquad ■$$

Suppose a principal of P dollars is invested at the simple interest rate of r per annum, then at the end of each year the investment earns rP dollars in interest. Therefore we have the following:

If A denotes the total value of the investment at the end of t years, then

$$A = \text{original principal} + \text{interest}$$
$$= P + Prt$$
$$= P(1 + rt)$$

We now have the simple interest formula, relating A and P:

> The *amount* A owed at the end of t years is
> $$A = P + Prt = P(1 + rt) \qquad (2)$$

P is often referred to as the *present value* and A as the *future value*.

Example 2

A person borrows $1000 for a period of 6 months. What simple interest rate is being charged if the amount A that must be repaid after 6 months is $1045?

Solution

The principal P is $1000 and the period is $\frac{1}{2}$ year (6 months), the amount A owed after 6 months is $1045. We seek r in the equation:

$$A = P + Prt$$

$$1045 = 1000 + 1000r\left(\frac{1}{2}\right)$$

$$45 = 500r$$

$$r = \frac{45}{500} = 0.09$$

The per annum rate of interest is 9%. ■

Example 3

A bank borrows $1,000,000 for 1 month at a simple interest rate of 9% per annum. How much must the bank pay back at the end of 1 month?

Solution

The principal P is $1,000,000, the period t is $\frac{1}{12}$ year and the rate r is 0.09.

$$A = P(1 + rt)$$

$$A = 1,000,000 \left[1 + 0.09 \left(\frac{1}{12} \right) \right]$$

$$= 1,000,000(1.0075)$$

$$= \$1,007,500 \qquad \blacksquare$$

SIMPLE DISCOUNT

If a lender deducts the interest at the time the loan is made, the loan is said to be *discounted*. The interest discounted from a loan is referred to as the *simple discount*. The amount the borrower receives is called the *proceeds,* and the amount to be repaid is called the *maturity value.*

The formula for the simple discount is

$$d = Art$$

and the formula for the proceeds is

$$P = A - d = A - Art = A(1 - rt)$$

where

$$d = \text{simple discount}$$
$$A = \text{maturity value}$$
$$r = \text{simple discount rate}$$
$$t = \text{time in years}$$
$$P = \text{proceeds, the amount the borrower receives}$$

Example 4

A borrower signs a note and agrees to pay $1000 in 9 months at 10% simple discount. How much does this borrower receive?

Solution

Here $A = 1000$, $r = .10$, and $t = \frac{9}{12}$. The discount d is given by

$$d = Art = \$1000(0.10) \left(\frac{9}{12} \right) = \$75$$

This amount is deducted from the maturity value of $1000; that is, the proceeds are

$$P = A - Art = 1000 - 75 = \$925 \qquad \blacksquare$$

Example 5

What simple interest rate is the borrower in Example 4 paying on the $925 that was borrowed for 9 months?

Solution

The principal P is $925, t is $\frac{3}{4}$ of a year, and the amount A is $1000.

$$A = P + Prt$$

$$1000 = 925 + 925r\left(\frac{3}{4}\right)$$

$$75 = 693.75r$$

$$r = \frac{75}{693.75} = 0.108108$$

The per annum rate of interest is 10.81%. ∎

Example 6

You wish to borrow $10,000 for 3 months. If the person you are borrowing from offers an 8% simple discount, how much must you repay at the end of 3 months?

Solution

The amount P you borrow is $10,000, the discount rate r is 0.08, and the time t is $\frac{1}{4}$ year.

$$P = A(1 - rt)$$

$$10,000 = A\left[1 - 0.08\left(\frac{1}{4}\right)\right]$$

$$10,000 = 0.98A$$

$$A = \frac{10,000}{0.98}$$

$$= \$10,204.08$$ ∎

TREASURY BILLS

Treasury bills (T-Bills) are short-term securities issued by the Federal Reserve. The bills do not specify a rate of interest. They are sold at public auction with financial institutions making competitive bids. For example, a financial institution may bid $982,400 for a 60-day $1 million treasury bill. At the end of 60 days the financial institution receives $1 million, which covers the interest earned and the cost of the T-bill. This is basically a simple discount transaction.

Example 7

How much should a bank bid to earn 7.65% simple discount interest on a 180-day $1 million treasury bill?

Solution

We are given the maturity value $A = 1,000,000$, the discount rate $r = 0.0765$, and $t = \frac{180}{360} = \frac{1}{2}$.

Note: Some financial institutions use a 365-day year and others a 360-day year. In this section we use a 360-day year.

Substituting these in

$$P = A(1 - rt)$$

We get

$$P = 1,000,000[1 - 0.0765(\tfrac{1}{2})]$$
$$= 1,000,000(0.96175)$$
$$= 961,750$$

The bank should bid $961,750.   ■

Exercise 5.1

Answers to Odd-Numbered Problems begin on page A-17.

A   In Problems 1–4 use Formula (1) to find the indicated quantity.

  1. $P = \$100$; $r = 10\%$; $t = 12$ months; $I = ?$

  2. $P = \$500$; $r = 12\%$; $t = 6$ months; $I = ?$

  3. $P = \$400$; $r = ?$; $t = 9$ months; $I = \$30$

  4. $P = \$600$; $r = ?$; $t = 18$ months; $I = \$108$

In Problems 5–8 use Formula (2) to find the indicated quantity.

  5. $P = \$100$; $r = 10\%$; $t = 12$ months; $A = ?$

  6. $P = \$500$; $r = 12\%$; $t = 6$ months; $A = ?$

  7. $P = \$400$; $r = ?$; $t = 9$ months; $A = \$430$

  8. $P = \$600$; $r = ?$; $t = 18$ months; $A = \$708$

B   In Problems 9–14 find the interest due on each loan.

  9. $1000 is borrowed for 3 months at 10% simple interest.

 10. $100 is borrowed for 6 months at 8% simple interest.

 11. $500 is borrowed for 9 months at 12% simple interest.

 12. $800 is borrowed for 8 months at 12% simple interest.

 13. $1000 is borrowed for 18 months at 10% simple interest.

 14. $100 is borrowed for 24 months at 12% simple interest.

In Problems 15–20 find the simple interest rate for each loan.

 15. $1000 is borrowed; the amount owed after 6 months is $1050.

 16. $500 is borrowed; the amount owed after 8 months is $600.

 17. $300 is borrowed; the amount owed after 12 months is $400.

18. $600 is borrowed; the amount owed after 9 months is $660.

19. $900 is borrowed; the amount owed after 10 months is $1000.

20. $800 is borrowed; the amount owed after 3 months is $900.

C   In Problems 21–26 use the given formula to solve for the indicated variable.

21. $I = Prt$; for P

22. $I = Prt$; for t

23. $A = P(1 + rt)$; for P

24. $A = P(1 + rt)$; for t

25. $P = A(1 - rt)$; for A

26. $P = A(1 - rt)$; for t

APPLICATIONS   In Problems 27–30 find the proceeds for the given amount repaid, time, and simple discount rate.

27. $1200 repaid in 6 months with a simple discount rate of 10%.

28. $500 repaid in 8 months with a simple discount rate of 9%.

29. $2000 repaid in 24 months with a simple discount rate of 8%.

30. $1500 repaid in 18 months with a simple discount rate of 10%.

In Problems 31–34 find the amount you must repay.

31. You borrow $1200 for 6 months at a simple discount rate of 10%.

32. You borrow $500 for 8 months at a simple discount rate of 9%.

33. You borrow $2000 for 24 months at a simple discount rate of 8%.

34. You borrow $1500 for 18 months at a simple discount rate of 10%.

In Problems 35 and 36 determine at which rate the borrower pays the least interest.

35. A loan for 6 months using a simple discount rate of 9% or using a simple interest rate of 10%.

36. A loan for 9 months using a simple discount rate of 8% or using a simple interest rate of $8\frac{1}{2}$%.

37. Wendy wants to buy a $500 stereo set in 9 months. How much should she invest at 8% simple interest to have the money then?

38. A sum of $10,000 is borrowed for a period of 3 years at the rate of 10%. Determine the total interest paid.

39. A sum of $20,000 is borrowed for a period of 2 years at simple interest. The rate is 8.5% per annum to be paid semiannually.

(a)   Determine the interest that is to be paid semiannually.

(b)   Compute the total interest that will be paid at the end of the 2 years.

40. Tami borrowed $600 at 8% interest. The amount of interest paid was $156. What was the length of the loan?

41. Marci took a $10,000 loan to buy a car, at an interest rate of 12% paid over 5 years. Her monthly payments are $266.67. How much of her payment goes for interest and how much for principal?

42. The owner of a restaurant would like to borrow $12,000 from a bank to buy some equipment. The bank will give the owner a simple discount loan at 11% discount rate for 9 months. What maturity value should be used so that the owner will receive $12,000?

43. Joan needs $850 to pay some of her bills. She has a choice of either a simple interest or a simple discount loan at 12% for 4 months. What should she do?

44. Jay would like to borrow $2,000 for one year from a bank. He is given a choice of a simple interest loan at 12.3% or a discount loan at 12.1%. What should he do?

45. To meet its expenditure a village sold $50,000 in municipal bonds. The bonds will mature in 10 years paying an interest of 6% per annum. Determine the total interest paid.

46. A bank wants to earn 8.5% simple discount interest on a 60-day $1 million treasury bill. How much should they bid?

47. A bank paid $979,000 for a 90-day $1 million treasury bill. What was the simple discount rate?

48. How much should a bank bid on a 60-day $3 million treasury bill to receive a 7.715% interest rate?

5.2

COMPOUND INTEREST

COMPOUND INTEREST FORMULA
EFFECTIVE RATE OF INTEREST
PRESENT VALUE

COMPOUND INTEREST FORMULA

If the interest due at the end of a unit payment period is added to the principal, so that the interest computed for the next unit payment period is based on this new principal amount (old principal plus interest), then the interest is said to have been *compounded.* That is, *compound interest* is interest paid on previously earned interest.

Suppose an amount of $1000 is deposited in a savings account where interest is compounded quarterly at an annual rate of 8%. What is the total amount accumulated by the end of 1 year, assuming that no deposits or withdrawals were made during the year? Compounded quarterly means that interest is added to the principal at the end of every 3 months, and therefore the interest itself earns interest. Thus, using the simple interest formula of the previous section, the amount saved at the end of the first quarter is

$$A = P(1 + rt)$$
$$= 1000 \left[1 + 0.08 \left(\frac{1}{4} \right) \right]$$
$$= 1000(1.02) = \$1020$$

Now, the new principal is $1020, so that at the end of the second quarter the amount of savings accumulated is

$$A = 1020 \left[1 + 0.08 \left(\frac{1}{4} \right) \right]$$
$$= 1020(1.02) = \$1,040.40$$

At the end of the third quarter the accumulated savings are

$$A = 1040.40 \left[1 + 0.08 \left(\frac{1}{4} \right) \right]$$
$$= 1040.40(1.02) = \$1061.21$$

Finally, at the end of the fourth quarter the accumulated savings are

$$A = 1061.21 \left[1 + 0.08 \left(\frac{1}{4} \right) \right]$$
$$= 1061.21(1.02) = \$1082.43$$

Based on the four quarters of computation, let us see if we can identify how the amount of savings can be computed for any length of time.

$$A = 1000 \left[1 + 0.08 \left(\frac{1}{4} \right) \right] \qquad \text{Savings available at the end of first quarter}$$

$$A = 1000 \left[1 + 0.08 \left(\frac{1}{4} \right) \right] \left[1 + 0.08 \left(\frac{1}{4} \right) \right]$$

$$= 1000 \left[1 + 0.08 \left(\frac{1}{4} \right) \right]^2 \qquad \text{Savings at the end of second quarter}$$

$$A = 1000 \left[1 + 0.08 \left(\frac{1}{4} \right) \right]^2 \left[1 + 0.08 \left(\frac{1}{4} \right) \right]$$

$$= 1000 \left[1 + 0.08 \left(\frac{1}{4} \right) \right]^3 \qquad \text{Savings at the end of third quarter}$$

$$A = 1000 \left[1 + 0.08 \left(\frac{1}{4} \right) \right]^3 \left[1 + 0.08 \left(\frac{1}{4} \right) \right]$$

$$= 1000 \left[1 + 0.08 \left(\frac{1}{4} \right) \right]^4 \qquad \text{Savings at the end of fourth quarter}$$

Based on the above analysis, it appears that

$$A = 1000 \left(1 + \frac{0.08}{4} \right)^n \qquad \text{savings at the end of } n\text{th quarter}$$

Here $i = \dfrac{0.08}{4} = 0.02$ is the interest rate per quarter. Since interest rates are stated on an annual basis, to obtain the interest rate per compounding period we divide the annual rate by the number of compounding periods. The annual rate is sometimes referred to as the *nominal interest rate*.

Next we develop a formula for computing the amount when interest is compounded. Let P denote the principal that earns interest at an annual rate r compounded m times a year, and let i denote the interest rate for each compounding period, that is

$$i = \frac{r}{m}$$

Then the amount A accumulated at the end of the first compounding period will be:

$$A = P + iP = P(1 + i)$$

at the end of the second compounding period we will have accumulated

$$A = P(1 + i) + iP(1 + i) = P(1 + i)(1 + i) = P(1 + i)^2$$

at the end of the third compounding period we will have accumulated

$$A = P(1 + i)^2 + iP(1 + i)^2 = P(1 + i)^2(1 + i) = P(1 + i)^3$$

at the end of the nth compounding period we will have accumulated

$$A = P(1 + i)(1 + i)^{n-1} = P(1 + i)^n$$

Thus we are led to the compound interest formula:

$$A = P(1 + i)^n \qquad (1)$$

$$i = \frac{r}{m} = \frac{\text{annual rate of interest}}{\text{number of payment periods per year}}$$

where

P = amount of principal (present value)

r = annual rate of interest (nominal rate)

i = rate per compounding period

m = number of payment periods per year

n = total number of payment periods

A = accumulated amount at the end of n payment periods (future value)

The values of $(1 + i)^n$ can be computed by using a calculator with a y^x key or by referring to Table 1 in Appendix II.

Example 1

A bank pays 6% per annum compounded quarterly. If $200 is placed in a savings account and the quarterly interest is left in the account, how much money is in the account after 1 year?

Solution

Here $P = 200$, $i = \dfrac{0.06}{4} = 0.015$, and $n = 4$ (four payments per year)

Using (1) we get

$A = P(1 + i)^n$

$\quad = 200(1 + 0.015)^4$ Use calculator (or Table 1 in Appendix II)

$\quad = 200(1.061363)$

$\quad = \$212.27$ ■

Example 2

If $1000 is invested at an annual rate of interest of 10%, what is the amount after 5 years if the compounding takes place (a) annually, (b) quarterly, (c) monthly, or (d) daily? Find the interest earned in each case.

Solution

(a) Compounding annually means that there is one payment period per year. Thus, $n = 5$, and $i = r = 0.1$

$$A = P(1 + i)^n$$
$$\quad = 1000(1 + 0.1)^5 \qquad \text{Use calculator}$$
$$\quad = 1000(1.61051)$$
$$\quad = \$1610.51$$

Interest earned $= A - P = 1610.51 - 1000 = \$610.51$.

(b) Compounding quarterly means that there are four payment periods per year. Thus $n = 4(5) = 20$ and $i = \dfrac{0.1}{4} = 0.025$

$$A = P(1 + i)^{20}$$
$$\quad = 1000(1 + 0.025)^{20} \qquad \text{Use calculator (or Table 1 in Appendix II)}$$
$$\quad = 1000(1.638616)$$
$$\quad = \$1638.62$$

Interest earned $= A - P = 1638.62 - 1000 = \$638.62$.

(c) Monthly compounding means there are 12 payment periods per year. Thus $n = 12(5) = 60$ and* $i = \dfrac{0.10}{12}$. Then,

$$A = P(1 + i)^n$$
$$\quad = 1000\left(1 + \frac{0.10}{12}\right)^{60} \qquad \text{Use calculator}$$
$$\quad = 1000(1.645309)$$
$$\quad = \$1645.31$$

* When you use a calculator to compute, do not round off your answer at any intermediate step.

Interest earned $= A - P = 1645.31 - 1000 = \$645.31$.

(d) Daily compounding means there are 365 payment periods per year. Thus $n = 365(5) = 1825$ and $i = \dfrac{0.10}{365}$. So

$$A = P(1 + i)^{1825}$$

$$= 1000 \left(1 + \frac{0.10}{365}\right)^{1825} \quad \text{Use calculator}$$

$$= 1000(1.648608)$$

$$= \$1648.61$$

Interest earned $= A - P = 1648.61 - 1000 = \$648.61$.

The results are summarized in the following table

Nominal Rate (%)	Computation Method	Interest Rate per Period (%)	Initial Principal	Accumulated Amount	Interest Earned
10	Annual ($m = 1$)	10	$1000	$1610.51	$610.51
10	Quarterly ($m = 4$)	2.5	$1000	$1638.61	$638.61
10	Monthly ($m = 12$)	0.833	$1000	$1645.31	$645.31
10	Daily ($m = 365$)	0.0274	$1000	$1648.61	$648.61

Notice the substantial increase in interest earned between annual compounding and quarterly compounding ($28.10), compare to the slight increase between monthly compounding and daily compounding ($3.30). In fact, it can be shown that as the number of compoundings per year gets larger and larger, the interest earned approaches a limit. This type of compounding is referred to as *continuous compounding.*

EFFECTIVE RATE OF INTEREST

In Example 2 we learned that, for a given annual rate, more frequent compounding of interest gives larger earnings at the end of the year. A 10% annual rate, compounded monthly, gives a larger amount at the end of the year than does 10% compounded quarterly. However, a lower rate, compounded more frequently, may or may not give a larger return. For example; which is a better investment, one that pays 10% compounded monthly or another that pays 10.2% compounded semiannually? We need a common basis for comparing various nominal rates. This can be done by computing the *effective rate.* The *effective rate* of an annual interest rate r compounded m times per year is the simple interest rate that produces, in 1 year, the same accumulated amount as the nominal rate compounded m times a year. We need therefore a formula that gives a relationship between the nominal interest rate r per year compounded m times a year and its corresponding effective rate r_e per year. Let us suppose that a principal P is invested for 1 year and earns interest at the stated rate. Then by definition

$$\begin{pmatrix} \text{amount at effective} \\ \text{rate compounded} \\ \text{annually} \\ \text{after 1 year} \end{pmatrix} = \begin{pmatrix} \text{amount of} \\ \text{compound interest} \\ \text{after 1 year} \end{pmatrix}$$

$$P(1 + r_e) = P\left(1 + \frac{r}{m}\right)^m$$

$$1 + r_e = \left(1 + \frac{r}{m}\right)^m \qquad \text{Divide both sides by } P$$

$$r_e = \left(1 + \frac{r}{m}\right)^m - 1 \qquad \text{Solve for } r_e$$

We now have a formula for computing the effective rate of interest.

Effective Rate If principal P is invested at an annual rate r and compounded m times a year, the effective interest rate r_e, is

$$r_e = (1 + i)^m - 1 \qquad (2)$$

where

$$i = \frac{r}{m}$$

Example 3
If an investment pays 8% compounded quarterly, what is the effective interest rate?

Solution
Here $m = 4$ and $i = \dfrac{0.08}{4}$

$$\begin{aligned} r_e &= (1 + i)^m - 1 \\ &= \left(1 + \frac{0.08}{4}\right)^4 - 1 \\ &= (1.02)^4 - 1 \qquad \text{Use calculator (or Table 1 in Appendix II)} \\ &= 1.082432 - 1 \\ &= 0.082432 \end{aligned}$$

In percent form, $r_e = 8.243\%$ is the effective rate of 8% compounded quarterly.

∎

Now let us answer the question stated earlier. Which is a better investment, one that pays 10% compounded monthly or another that pays 10.2% compounded semiannually?

Effective rate for 10% compounded monthly

$$r_e = \left(1 + \frac{r}{m}\right)^m - 1$$

$$= \left(1 + \frac{0.10}{12}\right)^{12} - 1 \qquad \text{Use calculator}$$

$$= 1.1047 - 1$$

$$= .1047 \qquad \text{or} \qquad 10.47\%$$

Effective rate for 10.2% compounded semiannually

$$r_e = \left(1 + \frac{r}{m}\right)^m - 1$$

$$r_e = \left(1 + \frac{0.102}{2}\right)^2 - 1$$

$$r_e = (1 + 0.051)^2 - 1$$

$$r_e = 1.1046 - 1$$

$$= .1046 \qquad \text{or} \qquad 10.46\%$$

Therefore, it would be a better investment to accept 10% interest compounded monthly.

Example 4

Laura wishes to invest in a money market deposit account. She has the following three choices:

(a) Beverly Bank   8.20%   Compounded quarterly
(b) First National   8.05%   Compounded semiannually
(c) Prairie State Bank 8.25%   Compounded annually

At which bank should she deposit her money?

Solution

We cannot compare the three interest rates directly since the periods over which they are compounded are different. To obtain comparable interest rates, we must compute their effective rates.

(a) Effective rate for Beverly Bank; since compounding is quarterly we have $m = 4$, $i = \dfrac{0.082}{4}$.

$$r_e = \left(1 + \frac{r}{m}\right)^m - 1$$

$$= \left(1 + \frac{0.082}{4}\right)^4 - 1$$

$$= (1.0205)^4 - 1 \qquad \text{Use calculator}$$

$$= 1.08456 - 1$$

$$= 0.08456 \qquad \text{or} \qquad 8.456\%$$

(b) Effective rate for First National; since compounding is done semiannually we have $m = 2$, $i = \dfrac{0.0805}{2}$.

$$r_e = \left(1 + \frac{r}{m}\right)^m - 1$$

$$= \left(1 + \frac{0.0805}{2}\right)^2 - 1$$

$$= (1.04025)^2 - 1 \qquad \text{Use calculator}$$

$$= 1.082120 - 1$$

$$= 0.082120 \qquad \text{or} \qquad 8.212\%$$

(c) Since the interest for Prairie State Bank is compounded annually, the nominal rate and the effective rate are the same, and hence the effective rate is 8.25%.

Comparing the three rates we see that Laura will do best by depositing her money in the Beverly Bank. ∎

Today, because of ease of calculation with computers, most banks, money market funds, and credit cards use daily compounding. For legal reasons, the interest rate per day and the effective annual interest rate are usually printed on every statement.

PRESENT VALUE
If we solve for P in the compound interest formula (1), we obtain

$$P = A(1 + i)^{-n}$$

> **Present Value** In this formula, P is called the *present value* of the amount A due at the end of n interest periods at a rate i per interest period. In other words, P is the amount that must be invested for n interest periods at a rate i per interest period in order to accumulate the amount A.

Values for $(1 + i)^{-n}$ can be found using a hand-held calculator with a y^x key, or the reciprocal of the values of $(1 + i)^n$ from Table 1 in Appendix II.

Example 5
How much money should be invested at 8% per annum so that after 2 years the amount will be $10,000 when the interest is compounded:
(a) Annually?   (b) Monthly?   (c) Daily?

Solution
In this problem, we want to find the principal P when we know that the amount A after 2 years is going to be $10,000. That is, we want to find the present value of $10,000. Using a calculator

(a)   Since compounding is once per year for 2 years, $n = 2$. We find

$$P = A(1 + i)^{-n}$$
$$= 10{,}000(1 + 0.08)^{-2} \qquad \text{Use calculator}$$
$$= 10{,}000(0.8573388)$$
$$= \$8573.39$$

(b)   Since compounding is 12 times per year for 2 years, $n = 24$. We find

$$P = A(1 + i)^{-n}$$
$$= 10{,}000 \left(1 + \frac{0.08}{12} \right)^{-24} \qquad \text{Use calculator}$$
$$= 10{,}000(0.852596)$$
$$= \$8525.96$$

(c)   Since compounding is 365 times per year for 2 years, $n = 730$. We find

$$P = A(1 + i)^{-n}$$
$$= 10{,}000 \left(1 + \frac{0.08}{365} \right)^{-730} \qquad \text{Use calculator}$$
$$= 10{,}000(0.8521587)$$
$$= \$8521.59$$

In the next example we solve for the variable i.

Example 6

What annual rate of interest compounded annually should you seek if you want to double your investment in 5 years?

Solution

If P is the principal and we want P to double, the amount, A will be $2P$. We use the compound interest formula with $n = 5$ to find i:

$$2P = P(1 + i)^5$$
$$2 = (1 + i)^5$$
$$1 + i = \sqrt[5]{2}$$
$$i = \sqrt[5]{2} - 1 \qquad \text{Use calculator}$$
$$= 1.148698 - 1 = 0.148698$$

The annual rate of interest needed to double the principal in 5 years is 14.87%.

In the next example we show how to compute the time it takes a given principal to reach a particular value.

Example 7

How long will it take an investment of \$100,000 to grow to \$120,000 at 9% compounded monthly?

Solution

$$A = P(1 + i)^n$$

$$120,000 = 100,000 \left(1 + \frac{0.09}{12}\right)^n$$

$$1.2 = \left(1 + \frac{0.09}{12}\right)^n$$

or

$$1.2 = (1.0075)^n$$

To find the time, we have to solve for n. Look at various values of n (integer) and find the smallest value of n such that the right side of the equation exceeds 1.2. If your first guess is $n = 15$ you will find it to be too small; so is $n = 20$. For $n = 25$, the value is larger than 1.2, and so on. In this case, it is $n = 24$ months. Thus it takes 2 years for the investment of \$100,000 to grow to \$120,000. By examining the values of $(1 + i)^n$ in Table 1 of Appendix II for various values of n, you could also obtain the answer. ∎

Exercise 5.2

Answers to Odd-Numbered Problems begin on page A-17.

A   In Problems 1–6 use a calculator to find the following.

1. $(1.02)^5$ **2.** $(1.04)^6$ **3.** $(1.076)^5$

4. $(1.0015)^{10}$ **5.** $(1.0125)^6$ **6.** $(1.025)^{10}$

In Problems 7–10 the annual and the compounding period are given. Find i, the interest rate per compounding period.

7. 12% compounded monthly   **8.** 11% compounded quarterly

9. 7% compounded semiannually   **10.** 9% compounded annually

In Problems 11–14 the rate per compounding period is given. Find r, the annual rate.

11. 0.01 compounded monthly   **12.** 2.5% per half year

13. 8% per year   **14.** 2.25% per quarter

In Problems 15–20 use the compound interest formula (1) to find the amount owed.

15. \$1000 is borrowed at 10% compounded monthly for 36 months.

16. \$100 is borrowed at 14% compounded monthly for 36 months.

17. \$500 is borrowed at 9% compounded annually for 1 year.

18. \$200 is borrowed at 10% compounded annually for 10 years.

19. \$800 is borrowed at 12% compounded daily for 200 days.

20. $400 is borrowed at 10% compounded daily for 180 days.

In Problems 21–24 find the principal needed now to get each amount.

21. To get $100 in 6 months at 10% compounded monthly

22. To get $500 in 1 year at 12% compounded annually

23. To get $500 in 1 year at 9% compounded daily

24. To get $800 in 2 years at 10% compounded monthly

B **25.** If $1000 is invested at 9% compounded

    (a) annually   (b) semiannually   (c) quarterly   (d) monthly:

    What is the amount after 3 years. How much interest is earned?

26. If $2000 is invested at 12% compounded

    (a) annually   (b) semiannually   (c) quarterly,   (d) monthly:

    What is the amount after 5 years. How much interest is earned?

27. If $1000 is invested at 12% compounded quarterly, what is the amount after

    (a) 2 years   (b) 3 years   (c) 4 years

28. If $2000 is invested at 8% compounded quarterly, what is the amount after

    (a) 2 years   (b) 3 years   (c) 4 years

29. If a bank pays 6% compounded semiannually, how much should be deposited now to have $5000

    (a) 4 years later   (b) 8 years later

30. If a bank pays 8% compounded quarterly, how much should be deposited now to have $10,000

    (a) 5 years later   (b) 10 years later

31. Use formula (2) to compute the effective rate of interest for money invested at

    (a) 8% compounded semiannually

    (b) 12% compounded monthly

32. Use formula (2) to compute the effective rate of interest for money invested at

    (a) 6% compounded monthly

    (b) 14% compounded semiannually

C **33.** What annual rate of interest is required to double an investment in 3 years?

34. What annual rate of interest is required to double an investment in 10 years?

35. Approximately how long will it take to triple an investment at 10% compounded annually?

36. Approximately how long will it take to triple an investment at 9% compounded annually?

37. What interest rate compounded quarterly will give an effective interest rate of 7%.

38. Repeat Problem 37 using 10%.

In Problems 39–42 which of the two rates would yield the larger amount in one year:

39. 6% compounded quarterly or $6\frac{1}{4}\%$ compounded annually?

40. 9% compounded quarterly or $9\frac{1}{4}$% compounded annually?

41. 9% compounded monthly or 8.8% compounded daily?

42. 8% compounded semiannually or 7.9% compounded daily?

APPLICATIONS

43. If the price of homes rises an average of 5% per year for the next 4 years, what will be the selling price of a home that is selling for $90,000 today 4 years from today?

44. Mr. Nielsen wants to borrow $1000 for 2 years. He is given the choice of (a) a simple interest loan of 12% or (b) a loan at 10% compounded monthly. Which loan results in less interest due?

45. Rework Problem 44 if the simple interest loan is 15% and the other loan is at 14% compounded daily.

46. A department store charges 1.25% per month on the unpaid balance for customers with charge accounts (interest is compounded monthly). A customer charges $200 and does not pay her bill for 6 months. What is the bill at that time?

47. A major credit card company has a finance charge of 1.5% per month on the outstanding indebtedness. Caryl charged $600 and did not pay her bill for 6 months. What is the bill at that time?

48. Laura wishes to have $8000 available to buy a car in 3 years. How much should she invest in a savings account now so that she will have enough if the bank pays 8% interest compounded quarterly?

49. Tami and Todd will need $40,000 for a down payment on a house in 4 years. How much should they invest in a savings account now so that they will be able to do this? The bank pays 8% compounded quarterly.

50. A newborn child receives a $3000 gift toward a college education. How much will the $3000 be worth in 17 years if it is invested at 10% compounded quarterly?

51. A child's grandparents have opened a $6000 savings account for the child on the day of her birth. The account pays 8% compounded semiannually. The child will be allowed to withdraw the money when she becomes 25 years old. What will the account be worth at that time?

52. What will a $90,000 house cost 5 years from now if the inflation rate over that period averages 5% compounded annually?

53. A town increased in population 2% per year for 8 years. If the population was 17,000 at the beginning, what is the size of the population at the end of 8 years?

54. Omega Company can invest its earnings at (a) 7% per year compounded monthly, at (b) 7.15% per year compounded semiannually or at (c) 7.20% per year compounded annually. Which rate should Omega choose?

55. Jack is considering buying 1000 shares of a stock that sells at $15 per share. The stock pays no dividends. From the history of the stock, Jack is certain that he will be able to sell it 4 years from now at $20 per share. Jack's goal is not to make any investment unless it returns at least 7% compounded quarterly. Should Jack buy the stock?

56. Repeat Problem 55 if Jack requires a return of at least 14% compounded quarterly.

57. An Individual Retirement Account (IRA) has $2000 in it, and the owner decides not to add any more money to the account other than the interest earned at 9% compounded quarterly. How much will be in the account 25 years from the day the account was opened?

58. If Jack sold a stock for $35,281.50 (net) that cost him $22,485.75 three years ago, what annual compound rate of return did Jack make on his investment?

For problems 59–61, zero coupon bonds are used. A *zero coupon bond* is a bond that is sold now at a discount and will pay its face value at some time in the future when it matures; no interest payments are made.

59. Tami's grandparents are considering buying a $40,000 face value zero coupon bond at birth so that she will have enough money for her college education 17 years later. If money is worth 8% compounded annually, what should they pay for the bond?

60. How much should a $10,000 face value zero coupon bond, maturing in 10 years, be sold for now if its rate of return is to be 8% compounded annually?

61. If you pay $12,485.52 for a $25,000 face value zero coupon bond that matures in 8 years, what is your annual compound rate of return?

62. A bank advertises that it pays interest on saving accounts at the rate of 6.25% compounded daily. Find the effective rate if the bank uses (a) 360 days, or (b) 365 days in determining the daily rate.

5.3

ANNUITY: SINKING FUND

FUTURE VALUE OF AN ANNUITY
SINKING FUND

FUTURE VALUE OF AN ANNUITY

In the previous sections we learned how to compute the future value of an investment when a fixed amount of money is deposited in an account that pays interest compounded periodically. Often, however, people and financial institutions do not deposit money and then sit back and watch it grow. Rather, money is invested in small amounts at periodic intervals. Examples of such investments are annual life insurance premiums, monthly deposits in a bank, and installment loan payments.

An *annuity* is a sequence of equal periodic payments. When the payments are made at the same time the interest is credited, the annuity is termed *ordinary*. We will only concern ourselves with *ordinary annuities* in this book.

The *payment period* can be annual, semiannual, quarterly, monthly or any fixed length of time.

Future Value of an Annuity **The *future value of an annuity* is the sum of all payments made plus all interest accumulated.**

We wish to develop a formula to compute the future value of an annuity. We start with an example. Suppose an individual deposits $1000 every 6 months in a savings annuity that pays 8% compounded semiannually. If six deposits are made, one at the *end* of each compounding period, what will be the value of this annuity at the end of 3 years? That is, how much will be in the savings account at the end of the 3 years? Figure 1 depicts the time sequence of deposits for this illustration.

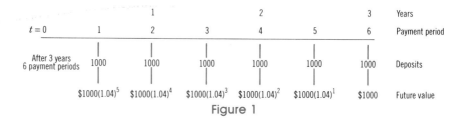

Figure 1

Note that we refer to the current time as $t = 0$, one payment period later as $t = 1$, two payment periods later as $t = 2$, and so on.

With the aid of the compound interest formula $A = P(1 + i)^n$, we list in Figure 1 the value of each deposit after it has earned compound interest up to the sixth deposit. The final future value of the annuity is obtained by adding the six payments and the interest accumulated on each one. Thus if A is the future value of the annuity, then

$$A = 1000 + 1000(1.04) + 1000(1.04)^2 + 1000(1.04)^3$$
$$+ 1000(1.04)^4 + 1000(1.04)^5$$

Clearly, computing this total period-by-period can be tedious if the annuity runs for a large number of payment periods. A formula for computing the future value of an annuity will help ease the computations. Suppose R is the payment for an annuity at an interest rate of $i\%$ per payment period and with a term of n payment periods. Since payments are made at the end of each period, the first payment of amount R will earn interest for $n - 1$ compounding periods and so will accumulate to

$$R(1 + i)^{n-1}$$

The second payment will earn interest for $n - 2$ compounding periods and so will accumulate

$$R(1 + i)^{n-2}$$

and so on. The next-to-last payment will earn interest for only one compounding period, so it will accumulate to

$$R(1 + i)^1$$

and the last payment will earn no interest, so that its "accumulated" amount is just R. Therefore, the total amount of the annuity after n payment periods, denoted by A, is

$$A = R(1 + i)^{n-1} + R(1 + i)^{n-2} + \cdots + R(1 + i) + R$$
$$A = R + R(1 + i) + \cdots + R(1 + i)^{n-2} + R(1 + i)^{n-1}$$

But this is the sum of the first n terms of a geometric sequence* with first term R and common ratio $1 + i$. Hence,

---

* See Appendix I for a definition of a geometric sequence.

$$A = \frac{R[1 - (1 + i)^n]}{1 - (1 + i)}$$

or   (1)

$$A = \frac{R[(1 + i)^n - 1]}{i}$$

It is customary to use the symbol $s_{\overline{n}|i}$, which reads "s angle n at i," for $\left[\dfrac{(1 + i)^n - 1}{i}\right]$.

Thus by setting $s_{\overline{n}|i} = \dfrac{(1 + i)^n - 1}{i}$ we obtain

$$A = Rs_{\overline{n}|1}$$

Table 1 of Appendix II gives the values of $s_{\overline{n}|i}$ for various values of n and i. Returning to our example above, the amount of the annuity is

$$A = 1000 \frac{(1.04)^6 - 1}{0.04} \qquad \text{Use calculator (or Table 1 in Appendix II)}$$

$$= 1000(6.632975)$$

$$= \$6632.98$$

We now state the general formula

Future Value of an Ordinary Annuity

$$A = R\frac{(1 + i)^n - 1}{i} \qquad\qquad (2)$$

where payments are made at the end of each period and

$R =$ **periodic payment**
$i =$ **rate per period**
$n =$ **number of payment periods**
$A =$ **future value**

Example 1
Find the amount of an annuity if a payment of $100 per year is made for 5 years at 8% compounded annually. How much of this value is interest?

Solution
To find the value of the annuity we use Formula (2) with $R = 100$, $i = 0.08$, and $n = 5$.

$$A = \frac{R[(1 + i)^n - 1]}{i} = Rs_{\overline{n}|i}$$

$$= \frac{100[(1 + 0.08)^5 - 1]}{.08} \qquad \text{Use calculator (or Table 1 in Appendix II)}$$

$$= 100(5.866601)$$

$$= \$586.66$$

To find the interest, we subtract the total amount deposited from the annuity:

$$\text{Deposits} \quad 5(100) = \$500$$
$$\text{Interest} = A - \text{deposits}$$
$$= 586.66 - 500$$
$$= \$86.66$$

Example 2
Caryl decides to put aside $100 every month in a money market fund that pays 9% compounded monthly. After making 12 deposits, how much money does Caryl have?

Solution
This is an annuity problem. We use formula (2) with $R = 100$, $i = \frac{0.09}{12}$, and $n = 12$.

$$A = \frac{R[(1 + i)^n - 1]}{i} = Rs_{\overline{n}|i}$$

$$= \frac{100 \left[(1 + \frac{0.09}{12})^{12} - 1\right]}{\frac{0.09}{12}} = 100s_{\overline{12}|0.75} \qquad \begin{array}{l}\text{Use calculator (or Table 1}\\\text{in Appendix II)}\end{array}$$

$$= \$100(12.507586) = \$1250.76$$

Example 3
To save for his son's college education, Mr. Graff decides to put $50 aside every month in a credit union account paying 10% interest compounded monthly. If he begins this saving program when his son is 3 years old, how much will he have saved by the time his son is 18 years old?

Solution
When his son is 18 years old, Mr. Graff will have made his 180th payment, so he will have

$$A = Rs_{\overline{n}|i} = R\frac{(1 + i)^n - 1}{i}$$

$$= \$50\frac{(1 + \frac{0.1}{12})^{180} - 1}{\frac{0.1}{12}} \qquad \text{Use calculator}$$

$$= \$50(414.4703) = \$20{,}723.52$$

Individual Retirement Accounts (IRA) are accounts in which people deposit money into an account and earnings from this account are tax-deferred until money is withdrawn. It is important, if possible, to open IRA accounts early in one's life.

Example 4

Joe, at age 35, decides to invest in an IRA. He will put aside $2000 per year for the next 30 years. How much will he have at age 65 if his rate of return is assumed to be 8% per annum compounded annually?

Solution

This is an annuity problem in which $R = \$2000$. Use Equation (2) with annual compounding at 8% and with $n = 30$. The amount A in Joe's IRA after 30 years is

$$A = R\frac{(1 + i)^n - 1}{i} = Rs_{\overline{n}|i} \qquad \text{Use calculator (or Table 1 in Appendix II)}$$

$$= 2000\frac{(1 + 0.08)^{30} - 1}{0.08} = 2000 \, s_{\overline{30}|0.08}$$

$$= 2000(113.283211)$$

$$= \$226,566.42 \qquad \blacksquare$$

Example 5

If Joe had begun his IRA at age 25, his IRA at age 65 would amount to

$$A = R\frac{(1 + i)^n - 1}{i} = Rs_{\overline{n}|i} \qquad \text{Use calculator (or Table 1 in Appendix II)}$$

$$= \$2000\frac{(1 + 0.08)^{40} - 1}{0.08} = 2000 \, s_{\overline{40}|0.08}$$

$$= \$2000(259.056519)$$

$$= \$518,113.04 \qquad \blacksquare$$

SINKING FUND

Quite often, a person with a debt decides to accumulate sufficient funds to pay off the debt by agreeing to set aside enough money each month (or quarter, or year) so that when the debt becomes payable, the money set aside each month plus the interest earned will equal the debt. The fund created by such a plan is called a *sinking fund*. Companies often use sinking funds to accumulate capital for the purpose of purchasing new equipment.

We limit our discussion of sinking funds to those in which equal payments are made at equal time intervals. Usually, the debtor agrees to pay interest on the debt as a separate item, so that the amount needed in a sinking fund is only the amount originally borrowed.

Example 6

A woman borrows \$3000 and agrees to pay interest monthly at an annual rate of 12%. At the same time, she sets up a sinking fund in order to repay the loan at the end of 5 years. If the sinking fund earns interest at the rate of 8% compounded annually, find the size of each annual sinking fund deposit. Construct a table showing the growth of the sinking fund.

Solution

Here we have an annuity whose value 5 years from now should be \$3000. We use Formula (2) with $A = 3000$, $n = 5$, $i = 0.08$ and our problem is to find R. Thus

$$A = \frac{R(1 + i)^n}{i} = Rs_{\overline{n}|i}$$

$$3000 = R \frac{(1.08)^5 - 1}{0.08} = Rs_{\overline{5}|0.08} \qquad \text{Use calculator (or Table 1 in Appendix II)}$$

$$3000 = R(5.866601)$$

If we solve for R we get

$$R = 3000(0.17045645)$$
$$= \$511.37$$

Thus, annual sinking fund payments of \$511.37 are needed.
The growth of the sinking fund is shown in Table 1.

Table 1

Payment	Interest	Sinking Fund Deposit	Total
0	—	—	—
1	0	511.37	511.37
2	40.91	511.37	1063.65
3	85.09	511.37	1660.11
4	132.81	511.37	2304.29
5	184.34	511.37	3000.00

Note: Because of round-off error, the final total will sometimes be slightly off. ∎

We can use the technique of Example 6 to obtain the general formula for a sinking fund payment. Since payments are made in the form of an ordinary annuity, we can use Formula (2) and solve for R.

$$A = R \frac{(1 + i)^n - 1}{i}$$

$$R = A \frac{i}{(1 + i)^n - 1}$$

We now have the general formula:

Sinking Fund Payment

$$R = A \frac{i}{(1+i)^n - 1} = \frac{A}{s_{\overline{n}|i}} \tag{3}$$

where payments are made at the end of each period, and

R = Sinking fund payment

A = Future value, (value of annuity after n payments)

n = Number of payment periods

i = rate per payment period

Example 7

The Board of Education received permission to issue a bond to build a new high school. The board is required to make payments every 6 months into a sinking fund paying 8% compounded semiannually. At the end of 12 years, the bond obligation will be retired at a cost of $4,000,000. What should each payment be?

Solution

We use Formula (3) with $A = 4,000,000$, $i = \dfrac{0.08}{2} = 0.04$, and $n = 2(12) = 24$:

$$R = A \frac{i}{(1+i)^n - 1}$$

$$R = 4,000,000 \frac{0.04}{(1+0.04)^{24} - 1} = 4,000,000 \frac{1}{s_{\overline{24}|0.04}} \qquad \text{Use calculator (or Table 1 in Appendix II)}$$

$$R = 4,000,000(0.0255868)$$

$$R = \$102,347.33$$

Thus a payment of 102,347.33 should be made every 6 months. ∎

Example 8

Depletion Investment A gold mine is expected to yield an annual net return of $200,000 for the next 10 years, after which it will be worthless. An investor wants an annual return on his investment of 18%. If he can establish a sinking fund earning 10% annually, how much should he be willing to pay for the mine?

Solution

Let p denote the purchase price. Then $0.18p$ represents an 18% return on investment. The annual sinking fund contribution needed to obtain the amount p in 10 years is $p/s_{\overline{n}|i}$, where $n = 10$ and the rate is 8%. The investor should be willing to pay an amount p so that

Return on investment + Sinking fund requirement = Annual return

$$0.18p + p \frac{1}{s_{\overline{10}|0.08}} = \$200{,}000$$

$$0.18p + p \left(\frac{1}{14.486562} \right) = \$200{,}000 \qquad \text{Use calculator (or Table 1 in Appendix II)}$$

$$0.18p + p(0.0690295) = \$200{,}000$$

$$p(0.249029) = \$200{,}000$$

Solve for p

$$p = \$803{,}117.74$$

Exercise 5.3

Answers to Odd-Numbered Problems begin on page A-17.

In Problems 1 – 10 use Formula (2) and a calculator to solve each problem

A
1. $A = ?$	$n = 10$	$i = .02$	$R = \$1000$
2. $A = ?$	$n = 15$	$i = .03$	$R = \$\ 500$
3. $A = ?$	$n = 20$	$i = .04$	$R = \$\ 100$
4. $A = ?$	$n = 25$	$i = .02$	$R = \$\ \ 50$

B
5. $A = \$2000$	$n = 10$	$i = .02$	$R = ?$
6. $A = \$3000$	$n = 15$	$i = .03$	$R = ?$
7. $A = \$1000$	$n = 20$	$i = .04$	$R = ?$
8. $A = \$5000$	$n = 25$	$i = .03$	$R = ?$

C
9. $A = \$5693$	$n = ?$	$i = .08$	$R = \$300$
10. $A = \$27432$	$n = ?$	$i = .06$	$R = \$500$

APPLICATIONS   In Problems 11 – 16 find the amount of each annuity.

11.   The payment is $100 annually for 10 years at 10% compounded annually.

12.   The payment is $200 monthly for 1 year at 8% compounded monthly.

13.   The payment is $400 monthly for 1 year at 12% compounded monthly.

14.   The payment is $1000 annually for 5 years at 10% compounded annually.

15.   The payment is $200 monthly for 3 years at 10% compounded monthly.

16.   The payment is $2000 annually for 20 years at 10% compounded annually.

17.   Caryl invests $2500 a year in a mutual fund for 15 years. If the market value of the fund increases on the average 7% per year, what will be the value of her share at the end of the 15th year?

18.   Laura wants to invest an amount at the end of every 3 months so that she will have $12,000 in 3 years to buy a new car. The account pays 8% compounded quarterly. How much should she deposit each quarter?

19.   Todd and Tami pay $300 at the end of each 3 months for 6 years into an ordinary annuity paying 8% compounded quarterly. What is the future value at the end of 6 years?

20.   In 4 years a couple would like to have $30,000 for a down payment on a house. How

much should they deposit each month into an account paying 9% compounded monthly?

21. A person wishes to have $350,000 in a pension fund 20 years from now. How much should he deposit each month in an account paying 9% compounded monthly?

22. A self-employed person has a Keogh retirement plan (this type of plan is tax-deferred until money is withdrawn). If deposits of $7500 are made each year into an account paying 8% compounded annually, how much will be in the account after 25 years?

23. A company establishes a sinking fund to provide for the payment of a $100,000 debt, maturing in 4 years. Contributions to the fund are to be made at the end of every year. Find the amount of each annual deposit if interest is 8% per annum.

24. A state has $5,000,000 worth of bonds that are due in 20 years. A sinking fund is established to pay off the debt. If the state can earn 7% annually on its money, what is the annual sinking fund deposit needed?

25. An investor wants to know the amount she should pay for an oil well expected to yield an annual return of $30,000 for the next 30 years, after which the well will be dry. Find the amount she should pay to yield a 14% annual return if a sinking fund earns 8% annually.

26. If you deposit $10,000 every year in an account paying 8% compounded annually, how long will it take to accumulate $1,000,000?

27. A city has issued bonds to finance a new library. The bonds have a total face value of $1,000,000 and are payable in 10 years. A sinking fund has been opened to retire the bonds. If the interest rate on the fund is 8% compounded quarterly, what will the quarterly payments be?

28. (a) Laura invested $2000 per year in an IRA each year for 10 years earning 8% compounded annually. At the end of 10 years she ceased the IRA payments, but continued to invest her accumulated amount at 8% compounded annually, for the next 30 years. What was the value of her IRA investment at the end of 10 years? What was the value of her investment at the end of the next 30 years?

(b) Tami started her IRA investment in the 11th year and invested $2000 per year for the next 30 years at 8% compounded annually. What was the value of her investment at the end of the 30 years?

(c) Who had more money at the end of the period?

5.4

PRESENT VALUE OF AN ANNUITY; AMORTIZATION

PRESENT VALUE OF AN ANNUITY
AMORTIZATION
AMORTIZATION SCHEDULES

PRESENT VALUE OF AN ANNUITY

Suppose an individual wants to withdraw $1000 every 6 months from a savings account for a period of 3 years. If the interest rate is 8% compounded semiannually, how much should be in the account initially if after the last withdrawal no money is to be left in the account? In fact, what we are looking for is the present value of each $1000 that is withdrawn during the 3 years. We therefore have the following definition.

Present Value of an Annuity **The *present value* of an annuity is the sum of the present values of the payments.**

In other words, the present value of an annuity is the amount of money needed now so that if it is invested at i percent, n equal payments can be withdrawn without any money left over.

Getting back to our illustration, to determine the present value of the six consecutive withdrawals we need to compute the sum of the present values corresponding to each withdrawal. We can do this by solving for P in the compound interest formula.

$$A = P(1 + i)^n$$

$$P = \frac{A}{(1 + i)^n} = A(1 + i)^{-n}$$

The rate per period is $i = \dfrac{0.08}{2} = 0.04$. The following table illustrates the time sequence of withdrawals for this example.

Years	Starting Time	1 Year Later		2 Years Later		3 Years Later
Number of payment	0	1	2	3	4	5	6
Withdrawal		1000	1000	1000	1000	1000	1000
Present value		$1000(1.04)^{-1}$	$1000(1.04)^{-2}$	$1000(1.04)^{-3}$	$1000(1.04)^{-4}$	$1000(1.04)^{-5}$	$1000(1.04)^{-6}$

$$\text{Sum of all present values} = 1000(1.04)^{-1} + 1000(1.04)^{-2} + 1000(1.04)^{-3}$$
$$+ 1000(1.04)^{-4} + 1000(1.04)^{-5} + 1000(1.04)^{-6}$$

Clearly, computing this total period by period can be tedious, especially when the number of withdrawals gets very large. We need, therefore, to develop a formula to compute the present value of an annuity. Suppose R is the periodic payment, i the rate per period, and n the number of periods, then the present value of all payments is given by

$$V = R(1 + i)^{-1} + R(1 + i)^{-2} + \cdots + R(1 + i)^{-n}$$

But V is the sum of the first n terms of a geometric sequence with the first term $R(1 + i)^{-1}$ and common ratio $(1 + i)^{-1}$. Therefore,

$$V = \frac{R(1 + i)^{-1}[1 - (1 + i)^{-n}]}{1 - (1 + i)^{-1}}$$

which simplifies to

$$V = \frac{R[1 - (1 + i)^{-n}]}{i} \tag{1}$$

Often, the symbol $a_{\overline{n}|i}$, which reads "a angle n at i," is used for

$$a_{\overline{n}|i} = \frac{1 - (1 + i)^{-n}}{i}$$

Then Equation (1) can be written as

$$V = Ra_{\overline{n}|i}$$

Table 1 in Appendix II lists various values of $a_{\overline{n}|i}$ for different values of n and i. However, you will find it very helpful to use a financial or scientific calculator. Such calculators can handle more cases regardless of how detailed they may be. Returning to our illustration, we can now complete the problem, when $R = 1000$, $i = 0.04$, $n = 6$.

$$P = \frac{1000[1 - (1 + 0.04)^{-6}]}{0.04} = 1000\, a_{\overline{6}|0.04} \qquad \text{Use calculator (or Table 1 in Appendix II)}$$

$$P = 1000(5.242137)$$

$$= \$5242.14$$

We now state the present value formula just derived:

Present Value of an Ordinary Annuity

$$V = R\,\frac{1 - (1 + i)^{-n}}{i} = Ra_{\overline{n}|i} \tag{2}$$

where payments are made at the end of each payment period and

R = periodic payment

i = rate per period

n = number of periods

V = present value of all payments

Example 1

Compute the amount of money required to pay out $100 per year for 5 years at 10% compounded annually.

Solution

We use Formula (2) to solve the problem, where $R = 100$, $i = 0.1$, and $n = 5$.

$$V = R \frac{1 - (1 + i)^{-n}}{i}$$

$$= 100 \frac{1 - (1 + 0.1)^{-5}}{0.1} \qquad \text{Use calculator}$$

$$= 100(3.790787)$$

$$= \$379.08$$

Thus, a person would need $379.08 now invested at 10% per annum in order to withdraw $100 per year for the next 5 years.

Table 2a summarizes these results. Table 2b lists the amount at the beginning of each subsequent year.

Table 2a

Payment	Present Value
1st	$100(1.10)^{-1} = \$\ 90.91$
2nd	$100(1.10)^{-2} = \$\ 82.64$
3rd	$100(1.10)^{-3} = \$\ 75.13$
4th	$100(1.10)^{-4} = \$\ 68.31$
5th	$100(1.10)^{-5} = \$\ 62.09$
Total	$379.08

Table 2b

Year	Amount at the Beginning of the Year	Interest
1	379.08	37.91
2	316.99	31.70
3	248.69	24.87
4	173.56	17.36
5	90.92	9.09
6	0	—

Example 2

A man agrees to pay $300 per month for 48 months to pay off a car loan. If interest of 12% per annum is charged monthly, how much did the car originally cost? How much interest was paid?

Solution

This is the same as asking for the present value of an annuity of $300 per month at 12% for 48 months. Using Formula (2) with $R = 300$, $i = \frac{0.12}{12} = 0.01$, $n = 48$ we get

$$V = R \frac{1 - (1 + i)^{-n}}{i} = R s_{\overline{n}|i}$$

$$= 300 \frac{1 - (1 + 0.01)^{-48}}{0.01} = 300\, a_{\overline{48}|0.01} \qquad \text{Use calculator (or Table 1 in Appendix II)}$$

$$= 300(37.973959)$$

$$= \$11,392.19$$

The total payment is ($300)(48) = $14,400. Thus, the interest paid is

$$\$14,400 - \$11,392.19 = \$3007.81$$

AMORTIZATION

We can look at the last example differently. What it also says is that if the man pays $300 per month for 48 months with an interest of 12% compounded monthly then the car is his. In other words, he *amortized* the debt in 48 equal monthly payments. (The Latin word *mort* means "death." Paying off a loan is regarded as "killing" it.) In general, a loan with a fixed rate of interest is said to be *amortized* if both principle and interest are paid by a sequence of equal payments made over equal periods of time.

When a loan of V dollars is amortized at a particular rate i of interest per payment period over n payment periods, the question is, "What is the payment R?" In other words, in amortization problems, we want to find the amount of payment R that, after n payment periods at a rate i of interest per payment period, gives us a present value equal to the amount of the loan. Thus, we need to find R in Formula (2).

Solving the present value formula for R, we obtain the following amortization formula:

$$R = V \frac{i}{1 - (1 + i)^{-n}} \tag{3}$$

$$= \frac{V}{a_{\overline{n}|i}}$$

where payments are made at the end of each period and

$$V = \text{amount of loan (present value)}$$
$$i = \text{rate per period}$$
$$n = \text{number of period payments}$$
$$R = \text{periodic payment}$$

Example 3

What monthly payment is necessary to pay off a loan of $800 at 12% per annum:
(a) In 2 years?   (b) In 3 years?

Solution
(a) Use Formula (3) with $V = \$800$, $i = \frac{0.12}{12} = 0.01$, and $n = 2(12) = 24$:

$$R = V \frac{i}{1 - (1 + i)^{-n}} = \frac{V}{a_{\overline{n}|i}}$$

$$= 800 \frac{0.01}{1 - (1 + 0.01)^{-24}} = \frac{800}{a_{\overline{24}|0.01}} \qquad \text{Use calculator (or Table 1 in Appendix II)}$$

$$= 800(0.047074)$$

$$= \$37.66$$

(b) For the 3-year loan $V = \$800$, $i = 0.01$, and $n = 3(12) = 36$, the monthly payment is

$$R = V \frac{i}{1-(1+i)^{-n}} = \frac{V}{a_{\overline{n}|i}}$$

$$= 800 \frac{0.01}{1-(1+0.01)^{-36}} = \frac{800}{a_{\overline{36}|0.01}}$$

Use calculator (or Table 1 in Appendix II)

$$= 800(0.033214)$$

$$= \$26.57$$

For the 2-year loan in Example 3, the total amount paid out is 37.66 (24) = $903.84. For the 3-year loan, the total amount paid is 26.57 (36) = $956.52. It should be clear that the longer the term of a debt, the more it costs the borrower to pay off the loan.

AMORTIZATION SCHEDULES

For amortization problems involving long-term mortgages with monthly payments, it is convenient to use a table of mortgages such as Table 3, in which appropriate values of $a_{\overline{n}|i}$ and $1/a_{\overline{n}|i}$ are listed.

Table 3
Mortgage Table

Monthly Payments n	8% per Annum $a_{\overline{n}|i}$	9% per Annum $a_{\overline{n}|i}$	10% per Annum $a_{\overline{n}|i}$	8% per Annum $i/a_{\overline{n}|i}$	9% per Annum $i/a_{\overline{n}|i}$	10% per Annum $i/a_{\overline{n}|i}$
60	49.3184	48.17337	47.0654	0.020276	0.020758	0.021247
120	82.42148	78.94169	75.6712	0.012133	0.012668	0.0132151
180	104.64059	98.5934	93.0574	0.009557	0.010143	0.0107461
240	119.55429	111.14495	103.625	0.008364	0.008997	0.0096502
300	129.5645	119.1616	110.047	0.007718	0.008392	0.00908701

Example 4
Mr. and Mrs. Corey have just purchased a $70,000 house and have made a down payment of $15,000. They can amortize the balance ($55,000) at 9% for 25 years.
(a) What are the monthly payments?
(b) What is their total interest payment?
(c) After 20 years, what equity do they have in their house?

Solution
We use Table 3, with $n = 25(12) = 300$, and $V = 55,000$.
(a) The monthly payment R needed to pay off the loan of $55,000 at 9% for 25 years (300 months) is

$$R = \$55,000 \left[\frac{1}{a_{\overline{n}|i}}\right]$$

$$= \$55,000(0.008392) = \$461.56$$

(b) The total paid out for the loan is

$$(\$461.56)(300) = \$138,468.00$$

Thus, the interest on this loan amounts to

$$\$138,468 - \$55,000 = \$83,468.00$$

(c) After 20 years (240 months) there remains 5 years (or 60 months) of payments. The present value of the loan is the present value of a monthly payment of $461.56 for 60 months at 9%, namely,

$$(\$461.56)a_{\overline{n}|i} = (\$461.56)(48.17337) = \$22,234.90$$

Thus, the equity after 20 years is

$$\text{Downpayment} + \text{Amount paid on loan} = \$15,000 + [\$55,000 - \$22,234.90]$$
$$= \$47,765.10 \qquad \blacksquare$$

Usually, the monthly payment is rounded up to the next dollar to ensure liquidation in the alloted term. The final payment is then not the same as the rest of the payments.

Table 4 gives a partial schedule of payments for the loan in Example 4. The Amount Paid on Loan column was obtained by the method of Example 4. However, the data in the Principal and Interest columns were obtained through a computer program. It is interesting to observe how slowly the amount paid on the loan increases early in the payment schedule, with very little of the payment used to reduce principal, and how quickly the amount paid on the loan increases during the last 5 years.

Table 4

Payment Number	Monthly Payment	Principal	Interest	Amount Paid on Loan
1	$461.56	$49.06	$412.50	$49.06
60	$461.56	$75.94	$385.62	$3,699.94
120	$461.56	$119.27	$342.29	$9,493.23
180	$461.56	$186.85	$274.71	$18,563.67
240	$461.56	$292.56	$169.00	$32,765.10
300	$461.56	$458.16	$3.40	$55,000.00

Example 5

When Mr. Nicholson died, he left an inheritance of $15,000 for his family to be paid to them over a 10 year period in equal amounts at the end of each year. If the $15,000 is invested at 8%, what is the annual payout to the family?

Solution

This example asks what annual payment is needed at 8% for 10 years to give a total of $15,000. That is, we can think of the $15,000 as a loan amortized at 8% for 10 years. The payment needed to pay off the loan is the yearly amount Mr. Nicholson's family will receive. Thus, the yearly payout R is

$$R = 15{,}000 \, \frac{0.08}{1 - (1 + 0.08)^{-10}} = 15{,}000 \, \frac{1}{a_{\overline{10}|0.08}} \qquad \text{Use calculator (or Table 1 in Appendix II)}$$

$$= 15{,}000(0.149030)$$

$$= \$2{,}235.44$$

Example 6

Laura is 20 years away from retiring and starts saving $100 a month in an account paying 6% compounded monthly. When she retires she wishes to withdraw a fixed amount each month for 25 years. What will this fixed amount be?

Solution

After 20 years, the amount accumulated in her account

$$A = R \cdot s_{\overline{n}|i} = 100 \cdot s_{\overline{240}|0.005} = 100 \cdot (462.0408951) = \$46{,}204.09$$

To find the amount she can withdraw for 300 payments (25 years) at 0.005% per payment (6% compounded monthly)

$$V = R \cdot a_{\overline{n}|i}$$

$$46{,}204.09 = R \cdot a_{\overline{300}|0.005}$$

$$R = \frac{46{,}204.09}{a_{\overline{300}|0.005}} = \frac{46{,}204.09}{155.206864} = \$297.69 \qquad \text{Use calculator}$$

Exercise 5.4   *Answers to Odd-Numbered Problems begin on page A-17.*

A   In Problems 1–10 use Formula (2) or (3) and a calculator to solve each problem.

1. $V = ?$,	$n = 20$,	$i = .03$,	$R = \$300$
2. $V = ?$,	$n = 30$,	$i = .04$,	$R = \$600$
3. $V = ?$,	$n = 35$,	$i = .025$,	$R = \$200$
4. $V = ?$,	$n = 40$,	$i = .00125$,	$R = \$300$

B
5. $V = \$\ 5000$,	$n = 24$,	$i = .02$,	$R = ?$
6. $V = \$\ 3000$,	$n = 30$,	$i = .025$,	$R = ?$
7. $V = \$20{,}000$,	$n = 80$,	$i = .00125$,	$R = ?$

   8. $V = \$10,000,$ $n = 48,$ $i = .0075,$ $R = ?$

C   9. $V = \$12,744,$ $n = ?,$ $i = .02,$ $R = \$500$

  10. $V = \$2251,$ $n = ?,$ $i = .01,$ $R = \$200$

In Problems 11 – 16 find the present value of each annuity.

11. The payment is to be $500 per month for 3 years at 9% compounded monthly.

12. The payment is to be $1000 per year for 3 years at 8% compounded annually.

13. The payment is to be $100 per month for 9 months at 12% compounded monthly.

14. The payment is to be $400 per month for 18 months at 9% compounded monthly.

15. The payment is to be $5,000 twice a year for 20 years at 10% compounded semiannually.

16. The payment is to be $2000 per month for 3 years at 12% compounded monthly.

17. A husband and wife contribute $4000 per year to an IRA paying 8% compounded annually for 20 years. What is the value of their IRA after 20 years?

18. Rework Problem 17 if the interest rate is 7%.

19. What annual payment is needed to pay off a loan of $10,000 amortized at 12% compounded monthly for 2 years?

20. What monthly payment is needed to pay off a loan of $500 amortized at 12% compounded monthly for 2 years?

21. In Example 4, if Mr. and Mrs. Corey amortize the $55,000 loan at 10% for 20 years, what is their monthly payment?

22. In Example 4, if Mr. and Mrs. Corey amortize their $55,000 loan at 12% for 15 years, what is their monthly payment?

23. In Example 5, if Mr. Nicholson left $15,000 to be paid over 20 years in equal yearly payments and if this amount were invested at 12%, what would the annual payout be?

APPLICATIONS   24. Joan has a sum of $30,000 that she invests at 10% compounded monthly. What equal monthly payments can she receive over a 10 year period? Over a 20 year period?

25. Mr. Doody, at age 65, can expect to live for 20 years. If he can invest at 10% per annum compounded monthly, how much does he need now to guarantee himself $250 every month for the next 20 years?

26. Ms. Joy, at age 65, can expect to live for 25 years. If she can invest at 10% per annum compounded monthly, how much does she need now to guarantee herself $300 every month for the next 25 years?

27. A couple wishes to purchase a house for $120,000 with a down payment of $25,000. They can amortize the balance either at 8% for 20 years or at 9% for 25 years. Which monthly payment is greater? For which loan is the total interest paid greater? After 10 years, which loan provides the greater equity?

28. A couple have decided to purchase a $100,000 house using a down payment of $20,000. If they can amortize the balance at 8% for 25 years, what is their monthly payment? What is the total interest paid? What is their equity after 5 years? What is the equity after 20 years?

29. John is 45 years old and wants to retire at 65. He wishes to make monthly deposits in an account paying 9% compounded monthly and when he retires withdraw $300 a month for 30 years. How much should John deposit each month?

30. The grand prize in the Illinois Lottery was $6,000,000 paid out in 20 equal yearly payments of $300,000 each. How much should they deposit in an account paying 8% compounded annually to achieve this goal?

31. Find the present value of an annuity that will pay $1200 every 3 months for 10 years from an account paying interest at a rate of 9% compounded quarterly.

32. Mr. Smith obtained a 25-year mortgage on a house. The monthly payments are $749.19 (principal and interest) and are based on a 7% interest rate. How much did Mr. Smith borrow? How much interest will be paid?

33. A car costs $12,000. You put 20% down and amortize the rest with equal monthly payments over a 3-year period at 15% to be compounded monthly. What will the monthly payment be?

34. Jay pays $320 per month for 36 months for furniture, making no down payment. If the interest charged is 0.5% per month on the unpaid balance, what was the original cost of the furniture? How much interest did he pay?

35. A restaurant owner buys equipment costing $20,000. If the owner pays 10% down and amortizes the rest with equal monthly payments over 4 years at 12% compounded monthly, what will be the monthly payments? How much interest is paid?

36. A house that was bought 8 years ago for $50,000 is now worth $100,000. Originally the house was financed by paying 20% down with the rest financed through a 25-year mortgage at 10.5% interest. The owner (after making 96 equal monthly payments) is in need of cash, and would like to refinance the house. The finance company is willing to loan 80% of the new value of the house amortized over 25 years with the same interest rate. How much cash will the owner receive after paying the balance of the original loan?

37. A home buyer is purchasing a $140,000 house. The down payment will be 20% of the price of the house, and the remainder will be financed by a 30-year mortgage at a rate of 9.8% interest compounded monthly. What will the monthly payment be? Compare the monthly payments and the total amounts of interest paid if a 15-year mortgage is chosen instead of a 30-year mortgage.

38. Mr. and Mrs. Hoch are interested in building a house that will cost $180,000. They intend to use the $60,000 equity in their present house as a down payment on the new one and will finance the rest with a 25-year mortgage at an interest rate of 10.2% compounded monthly. How large will their monthly payment be on the new house?

CHAPTER REVIEW

Important Terms and Formulas			
interest	proceeds	ordinary annuity
principal	maturity value	future value of an annuity
rate of interest	compound interest	payment period
per annum	effective rate of interest	sinking fund
simple interest	compound interest formula	present value of an annuity
amount	present value	amortization
simple discount	annuity	amount of annuity

$$I = Prt \qquad\qquad A = P(1 + rt) \qquad\qquad P = A - Art = A(1 - rt)$$

$$A = P(1 + i)^n \qquad r_e = \left(1 + \frac{r}{m}\right)^m - 1 \qquad V = R\,\frac{1 - (1 + i)^{-n}}{i}$$

$$a_{\overline{n}|i} = \frac{1 - (1 + i)^{-n}}{i} \qquad V = Ra_{\overline{n}|i} \qquad\qquad A = R\,\frac{(1 + i)^n - 1}{i}$$

$$s_{\overline{n}|i} = \frac{(1 + i)^n - 1}{i} \qquad A = Rs_{\overline{n}|i}$$

True-False Questions   *(Answers on page A-18)*

T   F   1. Compound interest is interest paid on previously earned interest.

T   F   2. Effective interest rate and nominal rate are the same.

T   F   3. The future value of an annuity is the sum of all payments made plus all interest accumulated.

T   F   4. The present value of an annuity is the sum of the present values of the payments.

T   F   5. To amortize a loan is to pay both interest and principal in a sequence of equal payments made over equal periods of time.

Fill in the Blanks   *(Answers on page A-18)*

1. In the formula $A = P + Prt$, P is referred to as the _____ and A as the _____.

2. After 2 years the compound amount of an initial investment of $1000 is $1217. The compound interest rate equals _____.

3. An annual rate of 10% compounded quarterly corresponds to an effective annual rate of _____.

4. The value today of a sum of money due in the future is called the _____ _____ of the sum.

5. A sequence of payments made at fixed periods of time is called an _____.

Review Exercises   *Answers to Odd-Numbered Problems begin on page A-18.*

A   In Problems 1–4 use the formula $A = P(1 + rt)$ to find the indicated quantity.

1. $A = ?,$ $P = \$1000,$ $r = 8\%,$ $t = 3$ months
2. $A = \$630,$ $P = ?,$ $r = 10\%,$ $t = 6$ months
3. $A = \$327,$ $P = \$300,$ $r = 9\%,$ $t = ?$
4. $A = \$1486,$ $P = \$1000,$ $r = ?,$ $t = 3$ months

B   In Problems 5 and 6, use the formulas $A = P(1 + i)^n$ and $P = \dfrac{A}{(1 + i)^n}$ to find the indicated quantities.

5. $A = ?$, $P = \$800$, $i = .03$, $n = 20$
6. $A = \$1500$, $P = ?$, $i = .0025$, $n = 30$

In Problems 7 and 8, use the formulas

$$A = R\,\frac{(1 + i)^n - 1}{i} \quad \text{and} \quad R = A\,\frac{i}{(1 + i)^n - 1}$$

to find the indicated quantity

7. $A = ?$, $R = \$100$, $i = .002$, $n = 40$
8. $A = \$5000$, $R = ?$, $i = .0175$, $n = 50$

In Problems 9 and 10, use the formulas

$$V = R\,\frac{1 - (1 + i)^{-n}}{i} \quad \text{and} \quad R = V\,\frac{i}{1 - (1 + i)^{-n}}$$

to find the indicated quantity

9. $V = ?$, $R = \$2000$, $i = .0125$, $n = 20$
10. $V = 2000$, $R = ?$, $i = .075$, $n = 30$

C   In Problems 11 and 12 use a calculator to solve for n to the nearest integer.

11. $3000 = 2000(1.075)^n$ 12. $1000 = 400\,\dfrac{(1.04)^n - 1}{.04}$

APPLICATIONS   13. Find the interest (I) and amount (A) if \$400 is borrowed for 9 months at 12% simple interest.

14. Dan borrows \$500 at 9% per annum simple interest for 1 year and 2 months. What is the interest charged and what is the amount of the loan?

15. Find the amount of an investment of \$100 after 2 years and 3 months at 10% compounded monthly.

16. Mike places \$200 in a savings account that pays 10% per annum compounded monthly. How much is in his account after 9 months?

17. A car dealer offers Mike the choice of two loans:
(a) \$3000 for 3 years at 12% per annum simple interest
(b) \$3000 for 3 years at 10% per annum compounded monthly
Which loan costs Mike the least?

18. A money market fund pays 9% per annum compounded monthly. How much should I invest now so that 2 years from now I will have \$100.00 in the account?

19. Katy wants to buy a bicycle that costs \$75 and will purchase it in 6 months. How much should she put in her savings account for this if she can get 10% per annum compounded monthly?

20. Mike decides he needs \$500.00 one year from now to buy a used car. If he can invest at 8% compounded monthly, how much should he save each month to buy the car?

21. Mr. and Mrs. Corey are newlyweds and want to purchase a home, but they need a

down payment of $10,000. If they want to buy their home in 2 years, how much should they save each month in their savings account that pays 10% per annum compounded monthly?

22. Mike has just purchased a used car on time and will make equal payments of $50 per month for 18 months at 12% per annum charged monthly. How much did the car actually cost?

23. Mr. and Mrs. Ostedt have just purchased an $80,000 home and made a 25% down payment. The balance can be amortized at 10% for 25 years. What are the monthly payments? How much interest will be paid? What is their equity after 5 years?

24. An inheritance of $25,000 is to be paid in equal amounts over a 5 year period at the end of each year. If the $25,000 can be invested at 10% per annum, what is the annual payment?

25. A mortgage of $125,000 is to be amortized at 9% per annum for 25 years. What are the monthly payments? What is the equity after 10 years?

26. A state has $8,000,000 worth of construction bonds that are due in 25 years. What annual sinking fund deposit is needed if the state can earn 10% per annum on its money?

27. How much should Mr. Graff pay for a gold mine expected to yield an annual return of $20,000 and to have a life expectancy of 20 years, if he wants to have a 15% annual return on his investment and he can set up a sinking fund that earns 10% a year?

28. Mr. Doody, at age 70, is expected to live for 15 years. If he can invest at 12% per annum compounded monthly, how much does he need now to guarantee himself $300 every month for the next 15 years?

29. An oil well is expected to yield an annual net return of $25,000 for the next 15 years, after which it will run dry. An investor wants a return on his investment of 20%. He can establish a sinking fund earning 10% annually. How much should he pay for the oil well?

30. Hal deposited $100 a month in an account paying 9% per annum compounded monthly for 25 years. What is the largest amount he may now withdraw monthly for the next 35 years?

31. Mr. Jones wants to save for his son's education. If he deposits $500 every 6 months at 6% compounded semiannually, how much will he have on hand at the end of 8 years?

32. How much money should be invested at 8% compounded quarterly in order to have $20,000 in 6 years?

33. Bill borrows $1000 at 5% compounded annually. He is able to establish a sinking fund to pay off the debt and interest in 7 years. The sinking fund will earn 8% compounded quarterly. What should be the size of the quarterly payments into the fund?

34. A man has $50,000 invested at 12% compounded quarterly at the time he retires. If he wishes to withdraw money every 3 months for the next 7 years, what will be the size of his withdrawals?

35. How large should monthly payments be to amortize a loan of $3000 borrowed at 12% compounded monthly for 2 years?

36. A school board issues bonds in the amount of $20,000,000 to be retired in 25 years. How much must be paid into a sinking fund at 6% compounded annually to pay off the total amount due?

37. What effective rate of interest corresponds to a nominal rate of 9% compounded monthly?

38. If an amount was borrowed 5 years ago at 6% compounded quarterly, and $6000 is owed now, what was the original amount borrowed?

39. John is the beneficiary of a trust fund set up for him by his grandparents. If the trust fund amounts to $20,000 earning 8% compounded semiannually and he is to receive the money in equal semiannual installments for the next 15 years, how much will he receive each 6 months?

40. For 20 years, a person has had on deposit $4000. At 6% annual interest compounded monthly, how much will have been saved at the end of the 20 years?

41. An employee gets paid at the end of each month and $60 is withheld from her paycheck for a retirement fund. The fund pays 1% per month (equivalent to 12% annually compounded monthly). What amount will be in the fund at the end of 30 months?

42. $6000 is borrowed at 10% compounded semiannually. The amount is to be paid back in 5 years. If a sinking fund is established to repay the loan and interest in 5 years, and the fund earns 8% compounded quarterly, how much will have to be paid into the fund every 3 months?

43. A student borrowed $4000 from a credit union toward purchasing a car. The interest rate on such a loan is 14% compounded quarterly, with payments due every quarter. The student wants to pay off the loan in 4 years. Find the quarterly payment.

Mathematical
Questions

From CPA Exams (Answers on page A-18)

1. Which of the following tables should be used to calculate the amount of the equal periodic payments that could be equivalent to an outlay of $3000 at the time of the last payment?
  (a) Amount of 1
  (b) Amount of an annuity of 1
  (c) Present value of an annuity of 1
  (d) Present value of 1

2. A businessman wants to withdraw $3000 (including principal) from an investment fund at the end of each year for 5 years. How should he compute his required initial investment at the beginning of the first year if the fund earns 6% compounded annually?
  (a) $3000 times the amount of an annuity of $1 at 6% at the end of each year for 5 years
  (b) $3000 divided by the amount of an annuity of $1 at 6% at the end of each year for 5 years
  (c) $3000 times the present value of an annuity of $1 at 6% at the end of each year for 5 years
  (d) $3000 divided by the present value of an annuity of $1 at 6% at the end of each year for 5 years

3. A businessman wants to invest a certain sum of money at the end of each year for 5 years. The investment will earn 6% compounded annually. At the end of 5 years, he will need a total of $30,000 accumulated. How should he compute his required annual investment?
  (a) $30,000 times the amount of an annuity of $1 at 6% at the end of each year for 5 years

(b) $30,000 divided by the amount of an annuity of $1 at 6% at the end of each year for 5 years

(c) $30,000 times the present value of an annuity of $1 at 6% at the end of each year for 5 years

(d) $30,000 divided by the present value of an annuity of $1 at 6% at the end of each year for 5 years

4. Shaid Corporation issued $2,000,000 of 6%, 10 year convertible bonds on June 1, 1973, at 98 plus accrued interest. The bonds were dated April 1, 1973, with interest payable April 1 and October 1. Bond discount is amortized semiannually on a straight-line basis.

   On April 1, 1974, $500,000 of these bonds were converted into 500 *shares of* $-20 par value common stock. Accrued interest was paid in cash at the time of conversion.

   What was the effective interest rate on the bonds when they were issued?

   (a) 6%
   (b) Above 6%
   (c) Below 6%
   (d) Cannot be determined from the information given.

Items 5 through 7 apply to the appropriate use of present-value tables. Given below are the present-value factors for $1.00 discounted at 8% for one to five periods. Each of the following items is based on 8% interest compounded annually from day of deposit to day of withdrawal.

Periods	Present Value of $1 Discounted at 8% per Period
1	0.926
2	0.857
3	0.794
4	0.735
5	0.681

5. What amounts should be deposited in a bank today to grow to $1000 three years from today?

   (a) $\dfrac{\$1000}{0.794}$

   (b) $1000 × 0.926 × 3

   (c) ($1000 × 0.926) + ($1000 × 0.857) + ($1000 × 0.794)

   (d) $1000 × 0.794

6. What amount should an individual have in his bank account today before withdrawal if he needs $2000 each year for 4 years with the first withdrawal to be made today and each subsequent withdrawal at 1-year intervals? (He is to have exactly a zero balance in his bank account after the fourth withdrawal.)

   (a) $2000 + ($2000 × 0.926) + ($2000 × 0.857) + ($2000 × 0.794)

   (b) $\dfrac{\$2000}{0.735} × 4$

   (c) ($2000 × 0.926) + ($2000 × 0.857) + ($2000 × 0.794) + ($2000 × 0.735)

   (d) $\dfrac{\$2000}{0.926} × 4$

7. If an individual put $3000 in a savings account today, what amount of cash would be available 2 years from today?

(a) $3000 $\times$ 0.857

(b) $3000 $\times$ 0.857 $\times$ 2

(c) $\dfrac{\$3000}{0.857}$

(d) $\dfrac{\$3000}{0.926} \times 2$

PART TWO

PROBABILITY

6

SETS; COUNTING TECHNIQUES

6.1

SETS

SET PROPERTIES AND SET NOTATION
OPERATIONS ON SETS
THE NUMBER OF ELEMENTS IN A SET
APPLICATIONS

SET PROPERTIES AND SET NOTATION

Recall from Chapter 1 that a *set* is a collection of objects considered as a whole. The objects of a set S are called *elements* of S, or *members* of S. A set that has no elements, called the *empty set* or *null set,* is denoted by the symbol $\varnothing$.

The elements of a set are not repeated. Thus, we never write {3, 2, 2} but rather write {3, 2}. Also, the order in which the elements of a set are listed does not make any difference. Thus, the three sets

$$\{3, 2, 4\} \qquad \{2, 3, 4\} \qquad \{4, 3, 2\}$$

are different listings of the same set. The *elements* of a set distinguish the set — not the order in which the elements are written.

Some other examples of sets are the following:

(a) Let

$$E = \{\text{All possible outcomes resulting from tossing a coin three times}\}$$

If we let H denote "heads" and T denote "tails," then the set E can also be written as

$$E = \{TTT, HTT, THT, TTH, HHT, HTH, THH, HHH\}$$

where, for instance, THT means the first toss resulted in tails, the second toss in heads, and the third toss in tails.

(b) Let

$$F = \{\text{Possible arrangements of all the digits}\}$$

Some typical elements of F are

$$1478906532, \qquad 4875326019, \qquad 3214569870$$

The number of elements in F is very large, so listing all of them is impractical.

GEORG F. L. P. CANTOR (1845–1918) was born in St. Petersburg, Russia. His father was a Danish merchant and his mother was a talented artist. The family, which was of Jewish descent converted to Christianity, moved to Frankfurt, Germany in 1856. Cantor was educated at Zurich and the University of Berlin. At Berlin his instructors were the famous mathematicians Kummer, Weierstrauss, and Kronecker. After receiving his Ph.D. degree in 1867, Cantor had an active professional career, but spent it at a mediocre university. At the age of 29 he published his revolutionary paper on the theory of infinite sets, a work that provided a common language for most of mathematics.

Later in this chapter we will study a technique to compute the number of elements in *F*.

Equality of Sets Let *A* and *B* be two sets. We say that *A is equal to B,* written as

$$A = B$$

if and only if *A* and *B* have the same elements. If two sets *A* and *B* are not equal, we write

$$A \neq B$$

Subset Let *A* and *B* be two sets. We say that *A is a subset of B* or that *A is contained in B,* written as

$$A \subseteq B$$

if and only if every element of *A* is also an element of *B*. If a set *A is not a subset* of a set *B*, we write

$$A \nsubseteq B$$

Of course, $A \subseteq B$ if and only if whenever $x \in A$, then $x \in B$. This latter way of interpreting the meaning of $A \subseteq B$ is useful for obtaining various laws that sets obey.

Proper Subset Let *A* and *B* be two sets. We say that *A is a proper subset of B* or that *A is properly contained in B,* written as

$$A \subset B$$

if and only if every element of the set *A* is also an element of set *B*, but there is at least one element in set *B* that is *not* in set *A*.

If a set *A* is *not* a proper subset of a set *B*, we write

$$A \not\subset B$$

The following example illustrates some uses of the three relationships, $=, \subseteq,$ and $\subset,$ just defined.

Example 1
Consider three sets *A*, *B*, and *C* given by

$$A = \{1, 2, 3\} \qquad B = \{1, 2, 3, 4, 5\} \qquad C = \{1, 2, 3\}$$

Some of the relationships between pairs of these sets are:
(a) $A = C$ (b) $A \subseteq B$ (c) $A \subseteq C$ (d) $A \subset B$ (e) $C \subseteq A$ ■

In comparing the two definitions of *subset* and *proper subset,* you should notice that if a set *A* is a subset of a set *B*, then either *A* is a proper subset of *B* or else *A* equals *B*. That is,

$$A \subseteq B \text{ if and only if either } A \subset B \text{ or } A = B$$

Also, if A is a proper subset of B, we can infer that A is a subset of B, but A does not equal B. That is,

$$A \subset B \text{ if and only if } A \subseteq B \text{ and } A \ne B$$

The distinction that is made between *subset* and *proper subset* is subtle, but important.

We can think of the relationship $\subset$ as a refinement of $\subseteq$. On the other hand, the relationship $\subseteq$ is an extension of $\subset$, in the sense that $\subseteq$ may include equality whereas with $\subset$, equality cannot be included.

Because of the way the relationship $\subseteq$ (is a subset of) has been defined, it is easy to see that for any set A, we have

$$\varnothing \subseteq A$$

Since the empty set $\varnothing$ has no elements, there is no element of the set $\varnothing$ that is not also in A.

Also, if A is any nonempty set, that is, any set having at least one element, then

$$\varnothing \subset A$$

In applications, the elements that may be considered are usually limited to some specific all-encompassing set. For example, in discussing students eligible to graduate from Midwestern University, the discussion would be limited to students enrolled at the university.

Universal Set **The *universal set* U is defined as the set consisting of all elements under consideration.**

Thus, if A is any set and if U is the universal set, then every element in A must be in U (since U consists of all elements under consideration). Hence, we may write

$$A \subseteq U$$

for *any* set A; and if A does not contain every element of U, then

$$A \subset U$$

It is convenient to represent a set as the interior of a circle. Pairs of sets are usually depicted as interlocking circles enclosed in a rectangle, which represents the universal set. Such diagrams of sets are called *Venn diagrams*. See Figure 1.

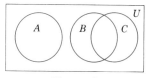

Figure 1

JOHN VENN (1834–1923), the son of a minister, graduated from Gonville and Caius College in Cambridge, England in 1853, after which he pursued theological interests as a curate in the parishes of London. In addition to his work in logic, he made important contributions to the mathematics of probability. He was an accomplished linguist, a botanist, and a noted mountaineer.

OPERATIONS ON SETS
Next, we introduce operations that may be performed on sets.

Union of Sets **Let A and B be any two sets. The *union of A with B*, written as**

$$A \cup B$$

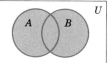

Figure 2

is defined to be the set consisting of those elements either in A or in B or in both A and B. That is,

$$A \cup B = \{x \mid x \in A \text{ or } x \in B\}$$

In the Venn diagram in Figure 2 the shaded area corresponds to $A \cup B$.

Intersection of Sets **Let A and B be any two sets. The *intersection of A with B*, written as**

$$A \cap B$$

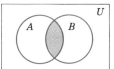

Figure 3

is defined as the set consisting of those elements that are both in A and in B. That is,

$$A \cap B = \{x \mid x \in A \text{ and } x \in B\}$$

In other words, to find the intersection of two sets A and B means to find the elements *common* to A and B. In the Venn diagram in Figure 3 the shaded region is $A \cap B$.

Example 2
For the sets

$$A = \{1, 3, 5\} \qquad B = \{3, 4, 5, 6\} \qquad C = \{6, 7\}$$

find: (a) $A \cup B$ (b) $A \cap B$ (c) $A \cap C$

Solution
(a) $A \cup B = \{1, 3, 5\} \cup \{3, 4, 5, 6\} = \{1, 3, 4, 5, 6\}$
(b) $A \cap B = \{1, 3, 5\} \cap \{3, 4, 5, 6\} = \{3, 5\}$
(c) $A \cap C = \{1, 3, 5\} \cap \{6, 7\} = \varnothing$ ∎

Example 3
Let T be the set of all taxpayers and let S be the set of all people over 65 years of age. Describe $T \cap S$.

Solution
$T \cap S$ is the set of all taxpayers who are also over 65 years of age. ∎

Disjoint Sets **If two sets A and B have no elements in common, that is, if**

$$A \cap B = \varnothing$$

then A and B are called *disjoint sets*.

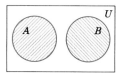

Figure 4

Two disjoint sets A and B are illustrated in the Venn diagram in Figure 4. Since the areas corresponding to A and B do not overlap anywhere, $A \cap B$ is empty.

Example 4

Let $U = \{1, 2, 3, 4, 5, 6, 7, 8\}$, $A = \{1, 3, 7\}$, and $B = \{2, 4, 6\}$. Then A and B are disjoint sets since

$$A \cap B = \varnothing$$ ∎

Example 5

Suppose that a die* is tossed. Let A be the set of outcomes in which an even number turns up; let B be the set of outcomes in which an odd number shows. Find $A \cap B$.

Solution

$$A = \{2, 4, 6\} \qquad B = \{1, 3, 5\}$$

We note that A and B have no elements in common, since an even number and an odd number cannot occur simultaneously on a single toss of a die. Therefore, $A \cap B = \varnothing$ and the sets A and B are disjoint sets.   ∎

Suppose we consider all the employees of some company as our universal set U. Let A be the subset of employees who smoke. Then all the nonsmokers will make up some subset of U that is called the *complement* of the set of smokers.

Complement   Let A be any set. The *complement of A*, written as

$$\overline{A} \quad (\text{or } A', \quad \text{or } -A)$$

is defined as the set consisting of all elements in the universal set U that are not in A. Thus,

$$\overline{A} = \{x \mid x \in U, x \notin A\}$$

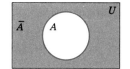

Figure 5

The shaded region in Figure 5 illustrates the complement, $\overline{A}$.

Example 6

Let

$$U = \{a, b, c, d, e, f\} \qquad A = \{a, b, c\} \qquad B = \{a, c, f\}$$

List the elements of the following sets:

(a) $\overline{A}$ (b) $\overline{B}$ (c) $\overline{A \cup B}$

(d) $\overline{A} \cap \overline{B}$ (e) $\overline{A \cap B}$ (f) $\overline{A} \cup \overline{B}$

---

* A *die* (plural *dice*) is a cube with the numbers 1, 2, 3, 4, 5, 6 showing on the six faces.

Solution
(a) $\bar{A}$ consists of all the elements in U that are not in A, namely, $\bar{A} = \{d, e, f\}$.
(b)   Similarly, $\bar{B} = \{b, d, e\}$.
(c)   To determine $\overline{A \cup B}$, we first determine the elements in $A \cup B$:

$$A \cup B = \{a, b, c, f\}$$

The complement of the set $A \cup B$ is then

$$\overline{A \cup B} = \{d, e\}$$

(d)   From parts (a) and (b) we find that

$$\bar{A} \cap \bar{B} = \{d, e\}$$

(e)   As in part (c), we first determine the elements in $A \cap B$:

$$A \cap B = \{a, c\}$$

Then,

$$\overline{A \cap B} = \{b, d, e, f\}$$

(f)   From parts (a) and (b) we find that

$$\bar{A} \cup \bar{B} = \{b, d, e, f\}$$ ■

The answers to parts (c) and (d) in Example 6 are the same, and so are the results from parts (e) and (f). This is no coincidence. There are two fundamental formulas involving intersections and unions of complements of sets. They are known as *De Morgan's laws.*

De Morgan's Laws Let A and B be any two sets.

(a) $\overline{A \cup B} = \bar{A} \cap \bar{B}$ (b) $\overline{A \cap B} = \bar{A} \cup \bar{B}$

De Morgan's laws state that all we need to do to form the complement of a union (or intersection) of sets is to form the complements of the individual sets and then change the union symbol to an intersection (or the intersection to a union). We will employ Venn diagrams to verify De Morgan's laws.

(a)   First, we draw two diagrams, as shown in Figure 6.

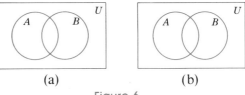

(a)   (b)

Figure 6

We will use the diagram on the left for $\bar{A} \cap \bar{B}$ and the one on the right for $\overline{A \cup B}$.

Figure 7 illustrates the completed Venn diagrams of these sets.

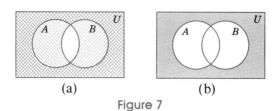

(a)   (b)

Figure 7

Thus, in Figure 7a $\overline{A} \cap \overline{B}$ is represented by the cross-hatched region and in Figure 7b $\overline{A \cup B}$ is represented by the shaded region. Since these regions correspond, this illustrates that the two sets $\overline{A} \cap \overline{B}$ and $\overline{A \cup B}$ are equal.

(b)   This verification is left to you. See Problem 30, part c.

Example 7
Use a Venn diagram to illustrate

$$A \cup B = (A \cap \overline{B}) \cup (A \cap B) \cup (\overline{A} \cap B)$$

Solution
First, we construct Figure 8.

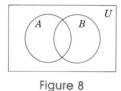

Figure 8

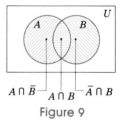

$A \cap \overline{B}$ $A \cap B$ $\overline{A} \cap B$

Figure 9

Now, shade the regions $A \cap \overline{B}$, $A \cap B$, and $\overline{A} \cap B$, as shown in Figure 9. The three regions together represent the set $A \cup B$.   ∎

The equation in Example 7 has a simple interpretation: it says that $A \cup B$ consists of

All elements in both A and B,
and all elements in A but not B,
and all elements in B but not A.

Example 8
Use a Venn diagram to illustrate

$$(A \cup B) \cap C$$

Solution
First we construct Figure 10a. Then we shade $A \cup B$ and C as in Figure 10b. The cross-hatched region of Figure 10b is the set $(A \cup B) \cap C$.

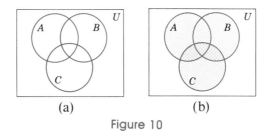

(a) (b)

Figure 10 ■

We list without any further discussion several results we will be using in later chapters:

(a) $A \cup \varnothing = A$ (b) $A \cap \varnothing = \varnothing$ (c) $A \cup \overline{A} = U$
(d) $A \cap \overline{A} = \varnothing$ (e) $\overline{U} = \varnothing$ (f) $\overline{\varnothing} = U$
(g) $\overline{(\overline{A})} = A$ (h) $A \cup U = U$ (i) $A \cap U = A$

THE NUMBER OF ELEMENTS IN A SET

When you count objects, what you are actually doing is taking each object to be counted and matching each of these objects exactly once to the counting numbers 1, 2, 3, and so on, until *no* objects remain. Even before numbers had names and symbols assigned to them, this method of counting was used. Early cavemen determined how many of their herd of cattle did not return from pasture by using rocks. As each cow left, a rock was placed aside. As each cow returned, a rock was removed from the pile. If rocks remained after all the cows returned, it was then known that some cows were missing. It is important to realize that cavemen were able to do this without developing a language or symbolism for numbers.

We will need some new notation. If A is any set, we will denote by $c(A)$ the number of elements in A. Thus, for example, for the set L of letters in the alphabet,

$$L = \{a, b, c, d, e, f, \ldots, x, y, z\}$$

we write $c(L) = 26$ and say "the number of elements in L is 26."

Also, for the set

$$N = \{1, 2, 3, 4, 5\}$$

we write $c(N) = 5$.

The empty set $\varnothing$ has no elements, and we write

$$c(\varnothing) = 0$$

If the number of elements in a set is zero or a positive integer, we say that the set is *finite*. Otherwise, the set is said to be *infinite*. The area of mathematics that deals with the study of finite sets is called *finite mathematics*.

Example 9

A survey of a group of people indicated there were 25 with brown eyes and 15 with black hair. If 10 people had both brown eyes and black hair and 23 people had neither, how many people were interviewed?

Solution

Let A denote the set of people with brown eyes and B the set of people with black hair. Then the data given tell us

$$c(A) = 25 \qquad c(B) = 15 \qquad c(A \cap B) = 10$$

Now, the number of people with either brown eyes or black hair cannot be $c(A) + c(B)$, since those with both would be counted twice. The correct procedure then would be to subtract those with both. That is,

$$c(A \cup B) = c(A) + c(B) - c(A \cap B) = 25 + 15 - 10 = 30$$

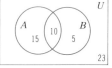

Figure 11

The sum of people found either in A or in B and those found neither in A nor in B is the total interviewed. Thus, the number of people interviewed is

$$30 + 23 = 53$$

See Figure 11.

In Example 9 we discovered the following important relationship:

> Let A and B be two finite sets. Then
>
> $$c(A \cup B) = c(A) + c(B) - c(A \cap B)$$

Example 10

Let $A = \{a, b, c, d, e\}$, $B = \{a, e, g, u, w, z\}$. Find $c(A)$, $c(B)$, $c(A \cap B)$, and $c(A \cup B)$.

Solution

$c(A) = 5$ and $c(B) = 6$. To find $c(A \cap B)$, we note that $A \cap B = \{a, e\}$ so that $c(A \cap B) = 2$. Since $A \cup B = \{a, b, c, d, e, g, u, w, z\}$, we have $c(A \cup B) = 9$. This checks with the formula

$$c(A \cup B) = c(A) + c(B) - c(A \cap B) = 5 + 6 - 2 = 9$$

APPLICATIONS

Example 11

Consumer Survey In a survey of 75 consumers, 12 indicated they were going to buy a new car, 18 said they were going to buy a new refrigerator, and 24 said they were going to buy a new stove. Of these, 6 were going to buy both a car and a refrigerator, 4 were going to buy a car and a stove, and 10 were going to buy a stove and refrigerator. One person indicated he was going to buy all three items.

(a) How many were going to buy none of these items?
(b) How many were going to buy only a car?

(c)   How many were going to buy only a stove?

(d)   How many were going to buy only a refrigerator?

(e)   How many were going to buy a car and a stove but not a refrigerator?

Solution

Denote the sets of people buying cars, refrigerators, and stoves by C, R, and S, respectively. Then we know from the data given that

$$c(C) = 12 \qquad c(R) = 18 \qquad c(S) = 24$$
$$c(C \cap R) = 6 \qquad c(C \cap S) = 4 \qquad c(S \cap R) = 10$$
$$c(C \cap R \cap S) = 1$$

We use the information given above in the reverse order and put it into a Venn diagram. Thus, beginning with the fact that $c(C \cap R \cap S) = 1$, we place a 1 in that set, as shown in Figure 12a. Now, $c(C \cap R) = 6$, $c(C \cap S) = 4$, and $c(S \cap R) = 10$.

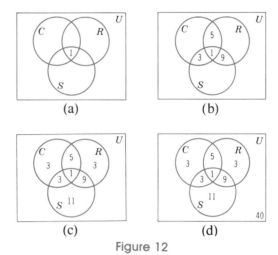

Figure 12

Thus, we place $6 - 1 = 5$ in the proper region (giving a total of 6 in the set $C \cap R$). Similarly, we place 3 and 9 in the proper regions for the sets $C \cap S$ and $S \cap R$. See Figure 12b. Now, $c(C) = 12$ and 9 of these 12 are already accounted for (in the three parts of circle C containing 5, 1, and 3 elements, respectively). Similarly $c(R) = 18$ with 15 accounted for $(5 + 1 + 9)$ and $c(S) = 24$ with 13 accounted for $(3 + 1 + 9)$. See Figure 12c. Finally, the number in $C \cup R \cup S$ is the total of 75 less those accounted for in C, R, and S, namely $3 + 5 + 1 + 3 + 3 + 9 + 11 = 35$. Thus,

$$c(\overline{C \cup R \cup S}) = 75 - 35 = 40$$

See Figure 12d. From this figure, we can see that: (a) 40 were going to buy none of the items, (b) 3 were going to buy only a car, (c) 11 were going to buy only a stove, (d) 3 were going to buy only a refrigerator, and (e) 3 were going to buy a car and a stove but not a refrigerator.   ■

Example 12

Demographic Survey In a survey of 10,281 people restricted to those who were either black or male or over 18 years of age, the following data were obtained:

Black: 3490
Male: 5822
Over 18: 4722
Black males: 1745
Over 18 and male: 859
Over 18 and black: 1341
Black male over 18: 239

The data are inconsistent. Why?

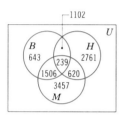

Figure 13

Solution
We denote the set of people who were black by B, male by M, and over 18 by H. Then we know that

$$c(B) = 3490 \qquad c(M) = 5822 \qquad c(H) = 4722$$
$$c(B \cap M) = 1745 \qquad c(H \cap M) = 859$$
$$c(H \cap B) = 1341 \qquad c(H \cap M \cap B) = 239$$

Since $H \cap M \cap B \neq \varnothing$, we use the Venn diagram shown in Figure 13. This means that

$$239 + 1102 + 620 + 1506 + 3457 + 643 + 2761 = 10,328$$

people were interviewed. However, it is given that only 10,281 were interviewed. This means the data are inconsistent. ■

Exercise 6.1

Answers to Odd-Numbered Problems begin on page A-18.

In Problems 1–10 replace the asterisk by the symbol(s) =, ⊂, and/or ⊆ to give a true statement. If none of these relationships holds true, write "None of these."

A
1. {1, 3, 7} * {1, 3}
2. {4, 9} * {9, 10, 4}
3. {5, 7} * {5, 8}
4. {0, 1, 4} * {0, 5, 8, 9}
5. ∅ * {1, 3}
6. {0} * {1, 3}
7. {1, 3} * {1, 3, 5, 8}
8. {5, 8, 9, 15} * {9}
9. {2, 3} * {2, 3, 6, 8, 0}
10. {2, 3} * {2, 3}

11. If $A \subseteq B$ and $B \subseteq C$, what do you conclude? Why? What if $A \subset B$ and $B \subset C$? If $A \subseteq B$ and $B \subset C$?

12. Write down all possible subsets of the set $\{a, b\}$. Which are proper subsets?

13. Write down all possible subsets of the set $\{a, b, c\}$. Which are proper subsets?

14. Write down all possible subsets of the set $\{a, b, c, d\}$. Which are proper subsets?

15. Based on your answers to Problems 12–14, can you guess the number of subsets of

$\{a, b, c, d, e\}$ without actually writing them down? Of $\{a, b, c, d, e, f\}$? Of $\{a, b, c, d, e, f, g, h, i, j\}$?

16. If the universal set is the set of people, let A denote the subset of fat people, let B denote the subset of bald people, and let C denote the subset of bald and fat people. Write down several correct relationships involving A, B, and C.

In Problems 17–24 use $A = \{1, 2, 3\}$, $B = \{3, 4, 5, 6\}$, $C = \{3, 5, 7\}$ to evaluate each set.

17. $A \cap B$ 18. $A \cap C$

19. $A \cup C$ 20. $B \cup C$

21. $(A \cup B) \cap C$ 22. $(A \cap B) \cap C$

23. $A \cup (B \cup C)$ 24. $(A \cap B) \cup C$

25. If $U =$ universal set $= \{0, 1, 2, 3, 4, 5, 6, 7, 8, 9\}$ and if $A = \{0, 1, 5, 7\}$, $B = \{2, 3, 5, 8\}$, $C = \{5, 6, 9\}$, find:

   (a) $A \cup B$ (b) $B \cap C$

   (c) $A \cap B$ (d) $\overline{A \cap B}$

   (e) $\overline{A} \cap \overline{B}$ (f) $A \cup (B \cap A)$

   (g) $(C \cap A) \cap (\overline{A})$ (h) $(A \cap B) \cup (B \cap C)$

26. If $U =$ universal set $= \{1, 2, 3, 4, 5\}$ and if $A = \{3, 5\}$, $B = \{1, 2, 3\}$, $C = \{2, 3, 4\}$, find:

   (a) $\overline{A} \cap \overline{C}$ (b) $(A \cup B) \cap C$

   (c) $A \cup (B \cap C)$ (d) $(A \cup B) \cap (A \cup C)$

   (e) $\overline{A \cap C}$ (f) $\overline{A \cup B}$

   (g) $\overline{A} \cap \overline{B}$ (h) $(A \cap B) \cup C$

27. Let
$$U = \{\text{All letters of the alphabet}\}$$
$$A = \{b, c, d\} \qquad B = \{c, e, f, g\}$$

List the elements of the sets:

   (a) $A \cup B$ (b) $A \cap B$ (c) $\overline{A} \cap \overline{B}$ (d) $\overline{A} \cup \overline{B}$

28. Let
$$U = \{a, b, c, d, e, f\} \qquad A = \{b, c\} \qquad B = \{c, d, e\}$$

List the elements of the sets:

   (a) $A \cup B$ (b) $A \cap B$ (c) $\overline{A}$

   (d) $\overline{B}$ (e) $\overline{A \cap B}$ (f) $\overline{A \cup B}$

29. Use Venn diagrams to illustrate the following sets:

   (a) $\overline{A} \cap B$ (b) $(\overline{A} \cap \overline{B}) \cup C$

   (c) $A \cap (A \cup B)$ (d) $A \cup (A \cap B)$

   (e) $(A \cup B) \cap (A \cup C)$ (f) $A \cup (B \cap C)$

   (g) $A = (A \cap B) \cup (A \cap \overline{B})$ (h) $B = (A \cap B) \cup (\overline{A} \cap B)$

30. Use Venn diagrams to illustrate the following laws:

   (a) $A \cap (B \cup C) = (A \cap B) \cup (A \cap C)$ (Distributive law)

(b) $A \cap (A \cup B) = A$ (Absorption law)

(c) $\overline{A \cap B} = \overline{A} \cup \overline{B}$ (De Morgan's law)

(d) $(A \cup B) \cup C = A \cup (B \cup C)$ (Associative law)

In Problems 31–34 use

$$A = \{x | x \text{ is a customer of IBM}\}$$
$$B = \{x | x \text{ is a secretary employed by IBM}\}$$
$$C = \{x | x \text{ is a computer operator at IBM}\}$$
$$D = \{x | x \text{ is a stockholder of IBM}\}$$
$$E = \{x | x \text{ is a member of the Board of Directors of IBM}\}$$

to describe each set.

31. $A \cap E$ **32.** $B \cap D$

33. $A \cup D$ **34.** $C \cap E$

In Problems 35–38 use

$$U = \{\text{All college students}\}$$
$$M = \{\text{All male students}\}$$
$$S = \{\text{All students who smoke}\}$$

to describe each set.

35. $M \cap S$ **36.** $\overline{M}$ **37.** $\overline{M} \cap \overline{S}$ **38.** $M \cup S$

In Problems 39–42 find the number of elements in each set.

39. $\{1, 3, 5, 7\}$ **40.** $\{0, 1, 2\}$

41. $\{0, 1, 2, 3, 4, 5, 6, 7, 8, 9\}$ **42.** $\{2, 4\}$

In Problems 43–48 use the sets $A = \{1, 2, 3, 5\}$ and $B = \{4, 6, 8\}$ to find the number of elements in each set.

43. A **44.** B **45.** $A \cap B$

46. $A \cup B$ **47.** $(A \cap B) \cup A$ **48.** $(B \cap A) \cup B$

B   In Problems 49–52 use the sets $A = \{1, 3, 6, 8\}$, $B = \{8\}$, and $C = \{8, 10\}$ to find the number of elements in each set.

49. $A \cup (B \cap C)$ **50.** $A \cup (B \cup C)$

51. $A \cap (B \cap C)$ **52.** $(A \cap B) \cup C$

53. Find $c(A \cup B)$, given that $c(A) = 4$, $c(B) = 3$, and $c(A \cap B) = 2$.

54. Find $c(A \cup B)$, given that $c(A) = 14$, $c(B) = 11$, and $c(A \cap B) = 6$.

55. Find $c(A \cap B)$, given that $c(A) = 5$, $c(B) = 4$, and $c(A \cup B) = 7$.

56. Find $c(A \cap B)$, given that $c(A) = 8$, $c(B) = 9$, and $c(A \cup B) = 16$.

57. Find $c(A)$, given that $c(B) = 8$, $c(A \cap B) = 4$, and $c(A \cup B) = 14$.

58. Find $c(B)$, given that $c(A) = 10$, $c(A \cap B) = 5$, and $c(A \cup B) = 29$.

59. Motors, Inc., manufactured 325 cars with automatic transmissions, 216 with power steering, and 89 with both these options. How many cars were manufactured if every car has at least one option?

60. Suppose that out of 1500 first-year students at a certain college, 350 are taking history, 300 are taking mathematics, and 270 are taking both history and mathematics. How many first-year students are taking history or mathematics or both? How many are taking neither?

In Problems 61–69 use the data in the figure to answer each question.

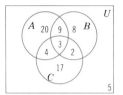

61. How many are in set A?

62. How many are in set B?

63. How many are in A or B?

64. How many are in B or C?

65. How many are in A but not B?

66. How many are in B but not C?

67. How many are in all three?

68. How many are in none?

69. How many are in A or B or C?

APPLICATIONS

70. Voting Patterns In 1948, according to a study made by Berelsa, Lazarfeld, and McPhee, the influence of religion and age on voting in Elmira, New York, was given by the following table:

	Age		
	Below 35	**35–54**	**Over 54**
Protestant Voting Republican	82	152	111
Protestant Voting Democratic	42	33	15
Catholic Voting Republican	27	33	7
Catholic Voting Democratic	44	47	33

Find:

(a) The number of voters who are Catholic or Republican or both.

(b) The number of voters who are Catholic or over 54 or both.

(c) The number of Democratic voters below 35 or over 54.

71. Enrollment Patterns The Venn diagram in the margin illustrates the number of seniors (S), female students (F), and students on the dean's list (D) at a small western college. Describe each number.

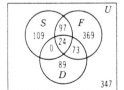

72. Enrollment Patterns At a small midwestern college:

31	female seniors were on the dean's list
62	females were on the dean's list who were not seniors
45	male seniors were on the dean's list
87	female seniors were not on the dean's list
96	male seniors were not on the dean's list
275	females were not seniors and were not on the dean's list
89	men were on the dean's list who were not seniors
227	men were not seniors and were not on the dean's list

(a)   How many were seniors?

(b)   How many were females?

(c)   How many were on the dean's list?

(d)   How many were seniors on the dean's list?

(e)   How many were female seniors?

(f)   How many were females on the dean's list?

(g)   How many were students at the college?

73. **Subscription Patterns** In a survey of 75 college students, it was found that of the three weekly news magazines *Time, Newsweek,* and *U.S. News and World Report:*

23	read *Time*
18	read *Newsweek*
14	read *U.S. News and World Report*
10	read *Time* and *Newsweek*
9	read *Time* and *U.S. News and World Report*
8	read *Newsweek* and *U.S. News and World Report*
5	read all three

(a)   How many read none of these three magazines?

(b)   How many read *Time* alone?

(c)   How many read *Newsweek* alone?

(d)   How many read *U.S. News and World Report* alone?

(e)   How many read neither *Time* nor *Newsweek?*

(f)   How many read *Time* or *Newsweek* or both?

74. **Purchasing Patterns** Of the cars sold during the month of July, 90 had air conditioning, 100 had automatic transmissions, and 75 had power steering. Five cars had all three of these extras. Twenty cars had none of these extras. Twenty cars had only air conditioning; 60 cars had only automatic transmissions; and 30 cars had only power steering. Ten cars had both automatic transmission and power steering.

(a)   How many cars had both power steering and air conditioning?

(b)   How many had both automatic transmission and air conditioning?

(c)   How many had neither power steering nor automatic transmission?

(d)   How many cars were sold in July?

(e)   How many had automatic transmission or air conditioning or both?

75. **Enrollment Patterns** A staff member at a large engineering school was presenting data to show that the students there received a liberal education as well as a scientific one. "Look at our record," she said. "Out of one senior class of 500 students, 281 are taking English, 196 are taking English and History, 87 are taking History and a foreign language, 143 are taking a foreign language and English, and 36 are taking all of these." She was fired. Why?

76. **Blood Classification** Blood is classified as being either Rh-positive or Rh-negative and according to type. If blood contains an A antigen, it is type A; if it has a B antigen, it is type B; if it has both A and B antigens, it is type AB; and if it has neither antigen, it is type O. Use a Venn diagram to illustrate these possibilities. How many different possibilities are there?

77. **Demographic Patterns** A survey of 52 families from a suburb of Chicago indi-

cated that there was a total of 241 children below the age of 18. Of these, 109 were boys; 132 were below the age of 11; 143 had played Little League; and 69 boys were below the age of 11. If 45 girls under 11 had played Little League and 30 boys under 11 had played Little League, how many children over 11 and under 18 had played Little League?

6.2

MULTIPLICATION PRINCIPLE

In this section we introduce a general principle of counting, the *multiplication principle.* We begin with three examples.

Example 1

In traveling from New York to Los Angeles, Mr. Doody wishes to stop over in Chicago. If he has five different routes to choose from in driving from New York to Chicago and has three routes to choose from in driving from Chicago to Los Angeles, in how many ways can Mr. Doody travel from New York to Los Angeles?

Solution

The task of traveling from New York to Los Angeles is composed of two consecutive operations:

Choose a route from New York to Chicago Task 1	Choose a route from Chicago to Los Angeles Task 2

In Figure 14, we see that following each of the five routes from New York to Chicago there are three routes from Chicago to Los Angeles. Thus, in all, there are $5 \cdot 3 = 15$ different routes.

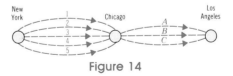

Figure 14

These 15 different routes can be enumerated as

1*A*, 1*B*, 1*C*   2*A*, 2*B*, 2*C*   3*A*, 3*B*, 3*C*   4*A*, 4*B*, 4*C*   5*A*, 5*B*, 5*C*   ■

Notice that the total number of ways the trip can be taken is simply the product of the number of ways of doing task 1 with the number of ways of doing task 2.

The different routes in Example 1 can also be depicted in a *tree diagram.* See Figure 15.

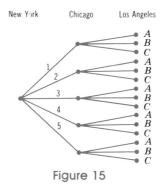

Figure 15

Example 2

In a city election, there are four candidates for mayor, three candidates for vice-mayor, six candidates for treasurer, and two for secretary. In how many ways can these four offices be filled?

Solution

The task of filling an office can be divided into four consecutive operations:

| Select a mayor | Select a vice-mayor | Select a treasurer | Select a secretary |

Corresponding to each of the 4 possible mayors, there are 3 vice-mayors. These two offices can be filled in $4 \cdot 3 = 12$ different ways. Also, corresponding to each of these 12 possibilities, we have 6 different choices for treasurer—giving $12 \cdot 6 = 72$ different possibilities. Finally, to each of these 72 possibilities there can correspond 2 choices for secretary. Thus, all told, these offices can be filled in $4 \cdot 3 \cdot 6 \cdot 2 = 144$ different ways. A partial illustration is given by the tree diagram in Figure 16.

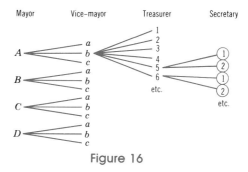

Figure 16

Example 3

A particular type of combination lock has 10 numbers on it. How many combinations of four different numbers can be formed to open the lock?

Solution
The first number can be chosen in 10 ways, then the second can be chosen in nine
ways, then the third in 8 ways, and then the fourth in 7 ways. Therefore, there are

$$10 \cdot 9 \cdot 8 \cdot 7 = 5040$$

different combinations. ■

The examples just solved demonstrate a general type of counting problem,
which can be solved by the multiplication principle.

> **Multiplication Principle** If we can perform a first task in p different ways,
> after that a second task in q different ways, after that a third task in r different
> ways, . . . , then the total act of performing the first task, followed by
> performing the second task, and so on, can be done in $p \cdot q \cdot r \cdot \ . \ . \ .$ differ-
> ent ways.

Example 4

Batting Orders

(a) How many possible batting orders can the manager of a baseball team construct from nine available players?
(b) If the manager adheres to the rule that the pitcher always bats last and the star
homerun hitter is always fourth (cleanup), then how many batting orders are
possible from nine given players?

Solution
(a) In the first slot, any of the 9 players can be chosen. In the second slot, any one
of the remaining 8 can be chosen; and so on, so that there are

$$9 \cdot 8 \cdot 7 \cdot 6 \cdot 5 \cdot 4 \cdot 3 \cdot 2 \cdot 1 = 362{,}880$$

possible batting orders.
(b) The first position can be filled in any one of 7 ways, the second in any of 6
ways, the third in any of 5 ways, the fourth has been designated, the fifth in
any of 4 ways, . . . , so that here there are

$$7 \cdot 6 \cdot 5 \cdot 4 \cdot 3 \cdot 2 \cdot 1 = 5040$$

different batting orders of 9 players after 2 have been designated. ■

Example 5
License plates in the state of Maryland consist of 3 letters of the alphabet followed
by 3 digits.

(a)   The Maryland system will allow how many possible license plates?

(b)   Of these, how many will have all their digits distinct?

Solution

(a)   There are 6 positions on the plate to be filled, the first 3 by letters and the last 3 by digits.

Positions 1, 2, and 3 can be filled in any one of 26 ways, while the remaining positions can each be filled in any of 10 ways. The total number of plates, by the multiplication principle, is then

$$26 \cdot 26 \cdot 26 \cdot 10 \cdot 10 \cdot 10 = 17,576,000$$

(b)   Here, the tasks involved in filling the digit positions are slightly different. The first digit can be any one of 10, but the second digit can be only any one of 9 (we cannot duplicate the first digit); there are the 8 choices for the third digit (we cannot duplicate either the first or the second). Thus there are

$$26 \cdot 26 \cdot 26 \cdot 10 \cdot 9 \cdot 8 = 12,654,720$$

plates with no repeated digit.   ■

Exercise 6.2

Answers to Odd-Numbered Problems begin on page A-19.

A   1.   There are two roads between towns A and B. There are 4 roads between towns B and C. How many different routes may one travel between towns A and C?

2.   A woman has 4 blouses and 5 skirts. How many different outfits can she wear?

3.   XYZ Company wants to build a complex consisting of a factory, office building, and warehouse. If the building contractor has 3 different kinds of factories, 2 different office buildings, and 4 different warehouses, how many models must be built to show all possibilities to XYZ Company?

4.   Cars, Inc., has 3 different car models and 6 color schemes. If you are one of the dealers, how many cars must you display to show each possibility?

5.   A man has 3 pairs of shoes, 8 pairs of socks, 4 pairs of slacks, and 9 sweaters. How many outfits can he wear?

6.   There are 14 teachers in a math department. A student is asked to indicate her favorite and her least favorite. In how many ways is this possible?

7.   A house has 3 doors and 12 windows. In how many ways can a burglar rob the house by entering through a window and exiting through a door?

8.   A man has 4 pairs of gloves. In how many ways can he select a right-hand glove and a left-hand glove that do not match?

9.   A corporation has a board of directors consisting of 12 members. The board must select from its members a chairman, a vice chairman, and a secretary. In how many ways can this be done?

10.   How many license plates consisting of 2 letters followed by 2 digits are possible?

B   11.   A restaurant offers 6 different salads, 5 different main courses, 10 different desserts, and 4 different drinks. How many different lunches—each consisting of a salad, a main course, a dessert, and a drink—are possible?

12. A quiz consists of 4 multiple-choice questions with 5 possible responses to each question. In how many different ways can the quiz be answered?

13. How many ways can 6 people be seated in a row of 6 seats? 8 people in a row of 8 seats?

14. How many different 5-letter words can be formed using the 5 letters of the word "CLIPS"?

15. How many four-letter code words are possible from the first of 6 letters of the alphabet with no letters repeated? Allowing letters to repeat?

16. How many outcomes are there for the experiment of flipping 2 coins?

17. How many outcomes are there for the experiment of flipping 4 coins?

18. Refer to Problem 17. In how many of the outcomes are the first and the last toss identical?

19. How many ways are there to rank 7 candidates who apply for a job?

20. (a) How many different ways are there to arrange the 7 letters in the word "PROB-LEM"? (b) If we insist that the letter P come first, how many ways are there? (c) If we insist that the letter P come first and the letter M be last, how many ways are there?

21. On a math test there are 10 multiple-choice questions with 4 possible answers and 15 true-false questions. In how many possible ways can the 25 questions be answered?

22. Using the digits 1, 2, 3, and 4, how many different 4-digit numbers can be formed?

23. How many different license plate numbers can be made using 2 letters followed by 4 digits selected from the digits 0 through 9, if

   (a) Letters and digits may be repeated?
   (b) Letters may be repeated, but digits may not be repeated?
   (c) Neither letters nor digits may be repeated?

C 24. Security A system has seven switches, each of which may be either open or closed. The state of the system is described by indicating for each switch whether it is open or closed. How many different states of the system are there?

25. Psychology Testing In an ESP experiment a person is asked to select and arrange 3 cards from a set of 6 cards labeled A, B, C, D, E, and F. Without seeing the card a second person is asked to give the arrangement he thinks he perceives. Determine the number of possible responses by the second person if he simply guesses.

26. Find the number of 7-digit telephone numbers

   (a) With no repeated digits (lead 0 is allowed).
   (b) With no repeated digits (lead 0 not allowed).
   (c) With repeated digits allowed including a lead 0.

27. Product Choice An automobile manufacturer produces 3 different models. Models A and B can come in any of 3 body styles; Model C can come in only 2 body styles. Each car also comes in either black or green. How many distinguishable car types are there? *Hint:* Use a tree diagram.

28. A combination lock is unlocked by dialing a sequence of numbers, first to the right, then to the left, and to the right again. If there are 10 digits on the dial, determine the number of possible combinations.

29. Opinion Polls An opinion poll is to be conducted among college students. Eight multiple-choice questions, each with 3 possible answers, will be asked. In how many different ways can a student complete the poll, if exactly one response is given to each question?

30. Path Selection in a Maze The maze below is constructed so that a rat must pass through a series of one-way doors. How many different paths are there from start to finish?

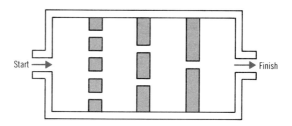

31. How many 3-letter code words are possible using the first 10 letters of the alphabet if:

(a) No letter can be repeated?

(b) Letters can be repeated?

(c) Adjacent letters cannot be the same?

6.3

PERMUTATIONS

FACTORIAL
PERMUTATION FORMULA

In the next two sections we use the multiplication principle to discuss two general types of counting problems, called *permutations* and *combinations*. These concepts arise often in applications, especially in probability.

FACTORIAL

Before discussing the nature of permutations, we introduce a useful shorthand notation — the *factorial symbol.*

Factorial **The symbol $n!$, read as "n factorial," means**

$$0! = 1$$
$$1! = 1$$
$$2! = 2 \cdot 1 \qquad = 2$$
$$3! = 3 \cdot 2 \cdot 1 \quad = 6$$
$$4! = 4 \cdot 3 \cdot 2 \cdot 1 = 24$$

and in general, for n an integer ≥ 1,

$$n! = n \cdot (n-1) \cdot (n-2) \, \cdots \, 3 \cdot 2 \cdot 1$$

Thus, to compute $n!$, we find the product of all consecutive integers from 1 to n inclusive.

A formula we will find useful is

$$n! = n \cdot (n - 1)! \qquad n \geq 1$$

Example 1

(a) $4! = (4)(3)(2)(1) = 24$

(b) $\dfrac{5!}{4!} = \dfrac{5 \cdot 4!}{4!} = 5$

(c) $\dfrac{52!}{5!47!} = \dfrac{52 \cdot 51 \cdot 50 \cdot 49 \cdot 48 \cdot 47!}{5 \cdot 4 \cdot 3 \cdot 2 \cdot 47!} = 2{,}598{,}960$

(d) $\dfrac{7!}{(7 - 5)!5!} = \dfrac{7!}{2!5!} = \dfrac{7 \cdot 6 \cdot 5!}{2!5!} = \dfrac{7 \cdot 6}{2} = 21$

(e) $\dfrac{50 \cdot 49 \cdot 48 \cdot 47 \cdot 46}{50!} = \dfrac{50 \cdot 49 \cdot 48 \cdot 47 \cdot 46}{50 \cdot 49 \cdot 48 \cdot 47 \cdot 46 \cdot 45!} = \dfrac{1}{45!}$ ■

It is interesting to note that $n!$ grows very quickly. Compare the following

$$5! = 120$$
$$10! = 3{,}628{,}800$$
$$15! = 1{,}307{,}674{,}368{,}000$$

PERMUTATION FORMULA

We start with an example.

Example 2

Suppose we are setting up a code of three-letter words and have six different letters, a, b, c, d, e, and f, from which to choose. If the code must not repeat any letter and if such words as abc and bac are considered different, how many different words can be formed?

Solution

We solve the problem by using the multiplication principle. In selecting a first letter, we have 6 choices. Since whatever letter is chosen cannot be repeated, we have 5 choices available for the second letter and 4 for the last letter. In all, then, there are $6 \cdot 5 \cdot 4 = 120$ words of three letters that can be formed. See Figure 17 for a partial tree diagram of this solution. ■

A way of rephrasing the question posed in Example 2 would be to ask: How many ordered arrangements using three distinct letters can be formed from the six letters a through f?

In general, we could ask: Given n distinct objects, how many ordered arrangements can be formed using r of the objects?

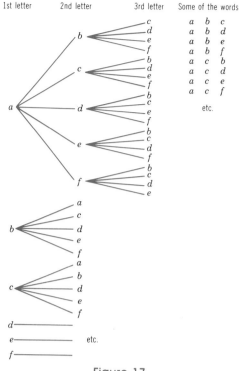

Figure 17

Permutation **Ordered arrangements of objects are called *permutations*. An *r*-permutation of a set of *n* objects is an ordered arrangement using *r* of the *n* objects. *P*(*n*, *r*) is defined to be the *number* of *r*-permutations of a set of *n* distinct objects.***

P(*n*, *r*) is also referred to as the number of *permutations of n different objects taken r at a time.*

Thus, *P*(*n*, *r*) is the number of ordered arrangements that can be formed using *r* objects chosen from a set of *n* distinct objects. To help you feel more comfortable, here is a list of short problems with their solutions given in *P*(*n*, *r*) notation.

Find the Number of	**Solution**
Ways of choosing 5 people from a group of 10 and arranging them in a line.	$P(10, 5)$
Six-letter "words" that can be formed with no letter repeated.	$P(26, 6)$
Seven-digit telephone numbers, with no repeated digit (allow 0 for a first digit).	$P(10, 7)$
Ways of arranging 8 people in a line.	$P(8, 8)$

Note that in all of the above examples, *order is important.*

---

* Sometimes the notation *nPr* is used for *P*(*n*, *r*).

We now proceed to find a formula for $P(n, r)$.

In computing $P(n, r)$, we want to find the number of possible different arrangements of r quantities that are chosen from n different quantities in which no item is repeated and order is important. The first entry can be filled by any one of the n possibilities, the second by any one of the remaining $(n - 1)$, the third by any one of the now remaining $(n - 2)$, and so on. Since there are r positions to be filled, the number of possibilities is

$$P(n, r) = \underbrace{n(n - 1)(n - 2) \cdots (n - r + 1)}_{r \text{ factors}}$$

For example, if $n = 6$ and $r = 2$,

$$P(6, 2) = \underbrace{6 \cdot 5}_{2 \text{ factors}} = 30$$

Other examples are

$$P(7, 3) = \underbrace{7 \cdot 6 \cdot 5}_{3 \text{ factors}} = 210 \qquad P(5, 5) = \underbrace{5 \cdot 4 \cdot 3 \cdot 2 \cdot 1}_{5 \text{ factors}} = 5! = 120$$

To obtain the last factor in the expression for $P(n, r)$, we observe the following pattern:

First factor is n

Second factor is $n - 1$

Third factor is $n - 2$

.

.

.

rth factor is $n - (r - 1) = n - r + 1$

The number of different arrangements using r objects chosen from n objects in which

1. The n objects are all different
2. No object is repeated more than once in an arrangement
3. Order is important

is given by the formula

$$P(n, r) = n(n - 1) \ldots (n - r + 1)$$

Multiplying the right side by 1 in the form of $(n - r)!/(n - r)!$, we obtain

$$P(n, r) = n(n - 1)(n - 2) \ldots (n - r + 1)\frac{(n - r)!}{(n - r)!}$$

Thus, since $n(n - 1)(n - 2) \ldots (n - r + 1)(n - r)! = n!$, we get the following useful formula for $P(n, r)$:

$$P(n, r) = \frac{n!}{(n - r)!}$$

Based on this formula we have

$$P(n, 0) = 1$$
$$P(n, 1) = n$$

Example 3
There are eight different mathematics books and six different computer science books. How many ways can a shelf arrangement using five of the mathematics books be formed?

Solution
We are seeking the number of arrangements using five of the eight mathematics books. The answer is given by

$$P(8, 5) = \frac{8!}{(8 - 5)!} = \frac{8!}{3!}$$
$$= 8 \cdot 7 \cdot 6 \cdot 5 \cdot 4 = 6720$$

Example 4
A student has six questions on an examination and is allowed to answer the questions in any order. In how many different orders could the student answer these questions?

Solution
The student wants the number of ordered arrangements of the six questions using all six of them. The number is given by

$$P(6, 6) = \frac{6!}{(6 - 6)!} = \frac{6!}{0!} = \frac{6!}{1} = 720$$

In general the number of permutations (arrangements) of n different objects using all n of them is given by

$$P(n, n) = n!$$

So, in a class of n students, there are $n!$ ways of coercing all the students into a straight line.

Example 5
From the eight mathematics books and the six computer science books of Example 3, seven positions on a shelf are to be filled. If the first four positions are to be occupied by math books and the last three by computer science books, in how many ways can this be done?

Solution

We think of the problem as consisting of two tasks. Task 1 is to fill the first four positions with four of the eight mathematics books. This can be done in $P(8, 4)$ ways. Task 2 is to fill the remaining three positions with three of six computer books. This can be done in $P(6, 3)$ ways. By the multiplication principle, the seven positions can be filled as specified in

$$P(8, 4) \cdot P(6, 3) = \frac{8!}{4!} \cdot \frac{6!}{3!} = 8 \cdot 7 \cdot 6 \cdot 5 \cdot 6 \cdot 5 \cdot 4 = 201,600 \text{ ways} \quad \blacksquare$$

Exercise 6.3

Answers to Odd-Numbered Problems begin on page A-19.

In Problems 1–20 evaluate each expression.

A

1. $\dfrac{5!}{2!}$ 2. $\dfrac{8!}{2!}$ 3. $\dfrac{6!}{3!}$ 4. $\dfrac{9!}{3!}$

5. $\dfrac{10!}{8!}$ 6. $\dfrac{11!}{9!}$ 7. $\dfrac{9!}{8!}$ 8. $\dfrac{10!}{9!}$

9. $\dfrac{8!}{2!6!}$ 10. $\dfrac{9!}{3!6!}$ 11. $P(7, 2)$ 12. $P(5, 1)$

13. $P(8, 1)$ 14. $P(6, 6)$ 15. $P(5, 0)$ 16. $P(6, 4)$

17. $\dfrac{8!}{(8 - 3)!3!}$ 18. $\dfrac{9!}{(9 - 5)!5!}$ 19. $\dfrac{6!}{(6 - 6)!6!}$ 20. $\dfrac{7!}{(0 - 0)!7!}$

21. (a) How many different ways are there to arrange the 6 letters in the word SUN-DAY?

    (b) If we insist that the letter S come first, now how many ways are there?

    (c) If we insist that the letter S come first and the letter Y be last, now how many ways are there?

22. How many ways are there to rank 8 candidates who apply for a job?

23. From a pool of 10 job applicants, a list ranking the top 4 must be made. How many such lists are possible?

24. How many different 5-letter "words" (sequences of letters) can be formed from the standard alphabet if repeated letters are not allowed? If repeated letters are allowed?

25. A station wagon has 9 seats. In how many different ways can 5 people be seated in it?

B 26. There are 5 different French books and 5 different Spanish books. How many ways are there to arrange them on a shelf if

    (a) Books of the same language must be grouped together?

    (b) French and Spanish books must alternate in the grouping?

27. In how many ways can 8 different books be distributed to 12 children if no child gets more than one book?

28. A computer must assign each of 4 outputs to one of 8 different printers. In how many ways can it do this provided no printer gets more than one output?

29. **Lottery Tickets** From the 1500 lottery tickets that are sold, 3 tickets are to be selected for first, second, and third prizes. How many possible outcomes are there?

30. Personnel Assignment A salesperson is needed in each of 7 different sales territories. If 10 equally qualified persons apply for the jobs, in how many ways can the jobs be filled?

31. Slate Assignment A club has 15 members. In how many ways can they choose a slate of four officers consisting of a president, vice president, secretary and treasurer?

32. Name Assignment A newborn child is to be given a first name and a middle name from a selection of 15 names. How many different possibilities are there?

33. Scheduling How many basketball games are played in the Big Ten, if every team hosts every other team once?

6.4

COMBINATIONS

| COMBINATION FORMULA
| PASCAL'S TRIANGLE

Permutations focus on the order in which objects are arranged. However, in many cases, order is not important. For example, in a draw poker hand, the order in which you receive the cards is not relevant — all that matters is what cards are received. That is, with poker hands, we are concerned only with what *combination* of cards we have — not the particular order of the cards.

COMBINATION FORMULA

To further emphasize the distinction between ordered and unordered selections, suppose we have four letters a, b, c, d, and wish to choose two of them without repeating any letter. If order is important, then we have $P(4, 2) = 4 \cdot 3 = 12$ possible arrangements, namely

$$ab, ac, ad \quad bc, bd, cd \quad ba, ca, da \quad cb, db, dc$$

If order is not a consideration, we have only six selections, namely,

$$ab, ac, ad, bc, bd, cd$$

Notice that in this example there are twice as many permutations (ordered selections) as unordered selections. The reason is that any selection can occur in one of two orderings — for example, the arrangements ab and ba both correspond to the unordered selection ab. Unordered selections are called *combinations*.

Combinations $C(n, r)$ is defined to be the *number* of ways of choosing r distinct objects from a set of n distinct objects, without regard to the order of the selection.*

---

* Sometimes the notation nCr or $\binom{n}{r}$ is used for $C(n, r)$.

$C(n, r)$ is also referred to as the number of *combinations of n objects taken r at a time.*

The following lists problems whose solutions are given in $C(n, r)$ notation.

Find the Number of	Solution
Ways of selecting four people from a group of six	$C(6, 4)$
Five-card unordered poker hands	$C(52, 5)$
Committees of six that can be formed from the U.S. Senate (100 members)	$C(100, 6)$
Ways of selecting five courses from a catalog containing 200	$C(200, 5)$

In order to obtain a formula for $C(n, r)$, we observe that each unordered selection of r objects from a set of n can be arranged or permuted in $r!$ different ways.

For example, the unordered selection a, b, c gives rise to the orderings listed below.

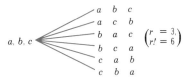

Thus, since each unordered selection yields $r!$ orderings or permutations, the number of permutations of r objects chosen from a set of n is $r!$ times as big as the number of unordered selections of r objects from a set of n.

That is,

$$r!C(n, r) = P(n, r)$$

and if we use our previously developed formula for $P(n, r)$,

$$C(n, r) = \frac{P(n, r)}{r!} = \frac{n!}{r!(n - r)!}$$

The number of different selections of r objects chosen from n objects in which

1. The n objects are all different
2. No object is repeated
3. Order is not important

is given by the formula

$$C(n, r) = \frac{P(n, r)}{r!} = \frac{n!}{r!(n - r)!}$$

Based on this formula we have

$$C(n, 0) = 1$$
$$C(n, 1) = n$$
$$C(n, n) = 1$$

Example 1

Compute the following numbers:

(a) $C(50, 2)$ (b) $C(7, 5)$ (c) $C(7, 7)$ (d) $C(7, 0)$

Solution

(a) $C(50, 2) = \dfrac{50!}{2!(50 - 2)!} = \dfrac{50!}{2!48!} = 1225$

(b) $C(7, 5) = \dfrac{7!}{5!(7 - 5)!} = \dfrac{7!}{5!2!} = 21$

(c) $C(7, 7) = \dfrac{7!}{7!(7 - 7)!} = \dfrac{7!}{7!0!} = 1$

(d) $C(7, 0) = \dfrac{7!}{0!(7 - 0)!} = \dfrac{7!}{0!7!} = 1$

Example 2

From a deck of 52 cards, a hand of 5 cards is dealt. How many different hands are possible?

Solution

Such a hand is an unordered selection of 5 cards from a deck of 52. So, the number of different hands is

$$C(52, 5) = \frac{52!}{5!47!} = \frac{52 \cdot 51 \cdot 50 \cdot 49 \cdot 48}{5 \cdot 4 \cdot 3 \cdot 2 \cdot 1}$$
$$= 2{,}598{,}960$$

Example 3

Computer Science A bit is a 0 or a 1. A 6-bit string is a sequence of length 6 consisting of 0's and 1's. How many 6-bit strings contain

(a) Exactly one 1
(b) Exactly two 1's?

Solution

(a) To form a 6-bit string having one 1, we need only specify where the single 1 is located (the other positions are 0's). The location for the 1 can be chosen in

$$C(6, 1) = 6 \text{ ways}$$

(b) Here, we must choose two of the 6 positions to contain 1's. Hence there are

$$C(6, 2) = 15 \text{ such strings}$$

Example 4

Sampling   A sociologist needs a sample of 12 welfare recipients located in a large metropolitan area. He divides the city into 4 areas — northwest, northeast, southwest, southeast. Each section contains 25 welfare recipients. The sociologist may select the 12 recipients in any way he wants — all from the same area, 2 from the southwest area and 10 from the northwest area, and so on. How many different groups of 12 recipients are there?

Solution
Since order of selection is not important and since the selection is of 12 things from a possible $4 \cdot 25 = 100$ things, there are $C(100, 12)$ different groups. That is,

$$C(100, 12) = \frac{100!}{12!88!}$$

∎

Example 5
From five faculty members and four students, a committee of four is to be chosen that includes two students and two faculty members. In how many ways can this be done?

Solution
The faculty members can be chosen in $C(5, 2)$ ways. The students can be chosen in $C(4, 2)$ ways. By the multiplication principle, there are then

$$C(5, 2) \cdot C(4, 2) = \frac{5!}{2!3!} \cdot \frac{4!}{2!2!} = 10 \cdot 6 = 60 \text{ different ways}$$

∎

PASCAL'S TRIANGLE
Sometimes the notation $\binom{n}{r}$, read as "n choose r," is used in place of $C(n, r)$. $\binom{n}{r}$ is called a *binomial coefficient* because of its connection with the binomial theorem (discussed in Section 6.6). A triangular display of $\binom{n}{r}$ for $n = 0$ to $n = 6$ is given in Figure 18. This triangular display is called *Pascal's triangle.*

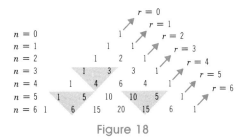

Figure 18

BLAISE PASCAL (1623–1662) is most well-known for his creation of a theory of probability. He constructed the first computer and later in life became interested in theology, contributing several literary masterpieces in this area. The computer language PASCAL is named for him.

For example, $\binom{5}{2} = 10$ is found in the row marked $n = 5$ and on the diagonal marked $r = 2$.

In Pascal's triangle, each entry can be obtained by adding the two nearest entries in the row above it. The shaded triangles in Figure 18 illustrate this. For example, $10 + 5 = 15$, etc.

Pascal's triangle, as the figure indicates, is symmetric. That is, when n is even, the largest entry occurs in the middle, and corresponding entries on either side are equal. When n is odd, there are two equal middle entries with corresponding equal entries on either side.

The reasons behind these properties of Pascal's triangle as well as other properties of binomial coefficients are discussed in Section 6.6.

Exercise 6.4

Answers to Odd-Numbered Problems begin on page A-19.

In Problems 1–8 find the value of each expression.

1. $C(6, 4)$ 2. $C(5, 4)$ 3. $C(7, 2)$ 4. $C(8, 7)$
5. $\binom{5}{1}$ 6. $\binom{8}{1}$ 7. $\binom{8}{6}$ 8. $\binom{8}{4}$

9. In how many ways can a committee of 3 senators be selected from a group of 8 senators?

10. In how many ways can a committee of 4 representatives be selected from a group of 9 representatives?

11. A math department is allowed to tenure 4 of 17 eligible teachers. In how many ways can the selection for tenure be made?

12. How many different hands are possible in a bridge game? (A bridge hand consists of 13 cards dealt from a deck of 52 cards.)

13. There are 25 students in the Math Club. How many ways can 3 officers be selected?

14. How many different relay teams of 4 persons can be chosen from a group of 10 runners?

15. A basketball team has 6 players who play at guard (2 of 5 starting positions). How many different teams are possible, assuming the remaining 3 positions are filled and it is not possible to distinguish a left guard from a right guard?

16. On a basketball team of 12 players, 2 play only at center, 3 play only at guard, and the rest play at forward (5 men on a team: 2 forwards, 2 guards, and 1 center). How many different teams are possible, assuming it is not possible to distinguish left and right guards and left and right forwards?

17. The Student Affairs Committee has 3 faculty, 2 administration members, and 5 students on it. In how many ways can a subcommittee of 1 faculty, 1 administration member, and 2 students be formed?

18. Of 1352 stocks traded in 1 day on the New York Stock Exchange, 641 advanced, 234 declined, and the remainder were unchanged. In how many ways can this happen?

19. How many different ways can an offensive football team be formed from a squad that consists of 20 linemen, 3 quarterbacks, 8 halfbacks, and 4 fullbacks? This football team must have 1 quarterback, 2 halfbacks, 1 fullback, and 7 linemen.

20. How many different ways can a baseball team be made up from a squad of 15 players,

if 3 players are used only as pitchers and the remaining players can be placed at any position except pitcher (9 players on a team)?

21. A little girl has 1 penny, 1 nickel, 1 dime, 1 quarter, and 1 half dollar in her purse. If she pulls out 3 coins, how many different sums are possible?

22. How many 8-bit strings contain

    (a) Exactly two 1's?
    (b) Exactly three 1's?

23. A state of Maryland million dollar lottery ticket consists of 6 distinct numbers chosen from the range 00 through 99. The order in which the numbers appear on the ticket is irrelevant. How many distinct lottery tickets can be issued?

24. How many poker hands contain all spades?

25. How many committees of 5 can be formed from members of the U. S. Senate?

26. A test has 3 parts. In part 1 a student must do 3 of 5 questions, in part 2 a student must choose 2 of 4 questions, and in part 3 a student must pick 3 of 6 questions. In how many different ways can a student complete the test?

27. A student has 7 books on a shelf. In how many different ways can she select a set of 3?

28. How many different committees can be formed from a group of 7 women and 9 men if a committee is composed of 3 women and 3 men?

29. A bag contains 14 balls: 8 red and 6 white. Suppose an experiment consists of drawing out 3 of the balls. How many possibilities are there?

30. In how many different ways can a panel of 12 jurors and 2 alternate jurors be chosen from a group of 40 prospective jurors?

31. **Test Panel Selection** A sample of 8 persons is selected for a test from a group containing 40 smokers and 15 nonsmokers. In how many ways can the 8 persons be selected?

32. **Resource Allocation** A trucking company has 8 trucks and 6 drivers available when requests for 4 trucks are received. How many different ways are there of selecting the trucks and the drivers to meet these requests.

33. **Group Selection** From a group of 5 people, we are required to select a different person to participate in each of 3 different tests. In how many ways can the selections be made?

34. **Congressional Committees** In the U.S. Congress, a conference committee is to be composed of 5 senators and 4 representatives. In how many ways can this be done? (There are 435 representatives and 100 senators.)

35. **Quality Control** A box contains 24 light bulbs. The quality control engineer will pick a sample of 4 light bulbs for inspection. How many different samples are possible?

36. **Mating** A horse stable has 12 mares and 4 stallions. How many different ways can they be mated?

37. **Rating** A sportswriter makes a preseason guess of the top 15 university basketball teams (in order) from among 50 major university teams. How many different possibilities are there?

38. **Packaging** A manufacturer produces 8 different items. He packages assortments of equal parts of 3 different items. How many different assortments can be packaged?

39. The digits 0 through 9 are written on 10 cards. Four different cards are drawn, and a 4-digit number is formed. How many different 4-digit numbers can be formed in this way?

40. Investment Selection An investor is going to invest $21,000 in 3 stocks from a list of 12 prepared by his broker. How many different investments are possible if:

(a) $7000 is to be invested in each stock?
(b) $10,000 is to be invested in one stock, $6000 in another, and $5000 in the third?
(c) $8000 is to be invested in each of 2 stocks and $5000 in a third stock?

6.5

MORE COUNTING TECHNIQUES

EXAMPLES
PERMUTATIONS WITH REPETITION

In this section we consider some counting problems that will be useful in our discussion of probability and that are also meant to sharpen your counting skills.

EXAMPLES

The first example deals with a coin-tossing experiment in which a coin is tossed a fixed number of times. There are exactly two possible outcomes on each trial or toss (heads, H, or tails, T). For instance, in tossing a coin three times, one possible outcome is HTH—heads on the first toss, tails on the second toss, and heads on the third toss.

Example 1
Suppose an experiment consists of tossing a fair coin eight times.

(a)   How many different outcomes are possible?
(b)   How many different outcomes have exactly 3 tails?
(c)   How many outcomes have at most 2 tails?
(d)   How many outcomes have at least 3 tails?

Solution
(a)   Each outcome of the experiment consists of a sequence of eight letters H or T, where the first letter records the result of the first toss, the second letter the result of the second toss, and so forth. Thus, the process can be visualized as filling an empty box at each toss with either an H or a T.

1st toss	T							

2nd toss	T	H						

3rd toss	T	H	H					

$\vdots$

8th toss	T	H	H	H	T	H	H	T

Since each box can be filled in two ways, by the multiplication principle, the sequence of eight boxes can be filled in

$$\underbrace{2 \cdot 2 \cdot 2 \cdot \ \cdot \ \cdot \ \cdot \ 2}_{8 \text{ factors}} = 2^8 = 256 \text{ ways}$$

Thus, there are $2^8 = 256$ different possible outcomes.

(b)   Any sequence that contains exactly 3 T's must contain 5 H's. A particular outcome is determined as soon as we decide where to place the T's in the eight boxes. The three boxes to receive the T's can be selected from the eight boxes in $C(8, 3)$ different ways. So the number of outcomes with exactly 3 tails is

$$C(8, 3) = \frac{8!}{3!5!} = \frac{8 \cdot 7 \cdot 6}{3!} = 8 \cdot 7 = 56$$

(c)   The outcomes with at most 2 tails correspond to the sequences with 0, 1, or 2 T's, and they are:

      0 T. Only one outcome is possible, namely *HHHHHHHH*.

      1 T. These outcomes, which include *THHHHHHH* and *HHHHHTHH*, are determined by selecting one box out of eight in which to place the single T. This can be done in $C(8, 1) = 8$ ways.

      2 T. These outcomes, which include *HTHHTHHH* and *HHHHHTTH*, are determined by selecting two boxes out of eight in which to place the two T's. This can be done in $C(8, 2) = 28$ ways.

Thus, the number of outcomes with at most 2 tails is just the sum of all these results, which is $1 + 8 + 28 = 37$.

(d)   The outcomes with at least 3 tails are the results with 3, 4, 5, 6, 7, or 8 tails. The total number of such outcomes is

$$C(8, 3) + C(8, 4) + C(8, 5) + C(8, 6) + C(8, 7) + C(8, 8)$$

But there is a simpler way of obtaining the answer. If we start with the total number of outcomes obtained in part (a) and subtract the number of outcomes of at most 2 tails obtained in part (c), we get the number of outcomes with at least 3 tails:

$$256 - 37 = 219 \text{ ways}$$

∎

Part (d) of Example 1 illustrates a counting technique that is often useful: Count the outcomes that are not favorable to you and subtract from the total number of outcomes. This "backdoor" approach can be easier at times than a direct attack. Here is another example of its use.

Example 2

Find the number of seven-digit telephone numbers that have at least one repeated digit. Lead 0's are allowed.

Solution

A direct application of the multiplication principle shows that there are 10^7 total possible seven-digit telephone numbers. The number of telephone numbers that have *no* repeated digits is

$$P(10, 7) = \frac{10!}{3!} = 604,800$$

Hence, the number that have at least one repeated digit is

$$10^7 - \frac{10!}{3!} = 9,395,200$$

∎

Example 3

An urn contains 8 white balls and 4 red balls. Four balls are selected. In how many ways can the 4 balls be drawn from the total of 12 balls:

(a) If the color is not considered?
(b) If 3 balls are white and 1 is red?
(c) If all 4 balls are white?
(d) If all 4 balls are red?

Solution

(a) Since order is not important, this experiment is simply a selection of 4 balls out of 12. There are $C(12, 4)$ possible ways to select the balls, that is,

$$C(12, 4) = \frac{12!}{4!8!} = \frac{12 \cdot 11 \cdot 10 \cdot 9}{4 \cdot 3 \cdot 2 \cdot 1} = 495 \text{ ways}$$

(b) The desired answer involves two operations: first, the selection of 3 white balls from 8; and second, the selection of 1 red ball from 4:

Select 3 white balls Operation 1	Select 1 red ball Operation 2

The first operation can be performed in $C(8, 3)$ ways; the second operation can be performed in $C(4, 1)$ ways. By the multiplication principle, the answer is

$$C(8, 3) \cdot C(4, 1) = \frac{8!}{3!5!} \cdot \frac{4!}{1!3!} = 224$$

(c) Since all 4 balls must be selected from the 8 that are white, the answer is

$$C(8, 4) = \frac{8!}{4!4!} = \frac{8 \cdot 7 \cdot 6 \cdot 5}{4 \cdot 3 \cdot 2 \cdot 1} = 70 \text{ ways}$$

(d) Since the 4 red balls must be selected from 4 red balls, the answer is 1 way. ∎

PERMUTATIONS WITH REPETITION

Our previous discussion of permutations required that the objects we were rearranging be distinct. We now examine what happens when repetitions are allowed. The following example shows that allowing repetition of some objects introduces modifications.

Example 4

How many three-letter words (real or imaginary) can be formed from the letters in the word

(a) MAD   (b) DAD?

Solution

(a) The three distinct letters in MAD can be rearranged in

$$P(3, 3) = 3! = 6 \text{ ways}$$

(b) Straightforward listing shows that there are only three ways of rearranging the letters in the word DAD:

DAD, DDA, and ADD ∎

The word DAD in Example 4 contains 2 D's, and it would seem that it is this duplication that results in fewer rearrangements for DAD than for MAD. In the next example we describe a way of dealing with the problem of duplication.

Example 5

How many distinct "words" can be formed using all the letters of the word

M A M M A L   ?

Solution

Any such word will have 6 letters formed from 3 M's, 2 A's, and 1 L. To form a word think of six blank positions that will have to be filled in by the above letters.

$$\overline{1}\ \overline{2}\ \overline{3}\ \overline{4}\ \overline{5}\ \overline{6}$$

We separate the construction of a word into three tasks.

> Task 1: Choose 3 of the positions for the M's.
>
> Task 2: Choose 2 of the remaining positions for the A's.
>
> Task 3: Choose the remaining position for the L.

Doing this sequence of tasks will result in a word and, conversely, every rearrangement of MAMMAL can be interpreted as resulting from this sequence of tasks.

Task 1 can be done in $C(6, 3)$ ways. There are now three positions left for the 2 A's, so task 2 can be done in $C(3, 2)$ ways. Five blanks have been filled, so that the L must go in the remaining blank. That is, task 3 can be done in $C(1, 1)$ ways. The multiplication principle says that the number of rearrangements is

$$C(6, 3) \cdot C(3, 2) \cdot C(1, 1) = \frac{6!}{3!3!} \cdot \frac{3!}{2!1!} \cdot \frac{1!}{1!}$$

$$= \frac{6!}{3!2!1!}$$

∎

The form of the answer in Example 5 is suggestive. Had the letters in MAMMAL been distinct, there would have been 6! rearrangements possible. The presence of 3 M's, 2 A's, and 1 L yielded the denominator above. The very same reasoning used in Example 5 can be used to derive the following general result.

The number of distinct permutations of n things of which n_1 are of one kind, n_2 of a second kind, . . . , n_k of a kth kind, is

$$\frac{n!}{n_1! \cdot n_2! \cdot \ \cdots \ \cdot n_k!}$$

Example 6

How many different vertical arrangements are possible for 10 flags, if 2 are white, 3 are red, and 5 are blue?

Solution

Here we want the different arrangements of 10 objects, which are not all different. Following the result above, we have

$$\frac{10!}{2!3!5!} = \frac{10 \cdot 9 \cdot 8 \cdot 7 \cdot 6 \cdot 5!}{2 \cdot 3 \cdot 2 \cdot 5!} = 2520 \text{ different arrangements}$$

∎

Example 7

How many different 11 letter words (real or imaginary) can be formed from the word below?

MISSISSIPPI

Solution

Here we want the number of distinct 11 letter words with 4 I's, 4 S's, 2 P's, and 1 M, so that the total number of 11 letter words is

$$\frac{11!}{2!4!4!} = \frac{39,916,800}{1152} = 34,650$$

The ideas above can be adapted to problems involving assignments of objects to locations.

Example 8

A sorority house has three bedrooms and 10 students. One bedroom has three beds, the second has two beds, and the third has five beds. In how many different ways can the students be assigned rooms?

Solution

Designate the bedrooms as A, B, and C. We think of the 10 students as standing in a row and we hand each student a letter corresponding to her assigned bedroom. An assignment of rooms can then be visualized as a sequence of length 10 (the number of students) containing 3 A's, 2 B's, and 5 C's (the capacity of the rooms). For example, the sequence

$$A \quad B \quad B \quad C \quad . \, . \, .$$

would have the first student in room A, students 2 and 3 in room B, and so on. There are

$$\frac{10!}{3!2!5!} = 2520$$

such sequences and, hence, room assignments.

Exercise 6.5

Answers to Odd-Numbered Problems begin on page A-20.

1. An experiment consists of tossing a coin 10 times.

    (a) How many different outcomes are possible?
    (b) How many different outcomes have exactly 4 heads?
    (c) How many different outcomes have at most 2 heads?
    (d) How many different outcomes have at least 3 heads?

2. An experiment consists of tossing a coin six times.

   (a) How many different outcomes are possible?
   (b) How many different outcomes have exactly 3 heads?
   (c) How many different outcomes have at least 2 heads?
   (d) How many different outcomes have 4 heads or 5 heads?

3. An urn contains 7 white balls and 3 red balls. Three balls are selected. In how many ways can the 3 balls be drawn from the total of 10 balls:

   (a) If the color is not considered?   (c) If all 3 balls are white?
   (b) If 2 balls are white and 1 is red?   (d) If all 3 balls are red?

4. An urn contains 15 red balls and 10 white balls. Five balls are selected. In how many ways can the 5 balls be drawn from the total of 25 balls:

   (a) If the color is not considered?   (c) If 3 balls are red and 2 are white?
   (b) If all balls are red?   (d) If at least 4 are red balls?

5. In the World Series the American League team (A) and the National League team (N) play until one team wins four games. If the sequence of winners is designated by letters (for example, $NAAAA$ means the National League team won the first game and the American League team won the next four), how many different sequences are possible?

6. How many different ways can 3 red, 4 yellow, and 5 blue bulbs be arranged in a string of Christmas tree lights with 12 sockets?

7. In how many ways can 3 apple trees, 4 peach trees, and 2 plum trees be arranged along a fence line if one does not distinguish between trees of the same kind?

8. How many different 9-letter words (real or imaginary) can be formed from the letters in the word ECONOMICS?

9. How many different 11-letter words (real or imaginary) can be formed from the letters in the word MATHEMATICS?

10. The U.S. Senate has 100 members. Suppose it is desired to place each senator on exactly 1 of 7 possible committees. The first committee has 22 members, the second has 13, the third has 10, the fourth has 5, the fifth has 16, and the sixth and seventh have 17 apiece. In how many ways can these committees be formed?

11. In how many ways can 10 children be placed on 3 distinct teams of 3, 3, and 4 members?

12. A group of 9 people is going to be formed into committees of 4, 3, and 2 people. How many committees can be formed if:

    (a) A person can serve on any number of committees?
    (b) No person can serve on more than one committee?

13. A group consists of 5 men and 8 women. A committee of 4 is to be formed from this group, and policy dictates that at least 1 woman be on this committee.

    (a) How many committees can be formed that contain exactly 1 man?
    (b) How many committees can be formed that contain exactly 2 women?
    (c) How many committees can be formed that contain at least 1 man?

14. How many distinct seven-digit telephone numbers can be formed (a) If the digits 1, 1, 2, 2, 5, 5, 5 are used? (b) If it is required that the first digit be a 5?

15. In how many ways can 30 diplomats be assigned to 5 countries, with each country receiving an equal number of diplomats?

16. How many rearrangements of the letters in the word SUCCESS have the U before the E?

17. An experiment consists of tossing a coin 8 times. How many outcomes have more heads than tails?

18. Eight couples (husband and wife) are present at a meeting where a committee of 3 is to be chosen. How many ways can this be done so that the committee

    (a) Contains a couple?
    (b) Contains no couple?

19. In how many ways can a committee of 4 be selected from 6 men and 8 women if the committee must contain at least 2 women?

20. A man wants to invite 1 or more of his 4 friends to dinner. In how many ways can he do this?

21. How many 8-bit strings contain an even number of 1's? An odd number of 1's?

22. An office manager must locate 12 secretaries into three offices that hold respectively 6, 4, and 2 secretaries. In how many ways can the three groups be chosen to occupy the three offices?

6.6

THE BINOMIAL THEOREM

INTRODUCTION
BINOMIAL IDENTITIES

INTRODUCTION

The *binomial theorem* deals with the problem of expanding an expression of the form $(x + y)^n$, where n is a positive integer.

Expressions such as $(x + y)^2$ and $(x + y)^3$ are not too difficult to expand. For example.

$$(x + y)^2 = x^2 + 2xy + y^2 \qquad (x + y)^3 = x^3 + 3x^2y + 3xy^2 + y^3$$

However, expanding expressions such as $(x + y)^6$ or $(x + y)^8$ by the normal process of multiplication would be tedious and time-consuming. It is here that the binomial theorem is especially useful.

Recall that

$$C(n, r) = \binom{n}{r} = \frac{n!}{r!(n - r)!}$$

Then, for example, the expression

$$(x + y)^2 = x^2 + 2xy + y^2$$

can be written as

$$(x + y)^2 = \binom{2}{0} x^2 + \binom{2}{1} xy + \binom{2}{2} y^2$$

The expansion of $(x + y)^3$ can be written as

$$(x + y)^3 = x^3 + 3x^2y + 3xy^2 + y^3$$

$$= \binom{3}{0} x^3 + \binom{3}{1} x^2y + \binom{3}{2} xy^2 + \binom{3}{3} y^3$$

The expansion of $(x + y)^4$ can be written as

$$(x + y)^4 = x^4 + 4x^3y + 6x^2y^2 + 4xy^3 + y^4$$

$$= \binom{4}{0} x^4 + \binom{4}{1} x^3y + \binom{4}{2} x^2y^2 + \binom{4}{3} xy^3 + \binom{4}{4} y^4$$

The binomial theorem generalizes this pattern.

Binomial Theorem If n is a positive integer,

$$(x + y)^n = \binom{n}{0} x^n + \binom{n}{1} x^{n-1}y + \binom{n}{2} x^{n-2}y^2$$

$$+ \cdots + \binom{n}{k} x^{n-k}y^k + \cdots + \binom{n}{n} y^n$$

(1)

Observe that the powers of x begin at n and decrease by 1, while the powers of y begin with 0 and increase by 1. Also, the coefficient of the term involving y^k is always $\binom{n}{k}$.

Let's get some practice in using the binomial theorem.

Example 1
Expand $(x + y)^6$ using the binomial theorem.

Solution

$$(x + y)^6 = \binom{6}{0} x^6 + \binom{6}{1} x^5y + \binom{6}{2} x^4y^2 + \binom{6}{3} x^3y^3$$

$$+ \binom{6}{4} x^2y^4 + \binom{6}{5} xy^5 + \binom{6}{6} y^6$$

$$= x^6 + 6x^5y + 15x^4y^2 + 20x^3y^3 + 15x^2y^4 + 6xy^5 + y^6 \qquad \blacksquare$$

Note that the coefficients in the expansion of $(x + y)^6$ are the entries in Pascal's triangle for $n = 6$. (See Figure 18, page 309.)

Example 2

Find the coefficient of x^3y^4 in the expansion of $(x + y)^7$, using the binomial theorem.

Solution

The coefficient of x^3y^4 is

$$\binom{7}{4} = \frac{7 \cdot 6 \cdot 5}{3 \cdot 2 \cdot 1} = 35$$

Example 3

Expand $(x + 2y)^4$ using the binomial theorem.

Solution

Here, we let "$2y$" play the role of "y" in the binomial theorem. We then get

$$(x + 2y)^4 = \binom{4}{0} x^4 + \binom{4}{1} x^3(2y) + \binom{4}{2} x^2(2y)^2$$
$$+ \binom{4}{3} x(2y)^3 + \binom{4}{4} (2y)^4$$
$$= x^4 + 8x^3y + 24x^2y^2 + 32xy^3 + 16y^4$$

To explain why the binomial theorem is true, we take a close look at what happens when we compute $(x + y)^3$. Think of $(x + y)^3$ as the product of three factors, namely,

$$(x + y)^3 = (x + y)(x + y)(x + y)$$

Were we to multiply these three factors together without any attempt at simplification or collecting of terms we would get

$$(x + y)\ (x + y)\ (x + y) = xxx + xyx + yxx + yyx + xxy + xyy + yxy + yyy$$
Factor Factor Factor
   1   2   3

Notice that the terms on the right yield all possible products that can be formed by picking either an x or y from each of the three factors on the left. Thus, for example.

xyx results from choosing an x from factor 1,
a y from factor 2, and an x from factor 3

Now, the number of terms on the right that will simplify to, say, xy^2, will be those terms that resulted from choosing y's from two of the factors and an x from the remaining factor. How many such terms are there? There are as many as there are ways of choosing two of the three factors to contribute y's—that is, there are $C(3, 2) = \binom{3}{2}$ such terms. This is why the coefficient of xy^2 in the expansion of $(x + y)^3$ is $\binom{3}{2}$.

In general, if we think of $(x + y)^n$ as the product of n factors,

$$(x + y)^n = \underbrace{(x + y) \cdot (x + y) \cdots (x + y)}_{n \text{ factors}}$$

then upon multiplying out and simplifying there will be as many terms of the form $x^{n-k} y^k$ as there are ways of choosing k of the n factors to contribute y's (and the remaining $n - k$ to contribute x's). There are $C(n, k)$ ways of making this choice. So, the coefficient of $x^{n-k} y^k$ is thus $\binom{n}{k}$, and this is the assertion of the binomial theorem.

BINOMIAL IDENTITIES

Binomial coefficients are involved in many relations that can be useful when applying them. We discuss some of these in the examples below.

Example 4
Show that

$$\binom{n}{k} = \binom{n}{n - k}$$

Solution
By definition,

$$\binom{n}{k} = \frac{n!}{k!(n - k)!}$$

while

$$\binom{n}{n - k} = \frac{n!}{(n - k)![n - (n - k)]!}$$

Since $n - (n - k) = k$, a comparison of the expressions shows that they are equal.

∎

Another way of explaining the equality would be to imagine that we wanted to pick a team of k players from n people. Then choosing those k who will play is the same as choosing those $n - k$ who will not. So the number of ways of choosing the players equals the number of ways of choosing the nonplayers. The players can be chosen in $C(n, k) = \binom{n}{k}$ ways, while those to be left out can be chosen in $\binom{n}{n-k}$ ways and the equality follows.
    Thus,

$$\binom{5}{3} = \binom{5}{2}, \qquad \binom{10}{2} = \binom{10}{8}$$

and so on. This identity accounts for the symmetry in the rows of Pascal's triangle.

Example 5
Show that

$$\binom{n}{k} = \binom{n - 1}{k} + \binom{n - 1}{k - 1}$$

Solution

We could expand both sides of the above identity using the definition of binomial coefficients and, after some algebra, demonstrate the equality. But we choose the route of posing a problem that we solve two different ways. Equating the two solutions will prove the identity.

A committee of k is to be chosen from n people. The total number of ways this can be done is $C(n, k) = \binom{n}{k}$.

We now count the total number of committees a different way. Assume that one of the n people is Mr. Martin. We compute

(1)   those not containing Mr. Martin

and

(2)   those containing Mr. Martin

The number of committees of type (1) is $\binom{n-1}{k}$ since the k people must be chosen from the $n - 1$ people who are not Mr. Martin. The number of committees of type (2) will correspond to the number of ways we can choose the $k - 1$ people other than Mr. Martin to be on the committee. So, the number of committees of type (2) is given by $\binom{n-1}{k-1}$. Since the number of committees of type (1) plus the number of committees of type (2) equals the total number of committees, our identity follows. ∎

For example, $\binom{8}{5} = \binom{7}{5} + \binom{7}{4}$. It is precisely this identity that explains the reason why an entry in Pascal's triangle can be obtained by adding the nearest two entries in the row above it. Due to a recursive character, the identity in Example 5 is sometimes used in computer programs that evaluate binomial coefficients.

Example 6

Explain why

$$\binom{6}{3} = \binom{2}{2} + \binom{3}{2} + \binom{4}{2} + \binom{5}{2}$$

Solution

We make repeated use of the identity in Example 5.

So,

$$\binom{6}{3} = \binom{5}{3} + \binom{5}{2}$$

$$= \binom{4}{3} + \binom{4}{2} + \binom{5}{2} \qquad \left[\text{Applying the identity to } \binom{5}{3}\right]$$

$$= \binom{3}{3} + \binom{3}{2} + \binom{4}{2} + \binom{5}{2} \qquad \left[\text{Applying the identity to } \binom{4}{3}\right]$$

$$= \binom{2}{2} + \binom{3}{2} + \binom{4}{2} + \binom{5}{2}. \qquad \left[\text{Since } \binom{2}{2} = \binom{3}{3} = 1\right]$$

∎

Example 7
Show that

$$\binom{n}{0} + \binom{n}{1} + \binom{n}{2} + \cdots + \binom{n}{n} = 2^n$$

Solution
We make use of the binomial theorem. Since the binomial theorem is valid for all x and y, we may set $x = y = 1$ in (1). This gives

$$2^n = (1+1)^n = \binom{n}{0} + \binom{n}{1} + \binom{n}{2} + \cdots + \binom{n}{n}$$ ∎

This shows, for example, that the sum of the elements in the row marked $n = 6$ of Pascal's triangle is $2^6 = 64$.

The result in Example 7 has an interpretation in terms of the number of subsets of a set with n elements. $\binom{n}{0}$ gives the number of subsets with 0 elements; $\binom{n}{1}$ the number of subsets with 1 element; $\binom{n}{2}$ the number of subsets with 2 elements; and so on. The sum $\binom{n}{0} + \binom{n}{1} + \cdots + \binom{n}{n}$ is thus the total number of subsets of a set with n elements. The result in Example 7 can then be rephrased as:

A set with n elements has 2^n subsets.

Thus, a set with five elements has $2^5 = 32$ subsets.

Example 8
Show that

$$\binom{n}{0} - \binom{n}{1} + \binom{n}{2} - \cdots + (-1)^n \binom{n}{n} = 0$$

(The last term will be preceded by a plus or minus sign depending on whether n is even or odd.)

Solution
We again make use of the binomial theorem. This time we let $x = 1$ and $y = -1$ in (1). This produces

$$0 = (1-1)^n$$
$$= \binom{n}{0} + \binom{n}{1}(-1) + \binom{n}{2}(-1)^2 + \binom{n}{3}(-1)^3 + \cdots + \binom{n}{n}(-1)^n$$
$$= \binom{n}{0} - \binom{n}{1} + \binom{n}{2} - \cdots + (-1)^n \binom{n}{n}$$ ∎

We mention an interpretation of this identity by examining the instance where $n = 5$. The identity gives

$$\binom{5}{0} - \binom{5}{1} + \binom{5}{2} - \binom{5}{3} + \binom{5}{4} - \binom{5}{5} = 0$$

Transposing this equality yields

$$\binom{5}{0} + \binom{5}{2} + \binom{5}{4} = \binom{5}{1} + \binom{5}{3} + \binom{5}{5}$$

This says that a set with five elements has as many subsets containing an even number of elements as it has subsets containing an odd number of elements.

Exercise 6.6   *Answers to Odd-Numbered Problems begin on page A-20.*

In Problems 1–6 use the binomial theorem to expand each expression.

A   1. $(x + y)^5$ 2. $(x + y)^4$ 3. $(x + 3y)^3$

    4. $(2x + y)^3$ 5. $(2x - y)^4$ 6. $(x - y)^4$

    7. What is the coefficient of x^2y^3 in the expansion of $(x + y)^5$?

    8. What is the coefficient of x^2y^6 in the expansion of $(x + y)^8$?

    9. What is the coefficient of x^8 in the expansion of $(x + 3)^{10}$?

    10. What is the coefficient of x^3 in the expansion of $(x + 2)^5$?

    11. How many different subsets can be chosen from a set with 5 elements?

    12. How many different subsets can be chosen from a set of 50 elements?

    13. How many nonempty subsets does a set with 10 elements have?

    14. How many subsets with an even number of elements does a set with 10 elements have?

    15. How many subsets with an odd number of elements does a set with 10 elements have?

B   16. Explain why

$$\binom{8}{5} = \binom{4}{4} + \binom{5}{4} + \binom{6}{4} + \binom{7}{4}$$

    17. Explain why

$$\binom{10}{7} = \binom{6}{6} + \binom{7}{6} + \binom{8}{6} + \binom{9}{6}$$

    18. Show that

$$\binom{7}{1} + \binom{7}{3} + \binom{7}{5} + \binom{7}{7} = 2^6$$

    19. Replace $\binom{11}{6} + \binom{11}{5}$ by a single binomial coefficient.

    20. Replace $\binom{8}{8} + \binom{9}{8} + \binom{10}{8}$ by a single binomial coefficient.

CHAPTER REVIEW

Important Terms and Formulas

set	union ($\cup$)	multiplication principle
empty set ($\varnothing$)	intersection ($\cap$)	factorial

equality of sets   disjoint sets   permutation

subset ($\subseteq$)   complement   combination

proper subset ($\subset$)   De Morgan's laws   binomial coefficient

universal set   finite sets   Pascal's triangle

Venn diagram   tree diagram

$$\overline{A \cup B} = \overline{A} \cap \overline{B}$$

$$\overline{A \cap B} = \overline{A} \cup \overline{B}$$

$$c(A \cup B) = c(A) + c(B) - c(A \cap B)$$

$$P(n, r) = n(n-1) \ldots (n-r+1) = \frac{n!}{(n-r)!}$$

$$C(n, r) = \binom{n}{r} = \frac{n!}{r!(n-r)!}$$

$$= \frac{P(n, r)}{r!}$$

$$(x + y)^n = \binom{n}{0} x^n + \binom{n}{1} x^{n-1}y + \binom{n}{2} x^{n-2}y^2 + \cdots + \binom{n}{k} x^{n-k}y^k$$

$$+ \cdots + \binom{n}{n} y^n$$

True–False Questions *(Answers on page A-20)*

T   F   1. If $A \cup B = A \cap B$, then $A = B$.

T   F   2. If A and B are disjoint sets, then $c(A \cup B) = c(A) + c(B)$.

T   F   3. The number of permutations of 4 different objects taken 4 at a time is 12.

T   F   4. $C(5, 3) = 20$

T   F   5. $5! = 120$

T   F   6. $\dfrac{7!}{6!} = 7$

T   F   7. $2^n = \binom{n}{0} + \binom{n}{1} + \binom{n}{2} + \cdots + \binom{n}{n}$

T   F   8. In the binomial expansion of $(x + 1)^7$, the coefficient of x^4 is 4.

Fill in the Blanks *(Answers on page A-20)*

1. Two sets that have no elements in common are called ———————.

2. The number of different arrangements of r objects from n objects in which (a) the n objects are different, (b) no object is repeated more than once in an arrangement, and (c) order is important is called a ———————.

3. If in 2 above, condition (c) is replaced by "order is not important," we have a ———————.

4. A triangular display of combinations is called the ——————— triangle.

5. The combinations $\binom{n}{r}$ are sometimes called ——————— ———————.

6. To expand an expression such as $(x + y)^n$ where n is a positive integer, we use the ——————— ———————.

Review Exercises

Answers to Odd-Numbered Problems begin on page A-20.

In Problems 1–16 replace the asterisk by the symbol(s) $\in$, $\subset$, $\subseteq$, and/or $=$ to give a true statement. If none of these relationships hold, write "None of these."

A

1. $0 * \varnothing$

2. $\{0\} * \{1, 0, 3\}$

3. $\{5, 6\} \cap \{2, 6\} * \{8\}$

4. $\{2, 3\} \cup \{3, 4\} * \{3\}$

5. $\{8, 9\} * \{9, 10, 11\}$

6. $1 * \{1, 3, 5\} \cup \{3, 4\}$

7. $5 * \{0, 5\}$

8. $\varnothing * \{1, 2, 3\}$

9. $\varnothing * \{1, 2\} \cap \{3, 4, 5\}$

10. $\{2, 3\} * \{3, 4\}$

11. $\{1, 2\} * \{1\} \cup \{3\}$

12. $5 * \{1\} \cup \{2, 3\}$

13. $\{4, 5\} \cap \{5, 6\} * \{4, 5\}$

14. $\{6, 8\} * \{8, 9, 10\}$

15. $\{6, 7, 8\} \cap \{8\} * \{6\}$

16. $4 * \{6, 8\} \cap \{4, 8\}$

17. For the sets

$$A = \{1, 3, 5, 6, 8\} \qquad B = \{2, 3, 6, 7\} \qquad C = \{6, 8, 9\}$$

find:

(a) $(A \cap B) \cup C$ (b) $(A \cap B) \cap C$ (c) $(A \cup B) \cap B$

18. For the sets U = universal set = $\{1, 2, 3, 4, 5, 6, 7\}$ and

$$A = \{1, 3, 5, 6\} \qquad B = \{2, 3, 6, 7\} \qquad C = \{4, 6, 7\}$$

find:

(a) $\overline{A \cap B}$ (b) $(B \cap C) \cap A$ (c) $\overline{B} \cup \overline{A}$

19. If A and B are sets and if $c(A) = 24$, $c(A \cup B) = 33$, $c(B) = 12$, find $c(A \cap B)$.

B

20. During June, Colleen's Motors sold 75 cars with air conditioning, 95 with power steering, and 100 with automatic transmission. Twenty cars had all three options, 10 cars had none of these options, and 10 cars were sold that had only air conditioning. In addition, 50 cars had both automatic transmission and power steering, and 60 cars had both automatic transmission and air conditioning.

(a) How many cars were sold in June?

(b) How many cars had only power steering?

21. In a survey of 125 college students, it was found that of three newspapers, the *Wall Street Journal*, *New York Times*, and *Chicago Tribune*:

60	read the *Chicago Tribune*
40	read the *New York Times*
15	read the *Wall Street Journal*
25	read the *Chicago Tribune* and *New York Times*
8	read the *New York Times* and *Wall Street Journal*
3	read the *Chicago Tribune* and *Wall Street Journal*
1	read all three

(a) How many read none of these papers?

(b) How many read only the *Chicago Tribune*?

(c) How many read neither the *Chicago Tribune* nor the *New York Times*?

22. If U = universal set = $\{1, 2, 3, 4, 5\}$ and $B = \{1, 4, 5\}$, find all sets A for which $A \cap B = \{1\}$.

23. Compute $P(6, 3)$.

24. Compute $C(6, 2)$.

25. In how many different ways can a committee of 3 people be formed from a group of 5 people?

26. In how many different ways can 4 people line up?

27. In how many different ways can 3 books be placed on a shelf?

28. In how many different ways can 3 people be seated in 4 chairs?

29. How many house styles are possible if a contractor offers 3 choices of roof designs, 4 choices of window designs, and 6 choices of brick?

30. How many different answers are possible in a true–false test consisting of 10 questions?

31. You are to set up a code of 2-digit words using the digits 1, 2, 3, 4 without using any digit more than once. What is the maximum number of words in such a language? If the words 12 and 21, for example, designate the same word, how many words are possible?

32. You are to set up a code of 3-digit words using the digits 1, 2, 3, 4, 5, 6 without using any digit more than once in the same word. What is the maximum number of words in such a language? If the words 124, 142, etc., designate the same word, how many different words are possible?

33. A small town consists of a north side and a south side. The north side has 16 houses and the south side has 10 houses. A pollster is asked to visit 4 houses on the north side and 3 on the south side. In how many ways can this be done?

34. A ceremony is to include 7 speeches and 6 musical selections.

    (a)   How many programs are possible?
    (b)   How many programs are possible if speeches and musical selections are to be alternated?

35. There are 7 boys and 6 girls willing to serve on a committee. How many 7-member committees are possible if a committee is to contain:

    (a)   3 boys and 4 girls?
    (b)   At least one member of each sex?

36. Laura's Ice Cream Parlor offers 31 different flavors to choose from, and specializes in double dip cones.

    (a)   How many different cones are there to choose from if you may select the same flavor for each dip?
    (b)   How many different cones are there to choose from if you cannot repeat any flavor? Assume that a cone with vanilla on top of chocolate is different from a cone with chocolate on top of vanilla.
    (c)   How many different cones are there if you consider any cone having chocolate on top and vanilla on the bottom the same as having vanilla on top and chocolate on the bottom?

37. A person has 4 History, 5 English, and 6 Mathematics books. How many ways can they be arranged on a shelf if books on the same subject must be together?

38. Five people are to line up for a group photograph. If 2 of them refuse to stand next to each other, in how many ways can the photograph be taken?

39. There are 5 rotten plums in a crate of 25 plums. How many samples of 4 of the 25 plums contain

(a) Only good plums?

(b) Three good plums and 1 rotten plum?

(c) One or more rotten plums?

40. An admissions test given by a university contains 10 true–false questions. Eight or more of the questions must be answered correctly in order to be admitted.

(a) How many different ways can the answer sheet be filled out?

(b) How many different ways can the answer sheet be filled out so that 8 or more questions are answered correctly?

41. The figure below indicates the locations of two houses, A and B, in a city, where the lines represent streets. A person at A wishes to reach B and can only travel in two directions, to the right and up. How many different paths are there from A to B?

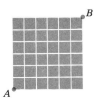

42. A cab driver picks up a passenger at point A (see the figure below) whose destination is point B. After completing the trip, the driver is to proceed to the garage at point C. If the cab must travel to the right or up, how many different routes are there from A to C?

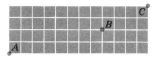

43. In how many ways can we choose three words, one each from five 3-letter words, six 4-letter words, and eight 5-letter words?

44. A newborn child can be given 1, 2, or 3 names. In how many ways can a child be named if we can choose from 100 names?

45. In how many ways can 5 girls and 3 boys be divided into 2 teams of 4 if each team is to include at least 1 boy?

46. A meeting is to be addressed by 5 speakers, A, B, C, D, E. In how many ways can the speakers be ordered if B must not precede A?

47. What is the answer to Problem 46 if B is to speak immediately after A?

48. An automobile license number contains 1 or 2 letters followed by a 4-digit number. Compute the maximum number of different licenses.

7

INTRODUCTION TO PROBABILITY

7.1

INTRODUCTION

Probability theory is a part of mathematics that is useful for discovering and investigating the *regular* features of *random events.* Although it is not really possible to give a precise and simple definition of what is meant by the words *random* and *regular,* we hope that the explanation and the examples given below will help you understand these concepts.

Certain phenomena in the real world may be considered *chance phenomena.* These phenomena do not always produce the same observed outcome, and the outcome of any given observation of the phenomena may not be predictable. But they have a long-range behavior known as *statistical regularity.*

In some cases, we are familiar enough with the phenomenon under investigation to feel justified in making *exact* predictions with respect to the result of each individual observation. For example, if you want to know the time and place of a solar eclipse, you may not hesitate to predict an exact time, based on astronomical data.

However, in many cases our knowledge is not precise enough to allow exact predictions in particular situations. Some examples of such cases, called *random events,* are:

1. Tossing a fair coin gives a result that is either a head or a tail. For any one throw, we cannot predict the result, although it is obvious that it is determined by definite causes (such as the initial velocity of the coin, the initial angle of throw, and the surface on which the coin rests). Even though some of these causes can be controlled, we cannot predetermine the result of any particular toss. Thus, the result of tossing a coin is a *random event.*

2. In a series of throws with an ordinary die, each throw yields as its result one of the numbers 1, 2, 3, 4, 5, or 6. Thus, the result of throwing a die is a *random event.*

3. The sex of a newborn baby is either male or female. However, the sex of a newborn baby cannot be predicted in any particular case. This, too, is an example of a *random event.*

The above examples demonstrate that in studying a sequence of random experiments, it is not possible to forecast individual results. These are subject to irregular, random fluctuations that cannot be exactly predicted. However, if the number of observations is large, that is, if we deal with a *mass phenomenon,* some regularity appears.

In the first example, we cannot predict the result of any particular toss of the coin. However, if we perform a long sequence of tosses, we notice that the number of times heads occurs is approximately equal to the number of times tails appears. That is, it seems *reasonable* to say that in any toss of this fair coin, a head or a tail is *equally likely* to occur. As a result, we might *assign a probability* of $\frac{1}{2}$ for obtaining a head (or tail) on a particular toss.

For the second example, the appearance of any particular face of the die is a random event. However, if we perform a long series of tosses, any face is as *equally likely* to occur as any other, provided the die is fair. Here we might *assign a probability* of $\frac{1}{6}$ for obtaining a particular face.

For the third example our intuition tells us that a boy baby and a girl baby are *equally likely* to occur. If we follow this reasoning, we might *assign a probability* of $\frac{1}{2}$ to having a girl baby. However, if we consult the data found in Table 1,* we see that it might be more accurate to *assign a probability* of .487 to having a girl baby.

Table 1

Year of Birth	Number of Births		Total Number of Births $b+g$	Ratio of Births	
	Boys b	Girls g		$\dfrac{b}{b+g}$	$\dfrac{g}{b+g}$
1970	1,915,378	1,816,008	3,731,386	.513	.487
1971	1,822,910	1,733,060	3,555,970	.513	.487
1972	1,669,927	1,588,484	3,258,411	.512	.488
1973	1,608,326	1,528,639	3,136,965	.513	.487
1974	1,622,114	1,537,844	3,159,958	.513	.487
1975	1,613,135	1,531,063	3,144,198	.513	.487
1976	1,624,436	1,543,352	3,167,788	.513	.487
1977	1,705,916	1,620,716	3,326,632	.513	.487
1978	1,709,394	1,623,885	3,333,279	.513	.487
1979	1,791,267	1,703,131	3,494,398	.513	.487
Total	17,082,803	16,226,182	33,308,985	.513	.487

The examples below illustrate some of the kinds of problems we will encounter and solve in this and the next chapter.

Example 1

A fair die is thrown. With what probability will the face 5 occur?   ■

Example 2

A fair coin is tossed. If it comes up heads (H), a fair die is rolled and the experiment is finished. If the coin comes up tails (T), the coin is tossed again and the experiment is finished. With what probability will the situation heads first, 5 second, occur?   ■

Example 3

A room contains 50 people. With what probability will at least 2 of them have the same birthday?   ■

---

* U.S. Census, 1970–1979.

Example 4

A factory produces light bulbs of which 80% are not defective. A sample of 6 bulbs is taken. With what probability will 3 or more of them be defective? ▪

Example 5

Two dice are rolled. If they are fair, with what probability will the total be 11? If one is "loaded" in a certain way, what is the probability of an 11? ▪

Example 6

A group of 1200 people includes 50 who qualify for an executive position and 500 females. Furthermore, suppose 35 females qualify. With what probability will a person chosen at random be both qualified and female? ▪

Other questions that can be answered by using probability theory are given below. They are listed here merely as illustrations of the power of probability theory.

Example 7

Consider a telephone exchange with a finite number of lines. Suppose that the exchange is constructed in such a way that an incoming call that finds all the lines busy does not wait, but is lost. This is called an *exchange without waiting lines.* The most important problem to be solved *before* the exchange is constructed is to determine for any time t, the probability of finding all the lines busy at time t.

▪

Example 8

People arrive at random times at a ticket counter to be served by an attendant, lining up in a queue if others are waiting. Given information about the rate of arrival and the length of time an attendant requires to serve each customer, how much of the time is the attendant idle? How much of the time is the line more than 15 persons long? What would be the effect of adding another attendant? If the people are not allowed to wait in line but must go elsewhere, what percentage of arrivals go unanswered? The same questions can be asked about gas stations, toll booths on roads, hospital beds, and so on. ▪

Example 9

The following problem occurs quite often in physics and biology and was formulated by Galton in 1874 in his study of the disappearance of family lines: Given that a man of a known family has a probability of $p_0, p_1, p_2, \ldots$ of producing 0, 1, 2 male offspring, what is the probability that the family will eventually die out?

▪

Example 10

Suppose that an infectious disease is spread by contact, that a susceptible person has a chance of catching it with each contact with an infected person, but that one becomes immune after having had the disease and can no longer transmit it. Some of the questions we would like answered are: How many susceptibles will be left when the number of infected is 0? How long will the epidemic last? For a given community, what is the probability that the disease will die out? ■

Exercise 7.1

1. Consult the United States census from 1980–1989. What probability would you assign to the birth of a boy baby based on these data?

2. Pick a page at random in your telephone directory and list the last digit of every number. Are all digits used equally often?

3. Toss a die six times and record the face shown. Denote this number by x (x can take the values 1, 2, 3, 4, 5, 6). Repeat the experiment 20 more times and tabulate the results. Do you observe any regularity?

4. **Buffon Needle Problem** Suppose a needle is dropped at random onto a floor that is marked with parallel lines 6 inches apart. Suppose the needle is 4 inches long. With what probability does the needle fall between the two lines? Perform the experiment 20 times and record the number of times the needle rests completely within the two lines. An illustration is provided in the margin and a solution may be found in *Mathematics in the Modern World.**

5. **Gambler's Ruin Problem** Conduct the following experiment with a classmate. Assume that you have $90 and your classmate has $10. Flip a fair coin. If you lose, you give your classmate $1, and if you win, your classmate gives you $1. The game ends when one of you loses all your money. Can you guess how long the game will last? Can you guess who has the better chance of winning?

6. **Birthday Problem** Conduct the following experiment in your class. Find out how many students would bet that no two students have the same birthday. List the birthdays of all the students and determine whether any two have the same birthday. Would you have won or lost the bet? (See page 359 for a table that provides a rationale for betting one way or the other.)

7. **Chevalier de Mere's Problem†** Which of the following random events do you think is more likely to occur?

   (a)   To obtain a 1 on at least one die in a simultaneous throw of four fair dice.

   (b)   To obtain at least one pair of 1's in a series of 24 throws of a pair of fair dice.

   (See Problem 29, page 388.)

8. Pick an editorial from your daily newspaper that contains at least 1000 words. List the

---

* *Mathematics in the Modern World,* Readings from *Scientific American,* W. H. Freeman, San Francisco, 1968, p. 169ff.

† CHEVALIER DE MERE (1607–1684), a philosopher and a man of letters, was a prominent figure at the court of Louis XIV. A French knight, de Mere was an ardent dice gambler and tried to become rich by this game. He was constantly thinking of various complicated rules that he hoped would help him reach his goal. He knew and corresponded with almost all leading mathematicians of his time, including Pascal, seeking their assistance.

number of times each letter of the alphabet is used. Based on this, what probability might you assign to the occurrence of a letter in an editorial?

7.2

SAMPLE SPACES AND ASSIGNMENT OF PROBABILITIES

SAMPLE SPACES
EVENTS AND SIMPLE EVENTS
PROBABILITY OF AN EVENT
CONSTRUCTING A PROBABILITY MODEL

In studying probability we are concerned with experiments, real or conceptual, and their outcomes. In this study we try to formulate in a precise manner a mathematical theory that closely resembles the experiment in question. The first stage of development of a mathematical theory is the building of what is termed a *mathematical model.* This model is then used as a predictor of outcomes of the experiment. The purpose of this section is to learn how a *probabilistic model* can be constructed.

SAMPLE SPACES

We begin by writing down the associated *sample space* of an experiment; that is, we write down all outcomes that can occur as a result of the experiment.

For example, if the experiment consists of flipping a coin, we would ordinarily agree that the only possible outcomes are heads, H, and tails, T. Therefore, a sample space for the experiment is the set $\{H, T\}$.

Example 1

Consider an experiment in which, for the sake of simplicity, one die is green and the other is red. When the two dice are rolled, the set of outcomes consists of all the different ways that the dice may come to rest. This is referred to as a set of all *logical possibilities.* This experiment can be displayed in two different ways. One way is to use a tree diagram, as shown in Figure 1.

Another way is to let g and r denote, respectively, the numbers that come up on the green die and the red die. Then an *outcome* can be represented by an ordered pair (g, r), where both g and r can assume any of the values 1, 2, 3, 4, 5, 6. Thus, a sample space S for this experiment is the set

$$S = \{(g, r) | 1 \le g \le 6, 1 \le r \le 6\}$$

Also, notice that the multiplication principle shows that the number of elements in S is 36; since there are 6 choices for g, 6 choices for r, and $6 \cdot 6 = 36$.

Figure 2 gives a graphic representation of the elements of S. ∎

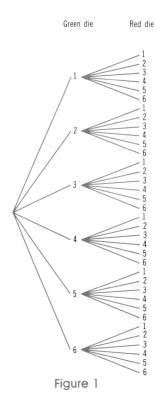

Figure 1

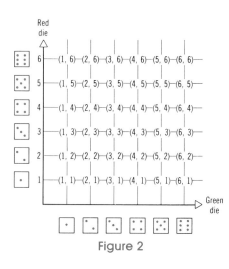

Figure 2

Sample Space

A *sample space S,* associated with a real or conceptual experiment, is the set of all logical possibilities that can occur as a result of the experiment. Each element of a sample space S is called an *outcome.*

In the table below we list some experiments and their sample spaces.

Experiment	Sample Space
(a) Tossing two coins, a penny and a nickel, and noting whether each coin falls heads (H) or tails (T).	$S = \{HH, HT, TH, TT\}$
(b) Same as (a) but we are interested only in the number of heads that appear on a single toss of the two coins.	$S = \{0, 1, 2\}$
(c) Same as (a) if we are interested in whether the coins match (M) or do not match (D).	$S = \{M, D\}$
(d) An experiment consists of selecting three manufactured parts from the production process and observing whether they are acceptable (A) or defective (D).	$S = \{AAA, AAD, ADA, ADD, DAA, DAD, DDA, DDD\}$
(e) Polling customers at a supermarket whether they *like* (L) a certain product or *dislike* (D) it and recording their response.	$S = \{L, D\}$

(f)   A spinner is marked from 1 to 8. An experi- $S = \{1, 2, 3, 4, 5, 6, 7, 8\}$
      ment consists of spinning the dial once.

(g)   Surveying whether a customer entering a $S = \{H, F, B, N\}$
      fast-food restaurant orders a hamburger
      (H), french fries (F), both (B), or neither
      (N).

The theory of probability begins as soon as a sample space has been specified. The sample space of an experiment plays the same role as the universal set in set theory for all questions concerning the experiment.

In this chapter, we confine our attention to those cases for which the sample space is finite, that is, to those situations in which it is possible to have only a finite number of outcomes.

Notice that in our definition we say a sample space, rather than *the* sample space, since an experiment can be described in many different ways. In general, it is a safe guide to include as much detail as possible in the description of the outcomes of the experiment in order to answer all pertinent questions concerning the result of the experiment.

The first three sample spaces in the last table all resulted from the same experiment, depending on the information we seek. The next example illustrates the same idea.

Example 2

Consider the set of all different types of families with three children. Describe the sample space for the experiment of drawing one family from the set of all possible three-child families.

Solution

One way of describing the sample space is by denoting the number of girls in the family. The only possibilities are members of the set

$$\{0, 1, 2, 3\}$$

That is, a three-child family can have 0, 1, 2, or 3 girls.

This sample space has four outcomes. A disadvantage of describing the experiment using this sample space is that a question such as "Was the second child a girl?" cannot be answered. Thus, this method of classifying the outcomes may be too coarse, since it may not provide enough wanted information.

Another way of describing the sample space is by first defining B and G as "Boy" and "Girl," respectively. Then the sample space might be given as

$$\{BBB, BBG, BGB, BGG, GBB, GBG, GGB, GGG\}$$

where BBB means first child is a boy, second child is a boy, third child is a boy; and so on. This experiment can be depicted by the tree diagram in Figure 3. Notice that the experiment has eight possible outcomes.

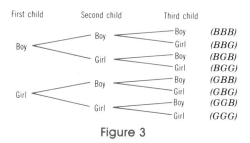

First child   Second child   Third child

Figure 3   ■

The advantage of the second type of classification of the sample space in Example 2 is that each outcome of the experiment corresponds to exactly one element in the sample space.

EVENTS AND SIMPLE EVENTS

Event; Simple Event An *event* is a subset of the sample space. If an event has exactly one element, that is, consists of only one outcome, it is called a *simple event*.

Every event can be written as the union of simple events. For example, consider the sample space of Example 2. The event E that the family consists of exactly two boys is

$$E = \{BBG, BGB, GBB\}$$

The event E is the union of the three simple events $\{BBG\}, \{BGB\}, \{GBB\}$. That is,

$$E = \{BBG\} \cup \{BGB\} \cup \{GBB\}$$

Since a sample space S is also an event, we can express a sample space as the union of simple events. Thus, if the sample space S consists of n outcomes,

$$S = \{e_1, e_2, \ldots, e_n\}$$

then

$$S = \{e_1\} \cup \{e_2\} \cup \ldots \cup \{e_n\}$$

Example 3
In the experiment of Example 1, involving a green die and a red die, let A be the event that the green die comes up less than or equal to 3 and the red die is 5. Let B be the event that the green die is 2 and the red die is 5 or 6. Describe the event A or B.

Solution

$$A = \{(g, r) | g \leq 3, \quad r = 5\} \qquad B = \{(g, r) | g = 2, \quad r = 5, 6\}$$
$$= \{(1, 5), (2, 5), (3, 5)\} \qquad\qquad = \{(2, 5), (2, 6)\}$$

Note that A has three members and B has two members. Now the event A or B is merely the union of A with B. Thus,

$$A \cup B = \{(1, 5), (2, 5), (3, 5), (2, 6)\}$$

We conclude the event A *or* B has four members. [The event A *and* B, $A \cap B$, has one member, namely, $(2, 5)$.] ∎

PROBABILITY OF AN EVENT

We are now in a position to formulate a definition for the probability of a simple event of a sample space.

> **Probability of a Simple Event** Let S denote a sample space. To each simple event $\{e\}$ of S, we assign a real number, $P(e)$, called the *probability of the simple event* $\{e\}$, which has the two properties:
>
> (I) $P(e) \geq 0$ for all simple events $\{e\}$ in S
> (II) The sum of the probabilities of all the simple events of S equals 1
>
> If the sample space S is given by
>
> $$S = \{e_1, e_2, \ldots, e_n\}$$
>
> then
>
> $$\text{(I)} \quad P(e_1) \geq 0, \quad P(e_2) \geq 0, \ldots, P(e_n) \geq 0$$
> $$\text{(II)} \quad P(e_1) + P(e_2) + \cdots + P(e_n) = 1 \qquad (1)$$

The real number assigned to the simple event $\{e\}$ is *completely arbitrary* within the framework of the above restrictions. For example, let a die be thrown. A sample space S is then

$$S = \{1, 2, 3, 4, 5, 6\}$$

There are six simple events in S: $\{1\}, \{2\}, \{3\}, \{4\}, \{5\}, \{6\}$. Either of the following two assignments of probabilities is acceptable.

(a)
$$P(1) = \frac{1}{6} \qquad P(2) = \frac{1}{6} \qquad P(3) = \frac{1}{6}$$
$$P(4) = \frac{1}{6} \qquad P(5) = \frac{1}{6} \qquad P(6) = \frac{1}{6}$$

This choice is in agreement with the definition, since the probability of each simple event is nonnegative and their sum is 1. This is an example of a "fair" die in which each simple event is equally likely to occur.

(b)
$$P(1) = 0 \qquad P(2) = 0 \qquad P(3) = \frac{1}{3}$$
$$P(4) = \frac{2}{3} \qquad P(5) = 0 \qquad P(6) = 0$$

This choice is also acceptable, even though it is unnatural. It implies that the die is "loaded," since only a 3 or a 4 appears and a 4 is twice as likely to occur as a 3.

Example 4

A coin is weighted so that heads $\{H\}$ is five times more likely to occur than tails $\{T\}$. What probability should we assign to heads? To tails?

Solution

Let x denote the probability that tails occurs. Then,

$$P(T) = x \quad \text{and} \quad P(H) = 5x$$

Since the sum of the probabilities of all the simple events equals 1, we must have

$$P(H) + P(T) = 5x + x = 1$$
$$6x = 1$$
$$x = \frac{1}{6}$$

Thus, we assign the probabilities

$$P(H) = \frac{5}{6} \quad P(T) = \frac{1}{6}$$

∎

Suppose probabilities have been assigned to each simple event of S. We now raise the question, "What is the probability of an event?" Let S be a sample space and let E be any event of S. It is clear that either $E = \varnothing$ or E is a simple event or E is the union of two or more simple events.

Probability of an Event If $E = \varnothing$, we define the *probability of $\varnothing$* to be

$$P(\varnothing) = 0 \tag{2}$$

In this case, the event $E = \varnothing$ is said to be *impossible*.
If E is simple, (1) is applicable.
If E is the union of r simple events $\{e_{i_1}\}, \{e_{i_2}\}, \ldots, \{e_{i_r}\}$, we define the *probability of E* to be

$$P(E) = P(e_{i_1}) + P(e_{i_2}) + \cdots + P(e_{i_r})$$

In particular, if the sample space S is given by

$$S = \{e_1, e_2, \ldots, e_n\}$$

we must have

$$P(S) = P(e_1) + \cdots + P(e_n) = 1$$

Thus, the probability of S, the sample space, is 1.

Example 5

Let two coins be tossed. A sample space S is

$$S = \{HH, TH, HT, TT\}$$

Let E be the event that they are both heads or both tails.
This can be depicted in two ways. See Figure 4.

E = same result on
both tosses = $\{HH, TT\}$

Figure 4

Compute the probability for event E with the following two assignments of probabilities:

(a) $P(HH) = P(TH) = P(HT) = P(TT) = \frac{1}{4}$
(b) $P(HH) = \frac{1}{9}$, $P(TH) = \frac{2}{9}$, $P(HT) = \frac{2}{9}$, $P(TT) = \frac{4}{9}$

Solution
The event E is $\{HH, TT\}$.

(a) $P(E) = P(HH) + P(TT) = \frac{1}{4} + \frac{1}{4} = \frac{1}{2}$
(b) $P(E) = P(HH) + P(TT) = \frac{1}{9} + \frac{4}{9} = \frac{5}{9}$

The fact that we obtained different probabilities for the same event in Example 5 is not unexpected, since they result from different assignments of probabilities to the simple events of the experiment. Any assignment that conforms to the restrictions given in (1) is mathematically correct. The question of which assignment to make is not a mathematical question, but is one that depends on the real-world situation to which the theory is applied. In this example the coins were fair in case (a) and were loaded in case (b).

CONSTRUCTING A PROBABILISTIC MODEL
Now that we have introduced the concepts of sample space, event, and probability of events, we introduce the idea of a *probabilistic model*.

To construct a *probabilistic model* we need to do the following:

Step 1. List all possible outcomes of the experiment under investigation; that is, give a sample space or, if this is not easy to do, determine the number of simple events in the sample space.

Step 2. Assign to each simple event e a probability $P(e)$ such that (1) is satisfied.

Example 6

A fair coin* is tossed. If it comes up heads, H, a fair die is rolled and the experiment is finished. If it comes up tails, T, the coin is tossed once more. Describe a probabilistic model for this experiment.

Solution

All the possible outcomes of this experiment are

$$S = \{H1, H2, H3, H4, H5, H6, TT, TH\}$$

where $H1$ indicates heads for the coin and then 1 for the die, and so on. See Figure 5 for the tree diagram.

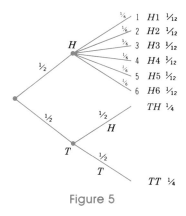

Figure 5

To these eight possible outcomes we assign the following probabilities:

$$P(H1) = P(H2) = P(H3) = P(H4) = P(H5) = P(H6) = \frac{1}{12}$$

$$P(TH) = P(TT) = \frac{1}{4}$$

The above discussion constitutes a model, called a *probabilistic* or *stochastic model*, for the experiment. ∎

Example 7

For the model described in Example 6, let E be the event

$$E = \{H2, H4, TH\}$$

The probability of the event E is

$$P(E) = P(H2) + P(H4) + P(TH)$$
$$= \frac{1}{12} + \frac{1}{12} + \frac{1}{4} = \frac{5}{12}$$

∎

* A fair coin is a coin in which the probability of heads is equal to the probability of tails.

Exercise 7.2   *Answers to Odd-Numbered Problems begin on page A-20.*

A   **1.** A nickel and a dime are tossed. List the elements of the sample space:

(a) If we are interested in whether the dime falls heads (*H*) or tails (*T*).

(b) If we are interested only in the number of heads that appear on a single toss of the two coins.

(c) If we are interested in whether the coins match (*M*) or do not match (*D*).

2. A card is selected from a standard deck of playing cards and its color—red (*R*) or black (*B*) is recorded. Describe an appropriate sample space for this experiment.

In Problems 3–6 describe the sample space associated with each random experiment. List the elements in each sample space.

3. Tossing a coin 3 times   **4.** Tossing 3 coins once

5. Tossing a coin 2 times and then a die   **6.** Tossing a coin and then a die

In Problems 7–14 use the spinners pictured below to list the elements in the sample space associated with each experiment described.

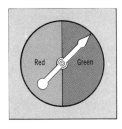 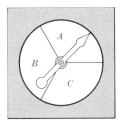 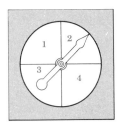

7. First Spinner 1 is spun and then Spinner 2 is spun.

8. First Spinner 3 is spun and then Spinner 1 is spun.

9. Spinner 1 is spun twice.

10. Spinner 3 is spun twice.

11. Spinner 2 is spun twice and then Spinner 3 is spun.

12. Spinner 3 is spun once and then Spinner 2 is spun twice.

13. Spinners 1, 2, 3 are each spun once in this order.

14. Spinners 3, 2, 1 are each spun once in this order.

In Problems 15–20 count the elements in the sample space associated with each random experiment.

15. Tossing a coin 4 times   **16.** Tossing a coin 5 times

17. Tossing 3 dice   **18.** Tossing 2 dice and then a coin

19. Selecting 2 cards (without replacement) from an ordinary deck of 52 cards*

20. Selecting 3 cards (without replacement) from an ordinary deck of 52 cards*

* An ordinary *deck of cards* has 52 cards. There are four suits of 13 cards each. The suits are called *clubs* (black), *diamonds* (red), *hearts* (red), and *spades* (black). In each suit the 13 cards are labeled A (ace), 2, 3, 4, 5, 6, 7, 8, 9, 10, J (jack), Q (queen), and K (king).

In Problems 21–24 consider the experiment of tossing a coin twice. The table lists seven possible assignments of probabilities for this experiment:

Simple Event	Sample Space			
	HH	**HT**	**TH**	**TT**
1	$\frac{1}{4}$	$\frac{1}{4}$	$\frac{1}{4}$	$\frac{1}{4}$
2	0	0	0	1
3	$\frac{3}{16}$	$\frac{5}{16}$	$\frac{5}{16}$	$\frac{3}{16}$
4	$\frac{1}{2}$	$\frac{1}{2}$	$\frac{-1}{2}$	$\frac{1}{2}$
5	$\frac{1}{8}$	$\frac{1}{4}$	$\frac{1}{4}$	$\frac{1}{8}$
6	$\frac{1}{9}$	$\frac{2}{9}$	$\frac{2}{9}$	$\frac{4}{9}$
7	.1	.5	.3	.1

(The left side of the table is labeled **Assignments**.)

Using this table, answer the following questions.

21. Which of the assignments of probabilities satisfy the restrictions in (1)?

22. Which of the assignments should be used if the coin is known to be fair?

23. Which of the assignments of probabilities should be used if the coin is known to always come up tails?

24. What assignment of probabilities should be used if tails is twice as likely as heads to occur?

In Problems 25–28 assign valid probabilities to the simple events of each random experiment.

25. Tossing a fair coin twice

26. Tossing a fair coin three times

27. Tossing a fair die and then a fair coin

28. Tossing a fair die and then 2 fair coins

In Problems 29–34 the random experiment consists of tossing a fair coin four times.

29. List the elements of the sample space and assign probabilities to each simple event.

30. Write the elements of the event, "The first two tosses are heads."

31. Write the elements of the event, "The last three tosses are tails."

32. Write the elements of the event, "Exactly three tosses come up tails."

33. Write the elements of the event, "The number of heads exceeds 1 but is fewer than 4."

34. Write the elements of the event, "The first two tosses are heads and the second two are tails."

In Problems 35–40 the random experiment consists of tossing two fair dice. Construct a probabilistic model for this experiment and find the probability of each given event.

35. $A = \{(1, 2), (2, 1)\}$

36. $B = \{(1, 5), (2, 4), (3, 3), (4, 2), (5, 1)\}$

37. $C = \{(1, 4), (2, 4), (3, 4), (4, 4)\}$

38. $D = \{(1, 2), (2, 1), (2, 4), (4, 2), (3, 6), (6, 3)\}$

39. $E = \{(1, 1), (2, 2), (3, 3), (4, 4), (5, 5), (6, 6)\}$

40. $F = \{(6, 6)\}$

41. A coin is loaded so that the probability of heads is three times the probability of tails. What is the probability of heads? What is the probability of tails?

42. A coin is loaded so that the probability of tails is $\frac{1}{4}$. What is the probability of heads?

In Problems 43–48 the random experiment consists of tossing a fair die and then a fair coin. Construct a probabilistic model for this experiment and find the probability of each given event.

43. *A*: The coin comes up heads   44. *B*: The die comes up 1

45. *C*: The die does not come up 1   46. *D*: The die comes up 5 or 6

47. *E*: The die comes up 3, 4, or 5

48. *F*: The coin comes up heads and the die comes up a number less than 4

49. In Example 5, if the events *E*, *F*, *G* are defined as

    E: At least 1 head   *F*: Exactly 1 tail   *G*: Tails on both tosses

    find $P(E)$, $P(F)$, $P(G)$ for the two given assignments of probabilities.

50. If the events *E* and *F* are defined as

    E: Common stocks are a good buy   *F*: Corporate bonds are a good buy

    state in words the meaning of:

    (a) $P(E \cup F)$ (b) $P(\overline{E})$ (c) $P(\overline{E} \cap \overline{F})$
    (d) $P(E \cup \overline{F})$ (e) $P(\overline{E} \cup F)$ (f) $P(\overline{E} \cap F)$

B 51. In a T-maze a mouse may turn to the right (R) and receive a mild shock, or to the left (L) and get a piece of cheese. Its behavior in making such "choices" is studied by psychologists. Suppose a mouse runs a T-maze three times. List the set of all possible outcomes and assign valid probabilities to each simple event. Find the probability of each of the following events:

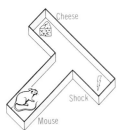

    (a) *E*: Run to the right two consecutive times
    (b) *F*: Never run to the right
    (c) *G*: Run to the left on the first trial
    (d) *H*: Run to the right on the second trial

52. A marketing study reveals two major rival companies, *G* and *H*, and four minor firms, *a, b, c, d*, who always ally themselves with one of the two major companies on prices. Free competition results if a major company is backed by two smaller ones. Construct a sample space and list the events that maintain free competition.

53. A red die and a green die are tossed. If *r* denotes the result on the red die and *g* the result on the green die, give verbal descriptions of the following algebraically described events:

    (a) $r = 3g$ (b) $r - g = 1$ (c) $r \le g$
    (d) $r + g = 8$ (e) $g = r^2$ (f) $r = g$

    Graph each of these events using Figure 2 as a backdrop.

C **54.** Let $S = \{e_1, e_2, e_3, e_4, e_5, e_6, e_7\}$ be a given sample space. Let the probabilities assigned to the simple events be given as follows:

$$P(e_1) = P(e_2) = P(e_6)$$

$$P(e_3) = 2P(e_4) = \frac{1}{2} P(e_1)$$

$$P(e_5) = \frac{1}{2} P(e_7) = \frac{1}{4} P(e_1)$$

(a) Find $P(e_1)$, $P(e_2)$, $P(e_3)$, $P(e_4)$, $P(e_5)$, $P(e_6)$, $P(e_7)$.

(b) If $A = \{e_1, e_2\}$, $B = \{e_2, e_3, e_4\}$, $C = \{e_5, e_6, e_7\}$, and $D = \{e_1, e_5, e_6\}$, find $P(A)$, $P(B)$, $P(C)$, $P(D)$, $P(A \cup B)$, $P(A \cap D)$, $P(D \cap B)$, and $P(A \cap \bar{B})$.

55. Three cars, C_1, C_2, C_3, are in a race. If the probability of C_1 winning is p, that is, $P(C_1) = p$, and $P(C_1) = \frac{1}{2}P(C_2)$, and $P(C_3) = \frac{1}{2}P(C_2)$, find $P(C_1)$, $P(C_2)$, and $P(C_3)$. Also, find $P(C_1 \cup C_2)$ and $P(C_1 \cup C_3)$.

56. Consider an experiment with a loaded die such that the probability of any of the faces appearing in a toss is equal to that face times the probability that a 1 will occur. That is, $P(6) = 6 \cdot P(1)$, and so on.

(a) Describe the sample space.

(b) Find $P(1)$, $P(2)$, $P(3)$, $P(4)$, $P(5)$, and $P(6)$.

(c) Let the events A, B, and C be described as

A: Even-numbered face

B: Odd-numbered face

C: Prime number on face (2, 3, 5 are prime)

Find $P(A)$, $P(B)$, $P(C)$, $P(A \cup B)$, and $P(A \cup \bar{C})$.

7.3

PROPERTIES OF THE PROBABILITY OF AN EVENT

MUTUALLY EXCLUSIVE EVENTS
ADDITIVE RULE
COMPLEMENT OF AN EVENT
ODDS

In this section, we state and prove results involving the probability of an event after the probabilistic model has been determined. Since set operations will enter into many of our later formulas, the following operations and symbols should be thoroughly understood:

$E \cup F$ Represents the event "either E occurs or F occurs (or both)," that is, the event "at least one of E or F occurs"

$E \cap F$ Represents the event "both E and F occur"

$\bar{E}$ Represents the event "E does not occur"

Example 1

In an experiment of tossing one fair die, consider the events:

E: Face is even

F: Face is above 2

Then

$$S = \{1, 2, 3, 4, 5, 6\}$$
$$E = \{2, 4, 6\} \qquad F = \{3, 4, 5, 6\}$$

So that

$$E \cup F = \{2, 3, 4, 5, 6\}$$
$$E \cap F = \{4, 6\}$$
$$\overline{E} = \{1, 3, 5\}$$

MUTUALLY EXCLUSIVE EVENTS

Mutually Exclusive Events   **Two or more events of a sample space S are said to be *mutually exclusive* if and only if they have no outcomes in common.**

That is, if we treat the events as sets, they are disjoint.

The following result gives us a way of computing probabilities for mutually exclusive events.

Let E and F be two events of a sample space S. If E and F are mutually exclusive, that is, if $E \cap F = \varnothing$, then the probability of the event E or F is the sum of their probabilities, namely,

$$P(E \cup F) = P(E) + P(F) \tag{1}$$

Since E and F can be written as a union of simple events in which no simple event of E appears in F and no simple event of F appears in E, the result follows.

Example 2

In the experiment of tossing two fair dice, what is the probability of obtaining either a sum of 7 or a sum of 11?

Solution

Let E and F be the events:

E: sum is 7
    $= \{(1, 6), (2, 5), (3, 4), (4, 3), (5, 2), (6, 1)\}$

F: sum is 11
    $= \{(5, 6), (6, 5)\}$

Since the dice are fair,

$$P(E) = \frac{6}{36} \qquad P(F) = \frac{2}{36}$$

The two events E and F are mutually exclusive. Therefore, by Equation (1), the probability that the sum is 7 or 11 is

$$P(E \cup F) = P(E) + P(F) = \frac{6}{36} + \frac{2}{36} = \frac{8}{36} = \frac{2}{9} \qquad \blacksquare$$

The probability of any simple event of a sample space S is nonnegative. Furthermore, since any event E of S is the union of simple events in S, and since $P(S) = 1$, it is easy to see that

$$0 \le P(E) \le 1$$

To summarize, the probability of an event E of a sample space S has the following three properties:

(I) **Positiveness.** $0 \le P(E) \le 1$ for every event E of S

(II) **Certainty.** $P(S) = 1$

(III) **Union.** $P(E \cup F) = P(E) + P(F)$ for any two events E and F of S for which $E \cap F = \emptyset$

ADDITIVE RULE

The following result, called the *additive rule,* provides a technique for finding the probability of the union of two events when they are not disjoint.

Additive Rule For any two events E and F of a sample space S,
$$P(E \cup F) = P(E) + P(F) - P(E \cap F)$$

Proof From Example 7 in Section 6.1 (page 286), we have

$$E \cup F = (E \cap \overline{F}) \cup (E \cap F) \cup (\overline{E} \cap F)$$

Since $E \cap \overline{F}$, $E \cap F$, and $\overline{E} \cap F$ are pairwise disjoint, we can use Property (III) to write

$$P(E \cup F) = P(E \cap \overline{F}) + P(E \cap F) + P(\overline{E} \cap F) \qquad (2)$$

We may write the sets E and F in the form

$$E = (E \cap F) \cup (E \cap \overline{F})$$
$$F = (E \cap F) \cup (\overline{E} \cap F)$$

Since $E \cap F$ and $E \cap \overline{F}$ are disjoint and $E \cap F$ and $\overline{E} \cap F$ are disjoint, we have

$$P(E) = P(E \cap F) + P(E \cap \overline{F})$$
$$P(F) = P(E \cap F) + P(\overline{E} \cap F) \qquad (3)$$

Combining the results in (2) and (3), we get

$$P(E \cup F) = P(E) + P(F) - P(E \cap F)$$

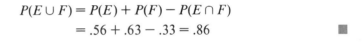

Example 3
Consider the two events

E: A shopper spends at least \$40 for food

F: A shopper spends at least \$15 for meat

Because of recent studies, we might assign

$$P(E) = .56 \qquad P(F) = .63$$

Suppose the probability that a shopper spends at least \$40 for food and \$15 for meat is .33. What is the probability that a shopper spends at least \$40 for food or at least \$15 for meat?

Solution
Since we are looking for the probability of $E \cup F$, we use the additive rule and find that

$$P(E \cup F) = P(E) + P(F) - P(E \cap F)$$
$$= .56 + .63 - .33 = .86$$

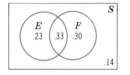

Figure 6

Often a Venn diagram is helpful in solving probability problems. A Venn diagram depicting the information of Example 3 is given in Figure 6. From this figure, we conclude that the probability of the event E, but not F, is .23 and that the probability of neither E nor F is .14.

Example 4
In an experiment with two fair dice, consider the events

E: The sum of the faces is 8

F: Doubles are thrown

What is the probability of obtaining E or F?

Solution
$$E = \{(2, 6), (3, 5), (4, 4), (5, 3), (6, 2)\}$$
$$F = \{(1, 1), (2, 2), (3, 3), (4, 4), (5, 5), (6, 6)\}$$
$$E \cap F = \{(4, 4)\}$$

Also,

$$P(E) = \frac{5}{36} \qquad P(F) = \frac{6}{36} \qquad P(E \cap F) = \frac{1}{36}$$

Thus, the probability of E or F is

$$P(E \cup F) = \frac{5}{36} + \frac{6}{36} - \frac{1}{36} = \frac{10}{36} = \frac{5}{18}$$

∎

Example 5
If $P(E) = .30$, $P(F) = .20$, and $P(E \cup F) = .40$, find $P(E \cap F)$.

Solution
By the additive rule we know that

$$P(E \cup F) = P(E) + P(F) - P(E \cap F)$$

or

$$P(E \cap F) = P(E) + P(F) - P(E \cup F)$$

Using the data given above we have

$$P(E \cap F) = .30 + .20 - .40 = .10$$

∎

COMPLEMENT OF AN EVENT

Let E be an event of a sample space. The complement of E is the event "Not E" in S. The next result gives a relationship between their probabilities.

Let E be an event of a sample space S. Then

$$P(\overline{E}) = 1 - P(E) \tag{4}$$

where $\overline{E}$ is the complement of E.

Proof   We know that

$$S = E \cup \overline{E} \qquad E \cap \overline{E} = \varnothing$$

Since E and $\overline{E}$ are mutually exclusive,

$$P(S) = P(E) + P(\overline{E})$$

Now, we apply Property II, setting $P(S) = 1$. It follows that

$$P(\overline{E}) = 1 - P(E)$$

∎

This result gives us a tool for finding the probability that an event does not occur if we know the probability that it does occur. Thus, the probability $P(\overline{E})$ that E does not occur is obtained by subtracting from 1 the probability $P(E)$ that E does occur.

Example 6
A study of people over 40 with an MBA degree shows that it is reasonable to assign a probability of .756 that such a person will have annual earnings in excess of

$30,000. The probability that such a person will earn $30,000 or less is then

$$1 - .756 = .244$$
■

Example 7
In an experiment using two fair dice find:

(a)   The probability that the sum of the faces is less than or equal to 3.
(b)   The probability that the sum of the faces is greater than 3. (Refer to Figure 2 on page 336).

Solution

(a)   The number of simple events in the event E described is 3. The number of simple events of the sample space S is 36. Thus, because the dice are fair,

$$P(E) = \frac{3}{36} = \frac{1}{12}$$

(b)   We need to find $P(\overline{E})$. From (4)

$$P(\overline{E}) = 1 - P(E) = 1 - \frac{1}{12} = \frac{11}{12}$$

That is, the probability that the sum of the faces is greater than 3 is $\frac{11}{12}$. ■

ODDS
In many instances, the probability of an event may be expressed as *odds* — either *odds for* an event or *odds against* an event.

If E is an event:

The *odds for E* are $\dfrac{P(E)}{P(\overline{E})}$ or $P(E)$ to $P(\overline{E})$

The *odds against E* are $\dfrac{P(\overline{E})}{P(E)}$ or $P(\overline{E})$ to $P(E)$

Example 8
The probability of the event

$$E: \quad \text{It will rain}$$

is .3. The odds for rain are

$$\frac{.3}{.7} = \frac{3}{7} \quad \text{or} \quad 3 \text{ to } 7$$

The odds against rain are

$$\frac{.7}{.3} = \frac{7}{3} \quad \text{or} \quad 7 \text{ to } 3$$
■

To obtain the probability of the event E when either the odds for E or the odds against E are known, we use the following formulas:

If the odds for E are a to b, then

$$P(E) = \frac{a}{a+b} \tag{5}$$

If the odds against E are a to b, then

$$P(E) = \frac{b}{a+b}$$

The proof of (5) is given after the following example.

Example 9

(a)   The odds for a Republican victory in the 1992 Presidential election are 7 to 5. What is the probability that a Republican victory occurs?

(b)   Suppose the odds against the Chicago Cubs winning the league pennant are 200 to 1. What is the probability that the Cubs win the pennant?

Solution

(a)   The event E is "A Republican victory occurs." The odds for E are 7 to 5. Thus,

$$P(E) = \frac{7}{7+5} = \frac{7}{12} \approx .583$$

(b)   The event F is "The Cubs win the pennant." The odds against F are 200 to 1. Thus,

$$P(F) = \frac{1}{200+1} = \frac{1}{201} \approx .00498$$ ∎

Proof of (5)   Suppose the odds for E are a to b. Then, by the definition of odds,

$$\frac{P(E)}{P(\overline{E})} = \frac{a}{b}$$

But, $P(\overline{E}) = 1 - P(E)$. So,

$$\frac{P(E)}{1 - P(E)} = \frac{a}{b}$$

Cross multiplying, we get

$$bP(E) = a[1 - P(E)] = a - aP(E)$$

or

$$aP(E) + bP(E) = a$$

Solving for $P(E)$ yields

$$P(E) = \frac{a}{a+b}$$

∎

Exercise 7.3

Answers to Odd-Numbered Problems begin on page A-21.

A   In Problems 1–6 find the probability of the indicated event when $P(A) = .20$ and $P(B) = .30$.

1. $P(\bar{A})$

2. $P(\bar{B})$

3. $P(A \cup B)$ if A, B are mutually exclusive

4. $P(A \cap B)$ if A, B are mutually exclusive

5. $P(A \cup B)$ if $P(A \cap B) = .15$

6. $P(A \cap B)$ if $P(A \cup B) = .40$

7. In tossing 2 fair dice, are the events "Sum is 2" and "Sum is 12" mutually exclusive? What is the probability of obtaining either a 2 or a 12?

8. In tossing 2 fair dice, are the events "Sum is 6" and "Sum is 8" mutually exclusive? What is the probability of obtaining either a 6 or an 8?

9. The Chicago Bulls basketball team has a probability of winning of .65. What is their probability of losing?

10. The Chicago Black Hawks hockey team has a probability of winning of .6 and a probability of losing of .25. What is the probability of a tie?

11. Jill needs to pass both mathematics and English in order to graduate. She estimates her probability of passing mathematics at .4 and English at .6, and she estimates her probability of passing at least one of them at .8. What is her probability of passing both courses?

12. After midterm exams, the student in Problem 11 reassesses her probability of passing mathematics to .7. She feels her probability of passing at least one of these courses is still .8, but she has only a probability of .1 of passing both courses. If her probability of passing English is less than .4, she will drop English. Should she drop English? Why?

13. Let A and B be events of a sample space S and let $P(A) = .5$, $P(B) = .3$, and $P(A \cap B) = .1$. Find the probabilities for each of the following events:

(a) A or B

(b) A but not B

(c) B but not A

(d) Neither A nor B

14. If A and B represent two mutually exclusive events such that $P(A) = .35$ and $P(B) = .50$, find each of the following:

(a) $P(A \cup B)$

(b) $P(\overline{A \cup B})$

(c) $P(\bar{B})$

(d) $P(\bar{A})$

(e) $P(A \cap B)$

15. At Milex, a tune-up and brake repair shop, the manager has found that a car will require a tune-up with probability .6, a brake job with probability .1, and both with probability .02.

(a) What is the probability that a car requires either a tune-up or a brake job?

(b) What is the probability that a car requires a tune-up but not a brake job?

(c) What is the probability that a car requires neither type of repair?

16. A factory needs two raw materials, say E and F. The probability of not having an

adequate supply of material E is .06, whereas the probability of not having an adequate supply of material F is .04. A study shows that the probability of a shortage of both E and F is .02. What is the probability of the factory being short of either material E or F?

B   **17.** In a survey of the number of TV sets in a house, the following probability table was constructed:

Number of TV sets	0	1	2	3	4 or more
Probability	.05	.24	.33	.21	.17

Find the probability of a house having:

(a)   1 or 2 TV sets   (b)   1 or more TV sets
(c)   3 or fewer TV sets   (d)   3 or more TV sets
(e)   Less than 2 TV sets   (f)   Less than 1 TV set
(g)   1, 2, or 3 TV sets   (h)   2 or more TV sets

18. Through observation it has been determined that the probability of a given number of people waiting in line at a particular checkout register of a supermarket is

Number waiting in line	0	1	2	3	4 or more
Probability	.10	.15	.20	.24	.31

Find the probability of:

(a)   At most 2 people in line
(b)   At least 2 people in line
(c)   At least 1 person in line

In Problems 19–24 determine the probability of E for the given odds.

19.   3 to 1 for E **20.**   4 to 1 against E
21.   7 to 5 against E **22.**   2 to 9 for E
23.   1 to 1 for E (even)   **24.**   50 to 1 for E

In Problems 25–28 determine the odds for and against each event for the given probability.

25. $P(E) = .7$ **26.** $P(H) = \frac{1}{3}$ **27.** $P(F) = \frac{4}{5}$ **28.** $P(G) = .01$

29. The probability of a person getting a job interview is .54; what are the odds against getting the interview?

30. If the probability of war is .6, what are the odds against war?

31. If 2 fair dice are thrown, what are the odds of obtaining a 7? An 11? A 7 or 11?

32. If the odds for event A are 1 to 5 and the odds for event B are 1 to 3, what are the odds for the event A or B, assuming the event A and B is impossible?

33. In a track contest, the odds that A will win are 1 to 2 and the odds that B will win are 2 to 3. Find the probability and the odds that A or B wins the race, assuming a tie is impossible.

34. It has been estimated that in 70% of the fatal accidents involving two cars, at least one of the drivers is drunk. If you hear of a two-car fatal accident, what odds should you give a friend for the event that at least one of the drivers is drunk?

C   **35.** Generalize the additive rule by showing that the probability of the occurrence of at least one of the three events A, B, C is given by

$$P(A \cup B \cup C) = P(A) + P(B) + P(C)$$
$$- P(A \cap B) - P(A \cap C) - P(B \cap C)$$
$$+ P(A \cap B \cap C)$$

7.4

PROBABILITY FOR THE CASE OF EQUALLY LIKELY EVENTS

Thus far, we have seen that in some circumstances it is reasonable to assign the same probability to each simple event of the sample space. Such events are termed *equally likely events.*

Equally likely events occur when items are selected randomly. For example, in randomly selecting 1 person from a group of 10, the probability of selecting a particular individual is $\frac{1}{10}$. If a card is chosen randomly from a deck of 52 cards, the probability of drawing a particular card is $\frac{1}{52}$. In general, if a sample space S has n equally likely simple events, the probability assigned to each simple event is $\frac{1}{n}$.

Let a sample space S be given by

$$S = \{e_1, e_2, \ldots, e_n\}$$

Suppose all of the simple events $\{e_1\}, \ldots, \{e_n\}$ are equally likely to occur. If E is the union of m of these n simple events, then

$$P(E) = \frac{m}{n} \tag{1}$$

Proof   Since the simple events $e_1, \ldots, e_n$ are equally likely, we know that

$$P(e_1) = \cdots = P(e_n) \tag{2}$$

Since

$$S = \{e_1\} \cup \cdots \cup \{e_n\}$$

we have

$$P(S) = P(e_1) + \cdots + P(e_n) = 1 \tag{3}$$

To satisfy both (2) and (3) we must have

$$P(e_1) = P(e_2) = \cdots = P(e_n) = \frac{1}{n}$$

Now, for the event E of S, in which the m events have been reordered for convenience, we have

$$E = \{e_1\} \cup \cdots \cup \{e_m\}$$

Thus,

$$P(E) = P(e_1) + \cdots + P(e_m) = \underbrace{\frac{1}{n} + \cdots + \frac{1}{n}}_{m \text{ times}} = \frac{m}{n}$$

■

The above result is sometimes stated in the following way:

If an experiment has n equally likely outcomes, among which the event E occurs m times, then the probability of event E, written as $P(E)$, is m/n. That is,

$$P(E) = \frac{\text{Number of possible ways the event } E \text{ can take place}}{\text{Number of outcomes in } S} = \frac{m}{n}$$

If the fact that the event E has occurred is termed a *success,* then

$$P(E) = \frac{\text{Number of successes}}{\text{Number of outcomes in } S} = \frac{m}{n} \tag{4}$$

Event E not occurring is referred to as a *failure.*

Thus, to compute the probability of an event E in which the outcomes are equally likely, count the number $c(E)$ of simple events in E, and divide by the total number $c(S)$ of simple events in the sample space. Then,

$$P(E) = \frac{c(E)}{c(S)}$$

Statement (4) is often used to define *probability.* Using this definition, Properties (I), (II), and (III) previously stated in Section 7.3 are still valid.

Suppose an experiment resulted in no successes at all. By (4), we have

$$P(\text{Success}) = \frac{0}{n} = 0 \tag{5}$$

If the experiment resulted in all events in the sample space being successes, then $m = n$, and

$$P(\text{Success}) = \frac{m}{n} = \frac{n}{n} = 1$$

From this result and (5), we see that

$$0 \le P(\text{Success}) \le 1$$

Suppose E is an event with m successes in a sample space S of n elements. Then

$$P(E) = \frac{m}{n}$$

Now $\bar{E}$ contains $n - m$ elements. Hence,

$$P(\bar{E}) = \frac{(n-m)}{n} = \frac{n}{n} - \frac{m}{n} = 1 - \frac{m}{n} = 1 - P(E)$$

Example 1

In an experiment with two dice, we present the following in terms of events:

(a) The sum of the faces is 3.
(b) The sum of the faces is 7.
(c) The sum of the faces is 7 or 3.
(d) The sum of the faces is 7 and 3.

Find the probability of these events, assuming the dice are fair.

Solution

(a) The sum of the faces is 3 if and only if the outcome is the event $A = \{(1, 2),$ $(2, 1)\}$. Since $c(A) = 2$ and $c(S) = 36$, we have

$$P(A) = \frac{2}{36} = \frac{1}{18}$$

(b) The sum of the faces is 7 if and only if the outcome is a member of the event $B = \{(1, 6), (2, 5), (3, 4), (4, 3), (5, 2), (6, 1)\}$. Since $c(B) = 6$, we have

$$P(B) = \frac{6}{36} = \frac{1}{6}$$

(c) The sum of the faces is 7 or 3 if and only if the outcome is a member of $A \cup B$, as defined in parts (a) and (b).

$$A \cup B = \{(2, 1), (1, 2), (1, 6), (2, 5), (3, 4), (4, 3), (5, 2), (6, 1)\}$$

Since $c(A \cup B) = 8$,

$$P(A \cup B) = \frac{8}{36} = \frac{2}{9}$$

(d) The sum of the faces is 3 and 7 if and only if the outcome is a member of $A \cap B$. Since $A \cap B = \varnothing$, the event is impossible. That is, $P(A \cap B) = 0$. ■

Example 2

What is the probability that a random four-digit telephone extension has one or more repeated digits?

Solution

There are $10^4 = 10,000$ distinct four-digit telephone extensions; this, therefore, is the number of simple events in the sample space.

We wish to find the probability that a random four-digit telephone extension has one or more repeated digits. Since it is difficult to count the elements in this event, we first compute the probability of the following event:

$$E: \quad \text{No repeated digits in four-digit extension}$$

Notice that the event $\overline{E}$ is one or more repeated digits. By the multiplication principle, the number of four-digit extensions that have *no* repeated digits is

$$10 \cdot 9 \cdot 8 \cdot 7 = 5040$$

Hence

$$P(E) = \frac{5040}{10,000} = .504$$

Therefore, the probability of one or more repeated digits is

$$P(\overline{E}) = 1 - .504 = .496$$

Example 3

Birthday Problem An interesting problem in which Formula (4) is used is the so-called *birthday problem.* In general, the problem is to find the probability that in a group of r people there are at least two people who have the same birthday (the same month and day of the year).

Solution

To solve the problem, let us first determine the number of simple events in the sample space. There are 365 possibilities for each person's birthday (we exclude February 29 for simplicity). Since there are r people in the group, there are 365^r possibilities for the birthdays. [For one person in the group, there are 365 days on which his or her birthday can fall; for two people, there are $(365)(365) = 365^2$ pairs of days; and, in general, using the multiplication principle, for r people there are 365^r possibilities.]

Next, we assume a person is no more likely to be born on one day than another, so that we assign the probability $1/365^r$ to each simple event.

We wish to find the probability that at least two people have the same birthday. It is difficult to count the elements in this set; it is much easier to count the elements of the event

$$E: \quad \text{No two people have the same birthday}$$

Notice that the event $\overline{E}$ is that at least two people have the same birthday. To find the probability of E, we proceed as follows: Choose one person at random. There are 365 possibilities for his or her birthday. Choose a second person. There are 364 possibilities for this birthday, if no two people are to have the same birthday. Choose a third person. There are 363 possibilities left for this birthday. Finally, we arrive at the rth person. There are $365 - (r - 1)$ possibilities left for this birthday. By the multiplication principle, the total number of possibilities is $365 \cdot 364 \cdot 363 \cdot \cdots \cdot (365 - r + 1)$.

Hence, the probability of event E is

$$P(E) = \frac{365 \cdot 364 \cdot 363 \cdot \cdots \cdot (365 - r + 1)}{365^r}$$

The probability of two or more people having the same birthday is then $P(\bar{E}) = 1 - P(E)$. ■

For example, in a group of eight people

$$P(\bar{E}) = 1 - \frac{365 \cdot 364 \cdot 363 \cdot 362 \cdot 361 \cdot 360 \cdot 359 \cdot 358}{365^8}$$

$$= 1 - .93$$

$$= .07$$

The table below gives the probabilities for two or more people having the same birthday for some values of r. Notice that the probability is better than $\frac{1}{2}$ for any group of 23 or more people.

	Number of People														
	5	10	15	20	21	22	23	24	25	30	40	50	60	70	80	90
Probability that 2 or more have same birthday	.027	.117	.253	.411	.444	.476	.507	.538	.569	.706	.891	.970	.994	.99916	.99991	.99999

Example 4

A box contains 12 light bulbs of which 5 are defective. All bulbs look alike and have equal probability of being chosen. Three light bulbs are picked at random.

(a) What is the probability that all 3 are defective?
(b) What is the probability that exactly 2 are defective?
(c) What is the probability that at least two are defective?

Solution

(a) The number of elements in the sample space S is equal to the number of combinations of 12 light bulbs taken 3 at a time, namely,

$$\binom{12}{3} = \frac{12!}{3!9!} = 220$$

Define E as the event, "3 bulbs are defective." Then E can occur in $\binom{5}{3}$ ways, that is, the number of ways in which 3 defective bulbs can be chosen from 5 defectives ones. The probability $P(E)$ is

$$P(E) = \frac{\binom{5}{3}}{\binom{12}{3}} = \frac{\frac{5!}{3!2!}}{220} = \frac{10}{220} = .04545$$

(b) Define F as the event, "2 bulbs are defective." Then F can occur in $\binom{5}{2}\binom{7}{1}$ ways. Thus the probability is given by

$$P(F) = \frac{\binom{5}{2}\binom{7}{1}}{\binom{12}{3}} = \frac{\frac{5!}{2!3!} \cdot 7}{220} = \frac{70}{220} = .31818$$

(c)   What is the probability of the event G: At least 2 are defective? The event G is equivalent to asking for the probability of selecting either 2 or 3 defective bulbs. Thus,

$$P(G) = P(E) + P(F) = \frac{10}{220} + \frac{70}{220} = \frac{80}{220} = .36364$$ ■

Example 5

Find the probability of obtaining (a) a straight and (b) a flush in a poker hand. (A poker hand is a set of 5 cards chosen at random from a deck of 52 cards.)

Solution

(a)   A straight consists of 5 consecutive cards, not all of the same suit. The sample space contains $\binom{52}{5}$ simple events, each equally likely to occur. Now, for the straight 4, 5, 6, 7, 8, the 4 can be drawn in 4 different ways, as can the 5, the 6, the 7, and the 8, for a total of 4^5 ways. There are a total of 10 different kinds of straights (A, 2, 3, 4, 5), (2, 3, 4, 5, 6), . . . , (9, 10, J, Q, K), and (10, J, Q, K, A). Thus, all together, there are $10 \cdot 4^5$ straights. However, among these are the straight flushes (36 straights all in one suit) and the four royal flushes (straight flushes consisting of 10, J, Q, K, A), which should not be included in the straight category. Thus, there are

$$10 \cdot 4^5 - 36 - 4 = 10{,}240 - 40 = 10{,}200 \text{ straights}$$

The probability of drawing a straight is therefore

$$\frac{10{,}200}{\binom{52}{5}} = .0039$$

(b)   A flush consists of 5 cards in a single suit. The number of ways of obtaining a flush in a given suit is $\binom{13}{5} = 1287$, and there are four different suits, for a total of $4(1287) = 5148$ flushes. However, straight flushes (36) and the four royal flushes should not be included in the flush category. Thus, there are

$$5148 - 36 - 4 = 5108 \text{ flushes}$$

The probability of drawing a flush is therefore

$$\frac{5108}{\binom{52}{5}} = .0020$$ ■

Exercise 7.4   *Answers to Odd-Numbered Problems begin on page A-21.*

In Problems 1–10 a card is drawn at random from a regular deck of 52 cards. Calculate the probability of each event.

A   **1.** The ace of hearts is drawn.   **2.** An ace is drawn.

    3. A spade is drawn.   **4.** A red card is drawn.

    5. A picture card (J, Q, K) is drawn.

    6. A number card (A, 2, 3, 4, 5, 6, 7, 8, 9, 10) is drawn.

    7. A card with a number less than 6 is drawn (count A as 1).

    8. A card with a value of 10 or higher is drawn.

    9. A card that is not an ace is drawn.

  10. A card that is either a queen or king of any suit is drawn.

In Problems 11–18 a ball is picked at random from a box containing 3 white, 5 red, 8 blue, and 7 green balls. Find the probability of each event.

  11. White ball is picked.   **12.** Blue ball is picked.

  13. Green ball is picked.   **14.** Red ball is picked.

  15. White or red ball is picked.   **16.** Green or blue ball is picked.

  17. Neither red nor green ball is picked.   **18.** Red or white or blue ball is picked.

  19. In a throw of 2 fair dice, what is the probability that the number on one die is double the number on the other?

  20. In a throw of 2 fair dice, what is the probability that one die gives a 5 and the other die a number less than 5?

  21. From a sales force of 150 people, 1 person will be chosen to attend a special sales meeting. If 52 are single, 72 are college graduates, and, of the 52 who are single, $\frac{3}{4}$ are college graduates, what is the probability that a salesperson selected at random will be neither single nor a college graduate?

  22. In an election, two amendments were proposed. The results indicate that of 850 people eligible to cast a ballot, 480 voted in favor of Amendment I, 390 voted for Amendment II, 120 voted for both, and 100 approved of neither. If an eligible voter is selected at random (that is, any one is as likely to be chosen as another), compute the following probabilities:

    (a)   The voter is in favor of I, but not II.

    (b)   The voter is in favor of II, but not I.

  23. What is the probability that, in a group of 3 people, at least 2 were born in the same month (disregard day and year)?

  24. What is the probability that, in a group of 6 people, at least 2 were born in the same month (disregard day and year)?

  25. A box contains 100 slips of paper numbered from 1 to 100. If 3 slips are drawn in succession with replacement, what is the probability that at least 2 of them have the same number?

  26. If, in Problem 25, 10 slips are drawn with replacement, what is the probability that at least 2 of them have the same number?

  27. Use a calculator and the idea behind Example 3 to find the approximate probability that 2 or more U.S. Senators have the same birthday. (There are 100 Senators.)

28. Follow the directions of Problem 27 for the House of Representatives. (There are more than 365 Representatives.)

B **29.** Through a mix-up on the production line, 6 defective refrigerators were shipped out with 44 good ones.

(a) If 5 are selected at random, what is the probability that all 5 are defective?

(b) What is the probability that at least 2 of them are defective?

30. In a shipment of 50 transformers, 10 are known to be defective. If 30 transformers are picked at random, what is the probability that all 30 are nondefective? Assume that all transformers look alike and have an equal probability of being chosen.

31. Five cards are dealt at random from a regular deck of 52 playing cards. Find the probability that:

(a) All are hearts.

(b) Exactly 4 are spades.

(c) Exactly 2 are clubs.

C **32.** Bridge   In a game of bridge, find the probability that a hand of 13 cards consists of 5 spades, 4 hearts, 3 diamonds, and 1 club.

33. Poker   Find the probability of obtaining each of the following poker hands:

(a) Royal flush (10, J, Q, K, A in a single suit).

(b) Straight flush (5 cards in sequence in a single suit, but not a royal flush).

(c) Four of a kind (4 cards of the same face value).

(d) Full house (one pair and one triple of the same face values).

(e) Straight or better.

34. Elevator Problem*   An elevator starts with 5 passengers and stops at 8 floors. Find the probability that no 2 passengers leave at the same floor. Assume that all arrangements of discharging the passengers have the same probability.

7.5

PROBABILITY PROBLEMS USING COUNTING TECHNIQUES

The following three examples utilize counting techniques to solve probability problems.

Example 1
A class contains 10 male students and 12 female students. Two students are selected at random. What is the probability that they are of opposite sex?

* William Feller, *An Introduction to Probability Theory and Its Applications,* 3rd ed., Wiley, New York, 1968.

Solution

Since there are 22 students in the class, there are $\binom{22}{2}$ ways of selecting two of them. If they are to be of opposite sex, we must obviously choose one male and one female. The first task can be accomplished in $\binom{10}{1}$ ways, while the second can be done in $\binom{12}{1}$ ways. The multiplication principle yields $\binom{10}{1} \cdot \binom{12}{1}$ ways of picking a male and a female. Consider the event

$$E: \text{"Students are of opposite sex"}$$

The desired probability is then

$$P(E) = \frac{\binom{10}{1} \cdot \binom{12}{1}}{\binom{22}{2}} = .519$$

∎

An alternate way of attacking the problem would be to consider the event

$$\overline{E}: \text{"Students chosen are of the same sex"}$$

There are $\binom{10}{2}$ ways of picking two males and $\binom{12}{2}$ ways of selecting two females. So the probability that both are of the same sex is

$$P(\overline{E}) = \frac{\binom{10}{2} + \binom{12}{2}}{\binom{22}{2}} = .481$$

Since $1 - .481 = .519$, the same answer results.

Example 2

A fair coin is tossed 10 times.

(a) What is the probability of obtaining exactly 5 heads and 5 tails?
(b) What is the probability of obtaining between 4 and 6 heads?

Solution

(a) The sample space consists of all possible outcomes resulting from tossing a coin 10 times and has 2^{10} elements (two possibilities, H or T, on each of 10 tosses). The number of tosses containing exactly 5 heads (and 5 tails) is $\binom{10}{5}$. Hence, the probability of an equal number of heads and tails is

$$\frac{\binom{10}{5}}{2^{10}} = .2461$$

(b) We are interested in how many outcomes result in 4, 5, or 6 heads. There are $\binom{10}{4}$ tosses with exactly 4 heads since counting them is the same as choosing which 4 of the 10 tosses should contain H's. Likewise, there are $\binom{10}{5}$ and $\binom{10}{6}$

tosses, respectively, with exactly 5 heads and 6 heads. Hence, the probability of obtaining between 4 and 6 heads is given by

$$\frac{\binom{10}{4} + \binom{10}{5} + \binom{10}{6}}{2^{10}} = .6563$$

∎

Example 3

If the letters in the word MISTER are randomly scrambled, what is the probability that the resulting rearrangement has the I preceding the E?

Solution

The total number of ways of rearranging the letters in MISTER is 6!, which forms the denominator of our desired probability. The numerator counts the number of arrangements having the I before the E. To compute this, we imagine any rearrangement as resulting from some placement of the letters in the following six positions

— — — — — —

We consider the formation of an arrangement having I preceding E as consisting of two tasks. Task 1 is to choose two of the six positions for the I and E — the I would then of necessity have to occupy the leftmost position of the two chosen.

Task 1 can be done in $\binom{6}{2}$ ways.

Task 2 would be to arrange the letters MSTR in the remaining four positions. This can be done in 4! ways. Hence, by the multiplication principle, the number of arrangements with the I preceding the E is

$$\binom{6}{2} \cdot 4!$$

So, the desired probability is

$$\frac{\binom{6}{2} \cdot 4!}{6!} = \frac{\frac{6!}{2!4!} \cdot 4!}{6!} = \frac{1}{2!} = .5$$

∎

The answer should be intuitively plausible — we would expect one half of the arrangements to have I before E and the other half to have E before I.

Exercise 7.5

Answers to Odd-Numbered Problems begin on page A-22.

1. A fair coin is tossed 10 times. Find the probability that 6 heads and 4 tails result.

2. A fair coin is tossed 8 times. Find the probability that

    (a) Four heads and 4 tails occur.

    (b) Between 3 and 5 heads occur.

    (c) More heads than tails occur.

3. Five distinct letters are randomly picked from the alphabet. What is the probability that at least one of them is a vowel (a, e, i, o, u)?

4. What is the probability that a 5-digit telephone extension has

   (a) No repeated digits?
   (b) One or more repeated digits?
   (c) Contains the number 2?

5. A room contains 10 married couples. Two people are chosen randomly from the room. What is the probability that they are husband and wife?

6. A room contains 5 males and 8 females. If two people are chosen at random, what is the probability that they are of opposite sex?

7. If the letters in EQUAL are randomly rearranged, what is the probability that the result has them in alphabetical order?

8. In a random rearrangement of the letters in the word CARPET, what is the probability that the R precedes the P?

9. If the letters in RANDOM are scrambled, what is the probability that the result has the M preceding the O?

10. A fair coin is tossed 10 times. What is the probability of obtaining an odd number of tails?

CHAPTER REVIEW

Important Terms and Important Formulas			
random event	simple event		equally likely events
probabilistic model	mutually exclusive events		success
sample space	additive rule		failure
outcome	odds for		
event	odds against		

$$P(E \cup F) = P(E) + P(F) \quad \text{if } E \cap F = \varnothing \qquad P(E) = \frac{a}{a + b}$$

$$P(E \cup F) = P(E) + P(F) - P(E \cap F) \qquad P(E) = \frac{m}{n}$$

$$P(\overline{E}) = 1 - P(E)$$

True–False Questions

(Answers on page A-22)

T   F   1. $P(E) + P(\overline{E}) = 1$

T   F   2. If $P(E) = .4$ and $P(F) = .3$, then $P(E \cup F)$ must be .7.

T   F   3. If $P(E \cap F) = 0$ then E and F are said to be mutually exclusive.

T   F   4. If $P(E \cup F) = .7$, $P(E) = .4$, and $P(F) = 3$, then $P(E \cap F) = 0$

T   F   5. If the odds for an event E are 2 to 1, then $P(E) = \frac{2}{3}$.

Fill in the Blanks   *(Answers on page A-22)*

1. If $S = \{a, b, c, d\}$ is a sample space and $E = \{a, b\}$, then $\overline{E} = $ _____.

2. If a coin is tossed five times, the number of simple events in the sample space of this experiment is _____.

3. If an event is certain to occur, then its probability is _____. If an event is impossible, its probability is _____.

4. If $P(E) = .6$, the odds _____ E are 3 to 2.

5. When each simple event in a sample space is assigned the same probability, the events are termed _____ _____.

6. If two events in a sample space have no simple events in common, they are said to be _____ _____.

7. If $P(E) = .6$, then $P(\overline{E}) = $ _____.

Review Exercises   *Answers to Odd-Numbered Problems begin on page A-22.*

A

1. A survey of families with 2 children is made, and the sexes of the children are recorded. Describe the sample space and draw a tree diagram of this random experiment.

2. A fair coin is tossed three times.

    (a) Construct a probabilistic model corresponding to this experiment.
    (b) Find the probabilities of the following events:

    (i)   The first toss is T.
    (ii)   The first toss is H.
    (iii) Either the first toss is T or the third toss is H.
    (iv)   At least one of the tosses is H.
    (v)   There are at least two T's.
    (vi)   No tosses are H.

3. An urn contains 3 white marbles, 2 yellow marbles, 4 red marbles, and 5 blue marbles. Two marbles are picked at random. What is the probability that:

    (a) Both are blue?
    (b) Exactly 1 is blue?
    (c) At least 1 is blue?

4. Let A and B be events with $P(A) = .3$, $P(B) = .5$, and $P(A \cap B) = .2$. Find the probability that:

    (a) A or B happens.
    (b) A does not happen.
    (c) Neither A nor B happens.
    (d) Either A does not happen or B does not happen.

5. A loaded die is rolled 400 times, and the following outcomes are recorded:

Face	1	2	3	4	5	6
No. times showing	32	45	84	74	92	73

Estimate the probability of rolling a

(a) 3   (b) 5   (c) 6

6. A survey of a group of criminals shows that 65% came from low-income families, 40% from broken homes, and 30% came from low-income and broken families. Define:

$$E: \quad \text{Criminal came from low income}$$
$$F: \quad \text{Criminal came from broken home}$$

If a criminal is selected at random find

(a) The probability that the criminal is not from a low-income family.
(b) The probability that the criminal comes from a broken home or a low-income family.
(c) Are E and F mutually exclusive?

B 7. Jones lives at O (see the figure). He owns 5 gas stations located 4 blocks away (dots). Each afternoon he checks on one of his gas stations. He starts at O. At each intersection he flips a fair coin. If it shows heads, he will head North (N); otherwise, he will head toward the East (E). What is the probability that he will end up at gas station G before coming to one of the other stations?

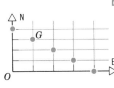

8. Consider the experiment of spinning the spinner in the figure three times. (Assume the spinner cannot fall on a line.)

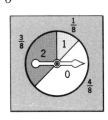

(a) Are all outcomes equally likely?
(b) If not, which of the outcomes has the highest probability?
(c) Let F be the event, "Each digit will occur exactly once." Find $P(F)$.

9. If E and F are events with $P(E \cup F) = \frac{5}{8}$, $P(E \cap F) = \frac{1}{3}$, and $P(E) = \frac{1}{2}$, find:

(a) $P(\overline{E})$ (b) $P(F)$ (c) $P(\overline{F})$

10. If E and F represent mutually exclusive events, $P(E) = .30$, and $P(F) = .45$, find each of the following probabilities:

(a) $P(\overline{E})$ (b) $P(\overline{F})$ (c) $P(E \cap F)$
(d) $P(E \cup F)$ (e) $P(\overline{E \cap F})$ (f) $P(\overline{E \cup F})$
(g) $P(\overline{E} \cup \overline{F})$ (h) $P(\overline{E} \cap \overline{F})$

11. Three envelopes are addressed for three secret letters written in invisible ink. A secretary randomly places each of the letters in an envelope and mails them. What is the probability that at least one person receives the correct letter?

12. What are the odds in favor of a 5 when a fair die is thrown?

13. A better is willing to give 7 to 6 odds that the Bears will win the NFL title. What is the probability of the Bears winning?

14. Roulette An American roulette table has 38 numbers on it:

$$00, 0, 1, 2, 3, \ldots 35, 36.$$

(a) If the bet is on a single number, what is the probability of winning? What are the odds against winning?
(b) If the bet is on "evens" coming up, that is on the number 2, 4, 6, ... , 34, 36, what is the probability of winning? What are the odds against winning?

Mathematical Questions

From Actuary Exams (Answers on page A-22)

1. *Actuary Exam — Part I*
   A box contains 12 varieties of candy and exactly 2 pieces of each variety. If 12 pieces of candy are selected at random, what is the probability that a given variety is represented?

   (a) $\dfrac{2^{12}}{(12!)^2}$ (b) $\dfrac{2^{12}}{24!}$ (c) $\dfrac{2^{12}}{\binom{24}{12}}$ (d) $\dfrac{11}{46}$ (e) $\dfrac{35}{46}$

2. *Actuary Exam — Part II*
   What is the probability that a 3-card hand drawn at random and without replacement from an ordinary deck consists entirely of black cards?

   (a) $\frac{1}{17}$ (b) $\frac{2}{17}$ (c) $\frac{1}{8}$ (d) $\frac{3}{17}$ (e) $\frac{4}{17}$

3. *Actuary Exam — Part II*
   The probability that both S and T occur, the probability that S occurs and T does not, and the probability that T occurs and S does not are all equal to p. What is the probability that either S or T occurs?

   (a) p (b) $2p$ (c) $3p$ (d) $3p^2$ (e) p^3

4. *Actuary Exam — Part II*
   What is the probability that a bridge hand contains 1 card of each denomination (i.e., 1 ace, 1 king, 1 queen, . . . , 1 three, 1 two)?

   (a) $\dfrac{13!}{13^{13}}$ (b) $\dfrac{4^{13}}{\binom{52}{13}}$ (c) $\dfrac{\binom{52}{4}}{\binom{52}{13}}$ (d) $\left(\dfrac{1}{13}\right)^{13}$ (e) $\dfrac{13^4}{\binom{52}{13}}$

8

ADDITIONAL TOPICS IN PROBABILITY

In this chapter we continue our study of probability theory by considering several of its most extensive applications.

8.1

CONDITIONAL PROBABILITY

DEFINITION OF CONDITIONAL PROBABILITY
PRODUCT RULE
PROBABILITY TREE

In this section we introduce *conditional probability.* Recall that whenever we compute the probability of an event we do it relative to the entire sample space in question. Thus, when we ask for the probability $P(E)$ of the event E, this probability $P(E)$ represents an appraisal of the likelihood that a chance experiment will produce an outcome in the set E relative to a sample space S.

However, sometimes we would like to compute the probability of an event E of a sample space relative to another event F of the same sample space. That is, if we have *prior* information that the outcome must be in a set F, this information should be used to reappraise the likelihood that the outcome will also be in E. This reappraised probability is denoted by $P(E|F)$, and is read as the *conditional probability of E given F.* It represents the answer to the question, how probable is E, given that F has occurred?

DEFINITION OF CONDITIONAL PROBABILITY
Let's discuss some examples to illustrate conditional probability and then state the definition.

Example 1

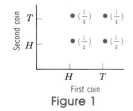

Second coin

Figure 1

Consider the experiment of flipping 2 fair coins. As we have previously seen, the sample space S is

$$S = \{HH, HT, TH, TT\}$$

Figure 1 illustrates the sample space and, for convenience, the probability of each event.

Suppose the experiment is performed by another person and we have no knowledge of the result, but we are informed that at least 1 tail was tossed. This information means the outcome HH could not have occurred. But the remaining outcomes HT, TH, TT are still possible. How does this alter the probabilities of the remaining outcomes?

For instance, we might be interested in calculating the probability of the event $\{TT\}$. The three simple events $\{TH\}$, $\{HT\}$, $\{TT\}$ were each assigned the probability $\frac{1}{4}$ *before* we knew the information that at least 1 tail occurred, but it is not reasonable to assign them this same probability now. Since only three outcomes are now possible, we assign to each of them the probability $\frac{1}{3}$. See Figure 2. ∎

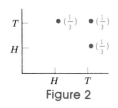

Figure 2

Example 2

Consider the experiment of drawing a single card from a deck of 52 playing cards. We are interested in the event E consisting of the outcome that a black ace is drawn. Since we may assume that there are 52 equally likely possible outcomes and there are 2 black aces in the deck, we have

$$P(E) = \frac{2}{52} = \frac{1}{26}$$

However, suppose a card is drawn and we are informed that it is a spade. How should this information be used to *reappraise* the likelihood of the event E?

Solution

Clearly, since the event F "A spade has been drawn" has occurred, the event "Not spade" is no longer possible. Hence, the sample space has changed from 52 playing cards to 13 spade cards, and the number of black aces that can be drawn has been reduced to 1. Therefore, we must compute the probability of event E relative to the new sample space F. This probability is denoted by $P(E|F)$ and has the value

$$P(E|F) = \frac{1}{13}$$

∎

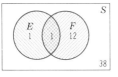

Figure 3

Let's analyze the situation in Example 2 more carefully. The event E is "A black ace is drawn." We have computed the probability of event E knowing event F has occurred. This means we are computing a probability relative to a *new sample space F*. That is, F is treated as the universal set. We should consider only that part of E that is included in F; that is, we consider $E \cap F$. See Figure 3.

Thus, the probability of E given F is the ratio of the number of entries in $E \cap F$ to the number of entries in F. Since $c(E \cap F) = 1$ and $c(F) = 13$, then

$$P(E|F) = \frac{c(E \cap F)}{c(F)} = \frac{1}{13}$$

Notice that

$$P(E|F) = \frac{1}{13} = \frac{\frac{1}{52}}{\frac{13}{52}} = \frac{P(E \cap F)}{P(F)}$$

In fact, for sample spaces involving equally likely events, this formula can be proved. Suppose E and F are two events for a particular experiment. Assume that the sample space S for this experiment has n possible equally likely outcomes. Suppose event F has m outcomes, while $E \cap F$ has k outcomes ($k \le m$). Since the events are equally likely we have

$$P(F) = \frac{[\text{Number of outcomes in } F]}{n} = \frac{m}{n}$$

and

$$P(E \cap F) = \frac{[\text{Number of outcomes in } E \cap F]}{n} = \frac{k}{n}$$

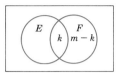

Figure 4

We wish to compute $P(E|F)$, the probability that E occurs given that F has occurred. Since we assume F has occurred, look only at the m outcomes in F. Of these m outcomes, there are k outcomes where E also occurs, since $E \cap F$ has k outcomes. See Figure 4. Thus

$$P(E|F) = \frac{k}{m}$$

Divide numerator and denominator by n to get

$$P(E|F) = \frac{\dfrac{k}{n}}{\dfrac{m}{n}} = \frac{P(E \cap F)}{P(F)}$$

With this result in mind we choose to define conditional probability as follows:

Conditional Probability **Let E and F be events of a sample space S and suppose $P(F) > 0$. The *conditional probability of event E assuming the event F*, denoted by $P(E|F)$, is defined as**

$$P(E|F) = \frac{P(E \cap F)}{P(F)} \qquad (1)$$

Thus, if E is any subset of the sample space S, then $P(E|F)$ provides the reappraisal of the likelihood that an outcome of the experiment will be in the set E if we have prior information that it must be in the set F. Thus we call $P(E|F)$ the conditional probability of E given F.

$P(E|F)$ satisfies the three Properties (I), (II), and (III) presented in Section 7.3, page 348. The student is asked to verify this in Problem 46.

Example 3

Suppose a population of 1000 people includes 70 accountants and 520 females. Let E be the event "A person is an accountant." Let F be the event "A person is female." Then

$$P(E) = \frac{70}{1000} = .07 \qquad P(F) = \frac{520}{1000} = .52$$

Instead of studying the entire population, we may want to investigate the female subpopulation and ask for the probability that a female chosen at random is also an accountant. If there are 40 females who are accountants, the ratio $\frac{40}{520}$ represents the conditional probability of the event E (accountant) assuming the event F (the person chosen is female). In symbols, we would write

$$P(E|F) = \frac{40}{520} = \frac{1}{13}$$

∎

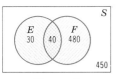

Figure 5

Figure 5 illustrates that in computing $P(E|F)$ in Example 3, we form the ratio of the numbers of those entries in E and in F with the numbers that are in F.

Alternately, using the definition of conditional probability, we see that

$$P(E \cap F) = \frac{40}{1000} \qquad P(F) = \frac{520}{1000}$$

$$P(E|F) = \frac{\frac{40}{1000}}{\frac{520}{1000}} = \frac{40}{520} = \frac{1}{13}$$

In the definition of conditional probability, if we replace F by S, the sample space, we get

$$P(E|S) = \frac{P(E \cap S)}{P(S)}$$

But $E \cap S = E$ and $P(S) = 1$. This reduces to

$$P(E|S) = P(E)$$

as expected.

The symbol $P(E|F)$ is usually read as "the probability of E given F."

Example 4

Consider a three-child family* for which the sample space S is

$$S = \{BBB, BBG, BGB, BGG, GBB, GBG, GGB, GGG\}$$

We assume each simple event is equally likely, so that each is assigned a probability of $\frac{1}{8}$. Let E be the event, "The family has exactly 2 boys" and let F be the event "The first child is a boy." What is the probability that the family has 2 boys, given that the first child is a boy?

Solution

We want to find $P(E|F)$. The events E and F are

$$E = \{BBG, BGB, GBB\} \qquad F = \{BBB, BBG, BGB, BGG\}$$

Clearly, $E \cap F = \{BBG, BGB\}$, so we have

$$P(E \cap F) = \frac{1}{4} \qquad P(F) = \frac{1}{2}$$

$$P(E|F) = \frac{P(E \cap F)}{P(F)} = \frac{\frac{1}{4}}{\frac{1}{2}} = \frac{1}{2}$$

∎

---

* Refer to Example 2 in Section 7.2 (page 337).

PRODUCT RULE

If in Formula (1), we multiply both sides of the equation by $P(F)$, we obtain the following useful relationship, which is referred to as the *product rule:*

$$P(E \cap F) = P(F) \cdot P(E|F)$$

Example 5

Motors, Inc., has two plants to manufacture cars. Plant I manufactures 80% of the cars and Plant II manufactures 20%. At Plant I, 85 out of every 100 cars are rated standard quality or better. At Plant II, only 65 out of every 100 cars are rated standard quality or better. We would like to find the answers to the following questions:

(a) What is the probability that a customer obtains a standard quality car if he buys a car from Motors, Inc.?

(b) What is the probability that the car came from Plant I if it is known that the car is of standard quality?

Solution

Let the events A, B, C, D, and E be defined as follows:

> A: The car purchased is of standard quality
> B: The car is of standard quality and came from Plant I
> C: The car is of standard quality and came from Plant II
> D: The car came from Plant I
> E: The car came from Plant II

We know that

$$P(D) = .80 \qquad P(E) = .20$$
$$P(A|D) = .85 \qquad P(A|E) = .65$$

Hence, by the product rule,

$$P(B) = P(D \cap A) = P(D) \cdot P(A|D) = (.80)(.85) = .68$$
$$P(C) = P(E \cap A) = P(E) \cdot P(A|E) = (.20)(.65) = .13$$

(a) We have

$$P(B) = .68 \qquad P(C) = .13$$

Since $B \cap C = \emptyset$,

$$P(A) = P(B \cup C) = P(B) + P(C) = .68 + .13 = .81$$

Thus, a total of 81% of the cars are of standard quality.

(b) We need to compute $P(D|A)$. Since $D \cap A = B$, we have

$$P(D|A) = \frac{P(D \cap A)}{P(A)} = \frac{P(B)}{P(A)} = \frac{.68}{.81} = .8395$$

■

The next example illustrates how the product rule is used to compute the probability of an event that is itself a sequence of two events.

Example 6

Two cards are drawn at random (without replacement) from a standard deck of 52 cards. What is the probability that the first card is a diamond and the second is red?

Solution

We seek the probability of an event that is a sequence of two events, namely,

E: The first card is a diamond

F: The second card is red

Since there are 52 cards in the deck, of which 13 are diamonds, it follows that

$$P(E) = \frac{13}{52} = \frac{1}{4}$$

If E occurred, that means that there are only 51 cards left in the deck, of which 25 are red, so

$$P(F|E) = \frac{25}{51}$$

By the product rule,

$$P(F \cap E) = P(E) \cdot P(F|E) = \frac{1}{4} \cdot \frac{25}{51} = \frac{25}{204}$$

■

PROBABILITY TREE

Many experiments consist of a sequence of two or more events. A *probability tree* provides a useful device for structurally defining the relationships within such experiments and for computing probabilities associated with possible outcomes.

The probability tree for Example 6 is given in Figure 6.

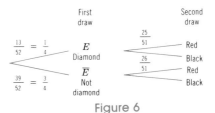

Figure 6

Example 7

From a box containing four white, three yellow, and one green ball, two balls are drawn one at a time without replacing the first before the second is drawn. Use the probability tree to find the probability that one white and one yellow ball are drawn.

Solution

We seek the probability of an event that is a sequence of two events, namely,

W: drawing a white ball

Y: drawing a yellow ball

Since four of the eight balls are white, on the first draw we have

$$P(W) = P(W \text{ on 1st}) = \frac{4}{8} = \frac{1}{2}$$

Since one white ball has been removed leaving seven balls in the box of which three are yellow, on the second draw we have

$$P(Y|W) = P(Y \text{ on 2nd}|W \text{ on 1st}) = \frac{3}{7}$$

The event drawing one white ball and one yellow ball can occur in two ways: drawing a white ball first and then a yellow ball (Path 2 of the probability tree in Figure 7), or drawing a yellow ball first and then a white ball (Path 4).

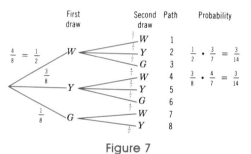

Figure 7

For Path 2 we have

$$P(W) \cdot P(Y|W) = P(W \text{ on 1st}) \cdot P(Y \text{ on 2nd}|W \text{ on 1st}) = \frac{1}{2} \cdot \frac{3}{7} = \frac{3}{14}$$

For Path 4 we have

$$P(Y) \cdot P(W|Y) = P(Y \text{ on 1st}) \cdot P(W \text{ on 2nd}|Y \text{ on 1st}) = \frac{3}{8} \cdot \frac{4}{7} = \frac{3}{14}$$

Since the two events are mutually exclusive, the final probability is the sum of these two probabilities

$$\frac{3}{14} + \frac{3}{14} = \frac{6}{14} = \frac{3}{7}$$
∎

Exercise 8.1   *Answers to Odd-Numbered Problems begin on page A-22.*

A   In Problems 1–6 find the indicated probabilities by referring to the following probability tree:

1. $P(C)$	**2.** $P(D)$
3. $P(C\|A)$	**4.** $P(D\|A)$
5. $P(C\|D)$	**6.** $P(D\|C)$

In Problems 7–14 use the table below to obtain probabilities for events in a sample space S. Read the following directly from the table:

	E	F	G	Totals
H	.10	.06	.08	.24
I	.30	.14	.32	.76
Totals	.40	.20	.40	1.00

7. $P(E)$	**8.** $P(G)$	**9.** $P(H)$	**10.** $P(I)$
11. $P(E \cap H)$	**12.** $P(E \cap I)$	**13.** $P(G \cap H)$	**14.** $P(G \cap I)$

In Problems 15–18 use Equation (1) and the appropriate values from the above table to compute the conditional probability.

15. $P(E\|H)$	**16.** $P(E\|I)$	**17.** $P(G\|H)$	**18.** $P(G\|I)$

19. If E and F are events with $P(E) = .4$, $P(F) = .2$, and $P(E \cap F) = .1$, find the probability of E given F. Find $P(F|E)$.

20. If E and F are events with $P(E) = .6$, $P(F) = .5$, and $P(E \cap F) = .3$, find the probability of E given F. Find $P(F|E)$.

21. If E and F are events with $P(E \cap F) = .1$ and $P(E|F) = .2$, find $P(F)$.

22. If E and F are events with $P(E \cap F) = .2$ and $P(E|F) = .5$, find $P(F)$.

23. If E and F are events with $P(F) = \frac{5}{13}$ and $P(E|F) = \frac{4}{5}$, find $P(E \cap F)$.

24. If E and F are events with $P(F) = .38$ and $P(E|F) = .46$, find $P(E \cap F)$.

25. If E and F are events with $P(E \cap F) = \frac{1}{3}$, $P(E|F) = \frac{1}{2}$, and $P(F|E) = \frac{2}{3}$, find:

(a) $P(E)$ (b) $P(F)$

26. If E and F are events with $P(E \cap F) = .1$, $P(E|F) = .25$, and $P(F|E) = .125$, find:

(a) $P(E)$ (b) $P(F)$

B

27. A card is drawn at random from a regular deck of 52 cards. What is the probability that:

(a) The card is a red ace?

(b) The card is a red ace if it is known an ace was picked?

(c) The card is a red ace if it is known a red card was picked?

28. A card is drawn at random from a regular deck of 52 cards. What is the probability that:

(a) The card is a black jack?

(b) The card is a black jack if it is known a jack was picked?

(c) The card is a black jack if it is known a black card was picked?

29. A recent poll of residents in a certain community revealed the following information about voting preferences:

	Democrat D	Republican R	Independent I	Totals
Male M	50	40	30	120
Female F	60	30	25	115
Totals	110	70	55	235

Find:

(a) $P(F|I)$ (b) $P(R|F)$ (c) $P(M|D)$

(d) $P(D|M)$ (e) $P(M|R \cup I)$ (f) $P(I|M)$

30. Let E be the event, "A person is an executive" and let F be the event "A person earns over \$25,000 per year." State in words what is expressed by each of the following probabilities:

(a) $P(E|F)$ (b) $P(F|E)$ (c) $P(\overline{E}|\overline{F})$ (d) $P(\overline{E}|F)$

31. The following table summarizes the graduating class of a midwestern university:

	Arts and Sciences A	Education E	Business B	Total
Male M	342	424	682	1448
Female F	324	102	144	570
Total	666	526	826	2018

A student is selected at random from the graduating class. Find the probability that the student:

(a) Is male.

(b) Is receiving an arts and sciences degree.

(c) Is a female receiving a business degree.

(d) Is a female, given that the student is receiving an education degree.

(e) Is receiving an arts and sciences degree, given that the student is a male.

(f) Is a female, given that the student is receiving an arts and sciences degree or an education degree.

32. The following data are the result of a survey conducted by a marketing company among students for their beer preferences.

	Do Not Drink Beer N	Light Beer L	Regular Beer R	Total
Female F	224	420	622	1266
Male M	196	512	484	1192
Total	420	932	1106	2458

A student is selected at random from this class. Find the probability that:

    (a)   The student does not drink beer.

    (b)   The student is a female.

    (c)   The student is a female who prefers regular beer.

    (d)   The student prefers regular beer, given the student is male.

    (e)   The student is male, given that the student prefers regular beer.

    (f)   The student is female, given that the student prefers regular beer or does not drink beer.

33. For a 3-child family, find the probability of exactly 2 girls, given that the first child is a girl.

34. For a 3-child family, find the probability of exactly 1 girl, given that the first child is a boy.

35. A fair coin is tossed four successive times. Find the probability of obtaining 4 heads. Does the probability change if we are told that the second throw resulted in a head?

36. A pair of fair dice is thrown and we are told that at least one of them shows a 2. If we know this, what is the probability that the total is 7?

37. In a small town, it is known that 25% of the families have no children, 25% have 1 child, 18% have 2 children, 16% have 3 children, 8% have 4 children, and 8% have 5 or more children. Find the probability that a family has more than 2 children if it is known that it has at least 1 child.

38. A sequence of 2 cards is drawn from an ordinary deck of 52 cards (without replacement). What is the probability that the first card is red and the second is black?

In Problems 39–42 use a probability tree to find the indicated probabilities.

39. Two cards are drawn at random (without replacement) from a standard deck of 52 cards. What is the probability that the first card is a heart and the second is red?

40. From a box containing 3 white, 2 green, and 1 yellow ball, two balls are drawn at a time without replacing the first before the second is drawn. Find the probability that one white and one yellow ball are drawn.

41. If two cards are drawn from a 52- card deck without replacement, what is the probability that the second card is a queen?

42. "Temp Help" uses a preemployment test to screen applicants for the job of a programmer. The test is passed by 70% of the applicants. Among those who pass the test 85% complete training successfully. In an experiment, a random sample of applicants who do not pass the test is also employed. Training is successfully completed by only 40% of this group. If no preemployment test is used, what percentage of applicants would you expect to complete training successfully?

C **43.** If E and F are two events with $P(E) > 0$ and $P(F) > 0$, show that

$$P(F) \cdot P(E|F) = P(E) \cdot P(F|E)$$

44. Show that $P(E|E) = 1$ when $P(E) \neq 0$.

45. Show that $P(E|F) + P(\overline{E}|F) = 1$.

46. Show that $P(E|F)$ satisfies the three Properties (I), (II), and (III) stated in Section 7.3.

47. Show that if $P(E|F) = P(E)$ and $P(E) \neq 0$, then $P(F|E) = P(F)$.

48. Show by means of numeric examples that $P(E|F) + P(E|\overline{F})$

    (a) May equal 1.   (b) May not equal 1.

APPLICATIONS

49. Records at the office of the dean of student affairs indicate that, for students who passed mathematics, the probability that they completed their assignments regularly is .861. Furthermore, the probability that students pass mathematics and complete their assignments regularly is .746. Find the probability that a student taking mathematics passes the course.

50. In a survey taken in a small town it was found that the probability that a person selected at random is married and has a college degree is .328. If the probability that a person selected at random has a college degree is .526, find the probability that the person is married given that the person has a college degree.

51. In a sample survey, it is found that 35% of the men and 70% of the women are under 160 pounds. Assume that 50% of the sample are men. If a person is selected at random and this person is under 160 pounds, what is the probability that this person is a woman?

52. If the probability that a married man will vote in a given election is .50, the probability that a married woman will vote in the election is .60, and the probability that a woman will vote in the election, given that her husband votes is .90, find:

(a) The probability that a husband and wife will both vote in the elections.

(b) The probability that a married man will vote in the election, given that at least one member of the married couple will vote.

53. Of the first-year students in a certain college, it is known that 40% attended private secondary schools and 60% attended public schools. The registrar reports that 30% of all students who attended private schools maintain an A average in their first year at college and that 24% of all first-year students had an A average. At the end of the year, one student is chosen at random from the class. If the student has an A average, what is the conditional probability that the student attended a private school? [*Hint:* Use a probability tree.]

54. In a rural area in the north, registered Republicans outnumber registered Democrats by 3 to 1. In a recent election, all Democrats voted for the Democratic candidate and enough Republicans also voted for the Democratic candidate so that the Democrat won by a ratio of 5 to 4. If a voter is selected at random, what is the probability he or she is Republican? What is the probability a voter is Republican, if it is known that he or she voted for the Democratic candidate?

8.2

INDEPENDENT EVENTS

DEFINITION OF INDEPENDENT EVENTS
TEST FOR INDEPENDENCE
STEPS TO CHECK FOR INDEPENDENCE
INDEPENDENCE FOR MORE THAN TWO EVENTS

One of the most important concepts in probability is that of independence. In this section we define what is meant by two events being *independent*. First, however, we try to develop an intuitive idea of the meaning of independent events.

Example 1

Consider a group of 36 students. Suppose that E and F are two properties that each student either has or does not have. For example, the events E and F might be

E: Student has blue eyes

F: Student is a male

With regard to these two properties, suppose it is found that the 36 students are distributed as follows:

	Blue Eyes E	Not Blue Eyes $\overline{E}$	Totals
Male, F	6	6	12
Female, $\overline{F}$	12	12	24
Totals	18	18	36

If we choose a student at random, the probabilities corresponding to the events E and F are

$$P(E) = \frac{18}{36} = \frac{1}{2}$$

$$P(F) = \frac{12}{36} = \frac{1}{3}$$

$$P(E \cap F) = \frac{6}{36} = \frac{1}{6}$$

$$P(E|F) = \frac{P(E \cap F)}{P(F)} = \frac{\frac{1}{6}}{\frac{1}{3}} = \frac{1}{2} = P(E)$$

■

DEFINITION OF INDEPENDENT EVENTS

In Example 1, the probability of E given F equals the probability of E. This situation can be described by saying that the information that the event F has occurred does not affect the probability of the event E. If this is the case, we say that E is independent of F.

Independent Events Let E and F be two events of a sample space S with $P(F) > 0$. The *event E is independent of the event F* if and only if

$$P(E|F) = P(E)$$

In Problem 43, Exercise 8.1, you were asked to show that

$$P(F) \cdot P(E|F) = P(E) \cdot P(F|E) \tag{1}$$

provided $P(E) > 0$ and $P(F) > 0$. If E is independent of F, then we know that $P(E|F) = P(E)$. Substituting this into (1), we find that

$$P(F|E) = P(F)$$

That is, the event F is independent of E.

Thus, if two events E and F have positive probabilities and if the event E is independent of F, then F is also independent of E. In this case, E and F are called *independent events.*

TEST FOR INDEPENDENCE

We also have the following result concerning independent events:

Test for Independence Two events E and F of a sample space S are independent events if and only if

$$P(E \cap F) = P(E) \cdot P(F) \tag{2}$$

That is, the probability of E *and* F is equal to the product of the probability of E and the probability of F.

Proof If E and F are independent events, then

$$P(E|F) = \frac{P(E \cap F)}{P(F)} \quad \text{and} \quad P(E|F) = P(E)$$

Thus,

$$\frac{P(E \cap F)}{P(F)} = P(E) \quad \text{or} \quad P(E \cap F) = P(E) \cdot P(F)$$

Conversely, if $P(E \cap F) = P(E) \cdot P(F)$, then

$$P(E|F) = \frac{P(E \cap F)}{P(F)} = \frac{P(E) \cdot P(F)}{P(F)} = P(E)$$

That is, E and F are independent events. ∎

The rule for independent events (2), gives us a method to determine if two events are independent when we know $P(E)$, $P(F)$, and $P(E \cap F)$.

Example 2

Suppose $P(E) = \frac{1}{4}$, $P(F) = \frac{2}{3}$, and $P(E \cap F) = \frac{1}{6}$. Show that E and F are independent.

Solution

$$P(E) \cdot P(F) = \frac{1}{4} \cdot \frac{2}{3}$$
$$= \frac{1}{6}$$
$$= P(E \cap F)$$

by the test of independence (2), E and F are independent. ∎

Example 3

Suppose E and F are independent events with $P(E) = .4$ and $P(F) = .3$. Find $P(E \cap F)$.

Solution

Since E and F are independent,

$$P(E \cap F) = P(E) \cdot P(F) \qquad \text{by (2)}$$
$$= (.4) \cdot (.3)$$
$$= .12 \qquad \blacksquare$$

Example 4

Suppose a red die and a green die are thrown. Let event E be "Throw a 5 with the red die," and let event F be "Throw a 6 with the green die." Show that E and F are independent.

Solution

In this experiment, the events E and F are

$$E = \{(5, 1), (5, 2), (5, 3), (5, 4), (5, 5), (5, 6)\}$$
$$F = \{(1, 6), (2, 6), (3, 6), (4, 6), (5, 6), (6, 6)\}$$

and

$$P(E) = \frac{1}{6} \qquad P(F) = \frac{1}{6}$$

Also, the event E and F is

$$E \cap F = \{(5, 6)\}$$

so that

$$P(E \cap F) = \frac{1}{36}$$

Since $P(E) \cdot P(F) = \frac{1}{6} \cdot \frac{1}{6} = \frac{1}{36} = P(E \cap F)$, E and F are independent events. $\blacksquare$

Example 5

For the data in the T-maze problem (Problem 51, Exercise 7.2), show that the events E and G are not independent, but that the events G and H are independent, where E, G, and H are defined as before:

> E: Run to the right two consecutive times
> G: Run to the left on the first trial
> H: Run to the right on the second trial

Solution

The events E and G are

$$E = \{RRL, LRR, RRR\}$$
$$G = \{LLL, LLR, LRL, LRR\}$$

The sample space S has eight elements so that

$$P(E) = \frac{3}{8} \qquad P(G) = \frac{1}{2}$$

Also, the event E and G is

$$E \cap G = \{LRR\}$$

so that

$$P(E \cap G) = \frac{1}{8}$$

Since $P(E \cap G) \neq P(E) \cdot P(G)$, the events E and G are not independent. Thus, running to the right two consecutive times and running to the left on the first trial are dependent.

Next, the event H is

$$H = \{RRL, RRR, LRL, LRR\}$$

so that

$$P(H) = \frac{1}{2}$$

The event G and H and its probability are

$$G \cap H = \{LRL, LRR\} \qquad P(G \cap H) = \frac{1}{4}$$

Since $P(G \cap H) = P(G) \cdot P(H)$, the events G and H are independent. Thus, running to the left on the first trial and running to the right on the second trial are independent events. ■

STEPS TO CHECK FOR INDEPENDENCE

Example 5 illustrates that the question of whether two events are independent can be answered simply by determining whether Formula (2) is satisfied. Although we may often suspect two events E and F of being independent, our intuition must be checked by computing $P(E)$, $P(F)$, and $P(E \cap F)$ and determining whether $P(E \cap F) = P(E) \cdot P(F)$.

The following steps summarize the procedure to use to check for events E and F being independent:

Step 1. Compute the probability of event E.

Step 2. Compute the probability of event F.

Step 3. Compute the probability of the event E and F, $P(E \cap F)$.

Step 4. If $P(E \cap F) = P(E) \cdot P(F)$, the events are independent.
If $P(E \cap F) \neq P(E) \cdot P(F)$, the events are not independent.

For some experiments it is clear that certain events are independent (repeated toss of a fair coin, repeated roll of a pair of fair dice, etc.). However, in many real-world experiments, we must rely on intuition and simply *assume* that the events in question are independent and apply Formula (2). The following example illustrates such a situation.

Example 6

In a group of seeds, $\frac{1}{4}$ of which should produce white plants, the best germination that can be obtained is 75%. If one seed is planted, what is the probability that it will grow into a white plant?

Solution
Let G and W be the events

$$G: \quad \text{The plant will grow}$$
$$W: \quad \text{The seed will produce a white plant}$$

Assume that W does not depend on G and vice versa, so that W and G are independent events.

Then, the probability that the plant grows and is white, namely, $P(W \cap G)$ is

$$P(W \cap G) = P(W) \cdot P(G) = \frac{1}{4} \cdot \frac{3}{4} = \frac{3}{16}$$

A white plant will grow 3 out of 16 times.   ■

There is a danger that mutually exclusive events and independent events may be confused. A source of this confusion is the common expression. "Events are independent if they have nothing to do with each other" This expression provides a description of independence when applied to everyday events; but when it is applied to sets, it suggests nonoverlapping. Non-overlapping sets are mutually exclusive, but in general are not independent. See Problem 23 in Exercise 8.2.

INDEPENDENCE FOR MORE THAN TWO EVENTS
The concept of independence can be applied to more than two events:

A set $E_1, E_2, \ldots, E_n$ of events is called *independent* if the occurrence of one or more of them does not change the probability of any of the others. It can be shown that, for such events,

$$P(E_1 \cap E_2 \cap \ldots \cap E_n) = P(E_1) \cdot P(E_2) \cdot \ldots \cdot P(E_n) \qquad (3)$$

Example 7

A new skin cream can cure skin infection 90% of the time. If five randomly selected people use this cream, assuming independence, what is the probability that:

(a) All five are cured?

(b) All five still have the infection?

Solution
(a)　Let

E_1:　first person does not have infection

E_2:　second person does not have infection

E_3:　third person does not have infection

E_4:　fourth person does not have infection

E_5:　fifth person does not have infection

then, since events E_1, E_2, E_3, E_4, E_5 are given to be independent

P(all 5 are cured)
$$= P(E_1 \cap E_2 \cap E_3 \cap E_4 \cap E_5) = P(E_1) \cdot P(E_2) \cdot P(E_3) \cdot P(E_4) \cdot P(E_5)$$
$$= (.9)^5$$
$$= .59$$

(b)　Let $\overline{E}_i$: i^{th} person has an infection $i = 1, 2, 3, 4, 5$
Then

$$P(\overline{E}_i) = 1 - .9 = .1$$
$$P(\text{all 5 are infected}) = P(\overline{E}_1 \cap \overline{E}_2 \cap \overline{E}_3 \cap \overline{E}_4 \cap \overline{E}_5)$$
$$= P(\overline{E}_1)P(\overline{E}_2)P(\overline{E}_3)P(\overline{E}_4)P(\overline{E}_5)$$
$$= (.1)^5$$
$$= .00001$$ ■

Exercise 8.2　　*Answers to Odd-Numbered Problems begin on page A-22.*

A　　**1.** If E and F are independent events and if $P(E) = .3$ and $P(F) = .5$, find $P(E \cap F)$.

2. If E and F are independent events and if $P(E) = .6$ and $P(E \cap F) = .3$, find $P(F)$.

3. If E and F are independent events, find $P(F)$ if $P(E) = .2$ and $P(E \cap F) = .3$.

4. If E and F are independent events, find $P(E)$ if $P(F) = .3$ and $P(E \cap F) = .6$.

5. Suppose E and F are two events such that $P(E) = \frac{4}{21}$, $P(F) = \frac{7}{12}$, and $P(E \cap F) = \frac{2}{9}$, are E and F independent?

6. If E and F are two events such that $P(E) = .25$, $P(F) = .36$, and $P(E \cap F) = .09$, are E and F independent?

7. If E and F are two independent events with $P(E) = .3$, and $P(F) = .5$, find:

(a) $P(E|F)$　　　　(b) $P(F|E)$　　　　(c) $P(E \cap F)$　　　　(d) $P(E \cup F)$

8. If E and F are independent events with $P(E) = .32$, $P(F) = .41$, find:

(a) $P(E|F)$　　　　(b) $P(F|E)$　　　　(c) $P(E \cap F)$　　　　(d) $P(E \cup F)$

9. If E, F, and G are three independent events with $P(E) = \frac{2}{3}$, $P(F) = \frac{3}{7}$, and $P(G) = \frac{2}{21}$, find $P(E \cap F \cap G)$.

10. If E_1, E_2, E_3, and E_4 are four independent events with $P(E_1) = .62$, $P(E_2) = .35$, $P(E_3) = .58$, and $P(E_4) = .41$, find $P(E_1 \cap E_2 \cap E_3 \cap E_4)$.

11. If $P(E) = .3$, $P(F) = .2$, and $P(E \cup F) = .4$, what is $P(E|F)$? Are E and F independent?

12. If $P(E) = .4$, $P(F) = .6$, and $P(E \cup F) = .7$, what is $P(E|F)$? Are E and F independent?

B 13. A fair die is rolled. Let E be the event "1, 2, or 3 is rolled" and let F be the event "3, 4, or 5 is rolled." Are E and F independent?

14. A loaded die is rolled. The probabilities for this die are $P(1) = P(2) = P(4) = P(5) = \frac{1}{8}$ and $P(3) = P(6) = \frac{1}{4}$. Are the events defined in Problem 13 independent in this case?

15. For a three-child family, let E be the event "The family has at most 1 boy" and let F be the event "The family has children of each sex." Are E and F independent events?

16. For a two-child family, are the events E and F as defined in Problem 15 independent?

17. A first card is drawn at random from a regular deck of 52 cards and is then put back in the deck. A second card is drawn. What is the probability that

    (a) The first card is a club?
    (b) The second card is a heart, given that the first is a club?
    (c) The first card is a club and the second is a heart?

18. For the situation described in Problem 17, what is the probability that

    (a) The first card is an ace?
    (b) The second card is a king, given that the first card is an ace?
    (c) The first card is an ace and the second is a king?

19. In the T-maze problem (Problem 51, Exercise 7.2), are the two events E and F independent?

20. A fair coin is tossed twice. Define the events E and F to be

    E: A head turns up on the first throw of a fair coin
    F: A tail turns up on the second throw of a fair coin

    Show that E and F are independent events.

21. A die is loaded so that

$$P(1) = P(2) = P(3) = \frac{1}{4} \qquad P(4) = P(5) = P(6) = \frac{1}{12}$$

    If $A = \{1, 2\}$, $B = \{2, 3\}$, $C = \{1, 3\}$, show that any pair of these events is independent.

22. If E and F are independent events with $P(E) = 4P(F)$, and $P(E \cap F) = \frac{1}{4}$, find $P(F)$.

C 23. Give an example of two events that are

    (a) Independent, but not mutually exclusive (disjoint)
    (b) Not independent, but mutually exclusive (disjoint)
    (c) Not independent and not mutually exclusive (disjoint)

24. Show that whenever two events are both independent and mutually exclusive, then at least one of them is impossible.

25. Let E be any event. If F is an impossible event, show that E and F are independent.

26. Show that if E and F are independent events, so are $\overline{E}$ and F. [*Hint:* Use De Morgan's law.]

27. Show that if E and F are independent events and if $P(E) \neq 0$, $P(F) \neq 0$, then E and F are not mutually exclusive.

28. Use Definition (3) to show that the events E, F, and G defined below for the experiment of tossing two fair dice are not independent.

$$E: \quad \text{The first die shows a 6}$$
$$F: \quad \text{The second die shows a 3}$$
$$G: \quad \text{The sum on the 2 dice is 7}$$

29. Chevalier de Mere's Problem Solve Problem 7 in Exercise 7.1 (page 334). [*Hint:* Part (a) $P(\text{No 1's are obtained}) = \frac{5^4}{6^4} = \frac{625}{1296} = .4823$. Part (b) The probability of not obtaining a double 1 on any given toss is $\frac{35}{36}$. Thus, $P(\text{No double 1's are obtained}) = \left(\frac{35}{36}\right)^{24} = .509$.]

30. A woman has 10 keys but only 1 fits her door. She tries them successively (without replacement). Find the probability that a key fits in exactly 5 tries.*

APPLICATIONS **31. Cardiovascular Disease** Records show that a child of parents with heart disease has a probability of $\frac{3}{4}$ of inheriting the disease. Assuming independence, what is the probability that, for a couple with heart disease that have two children:

(a)   Both children have heart disease.
(b)   Neither child has heart disease.
(c)   Exactly one child has heart disease.

32. Hitting a Target A marksman hits a target with probability $\frac{4}{5}$. Assuming independence for successive firings, find the probabilities of getting

(a)   One miss followed by two hits.
(b)   Two misses and one hit (in any order).

33. A coin is loaded so that tails is three times as likely as heads. If the coin is tossed three times, find the probability of getting

(a)   All tails.
(b)   Two heads and one tail (in any order).

34. Sex of Newborns The probability of a newborn baby being a girl is .49. Assuming that the sex of one baby is independent of the sex of all other babies, that is, the events are independent, what is the probability that all four babies born in a certain hospital on one day are girls?

35. Recovery Rate The recovery rate from a flu is .8. If 4 people have this flu, what is the probability (assume independence) that:

(a)   All will recover?
(b)   Exactly 2 will recover?
(c)   At least 2 will recover?

36. Germination In a group of seeds, $\frac{1}{3}$ of which should produce violets the best germination that can be obtained is 60%. If one seed is planted, what is the probability that it will grow into violets?

---

* William Feller, *An Introduction to Probability Theory and Its Applications,* 3rd ed., Wiley, New York, 1968.

37. A box has 9 marbles in it, 5 red and 4 white. Suppose we draw a marble from the box, replace it, and then draw another. Find the probability that:

   (a) Both marbles are red.

   (b) Just one of the two marbles is red.

38. **Survey** In a survey of 100 people, categorized as drinkers or nondrinkers, with or without a liver ailment, the following data were obtained:

	Liver Ailment F	No Liver Ailment $\overline{F}$
Drinkers, E	52	18
Nondrinkers, $\overline{E}$	8	22

   (a) Are the events E and F independent?

   (b) Are the events $\overline{E}$ and $\overline{F}$ independent?

   (c) Are the events E and $\overline{F}$ independent?

39. **Insurance** By examining the past driving records over a period of 1 year of 840 randomly selected drivers, the following data was obtained.

	Under 25 U	Over 25 $\overline{U}$	Totals
Accident A	40	5	45
No Accident $\overline{A}$	285	510	795
Totals	325	515	840

   (a) What is the probability of a driver having an accident, given the person is under 25?

   (b) What is the probability of a driver having an accident, given the person is over 25?

   (c) Are events U and A independent?

   (d) Are events U and $\overline{A}$ independent?

   (e) Are events $\overline{U}$ and A independent?

   (f) Are events $\overline{U}$ and $\overline{A}$ independent?

40. **Voting Patterns** The following data show the number of voters in a sample of 1000 from a large city categorized by religion and their voting preference.

	Democrat D	Republican R	Independent	Totals
Catholic C	160	150	90	400
Protestant P	220	220	60	500
Jewish	20	30	50	100
Totals	400	400	200	1000

    (a)   Find the probability a person is a Democrat.

    (b)   Find the probability a person is a Catholic.

    (c)   Find the probability a person is Catholic knowing the person is Democrat.

    (d)   Are the events R and D independent?

    (e)   Are the events P and R independent?

41. **Election** A candidate for office believes that $\frac{2}{3}$ of registered voters in her district will vote for her in the next election. If two registered voters are independently selected at random, what is the probability that:

    (a)   Both of them will vote for her in the next election?

    (b)   Neither will vote for her in the next election?

    (c)   Exactly one of them will vote for her in the next election?

8.3

BAYES' FORMULA

PARTITIONS AND PROBABILITIES
BAYES' FORMULA
A PRIORI AND *A POSTERIORI* PROBABILITIES

In this section we consider experiments with sample spaces that can be divided or partitioned into two (or more) mutually exclusive events. This study involves a further application of conditional probabilities and leads us to the famous formula of Thomas Bayes, first published in 1763.

We begin by considering the following example.

Example 1

Given two urns, I and II, suppose Urn I contains 4 black and 7 white balls. Urn II contains 3 black, 1 white, and 4 yellow balls. We select an urn at random and then draw a ball. What is the probability that we obtain a black ball?

Solution

Let U_I and U_{II} stand for the events "Urn I is chosen" and "Urn II is chosen," respectively. Similarly, let B, W, Y stand for the event that "A black," "A white," or "A yellow ball is chosen," respectively.

THOMAS BAYES (1702–1761), born in London, was the son of a Presbyterian minister. Bayes, who was also ordained and began his ministry by assisting his father, was elected a Fellow of the Royal Society in 1742. He published several theological papers, but is most famous for his paper on probability, which was published after his death by a friend who found it among his effects. This work is noteworthy because it is the first discussion of inductive inference in precise quantitive form.

A solution to Example 1 can be depicted using a probability tree, as shown in Figure 8.

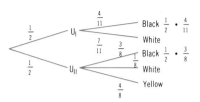 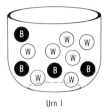 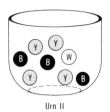

Figure 8

$$P(U_{\mathrm{I}}) = P(U_{\mathrm{II}}) = \frac{1}{2}$$

$$P(B|U_{\mathrm{I}}) = \frac{4}{11} \qquad P(B|U_{\mathrm{II}}) = \frac{3}{8}$$

The event B can be written as

$$B = (B \cap U_{\mathrm{I}}) \cup (B \cap U_{\mathrm{II}})$$

Since $B \cap U_{\mathrm{I}}$ and $B \cap U_{\mathrm{II}}$ are disjoint, we add their probabilities. Then

$$P(B) = P(B \cap U_{\mathrm{I}}) + P(B \cap U_{\mathrm{II}})$$

Using the definition of conditional probability, we have

$$P(B|U_{\mathrm{I}}) = \frac{P(B \cap U_{\mathrm{I}})}{P(U_{\mathrm{I}})} \qquad\qquad P(B|U_{\mathrm{II}}) = \frac{P(B \cap U_{\mathrm{II}})}{P(U_{\mathrm{II}})}$$

$$P(B \cap U_{\mathrm{I}}) = P(U_{\mathrm{I}}) \cdot P(B|U_{\mathrm{I}}) \qquad P(B \cap U_{\mathrm{II}}) = P(U_{\mathrm{II}}) \cdot P(B|U_{\mathrm{II}})$$

Thus,

$$P(B) = P(U_{\mathrm{I}}) \cdot P(B|U_{\mathrm{I}}) + P(U_{\mathrm{II}}) \cdot P(B|U_{\mathrm{II}})$$

$$= \frac{1}{2} \cdot \frac{4}{11} + \frac{1}{2} \cdot \frac{3}{8} = \frac{65}{176} = .369$$

PARTITIONS AND PROBABILITIES

Suppose A_1 and A_2 are two nonempty, mutually exclusive events of a sample space S and the union of A_1 and A_2 is S; that is,

$$A_1 \neq \varnothing \qquad A_2 \neq \varnothing \qquad A_1 \cap A_2 = \varnothing \qquad S = A_1 \cup A_2$$

In this case, we say that A_1 and A_2 form a *partition of S*. See Figure 9. Now, if we let E be any event in S, we may write the set E in the form

$$E = (E \cap A_1) \cup (E \cap A_2)$$

The sets $E \cap A_1$ and $E \cap A_2$ are disjoint, since

$$(E \cap A_1) \cap (E \cap A_2) = (E \cap E) \cap (A_1 \cap A_2) = E \cap \varnothing = \varnothing$$

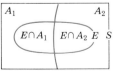

Figure 9

Using the definition for conditional probability, the probability of E is therefore

$$P(E) = P(E \cap A_1) + P(E \cap A_2)$$
$$= P(A_1) \cdot P(E|A_1) + P(A_2) \cdot P(E|A_2) \tag{1}$$

The above formula is used to find the probability of an event E of a sample space when the sample space is partitioned into two sets A_1 and A_2.

Example 2

Of the applicants to a medical school, it is felt that 80% are eligible to enter and 20% are not. To aid in the selection process, an admissions test is administered which is designed so that an eligible candidate will pass 90% of the time, while an ineligible candidate will pass only 30% of the time. What is the probability that an applicant for admission will pass the admissions test?

Solution

The sample space S consists of the applicants for admission, and S can be partitioned into the following two events:

$$A_1: \quad \text{Eligible applicant} \qquad A_2: \quad \text{Ineligible applicant}$$

These two events are disjoint, and their union is S. See Figure 10.
    The event E is

$$E: \quad \text{Applicant passes admissions test}$$

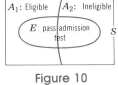

Figure 10

Now,

$$P(A_1) = .8 \qquad P(A_2) = .2$$
$$P(E|A_1) = .9 \qquad P(E|A_2) = .3$$

Using Formula (1), we have

$$P(E) = P(A_1) \cdot P(E|A_1) + P(A_2) \cdot P(E|A_2) = (.8)(.9) + (.2)(.3) = .78$$

Thus, the probability that an applicant will pass the admissions test is .78. ∎

If we partition a sample space S into three sets A_1, A_2, and A_3 so that

$$S = A_1 \cup A_2 \cup A_3$$
$$A_1 \cap A_2 = \varnothing \qquad A_2 \cap A_3 = \varnothing \qquad A_1 \cap A_3 = \varnothing$$
$$A_1 \neq \varnothing \qquad A_2 \neq \varnothing \qquad A_3 \neq \varnothing$$

we may write any set E in S in the form

$$E = (E \cap A_1) \cup (E \cap A_2) \cup (E \cap A_3)$$

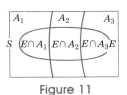

Figure 11

Since $E \cap A_1$, $E \cap A_2$, and $E \cap A_3$ are disjoint, the probability of event E is

$$P(E) = P(E \cap A_1) + P(E \cap A_2) + P(E \cap A_3)$$
$$= P(A_1) \cdot P(E|A_1) + P(A_2) \cdot P(E|A_2) + P(A_3) \cdot P(E|A_3) \tag{2}$$

See Figure 11.

Formula (2) is used to find the probability of an event E of a sample space when the sample space is partitioned into three sets A_1, A_2, and A_3.

Example 3
Three machines, I, II, and III, manufacture .4, .5, and .1 of the total production in a plant, respectively. The percentage of defective items produced by I, II, and III is 2%, 4%, and 1%, respectively. For an item chosen at random, what is the probability that it is defective?

Solution
In this example, the sample space S is partitioned into three events A_1, A_2, and A_3 defined as follows:

$$A_1: \quad \text{Item produced by Machine I}$$
$$A_2: \quad \text{Item produced by Machine II}$$
$$A_3: \quad \text{Item produced by Machine III}$$

Clearly, the events A_1, A_2, and A_3 are mutually exclusive, and their union is S. Define the event E in S to be

$$E: \quad \text{Item is defective}$$

Now,

$$P(A_1) = .4 \qquad P(A_2) = .5 \qquad P(A_3) = .1$$
$$P(E|A_1) = .02 \qquad P(E|A_2) = .04 \qquad P(E|A_3) = .01$$

Thus, using Formula (2), we see that

$$P(E) = (.4)(.02) + (.5)(.04) + (.1)(.01)$$
$$= .008 + .020 + .001 = .029$$ ∎

Figure 12 gives a probability tree solution to Example 3.

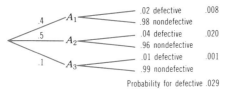

Probability for defective .029

Figure 12

To generalize Formulas (1) and (2) to a sample space S partitioned into n subsets, we first introduce the following definition:

Partition A sample space S is *partitioned* into n subsets $A_1, A_2, \ldots, A_n$, provided:

(a) The intersection of any two of the subsets is empty.
(b) Each subset is nonempty.
(c) $A_1 \cup A_2 \cup \cdots \cup A_n = S$

Let S be a sample space and let $A_1, A_2, A_3, \ldots, A_n$ be n events that form a partition of the set S. If E is any event in S, then

$$E = (E \cap A_1) \cup (E \cap A_2) \cup \cdots \cup (E \cap A_n)$$

Clearly, $E \cap A_1, E \cap A_2, \ldots, E \cap A_n$ are mutually exclusive events. Hence,

$$P(E) = P(E \cap A_1) + P(E \cap A_2) + \cdots + P(E \cap A_n) \qquad (3)$$

In (3), replace $P(E \cap A_1), P(E \cap A_2), \ldots, P(E \cap A_n)$ using the definition of conditional probability. Then we obtain the formula

$$P(E) = P(A_1) \cdot P(E|A_1) + P(A_2) \cdot P(E|A_2) + \cdots + P(A_n) \cdot P(E|A_n) \quad (4)$$

Example 4

In Example 2, suppose an applicant passes the admissions test. What is the probability that he or she was among those eligible; that is, what is the probability $P(A_1|E)$?

Solution

By the definition of conditional probability,

$$P(A_1|E) = \frac{P(A_1 \cap E)}{P(E)} = \frac{P(A_1) \cdot P(E|A_1)}{P(E)}$$

But $P(E)$ is given by Formula (4) when $n = 2$, or, by Formula (1). Thus,

$$P(A_1|E) = \frac{P(A_1) \cdot P(E|A_1)}{P(A_1) \cdot P(E|A_1) + P(A_2) \cdot P(E|A_2)} \qquad (5)$$

Using the information supplied in Example 2, we find

$$P(A_1|E) = \frac{(.8)(.9)}{.78} = \frac{.72}{.78} = .923$$

The admissions test is a reasonably effective device. Less than 8% of the students passing the test are ineligible. However, the test has one weakness: over 36% of those who fail it are actually eligible to enter the school.   ■

BAYES' FORMULA

Equation (5) is a special case of *Bayes' formula* when the sample space is partitioned into two sets A_1 and A_2. The general formula is given next.

Bayes' Formula Let S be a sample space partitioned into n events, $A_1, \ldots, A_n$. Let E be any event of S for which $P(E) > 0$. The probability of the event A_j ($j = 1, 2, \ldots, n$), given the event E, is

$$P(A_j|E) = \frac{P(A_j) \cdot P(E|A_j)}{P(E)}$$

$$= \frac{P(A_j) \cdot P(E|A_j)}{P(A_1) \cdot P(E|A_1) + P(A_2) \cdot P(E|A_2) + \cdots + P(A_n) \cdot P(E|A_n)} \qquad (6)$$

The proof is left as an exercise (see Problem 20, Exercise 8.3).

The following example illustrates a use for Bayes' formula when the sample space S is partitioned into three events.

Example 5

Motors, Inc., has three plants, I, II, and III. Plant I produces 35% of the car output, Plant II produces 20%, and Plant III produces the remaining 45%. One percent of the output of Plant I is defective, as is 1.8% of the output of Plant II, and 2% of the output of Plant III. The annual total output of Motors, Inc., is 1,000,000 cars. A car is chosen at random from the annual output and it is found to be defective. What is the probability that it came from Plant I? Plant II? Plant III?

Solution

To answer these questions, let's define the following events:

$$E: \quad \text{Car is defective}$$
$$A_1: \quad \text{Car produced by Plant I}$$
$$A_2: \quad \text{Car produced by Plant II}$$
$$A_3: \quad \text{Car produced by Plant III}$$

Also, $P(A_1|E)$ indicates the probability that a car is produced by Plant I, given that it was defective; $P(A_2|E)$ and $P(A_3|E)$ are similarly defined. To find these probabilities, we proceed as follows.

From the data given in the problem, we can determine the following:

$$P(A_1) = .35 \qquad P(E|A_1) = .010$$
$$P(A_2) = .20 \qquad P(E|A_2) = .018$$
$$P(A_3) = .45 \qquad P(E|A_3) = .020$$

Now,

$A_1 \cap E$ is the event "Produced by Plant I and is defective"

$A_2 \cap E$ is the event "Produced by Plant II and is defective"

$A_3 \cap E$ is the event "Produced by Plant III and is defective"

By the product rule, we find

$$P(A_1 \cap E) = P(A_1) \cdot P(E|A_1) = (.35)(.010) = .0035$$
$$P(A_2 \cap E) = P(A_2) \cdot P(E|A_2) = (.20)(.018) = .0036$$
$$P(A_3 \cap E) = P(A_3) \cdot P(E|A_3) = (.45)(.020) = .0090$$

Since $E = (A_1 \cap E) \cup (A_2 \cap E) \cup (A_3 \cap E)$, we have

$$P(E) = P(A_1 \cap E) + P(A_2 \cap E) + P(A_3 \cap E)$$
$$= .0035 + .0036 + .0090$$
$$= .0161$$

Thus, the probability that a defective car is chosen is .0161.

Given that the car chosen is defective, the probability that it came from Plant I is

$P(A_1|E)$, from Plant II is $P(A_2|E)$, and from Plant III is $P(A_3|E)$. To compute these probabilities, we use Bayes' formula:

$$P(A_1|E) = \frac{P(A_1) \cdot P(E|A_1)}{P(A_1) \cdot P(E|A_1) + P(A_2) \cdot P(E|A_2) + P(A_3) \cdot P(E|A_3)}$$

$$= \frac{P(A_1) \cdot P(E|A_1)}{P(E)} = \frac{(.35)(.01)}{.0161} = .217$$

$$P(A_2|E) = \frac{P(A_2) \cdot P(E|A_2)}{P(E)} = \frac{.0036}{.0161} = .224$$

$$P(A_3|E) = \frac{P(A_3) \cdot P(E|A_3)}{P(E)} = \frac{.0090}{.0161} = .559$$

∎

A PRIORI, A POSTERIORI PROBABILITIES

In Bayes' formula, the probabilities $P(A_j)$ are referred to as *a priori* probabilities, while the $P(A_j|E)$ are called *a posteriori* probabilities. We use Example 5 to explain the reason for this terminology. Knowing nothing else about a car, the probability that it was produced by Plant I is given by $P(A_1)$. Thus, $P(A_1)$ can be regarded as a "before the fact" or *a priori* probability. With the additional information that the car is defective, we reassess the likelihood of whether it came from Plant I and compute $P(A_1|E)$. Thus, $P(A_1|E)$ can be viewed as an "after the fact" or *a posteriori* probability.

Note that $P(A_1) = .35$, while $P(A_1|E) = .217$. So, the knowledge that a car is defective decreases the chances that it came from Plant I.

Example 6

The residents of a community are examined for cancer. The examination results are classified as positive (+), if a malignancy is suspected, and as negative (−), if there are no indications of a malignancy. If a person has cancer, the probability of a suspected malignancy is .98; and the probability of reporting cancer where none existed is .15. If 5% of the community has cancer, what is the probability of a person not having cancer if the examination is positive?

Solution

Let us define the following events:

A_1:   Person has cancer

A_2:   Person does not have cancer

E: Examination is positive

We want to know the probability of a person not having cancer if it is known that the examination is positive; that is, we wish to find $P(A_2|E)$. Now,

$$P(A_1) = .05 \qquad P(A_2) = .95$$
$$P(E|A_1) = .98 \qquad P(E|A_2) = .15$$

Using Bayes' formula, we get

$$P(A_2|E) = \frac{P(A_2) \cdot P(E|A_2)}{P(A_1) \cdot P(E|A_1) + P(A_2) \cdot P(E|A_2)}$$

$$= \frac{(.95)(.15)}{(.05)(.98) + (.95)(.15)} = .744$$

Thus, even if the examination is positive, the person examined is more likely not to have cancer than to have cancer. The reason the test is designed this way is that it is better for a healthy person to be examined more thoroughly, than for someone with cancer to go undetected. Simply stated, the test is useful because of the high probability that a person with cancer will not go undetected. ■

In the next example, the sample space is partitioned into five events.

Example 7

The manager of a car repair shop knows from experience that when a call is received from a person whose car will not start, the probabilities for various troubles (assuming no two can occur simultaneously) are as follows:

Event	Trouble	Probability
A_1:	Flooded	.3
A_2:	Battery cable loose	.2
A_3:	Points bad	.1
A_4:	Out of gas	.3
A_5:	Something else	.1

The manager also knows that if the person will hold the gas pedal down and try to start the car, the probability that it will start (E) is

$$P(E|A_1) = .9 \qquad P(E|A_2) = 0 \qquad P(E|A_3) = .2$$
$$P(E|A_4) = 0 \qquad P(E|A_5) = .2$$

(a) If a person has called and is instructed to "hold the pedal down . . . ," what is the probability that the car will start?

(b) If the car does start after holding the pedal down, what is the probability that the car was flooded?

Solution

(a) We need to compute $P(E)$. Using Formula (4) for $n = 5$ (the sample space is partitioned into five disjoint sets), we have

$$P(E) = P(A_1) \cdot P(E|A_1) + P(A_2) \cdot P(E|A_2) + P(A_3) \cdot P(E|A_3)$$
$$+ P(A_4) \cdot P(E|A_4) + P(A_5) \cdot (P(E|A_5)$$
$$= (.3)(.9) + (.2)(0) + (.1)(.2) + (.3)(0) + (.1)(.2)$$
$$= .27 + .02 + .02 = .31$$

(b) We use Bayes' formula to compute the *a posteriori* probability $P(A_1|E)$:

$$P(A_1|E) = \frac{P(A_1) \cdot P(E|A_1)}{P(E)} = \frac{(.3)(.9)}{.31} = \frac{.27}{.31} = .87$$

Thus, the probability that the car is flooded, after it is known that holding down the pedal started the car, is .87. ∎

When the probability of each event of the partition is equally likely,

$$P(A_1) = P(A_2) = \cdots = P(A_n)$$

we obtain a special case of Bayes' formula:

$$P(A_j|E) = \frac{P(E|A_j)}{P(E|A_1) + P(E|A_2) + \cdots + P(E|A_n)} \tag{7}$$

Let's return to Example 5 and assume that $P(A_1) = P(A_2) = P(A_3)$, so that the three plants' share of the total production is the same. In this case, it is easier to use (7) to obtain the probability that a car is produced at Plant I, given that it was defective:

$$P(A_1|E) = \frac{P(E|A_1)}{P(E|A_1) + P(E|A_2) + P(E|A_3)}$$

$$= \frac{.010}{.010 + .018 + .020} = \frac{.010}{.048} = \frac{5}{24} = .208$$

Exercise 8.3

Answers to Odd-Numbered Problems begin on page A-23.

A   In Problems 1–6 find the indicated probabilities by referring to the following probability tree and using Bayes' formula:

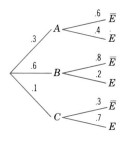

1. $P(A|E)$
2. $P(B|\overline{E})$
3. $P(C|E)$
4. $P(A|\overline{E})$
5. $P(B|E)$
6. $P(C|\overline{E})$

7. Events A_1 and A_2 form a partition of a sample space S with $P(A_1) = .3$ and $P(A_2) = .7$. If E is an event in S with $P(E|A_1) = .01$ and $P(E|A_2) = .02$, compute $P(E)$.

8. Events A_1 and A_2 form a partition of a sample space S with $P(A_1) = .4$ and $P(A_2) = .6$. If E is an event in S with $P(E|A_1) = .03$ and $P(E|A_2) = .01$, compute $P(E)$.

9. Events A_1, A_2, and A_3 form a partition of a sample space S with $P(A_1) = .5$, $P(A_2) = .3$, and $P(A_3) = .2$. If E is an event in S with $P(E|A_1) = .01$, $P(E|A_2) = .03$, and $P(E|A_3) = .02$, compute $P(E)$.

10. Events A_1, A_2, and A_3 form a partition of a sample space S with $P(A_1) = .3$, $P(A_2) = .3$, and $P(A_3) = .4$. If E is an event in S with $P(E|A_1) = .01$, $P(E|A_2) = .02$, and $P(E|A_3) = .02$, compute $P(E)$.

11. Use the information in Problem 7 to find $P(A_1|E)$ and $P(A_2|E)$.

12. Use the information in Problem 8 to find $P(A_1|E)$ and $P(A_2|E)$.

13. Use the information in Problem 9 to find $P(A_1|E)$, $P(A_2|E)$, and $P(A_3|E)$.

14. Use the information in Problem 10 to find $P(A_1|E)$, $P(A_2|E)$, and $P(A_3|E)$.

B   15. In Example 3 (page 393), suppose it is known that a defective item was produced. Find the probability that it came from Machine I; from Machine II; from Machine III.

16. In Example 5 (page 395), suppose $P(A_1) = P(A_2) = P(A_3) = \frac{1}{3}$. Find $P(A_2|E)$ and $P(A_3|E)$.

17. In Example 7 (page 397), compute the *a posteriori* probabilities $P(A_2|E)$, $P(A_3|E)$, $P(A_4|E)$, and $P(A_5|E)$.

18. In Example 6 (page 396), compute $P(A_1|E)$.

19. Three urns contain colored balls as follows:

Urn	Red, R	White, W	Blue, B
I	5	6	5
II	3	4	9
III	7	5	4

One urn is chosen at random and a ball is withdrawn. The ball is red. What is the probability that it came from Urn I? From Urn II? From Urn III? [*Hint:* Define the events E: Ball selected is red, U_I: Urn I selected, U_{II}: Urn II selected, and U_{III}: Urn III selected. Determine $P(U_I|E)$, $P(U_{II}|E)$, and $P(U_{III}|E)$ by using Bayes' formula.]

C   20. Prove Bayes' formula (6).

21. Show that $P(E|F) = 1$ if F is a subset of E and $P(F) \neq 0$.

APPLICATIONS   22. **Medical Diagnosis**   Suppose that if a person with tuberculosis is given a TB screening, the probability that his or her condition will be detected is .90. If a person without tuberculosis is given a TB screening, the probability that he or she will be diagnosed incorrectly as having tuberculosis is .3. Suppose, further, that 11% of the adult residents of a certain city have tuberculosis. If one of these adults is diagnosed as having tuberculosis based on the screening, what is the probability that he or she actually has tuberculosis? Interpret your result.

23. **Quality Control**   Cars are being produced by two factories, I and II, but Factory I produces twice as many cars as Factory II in a given time. Factory I is known to produce 2% defectives and Factory II produces 1% defectives. A car is examined and found to be defective. What are the *a priori* and *a posteriori* probabilities that the car was produced by Factory I?

24. **Malpractice**   An absent-minded nurse is to give Mr. Brown a pill each day. The probability that the nurse forgets to administer the pill is $\frac{2}{3}$. If he receives the pill, the probability that Brown will die is $\frac{1}{3}$. If he does not get his pill, the probability that he will die is $\frac{3}{4}$. Mr. Brown died. What is the probability that the nurse forgot to give Brown the pill?

25. **Color-blind**   In a human population, 51% are male and 49% are female. 5% of the males and .3% of the females are color-blind. If a person randomly chosen from the population is found to be color-blind, what is the probability that the person is a male?

26. **Medical Diagnosis**   In a certain small town, 16% of the population developed lung cancer. If 45% of the population are smokers, and 85% of those developing lung cancer are smokers, what is the probability that a smoker in this population will develop lung cancer?

27. **Voting Pattern**   In Cook County 55% of the registered voters are Democrats, 30% are Republicans, and 15% are independents. During a recent election, 35% of the Democrats voted, 65% of the Republicans voted, and 75% of the independents voted. What is the probability that someone who voted is a Democrat? Republican? Independent?

28. **Quality Control**   A computer manufacturer has three assembly plants. Records

show that 2% of the sets shipped from plant A turn out to be defective, as compared to 3% of those that come from plant B and 4% of those that come from plant C. In all, 30% of the manufacturer's total production comes from plant A, 50% from plant B, and 20% from plant C. If a customer finds that his computer is defective, what is the probability it came from plant B?

29. **Oil Drilling** An oil well is to be drilled in a certain location. The soil there is either rock (probability .53), clay (probability .21), or sand. If it is rock, a geological test gives a positive result with 35% accuracy; if it is clay, this test gives a positive result with 48% accuracy; and if it is sand, the test gives a positive result with 75% accuracy. Given that the test is positive, what is the probability that the soil is rock? What is the probability that the soil is clay? What is the probability that the soil is sand?

30. **Oil Drilling** A geologist is using seismographs to test for oil. It is found that if oil is present, the test gives a positive result 95% of the time, and if oil is not present, the test gives a positive result 2% of the time. Finally oil is present in 1% of the cases tested. If the test shows positive, what is the probability that oil is present?

31. **Political Polls** In conducting a political poll, a pollster divides the United States into four sections: Northeast (N), containing 40% of the population; South (S), containing 10% of the population; Midwest (M), containing 25% of the population; and West (W), containing 25% of the population. From the poll, it is found that in the next election 40% of the people in the Northeast say they will vote for Republicans, in the South 56% will vote Republican, in the Midwest 48% will vote Republican, and in the West 52% will vote Republican. What is the probability that a person chosen at random will vote Republican? Assuming a person votes Republican, what is the probability that he or she is from the Northeast?

32. **Marketing** To introduce a new beer, a company conducted a survey. It divided the United States into four regions, eastern, northern, southern, and western. The company estimates that 35% of the potential customers for the beer are in the eastern region, 30% are in the northern region, 20% are in the southern region, and 15% are in the western region. The survey indicates that 50% of the potential customers in the eastern region, 40% of the potential customers in the northern region, 36% of the potential customers in the southern region, and 42% of those in the western region will buy the beer. If a potential customer chosen at random indicates that he or she will buy the beer, what is the probability that the customer is from the southern region?

33. **College Majors** Data collected by the Office of Admissions of a large midwestern university indicate the following choices made by the members of the freshman class regarding their majors.

Major	Percentage of Freshman Choosing Major	Female (in percent)	Male (in percent)
Engineering	26	40	60
Business	30	35	65
Education	9	80	20
Social science	12	52	48
Natural science	12	56	44
Humanities	9	65	35
Other	2	51	49

What is the probability that a female student selected at random from the freshman class is majoring in engineering?

8.4

BINOMIAL PROBABILITY MODEL

BERNOULLI TRIALS
BINOMIAL PROBABILITIES
MODEL: TESTING A SERUM OR VACCINE
MODEL: ERROR CORRECTION IN ELECTRONIC TRANSMISSION OF DATA

In this section we study practical situations that can be interpreted by using a simple probabilistic model, called the *binomial probability model.* The model was first studied by J. Bernoulli about 1700 and, for this reason, the model is sometimes referred to as a *Bernoulli trial.*

BERNOULLI TRIALS

The binomial probability model is a sequence of trials, each of which consists of repetition of a single experiment. We assume the outcome of one experiment does not affect the outcome of any other one; that is, we assume the trials to be independent. Furthermore, we assume that there are only two possible outcomes for each trial and label them S, for *Success,* and F, for *Failure.* The probability of success, $p = P(S)$, remains the same from trial to trial. In addition, since there are only two outcomes in each trial, the probability of failure must be $1 - p$, and we write

$$q = 1 - p = P(F)$$

Thus, $p + q = 1$.

Any random experiment for which the binomial probability model is appropriate is called a *Bernoulli trial.*

Bernoulli Trial **Random experiments are called *Bernoulli trials* if:**

1. **The same experiment is repeated several times.**
2. **There are only two possible outcomes, success and failure.**

JAMES (JACQUES) BERNOULLI (1654–1705) was a member of a family of famous Swiss mathematicians. At the insistence of his father, he studied theology. Later, he refused a church appointment and began lecturing on experimental physics at the University of Basel. Thanks to Bernoulli's contributions, probability theory was raised to the status of a science. In 1713, his book *Ars Conjectardi* was published in Latin by his nephew Nicholas Bernoulli, who also was a mathematician.

3. The repeated trials are independent.

4. The probability of each outcome remains the same for each trial.

Many real-world situations have the characteristics of the binomial probability model. For example, in repeatedly running a subject through a T-maze, we may label a turn to the left by S and a turn to the right by F. The assumption of independence of each trial is equivalent to presuming the subject has no memory.

In opinion polls, one person's response is independent of any other person's response, and we may designate the answer "Yes" by an S and any other answer ("No" or "Don't know") by an F.

In testing TV's, we have a sequence of independent trials (each test of a particular TV is a trial) and we label a nondefective TV with an S and a defective one with an F.

In determining whether 9 out of 12 persons will recover from a tropical disease, we assume that each of the 12 persons has the same chance of recovery from the disease and that their recoveries are independent. We may designate "recovery" by S and "nonrecovery" by F.

BINOMIAL PROBABILITIES

Next we consider an experiment that will lead us to formulate a general expression for the probability of obtaining exactly k successes in a sequence of n Bernoulli trials ($k \leq n$).

The experiment is to test a drug to be administered to a group of people infected by a tropical disease. In such a case, we can define a success to be if a person recovers and a failure to occur if he or she does not recover. Then p is the probability of recovery and q is the probability of not recovering.

The probability tree in Figure 13 lists all the possible outcomes and their respective probabilities for a group of three persons.

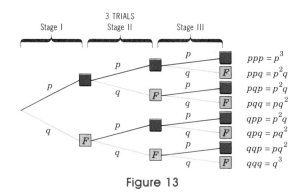

Figure 13

The set of outcomes is

$$SSS, SSF, SFS, SFF, FSS, FSF, FFS, FFF$$

Here for example, SSF means that the first two people recover (successes) and the third does not (failure).

The outcome *SSS*, in which all three recover, has a probability $ppp = p^3$, since each Bernoulli trial has a probability p of resulting in success.

If we wish to calculate the probability that exactly two people will recover, we consider only the outcomes

$$SSF, \ SFS, \ FSS$$

The outcome *SSF* has probability $ppq = p^2q$ since the two successes each have probability p and the failure has probability q. In the same way, the other two outcomes *SFS* and *FSS*, in which there are two successes and one failure, also have probabilities p^2q. Therefore the probability that exactly two people recover and one does not is equal to the sum of the probabilities of the three outcomes and, hence, is given by $3p^2q$.

In a similar way we compute the following probabilities:

$$P \text{ (exactly one person recovering and two not recovering)} = 3pq^2$$
$$P \text{ (all three recovered)} = p^3$$
$$P \text{ (all three not recovering)} = q^3$$

When considering a group larger than three it would be extremely tedious, to say the least, to solve problems of this type using a probability tree. This is why we want a general expression.

Suppose the probability of a success in a Bernoulli trial is p, and suppose we wish to find the probability of exactly k successes in n repeated trials. One possible outcome is

$$\underbrace{SSS\text{---}S}_{k \text{ successes}} \ \underbrace{FFF\text{---}F}_{n-k \text{ failures}} \qquad (1)$$

where k successes come first, followed by $n-k$ failures. The probability of this outcome is

$$= \underbrace{ppp\text{---}p}_{k \text{ factors}} \ \underbrace{qq\text{---}q}_{n-k \text{ factors}}$$
$$= p^k q^{n-k}$$

The k successes could also be obtained by rearranging the letters in (1) above. The number of such sequences must be equal to the number of ways of choosing k of the n trials to contain successes — namely, $\binom{n}{k}$. If we multiply this number by the probability of obtaining any one such sequence, we arrive at the following result:

In a Bernoulli trial the probability of exactly k successes in n trials is given by

$$b(n, k; p) = \binom{n}{k} p^k \cdot q^{n-k} = \frac{n!}{k!(n-k)!} p^k \cdot q^{n-k} \qquad (2)$$

where $q = 1 - p$.

The symbol $b(n, k; p)$, which represents the probability of exactly k successes in n trials, is called a *binomial probability.*

Example 1

A common example of a Bernoulli trial is a coin-flipping experiment:

1. There are exactly two possible mutually exclusive outcomes on each trial or toss (heads or tails).
2. The probability of a particular outcome (say, H) remains constant from trial to trial (toss to toss).
3. The outcome on any trial (toss) is independent of the outcome on any other trial (toss).

We would like to compute the probability of obtaining exactly 1 tail in 6 tosses of a fair coin.

Solution

Let S denote the simple event "Tail shows" and let F denote the simple event "Head shows." Using Formula (2) in which $k = 1$, $n = 6$, and $p = \frac{1}{2} = P(S)$, and since $q = 1 - p = 1 - \frac{1}{2} = \frac{1}{2}$ we obtain

$$P(\text{Exactly one success}) = b\left(6, 1; \frac{1}{2}\right) = \binom{6}{1}\left(\frac{1}{2}\right)^1\left(\frac{1}{2}\right)^{6-1}$$

$$= \frac{6}{64} \approx .0938$$

Example 2

Typist Error   A typist makes one error on the average in every fifth typed page. If nine pages are typed in one day, what is the probability that

(a)   Exactly three of them have errors.
(b)   None of them has an error.
(c)   No more than 7 pages have errors.

Solution

In this example, $P(\text{success}) = P(\text{making an error}) = \frac{1}{5} = .2$. The number of trials n is the number of typed pages and k is the number of pages containing errors. In all three parts $n = 9$, $q = 1 - .2 = .8$

(a)   For $k = 3$,

$$P(\text{exactly 3}) = b(9, 3; .2) = \binom{9}{3}(.2)^3(.8)^6 = .1762$$

(b)   If none of the 9 pages had any errors, $k = 0$.

$$P(\text{exactly 0}) = b(9, 0; .2) = \binom{9}{0}(.2)^0(.8)^9 = .1342$$

For part (c), as in many other applications, it is necessary to compute the probability not of exactly k successes, but of *at least* or *at most* k successes. To obtain such probabilities we have to compute all the individual probabilities and add them.

(c)   "No more than 7 pages have errors" means 0, 1, 2, 3, 4, 5, 6, or 7 pages with errors. We could add $b(9, 0; .2)$, $b(9, 1; .2)$, and so on, but it is easier to use the formula $P(E) = 1 - P(\overline{E})$. The complement of "no more than 7" is "8 or more."

$$P \text{ (no more than 7)} = 1 - P \text{ (8 or more)}$$
$$= 1 - [b(9, 8; .2) + b(9, 9; .2)] = .99998 \quad \blacksquare$$

Example 3

Quality Control A machine produces light bulbs to meet certain specifications, and 80% of the bulbs produced meet these specifications. A sample of 6 bulbs is taken from the machine's production. What is the probability that 3 or more of them fail to meet the specifications?

Solution
In this example we are looking for the probability of the event

E: At least 3 fail to meet specifications

But this event is just the union of the mutually exclusive events "Exactly 3 failures," "Exactly 4 failures," "Exactly 5 failures," and "Exactly 6 failures." Hence, we use Formula (2) for $n = 6$ and $k = 3, 4, 5,$ and 6. Since the probability of failure is .20, we have

$$P(\text{Exactly 3 failures}) = b(6, 3; .20) = .0819$$
$$P(\text{Exactly 4 failures}) = b(6, 4; .20) = .0154$$
$$P(\text{Exactly 5 failures}) = b(6, 5; .20) = .0015$$
$$P(\text{Exactly 6 failures}) = b(6, 6; .20) = .0001$$

Therefore,

$$P(\text{At least 3 failures}) = P(E) = .0819 + .0154 + .0015 + .0001 = .0989 \quad \blacksquare$$

Another way of getting this answer is to compute the probability of the complementary event

$\overline{E}$: Less than 3 failures

Then,

$$P(\overline{E}) = P(\text{Exactly 2 failures}) + P(\text{Exactly 1 failure}) + P(\text{Exactly 0 failures})$$
$$= b(6, 2; .20) + b(6, 1; .20) + b(6, 0; .20)$$
$$= .2458 + .3932 + .2621 = .9011$$

As a result,

$$P(\text{At least 3 failures}) = 1 - P(\overline{E}) = 1 - .9011 = .0989 \quad \blacksquare$$

Example 4

Product Promotion A man claims to be able to distinguish between two kinds of wine with 90% accuracy and presents his claim to an agency interested in promoting the consumption of one of the two kinds of wine. The following experiment is conducted to check his claim. The man is to taste the two types of wine and distinguish between them. This is to be done 9 times with a 3 minute break after each taste. It is agreed that if the man is correct at least 6 out of the 9 times, he will be hired.

The main questions to be asked are, on the one hand, whether the above procedure gives sufficient protection to the hiring agency against a person guessing and, on the other hand, whether the man is given sufficient chance to be hired if he is really a wine connoisseur.

Solution

To answer the first question, let's assume that the man is guessing. Then in each trial he has probability $\frac{1}{2}$ of identifying the wine correctly. Let k be the number of correct identifications. Let's compute the binomial probability for $k = 6, 7, 8, 9$, to find the likelihood of the man being hired while guessing:

$$b\left(9, 6; \frac{1}{2}\right) + b\left(9, 7; \frac{1}{2}\right) + b\left(9, 8; \frac{1}{2}\right) + b\left(9, 9; \frac{1}{2}\right)$$
$$= .1641 + .0703 + .0176 + .0020 = .2540$$

Thus, there is a likelihood of .254 that he will pass if he is just guessing.

To answer the second question in the case where the claim is true, we need to find the sum of the probabilities $b(9, k; .90)$ for $k = 6, 7, 8, 9$:

$$b(9, 6; .90) + b(9, 7; .90) + b(9, 8; .90) + b(9, 9; .90)$$
$$= .0446 + .1722 + .3874 + .3874 = .9916 \quad \blacksquare$$

You may obtain slightly different answers due to intermediate round off errors.

Notice that the test in Example 4 is fair to the man, since it practically assures him the position if his claim is true. However, the company may not like the test because 25% of the time a person who guesses will pass the test.

MODEL: TESTING A SERUM OR VACCINE*

Suppose that the normal rate of infection of a certain disease in cattle is 25%. To test a newly discovered serum, healthy animals are injected with it. How can we evaluate the result of the experiment?

For an absolutely worthless serum, the probability that exactly k of the n test animals catch infection may be equated to $b(n, k; .25)$. For $k = 0$ and $n = 10$, this probability is about $b(10, 0; .25) = .056$. Thus, if out of 10 test animals none catch infection, this may be taken as an indication that the serum has had an effect,

*P. V. Sukhatme and V. G. Panse, "Size of Experiments for Testing Sera or Vaccines," *Indiana Journal of Veterinary Science and Animal Husbandry,* **13** (1943), pp. 75–82.

although it is not conclusive proof. Notice that, without serum, the probability that out of 17 animals at most 1 catches infection is $b(17, 0; .25) + b(17, 1; .25) = .0501$. Therefore, there is *stronger evidence* in favor of the serum if out of 17 test animals at most 1 gets infected than if out of 10 all remain healthy. For $n = 23$ the probability of at most 2 animals catching infection is about .0492 and, thus, at most 2 failures out of 23 is again better evidence for the serum than at most 1 out of 17 or 0 out of 10.

MODEL: ERROR CORRECTION IN ELECTRONIC TRANSMISSION OF DATA

Electronically transmitted data, be it from computer to computer or from a satellite to a ground station, is normally in the form of strings of 0's and 1's—that is, in binary form. Bursts of noise or faults in relays, for example, may at times garble the transmission and produce errors so that the message received is not the same as the one originally sent. For example,

$$001 \quad \longmapsto\!\!\!\sim\!\!\sim\!\!\sim\!\!\longrightarrow \quad 101 \qquad\qquad (3)$$

Message sent	Noisy channel	Message received

is a transmission where the message received is in error since the initial 0 has been changed to a 1.

A naive way of trying to protect against such error would be to repeat the message twice. So, instead of transmitting 001, we would send 001001. Then, were the exact same error to creep in as did in (3), the received message would be

$$101001$$

The receiver would certainly know an error has occurred since the last half of the message is not a duplicate of the first half. But, he would have no way of recovering the original message, since he would not know where the error happened. For example, he would not be able to distinguish between the two messages

$$001001 \quad \text{and} \quad 101101$$

There are more sophisticated ways of coding binary data with redundancy that allow not only the detection of errors, but simultaneously permit their location and correction so that the original message can be recovered. These are referred to as *error-correcting codes* and are commonly used today in computer-implemented transmissions. One such is the (7, 4) Hamming code named after Richard Hamming, a former researcher at AT&T Bell Laboratories. It is a code of length 7, meaning that an individual message is a string consisting of seven items, each of which is either a 0 or 1. (The 4 refers to the fact that the first four elements in the string can be freely chosen by the sender, while the remaining three are determined by a fixed rule and constitute the redundancy that gives the code its error correction capability.) The (7, 4) Hamming code is capable of locating and correcting a single error. That is, if during transmission a 1 has been changed to a 0 or vice versa in one of the seven locations, the Hamming code is capable of detecting and correcting this. While we will not explain how or why the Hamming code works, we will analyze the benefit obtained by its use.

By an error we mean that an individual 1 has been changed to a 0 or that a 0 has been changed to a 1. We assume that the probability of an error happening remains constant during transmission, and we designate this probability by q. Thus, $p = 1 - q$ is the probability that an individual symbol remains unchanged. This is summarized in the diagram.

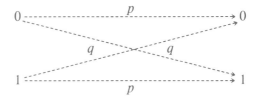

(In practice, values of q are normally small and values of p close to 1, since we would normally be using a relatively reliable channel.)

We also assume that errors occur randomly and independently. In short, we can think of the transmission of a binary string of length 7 as a Bernoulli trial, with failure corresponding to a symbol being received in error.

For a message of length 7, if *no* coding were used, the probability that the receiver would get the correct message would be

$$b(7, 7; p) = p^7$$

since none of the seven symbols could have been altered. For $p = .98$, this gives $(.98)^7 = .8681$.

Using the Hamming code, the receiver will get the correct message even if one error has occurred. Hence, the corresponding probability of correct reception would be

$$\underbrace{b(7,7; p)}_{\text{No errors}} + \underbrace{b(7,6; p)}_{\text{One error}} = p^7 + 7p^6q$$

For $p = .98$ this now gives $.8681 + .1240 = .9921$, which shows a considerable improvement.

There are codes in use that correct more than a single error. One such code, known by the initials of its originators as a BCH code, is a code of length 15 (messages are binary strings of length 15) that corrects up to two errors. Sending a message of length 15 with no attempt at coding would result in a probability of $b(15, 15; p) = p^{15}$ of the correct message being received. For $p = .98$ this gives $b(15, 15; .98) = .7386$. Using the BCH code that can correct two or fewer errors, the probability that a message will be correctly received becomes

$$\underbrace{b(15,15; p)}_{\text{No errors}} + \underbrace{b(15,14; p)}_{\text{One error}} + \underbrace{b(15,13; p)}_{\text{Two errors}}$$

Evaluating this for $p = .98$ we get

$$.7386 + .2261 + .0323 = .9970$$

Codes that can correct a high number of errors are clearly very desirable. Yet a basic result in the theory of codes states that as the error-correcting capability of a code increases, so, of necessity, must its length. But lengthier codes require more time for transmission and are clumsier to use. Thus, here, speed and correctness are at odds.

Exercise 8.4

Answers to Odd-Numbered Problems begin on page A-23.

A   In Problems 1–14 use Formula (2), page 403 and a calculator to compute each binomial probability.

1. $b(7, 5; .30)$ 2. $b(8, 6; .40)$
3. $b(15, 8; .70)$ 4. $b(8, 5; .60)$
5. $b(15, 10; .5)$ 6. $b(12, 6; .90)$
7. $b(15, 3; .3) + b(15, 2; .3) + b(15, 1; .3) + b(15, 0; .3)$
8. $b(8, 6; .4) + b(8, 7; .4) + b(8, 8; .4)$
9. $n = 3,\ k = 2,\ p = \frac{1}{3}$ 10. $n = 3,\ k = 1,\ p = \frac{1}{3}$
11. $n = 3,\ k = 0,\ p = \frac{1}{6}$ 12. $n = 3,\ k = 3,\ p = \frac{1}{6}$
13. $n = 5,\ k = 3,\ p = \frac{2}{3}$ 14. $n = 5,\ k = 0,\ p = \frac{2}{3}$
15. Find the probability of obtaining exactly 6 successes in 10 trials when the probability of success is .3.
16. Find the probability of obtaining exactly 5 successes in 9 trials when the probability of success is .2.
17. Find the probability of obtaining exactly 9 successes in 12 trials when the probability of success is .8.
18. Find the probability of obtaining exactly 8 successes in 15 trials when the probability of success is .75.
19. Find the probability of obtaining at least 5 successes in 8 trials when the probability of success is .25.
20. Find the probability of obtaining at most 3 successes in 7 trials when the probability of success is .15.

In Problems 21–25 a fair coin is tossed eight times.

21. What is the probability of obtaining exactly 1 head?
22. What is the probability of obtaining exactly 2 heads?
23. What is the probability of obtaining at least 5 tails?
24. What is the probability of obtaining at most 2 tails?
25. What is the probability of obtaining exactly 2 heads if it is known that at least 1 head appeared?

B 26. What is the probability of obtaining 11's exactly three times in 7 rolls of 2 fair dice?
27. What is the probability of obtaining 7's exactly two times in 5 rolls of 2 fair dice?
28. What is the probability that in a family of 7 children:
    (a) 4 will be girls?
    (b) At least 2 are girls?
    (c) At least 2 and not more than 4 are girls?

29. Assuming male and female birth rates are equally probable, what is the probability that a family with exactly 6 children will have 3 boys and 3 girls?

30. What is the probability of obtaining exactly 3 tails in three tosses if it is known that at least 1 tail appeared?

31. An experiment is performed four times, with 2 possible outcomes: F (Failure) and S (Success) with probabilities $\frac{1}{4}$ and $\frac{3}{4}$, respectively.

    (a) Draw the tree diagram describing the experiment.
    (b) Calculate the probability of exactly 2 successes and 2 failures by using the tree diagram from part (a).
    (c) Verify your answer to part (b) by using Formula (2).

32. In a Bernoulli trial with $p = \frac{1}{3}$, for what least value of n does the probability of exactly 2 successes have its maximum value?

33. What is the probability that the birthdays of 6 people fall in 2 calendar months, leaving exactly 10 months free? (Assume independent and equal probabilities for all months.)

34. How many times should a fair coin be flipped in order to have the probability of at least 1 head appear to be greater than .98?

C   35. Prove the identity

$$b(n, k; p) = b(n, n - k; 1 - p)$$

[*Hint:* Refer to Example 4, page 322].

APPLICATIONS   36. **Employment Screening** To screen prospective employees, a company gives a 10-question multiple-choice test. Each question has 4 possible answers, of which one is correct. The chance of answering the questions correctly by just guessing is $\frac{1}{4}$ or 25%. Find the probability of answering, by chance:

    (a) Exactly 3 questions correctly   (b) No questions correctly
    (c) At least 8 correctly   (d) No more than 7 questions correctly

37. **Opinion Poll** Mr. Austin and Mr. Moran are running for public office. A survey conducted just before the day of election indicates that 60% of the voters prefer Mr. Austin and 40% prefer Mr. Moran. If 8 people are chosen at random and asked their preference, find the probability that all 8 people will express a preference for Mr. Moran.

38. **Working Habits** If 40% of the workers at a large factory bring their lunch each day, what is the probability that in a randomly selected sample of eight workers

    (a) Exactly two bring their lunch each day?
    (b) At least two bring their lunch each day?
    (c) No one brings lunch?
    (d) No more than three bring lunch each day?

39. **Quality Control** Suppose that 5% of the items produced by a factory are defective. If 8 items are chosen at random, what is the probability that:

    (a) Exactly 1 is defective?   (b) Exactly 2 are defective?
    (c) At least 1 is defective?   (d) Less than 3 are defective?

40. **Opinion Polls** Suppose that 60% of the voters intend to vote for a conservative

candidate. What is the probability that a survey polling 8 people reveals that 3 or fewer intend to vote for a conservative candidate?

41. Heart Attack Approximately 23% of North American deaths are due to heart attacks. What is the probability that 4 of the next 10 unrelated deaths reported in a certain community will be due to heart attacks?

42. Batting Averages For a baseball player with a .250 batting average, what is the probability that the player will have at least 2 hits in four times at bat? What is the probability of at least 1 hit in four times at bat?

43. Targeting If the probability of hitting a target is $\frac{1}{3}$ and 10 shots are fired independently, what is the probability of the target being hit at least twice?

44. Quality Control A television manufacturer tests a random sample of 15 picture tubes to determine whether any are defective. The probability that a picture tube is defective has been found from past experience to be .05.

  (a)   What is the probability that there are no defective tubes in the sample?

  (b)   What is the probability that more than 2 of the tubes are defective?

45. True and False Tests In a 15 item true–false examination, what is the probability that a student who guesses on each question will get at least 10 correct answers? If another student has .8 probability of correctly answering each question, what is the probability that this student will answer at least 12 questions correctly?

46. Marriage The probability that an American man who has not married by the age of 45 will ever marry is approximately .44. In a group of 10 randomly selected unmarried 45-year-old men:

  (a)   What is the probability that exactly 2 of the 10 will ever marry?

  (b)   What is the probability that no more than 2 of the 10 will ever marry?

47. Product Testing A supposed coffee connoisseur claims she can distinguish between a cup of instant coffee and a cup of percolator coffee 75% of the time. You give her 6 cups of coffee and tell her that you will grant her claim if she correctly identifies at least 5 of the 6 cups.

  (a)   What are her chances of having her claim granted if she is in fact only guessing?

  (b)   What are her chances of having her claim rejected when in fact she really does have the ability she claims?

48. Opinion Polls Opinion polls based on small samples often yield misleading results. Suppose 65% of the people in a city are opposed to a bond issue and the others favor it. If 7 people are asked for their opinion, what is the probability that a majority of them will favor the bond issue?

49. Coding There is a Hamming code of length 15 that corrects a single error. Assuming $p = .98$, find the probability that a message transmitted using this code will be correctly received.

50. Coding There is a binary code of length 23 (called the Golay code) that can correct up to 3 errors. With $p = .98$, find the probability that a message transmitted using the Golay code will be correctly received.

8.5

RANDOM VARIABLES

RANDOM VARIABLE

When we perform an experiment, we are often interested not in a particular outcome, but rather in some number associated with that outcome. For example, in tossing a coin three times we may be interested in the number of heads obtained regardless of the particular sequence in which the heads appear. Similarly, the gambler throwing a pair of dice in a crap game is interested in the sum of the faces rather than the particular number on each face. When we sample mass-produced microchips we may be interested in the number of defective microchips, but not their price. When an experiment involves determining the time between arrivals of customers in a gas station, we may be interested only in the time between arrivals, but not in the makes of their cars.

In each of these examples we are interested in numbers that are associated with experimental outcomes. This process of assigning a number to each outcome is called *random variable* assignment.

A *random variable* is a rule that assigns a number to each outcome of an experiment.

It is customary to denote the random variable by a capital letter, such as X or Y.

Table 1 summarizes the results of the experiment of flipping a fair coin three times. The first column in the table gives the sample space for this experiment. The second column shows the number of heads for each simple event.

Table 1

Sample Space	Number of Heads $X(e)$
e_1: HHH	3
e_2: HHT	2
e_3: HTH	2
e_4: THH	2
e_5: HTT	1
e_6: THT	1
e_7: TTH	1
e_8: TTT	0

Table 2

Number of Heads Obtained in Three Flips of a Coin	Probability
0	$\frac{1}{8}$
1	$\frac{3}{8}$
2	$\frac{3}{8}$
3	$\frac{1}{8}$

Suppose that in this experiment we are interested only in the total number of heads. This information is given in Table 2.

The role of the random variable is to transform the original sample space $\{HHH, HHT, HTH, HTT, THH, THT, TTH, TTT\}$ into a new sample space that consists of the number of heads that occur: $\{0, 1, 2, 3\}$. If X denotes the random variable, then:

$$X(e_1) = X(HHH) = 3, \ X(e_2) = X(HHT) = 2, \ X(e_3) = (HTH) = 2, \ X(e_4) = X(THH) = 2$$
$$X(e_5) = X(HTT) = 1, \ X(e_6) = X(THT) = 1, \ X(e_7) = X(TTH) = 1, \ X(e_8) = X(TTT) = 0$$

We are interested in the probabilities that the random variable X assumes the values 0, 1, 2, and 3. Table 2 gives us the probabilities, so we may write

$$\text{probability } (X = 0) = \frac{1}{8}, \qquad \text{probability } (X = 1) = \frac{3}{8},$$

$$\text{probability } (X = 2) = \frac{3}{8}, \qquad \text{probability } (X = 3) = \frac{1}{8}.$$

PROBABILITY DISTRIBUTION

The discussion thus far illustrates two of the most important properties of a random variable:

1. The values it can assume.
2. The probabilities that are associated with each of these possible values.

Because values assumed by a random variable can be used to symbolize all outcomes associated with a given experiment, the probability assigned to a simple event can now be assigned as the likelihood that the random variable takes on the corresponding value.

If a random variable X has the values

$$x_1, x_2, x_3, \ldots, x_n \tag{1}$$

then the rule given by

$$p(x) = P(X = x)$$

where x assumes the values in (1) is called the *probability distribution* of X, or distribution of X.

It follows that $p(x)$ is a probability distribution of X if it satisfies the following two conditions:

1. $0 \leq p(x) \leq 1$ for $x = x_1, x = x_2, \ldots, x = x_n$
2. $p(x_1) + p(x_2) + \cdots + p(x_n) = 1$

where $x_1, x_2, \ldots, x_n$ are the values of X.

If we consider again the experiment of tossing a fair coin three times, we denote the probability distribution by

$$p(x) \quad \text{where } x = 0, 1, 2, \text{ or } 3$$

For instance, $p(3)$ is the probability of getting exactly three heads, that is,

$$p(3) = P(X = 3) = \frac{1}{8}$$

In a similar way we define $p(0)$, $p(1)$, and $p(2)$. Probability distributions are also represented graphically, as shown in Figure 14. The graph of a probability distribution is often called a *histogram*.

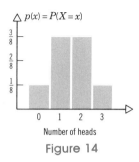

Figure 14

BINOMIAL PROBABILITY USING PROBABILITY DISTRIBUTION

We can now state the formula developed for the binomial probability model in terms of its probability distribution.

The binomial probability that assigns probabilities to the number of successes in n trials is an example of a probability distribution. For this distribution, X is the random variable whose value for any outcome of the experiment is the number of successes obtained. For this distribution, we may write

$$P(X = k) = b(n, k;p) = \binom{n}{k} p^k q^{n-k}$$

where $P(X = k)$ denotes the probability that the random variable equals k, that is, that exactly k successes are obtained.

Exercise 8.5

Answers to Odd-Numbered Problems begin on page A-24.

In Problems 1–6 list the values of the given random variable X together with the probability distributions.

1. A fair coin is tossed two times and X is the random variable whose value for an element in the sample space is the number of heads obtained.

2. A fair die is tossed once. The random variable X is the number showing on the top face.

3. The random variable X is the number of female children in a family with 3 children. (Assume the probability of a female birth is $\frac{1}{2}$).

4. A job applicant takes a 3-question true–false examination and guesses on each question. Let X be the number of right answers minus the number of wrong answers.

5. An urn contains 4 red balls and 6 white balls. Three balls are drawn with replacement. The random variable X is the number of red balls.

6. A couple getting married will have 3 children, and the random variable X denotes the number of boys they will have.

8.6

EXPECTATION

| EXPECTED VALUE
| STEPS TO COMPUTE EXPECTATION
| EXPECTED VALUE OF BERNOULLI TRIALS
| DERIVATION OF $E = np$

EXPECTED VALUE

An important concept, which has its origin in gambling and which uses probability, is *expected value.* For instance, gamblers are quite concerned with the *expectation,* or *expected value,* of a game. Suppose, for example, that 1000 tickets are printed and raffled off. Out of all these tickets one ticket has a cash value of $300, two are worth $100, 100 are worth $1, and the remaining are worth $0. The average *value* of a ticket is then

$$\frac{\$300 + \$100 + \$100 + \$1 + \$1 + \cdots + \$1 + \$0 + \$0 + \cdots + \$0}{1000}$$

$$= \frac{300 + 100(2) + 1(100) + 0(897)}{1000} = \frac{600}{1000} = \$0.60 \qquad (1)$$

Thus, if the raffle is to be nonprofit to all, the charge for each ticket should be $0.60. The result can also be viewed as saying that if we entered such a raffle many times, $\frac{1}{1000}$ of the time we would win $300, $\frac{2}{1000}$ of the time we would win $100, and so on, with our winnings in the long run averaging $0.60 per ticket.

Let's write (1) in another way.

$$\$300 \cdot \frac{1}{1000} + 100 \cdot \frac{2}{1000} + 1 \cdot \frac{100}{1000} + 0 \cdot \frac{897}{1000}$$

$$= 300(.001) + 100(.002) + 1(.1) + 0(.897)$$

This last expression is simply the sum obtained by adding each cash value (300, 100, 1, 0) times the probability that it occurred (.001, .002, .1, .897); that is,

$$300p(300) + 100p(100) + 1p(1) + 0p(897)$$

where $p(300)$ denotes the probability of winning $300, $p(100)$ the probability of winning $100, and so on.

This illustrates the procedure used to compute the average when each value occurs with a specified probability. This sum is called the *expected value;* it represents the long-term average of numerous trials (numerous trials, since a few trials may not average close to the expected value, while a large number of trials will tend to give an average closer to the expected value).

The above discussion leads us to the definition of expected value in terms of a random variable and probability distribution.

Given the probability distribution for the random variable X,

x_i	$x_1\ x_2\ .\ .\ .\ x_n$
p_i	$p_1\ p_2\ .\ .\ .\ p_n$

where $p_i = p(x_i)$, we define the *expected value of X*, denoted by $E(X)$, as the sum

$$E(X) = x_1 p_1 + x_2 p_2 + \cdots + x_n p_n$$

Again, the term *expected value* should not be interpreted as a value that actually occurs in the experiment. For example, there was no raffle ticket paying $.60. Rather, it represents the anticipated average value per experiment were we to repeat the experiment numerous times.

In gambling, for instance, $E(X)$ is interpreted as the average winnings expected for the player in the long run. If $E(X)$ is positive, we say that the game is *favorable* to the player; if $E(X) = 0$, we say the game is *fair*, and if $E(X)$ is negative, we say the game is *unfavorable* to the player.

When the value assigned to an outcome of an experiment is positive, it can be interpreted as profits, winnings, or gains. When it is negative, it represents losses, penalties, or deficits.

STEPS TO COMPUTE EXPECTATION

The following steps outline the general procedure involved in determining the expected value.

Step 1. Determine the sample space S, which describes the possible outcomes when the experiment is performed, and assign an appropriate probability to each simple event in S.

Step 2. Determine the values $x_1, x_2 \ldots , x_n$ of the random variable X, and for each value of X determine the corresponding event in S.

Step 3. Determine $p(x_i)$ for each value x_i of X.

Step 4. Calculate $E(X) = x_1 p_1 + x_2 p_2 + \cdots + x_n p_n$

Example 1
What is the expected number of heads in tossing a fair coin three times?

Solution

Step 1. The sample space is

$$S = \{HHH, HHT, HTH, HTT, THH, THT, TTH, TTT\}$$

Step 2. A coin tossed three times can have 0, 1, 2, or 3 heads. We treat these as the values of the random variable.

Step 3. The probabilities corresponding to 0, 1, 2, or 3 heads in Step 2 are $\frac{1}{8}, \frac{3}{8}, \frac{3}{8}$, and $\frac{1}{8}$, respectively.

Step 4. The expected number of heads can now be found by multiplying each value by its corresponding probability and finding the sum of these values.

Expected number of heads =

$$E(X) = 0 \cdot \frac{1}{8} + 1 \cdot \frac{3}{8} + 2 \cdot \frac{3}{8} + 3 \cdot \frac{1}{8} = \frac{3}{2} = 1.5$$

Thus, on the average, tossing a coin three times will result in 1.5 heads.   ▪

Example 2

Consider the experiment of rolling a fair die. The player recovers an amount of dollars equal to the number of dots on the face that turns up, except when face 5 or 6 turns up, in which case the player will lose $5 or $6, respectively. What is the expected value of the game?

Solution

The sample space is

$$\{1, 2, 3, 4, 5, 6\}$$

We assign a probability of $\frac{1}{6}$ to each of them, since all faces are equally likely to occur.

The random variable X assigns $1, $2, $3, $4 to the winning numbers and $-$5 and $-$6 to the remaining outcomes. The probability distribution for X is summarized in the following probability distribution table.

x_i	$1	$2	$3	$4	$-$5	$-$6
p_i	$\frac{1}{6}$	$\frac{1}{6}$	$\frac{1}{6}$	$\frac{1}{6}$	$\frac{1}{6}$	$\frac{1}{6}$

The expected value of the game is

$$E(X) = \$1 \cdot \frac{1}{6} + \$2 \cdot \frac{1}{6} + \$3 \cdot \frac{1}{6} + \$4 \cdot \frac{1}{6} + (-\$5) \cdot \frac{1}{6} + (-\$6) \cdot \frac{1}{6}$$

$$= -\$\frac{1}{6} = -16.7 \text{ cents}$$

The player would expect to lose an average of 16.7 cents on each throw.   ▪

Example 3

A laboratory contains 10 electron microscopes, of which 2 are defective. If all microscopes are equally likely to be chosen, and if 4 are chosen, what is the expected number of defective microscopes?

Solution

Since we are interested in determining the expected number of defective microscopes, we assign a value of 0 to the outcome "0 defectives are selected," a value of

1 to the outcome "1 defective is chosen," and a value of 2 to the outcome "2 defectives are chosen." These are the values of the random variable X.

The sample of four microscopes can contain 0, 1, or 2 defective microscopes. The probability p_0 that none in the sample is defective is

$$p(0) = p_0 = \frac{\binom{2}{0}\binom{8}{4}}{\binom{10}{4}} = \frac{1}{3}$$

Similarly, the probabilities p_1 and p_2 for 1 and 2 defective microscopes are

$$P(1) = p_1 = \frac{\binom{2}{1}\binom{8}{3}}{\binom{10}{4}} = \frac{8}{15} \quad \text{and} \quad P(2) = p_2 = \frac{\binom{2}{2}\binom{8}{2}}{\binom{10}{4}} = \frac{2}{15}$$

The following table summarize the above results

x_i	0	1	2
p_i	$\frac{1}{3}$	$\frac{8}{15}$	$\frac{2}{15}$

Note $\dfrac{1}{3} + \dfrac{8}{15} + \dfrac{2}{15} = 1$

The expected value is

$$E(X) = 0 \cdot p_0 + 1 \cdot p_1 + 2 \cdot p_2 = \frac{8}{15} + \frac{4}{15} = \frac{4}{5}$$

Of course, we cannot have four-fifths of a defective microscope. However, we can interpret this to mean that in the long run such a sample will average just under one defective microscope.   ■

We point out that $\frac{4}{5}$ is a reasonable answer for the expected number of defective microscopes, since one-fifth of the microscopes in the laboratory are defective and we are selecting a random sample consisting of four of these microscopes.

The last example illustrates that an expected value need not be an amount of money. It can be any "payoff" associated with any experiment.

Example 4

Mr. Richmond is producing an outdoor concert. He estimates that he will make $300,000 if it does not rain and make $60,000 if it does rain. The weather bureau predicts that the chance of rain is .34 for the day of the concert.

(a) What are Mr. Richmond's expected earnings from the concert?

(b) An insurance company is willing to insure the concert for $150,000 against rain for a premium of $30,000. If he buys this policy, what are his expected earnings from the concert?

(c) Based on the expected earnings, should Mr. Richmond buy an insurance policy?

Solution

(a) We compute the expected value of the random variable X using the following probability distribution table:

	Rain	Not Rain
x_i	60,000	300,000
p_i	.34	.66

$$
\begin{aligned}
E(X) &= x_1 p_1 + x_2 p_2 \\
&= 60{,}000(.34) + 360{,}000(.66) \\
&= \$218{,}400
\end{aligned}
$$

(b) In the event Mr. Richmond buys the policy, we obtain the following distribution table:

	Rain	No Rain
x_i	180,000	270,000
p_i	.34	.66

We obtain the entries in the table in the following way. If it does not rain, the insurance company has a net gain of $30,000, the premium of the policy, and Mr. Richmond's income is $300,000 − $30,000 = $270,000. If it does rain, however, the company keeps the premium ($30,000) but pays out the $150,000, and the income to Mr. Richmond is the insurance company payoff minus the premium plus the gate receipts; that is,

$$\$150{,}000 - \$30{,}000 + \$60{,}000 = \$180{,}000$$

The expected value for this course of action is

$$
\begin{aligned}
E(X) &= \$180{,}000(0.34) + \$270{,}000(0.66) \\
&= \$239{,}400
\end{aligned}
$$

(c) Yes, he should take the insurance since his expected earnings will be higher.

∎

Example 5

An oil company may bid on only one of two contracts for oil drilling in two different areas, I and II. It is estimated that a profit of $300,000 would be realized from the first field and $400,000 from the second field if oil is discovered. Legal and other costs of bidding for the first oil field are $25,000 and for the second are $50,000. The probability of discovering oil in the first field is .60 and in the second is .70. The question is which oil field should the company bid for; that is, for which oil field is the expectation larger?

Solution

The company has a choice between two courses of action. A_1: Bid on Area I and A_2: Bid on Area II. To be able to make a decision, we compute the expected value for each course of action. We do this with the aid of the following tables.

Course I		
x_i	300,000	$-25,000$
p_i	.6	.4

Course II		
x_i	400,000	$-50,000$
p_i	.7	.3

The expected value for each course of action is

$$A_1: \text{Area I}$$
$$E(X) = (300{,}000)(0.6) + (-25{,}000)(0.4)$$
$$= \$170{,}000$$

$$A_2: \text{Area II}$$
$$E(X) = 400{,}000(0.7) + (-50{,}000)(0.3)$$
$$= \$265{,}000$$

Since the expected value for the second field exceeds that for the first, the oil company should bid on the second field. ∎

EXPECTED VALUE OF BERNOULLI TRIALS

In 100 tosses of a coin, what is the expected number of heads? If a student guesses at random on a true–false exam with 50 questions, what is her expected grade? These are specific instances of the following more general question:

In n trials of a Bernoulli process, what is the expected number of successes?

We now compute this expected value. As before, p denotes the probability of success on any individual trial. We define the random variable X as follows: X assumes the value 1 for success and 0 for failure.

If $n = 1$ (there is but one trial), then the expected number of successes is

$$E(X) = 1 \cdot p + 0 \cdot (1 - p) = p$$

If $n = 2$ (two trials), then 0, 1, or 2 successes can occur. Computing the probability of each, we get

$$E(X) = 2 \cdot p^2 + 1 \cdot 2p(1 - p) + 0 \cdot (1 - p)^2 = 2p$$

This would seem to suggest that with n trials the expected value E would be given by $E = np$. This is indeed the case and we have the following result:

In a Bernoulli process with n trials, the expected number of successes is

$$E(X) = np$$

where p is the probability of success on any single trial.

A derivation of this result is included at the end of this section. The intuitive idea behind the result is fairly simple. Thinking of probabilities as percentages, if success results p percent of the time then out of n attempts p percent of them, namely np, should be successful.

Example 6

In flipping a fair coin five times, there are 6 possible outcomes: 0 tails, 1 tail, 2 tails, 3 tails, 4 tails, or 5 tails, each with the respective probabilities

$$\binom{5}{0}\left(\frac{1}{2}\right)^5, \binom{5}{1}\left(\frac{1}{2}\right)^5, \binom{5}{2}\left(\frac{1}{2}\right)^5, \binom{5}{3}\left(\frac{1}{2}\right)^5, \binom{5}{4}\left(\frac{1}{2}\right)^5, \binom{5}{5}\left(\frac{1}{2}\right)^5$$

The expected number of tails is

$$E = 0 \cdot \binom{5}{0}\left(\frac{1}{2}\right)^5 + 1 \cdot \binom{5}{1}\left(\frac{1}{2}\right)^5 + 2 \cdot \binom{5}{2}\left(\frac{1}{2}\right)^5$$

$$+ 3 \cdot \binom{5}{3}\left(\frac{1}{2}\right)^5 + 4 \cdot \binom{5}{4}\left(\frac{1}{2}\right)^5 + 5 \cdot \binom{5}{5}\left(\frac{1}{2}\right)^5 = \frac{5}{2}$$

Clearly, using the result $E = np$ is much easier, since for $n = 5$ and $p = \frac{1}{2}$, we obtain $E = (5)(\frac{1}{2}) = \frac{5}{2}$. ■

Example 7

In a multiple-choice test there are 100 questions; each question consists of four answers with only one choice being correct. What is the expected number of correct answers if a person guesses on each question?

Solution

This is an example of a Bernoulli trial. The probability for success (a correct answer) when guessing is $p = \frac{1}{4}$. Since there are $n = 100$ questions, the expected number of correct answers is

$$E = np = (100)(\tfrac{1}{4}) = 25$$ ■

DERIVATION OF $E = np$

The derivation is an exercise in handling binomial coefficients and using the binomial theorem. We will make use of the following identity:

$$k\binom{n}{k} = n\binom{n-1}{k-1} \tag{2}$$

This can be established by expanding both sides using the definition of a binomial coefficient. (See Problem 26, Exercise 8.6, page 424.)

Recall that the probability of obtaining exactly k successes in n trials is given by $b(n, k; p) = \binom{n}{k}p^k q^{n-k}$. Thus the expected number of successes is

$$E = 0 \cdot \binom{n}{0} p^0 q^n + 1 \cdot \binom{n}{1} p^1 q^{n-1} + 2 \cdot \binom{n}{2} p^2 q^{n-2} +$$

$$\cdots + \underbrace{k}_{\substack{\text{No. of}\\\text{successes}}} \underbrace{\binom{n}{k} p^k q^{n-k}}_{\substack{\text{Corresponding}\\\text{probability}}} + \cdots + n \binom{n}{n} p^n$$

Using (2) above,

$$E = n \binom{n-1}{0} pq^{n-1} + n \binom{n-1}{1} p^2 q^{n-2} + \cdots$$

$$+ n \binom{n-1}{k-1} p^k q^{n-k} + \cdots + n \binom{n-1}{n-1} p^n$$

Now, factor out an n and a p from each of the terms on the right to get

$$E = np \left[\binom{n-1}{0} q^{n-1} + \binom{n-1}{1} pq^{n-2} + \cdots \right.$$

$$\left. + \binom{n-1}{k-1} p^{k-1} q^{(n-1)-(k-1)} + \cdots + \binom{n-1}{n-1} p^{n-1} \right]$$

The expression in brackets is $(p + q)^{n-1}$. To see why, use the binomial theorem. Thus,

$$E = np(p + q)^{n-1}$$

Since $p + q = 1$, the result $E = np$ follows.

Exercise 8.6

Answers to Odd-Numbered Problems begin on page A-24.

A   1. For the data given below, compute the expected value.

Outcome	e_1	e_2	e_3	e_4
Probability	.4	.2	.1	.3
x_i	2	3	-2	0

2. For the data below, compute the expected value.

Outcome	e_1	e_2	e_3	e_4
Probability	$\frac{1}{3}$	$\frac{1}{6}$	$\frac{1}{4}$	$\frac{1}{4}$
x_i	1	0	4	-2

3. Attendance at a football game in a certain city results in the following pattern. If it is extremely cold, the attendance will be 35,000; if it is cold, it will be 40,000; if it is moderate, 48,000; and if it is warm, 60,000. If the probabilities for extremely cold, cold, moderate, and warm are .08, .42, .42, and .08, respectively, how many fans are expected to attend the game?

4. A player rolls a fair die and receives a number of dollars equal to the number of dots appearing on the face of the die. What is the least the player should expect to pay in order to play the game?

5. Mary will win $8 if she draws an ace from a set of 10 different cards numbered from ace to 10. How much should she pay for one draw?

6. Thirteen playing cards, ace through king, are placed randomly with faces down on a table. The prize for guessing correctly the value of any given card is $1. What would be a fair price to pay for a guess?

7. David gets $10 if he throws a double on a single throw of a pair of dice. How much should he pay for a throw?

8. You pay $1 to toss 2 coins. If you toss 2 heads, you get $2 (including your $1); if you toss only 1 head, you get back your $1; and if you toss no heads, you lose your $1. Is this a fair game to play?

9. In a raffle, 1000 tickets are being sold at 60¢ each. The first prize is $100, and there are three second prizes of $50 each. By how much does the price of a ticket exceed its expected value?

10. In a raffle, 1000 tickets are being sold at 60¢ each. The first prize is $100. There are two second prizes of $50 each, and five third prizes of $10 each (there are eight prizes in all). Laura buys one ticket. How much more than the expected value of the ticket does she pay?

B 11. A fair coin is tossed three times, and a player wins $3 if 3 tails occur, wins $2 if 2 tails occur, and loses $3 if no tails occur. If 1 tail occurs, no one wins.

   (a)   What is the expected value of the game?
   (b)   Is the game fair?
   (c)   If the answer to part (b) is "No," how much should the player win or lose for a toss of exactly 1 tail to make the game fair?

12. Colleen bets $1 on a 2-digit number. She wins $75 if she draws her number from the set of all 2-digit numbers, {00, 01, 02, . . . , 99}; otherwise, she loses her $1.

   (a)   Is this game fair to the player?
   (b)   How much is Colleen expected to lose in a game?

13. Two teams, A and B, have played each other 14 times. Team A won 9 games, and team B won 5 games. They will play again next week. Bob offers to bet $6 on team A while you bet $4 on team B. The winner gets the $10. Is the bet fair to you in view of the past records of the two teams? Explain your answer.

14. A department store wants to sell 11 purses that cost them $41 each and 32 purses that cost them $9 each. If all purses are wrapped in 43 identical boxes and if each customer picks a box randomly, find:

   (a)   Each customer's expectation.
   (b)   The department store's expected profit if it charges $13 for each box.

15. Caryl draws a card from a deck of 52 cards. She receives 40 cents for a heart, 50 cents for an ace, and 90 cents for the ace of hearts. If the cost of a draw is 15 cents, should she play the game? Explain.

16. The following data give information about family size in the United States for a household in which the wife resides and the male head of household is in the 30–34 age bracket:

Number of Children	0	1	2	3
Proportion of Families		10.2%	15.9%	31.8%	42.1%

A family is chosen at random. Find the expected number of children in the family.

17. Assume that the odds for a certain racehorse to win are 7 to 5. If a bettor receives $5 when the horse wins, how much should he bet when the horse loses to make the game fair?

18. **Roulette** In roulette, there are 38 equally likely possibilities: the numbers $1-36$, 0, and 00 (double zero). See the figure. What is the expected value for a gambler who bets $\$1$ on number 15 if she wins $\$35$ each time the number 15 turns up and loses $\$1$ if any other number turns up? If the gambler plays the number 15 for 200 consecutive times, what is the total expected gain?

19. Find the number of times the face 5 is expected to occur in a sequence of 2000 throws of a fair die.

20. What is the expected number of tails that will turn up if a fair coin is tossed 582 times?

21. A certain kind of light bulb has been found to have .02 probability of being defective. A shop owner receives 500 light bulbs of this kind. How many of these bulbs are expected to be defective?

22. A student enrolled in a math course has .9 probability of passing the course. In a class of 20 students, how many would you expect to fail the math course?

23. A coin is weighted so that $P(H) = \frac{1}{4}$ and $P(T) = \frac{3}{4}$. Find the expected number of tosses of the coin required in order to obtain either a head or 4 tails.

24. A true–false test consisting of 30 questions is scored by subtracting the number of wrong answers from the number of right ones. Find the expected number of correct answers of a student who just guesses on each question. What will the expected test score be?

25. A box contains 3 defective bulbs and 9 good bulbs. If 5 bulbs are drawn from the box without replacement, what is the expected number of defective bulbs?

C 26. Verify that $k\binom{n}{k} = n\binom{n-1}{k-1}$

27. Prove that if the numerical values assigned to the outcomes of an experiment that has expected value E are all multiplied by the constant k, then the expected value of the new experiment is $k \cdot E$. Similarly, if to all the numerical values we add the same constant k, prove that the expected value of the new experiment is $E + k$.

APPLICATIONS 28. **Drug Reaction** A doctor has found that the probability that a patient who is given a certain drug will have unfavorable reactions to the drug is .002. If a group of 500 patients is going to be given the drug, how many of them does the doctor expect to have unfavorable reactions?

29. **Site Selection** A company operating a chain of supermarkets plans to open a new

store in one of two locations. They conduct a survey of the two locations and estimate that the first location will show an annual profit of $15,000 if it is successful and a $3000 loss otherwise. For the second location, the estimated annual profit is $20,000 if successful and a $6000 loss results otherwise. The probability of success at each location is $\frac{1}{2}$. What location should the management decide on in order to maximize its expected profit?

30. Site Selection For Problem 29 assume the probability of success at the first location is $\frac{2}{3}$ and at the second location is $\frac{1}{2}$. What location should be chosen?

8.7

FURTHER APPLICATIONS

OPERATIONS RESEARCH

The field of *operations research,* the science of making optimal or best decisions, has experienced remarkable growth and development since the 1940s. The purpose of this section is to introduce you to some examples from operations research that utilize the concept of expectation.

Example 1

Market Assessment A national car rental agency rents cars for $16 per day (gasoline and mileage are additional expenses to the customer). The daily cost per car (for example, lease costs and overhead) is $6 per day. The daily profit to the company is $10 per car if the car is rented, and the company incurs a daily loss of $6 per car if the car is not rented. The daily profit depends on two factors: the demand for cars and the number of cars the company has available to rent. Previous rental records show that the daily demand is:

Number of customers	8	9	10	11	12
Probability	.10	.10	.30	.30	.20

Find the expected number of customers and determine the optimal number of cars the company should have available for rental. (This is the number that yields the largest expected profit.)

Solution

The expected number of customers is

$$8(.1) + 9(.1) + 10(.3) + 11(.3) + 12(.2) = 10.4$$

If 10.4 customers are expected, how many cars should be on hand? The number may not be the integer closest to 10.4, since costs play a major role in the determination of profit. We need to compute the expected profit for each possible number of cars. The largest expected profit will tell us how many cars to have on hand.

For example, if there are 10 cars available, the expected profit for 8, 9, or 10 customers is

$$68(.1) + 84(.1) + 100(.8) = \$95.20$$

We obtain the entry $68(.1)$ by noting that the 10 cars cost the company \$60, and 8 cars rented with probability .10 bring in \$128, for a profit of \$68. Similarly, we obtain the entry $84(.1)$ by noting that the 10 cars cost the company \$60, and 9 cars rented with probability .10 bring in \$144, for a profit of \$84. The entry $100(.8)$ is obtained since for 10 or more customers (probability $.3 + .3 + .2 = .8$) the profit is $10 \times \$16 - \$60 = \$100$.

The table lists the expected profit for 8 to 12 cars. Clearly, the optimal stock size is 11 cars, since this number of cars maximizes expected profit.

Number of cars	8	9	10	11	12
Expected profit	\$80.00	\$88.40	\$95.20	\$97.20	\$94.40

■

Example 2

Quality Control A factory produces electronic components, and each component must be tested. If the component is good, it will allow the passage of current; if the component is defective, it will block the passage of current. Let p denote the probability that a component is good. See Figure 15.

Figure 15

With this system of testing, a large number of components requires an equal number of tests. This increases the production cost of the electronic components since it requires one test per component. To reduce the number of tests, a quality control engineer proposes, instead, a new testing procedure: Connect the components pairwise in series, as shown in Figure 16.

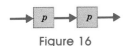

Figure 16

If the current passes two components in series, then both components are good and only one test is required. The probability that two components are good is p^2. If the current does not pass, they must be sent individually to the quality control department, where each component is tested separately. In this case, three tests are required. The probability that three tests are needed is $1 - p^2$ (1 minus probability of success p^2). The expected number of tests for a pair of components is

$$E = 1 \cdot p^2 + 3 \cdot (1 - p^2) = p^2 + 3 - 3p^2 = 3 - 2p^2$$

The number of tests saved for a pair is

$$2 - (3 - 2p^2) = 2p^2 - 1$$

The number of tests saved per component is

$$\frac{2p^2 - 1}{2} = p^2 - \frac{1}{2} \text{ tests}$$

The greater the probability p that the component is good, the greater the saving.

For example, if p is almost 1, we have a saving of almost $1 - \frac{1}{2}$ or $\frac{1}{2}$, which is 50% of the original number of tests needed. Of course, if p is small, say less than .7, we do not save anything since $(.7)^2 - \frac{1}{2}$ is less than 0, and we are wasting tests. ■

If the reliability of the components manufactured in Example 2 is very high, it might even be advisable to make larger groups. Suppose three components are connected in series. See Figure 17.

Figure 17

For individual testing, we need three tests. For group testing, we have

1   test needed with probability p^3

4   tests needed with probability $1 - p^3$

The expected number of tests is

$$E = 1 \cdot p^3 + 4 \cdot (1 - p^3) = 4 - 3p^3 \text{ tests}$$

The number of tests saved per component is

$$\frac{3p^3 - 1}{3} = p^3 - \frac{1}{3} \text{ tests}$$

In a similar way we can show that if the components are arranged in groups of four connected in series, then the number of tests saved per component is

$$p^4 - \frac{1}{4} \text{ tests}$$

In general, for groups of n, the number of tests saved per component is

$$p^n - \frac{1}{n} \text{ tests}$$

Notice from the above formula that as n, the group size, gets very large, the number of tests saved per component gets very, very small.

To determine the optimal group size for $p = .9$, we refer to Table 3. From the

Table 3

Group Size	Expected Tests Saved per Component $p = .9$	Percent Saving
2	$p^2 - \frac{1}{2} = .81 - .50 = .31$	31
3	$p^3 - \frac{1}{3} = .729 - .333 = .396$	39.6
4	$p^4 - \frac{1}{4} = .6561 - .25 = .4061$	40.61
5	$p^5 - \frac{1}{5} = .59049 - .2 = .39049$	39.05
6	$p^6 - \frac{1}{6} = .531 - .167 = .364$	36.4
7	$p^7 - \frac{1}{7} = .478 - .143 = .335$	33.5
8	$p^8 - \frac{1}{8} = .430 - .125 = .305$	30.5

428 8 / ADDITIONAL TOPICS IN PROBABILITY

table, we can see that the optimal group size is 4, resulting in a substantial saving of approximately 41%.

We also note that larger group sizes do not increase savings.

Example 3

A $75,000 oil detector is lowered under the sea to detect oil fields, and it becomes detached from the ship. If the instrument is not found within 24 hours, it will crack under the pressure of the sea. It is assumed that a skin diver will find it with probability .85, but it costs $500 to hire him. How many skin divers should be hired? (We assume that the chances of any diver's finding the detector are unaffected by how many other divers are looking.)

Solution

Let's assume that x skin divers are hired. The probability that they will fail to discover the instrument is $.15^x$. Thus, the instrument will be found with probability $1 - .15^x$.

The expected gain from hiring the skin divers is

$$\$75,000(1 - .15^x) = \$75,000 - \$75,000(.15^x)$$

while the cost for hiring them is

$$\$500 \cdot x$$

Thus, the expected net gain, denoted by $E(x)$, is

$$E(x) = \$75,000 - \$75,000(.15^x) - \$500x$$

The problem is then to choose x so that $E(x)$ is maximum.

We begin by evaluating $E(x)$ for various values of x:

$$E(1) = \$75,000 - \$75,000(.15^1) - \$500(1) = \$63,250.00$$
$$E(2) = \$75,000 - \$75,000(.15^2) - \$500(2) = \$72,312.50$$
$$E(3) = \$75,000 - \$75,000(.15^3) - \$500(3) = \$73,246.88$$
$$E(4) = \$75,000 - \$75,000(.15^4) - \$500(4) = \$72,962.03$$
$$E(5) = \$75,000 - \$75,000(.15^5) - \$500(5) = \$72,494.30$$
$$E(6) = \$75,000 - \$75,000(.15^6) - \$500(6) = \$71,999.15$$

Thus, the expected net gain is optimal when three divers are hired. Notice that hiring additional skin divers does not necessarily increase expected net gain. In fact, the expected net gain declines if more than three divers are hired. ▪

Exercise 8.7   *Answers to Odd-Numbered Problems begin on page A-24.*

1. Market Assessment A car agency has fixed costs of $8 per car per day and the revenue for each car rented is $14 per day. The daily demand is given in the table:

Number of customers	7	8	9	10	11
Probability	.10	.20	.40	.20	.10

Find the expected number of customers. Determine the optimal number of cars the company should have on hand each day. What is the expected profit in this case?

2. In Example 2 in this section, suppose $p = .8$. Show that the optimal group size is 3.

3. In Example 2 in this section, suppose $p = .95$. Show that the optimal group size is 5.

4. In Example 2 in this section, suppose $p = .99$. Compute savings for group sizes 10, 11, and 12, and thus show that 11 is the optimal group size. Determine the percent saving.

5. In Example 3 in this section, suppose the probability of any skin diver discovering the instrument is .95. Find:

   (a) An equation expressing the net expected gain.

   (b) The number x of skin divers that maximizes the expected net gain.

6. An airline must decide which of two aircraft it will use on a flight from New York to Los Angeles. Aircraft A has a seating capacity of 200, while aircraft B has a capacity of 300. Previous experience has allowed the airline to estimate the number of passengers on the flight as follows:

No. of passengers	150	180	200	250	300
Probability	.2	.3	.2	.2	.1

Regardless of aircraft used, the cost of a ticket is $300, but there are different operating costs attached to each aircraft. There is a fixed cost (fuel, crew, etc.) of $6000 attached to using aircraft A, while aircraft B has a fixed cost of $8000. There is also a per passenger cost (meals, luggage, added fuel) of $120 for aircraft A and $130 for aircraft B.

   Which aircraft should the airline schedule so that it maximizes its expected profit on the flight?

7. In Example 2, compute the expected number of tests saved per component if on the first test the current does not pass through 2 components in series, but on the second test the current does pass through 1 of them. A third test is not made (since the other component is obviously defective).

CHAPTER REVIEW

Important Terms
and Formulas

conditional probability	Bayes' formula	probability distribution
product rule	Bernoulli trials	histogram
probability tree	binomial probability	expected value
independent events	expectation	operations research
partition	random variable	

$$P(E|F) = \frac{P(E \cap F)}{P(F)}$$

$$P(E \cap F) = P(F) \cdot P(E|F)$$

$$P(E_1 \cap E_2 \cap \ldots \cap E_n) = P(E_1) \cdot P(E_2) \cdot \ldots \cdot P(E_n) \text{ for independent events}$$

$$P(E) = P(A_1) \cdot P(E|A_1) + P(A_2) \cdot P(E|A_2) + \cdots + P(A_n) \cdot P(E|A_n)$$

$$P(A_i|E) = \frac{P(A_i) \cdot P(E|A_i)}{P(E)}$$

$$b(n, k; p) = \binom{n}{k} p^k \cdot q^{n-k} = \frac{n!}{k!(n-k)!} p^k \cdot q^{n-k}$$

$$\text{If } P(E \cap F) = P(E) \cdot P(F) \text{ then } P(E|F) = P(E)$$

$$E(X) = x_1 p_1 + x_2 p_2 + \cdots + x_n p_n$$

True-False Questions

(Answers on page A-25)

T   F   1. The conditional probability of E given F is

$$P(E|F) = \frac{P(E \cap F)}{P(F)}$$

T   F   2. If two events in a sample space have no simple events in common, they are said to be *independent*.

T   F   3. $P(E|F) = P(F|E)$

T   F   4. Bayes' formula is useful for computing *a posteriori* probability.

T   F   5. $b(n, k; p)$ gives the probability of exactly n trials in k successes.

T   F   6. In flipping a fair coin 10 times, the expected number of tails is 5.

Fill in the Blanks

(Answers on page A-25)

1. If $P(E \cap F) = P(E) \cdot P(F)$ then E and F are said to be _____.

2. The formula $P(E|F) = \dfrac{P(E \cap P)}{P(F)}$ is called _____ _____.

3. A sample space S is partitioned into n subsets $A_1 A_2, \ldots A_n$ if

   (a) _____
   (b) _____
   (c) _____

4. The formula

$$P(A_1|E) = \frac{P(A_1) \cdot P(E|A_1)}{P(E)}$$

   is called _____ _____.

5. Random experiments are called Bernoulli trials if:

   (a) The same experiment is repeated several times.
   (b) There are only two possible outcomes: success and failure.
   (c) The repeated trials are _____.
   (d) The probability of each outcome remains the _____ for each trial.

6. A random variable on a sample space S is a rule that assigns _____ _____ to each element in S.

7. If X is a random variable assuming the values x_1, x_2. . . . x_n, then $E(X) = x_1 p(x_1) + x_2 p(x_2) + \cdots + x_n p(x_n)$ is the _____ _____ of X.

Review Exercises

A

(Answers on page A-25)

1. If $P(E) = .75$, $P(E|F) = .60$, $P(F|E) = .10$, find $P(E \cap F)$. Find $P(F)$.

2. If $P(E) = .30$, $P(E|F) = .35$, $P(F|E) = .90$, find $P(E \cap F)$. Find $P(F)$.

3. If E and F are mutually exclusive events with $P(E) = .35$ and $P(F) = .40$, find:

    (a) $P(E|F)$ (b) $P(F|E)$ (c) $P(E \cap F)$ (d) $P(E \cup F)$

4. If E and F are mutually exclusive events with $P(E) = .18$ and $P(F) = .27$, find:

    (a) $P(E|F)$ (b) $P(F|E)$ (c) $P(E \cap F)$ (d) $P(E \cup F)$

5. Let M be the event, "Student takes math course" and let C be the event, "Student takes computer science." If $P(M) = .30$, $P(C) = .15$, and $P(C|M) = .25$, find:

    (a) $P(M \cap C)$ (b) $P(M|C)$ (c) $P(M \cup C)$
    (d) $P(M \cap \overline{C})$ (e) $P(M \cup \overline{C})$

6. Let B be the event "Customer at a fast-food chain orders a cheeseburger," and let C be the event "Customer orders cola." If $P(B) = .70$, $P(C) = .40$, and $P(B|C) = .30$, find:

    (a) $P(B \cap C)$ (b) $P(C|B)$ (c) $P(B \cup C)$ (d) $P(\overline{B})$
    (e) $P(\overline{B} \cap C)$ (f) $P(\overline{B} \cup C)$ (g) are B and C mutually exclusive?
    (h) are B and C independent?

7. If E and F are independent events with probabilities $P(E) = .7$ and $P(F) = .9$, find:

    (a) $P(E|F)$ (b) $P(F|E)$ (c) $P(E \cap F)$ (d) $P(E \cup F)$

B

8. A biased coin is such that the probability of heads (H) is $\frac{1}{4}$ and the probability of tails (T) is $\frac{3}{4}$. Show that in flipping this coin twice the events E and F defined below are independent.

                 E: A head turns up in the first throw
                 F: A tail turns up in the second throw

9. The records of Midwestern University show that in one semester, 38% of the students failed Mathematics, 27% of the students failed Physics, and 9% of the students failed Mathematics and Physics. A student is selected at random.

    (a) If a student failed Physics, what is the probability that he or she failed Mathematics?

    (b) If a student failed Mathematics, what is the probability that he or she failed Physics?

    (c) What is the probability that he or she failed Mathematics or Physics?

10. A pair of fair dice is thrown three times. What is the probability that on the first toss the sum of the 2 dice is even, on the second toss the sum is less than 6, and on the third toss the sum is 7?

11. Management believes that 1 out of 5 people watching a television advertisement about their new product will purchase the product. Five people who watched the advertisement are picked at random. What is the probability that 0, 1, 2, 3, 4, or 5 of these people will purchase the product?

12. Suppose that the probability of a player hitting a home run is $\frac{1}{20}$. In five tries, what is the probability that the player hits at least 1 home run?

13. In a 12 item true–false examination:

(a) What is the probability that a student will obtain all correct answers by chance if he or she is guessing?

(b) If 7 correct answers constitute a passing grade, what is the probability that he or she will pass?

(c) What are the odds in favor of passing?

14. In a 20 item true–false examination:

(a) What is the probability that a student will obtain all correct answers by chance if he or she is guessing?

(b) If 12 correct answers constitute a passing grade, what is the probability that he or she will pass?

(c) What are the odds in favor of passing?

15. In a certain game, a player has the probability $\frac{1}{7}$ of winning a prize worth $89.99 and the probability $\frac{1}{3}$ of winning another prize worth $49.99. What is the expected value of the game for the player?

16. Frank pays 70¢ to play a certain game. He draws 2 balls (together) from a bag containing 2 red balls and 4 green balls. He receives $1 for each red ball that he draws. If he draws no red balls, he loses his 70¢. Has he paid too much? By how much?

17. In a lottery, 1000 tickets are sold at 25¢ each. There are three cash prizes: $100, $50, and $30. Alice buys five tickets.

(a) What would have been a fair price for a ticket?

(b) How much extra did Alice pay?

18. The figure here shows a spinning game for which a person pays $0.30 to purchase an opportunity to spin the dial. The numbers in the figure indicate the amount of payoff and its corresponding probability. Find the expected value of this game. Is the game fair?

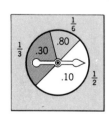

19. Consider the three boxes in the figure below. The game is played in two stages. The first stage is to choose a ball from Box A. If the result is a ball marked I, then we go to Box I, and select a ball from there. If the ball is marked II, then we select a ball from Box II. The number drawn on the second stage is the gain. Find the expected value of this game.

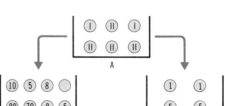

20. What is the expected number of heads that will turn up if a biased coin, $P(H) = \frac{1}{4}$, is tossed 200 times?

21. European Roulette A European roulette wheel has only 37 compartments, 18 red, 18 black, and 1 green. A player will be paid $2 (including his $1 bet) if he picks correctly the color of the compartment in which the ball finally rests. Otherwise, he loses $1. Is the game fair to the player? Compare this answer to the answer obtained in Problem 18, Exercise 8.6.

22. Find the probability of throwing an 11 at least three times in five throws of a pair of fair dice.

23. What is the expected number of girls in families having exactly three children?

24. Marketing Survey The table below indicates a survey conducted by a deodorant producer:

	Like the Deodorant	Did Not Like the Deodorant	No Opinion
Group I	180	60	20
Group II	110	85	12
Group III	55	65	7

Let the events E, F, G, H, and K be defined as follows:

E: Customer likes the deodorant

F: Customer does not like the deodorant

G: Customer is from Group I

H: Customer is from Group II

K: Customer is from Group III

Find:

(a) $P(E|G)$ (b) $P(G|E)$ (c) $P(H|E)$

(d) $P(K|E)$ (e) $P(F|G)$ (f) $P(G|F)$

(g) $P(H|F)$ (h) $P(K|F)$

25. ACT Scores The following data compare ACT scores of students with their performance in the classroom (based on a 4.0).

Average	Below 21	22–27	Above 28	Total
3.6–4	8	56	104	168
3.0–3.5	47	70	30	147
Below 3	47	34	4	85
Totals	102	160	138	400

A graduating student is selected at random. Find the probability that:

(a) The student scored above 28.

(b) The student point average is 3.6–4.

(c) The student scored above 28 with an average 3.6–4.

(d)   The student ACT score was in the 22–27 range.

(e)   The student had a 3.0–3.5 average.

(f)   Show that "ACT above 28" and "average below 3" are not independent.

26.   **Score Distribution**   Two forms of a standardized math exam were given to 100 students. The following are the results.

	Form A	Form B	Total
Over 80%	8	12	20
Under 80%	32	48	80
Totals	40	60	100

(a)   What is the probability that a student who scored over 80% took form A?

(b)   What is the probability that a student who took form A scored over 80%?

(c)   Show that the events "scored over 80%" and "took form A" are independent.

(d)   Are the events "scored over 80%" and "took form B" independent?

27.   **Genetics**   In a certain population of people, 25% are blue-eyed and 75% are brown-eyed. Also, 10% of the blue-eyed people are left-handed and 5% of the brown-eyed people are left-handed.

(a)   What is the probability that a person chosen at random is blue-eyed and left-handed?

(b)   What is the probability that a person chosen at random is left-handed?

(c)   What is the probability that a person is blue-eyed, given that the person is left-handed?

28.   **Quality Control**   Three machines in a factory, A_1, A_2, A_3, produce 55%, 30%, and 15% of total production, respectively. The percentage of defective output of these machines is 1%, 2% and 3%, respectively. An item is chosen at random and it is defective. What is the probability that it came from machine A_1? From A_2? From A_3?

29.   **Cancer Detection**   A lung cancer test has been found to have the following reliability. The test can detect 85% of the people who have cancer and does not detect 15% of these people. Among the noncancerous group it detects 92% of the people not having cancer, whereas 8% of this group are detected erroneously as having lung cancer. Statistics show that about 1.8% of the population have cancer. Suppose an individual is given the test for lung cancer and it detects the disease. What is the probability that the person actually has cancer?

30.   **The Blood Testing Problem***   A group of 1000 people are subjected to a blood test that can be administered in two ways: (1) each person can be tested separately (in this case 1000 tests are required) or (2) the blood samples of 30 people can be pooled

---

* William Feller, *An Introduction to Probability Theory and Its Applications*, 3rd ed., Wiley, New York, 1968, pp. 239–240.

and analyzed together. If we use the second way and the test is negative, then one test suffices for 30 people. If the test is positive, each of the 30 people can then be tested separately, and, in all, $30 + 1$ tests are required for the 30 people. Assume the probability p that the test is positive is the same for all people and that the people to be tested are independent.

(a) What is the probability that the test for a pooled sample of 30 people will be positive?

(b) What is the expected number of tests necessary under plan (2)?

31. Survival The probability that a recently transplanted shrub will survive is .75 if it rains the next 2 days and .25 otherwise. If the probability of rain the next 2 days is .15, what is the probability that the shrub will survive?

32. Quality Control A quality control inspector rejects 9% of all produced toys. Experience has shown that this inspector rejects 95% of all defective toys and that 8% of all produced toys have ultimately been proven to be defective. What is the probability that one of the inspector's rejected toys is actually defective?

Mathematical Questions

From CPA and Actuary Exams (Answers on page A-25.)

1. *Actuary Exam – Part I*
If P and Q are events having positive probability in the sample space S such that $P \cap Q = \varnothing$, then all of the following pairs are independent EXCEPT:

(a) $\varnothing$ and P (b) P and Q (c) P and S
(d) P and $P \cap Q$ (e) $\varnothing$ and the complement of P

2. *Actuary Exam – Part II*
Events S and T are independent with $Pr(S) < Pr(T)$, $Pr(S \cap T) = \frac{6}{25}$, and $Pr(S|T) + Pr(T|S) = 1$. What is $Pr(S)$?

(a) $\frac{1}{25}$ (b) $\frac{1}{5}$ (c) $\frac{5}{25}$ (d) $\frac{2}{5}$ (e) $\frac{3}{5}$

3. *Actuary Exam – Part II*
In a group of 20,000 men and 10,000 women, 6% of the men and 3% of the women have a certain affliction. What is the probability that an afflicted member of the group is a man?

(a) $\frac{3}{5}$ (b) $\frac{2}{3}$ (c) $\frac{3}{4}$ (d) $\frac{4}{5}$ (e) $\frac{8}{9}$

4. *Actuary Exam – Part II*
An unbiased die is thrown two independent times. Given that the first throw resulted in an even number, what is the probability that the sum obtained is 8?

(a) $\frac{5}{36}$ (b) $\frac{1}{6}$ (c) $\frac{4}{21}$ (d) $\frac{7}{36}$ (e) $\frac{1}{3}$

5. *Actuary Exam – Part II*
If the events S and T have equal probability and are independent with $Pr(S \cap T) = p > 0$, then $Pr(S) =$

(a) $\sqrt{p}$ (b) p^2 (c) $\frac{p}{2}$ (d) p (e) $2p$

6. *CPA Exam*
The Stat Company wants more information on the demand for its products. The following data are relevant:

Units Demanded	Probability of Unit Demand	Total Cost of Units Demanded
0	.10	$0
1	.15	1.00
2	.20	2.00
3	.40	3.00
4	.10	4.00
5	.05	5.00

What is the total expected value or payoff with perfect information?

(a) $2.40   (b) $7.40   (c) $9.00   (d) $9.15

7. *CPA Exam*

Your client wants your advice on which of two alternatives he should choose. One alternative is to sell an investment now for $10,000. Another alternative is to hold the investment 3 days after which he can sell it for a certain selling price based on the following probabilities:

Selling Price	Probability
$5,000	.4
$8,000	.2
$12,000	.3
$30,000	.1

Using probability theory, which of the following is the most reasonable statement?

(a) Hold the investment 3 days because the expected value of holding exceeds the current selling price.

(b) Hold the investment 3 days because of the chance of getting $30,000 for it.

(c) Sell the investment now because the current selling price exceeds the expected value of holding.

(d) Sell the investment now because there is a 60% chance that the selling price will fall in 3 days.

8. *CPA Exam*

The Polly Company wishes to determine the amount of safety stock that it should maintain for Product D that will result in the lowest cost.

The following information is available:

Stockout cost	$80 per occurrence
Carrying cost of safety stock	$2 per unit
Number of purchase orders	5 per year

The available options open to Polly are as follows:

Units of Safety Stock	10	20	30	40	50	55
Probability		50%	40%	30%	20%	10%	5%

The number of units of safety stock that will result in the lowest cost are:

(a) 20   (b) 40   (c) 50   (d) 55

9. *CPA Exam*

The ARC Radio Company is trying to decide whether to introduce as a new product a wrist "radiowatch" designed for shortwave reception of exact time as broadcast by the National Bureau of Standards. The "radiowatch" would be priced at $60, which is exactly twice the variable cost per unit to manufacture and sell it. The incremental fixed costs necessitated by introducing this new product would amount to $240,000 per year. Subjective estimates of the probable demand for the product are shown in the following probability distribution:

Annual Demand	6,000	8,000	10,000	12,000	14,000	16,000
Probability	.2	.2	.2	.2	.1	.1

The expected value of demand for new product is

(a) 11,000 units   (b)   10,200 units

(c) 9,000 units   (d)   10,600 units

(e) 9,800 units

10. *CPA Exam*
In planning its budget for the coming year, King Company prepared the following payoff probability distribution describing the relative likelihood of monthly sales volume levels and related contribution margins for product A:

Monthly Sales Volume	Contribution Margin	Probability
4,000	$ 80,000	.20
6,000	120,000	.25
8,000	160,000	.30
10,000	200,000	.15
12,000	240,000	.10

What is the expected value of the monthly contribution margin for product A?

(a) $140,000   (b) $148,000

(c) $160,000   (d) $180,000

11. *CPA Exam*
A decision tree has been formulated for the possible outcomes of introducing a new product line.

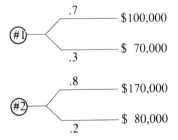

Branches related to Alternative 1 reflect the possible payoffs from introducing the product without an advertising campaign. The branches for Alternative 2 reflect the possible payoffs with an advertising campaign costing $40,000. The expected values of Alternatives 1 and 2, respectively, are

(a)   #1: $(.7 \times \$100,000) + (.3 \times \$70,000)$
      #2: $(.8 \times \$170,000) + (.2 \times \$80,000)$

(b)   #1: $(.7 \times \$100,000) + (.3 \times \$70,000)$
      #2: $(.8 \times \$130,000) + (.2 \times \$40,000)$

(c)   #1: $(.7 \times \$100,000) + (.3 \times \$70,000)$
      #2: $(.8 \times \$170,000) + (.2 \times \$80,000) - \$40,000$

(d)   #1: $(.7 \times \$100,000) + (.3 \times \$70,000) - \$40,000$
      #2: $(.8 \times \$170,000) + (.2 \times \$80,000) - \$40,000$

12. *CPA Exam*

A battery manufacturer warrants its automobile batteries to perform satisfactorily for as long as the owner keeps the car. Auto industry data show that only 20% of car buyers retain their cars for three years or more. Historical data suggest

Number of Years Owned	Probability of Battery Failure	Battery Exchange Costs	Percentage of Failed Batteries Returned
Less than 3 years	0.4	$50	75%
3 years or more	0.6	$20	50%

If 50,000 batteries were sold this year, what is the estimated warranty cost?

(a) $375,000   (b) $435,000
(c) $500,000   (d) $660,000

13. *Actuary Exam*

What is the probability that 10 independent tosses of an unbiased coin result in no fewer than 1 head and no more than 9 heads?

(a) $(\tfrac{1}{2})^9$ (b) $1 - 11(\tfrac{1}{2})^9$ (c) $1 - 11(\tfrac{1}{2})^{10}$ (d) $1 - (\tfrac{1}{2})^9$
(e) $1 - (\tfrac{1}{2})^{10}$

9

STATISTICS

9.1

INTRODUCTORY REMARKS

Statistics is the science of collecting, organizing, analyzing, and interpreting numerical facts. By making observations, statisticians obtain *data* in the form of measurements or counts. The *organization of data* involves the presentation of the collected measurements or counts in a form suitable for determining logical conclusions. Usually, tables or graphs are used to represent the collected data. The *analysis of data* is the process of extracting, from given measurements or counts, related and relevant information from which a brief numerical description can be formulated. In this process we use concepts known as the *mean, median, range, variance,* and *standard deviation.* By *interpretation of data* we mean the art of drawing conclusions from the analysis of the data. This involves the formulation of predictions concerning a large collection of objects based on the information available from a small collection of similar objects.

In collecting data concerning varied characteristics, it is often impossible or impractical to observe an entire group. Instead of examining the entire group, called the *population,* a small segment, called the *sample,* is chosen. It would be difficult, for example, to question all cigarette smokers in order to study the effects of smoking. Therefore, an appropriate sample of smokers is usually selected for questioning.

The method of selecting the sample is extremely important if we want the conclusions to be reliable. All members of the population under investigation should have an equal probability of being selected; otherwise, a *biased sample* could result. For example, if we want to study the relationship between smoking cigarettes and lung cancer, we cannot choose a sample of smokers who all live in the same location. The individuals chosen might have dozens of characteristics peculiar to their area, which would give a false impression with regard to all smokers.

As an example, suppose we want to study the effects of drugs on youths. If we decide to choose a sample from a group of students at a specific university or from a group of youths in a ghetto, we will get a biased sample, since university students and ghetto youths may be heavy users of drugs. It would be more appropriate to arrange the sample so that all members of the population under investigation have an equal probability of being selected.

Samples collected in such a way that each selection is equally likely to be chosen are called *random samples.* Of course, there are many random samples that can be chosen from a population. By combining the results of more than one random sample of a population, it is possible to obtain a more accurate representation of the population.

If a sample is representative of a population, important conclusions about the population can often be inferred from analysis of the sample. The phase of statistics dealing with conditions under which such inference is valid is called *inductive statistics* or *statistical inference.* Since such inference cannot be absolutely certain, the language of probability is often used in stating conclusions. Thus, when a meteorologist makes a forecast, weather data collected over a large region are studied, and based on this study, the weather forecast is given in terms of chances. A typical forecast might be "There is a 20% possibility for rain tomorrow."

To summarize, in statistics we are interested in four principles: gathering data or information, organizing it, analyzing it, and interpreting it.

In gathering data or in choosing a random sample it is important to:

1. Describe the method for choosing the sample clearly and carefully.

2. Choose the sample so that it is random, that is, so that it is dependable and not subject to personal choice or bias.

Example 1

Suppose television tubes on an assembly line pass an inspection, and suppose it is desired to test on the average one out of four tubes. If the test is to be performed in a random fashion, how should the inspection proceed?

Solution

To remove any personal choice on the part of the inspector, 2 fair coins can be flipped. Then, whenever 2 tails appear (probability $\frac{1}{4}$), the inspector can select a tube for testing.

Exercise 9.1

Answers to Odd-Numbered Problems begin on page A-25.

In Problems 1–6 list some possible ways to choose random samples for each study.

1. A study to determine opinion about a certain television program.

2. A study to detect defective radio resistors.

3. A study of the opinions of people toward Medicare.

4. A study to determine opinions about an election of a U.S. president.

5. A study of the number of savings accounts per family in the United States.

6. A national study of the monthly budget for a family of four.

7. The following is an example of a biased sample: In a study of political party preferences, poorly dressed interviewers obtained a significantly greater proportion of answers favoring Democratic party candidates in their samples than did their well-dressed and wealthier-looking counterparts. Give two more examples of biased samples.

8. In a study of the number of savings accounts per family, a sample of accounts totaling less than $10,000 was taken and, from the owners of these accounts, information about the total number of accounts owned by all family members was obtained. Criticize this sample.

9. It is customary for newsreporters to sample the opinions of a few people to find out how the population at large feels about the events of the day. A reporter questions people on a downtown street corner. Is there anything wrong with such an approach?

10. In 1936 the *Literary Digest* conducted a poll to predict the presidential election. Based

on its poll it predicted the election of Landon over Roosevelt. In the actual election, Roosevelt won. The sample was taken by drawing the mailing list from telephone directories and lists of car owners. What was wrong with the sample?

9.2

ORGANIZATION OF DATA

FREQUENCY TABLE
HISTOGRAM
FREQUENCY POLYGON
CUMULATIVE FREQUENCY DISTRIBUTION

Quite often a study results in a collection of large masses of data. If the data are to be understood and, at the same time, effective, they must be summarized in some manner. Two methods of presenting data are in common use. One method involves a summarized presentation of the numbers themselves according to order in a tabular form; the other involves presenting the quantitative data in pictorial form, such as by using graphs or diagrams.

Example 1

Suppose a random sample of 71 children from a group of 10,000 indicate their weight measurements, as shown in Table 1:

Table 1
Weight Measurements of 71 Students, in Pounds

69	71	71	55	52	55	58	58	58	62	67	94
82	94	95	89	89	104	93	93	58	62	67	62
94	85	92	75	75	79	75	82	94	105	115	104
105	109	94	92	89	85	85	89	95	92	105	71
72	72	79	79	85	72	79	119	89	72	72	69
79	79	69	93	85	93	79	85	85	69	79	

Certain information available from the sample becomes more evident once the data are ordered according to some scheme. If the 71 measurements are written in order of magnitude, we obtain Table 2:

Table 2

52	55	55	58	58	58	58	62	62	62	67	67
69	69	69	69	71	71	71	72	72	72	72	72
75	75	75	79	79	79	79	79	79	79	79	82
82	85	85	85	85	85	85	85	89	89	89	89
89	92	92	92	93	93	93	93	94	94	94	94
94	95	95	104	104	105	105	105	109	115	119	

FREQUENCY TABLE

The data in Table 2 can be presented in a so-called *frequency table.* This is done as follows: Tally marks are used to record the occurrence of the respective scores. Then the *frequency f* with which each score occurs can be determined. In doing this, further information may become evident. See Table 3.

Table 3

Score	Tally	Frequency, f	Score	Tally	Frequency, f
119	/	1	82	//	2
115	/	1	79	//// ///	8
109	/	1	75	///	3
105	///	3	72	////	5
104	//	2	71	///	3
95	//	2	69	////	4
94	////	5	67	//	2
93	////	4	62	///	3
92	///	3	58	////	4
89	////	5	55	//	2
85	//// //	7	52	/	1

A graph representation of the same data may be presented in a *line chart,* which is obtained in the following way: If we let the vertical axis denote the frequency *f* and the horizontal axis denote the score data, we obtain the graph shown in Figure 1.

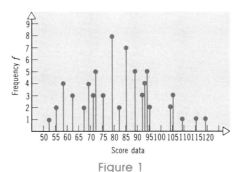

Figure 1

In studying data, a distinction should be made as to whether the data are *discrete* or *continuous.*

Variable; Continuous Variable; Discrete Variable **A measurable characteristic is called a *variable.* If a variable can assume any real value between certain limits, it is called a *continuous variable.* It is called a *discrete variable* if it can assume only a finite set of values or as many values as there are whole numbers.**

Examples of continuous variables are weight, height, length, time, etc. Examples of discrete variables are the number of votes a candidate gets, number of cars sold, etc.

Example 2

A random sample of 71 children from a group of 10,000 indicate their weight measurements, as shown in Table 4 (these are the data from Table 2, but measured more accurately).

Table 4

52.30	55.61	55.71	58.01	58.41	58.51	58.91	62.33	62.50	62.71
67.13	67.23	69.51	69.67	69.80	69.82	71.34	71.65	71.83	72.15
72.22	72.41	72.59	72.67	75.11	75.71	75.82	79.03	79.06	79.09
79.15	79.28	79.32	79.51	79.62	82.32	82.61	85.09	85.13	85.25
85.31	85.41	85.51	85.58	89.21	89.32	89.49	89.61	89.78	92.41
92.63	92.89	93.05	93.19	93.28	93.91	94.17	94.28	94.31	94.52
94.71	95.32	95.51	104.31	104.71	105.21	105.37	105.71	109.34	115.71
119.38									

The data in Table 4 illustrate an example of a continuous variable, whereas the data given in Table 1 are discrete. The difference is in the accuracy of the measuring device. ■

The data in Table 4 are ordered, but are still considered to be *raw data* because they have not yet been subjected to any kind of statistical treatment. To begin to classify the raw data, two decisions must be made:

1. We have to decide on the *number of classes* into which the data are to be grouped.
2. We must decide on the *range of values* each class is to cover.

In grouping any data, experience indicates that we should seldom use fewer than 6 classes or more than 20 classes. This is, of course, not a firm rule and there are exceptions to it.

The size of each class depends to a large extent on the nature of the data, and above all, on the actual number of items within each class. For the data in Table 4, the smallest value is 52.30 and the largest value is 119.38. To use all the data, we have to cover the interval from 52.30 to 119.38.

Range **The *range* of a set of numbers is the difference between the largest and the smallest value in the data under consideration. Thus,**

$$\text{Range} = (\text{Largest value}) - (\text{Smallest value})$$

For the weight measurements of Example 2, the range is

$$119.38 - 52.30 = 67.08$$

HISTOGRAM

Next we would like to represent data in the form of a graph called a *histogram*. To construct a histogram, we first determine the class intervals by dividing the range into equal intervals. We then construct a frequency table. Then we represent each interval by a bar or a rectangle. The height of the rectangle represents the frequency of the interval. We label the bottom of each rectangle by the interval the rectangle represents or the middle of the interval. Usually the rectangles touch, leaving no gaps in the histogram. See Figure 2, representing the histogram for the data given in Table 3 where the classes are 50–59, 60–69, 70–79, and so on.

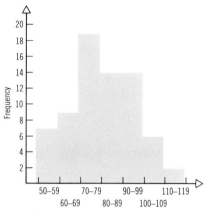

Class	Tally	Frequency
50–59	LHT II	7
60–69	LHT IIII	9
70–79	LHT LHT LHT IIII	19
80–89	LHT LHT IIII	14
90–99	LHT LHT IIII	14
100–109	LHT I	6
110–119	II	2

Figure 2

Next we show how to obtain the histogram for continuous data. We use the data in Table 4. To do this, we must first determine the *class intervals* and the *class limits.* The class intervals for our data will be obtained by dividing the range into equal intervals. Tables 5 and 6 show the use of two different class intervals — one of size 5 and the other of size 10. In choosing intervals of size 5 and 10, we are careful to cover all the scores.

The intervals given in column 2 of Tables 5 and 6 are called *class intervals.*

Table 5

Class	Class Interval	Tally	Frequency
14	114.995–119.995	//	2
13	109.995–114.995		0
12	104.995–109.995	////	4
11	99.995–104.995	//	2
10	94.995– 99.995	//	2
9	89.995– 94.995	⟋⟋⟋ ⟋⟋⟋ //	12
8	84.995– 89.995	⟋⟋⟋ ⟋⟋⟋ //	12
7	79.995– 84.995	//	2
6	74.995– 79.995	⟋⟋⟋ ⟋⟋⟋ /	11
5	69.995– 74.995	⟋⟋⟋ ///	8
4	64.995– 69.995	⟋⟋⟋ /	6
3	59.995– 64.995	///	3
2	54.995– 59.995	⟋⟋⟋ /	6
1	49.995– 54.995	/	1

Table 6

Class	Class Interval	Tally	Frequency
7	109.995–119.995	//	2
6	99.995–109.995	𝖳𝖧𝖫 /	6
5	89.995– 99.995	𝖳𝖧𝖫 𝖳𝖧𝖫 ////	14
4	79.995– 89.995	𝖳𝖧𝖫 𝖳𝖧𝖫 ////	14
3	69.995– 79.995	𝖳𝖧𝖫 𝖳𝖧𝖫 𝖳𝖧𝖫 ////	19
2	59.995– 69.995	𝖳𝖧𝖫 ////	9
1	49.995– 59.995	𝖳𝖧𝖫 //	7

Numbers such as 49.995–59.995 are called *class limits:* 49.995 is the *lower class limit* and 59.995 is the *upper class limit* for the class interval. To avoid confusion, we will always use one decimal place more for class limits than appears in the raw data. Thus, for our data, we choose the class intervals 114.995–119.995, 109.995–114.995, and so on, so that each score could be assigned to one and only one class interval. In performing arithmetic computations with the class limits, we will use the nearest integer value instead of the actual value.

Notice that once raw data are converted to grouped data, it is impossible to retrieve or recover the original data from the frequency table. The best we can do is to choose the midpoint of each class interval as a representative for each class. In Table 5, for example, the actual scores of 105.21, 105.37, 105.71, and 109.34 are viewed as being represented by the midpoint of the class interval 104.995–109.995, namely, $(105 + 110)/2 = 107.500$.

To build a histogram for the data in Table 6, we construct a set of rectangles having as a base the class interval and as height the frequency of occurrence of data in that particular interval. The center of the base is the midpoint of each class interval. See Figure 3.

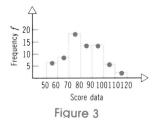

Figure 3

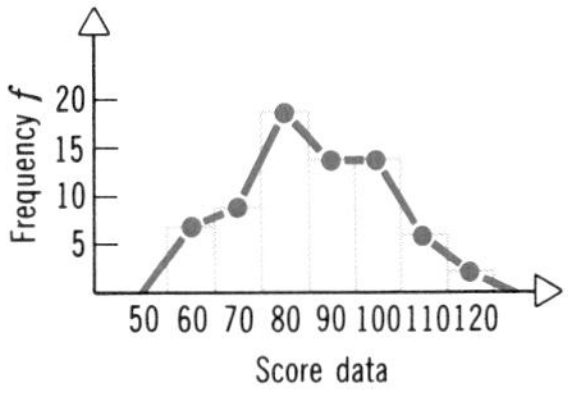

Figure 4

FREQUENCY POLYGON
If we connect all the midpoints of the tops of the rectangles in Figure 3, we obtain a line graph called a *frequency polygon.* (In order not to leave the graph hanging, we always connect it to the horizontal axis on both sides.) See Figure 4.

Sometimes it is useful to learn how many cases fall below (or above) a certain value. For the data of Table 6, we may want to know how many students had weights less than 99.995 or less than 69.995 (or how many had weights more than

59.995 or more than 99.995). If this is the case, we can convert the data as follows: Start at the lowest class interval (49.995–59.995) and note how many scores are below the upper limit of this interval. The number is 7. So we write 7 in the column labeled *cf* of Table 7 in the row for 49.995–59.995. Next, we ask how many scores fall below the upper limit of the next class interval (59.995–69.995); that is, how many scores are below 69.995? The answer is 7 + 9 = 16. The process is continued upward. The top entry of the last column should agree with the total number of scores in the sample. The numbers in this column are called the *cumulative (less than) frequencies.*

Table 7

Class Interval	Tally	f	cf
109.995–119.995	//	2	71
99.995–109.995	7#/ /	6	69
89.995– 99.995	7#/ 7#/ ////	14	63
79.995– 89.995	7#/ 7#/ ////	14	49
69.995– 79.995	7#/ 7#/ 7#/ ////	19	35
59.995– 69.995	7#/ ////	9	16
49.995– 59.995	7#/ //	7	7

CUMULATIVE FREQUENCY DISTRIBUTION

The graph in which the horizontal axis represents class intervals and the vertical axis represents cumulative frequencies is called the *cumulative (less than) frequency distribution.* See Figure 5 for the cumulative frequency distribution for the

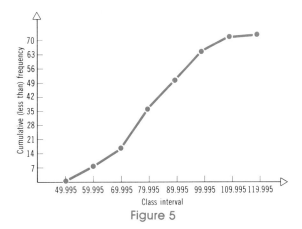

Figure 5

data from Table 7. Notice that the points obtained are connected by lines to aid the interpretation of the graph.

In a similar manner, the *cumulative (more than) frequency distribution* can be obtained.

Exercise 9.2   *Answers to Odd-Numbered Problems begin on page A-25.*

A   1. Consider the data given in the table.

Votes Cast	Number of Precincts
600–649	1
550–599	9
500–549	26
450–499	48
400–449	67
350–399	104
300–349	150
250–299	190
200–249	120
150–199	33
100–149	4
50–99	1
	Total: 753

Distribution of Cleveland Voting Precincts according to total vote cast for governor.
SOURCE: *Ohio Election Statistics,* 1932, pp. 218–242.

With reference to this table, determine the following:

(a) The lower limit of the fifth class
(b) The upper limit of the fourth class
(c) The midpoint of the fifth class
(d) The size of the fifth interval
(e) The frequency of the third class
(f) The class interval having the largest frequency
(g) The number of precincts with less than 600 votes
(h) Construct the histogram
(i) Construct the frequency polygon

2. The following scores were made on a 60-item test:

25	30	34	37	41	42	46	49	53
26	31	34	37	41	42	46	50	53
28	31	35	37	41	43	47	51	54
29	32	36	38	41	44	48	52	54
30	33	36	39	41	44	48	52	55
30	33	37	40	42	45	48	52	

(a) Set up a frequency table for the above data. What is the range?
(b) Draw a line chart for the data.
(c) Draw a histogram for the data using a class interval of size 2.
(d) Draw the frequency polygon for this histogram.
(e) Find the cumulative (less than) frequencies.
(f) Draw the cumulative (less than) frequency distribution.
(g) Find the cumulative (more than) frequencies.
(h) Draw the cumulative (more than) frequency distribution.

3. For Table 5 in the text:

(a) Draw a line chart.
(b) Draw a histogram.
(c) Draw the frequency polygon.
(d) Find the cumulative (less than) frequencies.
(e) Draw the cumulative (less than) frequency distribution.

4. Commercial Bank Earnings According to the *Fortune Directory* (June 15, 1967), the following are the earnings of the 50 largest commercial banks in the United States (as percentage of capital funds for the year 1966):

12.2	9.9	11.2	12.5	9.8
11.5	11.8	11.1	12.3	10.1
11.4	9.2	12.8	9.8	12.6
9.9	10.2	12.6	14.4	10.9
10.2	10.3	11.6	10.2	13.1
10.4	10.9	8.4	14.6	13.4
12.3	11.4	9.2	12.8	11.0
11.2	10.9	10.1	10.9	12.9
11.2	13.2	10.2	16.0	13.6
10.9	11.4	11.6	11.7	13.0

Answer the same questions as in Problem 2, using a class interval of 0.5 beginning with 8.05.

5. Number of Physicians The following are the number of physicians per 100,000 population in 110 selected large American cities in 1962 (*Statistical Abstract of the United States,* 1967):

131	245	145	129	155	232	256	204	296	222
185	166	198	127	153	230	175	161	240	169
169	158	116	171	111	152	126	140	218	142
141	116	176	127	156	185	207	218	153	128
176	162	100	138	129	211	178	198	132	289
165	129	137	78	146	148	145	146	161	119
119	116	245	137	95	169	131	156	136	122
194	113	184	132	172	91	110	188	185	144
105	166	154	108	144	202	212	190	165	128
131	157	115	153	127	224	171	154	149	112
134	190	130	192	123	224	131	190	136	123

Answer the same questions as in Problem 2, using a class interval of 10 beginning with 70.5.

6. High School Dropouts The high school dropout rates in 1986 for the 50 states and the District of Columbia are given below. (Data are from the *1989 World Almanac,* Pharos Books, 1989.) The figures are the percentages of high school students who dropped out during the year. Make a frequency distribution for the data using eight classes of equal width and draw the frequency histogram.

32.7	31.7	37.0	22.0	33.3	26.9	10.2	29.3	43.2
38.0	37.3	29.2	21.0	24.2	28.3	12.5	18.5	31.4
37.3	23.5	23.4	23.3	32.2	8.6	36.7	24.4	12.8
11.9	34.8	26.7	22.4	27.7	35.8	30.0	10.3	19.6
28.4	25.9	21.5	32.7	35.5	18.5	32.6	35.7	19.7
22.4	26.1	24.8	24.8	13.7	18.8			

7. The fiscal year 1986 per capita tax burdens of the 50 states, in dollars, are given on the next page.

(Data are from the *1989 World Almanac*, Pharos Books, 1989.) Make a frequency distribution for the data.

740	3490	975	770	1144	718	1202	1343	780	806
1400	743	848	810	863	777	863	807	940	1047
1314	1019	1163	731	712	755	700	1084	472	1096
989	1278	881	907	843	895	715	898	908	863
570	682	667	820	923	836	1169	964	1148	1569

9.3

OTHER GRAPHICAL TECHNIQUES

PIE CHARTS
BAR GRAPHS

In the preceding section we discussed how frequency distribution line charts and histograms are often used to communicate an overall impression of data. In this section we discuss some other forms of graphs that are of great help in communicating information contained in data.

PIE CHARTS

The *pie chart* is another form of graphical representation of data. The total data is represented by a circle with radii drawn to divide the pie in the same proportions as the categories divide the total data. Each section is labeled with the category name, accompanied by the corresponding percentage.

Since there are 360° in a complete circle, each category is allocated a sector bounded by radii whose angle at the center of the circle is the appropriate percentage of 360°. Each percentage point is equal to an angle of 3.6° (that is, 1% of 360°).

Figure 6 illustrates through two pie charts the changes in the U.S. auto market from 1979 to 1989.

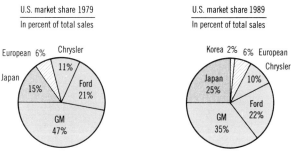

Figure 6

The two pie charts communicate very effectively the growth of the Japanese share of the total auto market (from 15% to 25%) and the decline of GM's share of the market (from 47% to 35%).

To illustrate the procedure of drawing pie charts, consider the data given in Table 8, which indicates the prerecorded music sales in the United States during 1988.

Table 8

Type	Amount (in Billions of Dollars)
Cassette tapes	3.8125
Cassette tapes singles	0.0625
Compact discs	1.1875
12-inch singles	0.125
Long playing	0.875
Singles	0.1875
Total	6.25

It is more convenient to work with percentages than with amounts of money. Thus, we may want to convert the amount of money spent in each category into percentages. We do this as follows:

Relative Frequency

$$\frac{3.8125}{6.25} = 0.61 \text{ or } 61\% \text{ of cassette tapes} \qquad \frac{0.125}{6.25} = 0.02 \text{ or } 2\% \text{ of 12-inch singles}$$

$$\frac{0.0625}{6.25} = 0.01 \text{ or } 1\% \text{ of tapes singles} \qquad \frac{0.875}{6.25} = 0.14 \text{ or } 14\% \text{ of long-playing}$$

$$\frac{1.1875}{6.25} = 0.19 \text{ or } 19\% \text{ of compact discs} \qquad \frac{0.1875}{6.25} = 0.03 \text{ or } 3\% \text{ of singles}$$

Figure 7 illustrates the above data.

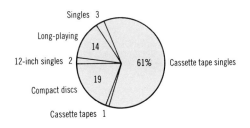

(Source: Chilton Research Services.)

Figure 7

BAR GRAPHS

The *bar graph* is also commonly used to graphically describe data. In such graphs *vertical bars* are usually used. The height of each bar represents the frequency of

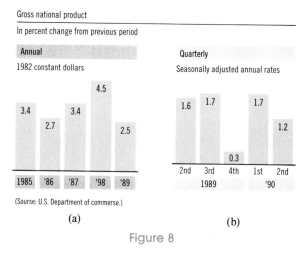

Figure 8

that category. Look at Figure 8*a* illustrating the gross national product (GNP) in percent change from the previous period first on an annual basis and then Figure 8*b* on a quarterly basis.

Sometimes the bars are also drawn horizontally. (See Figure 9). Note that in Figure 9, two bar graphs are drawn on the same scale. Such a graph will enable the U.S. Bureau of Labor Statistics to determine the relationship between working-class men and women.

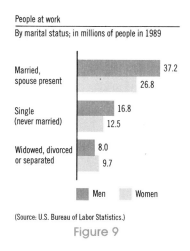

Figure 9

Exercise 9.3   *Answers to Odd-Numbered Problems begin on page A-29.*

A   In Problems 1–2 graph the data in the following table using a bar graph.

1. Average home mortgage rates in percentage (*Source:* Supervision, Resolution Trust Corporation).

1983	1984	1985	1986	1987	1988	1989
12.25	12.1	11.8	10.1	9	9.2	10

2. The table below gives the number of days per year the average American works to pay all federal, state and local taxes (*Source:* Tax Foundation).

1984	1985	1986	1987	1988	1989	1990
117	121	122	124	123	123	125

3. The table below gives the average income of young families for the years 1973 and 1987 headed by a person younger than 30 years old, adjusted for inflation in thousands of 1986 dollars (*Source:* Children's Defense Fund, U.S. Census Bureau). Use a double bar graph to graph the data, that is, a bar graph for each category on the same coordinate system.

1973	White	Black	Hispanic
	21	12	14

1987	White	Black	Hispanic
	16	6	9

4. Graph the data given below by using a double bar graph, that is, a bar graph for each category on the same coordinate system (*Source:* U.S. Department of Commerce).

Top Bank Card Issuers
Outstanding receivables by institution (in billions of dollars)

	1988	1989
Citibank	$21.8	$28.5
Chase Manhattan	7.2	9.3
Sears Discover Card	5.9	8.5
Bank of America	6.0	7.5
First Chicago Corp.	5.5	7.2
American Express Optima	3.8	5.2

5. Graph the data in the table below by using a pie graph: sources of U.S. personal income in percent of $4.06 trillion for 1988 (*Source:* U.S. Department of Commerce).

Wages and Salaries	Interest and Dividends	Social Security Pensions	Self-Employed	Employee Contributions	Small Farm
60	17	9.5	7	5	1

6. The following data show how office workers in Chicago get to work.

Means of Transportation	Percentage
Ride alone	64
Car pool	5
Ride bus	30
Other	1

Construct a pie chart and a bar chart, and compare them to see which one seems more informative to you.

7. To study their attitudes toward a new product, 1000 people were interviewed. Their response is given in the following table.

Attitude	No. of Responses
Do not like	420
Like	360
Like very much	220

Construct a pie chart and a bar chart, and compare them to see which one seems more informative to you.

9.4

MEASURES OF CENTRAL TENDENCY

MEAN
MEDIAN
MODE

The idea of taking an *average* is familiar to practically everyone. Quite often we hear people talk about average salary, average height, average grade, average family, and so on. However, the idea of averages is so commonly used that it should not surprise you to learn that several kinds of averages have been introduced in statistics.

MEAN
Averages are called *measures of central tendency*. The three most common measures of central tendency are the *arithmetic mean, median,* and *mode.*

Mean **The *arithmetic mean*, or *mean*, of a set of real numbers $x_1, x_2, \ldots, x_n$ is denoted by $\overline{X}$ and is defined as**

$$\overline{X} = \frac{x_1 + x_2 + \cdots + x_n}{n} \tag{1}$$

Example 1
The grades of a student on eight examinations were 70, 65, 69, 85, 94, 62, 79, and 100. Find the mean.

Solution
In this example, $n = 8$. The mean of this set of grades is

$$\overline{X} = \frac{70 + 65 + 69 + 85 + 94 + 62 + 79 + 100}{8} = 78$$

∎

An interesting fact about the mean is that the sum of deviations of each item from the mean is zero. In Example 1, the deviation of each score from the mean

$\overline{X} = 78$ is $(100 - 78)$, $(94 - 78)$, $(85 - 78)$, $(79 - 78)$, $(70 - 78)$, $(69 - 78)$, $(65 - 78)$, and $(62 - 78)$. Table 9 lists each score, the mean, and the deviation from the mean. If we add up the deviations from the mean, we obtain a sum of zero.

Table 9

Score	Mean	Deviation from Mean
100	78	22
94	78	16
85	78	7
79	78	1
70	78	-8
69	78	-9
65	78	-13
62	78	-16
		Sum of Deviations: 0

For any group of data, the following result is true:

The sum of the deviations from the mean is zero.

In fact, we could have defined the mean as that real number for which the sum of the deviations is zero.

Another interesting fact about the mean is that if Y is any guessed or assumed mean (which may be any real number) and if we let d_j denote the deviation of each item of the data from the assumed mean, $(d_j = x_j - Y)$, then the actual mean is

$$\overline{X} = Y + \frac{d_1 + d_2 + \cdots d_n}{n} \tag{2}$$

The purpose of introducing equation (2) to find the mean is that, if the numbers to be added in finding the mean are large, (2) can simplify the computation.

Example 2
Find the mean of the following data: 825, 856, 863, 869, 885, and 889.

Solution
Suppose we guess the mean to be 840, then by (2) we have

$$\overline{X} = 840 + \frac{\begin{array}{c}(889 - 840) + (885 - 840) + (869 - 840) \\ + (863 - 840) + (856 - 840) + (825 - 840)\end{array}}{6}$$

$$= 840 + 24.5 = 864.5 \qquad \blacksquare$$

Consider Example 1. We know that the actual mean is 78. Suppose we had guessed the mean to be 52. Then, using formula (2), we obtain

$$\overline{X} = 52 + \frac{\begin{array}{c}(100 - 52) + (94 - 52) + (85 - 52) + (79 - 52) \\ + (70 - 52) + (69 - 52) + (65 - 52) + (62 - 52)\end{array}}{8}$$

$$= 52 + 26 = 78$$

which agrees with what we have already found.

A method for computing the mean for grouped data given in a frequency table is illustrated here by the use of an example.

Example 3
Find the mean for the grouped data given in Table 6.

Solution

1. Take the midpoint (m_i) of each of the intervals as a reference point and enter the result in column 3 of Table 10. For example, the midpoint of the interval 79.995–89.995 is 85.
2. Next, multiply the entry in column 3 by the frequency f_i for that class interval and enter the product in column 4, which is labeled $f_i m_i$.
3. Add the entries in column 4.

The mean is then computed by dividing the sum by the number of entries. That is,

$$\overline{X} = \frac{\Sigma f_i m_i}{n} \tag{3}$$

where

Σ means to add up the entries

f_i = Number of entries in the ith class interval

m_i = Midpoint of ith class interval

n = Number of items

Table 10 displays the information needed to complete the example.

Table 10

Class Interval	f_i	m_i	$f_i m_i$
109.995–119.995	2	115	230
99.995–109.995	6	105	630
89.995– 99.995	14	95	1330
79.995– 89.995	14	85	1190
69.995– 79.995	19	75	1425
59.995– 69.995	9	65	585
49.995– 59.995	7	55	385
$n = 71$			5775 = Sum of $f_i m_i$

Now we can use the data in Table 10 and equation (3) to find the mean:

$$\overline{X} = \frac{5775}{71} = 81.34$$

When data is grouped, the original data are lost due to grouping. As a result, the number obtained by using equation (3) is only an approximation to the actual mean. The reason for this is that using (3) amounts to computing the weighted average midpoint of a class interval (weighted by the frequency of scores in that interval) and therefore cannot be a computation for $\overline{X}$ exactly.

MEDIAN

Median The *median* of a set of real numbers arranged in order of magnitude is the **middle value if the number of items is odd, and it is the mean of the two middle values if the number of items is even.**

Example 3
(a)   The group of data 2, 2, 3, 4, 5, 7, 7, 7, 11 has median 5.
(b)   The group of data 2, 2, 3, 3, 4, 5, 7, 7, 7, 11 has median 4.5, since

$$\frac{4 + 5}{2} = 4.5$$

To find the median for the grouped data in Table 6, page 446, we proceed as follows: The median is that point in the data that will have 50% of the cases above it and 50% below it. Now, 50% of 71 cases is 35.5 cases, so we are interested in finding the point in the distribution with 35 cases above it and 35 below it.

We start by counting up from the bottom until we come as close to 35 cases as possible, but not exceeding it. This brings us through the interval 69.995–79.995. Thus, the median must lie in the interval 79.995–89.995. Now, the median will equal the lower limit of the interval 79.995–89.995, namely, 79.995, plus an *interpolation factor*. The interpolation factor is determined as follows:

---

1. Count the number p of entries or fractional entries needed to reach the median (in our example, the number is 0.5).

2. If the frequency for the interval is q, divide the interval into q parts.

3. The interpolation factor I is

$$I = \frac{p}{q} \cdot i$$

where

p = Number of entries needed to reach the median
q = Number of entries in the interval
i = Size of the interval

The median M is then given by

$$M = (\text{Lower limit of interval}) + (\text{Interpolation factor})$$

For the data of Table 6, the median M is

$$M = 79.995 + \frac{0.5}{14} \cdot 10 = 80.352$$

Again, keep in mind that this median is an approximation to the actual median, since it is obtained from grouped data. If we go back to the original data listed in Table 4, we obtain $M = 82.32$.

The median is sometimes called the *centile point* and is denoted by C_{50} to indicate that 50% of the data are below it and 50% are above it. Similarly, we can find C_{25}, or the first quartile, and C_{75}, or the third quartile.

For the data in Table 10, C_{25} is formed by first finding the class interval containing the tally equal to 25% of all the tallies. Thus, the tally corresponding to C_{25} is found in the class interval 69.995 – 79.995, since

$$25\% \text{ of } 71 = 17.75$$

and 16 tallies lie in the first two class intervals. Using the interpolation factor, we find that

$$C_{25} = 69.995 + \frac{1.75}{19}(10) = 69.995 + 0.921 = 70.916$$

MODE

Mode   **The *mode* of a set of real numbers is the value that occurs with the greatest frequency exceeding a frequency of 1.**

The mode does not necessarily exist, and if it does, it is not always unique. For the data in Table 3, page 443, the mode is 79 (8 is the highest frequency).

Example 5
The group of data 2, 3, 4, 5, 7, 15 has no mode. ■

Example 6
The group of data 2, 2, 2, 3, 3, 7, 7, 7, 11, 15 has two modes 2 and 7, and is called *bimodal*. ■

When data have been listed in a frequency table, the mode is defined as the midpoint of the interval consisting of the largest number of cases. For example, the mode for the data in Table 6, page 446, is 75 (the midpoint of the interval 69.995 – 79.995).

Of the three measures of central tendency considered thus far, the mean is the most important, the most reliable, and the one most frequently used. The reason for this is that it is easy to understand, easy to compute, and uses all the data in the collection. If two samples are chosen from the same population, the two means corresponding to the two samples will not generally differ by as much as the two medians of these samples.

The second most reliable measure is the median. It, too, is easy to understand and usually easy to compute. One advantage of the median over the mean is that it is independent of extreme values.

Exercise 9.4

Answers to Odd-Numbered Problems begin on page A-29.

In Problems 1–6 compute the mean, median, and mode of the given raw data.

A
1. 21, 25, 43, 36

2. 16, 18, 24, 30

3. 55, 55, 80, 92, 70

4. 65, 82, 82, 95, 70

5. 62, 71, 83, 90, 75

6. 48, 65, 80, 92, 80

In Problems 7–8 use the given histograms to determine the mean and median.

7.

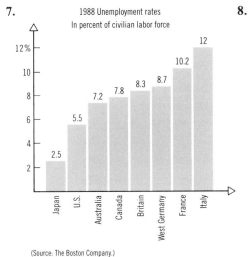

1988 Unemployment rates
In percent of civilian labor force

(Source: The Boston Company.)

8.

Direct investment in the United States.

Top countries investing in billions of dollars for 1988
(includes ownership of at least 10% of a company)

(Source: U.S. Department of Commerce.)

9. If an investor purchased 50 shares of IBM stock at $155 per share, 90 shares at $190 per share, 120 shares at $210 per share, and another 75 shares at $255 per share, what is the average cost per share?

10. If a farmer sells 120 bushels of corn at $2 per bushel, 80 bushels at $2.10 per bushel, 150 bushels at $1.90 per bushel, and 120 bushels at $2.20 per bushel, what is the average income per bushel?

11. The annual salaries of five faculty members in the Mathematics Department at a large university are $34,000, $35,000, $36,000, $36,500, and $55,000. Compute the mean and median. Which measure describes the situation more realistically? If you were among the four lowest-paid members, which measure would you use to describe the situation? What if you were the one making $55,000?

12. For the grouped data in Table 5, page 445, compute the mean, median, and mode.

B
13. The distribution of the monthly earnings of 1155 secretaries in May 1987 in the Chicago metropolitan area is summarized in the table. Find the median salary.

Monthly Earnings	Number of Secretaries
Under $1000	25
$1000–$1250	55
$1250–$1500	325
$1500–$1750	410
$1750–$2000	215
$2000–$2250	75
Over $2250	50

14. According to an article in the October 4, 1988, edition of *The Wall Street Journal,* for companies having fewer than 5000 employees the average sales per employee were as follows:

Size of Company (Number of Employees)	Sales per Employee (Thousands of Dollars)
1–4	112
5–19	128
20–99	127
100–499	118
500–4999	120

Estimate the mean sales per employee for all firms having fewer than 5000 employees.

15. An article by Lester Thurow ("A Surge in Inequality," *Scientific American,* May 1987) presented the following data about the net worth of U.S. families:

	Percentage of Families		
Net Worth ($)	1970	1977	1983
0–4,999	38	35	33
5,000–9,999	6	5	5
10,000–24,999	14	13	12
25,000–49,999	17	17	16
50,000–99,999	13	15	17
100,000–249,999	9	12	12
250,000–499,999	2	2	3
500,000 or more	1	1	2

Estimate the mean net worth of U.S. families in each year, using a class mark of $1 million for the last class.

16. For the data given in Problems 2, 4, and 5 in Exercise 9.2 (page 448), find the mean, median, and mode.

17. Find C_{75} and C_{40} for the grouped data in Tables 5 and 6.

18. In a labor–management wage negotiation in which the laborers are the lowest paid of the workers in the company, which measure of central tendency would labor tend to use as an argument for more pay? Which would management use? Why?

19. For the data given in Problem 1 in Exercise 9.2 (page 448), find the mean, using an assumed mean.

C   **20.** In a frequency table, the score x_1 appears f_1 times, the score x_2 appears f_2 times, . . . , and the score x_n appears f_n times. Show that the mean $\overline{X}$ is given by the formula

$$\overline{X} = \frac{x_1 \cdot f_1 + x_2 \cdot f_2 + \cdots + x_n \cdot f_n}{f_1 + f_2 + \cdots + f_n}$$

9.5

MEASURES OF DISPERSION

RANGE, VARIANCE
STANDARD DEVIATION
CHEBYCHEV'S THEOREM
STANDARD DEVIATION FOR GROUPED DATA

We begin with an example.

Example 1

Consider the following sets of scores:

$$S_1: \quad 4, 6, 8, 10, 12, 14, 16$$
$$S_2: \quad 4, 7, 9, 10, 11, 13, 16$$

Notice that the mean of S_1 and S_2 is 10, and the median of S_1 and S_2 is 10. The scores in each set are different, but those in S_2 seem to be more closely clustered around 10 than those in S_1.

We need a statistical measure to indicate the extent to which the scores in Example 1 are spread out. Such measures are called *measures of dispersion.*

RANGE, VARIANCE

The simplest measure of dispersion is the *range,* which we have already defined as the difference between the largest value and the smallest value. For both S_1 and S_2 the range is $16 - 4 = 12$. We can see that the range is a poor measure of dispersion, since it depends on only the two extreme measures and tells us nothing about the rest of the scores.

Another measure of dispersion is the *deviation from the mean.* Recall that this measure is characterized by the fact that if the deviations from the mean of each score are all added up, the result is zero. Because of this, it is not widely used as a measure of dispersion.

We need a measure that will give us an idea of how much deviation is involved without having these deviations add up to zero. By squaring each deviation from the mean, adding them, and dividing by the number of scores, we obtain an average squared deviation, which is called the *variance* of the set of scores. The formula for the variance, which is denoted by σ^2, is*

$$\sigma^2 = \frac{(x_1 - \overline{X})^2 + (x_2 - \overline{X})^2 + \cdots + (x_n - \overline{X})^2}{n}$$

where $\overline{X}$ is the mean of the scores $x_1, x_2, \ldots, x_n$ and n is the number of scores.

Example 2
Calculate the variance for sets S_1 and S_2 of Example 1

Solution
For S_1, $\overline{X} = 10$ so that

$$\sigma^2 = \frac{\begin{array}{c}(4-10)^2 + (6-10)^2 + (8-10)^2 + (10-10)^2 \\ + (12-10)^2 + (14-10)^2 + (16-10)^2\end{array}}{7}$$

$$= 16$$

For S_2, $\overline{X} = 10$ so that

$$\sigma^2 = \frac{\begin{array}{c}(4-10)^2 + (7-10)^2 + (9-10)^2 + (10-10)^2 \\ + (11-10)^2 + (13-10)^2 + (16-10)^2\end{array}}{7}$$

$$= 13.14$$

∎

STANDARD DEVIATION
In order to apply this measure in practical situations (for instance, if our data represent dollars, we cannot talk about "squared dollars"), we use the square root of the variance. This is called the *standard deviation* of a set of scores. The standard deviation is denoted by σ and is given by the formula

$$\sigma = \sqrt{\frac{(x_1 - \overline{X})^2 + (x_2 - \overline{X})^2 + \cdots + (x_n - \overline{X})^2}{n}}$$

where $\overline{X}$ and the x_i's are defined the same way as for the variance.
For the data in Example 1, the standard deviation for S_1 is

$$\sigma = \sqrt{\frac{36 + 16 + 4 + 0 + 4 + 16 + 36}{7}} = \sqrt{\frac{112}{7}} = \sqrt{16} = 4$$

and the standard deviation for S_2 is

$$\sigma = \sqrt{\frac{36 + 9 + 1 + 0 + 1 + 9 + 36}{7}} = \sqrt{\frac{92}{7}} = \sqrt{13.14} = 3.625$$

The fact that the standard deviation of the set S_2 is less than that for the set S_1 is

---

* The Greek letter sigma.

an indication that the scores of S_2 are more clustered around the mean than those of S_1.

___

Example 3
Find the standard deviation for the data

$$100, 90, 90, 85, 80, 75, 75, 75, 70, 70, 65, 65, 60, 40, 40, 40$$

Solution
The mean is

$$\overline{X} = \frac{100 + 2 \cdot 90 + 85 + 80 + 3 \cdot 75 + 2 \cdot 70 + 2 \cdot 65 + 60 + 3 \cdot 40}{16} = 70$$

The deviations from the mean and their squares are given in Table 11. The standard deviation is

$$\sigma = \sqrt{\frac{4950}{16}} = \frac{70.4}{4} = 17.6$$

Table 11

Scores x	Deviation from the Mean $x - \overline{X}$	Deviation Squared $(x - \overline{X})^2$
100	30	900
90	20	400
90	20	400
85	15	225
80	10	100
75	5	25
75	5	25
75	5	25
70	0	0
70	0	0
65	−5	25
65	−5	25
60	−10	100
40	−30	900
40	−30	900
40	−30	900
Mean = 70 $n = 16$	Sum = 0	Sum = 4950

Table 12

Scores x	Deviation from the Mean $x - \overline{X}$	Deviation Squared $(x - \overline{X})^2$
80	10	100
80	10	100
80	10	100
80	10	100
75	5	25
75	5	25
75	5	25
75	5	25
70	0	0
70	0	0
65	−5	25
65	−5	25
60	−10	100
60	−10	100
55	−15	225
55	−15	225
Mean = 70 $n = 16$	Sum = 0	Sum = 1200

___

Example 4
Find the standard deviation for the data

$$80, 80, 80, 80, 75, 75, 75, 75, 70, 70, 65, 65, 60, 60, 55, 55$$

Solution

Here the mean is $\overline{X} = 70$ for the 16 scores. Table 12 gives the deviations from the mean and their squares. The standard deviation is

$$\sigma = \sqrt{\frac{1200}{16}} = \sqrt{75} = 8.7$$

■

These two examples show that although the samples have the same mean, 70, and the same sample size, 16, the scores in the first set deviate further from the mean than do the scores in the second set.

> In general, a relatively small standard deviation indicates that the measures tend to cluster close to the mean, and a relatively high standard deviation shows that the measures are widely scattered from the mean.

CHEBYCHEV'S THEOREM

Suppose we are observing an experiment with numerical outcomes and that the experiment has mean $\overline{X}$ and standard deviation σ. We wish to estimate the fractions of outcomes that lie within k units of the mean.

> Chebychev's Theorem* For any distribution of numbers with mean $\overline{X}$ and standard deviation σ the probability that a randomly chosen outcome lies between $\overline{X} - k$ and $\overline{X} + k$ is at least $1 - \dfrac{\sigma^2}{k^2}$.

Example 5

Suppose that an experiment with numerical outcomes has mean 4 and standard deviation 1. Use Chebychev's theorem to estimate the probability that an outcome lies between 2 and 6.

Solution

Here $\overline{X} = 4$, $\sigma = 1$. Since we wish to estimate the probability that an outcome lies between 2 and 6, the value of k is $k = 6 - \overline{X} = 6 - 4 = 2$ (or $k = \overline{X} - 2 = 4 - 2 = 2$). Then by Chebychev's theorem the desired probability is at least

$$1 - \frac{\sigma^2}{k^2} = 1 - \frac{1}{2^2} = 1 - \frac{1}{4} = .75$$

That is, we expect at least 75% of the outcomes of this experiment will lie between 2 and 6.   ■

---

* Named after the nineteenth-century Russian mathematician P. L. Chebychev.

Example 6

An office supply company sells boxes containing 100 paper clips. Because of the packaging procedure not every box contains exactly 100 clips. From previous data it is known that the average number of clips in a box is indeed 100 and the standard deviation is 3. If the company ships 10,000 boxes, estimate the number of boxes having between 94 and 106 clips inclusive.

Solution

Our experiment involves counting the number of clips in the box. For this experiment we have $\overline{X} = 100$, and $\sigma = 3$. Therefore, by Chebychev's theorem the fraction of boxes having between $100 - 6$ and $100 + 6$ clips ($k = 6$) should be, at least,

$$1 - \frac{3^2}{6^2} = 1 - \frac{9}{36} = \frac{27}{36} = .75$$

That is, we expect at least 75% of 10,000 boxes, or 7500 boxes to have between 94 and 106 clips. ∎

The importance of Chebychev's theorem stems from the fact that it applies to *any* data—only the mean and standard deviation must be known. However the estimate is a crude one. Other results (such as the *normal distribution* given later) produce more accurate estimates about the probability of falling within k units of the mean.

STANDARD DEVIATION FOR GROUPED DATA

To find the standard deviation for grouped data we use the formula

$$\sigma = \sqrt{\frac{(x_1 - \overline{X})^2 \cdot f_1 + (x_2 - \overline{X})^2 \cdot f_2 + \cdots + (x_n - \overline{X})^2 \cdot f_n}{n}}$$

where $x_1, x_2, \ldots, x_n$ are the class midpoints; $f_1, f_2, \ldots, f_n$ are the respective frequencies; n is the sum of the frequencies, that is, $n = f_1 + f_2 + \cdots + f_n$; and $\overline{X}$ is the mean.

Example 7

Find the standard deviation for the grouped data given in Table 10, page 456.

Solution

We have already found that the mean for the grouped data is

$$\overline{X} = 81.3$$

The class midpoints are 55, 65, 75, 85, 95, 105, and 115. The deviations of the mean from the class midpoints, their squares, and the products of the squares by the respective frequencies are listed in Table 13. The standard deviation is

$$\sigma = \sqrt{\frac{16{,}447.99}{71}} = \sqrt{231.66} = 15.22$$

∎

Table 13

Class Midpoint	f_i	$x_i - \overline{X}$	$(x_i - \overline{X})^2$	$(x_i - \overline{X})^2 \cdot f_i$
115	2	33.7	1,135.69	2,271.38
105	6	23.7	561.69	3,370.14
95	14	13.7	187.69	2,627.66
85	14	3.7	13.69	191.66
75	19	−6.3	39.69	754.11
65	9	−16.3	265.69	2,391.21
55	7	−26.3	691.69	4,841.83
Sum	71			16,447.99

Exercise 9.5

Answers to Odd-Numbered Problems begin on page A-30.

A   1. Use histograms (*a*) and (*b*) to determine by inspection which distribution has the largest variance.

2. Use histograms (*b*) and (*c*) to determine by inspection which distribution has the largest variance.

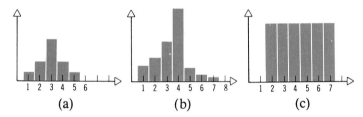

(a)   (b)   (c)

In Problems 3–8 compute the standard deviation for the given raw data.

3. 4, 5, 9, 9, 10, 14, 25   4. 6, 8, 10, 10, 11, 12, 18
5. 62, 58, 70, 70   6. 55, 65, 80, 80, 90
7. 85, 75, 62, 78, 100   8. 92, 82, 75, 75, 82
9. The lifetimes of six light bulbs are 968, 893, 769, 845, 922, and 815 hours. Calculate the mean lifetime and the standard deviation.

B   In Problems 10–11, calculate the mean and the standard deviation for all given data.

10.	Class	Frequency		11.	Class	Frequency
	0–3	2			10–16	1
	4–7	5			17–23	3
	8–11	8			24–30	10
	12–15	6			31–37	12
	16–19	3			38–44	5
					45–51	2

12. **Charge Accounts** A department store takes a sample of its customer charge accounts and finds the following:

Outstanding Balance	Number of Accounts
0–49	15
50–99	41
100–149	80
150–199	60
200–249	8

Find the mean and the standard deviations of the outstanding balances.

13. **Fishing** The number of salmon caught in each of two rivers over the past 15 years is as follows:

River I		River II	
Number Caught	Years	Number Caught	Years
500–1499	4	750–1249	2
1500–2499	8	1350–1799	3
2500–3499	2	1800–2249	4
3500–4499	1	2250–2699	4
		2700–3149	2

Which river should be preferred for fishing?

14. **Test Scores** A group of 25 applicants for admission to Midwestern University made the following scores on the quantitative part of an aptitude test:

591	570	425	472	555
490	415	479	517	570
606	614	542	607	441
502	506	603	488	460
550	551	420	590	482

Find the mean and standard deviation of these scores.

15. Find the standard deviation for the data given in Problem 1, Exercise 9.2.

16. Find the standard deviation for the grouped data given in Table 5 (page 445).

C 17. Suppose that an experiment with numerical outcomes has mean 25 and standard deviation 3. Use Chebychev's theorem to tell what percent of outcomes must lie:

(a) between 19 and 31   (b) between 20 and 30

(c) between 16 and 34   (d) less than 19 or more than 31

(e) less than 16 or more than 34

18. **Quality Control** A watch company determines that the number of defective watches in each box averages 6 with standard deviation 2. Suppose that 1000 boxes are produced. Estimate the number of boxes having between 0 and 12 defective watches.

19. **Sales** The average sale at a department store is $51.25, with a standard deviation of $8.50. Find the smallest interval such that by Chebychev's theorem at least 90% of the store's sales fall within it.

9.6

NORMAL DISTRIBUTION

NORMAL CURVE
STANDARD NORMAL CURVE
NORMAL CURVE TABLE
THE NORMAL CURVE AS AN APPROXIMATION TO THE BINOMIAL DISTRIBUTION

Frequency polygons or frequency distributions can assume almost any shape or form, depending on the data. However, the data obtained from many experiments often follow a common pattern. For example, heights of people, weights of people, test scores, and coin tossing all lead to data which have the same kind of frequency distribution. This distribution is referred to as the *normal distribution* or the *Gaussian distribution.* Because it occurs so often in practical situations, it is generally regarded as the most important distribution, and much statistical theory is based on it. The graph of the normal distribution, called the *normal curve,* is the bell-shaped curve shown in Figure 10.

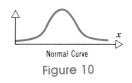

Normal Curve

Figure 10

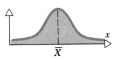

$\overline{X}$

Figure 11

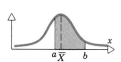

$a\,\overline{X}\ \ b$

Probability between a and b
= area of the shaded region.

Figure 12

NORMAL CURVE

Some properties of the normal curve are listed below.

1. Normal curves are bell-shaped and are symmetrical with respect to a vertical line. See Figure 11.

2. The mean is at the center. See Figure 11.

3. Irrespective of the shape, the area enclosed by the curve and the x-axis is always equal to 1. See the shaded region in Figure 11.

4. The probability that an outcome of a normally distributed experiment is between a and b equals the area under the associated normal curve from $x = a$ to $x = b$. See the shaded region in Figure 12.

5. The standard deviation of a normal distribution plays a major role in describing the area under the normal curve. As shown in Figure 13, the standard deviation is related to the area under the normal curve as follows:

(a) Within 1 standard deviation (from $\overline{X} - \sigma$ to $\overline{X} + \sigma$) is about 68.27% of the total area under the curve.

(b) Within 2 standard deviations (from $\overline{X} - 2\sigma$ to $\overline{X} + 2\sigma$) is about 95.45% of the total area under the curve.

(c) Within 3 standard deviations (from $\overline{X} - 3\sigma$ to $\overline{X} + 3\sigma$) is about 99.73% of the total area under the curve.

CARL FRIEDRICH GAUSS (1777–1855), sometimes called the "prince of mathematicians," made profound contributions to number theory, the theory of functions, probability and statistics. He discovered a way to calculate the orbits of asteroids, made basic discoveries in electromagnetic theory, and invented a telegraph.

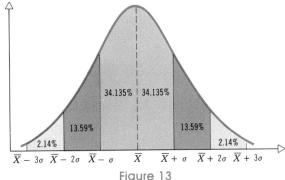

Figure 13

It is also important to recognize that, in theory, the normal curve will never touch the *x*-axis, but will extend to infinity in either direction. In addition, every normal distribution has its mean, median, and mode at the same point.

Example 1

At Jefferson High School, the average IQ score of the 1200 students is 100, with a standard deviation of 15. The IQ scores have a normal distribution.

(a) How many students will have an IQ between 85 and 115?
(b) How many students will have an IQ between 70 and 130?
(c) How many students will have an IQ between 55 and 145?
(d) How many students will have an IQ under 55 or over 145?
(e) How many students will have an IQ over 145?

Solution

(a) Since we are assuming that the IQ scores have a normal distribution, we know that the mean is 100. Since the standard deviation σ is 15, 1σ on either side of the mean is from 85 to 115. By property 5(a) we know that about 68.27% of 1200, or

$$(0.6827)(1200) = 819 \text{ students}$$

have IQ's between 85 and 115.

(b) The scores from 70 to 130 extend 2σ (= 30) on either side of the mean. By property 5(b) we know that about 95.45% of 1200, or

$$(0.9545)(1200) = 1145 \text{ students}$$

have IQ's between 70 and 130.

(c) The scores from 55 to 145 extend 3σ (= 45) on either side of the mean. By property 5(c) we know that about 99.73% of 1200, or

$$(0.9973)(1200) = 1197 \text{ students}$$

have IQ's between 55 and 145.

(d) There are about 3 students (1200 − 1197) who have scores that are not between 55 and 145.

(e)   Half of these 3 students (about 1 or 2 of them) are above 145.

See Figure 14.

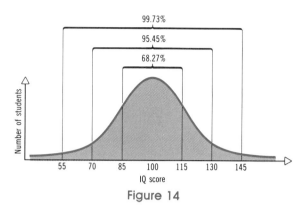

Figure 14   ◼

A normal distribution curve is completely determined by $\overline{X}$ and σ. Hence, different normal distributions of data with different means and standard deviations give rise to different shapes of the normal curve. Figure 15 indicates how the normal curve changes when the standard deviation changes. For the sake of clarity, we assume all data have mean 0.

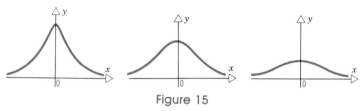

Figure 15

As the standard deviation increases, the normal curve flattens out. A flatter curve indicates a greater likelihood for the outcomes to be spread out. A sharper curve indicates that the outcomes are more likely to be close to the mean.

STANDARD NORMAL CURVE

It would be a hopeless task to attempt to set up separate tables of normal curve areas for every conceivable value of $\overline{X}$ and σ. Fortunately, we are able to transform all the observations to one table — the table corresponding to the so-called *standard normal curve,* which is the normal curve for which $\overline{X} = 0$ and $\sigma = 1$. This can be achieved by introducing new score data, called *Z-scores,* defined as

$$Z = \frac{\text{Distance between } x \text{ and } \overline{X}}{\text{Standard deviation}} = \frac{x - \overline{X}}{\sigma} \tag{1}$$

where

$x =$ Old score data

$\overline{X} =$ Mean of the old data

$\sigma =$ Standard deviation of the old data

The new score data defined by (1) will always have a *zero mean* and a *unit standard deviation*. Such data are said to be expressed in *standard units* or *standard scores*. By expressing data in terms of standard units, it becomes possible to make a comparison of distributions. Furthermore, as for all normal curves, the total area under a standard normal curve is equal to 1.

Example 2

On a test, 80 is the mean and 7 is the standard deviation. What is the Z-score of a score of:

(a) 88?   (b)   62?

Interpret your results.

Solution

(a)   Here, 88 is the regular score. Using (1) with $x = 88$, $\overline{X} = 80$, $\sigma = 7$, we get

$$Z = \frac{x - \overline{X}}{\sigma} = \frac{88 - 80}{7} = \frac{8}{7} = 1.1429$$

(b)   Here, 62 is the regular score. Using (1) with $x = 62$, $\overline{X} = 80$, and $\sigma = 7$, we get

$$Z = \frac{62 - 80}{7} = \frac{-18}{7} = -2.5714$$

The Z-score of 1.1429 tells us that the original score of 88 is 1.1429 standard deviations *above* the mean. See Figure 16. The Z-score of -2.5714 tells us that the original score of 62 is 2.5714 standard deviations *below* the mean.

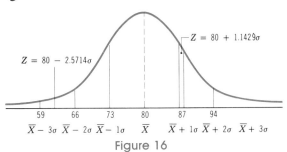

Figure 16

A negative Z-score always means that the score is below the mean. See Figure 16.   ∎

The curve in Figure 16 with mean $\overline{X} = 0$ and standard deviation $\sigma = 1$ is the standard normal curve. For this curve, the areas between $Z = -1$ and 1, $Z = -2$ and 2, and $Z = -3$ and 3 are equal, respectively, to 68.27%, 95.45%, and 99.73% of the total area under the curve, which is 1. To find the areas cut off between other points, we proceed as in the following example.

NORMAL CURVE TABLE

Figure 17

Example 3

Suppose, to begin with, we consider the standard normal curve illustrated in Figure 17. We wish to find the proportion of the area, or the proportion of cases, included between the two points 0.6 and 1.86 units from the mean.

Solution

This problem is worked by using a *normal curve table* (refer to Table 2 in Appendix 2). We begin by checking the table to find the area of the curve cut off between the mean and a point equivalent to a standard score of 0.6 from the mean. This value appears in the second column of the table next to 0.6 and is found to be 0.2257. Next, we continue down the table in the left-hand column until we come to a standard score of 1.8. By looking across the row to the column below 0.06 (column 8), we find that 0.4686 of the area is included between the mean and this point. Then the area of the curve between these two points is the difference between the two areas, $0.4686 - 0.2257$, which is 0.2429. We can then state that approximately 24.29% of the cases fall between 0.6 and 1.86, or that the probability of a score falling between these two points is about .2429.   ■

In the next example, we take two points that are on different sides of the mean.

Figure 18

Example 4

We want to determine what proportion of the area of the normal curve falls between a standard score of -0.39 and one of 1.86 for the standard normal curve given in. Figure 18.

Solution

There are no values for negative standard scores in Table 2 in Appendix 2. Because of the symmetry of normal curves, equal standard scores, whether positive or negative, give equal areas when taken from the mean. From Table 2 we find that a standard score of -0.39 cuts off an area of 0.1517 between it and the mean. A standard score of 1.86 includes 0.4686 of the area of the curve between it and the mean. The area included between both points is then equal to the sum of these two areas, $0.1517 + 0.4686$, which is 0.6203. Thus, approximately 62.03% of the area is between -0.39 and 1.86. In other words, the probability of a score falling between these two points is about 0.6203.   ■

Example 5

A student receives a grade of 82 on a final examination in Biology for which the mean is 73 and the standard deviation is 9. In his final examination in Sociology for which the mean grade is 81 and the standard deviation is 15, he receives an 89. In which examination is his relative standing higher?

Solution

In their present form, these distributions are not comparable, since they have different means and, more important, different standard deviations. To compare the data, we transform the data to Z-scores. For the Biology test data, the Z-score data for the student's examination score is

$$Z = \frac{82 - 73}{9} = \frac{9}{9} = 1$$

For the Sociology test data, the Z-score data for the student's examination score is

$$Z = \frac{89 - 81}{15} = \frac{8}{15} = 0.533$$

This means the student's score in the Biology exam is 1 standard unit above the mean, while his score in the Sociology exam is 0.533 standard units above the mean. Hence, his *relative standing* is higher in Biology.   ■

THE NORMAL CURVE AS AN APPROXIMATION TO THE BINOMIAL DISTRIBUTION

We start with an example.

Example 6

Consider an experiment in which a fair coin is tossed 10 times. Find the frequency distribution for the probability of tossing a head.

Solution

The probability for obtaining exactly k heads is given by a binomial distribution $b(10, k; \frac{1}{2})$. Thus, we obtain the distribution given in Table 14. If we graph this frequency distribution, we obtain the line chart shown in Figure 19. When we connect the tops of the lines of the line chart, we obtain a *normal curve,* as shown.

Table 14

No. of Heads	Probability $b(10, k; \frac{1}{2})$
0	.0010
1	.0098
2	.0439
3	.1172
4	.2051
5	.2461
6	.2051
7	.1172
8	.0439
9	.0098
10	.0010

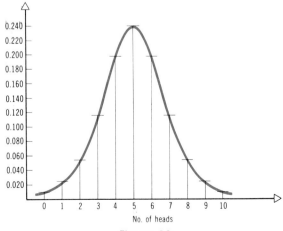

Figure 19   ■

This particular distribution for $n = 10$ and $p = \frac{1}{2}$ is not a result of the choice of n or p. As a matter of fact, the line chart for any binomial probability $b(n, k; p)$ will give an approximation to a normal curve. You should verify this for the cases in which $n = 15$, $p = .3$, and $n = 8$, $p = \frac{3}{4}$.

Probabilities associated with binomial experiments are readily obtainable from the formula $b(n, k; p)$ when n is small. If n is large, we can compute the binomial probabilities by an approximating procedure using a normal curve. It turns out that the normal distribution provides a very good approximation to the binomial distribution when n is large or p is close to $\frac{1}{2}$.

In Chapter 8, Section 6, we have shown that the mean $\bar{X}$ for the binomial distribution is given by $\bar{X} = np$. Moreover, it can be shown that the standard deviation is $\sigma = \sqrt{npq}$.

Example 7

A company manufactures 60,000 pencils each day. Quality control studies have shown that, on the average, 4% of the pencils are defective. A random sample of 500 pencils is selected from each day's production and tested. What is the probability that in the sample there are:

(a) At least 12 and no more than 24 defective pencils?

(b) 32 or more defective pencils?

Solution

(a) Since $n = 500$ is very large, it is appropriate to use a normal curve approximation for the binomial distribution. Thus, with $n = 500$ and $p = .04$,

$$\bar{X} = np = 500(.04) = 20 \qquad \sigma = \sqrt{npq} = \sqrt{500(.04)(.96)} \approx 4.38$$

To find the approximate probability of the number of defective pencils in a sample being at least 12 and no more than 24, we find the area under a normal curve from $x = 12$ to $x = 24$. See Figure 20.

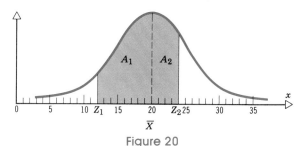

Figure 20

Areas A_1 and A_2 are found as follows:

$$Z_1 = \frac{x - \bar{X}}{\sigma} = \frac{12 - 20}{4.38} = -1.83 \qquad A_1 = 0.4664$$

$$Z_2 = \frac{x - \bar{X}}{\sigma} = \frac{24 - 20}{4.38} = 0.91 \qquad A_2 = 0.3186$$

Total area $= A_1 + A_2 = 0.4664 + 0.3186 = 0.785$

Thus, the approximate probability of the number of defective pencils in the sample being at least 12 and not more than 24 is 0.785.

(b) We want to find the area A_2 indicated in Figure 21. We know that the area to

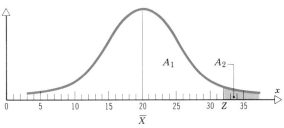

Figure 21

the right of the mean is 0.5, and if we subtract the area A_1 from 0.5, we will obtain A_2. Therefore, we find the area A_1:

$$Z = \frac{x - \overline{X}}{\sigma} = \frac{32 - 20}{4.38} = 2.74 \qquad A_1 = 0.4969$$

Then,

$$A_2 = 0.5 - A_1 = 0.5 - 0.4969 = 0.0031$$

Thus, the approximate probability of finding 32 or more defective pencils in the sample is .0031. ∎

Exercise 9.6

Answers to Odd-Numbered Problems begin on page A-30.

A In Problems 1–4 determine $\overline{X}$ and σ by inspection.

1.
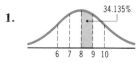

2.
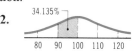

3.
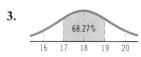

4.
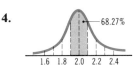

5. Given a normal distribution with a mean of 13.1 and a standard deviation of 9.3, find the Z-score equivalent of the following scores in the distribution:

$$7, 9, 13, 15, 29, 37, 41$$

6. Given a normal distribution with a mean of 15.2 and a standard deviation of 5.1, find the Z-score equivalent of the following scores in the distribution:

$$8, 9, 15, 16, 22, 23, 25$$

In Problems 7–10 use Table 2 in Appendix 2, to find each area of the shaded region under the standard normal curve.

7.

8.

9.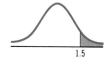

   1.5

10.

   $-0.5\ 0\ 0.5$

11. Given the following Z-scores on a standard normal distribution, find the area from the mean to each score.

   (a) 0.89   (b) 1.10   (c) 2.50   (d) 3.00
   (e) −0.75   (f) −2.31   (g) 0.80   (h) 3.03

B 12. An instructor assigns grades in an examination according to the following procedure:

   A   if score exceeds $\overline{X} + 1.6\sigma$
   B   if score is between $\overline{X} + 0.6\sigma$ and $\overline{X} + 1.6\sigma$
   C   if score is between $\overline{X} - 0.3\sigma$ and $\overline{X} + 0.6\sigma$
   D   if score is between $\overline{X} - 1.4\sigma$ and $\overline{X} - 0.3\sigma$
   F   if score is below $\overline{X} - 1.4\sigma$

   What percentage of each grade does this instructor give, assuming that the scores are normally distributed?

13. The average height of 2000 women in a random sample is 64 inches. The standard deviation is 2 inches. The heights have a normal distribution.

   (a) How many women are between 62 and 66 inches tall?
   (b) How many women are between 60 and 68 inches tall?
   (c) How many women are between 58 and 70 inches tall?

14. Corn flakes come in a box that says it holds a mean weight of 16 ounces of cereal. The standard deviation is 0.1 ounce. Suppose that the manufacturer packages 600,000 boxes with weights that have a normal distribution.

   (a) How many boxes weigh between 15.9 and 16.1 ounces?
   (b) How many boxes weigh between 15.8 and 16.2 ounces?
   (c) How many boxes weigh between 15.7 and 16.3 ounces?
   (d) How many boxes weigh under 15.7 or over 16.3 ounces?
   (e) How many boxes weigh under 15.7 ounces?

15. The weight of 100 college students closely follows a normal distribution with a mean of 130 pounds and a standard deviation of 5.2 pounds.

   (a) How many of these students would you expect to weigh at least 142 pounds?
   (b) What range of weights would you expect to include the middle 70% of the students in this group?

16. In Mathematics 135, the average final grade is 75.0 and the standard deviation is 10.0. The professor's grade distribution shows that 15 students with grades from 68.0 to 82.0 received C's. Assuming the grades follow a normal distribution, how many students are in Mathematics 135?

C 17. Draw the line chart and frequency curve for the probability of k heads in an experiment in which a biased coin is tossed 15 times and the probability that a head occurs is .3 on each toss. [*Hint:* Find $b(15, k; .30)$ for $k = 0, 1, \ldots , 15$.]

18. Follow the same directions as in Problem 17 for an experiment in which a biased coin is tossed 8 times and the probability that heads appears is $\frac{3}{4}$ on each toss.

In Problems 19–24 suppose a binomial experiment consists of 750 trials and the probability of success for each trial is .4. Then

$$\overline{X} = np = 300 \quad \text{and} \quad \sigma = \sqrt{npq} = \sqrt{(750)(.4)(.6)} \approx 13$$

Approximate the probability of obtaining the number of successes indicated by using a normal curve.

19. 285–315

20. 280–320

21. 300 or more

22. 300 or less

23. 325 or more

24. 275 or less

APPLICATIONS

25. Job Screening Caryl, Mary, and Kathleen are vying for a position as secretary. Caryl, who is tested with Group I, gets a score of 76 on her test; Mary, who is tested with Group II, gets a score of 89; and Kathleen, who is tested with Group III, gets a score of 21. If the average score for Group I is 82, for Group II is 93, and for Group III is 24, and if the standard deviations for each group are 7, 2, and 9, respectively, which person has the highest relative standing?

26. Life Expectancy of Clothing If the average life of a certain make of clothing is 40 months with standard deviation of 7 months, what percentage of these clothes can be expected to last from 28 months to 42 months? Assume that clothing lifetime follows a normal distribution.

27. Life Expectancy of Shoes Records show that the average life expectancy of a pair of shoes is 2.2 years with a standard deviation of 1.7 years. A manufacturer guarantees that shoes lasting less than a year are replaced free. For every 1000 shoes sold, how many shoes should the manufacturer expect to replace free?

28. Theater Attendance The attendance over a weekly period of time at a movie theater is normally distributed with a mean of 10,000 and a standard deviation of 1000 persons. Find:

(a) The number in the lowest 70% of the attendance figures

(b) The percentage of attendance figures that falls between 8500 and 11,000 persons

(c) The percentage of attendance figures that differs from the mean by 1500 persons or more

CHAPTER REVIEW

Important Terms and Formulas

population	frequency polygon	measure of dispersion
sample	cumulative (less than) frequency	range
biased sample		variance
random sample	cumulative (more than) frequency	standard deviation
inductive statistics		Chebychev's theorem
frequency table	measure of central tendency	normal distribution
line chart	mean	normal curve, bell-shaped curve
continuous variable	median	
discrete variable	mode	standard normal curve
histogram	deviation from the mean	Z-score
class interval	interpolation factor	standard score
upper class limit	centile point	normal curve table
lower class limit	bimodal	

$$\overline{X} = \frac{x_1 + x_2 + \cdots + x_n}{n}$$

$$\overline{X} = \frac{\Sigma f_i m_i}{n}$$

$$\sigma = \sqrt{\frac{(x_1 - \overline{X})^2 + (x_2 - \overline{X})^2 + \cdots + (x_n - \overline{X})^2}{n}}$$

$$\sigma = \sqrt{\frac{(x_1 - \overline{X})^2 \cdot f_1 + (x_2 - \overline{X})^2 \cdot f_2 + \cdots + (x_n - \overline{X})^2 \cdot f_n}{n}}$$

$$Z = \frac{x - \overline{X}}{\sigma}$$

True–False Questions *(Answers are on page A-31)*

T F 1. The range of a set of numbers is the difference between the standard deviation and the mean.

T F 2. Two sets of scores can have the same mean and median yet be different.

T F 3. A relatively small standard deviation indicates that measures are widely scattered from the mean.

T F 4. The sum of the deviations from the mean is zero.

T F 5. For the normal distribution 68.27% of the total area under the curve is within 2 standard deviations of the mean.

Fill in the Blanks *(Answers are on page A-31)*

1. The three most common measures of central tendency are (a) _____ (b) _____ (c) _____.

2. The square root of the variance is called _____ _____.

3. The graph of the normal distribution has a _____ shape.

4. The formula $\dfrac{x - \overline{X}}{\sigma}$ is called the _____.

5. The formula $1 - \dfrac{\sigma^2}{k^2}$ measures the probability that a randomly chosen variable lies between _____ and _____.

Review Exercises   Answers to Odd-Numbered Problems begin on page A-31.

1. The following scores were made on a math exam:

80	99	82	21	100	55	80	26	78	52
12	73	20	44	72	63	19	85	33	66
78	42	87	90	30	10	48	75	83	77
63	85	69	80	14	87	66	52	17	60
74	70	73	95	89	14	92	8	100	72

    (a) Set up a frequency table for the above data. What is the range?

    (b) Draw a line chart for the data.

    (c) Draw a histogram for the data using a class interval of size 5 beginning with 4.5.

    (d) Draw the frequency polygon for the histogram.

    (e) Find the cumulative (more than) frequencies.

    (f) Find the cumulative (less than) frequencies.

2. The following table gives the percentage of marginal tax rates for married couples filing jointly; rates shown apply to taxable income for 1990; each portion of income is taxed at its own rate; the maximum income subject to the 33% current rate varies by exemptions claimed (*Source:* U.S. Internal Revenue Service).

Income	Current Rates in Percent
0–32,450	15
32,450–78,400	28
78,400–162,770	33
162,770+	28

Graph the data using a bar graph.

3. Use the information in Problem 2 to obtain a pie graph.

4. Find the mean, the median, and the mode for each of the following sets of measurements:

    (a) 12, 10, 8, 2, 0, 4, 10, 5, 4, 4, 8, 0

    (b) 195, 5, 2, 2, 2, 2, 1, 0

    (c) 2, 5, 5, 7, 7, 7, 9, 9, 11

5. In which of the sets of data in Problem 4 is the mean a poor measure of central tendency? Why?

6. Give an example of data in which the preferred measure of central tendency would be the:

    (a) Mean   (b) Mode   (c) Median

7. Give an example of two sets of scores for which the mean is the same and the standard deviation is different.

8. Give one advantage of the standard deviation over the variance. Give an example.

9. In seven different rounds of golf, Joe scores 74, 72, 76, 81, 77, 76, and 73. What is the standard deviation of his scores?

10. A normal distribution has a mean of 25 and a standard deviation of 5.

    (a) What proportion of the scores fall between 20 and 30?

    (b) What proportion of the scores will lie above 35?

11. A set of 600 scores is normally distributed. How many scores would you expect to find:

    (a) Between $\pm 1\sigma$ above the mean?

    (b) Between 1σ and 3σ above the mean?

    (c) Between $\pm \frac{2}{3}\sigma$ of the mean?

12. The average life expectancy of a dog is 14 years, with a standard deviation of about

1.25 years. Assuming that the life spans of dogs are normally distributed, approximately how many dogs will die before reaching the age of 10 years, 4 months?

13. Use Table 2, in Appendix 2, to calculate the area under the normal curve between:

   (a) $Z = -1.35$ and $Z = -2.75$
   (b) $Z = 1.2$ and $Z = 1.75$

14. Bob got an 89 on the final exam in Mathematics and a 79 on the Sociology exam. In the Mathematics class the average grade was 79 with a standard deviation of 5, and in the Sociology class the average grade was 72 with a standard deviation of 3.5. Assuming that the grades in both subjects were normally distributed, in which class did Bob rank higher?

15. Suppose it is known that the number of items produced in a factory has a mean of 40. If the variance of a week's production is known to equal 25, then what can be said about the probability that this week's production will be between 30 and 50?

16. From past experience a teacher knows that the test scores of students taking an examination have a mean of 75 and a variance of 25.

   (a)   What can be said about the probability that a student will score between 65 and 85?

   (b)   How many students would have to take the examination so as to ensure, with probability of at least .9, that the class average would be within 5 of 75?

Mathematical Questions

From Actuary Exams (Answers on page A-33.)

1. *Actuary Exam—Part II*
   Under the hypothesis that a pair of dice are fair, the probability is approximately 0.95 that the number of 7's appearing in 180 throws of the dice will lie within $30 \pm K$. What is the value of K?
   (a) 2   (b) 4   (c) 6   (d) 8   (e) 10

2. *Actuary Exam—Part II*
   If X is normally distributed with mean μ and variance μ^2 and if $\Pr(-4 < X < 8) = 0.9974$, then $\mu =$
   (a) 1   (b) 2   (c) 4   (d) 6   (e) 8

3. *Actuary Exam—Part II*
   A manufacturer makes golf balls whose weights average 1.62 ounces, with a standard deviation of 0.05 ounce. What is the probability that the weight of a group of 100 balls will lie in the interval 162 ± 0.5 ounces?
   (a) 0.18   (b) 0.34   (c) 0.68   (d) 0.84   (e) 0.96

10

APPLICATIONS TO GAMES OF STRATEGY

10.1

INTRODUCTION

TWO PERSON GAMES
STRICTLY DETERMINED GAMES

Game theory, as a branch of mathematics, is a relatively new field. It is concerned with the analysis of human behavior in conflicts of interest. In other words, game theory gives mathematical expression to the strategies of opposing players and offers techniques for choosing the best possible strategy. In most parlor games, it is relatively easy to define winning and losing and, on this basis, to quantify the best strategy for each player. However, game theory is not merely a tool for the gambler so that he or she can take advantage of the odds; nor is it merely a method for winning games like tic-tac-toe, matching pennies, or the Italian game called *Morra* (described in Section 10.3).

TWO PERSON GAMES

Gottfried Wilhelm von Leibniz is generally recognized as being the first to see the relationship between games of strategy and the theory of social behavior. For example, when union and management sit down at the bargaining table to discuss contracts, each has definite strategies open to him. Each will bluff, persuade, and try to discover the other's strategy, while at the same time trying to prevent the discovery of his own. If enough information is known, results from the theory of games can determine what is the best possible rational behavior or the best possible strategy for each player. Another application of game theory can be made to politics. If two people are vying for the same political office, each has open to him or her various campaign strategies. If it is possible to determine the impact of alternate strategies on the voters, the theory of games can be used to find the best strategy (usually the one that gains the most votes, while losing the least votes). Thus, game theory can be used in certain situations of conflict to indicate how people should behave to achieve certain goals. Of course, game theory does not tell us how people actually behave. Game theory is the study, then, of rational behavior in conflict situations.

Two-Person Game **Any conflict or competition between two people is called a *two-person game.***

Let's consider some examples of two-person games.

GOTTFRIED WILHELM von LEIBNIZ (1646–1716), along with Newton is credited with the development of differential and integral calculus. He was a philosopher, lawyer, theologian, and historian and wrote in several languages. His later years were clouded by a controversy over who first discovered calculus, Leibniz or Newton. In fact, both men should be credited since each discovered calculus by different means.

Example 1

In a game similar to matching pennies, Player I picks heads or tails and Player II attempts to guess the choice. Player I will pay Player II $3 if both choose heads; Player I will pay Player II $2 if both choose tails. If Player II guesses incorrectly, he will pay Player I $5. ∎

We use Example 1 to illustrate some terminology. First, since two players are involved, this is a two-person game. Next, notice that no matter what outcome occurs (*HH, HT, TH, TT*), whatever is lost (or gained) by Player I is gained (or lost) by player II. Such games are called *two-person zero-sum games.*

If we denote the gains of Player I by positive entries and his losses by negative entries, we can display this game in a 2×2 matrix as

$$\begin{array}{c} \\ H \\ T \end{array} \begin{array}{cc} H & T \\ \begin{bmatrix} -3 & 5 \\ 5 & -2 \end{bmatrix} \end{array}$$

Each entry a_{ij} of a matrix game is termed a *payoff* and the matrix is called the *game* or *payoff matrix.*

Conversely, any $m \times n$ matrix $A = [a_{ij}]$ can be regarded as the game matrix for a two-person zero-sum game in which Player I chooses any one of the m rows of A and simultaneously Player II chooses any one of the n columns of A. The entry in the row and column chosen is the payoff to player I.

We will assume that the game is played repeatedly, and that the problem facing each player is what choice he should make so that he gains the most benefit. Thus, Player I wishes to maximize his winnings and Player II wishes to minimize his losses. By a strategy of Player I for a given matrix game A, we mean the decision by Player I to select rows of A in some manner.

Example 2

Consider a two-person zero-sum game given by the matrix

$$\begin{bmatrix} 3 & 6 \\ -2 & -3 \end{bmatrix}$$

in which the entries denote the winnings of Player I. The game consists of Player I choosing a row and Player II simultaneously choosing a column, with the intersection of row and column giving the payoff for this play in the game. For example, if Player I chooses row 1 and Player II chooses column 2, then Player I wins $6; if Player I chooses row 2 and Player II chooses column 1, then Player I loses $2.

It is immediately evident from the matrix that this particular game is biased in favor of Player I, who will always choose row 1 since he cannot lose by doing so. Similarly, Player II, recognizing that Player I will choose row 1 will always choose column 1, since his losses are then minimized.

Thus, the *best strategy* for Player I is row 1 and the *best strategy* for Player II is column 1. When both players employ their best strategy, the result is that Player I wins $3. This amount is called the *value* of the game. Notice that the payoff $3 is the minimum of the entries in its row and is the maximum of the entries in its column.   ■

STRICTLY DETERMINED GAMES

A game defined by a matrix is said to be *strictly determined* if and only if there is an entry of the matrix that is the smallest element in its row and is also the largest element in its column. This entry is then called a *saddle point* and is the *value* of the game.

If a game has a positive value, the game favors Player I. If a game has a negative value, the game favors Player II. Any game with a value of 0 is termed a *fair game.*

If a matrix game has a saddle point, it can be shown that the row containing the saddle point is the best strategy for Player I and the column containing the saddle point is the best strategy for Player II. (We will prove this in Section 10.5.) This is why such games are called *strictly determined games.* Such games are also called games of *pure strategy.*

Of course, a matrix may have more than one saddle point, in which case each player has more than one best strategy available. However, the value of the game is always the same no matter how many saddle points the matrix may have. See Problem 12 in Exercise 10.1.

Example 3

Determine whether the game defined by the matrix below is strictly determined.

$$\begin{bmatrix} 3 & 0 & -2 & -1 \\ 2 & -3 & 0 & -1 \\ 4 & 2 & 1 & 0 \end{bmatrix}$$

Solution

First, we look at each row and find the smallest entry in each row:

$$\text{Row 1: } -2 \qquad \text{Row 2: } -3 \qquad \text{Row 3: } 0$$

Next, we check to see if any of the above elements are also the largest in their column. The element -2 in row 1 is not the largest entry in column 3; the element -3 in row 2 is not the largest entry in column 2; however, the element 0 in row 3 is the largest entry in column 4. Thus, this game is strictly determined. Its value is 0, so the game is fair.   ■

The game of Example 3 is represented by a 3×4 matrix. This means that Player I has 3 strategies open to him, while Player II can choose from 4 strategies.

Example 4

Two franchising firms, Alpha Products and Omega Industries, are each planning to add an outlet in a certain city. It is possible for the site to be located either in the center of the city or in a large suburb of the city. If both firms decide to build in the center of the city, Alpha Products will show an annual profit of $1000 more than the profit of Omega Industries. If both firms decide to locate their outlet in the suburb, then it is determined that Alpha Products' profit will be $2000 less than the profit of Omega Industries. If Alpha locates in the suburb and Omega in the city, then Alpha will show a profit of $4000 more than Omega. Finally, if Alpha locates in the city and Omega in the suburb, then Alpha will have a profit of $3000 less than Omega's. Is there a best site for each firm to locate its franchise? In this case, by *best site* we mean the one that produces the most competition against the other firm — not the site that produces the highest gross sales. Of course, someone else may well have a different interpretation of what constitutes the best site.

Solution

If we assign rows as Alpha strategies and columns as Omega strategies and if we use positive entries to denote the gain of Alpha over Omega and negative entries for the gain of Omega over Alpha, then the matrix game for this situation is

$$\text{Alpha} \begin{array}{c} \\ \text{City} \\ \text{Suburb} \end{array} \overset{\overset{\text{Omega}}{\text{City}\quad\text{Suburb}}}{\begin{bmatrix} 1 & -3 \\ 4 & -2 \end{bmatrix}}$$

where the entries are in thousands of dollars.

This game is strictly determined and the saddle point is -2, which is the value of the game. Thus, if both firms locate in the suburb, this results in the best competition. This is so since Omega will always choose to locate in the suburb, guaranteeing a larger profit than Alpha. This being the case, Alpha, to minimize this larger profit of Omega, must always choose the suburb. Of course, the game is not fair, since it is favorable to Omega. ∎

Exercise 10.1   *Answers to Odd-Numbered Problems begin on page A-33.*

A   In Problems 1–4 write the matrix game that corresponds to each two-person conflict situation.

1. Tami and Laura simultaneously each show one or two fingers. If they show the same number of fingers, Tami pays Laura one dime. If they show a different number of fingers, Laura pays Tami one dime.

2. Tami and Laura simultaneously each show one or two fingers. If the total number of fingers shown is even, Tami pays Laura that number of dimes. If the total number of fingers shown is odd, Laura pays Tami that number of dimes.

3. Tami and Laura, simultaneously and independently, each write down one of the numbers 1, 4, or 7. If the sum of the numbers is even, Tami pays Laura that number of dimes. If the sum of the numbers is odd, Laura pays Tami that number of dimes.

4. Tami and Laura, simultaneously and independently, each write down one of the numbers 3, 6, or 8. If the sum of the numbers is even, Tami pays Laura that number of dimes. If the sum of the numbers is odd, Laura pays Tami that number of dimes.

B   In Problems 5–14 determine which of the two-person, zero-sum games are strictly determined. For those that are, find the value of the game. All entries are the winnings of Player I, who plays rows.

5. $\begin{bmatrix} -1 & 2 \\ -3 & 6 \end{bmatrix}$
　　　　　　　　　6. $\begin{bmatrix} 4 & 0 \\ 0 & -1 \end{bmatrix}$

7. $\begin{bmatrix} 4 & 2 \\ 3 & 1 \end{bmatrix}$
　　　　　　　　　8. $\begin{bmatrix} -6 & -1 \\ 0 & 0 \end{bmatrix}$

9. $\begin{bmatrix} 2 & 0 & -1 \\ 3 & 6 & 0 \\ 1 & 3 & 7 \end{bmatrix}$
　　　　　　10. $\begin{bmatrix} 2 & 3 & -2 \\ -2 & 0 & 4 \\ 0 & -3 & -2 \end{bmatrix}$

11. $\begin{bmatrix} 1 & 0 & 3 \\ -1 & 2 & 1 \\ 2 & 2 & 3 \end{bmatrix}$
　　　　　　12. $\begin{bmatrix} 1 & -3 & -2 \\ 2 & 5 & 4 \\ 2 & 3 & 2 \end{bmatrix}$

13. $\begin{bmatrix} 6 & 4 & -2 & 0 \\ -1 & 7 & 5 & 2 \\ 1 & 0 & 4 & 4 \end{bmatrix}$
　　　14. $\begin{bmatrix} 8 & 6 & 4 & 0 \\ -1 & 6 & 5 & -2 \\ 0 & 1 & 3 & 3 \end{bmatrix}$

C　**15.** For what values of a is the matrix below strictly determined?

$$\begin{bmatrix} a & 8 & 3 \\ 0 & a & -9 \\ -5 & 5 & a \end{bmatrix}$$

16. Show that the matrix below is strictly determined for any choice of a, b, or c.

$$\begin{bmatrix} a & a \\ b & c \end{bmatrix}$$

17. Find necessary and sufficient conditions for the matrix below to be strictly determined.

$$\begin{bmatrix} a & 0 \\ 0 & b \end{bmatrix}$$

10.2

MIXED STRATEGIES

MIXED STRATEGY GAMES
EXPECTED PAYOFF

Example 1
Consider a two-person zero-sum game given by the matrix

$$\begin{bmatrix} 6 & 0 \\ -2 & 3 \end{bmatrix}$$

in which the entries denote the winnings of Player I. Is this game strictly determined? If so, find its value.

Solution

We find that the smallest entry in each row is

$$\text{Row 1:} \quad 0 \qquad \text{Row 2:} \quad -2$$

The entry 0 in row 1 is not the largest element in its column; similarly, the entry -2 in row 2 is not the largest element in its column. Thus, this game is not strictly determined. ∎

At this stage, we would like to stress the point that a matrix game is not usually played just once. With this in mind, Player I in Example 1 might decide always to play row 1, since she may win $6 at best and win $0 at worst. Does this mean she should always employ this strategy? If she does, Player II would catch on and begin to choose column 2, since this strategy limits her losses to $0. However, after awhile, Player I would probably start choosing row 2 to obtain a payoff of $3. Thus, in a nonstrictly determined game, it would be advisable for the players to *mix* their strategies rather than to use the same one all the time. That is, a random selection is desirable. Indeed, to make certain that the other player does not discover the pattern of moves, it may be best not to have any pattern at all. For instance, Player I may elect to play row 1 in 40% of the plays (that is, with probability .4), while Player II elects to play column 2 in 80% of the plays (that is, with probability .8). This idea of mixing strategies is important and useful in game theory. Games in which each player's strategies are *mixed* are termed *mixed-strategy games*.

MIXED-STRATEGY GAMES

Suppose we know the probability for each player to choose a certain strategy. What meaning can be given to the term *payoff of a game* if mixed strategies are used? Since the payoff has been defined for a pair of pure strategies and in a mixed-strategy situation we do not know which strategy is being used, it is not possible to define a payoff for a single game. However, in the long run, we do know how often each strategy is being used, and we can use this information to compute the *expected payoff* of the game.

In Example 1, if Player I chooses row 1 in 50% of the plays and row 2 in 50% of the plays, and if Player II chooses column 1 in 30% of the plays and column 2 in 70% of the plays, the expected payoff of the game can be computed. For example, the strategy of row 1, column 1, is chosen $(.5)(.3) = .15$ of the time. This strategy has a payoff of $6, so that the expected payoff will be $(\$6)(.15) = \$0.90$. Table 1 summarizes the entire process. Thus, the expected payoff E of this game, when

Table 1

Strategy	Payoff	Probability	Expected Payoff
Row 1—Column 1	6	.15	$0.90
Row 2—Column 1	−2	.15	−0.30
Row 1—Column 2	0	.35	0.00
Row 2—Column 2	3	.35	1.05
Totals		1.00	$1.65

the given strategies are employed, is $1.65, which makes the game favorable to Player I.

If we look very carefully at the above derivation, we get a clue as to how the expected payoff of a game that is not strictly determined should be defined.

EXPECTED PAYOFF

Let's consider a game defined by the 2×2 matrix

$$A = \begin{bmatrix} a_{11} & a_{12} \\ a_{21} & a_{22} \end{bmatrix}$$

Let the strategy for Player I, who plays rows, be denoted by the row vector

$$P = [p_1 \quad p_2]$$

where p_1 is the probability that he chooses row 1 and p_2 is the probability that he chooses row 2; and the strategy for Player II, who plays columns, be denoted by the column vector

$$Q = \begin{bmatrix} q_1 \\ q_2 \end{bmatrix}$$

where q_1 is the probability that he chooses column 1 and q_2 is the probability that he chooses column 2. The probability that Player I wins the amount a_{11} is $p_1 q_1$. Similarly, the probabilities that he wins the amounts a_{12}, a_{21}, and a_{22} are $p_1 q_2$, $p_2 q_1$, and $p_2 q_2$, respectively. If we denote by $E(P, Q)$ the expectation of Player I, that is, the expected value of the amount he wins when he uses strategy P and Player II uses strategy Q, then

$$E(P, Q) = p_1 a_{11} q_1 + p_1 a_{12} q_2 + p_2 a_{21} q_1 + p_2 a_{22} q_2$$

By using matrix notation, the above can be expressed as

$$E(P, Q) = PAQ$$

In general, if A is an $m \times n$ matrix game, we are led to the following definition:

Expected Payoff **The *expected payoff* E of a two-person zero-sum game, defined by the matrix A, in which the row vector P and column vector Q define the respective strategy probabilities of Player I and Player II is**

$$E = PAQ$$

If a matrix game $A = [a_{ij}]$ of dimension $m \times n$ is strictly determined, then one of the entries is a saddle point. This saddle point can always be placed in the first row and first column by simply rearranging and renumbering the rows and columns of A. The value of the game is then a_{11}, and P and Q are vectors of the form

$$P = [1 \quad 0 \quad 0 \quad 0 \quad 0 \quad \cdots \quad 0] \qquad Q = \begin{bmatrix} 1 \\ 0 \\ \cdot \\ \cdot \\ \cdot \\ 0 \end{bmatrix}$$

where P is of dimension $1 \times m$ and Q is of dimension $n \times 1$.

Example 2

Find the expected payoff of the matrix game

$$A = \begin{bmatrix} 3 & -1 \\ -2 & 1 \\ 1 & 0 \end{bmatrix}$$

if Player I and Player II decide on the strategies

$$P = \begin{bmatrix} \frac{1}{3} & \frac{1}{3} & \frac{1}{3} \end{bmatrix} \qquad Q = \begin{bmatrix} \frac{1}{3} \\ \frac{2}{3} \end{bmatrix}$$

Solution

The expected payoff E of this game is

$$E = PAQ = \begin{bmatrix} \frac{1}{3} & \frac{1}{3} & \frac{1}{3} \end{bmatrix} \begin{bmatrix} 3 & -1 \\ -2 & 1 \\ 1 & 0 \end{bmatrix} \begin{bmatrix} \frac{1}{3} \\ \frac{2}{3} \end{bmatrix}$$

$$= \begin{bmatrix} \frac{2}{3} & 0 \end{bmatrix} \begin{bmatrix} \frac{1}{3} \\ \frac{2}{3} \end{bmatrix} = \frac{2}{9}$$

Thus, the game is biased in favor of Player I and has an expected payoff of $\frac{2}{9}$. ■

Most games are not strictly determined. That is, most games do not give rise to best pure strategies for each player. Examples of games that are not strictly determined are matching pennies (see Example 1, Section 10.1), bridge, poker, and so on. In the next two sections, we discuss techniques for finding optimal strategies for games that are not strictly determined.

Exercise 10.2   *Answers to Odd-Numbered Problems begin on page A-34.*

A   **1.** For the game of Example 1, find the expected payoff E if Player I chooses row 1 in 30% of the plays and Player II chooses column 1 in 40% of the plays.

2. For the game of Example 2, find the expected payoff E if Player I chooses row 1 with probability .3 and row 2 with probability .4, while Player II chooses column one half the time.

In Problems 3–6 find the expected payoff of the game $\begin{bmatrix} 4 & 0 \\ 2 & 3 \end{bmatrix}$ for the given strategies.

3. $P = \begin{bmatrix} \frac{1}{2} & \frac{1}{2} \end{bmatrix}; \quad Q = \begin{bmatrix} \frac{1}{2} \\ \frac{1}{2} \end{bmatrix}$ **4.** $P = \begin{bmatrix} \frac{1}{2} & \frac{1}{2} \end{bmatrix}; \quad Q = \begin{bmatrix} \frac{3}{4} \\ \frac{1}{4} \end{bmatrix}$

5. $P = \begin{bmatrix} \frac{1}{4} & \frac{3}{4} \end{bmatrix}; \quad Q = \begin{bmatrix} \frac{1}{2} \\ \frac{1}{2} \end{bmatrix}$ **6.** $P = \begin{bmatrix} 0 & 1 \end{bmatrix}; \quad Q = \begin{bmatrix} 0 \\ 1 \end{bmatrix}$

B   In Problems 7–10 find the expected payoff of each game.

7. $\begin{bmatrix} 4 & 0 \\ -3 & 6 \end{bmatrix}; \quad P = \begin{bmatrix} \frac{2}{3} & \frac{1}{3} \end{bmatrix}; \quad Q = \begin{bmatrix} \frac{1}{3} \\ \frac{2}{3} \end{bmatrix}$

8. $\begin{bmatrix} 1 & -1 \\ -2 & 3 \end{bmatrix}; \quad P = \begin{bmatrix} \frac{1}{4} & \frac{3}{4} \end{bmatrix}; \quad Q = \begin{bmatrix} \frac{1}{3} \\ \frac{2}{3} \end{bmatrix}$

9. $\begin{bmatrix} 1 & 0 & 0 \\ 0 & 1 & 0 \\ 0 & 0 & 1 \end{bmatrix}$; $P = [\frac{1}{3} \quad \frac{1}{3} \quad \frac{1}{3}]$; $Q = \begin{bmatrix} \frac{1}{3} \\ \frac{1}{3} \\ \frac{1}{3} \end{bmatrix}$

10. $\begin{bmatrix} 4 & -1 & 0 \\ 2 & 3 & 1 \end{bmatrix}$; $P = [\frac{1}{3} \quad \frac{2}{3}]$; $Q = \begin{bmatrix} \frac{2}{3} \\ \frac{1}{6} \\ \frac{1}{6} \end{bmatrix}$

C   **11.** Show that in a 2×2 game

$$\begin{bmatrix} a_{11} & a_{12} \\ a_{21} & a_{22} \end{bmatrix}$$

the only games that are not strictly determined are those for which either

(a) $a_{11} > a_{12}, \quad a_{11} > a_{21}, \quad a_{21} < a_{22}, \quad a_{12} < a_{22}$

or

(b) $a_{11} < a_{12}, \quad a_{11} < a_{21}, \quad a_{21} > a_{22}, \quad a_{12} > a_{22}$

Also show that all others are strictly determined.

10.3

OPTIMAL STRATEGY IN TWO-PERSON ZERO-SUM GAMES WITH 2 × 2 MATRICES

VALUE OF A GAME
APPLICATIONS

We have already seen that the best strategy for two-person zero-sum games that are strictly determined is found in the row and column containing the saddle point. Suppose the game is not strictly determined so that the conditions given in Problem 11, Exercise 10.2, are satisfied.

In 1927, John von Neumann, along with E. Borel, initiated research in the theory of games and proved that, even in nonstrictly determined games, there is a single course of action that represents the best strategy. In practice, this means that to avoid always using a single strategy, a player in a game may instead choose plays randomly according to a fixed probability pattern. This has the effect of making it impossible for the opponent to know what the player will do, since even the player will not know until the final moment. That is, by selecting a strategy randomly according to the laws of probability, the actual strategy chosen at any one time cannot be known even to the one choosing it.

EMIL BOREL (1871–1956) was a prominent French mathematician. In his book, *Le Hasard,* he described the penetration of probabilistic methods into physics, biology, and other branches of science as well as the relationship between probability theory and other branches of mathematics. His pioneering work helped launch the field of measure theory on which the modern notions of length, area, and probability rest. He was a member of the Chamber of Deputies and served as the Minister of the Navy.

For example, in the Italian game of *Morra* each player shows 1, 2, or 3 fingers and simultaneously calls out his guess as to what the sum of his and his opponent's fingers is. It can be shown that if he guesses 4 fingers each time and varies his own moves so that every twelve times he shows 1 finger five times, 2 fingers four times, and 3 fingers three times, he will, at worst, break even (in the long run).

Example 1

Consider the non-strictly determined game

$$A = \begin{bmatrix} 1 & -1 \\ -2 & 3 \end{bmatrix}$$

in which Player I plays rows and Player II plays columns. Determine the optimal strategy for each player.

Solution

If Player I chooses row 1 with probability p, then she must choose row 2 with probability $1 - p$. If Player II chooses column 1, Player I then expects to earn

$$E_\mathrm{I} = (1)p + (-2)(1 - p) = 3p - 2 \qquad (1)$$

Similarly, if Player II chooses column 2, Player I expects to earn

$$E_\mathrm{I} = (-1)p + 3(1 - p) = -4p + 3 \qquad (2)$$

We graph these using E_I as the vertical axis and p as the horizontal axis. See Figure 1a.

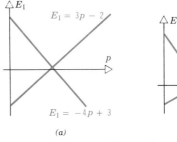

(a)   (b)

Figure 1

Player I wants to maximize her expected earning so she should maximize the minimum expected gain. This occurs when the two lines intersect, since for any other choice of p one or the other of the two expected earnings is less. Thus, solving equations (1) and (2) simultaneously, we obtain

$$3p - 2 = -4p + 3$$
$$7p = 5$$
$$p = \frac{5}{7}$$

The optimal strategy for Player I is therefore

$$P = [\tfrac{5}{7} \quad \tfrac{2}{7}]$$

Similarly, suppose Player II chooses column 1 with probability q (and therefore column 2 with probability $1 - q$. If Player I chooses row 1, Player II's expected earnings are

$$E_{II} = (1)q + (-1)(1 - q) = 2q - 1$$

If Player I chooses row 2, Player II's expected earnings are

$$E_{II} = (-2)(q) + 3(1 - q) = -5q + 3$$

The optimal strategy for Player II occurs when

$$2q - 1 = -5q + 3$$
$$7q = 4$$
$$q = \frac{4}{7}$$

See Figure 1*b*. The optimal strategy for Player II is

$$Q = \begin{bmatrix} \frac{4}{7} \\ \frac{3}{7} \end{bmatrix}$$

The expected payoff E corresponding to these optimal strategies is

$$E = PAQ = \begin{bmatrix} \frac{5}{7} & \frac{2}{7} \end{bmatrix} \begin{bmatrix} 1 & -1 \\ -2 & 3 \end{bmatrix} \begin{bmatrix} \frac{4}{7} \\ \frac{3}{7} \end{bmatrix} = \frac{1}{7}$$ ■

VALUE OF A GAME

Now, consider a two-person zero-sum game given by the 2×2 matrix

$$A = \begin{bmatrix} a_{11} & a_{12} \\ a_{21} & a_{22} \end{bmatrix}$$

in which Player I chooses row strategies and Player II chooses column strategies.

By using the method illustrated above, it can be shown that the optimal strategy for Player I is given by $P = [p_1 \quad p_2]$, where

$$p_1 = \frac{a_{22} - a_{21}}{a_{11} + a_{22} - a_{12} - a_{21}} \qquad p_2 = \frac{a_{11} - a_{12}}{a_{11} + a_{22} - a_{12} - a_{21}} \qquad (3)$$

with $a_{11} + a_{22} - a_{12} - a_{21} \neq 0$. Notice that $p_1 + p_2 = 1$, as expected. Similarly, the optimal strategy for Player II is given by $Q = \begin{bmatrix} q_1 \\ q_2 \end{bmatrix}$, where

$$q_1 = \frac{a_{22} - a_{12}}{a_{11} + a_{22} - a_{12} - a_{21}} \qquad q_2 = \frac{a_{11} - a_{21}}{a_{11} + a_{22} - a_{12} - a_{21}} \qquad (4)$$

with $a_{11} + a_{22} - a_{12} - a_{21} \neq 0$, and $q_1 + q_2 = 1$. The expected payoff E of the game corresponding to these optimal strategies is

$$E = PAQ = \frac{a_{11} \cdot a_{22} - a_{12} \cdot a_{21}}{a_{11} + a_{22} - a_{12} - a_{21}}$$

When optimal strategies are used, the expected payoff E of the game is called the *value V of the game.* (We alert the reader that the formulas for $p_1, p_2, q_1, q_2,$ and V are generally not valid in a game with a saddle point.)

Example 2

For the game matrix

$$\begin{bmatrix} 1 & -1 \\ -2 & 3 \end{bmatrix}$$

determine the optimal strategies for Player I and Player II, and find the value of the game.

Solution

Using formula (3), we have

$$p_1 = \frac{3-(-2)}{1+3-(-1)-(-2)} = \frac{5}{7}$$

$$p_2 = \frac{1-(-1)}{1+3-(-1)-(-2)} = \frac{2}{7}$$

Thus, Player I's optimal strategy is to select row 1 with probability $\frac{5}{7}$ and row 2 with probability $\frac{2}{7}$. Also, by (4), Player II's optimal strategy is

$$q_1 = \frac{3-(-1)}{1+3-(-1)-(-2)} = \frac{4}{7}$$

$$q_2 = \frac{1-(-2)}{1+3-(-1)-(-2)} = \frac{3}{7}$$

Player II's optimal strategy is to select column 1 with probability $\frac{4}{7}$ and column 2 with probability $\frac{3}{7}$. The value V of the game is

$$V = \frac{1 \cdot 3 - (-1)(-2)}{1+3-(-1)-(-2)} = \frac{1}{7}$$

Thus, in the long run, the game is favorable to Player I. ∎

The results obtained in Example 2 are in agreement with those obtained earlier using the graphical technique.

Example 3

Find the optimal strategy for each player, and determine the value of the game given by the matrix

$$\begin{bmatrix} 6 & 0 \\ -2 & 3 \end{bmatrix}$$

Solution

Using the graphical technique or formulas (3) and (4), we find Player I's optimal strategy to be

$$p_1 = \frac{5}{11} \qquad p_2 = \frac{6}{11}$$

Player II's optimal strategy is

$$q_1 = \frac{3}{11} \qquad q_2 = \frac{8}{11}$$

The value V of the game is

$$V = \frac{18}{11} = 1.64$$

Thus, the game favors Player I, whose optimal strategy is $[\frac{5}{11} \quad \frac{6}{11}]$. ∎

APPLICATIONS

Example 4

Election Strategy In a presidential campaign, there are two candidates, a Democrat (D) and a Republican (R), and two types of issues, domestic issues and foreign issues. The units assigned to each candidate's strategy are given in the table. We assume that positive entries indicate a strength for the Democratic candidate, whereas negative entries indicate a weakness. We also assume that a strength of one candidate equals a weakness of the other so that the game is zero-sum. The question is, what is the best strategy for each candidate? What is the value of the game?

		Republican	
		Domestic	**Foreign**
Democrat	Domestic	4	−2
	Foreign	−1	3

Solution

Notice first that this game is not strictly determined. If D chooses to always play foreign issues, then R will counter with domestic issues, in which case D would also talk about domestic issues, in which case, etc., etc. There is no *single* strategy either can use. We compute that the optimal strategy for the Democrat is

$$p_1 = \frac{3 - (-1)}{4 + 3 - (-2) - (-1)} = \frac{4}{10} = .4 \qquad p_2 = \frac{4 - (-2)}{10} = .6$$

The optimal strategy for the Republican is

$$q_1 = \frac{3 - (-2)}{10} = .5 \qquad q_2 = \frac{4 - (-1)}{10} = .5$$

Thus, the best strategy for the Democrat is to spend 40% of her time on domestic issues and 60% on foreign issues, whereas the Republican should divide her time evenly between the two issues.

The value of the game is

$$V = \frac{3 \cdot 4 - (-1)(-2)}{10} = \frac{10}{10} = 1.0$$

Thus, no matter what the Republican does, the Democrat gains at least 1.0 unit by employing her best strategy.   ■

Example 5

War Game In a naval battle, attacking bomber planes are trying to sink ships in a fleet protected by an aircraft carrier with fighter planes. The bombers can attack either high or low, with a low attack giving more accurate results. Similarly, the aircraft carrier can send its fighters at high altitudes or low altitudes to search for the bombers. If the bombers avoid the fighters, credit the bombers with 8 points; if the bombers and fighters meet, credit the bombers with -2 points. Also, credit the bombers with 3 additional points for flying low (since this results in more accurate bombing). Find optimal strategies for the bombers and the fighters. What is the value of the game?

Solution
First, we set up the game matrix. Designate the bombers as playing rows and the fighters as playing columns. Also, each entry of the matrix will denote winnings of the bombers. Then the game matrix is

$$\begin{array}{cc} & \text{Fighters} \\ & \begin{array}{cc} \text{Low} & \text{High} \end{array} \\ \text{Bombers} \begin{array}{c} \text{Low} \\ \text{High} \end{array} & \begin{bmatrix} 1 & 11 \\ 8 & -2 \end{bmatrix} \end{array}$$

The reason for a 1 in row 1, column 1, is that -2 points are credited for the planes meeting, but 3 additional points are credited to the bombers for a low flight.

Next, using formulas (3) and (4), the optimal strategies for the bombers $[p_1 \quad p_2]$ and for the fighters $[\begin{smallmatrix} q_1 \\ q_2 \end{smallmatrix}]$ are

$$p_1 = \frac{-10}{-20} = \frac{1}{2} \qquad p_2 = \frac{-10}{-20} = \frac{1}{2}$$

$$q_1 = \frac{-13}{-20} = \frac{13}{20} \qquad q_2 = \frac{-7}{-20} = \frac{7}{20}$$

The value V of the game is

$$V = \frac{-2 - 88}{-20} = \frac{-90}{-20} = 4.5$$

Thus, the game is favorable to the bombers, if both players employ their optimal strategies.

The bombers can decide whether to fly high or low by flipping a fair coin and flying high whenever heads appear. The fighters can decide whether to fly high or low by using an urn with 13 black balls and 7 white balls. Each day, a ball should be selected at random and then replaced. If the ball is black, they will go low; if it is white, they will go high. ■

Exercise 10.3   *Answers to Odd-Numbered Problems begin on page A-34.*

A   In Problems 1–6 find the optimal strategy for each player and determine the value of each 2×2 game by using graphical techniques. Check your answers by using formulas (3) and (4).

1. $\begin{bmatrix} 1 & 2 \\ 4 & 1 \end{bmatrix}$

2. $\begin{bmatrix} 2 & 4 \\ 3 & -2 \end{bmatrix}$

3. $\begin{bmatrix} -3 & 2 \\ 1 & 0 \end{bmatrix}$

4. $\begin{bmatrix} 3 & -2 \\ -1 & 2 \end{bmatrix}$

5. $\begin{bmatrix} 2 & -1 \\ -1 & 4 \end{bmatrix}$

6. $\begin{bmatrix} 5 & 4 \\ -3 & 7 \end{bmatrix}$

B   7. In Example 4, suppose the candidates are assigned the following weights for each issue:

		Republican	
		Domestic	**Foreign**
Democrat	Domestic	4	-1
	Foreign	0	3

What is each candidate's best strategy? What is the value of the game and whom does it favor?

C   8. Prove formulas (3) and (4).

9. In the matrix game

$$\begin{bmatrix} a_{11} & a_{12} \\ a_{21} & a_{22} \end{bmatrix}$$

what can be said if $a_{11} + a_{22} - a_{12} - a_{21} = 0$?

APPLICATIONS   10. **War Game** For the situation described in Example 5, credit the bomber with 4 points for avoiding the fighters and with -6 points for meeting the fighters. Also, grant the bombers 2 additional points for flying low. What are the optimal strategies and the value of the game? Give instructions to the fighters and bombers as to how they should decide whether to fly high or low.

11. **Spy Game** A spy can leave an airport through two exits, one a relatively deserted exit and the other an exit heavily used by the public. His opponent, having been notified of the spy's presence in the airport, must guess which exit he will use. If the spy and opponent meet at the deserted exit, the spy will be killed; if the two meet at the heavily used exit, the spy will be arrested. Assign a payoff of 30 points to the spy if he avoids his opponent by using the deserted exit and of 10 points to the spy if he avoids his opponent by using the busy exit. Assign a payoff of -100 points to the spy if he is killed and -2 points if he is arrested. What are the optimal strategies and the value of the game?

10.4

OPTIMAL STRATEGY IN OTHER TWO-PERSON ZERO-SUM GAMES USING GEOMETRIC METHODS

DOMINANT/RECESSIVE ROWS AND COLUMNS
$2 \times m$ OR $m \times 2$ MATRIX GAMES
MODEL: CULTURAL ANTHROPOLOGY

DOMINANT/RECESSIVE ROWS AND COLUMNS

Thus far we have discussed how to find optimal strategies for two-person zero-sum games that can be represented only by 2×2 matrices. In this section, we give techniques for finding optimal strategies when the matrix is not 2×2.

We begin with the following definition.

Dominant Row; Recessive Row **If a matrix A contains a row r^* with entries that are all less than or equal to the corresponding entries in some other row r, then row r is said to *dominate* row r^* and r^* is said to be *recessive*.**

Example 1

In the matrix

$$A = \begin{bmatrix} -6 & -3 & 2 & 2 \\ -2 & 0 & 3 & 2 \\ 5 & -2 & 4 & 0 \end{bmatrix}$$

row 1 is dominated by row 2, since each entry in row 1 is less than or equal to its corresponding entry in row 2; that is,

$$-6 < -2 \qquad -3 < 0 \qquad 2 < 3 \qquad 2 = 2$$

If the matrix A of Example 1 were a game in which the entries represent winnings for Player I and if Player I chooses rows, it is clear that Player I would always choose row 2 over row 1, since the values in row 2 always give greater benefit to him than those in row 1. Thus, as far as the matrix representation of this game is concerned, we could represent it by the *reduced matrix*

$$\begin{bmatrix} -2 & 0 & 3 & 2 \\ 5 & -2 & 4 & 0 \end{bmatrix}$$

in which row 1 of matrix A is eliminated, since it would never be chosen.

Dominant Column; Recessive Column **If a matrix A contains a column c^* with entries that are all greater than or equal to the corresponding entries in some other column c, then column c is said to *dominate* column c^* and c^* is said to be *recessive*.**

Example 2
In the matrix

$$A = \begin{bmatrix} -6 & 2 & 4 \\ 4 & 4 & 2 \\ 1 & 3 & -1 \end{bmatrix}$$

column 1 dominates column 2, since each entry in column 2 is greater than or equal to its corresponding entry in column 1; that is,

$$2 > -6 \qquad 4 = 4 \qquad 3 > 1$$ ∎

If the matrix A in Example 2 were a game in which the entries denote winnings for Player I, and if Player II chooses columns, it is clear that Player II would always prefer column 1 over column 2, since the smaller entries indicate lower losses. For this reason, column 2 can be eliminated from matrix A, and instead the reduced matrix below may be used:

$$\begin{bmatrix} -6 & 4 \\ 4 & 2 \\ 1 & -1 \end{bmatrix}$$

Example 3
By eliminating recessive rows and columns, find the reduced form of the matrix

$$A = \begin{bmatrix} -6 & -4 & 2 \\ 2 & -1 & 2 \\ -3 & 4 & 4 \end{bmatrix}$$

Solution

First we look at the rows of the matrix A. Notice that each entry in row 3 (and row 2) is greater than or equal to the corresponding entry in row 1. Thus, row 1 is recessive and can be eliminated. The reduced matrix is

$$\begin{bmatrix} 2 & -1 & 2 \\ -3 & 4 & 4 \end{bmatrix}$$

Neither row 1 nor row 2 in the new matrix is recessive, so we now consider the columns. Notice that each entry in column 3 is greater than or equal to the corresponding entry in column 1 (or column 2). Thus, column 3 is recessive and can be eliminated. The reduced matrix is

$$\begin{bmatrix} 2 & -1 \\ -3 & 4 \end{bmatrix}$$ ∎

The above example shows how a 3×3 matrix game can sometimes be reduced to a 2×2 matrix by eliminating recessive rows and recessive columns.

Example 4
Find the optimal strategy for each player, and find the value of the two-person, zero-sum game

$$\begin{bmatrix} -6 & -4 & 2 \\ 2 & -1 & 2 \\ -3 & 4 & 4 \end{bmatrix}.$$

in which the entries denote the winnings of Player I, who chooses rows, and in which each player has three possible strategies.

Solution
By eliminating recessive rows and columns, this matrix reduces to

$$\begin{bmatrix} 2 & -1 \\ -3 & 4 \end{bmatrix}$$

Using formulas (3) and (4) on page 492, we find that the optimal strategy for Player I is

$$p_1 = \frac{7}{10} \qquad p_2 = \frac{3}{10}$$

and the optimal strategy for Player II is

$$q_1 = \frac{5}{10} \qquad q_2 = \frac{5}{10}$$

The value of the game is

$$V = \frac{5}{10}$$

Thus, the game given in this example is favorable to Player I, and his best strategy is to choose row 2 in 70% of the plays and row 3 in 30% of the plays (row 1 is recessive). Player II's best strategy is to choose column 1 in 50% of the plays and column 2 in 50% of the plays (column 3 is recessive). ■

2 × m OR m × 2 MATRIX GAMES
Suppose we now consider two-person zero-sum games with matrix representations that are $2 \times m$ or $m \times 2$ ($m > 2$) matrices, that are not strictly determined, and that contain no recessive rows or columns. For a $2 \times m$ matrix game, Player I has two strategies and Player II has m strategies; for an $m \times 2$ matrix game, Player I has m strategies and Player II has 2 strategies.

We begin with the following example to illustrate how to find optimal strategies.

Example 5
Find the optimal strategy for each player in the 2×3 game

$$\begin{bmatrix} 4 & -1 & 0 \\ -1 & 4 & 2 \end{bmatrix}$$

in which entries denote winnings for Player I. What is the value of this game?

Solution

In the above game, Player I has two strategies and Player II has three strategies. Suppose p is the probability that Player I plays row 1. Then $1 - p$ is the probability that row 2 is played. Now let's compute the expected earnings of Player I in terms of p.

If Player II elects to play column 1, then the expected earnings E_I of Player I are equal to $4p - 1 \cdot (1 - p)$, or

$$\text{(1)} \ E_I = 5p - 1$$

Similarly, if Player II selects column 2 or column 3, the expected earnings for Player I are, respectively,

$$\text{(2)} \ E_I = 4 - 5p \qquad \text{(3)} \ E_I = 2 - 2p$$

Next, we graph each of these three straight lines measuring E_I along the y-axis and p along the x-axis, and we look at the situation from Player II's point of view. Player II wants to make Player I's earnings as small as possible, since then he maximizes his own earnings. Thus, Player II will always choose the lowest strategy (line), since the height of each line measures winnings of Player I. In other words, Player II's best strategy lies along the darkened line segments in Figure 2.

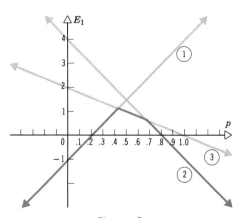

Figure 2

Player I, realizing this, will choose the value of p that yields the most earnings for him. This value occurs at the intersection of the lines

$$\text{(1)} \ E_I = 5p - 1 \qquad \text{(3)} \ E_I = 2 - 2p$$

Their intersection is where

$$p = \frac{3}{7} \qquad E_I = \frac{8}{7}$$

Thus, the optimal strategy for Player I is to choose row 1 in $\frac{3}{7}$ of the plays and row 2 in $\frac{4}{7}$ of the plays. The value of the game in this case is $\frac{8}{7}$.

To find the optimal strategy for Player II, notice that Player I's optimal strategy comes from earnings calculated by using columns 1 and 3 of the matrix game. The matrix that results by eliminating column 2 from the matrix is

$$\begin{bmatrix} 4 & 0 \\ -1 & 2 \end{bmatrix}$$

The optimal strategy for Player II can now be found by formula (4), page 492. It is

$$q_1 = \frac{2}{7} \qquad q_2 = 0 \qquad q_3 = \frac{5}{7}$$

Thus, Player II's strategy is to play column 1 in $\frac{2}{7}$ of the plays and column 3 in $\frac{5}{7}$ of the plays. Since column 2 is eliminated, it is never played. ∎

Example 6

Find the optimal strategy for each player in the 5×2 matrix game

$$\begin{bmatrix} -2 & 2 \\ -1 & 1 \\ 2 & 0 \\ 3 & -1 \\ 4 & -2 \end{bmatrix}$$

in which the entries denote winnings for Player I. What is the value of this game?

Solution

Here, Player II has two strategies. Let q be the probability that he chooses column 1, so that $1 - q$ is the probability that he chooses column 2. Player I's earnings E_I are then

① $E_I = -2q + 2(1 - q)$ ② $E_I = -q + (1 - q)$ ③ $E_I = 2q$
  $\quad = -4q + 2$ $\quad = -2q + 1$
④ $E_I = 3q - (1 - q)$ ⑤ $E_I = 4q - 2(1 - q)$
  $\quad = 4q - 1$ $\quad = 6q - 2$

for rows 1–5, respectively. We graph these five linear equations in Figure 3.

Player I may select any of the five strategies represented by the lines in Figure 3. Since the height of each line represents his earnings, he will employ strategies that carry him along the darkened line segments in Figure 3.

But Player II wants the earnings of Player I to be as small as possible. This occurs at the intersection of ① and ③, where $q = \frac{1}{3}$, $E_I = \frac{2}{3}$. Thus, the optimal strategy for Player II is to choose column 1 in $\frac{1}{3}$ of the plays and column 2 in $\frac{2}{3}$ of the plays. The value of the game is $\frac{2}{3}$, and it is favorable to Player I.

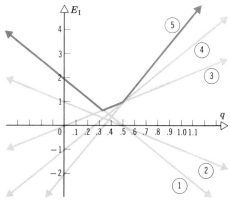

Figure 3

Now, to find Player I's optimal strategy, we notice that Player II's optimal strategy comes from lines ① and ③. If we eliminate rows 2, 4, and 5 from the matrix above, we obtain the matrix

$$\begin{bmatrix} -2 & 2 \\ 2 & 0 \end{bmatrix}$$

Applying formula (3), page 492, Player I's optimal strategy is

$$p_1 = \frac{-2}{-6} = \frac{1}{3} \qquad p_3 = \frac{-4}{-6} = \frac{2}{3}$$

The following is an example from a paper by J. D. Williams.* In this example, unlike those previously given, the optimal solution calls for a diversification of investments rather than a random choice among them.

Example 7

Investment Strategy An investor plans to invest $10,000 during a period of international uncertainty as to whether there will be peace, a continuation of the cold war, or an actual war. Her investment can be made in government bonds, armament stocks, or industrial stocks. The game is a struggle between the investor and nature. The matrix below gives the rate of interest for each player's strategy.

		Hot war	Cold war	Peace
	Government bonds	2	3	3.2
Investor	Armament stocks	18	6	−2
	Industrial stocks	2	7	12

Calculate the investor's optimal strategy.

---

* J. D. Williams, *La Stratégie dans les Actions Humaines,* Dunod, Paris, 1956.

Solution

First we look at the matrix to see if there is any row dominance or column dominance. Notice that row 3 dominates row 1 so that the reduced matrix for this game is

$$
\begin{array}{c}
\\
\text{Armament stocks} \\
\text{Industrial stocks}
\end{array}
\begin{array}{ccc}
\text{Hot war} & \text{Cold war} & \text{Peace} \\
\begin{bmatrix} 18 & 6 & -2 \\ 2 & 7 & 12 \end{bmatrix}
\end{array}
$$

This is a 2×3 matrix that can be solved by the graphing method. The optimal strategy for the investor is

$$p_1 = 0 \qquad p_2 = \frac{5}{17} \qquad p_3 = \frac{12}{17}$$

The value of the game is $V = 6.7$.

Thus, the investor is assured of a return of at least 6.7% when she invests $\frac{5}{17} = 29\%$ in armament stocks and $\frac{12}{17} = 71\%$ in industrial stocks. In the event of a hot war, the return is

$$18\left(\frac{5}{17}\right) + 2\left(\frac{12}{17}\right) = 6.7\%$$

In the event of a cold war, the return is

$$6\left(\frac{5}{17}\right) + 7\left(\frac{12}{17}\right) = 6.7\%$$

In the event of peace, the return is

$$(-2)\left(\frac{5}{17}\right) + 12\left(\frac{12}{17}\right) = 7.9\%$$

∎

Example 8

War Game General White's army and the enemy are each trying to occupy three hills. General White has three regiments and the enemy has two regiments. A hill is occupied when one force has more regiments present than the other force; if both try to occupy a hill with the same number of regiments, a draw results. How should the troops be deployed to gain maximum advantage?

Solution

We denote the three strategies available to General White as follows:

   3: All three regiments used together to attack one hill.

   2,1: Two regiments used together and one used by itself to attack two hills.

   1,1,1: All three regiments used separately to attack three hills.

White's opponent has two strategies available, namely:

2:   The two regiments used together to defend one hill.

1,1:   The two regiments used separately to defend two of the hills.

White will play rows, and the entries in the game matrix will denote White's expected winnings based on the rule that when White takes a hill, he earns 1 point; when a draw results, he earns 0 points; and when he is defeated, he loses 1 point. Also, for each division that is overpowered, 1 point is earned. Table 2 shows the points won (or lost) by General White for all possible deployments of his regiments. Notice that the order of deployment is quite important.

For example, if White deploys his regiments as 3,0,0 and his opponent uses the deployment 2,0,0, then White captures Hill I, winning 1 point, and overpowers two regiments, winning 2 points, for a total score of 3 points. If White uses 0,2,1 and his opponent uses 0,1,1, then Hill I is a standoff, White wins Hill II and overpowers one regiment, and Hill III is a draw. Here, White has a total score of 2 points.

Table 2

	2,0,0	0,2,0	0,0,2	1,1,0	1,0,1	0,1,1
3,0,0	3	0	0	1	1	-1
0,3,0	0	3	0	1	-1	1
0,0,3	0	0	3	-1	1	1
2,1,0	1	-1	1	2	2	0
2,0,1	1	1	-1	2	2	0
1,2,0	-1	1	1	2	0	2
0,2,1	1	1	-1	2	0	2
1,0,2	-1	1	1	0	2	2
0,1,2	1	-1	1	0	2	2
1,1,1	0	0	0	1	1	1

To determine the game matrix, we proceed as follows: If White uses a 3 deployment and his enemy uses a 2 deployment, then White expects to score 3 points $\frac{1}{3}$ of the time and score 0 points $\frac{2}{3}$ of the time. We assign an expected payoff to White of $3 \cdot \frac{1}{3} + 0 \cdot \frac{2}{3} = 1$ point in this case. If White uses a 2,1 deployment and his enemy uses a 1,1 deployment, then White expects to gain 2 points $\frac{2}{3}$ of the time and 0 points $\frac{1}{3}$ of the time for an expected payoff of $\frac{4}{3}$ points. The game matrix can be written as

$$
\begin{array}{cc}
 & \text{Enemy} \\
 & \begin{array}{cc} 2 & 1,1 \end{array} \\
\text{White}\ \begin{array}{c} 3 \\ 2,1 \\ 1,1,1 \end{array} &
\begin{bmatrix} 1 & \frac{1}{3} \\ \frac{1}{3} & \frac{4}{3} \\ 0 & 1 \end{bmatrix}
\end{array}
$$

Notice that this matrix can be reduced, since row 2 dominates row 3. The reduced matrix is

$$
\begin{array}{cc}
 & \text{Enemy} \\
 & \begin{array}{cc} 2 & 1,1 \end{array} \\
\text{White} \quad \begin{array}{c} 3 \\ 2,1 \end{array} &
\begin{bmatrix} 1 & \frac{1}{3} \\ \frac{1}{3} & \frac{4}{3} \end{bmatrix}
\end{array}
$$

This matrix is not strictly determined. The optimal (mixed) strategy for General White is

$$
p_1 = \frac{1}{\frac{7}{3} - \frac{2}{3}} = .6 \qquad p_2 = \frac{\frac{2}{3}}{\frac{5}{3}} = .4 \qquad p_3 = 0
$$

The optimal strategy for the enemy is

$$
q_1 = \frac{1}{\frac{5}{3}} = .6 \qquad q_2 = \frac{\frac{2}{3}}{\frac{5}{3}} = .4
$$

The value of the game is

$$
V = \frac{\frac{4}{3} - \frac{1}{9}}{\frac{5}{3}} = \frac{\frac{11}{9}}{\frac{5}{3}} = \frac{11}{15}
$$

The game is favorable to General White, who should deploy his troops in a 3 strategy 60% of the time and in a 2,1 strategy 40% of the time. Since no one hill is more likely to be chosen for attack than any other, it follows that each hill should be attacked by all three regiments 20% of the time. Furthermore, for the 2,1 deployment, each possible selection of the hills to receive 0,1, or 2 regiments (6 in all) will be used $\frac{40}{6} = 6.67\%$ of the time. ∎

MODEL: CULTURAL ANTHROPOLOGY

In 1960, Davenport* published an analysis of the behavior of Jamaican fishermen. Each fishing crew is confronted with a three-choice decision of fishing in the inside banks, the outside banks, or a combination of inside–outside banks. Fairly reliable estimates can be made of the quantity and quality of fish caught in these three areas under the two conditions that current is present, or not present.

If we take the village as a whole as one player and the environment as another player, we have the components for a two-person zero-sum game, in which the village has three strategies (inside, inside–outside, outside) and the environment has two strategies (current, no current). Davenport computed an estimate of income claimed by the fishermen using each of the alternatives. This estimate is given in matrix form by

$$
\begin{array}{ccc}
 & & \text{Environment} \\
 & & \begin{array}{cc} \text{Current} & \text{No current} \end{array} \\
\begin{array}{c} \\ \text{Village} \\ \\ \end{array}
\begin{array}{c} \text{Inside} \\ \text{Inside–Outside} \\ \text{Outside} \end{array}
& \begin{bmatrix} 17.3 & 11.5 \\ 5.2 & 17.0 \\ -4.4 & 20.6 \end{bmatrix} & \quad (1)
\end{array}
$$

* E. Davenport, "Jamaican Fishing: A Game Theory Analysis in Papers on Caribbean Anthropology," Yale University Publication in Anthropology, Nos. 57–64, 1960.

Here the environment has two strategies. Let q be the probability of current, so that $1 - q$ is the probability of no current. If E_I represents the villagers' expected earnings, then

① $E_I = 17.3q + 11.5(1 - q)$ ② $E_I = 5.2q + 17(1 - q)$
   $= 5.8q + 11.5$ $= -11.8q + 17$

③ $E_I = -4.4q + 20.6(1 - q)$
   $= -25q + 20.6$

Figure 4 shows that the optimal strategy of the environment comes from the intersection of lines ① and ②.

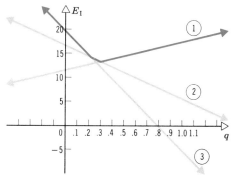

Figure 4

Computing this intersection, we obtain

$$q = .31 \quad \text{and} \quad 1 - q = .69$$

To obtain the optimal strategy of the village, we note that the optimal strategy of the environment comes from lines ① and ②. If we eliminate row 3 from the matrix in (1), we find

$$\begin{bmatrix} 17.3 & 11.5 \\ 5.2 & 17.0 \end{bmatrix}$$

Applying formula (3), page 492, the village's optimal strategy is

Table 3

	Observed	Predicted
Outside	0	0
Inside	.69	.67
Inside–Outside	.31	.33
Current	.25	.31
No Current	.75	.69

$$p_1 = \frac{17.0 - 5.2}{34.3 - 16.7} = \frac{11.8}{17.6} = .67 \qquad p_2 = \frac{5.8}{17.6} = .33$$

Table 3 compares the observed frequency of strategy usage as compared with the optimal usage as predicted by the game.

Exercise 10.4   *Answers to Odd-Numbered Problems begin on page A-35.*

A   In Problems 1 – 12, find the optimal strategy for each player, where Player I plays rows and entries denote winnings of Player I. What is the value of each game?

1. $\begin{bmatrix} 8 & 3 & 8 \\ 6 & 5 & 4 \\ -2 & 4 & 1 \end{bmatrix}$

2. $\begin{bmatrix} 3 & -1 & 0 \\ -2 & 1 & -1 \end{bmatrix}$

3. $\begin{bmatrix} 2 & 1 & 0 & 6 \\ 3 & -2 & 1 & 2 \end{bmatrix}$

4. $\begin{bmatrix} -1 & 1 \\ 5 & -3 \\ 1 & -2 \\ -2 & 5 \end{bmatrix}$

5. $\begin{bmatrix} 6 & -4 & 2 & -3 \\ -4 & 6 & -5 & 7 \end{bmatrix}$

6. $\begin{bmatrix} 3 & -2 & 2 \\ -1 & 1 & 0 \end{bmatrix}$

7. $\begin{bmatrix} 4 & -5 & 5 \\ -6 & 3 & 3 \\ 2 & -6 & 3 \end{bmatrix}$

8. $\begin{bmatrix} -5 & -4 & -3 & 2 & 3 \\ 3 & 2 & 1 & -2 & -4 \end{bmatrix}$

9. $\begin{bmatrix} 1 & 3 & 0 \\ 0 & -3 & 1 \\ 0 & 4 & 1 \\ -2 & 1 & 1 \end{bmatrix}$

10. $\begin{bmatrix} 6 & -4 \\ 4 & -3 \\ 1 & 0 \\ -3 & 2 \\ -5 & 4 \end{bmatrix}$

11. $\begin{bmatrix} 4 & 3 & -1 \\ 1 & 1 & 4 \\ 1 & 0 & 2 \end{bmatrix}$

12. $\begin{bmatrix} 3 & 2 & 0 \\ 1 & 5 & -2 \\ 0 & 1 & 1 \end{bmatrix}$

APPLICATIONS   13. **Surveillance** In a department store, one area (A) is usually very crowded and the other area (B) is usually relatively empty. The store employs two detectives and has closed-circuit television (T) to control pilferage. The television covers A and B and the detectives can be in either area A or area B or watching the television (T). The matrix below gives an estimate of the probability of the detectives finding and arresting a thief:

		Thief	
		A	B
Detectives	TT	.51	.75
	AA	.64	.36
	BB	.19	.91
	TA	.58	.60
	TB	.37	.85
	AB	.56	.76

Here, TT means both detectives are at the television, TA means the first detective is at the television and the second is in area A, and so on. Find the optimal strategy for the thief and the detectives. What is the value of the game?

14. Effectiveness of Antibiotics This problem is adapted from J. D. Williams.* Three antibiotics, A_1, A_2, and A_3, and five types of bacilli, M_1, M_1, M_3, M_4, and M_5, are involved in a study of the effectiveness of antibiotics on bacilli, with A_1 having a probability .3 of destroying M_1, and so on, as given below. Without knowing the proportion in which these germs are distributed during an epidemic, in what ratio should the antibiotics be mixed to have the greatest probability of being effective?

$$
\begin{array}{cc}
 & \text{Bacilli} \\
 & \begin{array}{ccccc} M_1 & M_2 & M_3 & M_4 & M_5 \end{array} \\
\text{Antibiotics} \quad \begin{array}{c} A_1 \\ A_2 \\ A_3 \end{array} & \left[\begin{array}{ccccc} .3 & .4 & .5 & 1 & 0 \\ .2 & .3 & .6 & 0 & 1 \\ .1 & .5 & .3 & .1 & 0 \end{array} \right]
\end{array}
$$

10.5

SOLVING NONSTRICTLY DETERMINED MATRIX GAMES USING LINEAR PROGRAMMING

LINEAR PROGRAMMING AND 2 × 2 GAMES
SIMPLEX SOLUTION FOR m × n GAMES

In this section we introduce a method for solving a nonstrictly determined matrix game without dominant row or recessive column by converting the game into a linear programming problem, which in turn can be solved by using the methods of Chapters 3 and 4. (Reviewing these chapters may be very helpful). We start with a 2 × 2 game; later on we will develop procedures to solve arbitrary m × n matrix games.

Consider a nonstrictly determined matrix game with all positive entries:

$$
A = \text{Player I} \quad \begin{array}{c} \text{Player II} \\ \left[\begin{array}{cc} a & b \\ c & d \end{array} \right] \end{array} \tag{1}
$$

We require that all the entries of matrix A be positive or zero (we will use this fact later on in our discussion). If some, or all of the entries of A are negative, we can make them positive or zero by making use of the following theorem:

> If every entry in a matrix game A is increased by an amount k, then the value of the game is also increased by k and the optimum strategies remain the same.

For example, consider the matrix game

---

* J. D. Williams, *La Stratégie dans les Actions Humaines,* Dunod, Paris, 1956.

$$A = \begin{bmatrix} -2 & 1 \\ 3 & -4 \end{bmatrix}$$

if we add 5 to each entry of A we obtain

$$A_1 = \begin{bmatrix} 3 & 6 \\ 8 & 1 \end{bmatrix}$$

a matrix game with positive entries only. Now if the value of this new game is V_1 then the value of the original game A is $V = V_1 - 5$, and the strategies for P and Q stay the same.

From Section 10.2 we know that if $P = [p_1 \, p_2]$ and $Q = \begin{bmatrix} q_1 \\ q_2 \end{bmatrix}$ are the respective strategy probabilities of Player I and Player II, then the expected payoff of the game is

$$E = PAQ \tag{2}$$

Now Player I is interested in finding strategy P^*, together with the largest possible value of V such that

$$P^*AQ \geq V \tag{3}$$

for any choice of strategy Q by Player II.

Now if Player II uses strategy $Q_1 = \begin{bmatrix} 1 \\ 0 \end{bmatrix}$ or $Q_2 = \begin{bmatrix} 0 \\ 1 \end{bmatrix}$, then (3) can be written as two inequalities.

$$[p_1 \, p_2] \begin{bmatrix} a & b \\ c & d \end{bmatrix} \begin{bmatrix} 1 \\ 0 \end{bmatrix} \geq V \quad \text{and} \quad [p_1 \, p_2] \begin{bmatrix} a & b \\ c & d \end{bmatrix} \begin{bmatrix} 0 \\ 1 \end{bmatrix} \geq V$$

If we carry out the matrix multiplication we obtain

$$ap_1 + cp_2 \geq V \qquad bp_1 + dp_2 \geq V$$

Thus the value of the game is the largest number V for which

$$\begin{aligned} ap_1 + cp_2 &\geq V \\ bp_1 + dp_2 &\geq V \end{aligned} \tag{4}$$

In other words, an optimal strategy P^* for Player I is a solution of the following linear programming problem.

Maximize V

subject to
$$\begin{aligned} ap_1 + cp_2 &\geq V & p_1 + p_2 &= 1 \\ bp_1 + dp_2 &\geq V & p_1 &\geq 0; \ p_2 \geq 0 \end{aligned} \tag{5}$$

In order for the linear programming system to conform to the form we used in Chapters 3 and 4, we perform the following operations.

First we divide the inequalities by V (the sense of the inequality does not change, since we know that V is positive because the entries in the matrix game are positive) to obtain

$$a\frac{p_1}{V} + c\frac{p_2}{V} \geq 1$$

$$b\frac{p_1}{V} + d\frac{p_2}{V} \geq 1 \tag{6}$$

To further simplify the inequalities in (6) we make the following substitution:

$$x_1 = \frac{p_1}{V} \qquad x_2 = \frac{p_2}{V} \quad x_1 \geq 0 \qquad x_2 \geq 0 \tag{7}$$

The system of inequalities in (6) takes the form

$$ax_1 + cx_2 \geq 1$$

$$bx_1 + dx_2 \geq 1 \tag{8}$$

Now, because $p_1 + p_2 = 1$ and $p_1, p_2, \geq 0$, we see that

$$x_1 + x_2 = \frac{p_1}{V} + \frac{p_2}{V} = \frac{p_1 + p_2}{V} = \frac{1}{V}$$

Thus

$$x_1 + x_2 = \frac{1}{V} \qquad \text{or} \qquad V = \frac{1}{x_1 + x_2} \tag{9}$$

Observe that maximizing V is the same as minimizing $\frac{1}{V}$.

Let

$$x = \frac{1}{V}$$

The linear programming in problem (5) becomes

$$\text{Minimize} \quad x = x_1 + x_2 \tag{10}$$

subject to

$$ax_1 + cx_2 \geq 1$$

$$bx_1 + dx_2 \geq 1$$

$$x_1 \geq 0, x_2 \geq 0$$

Note the coefficients of x_1 and x_2 in (10) are simply the entries of the matrix game A. This linear programming problem can be solved by using the techniques of Chapter 3. Once the values for x_1, x_2, and x are found, we can obtain the values of V, p_1, and p_2 from the relations

$$V = \frac{1}{x} \qquad p_1 = x_1 V \qquad \text{and} \qquad p_2 = x_2 V$$

Now we turn our attention to Player II who is interested in coming up with a strategy Q^* and the smallest value V such that

$$PAQ^* \leq V \tag{11}$$

for any choice of strategy P by Player I. In particular, inequality (11) is true if Player II chose the strategy

$$p_1 = [1 \quad 0] \quad \text{or} \quad p_2 = [0 \quad 1]$$

Substituting p_1 and p_2 into (11) and carrying out the matrix multiplication we obtain

$$\begin{aligned} aq_1 + bq_2 &\leq V \\ cq_1 + dq_2 &\leq V \end{aligned} \qquad (12)$$

Divide through by V (which is positive) to obtain

$$a\frac{q_1}{V} + b\frac{q_2}{V} \leq 1$$

$$c\frac{q_1}{V} + d\frac{q_2}{V} \leq 1$$

Introduce the new variables

$$y_1 = \frac{q_1}{V} \quad \text{and} \quad y_2 = \frac{q_2}{V} \quad y_1, y_2 \geq 0 \qquad (13)$$

to obtain

$$\begin{aligned} ay_1 + by_2 &\leq 1 \\ cy_1 + dy_2 &\leq 1 \quad y_1, y_2 \geq 0 \end{aligned} \qquad (14)$$

Since $q_1 + q_2 = 1$, we get

$$\begin{aligned} y_1 + y_2 &= \frac{q_1}{V} + \frac{q_2}{V} \\ &= \frac{q_1 + q_2}{V} \\ &= \frac{1}{V} \end{aligned}$$

Thus,

$$y_1 + y_2 = \frac{1}{V} \text{ or } V = \frac{1}{y_1 + y_2}$$

Note that minimizing V is the same as maximizing $\frac{1}{V}$. Thus, we have the following linear programming problem:

$$\text{Maximize} \quad y = y_1 + y_2 \qquad (15)$$

subject to

$$\begin{aligned} ay_1 + by_2 &\leq 1 \\ cy_1 + dy_2 &\leq 1 \\ y_1, y_2 &\geq 0 \end{aligned}$$

This linear programming problem can be solved by using geometric techniques for y_1, y_2, and y. Then the values q_1, q and V can be obtained by using (13).

If you compare the system in (10) with the one given in (15) you will notice that one is the dual of the other and, therefore, by the duality property stated in Chapter 4 the optimal solution V found by using (10) is the same as the one found for (15).

Example 1

Use linear programming techniques to solve the following matrix game

$$A = \begin{bmatrix} 1 & -2 \\ -3 & 4 \end{bmatrix}$$

Solution

The sequence of steps leading to the solution is as follows:

Step 1. Since A contains negative entries, we convert it by adding 4 to each entry in A:

$$A_1 = \begin{bmatrix} 5 & 2 \\ 1 & 8 \end{bmatrix}$$

Step 2. The two linear programming problems corresponding to (10) and (15) are

Minimize $x = x_1 + x_2$ Maximize $y = y_1 + y_2$

subject to $5x_1 + 1x_2 \geq 1$ subject to $5y_1 + 2y_2 \leq 1$
$2x_1 + 8x_2 \geq 1$ $1y_1 + 8y_2 \leq 1$
$x_1, x_2 \geq 0$ $y_1, y_2 \geq 0$

Step 3. We will use the geometric method to solve each linear programming system. See Figure 5.

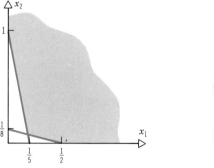 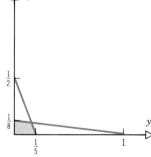

Figure 5

The solution for each problem occurs at the

Vertex	Minimize $x = x_1 + x_2$	Vertex	Maximize $y = y_1 + y_2$
$(0, 1)$	1	$(0, 0)$	0
$(\frac{7}{38}, \frac{3}{38})$	$\frac{5}{19}$	$(0, \frac{1}{8})$	$\frac{1}{8}$
$(\frac{1}{2}, 0)$	$\frac{1}{2}$	$(\frac{3}{19}, \frac{2}{19})$	$\frac{5}{19}$
		$(\frac{1}{5}, 0)$	$\frac{1}{5}$

Minimum occurs at $\qquad$ Maximum occurs at

$$x_1 = \tfrac{7}{38} \quad x_2 = \tfrac{3}{38} \qquad\qquad y_1 = \tfrac{3}{19} \quad y_2 = \tfrac{2}{19}$$

Step 4. To find the value V_1 for the matrix game A_1, we use the solution just obtained:

$$V_1 = \frac{1}{x_1 + x_2} \qquad\qquad V_1 = \frac{1}{y_1 + y_2}$$

$$= \frac{1}{\dfrac{7}{38} + \dfrac{3}{38}} \qquad\qquad = \frac{1}{\dfrac{3}{19} + \dfrac{2}{19}}$$

$$= \frac{19}{5} \qquad\qquad = \frac{19}{5}$$

$$p_1 = V_1 x_1 = \frac{19}{5}\frac{7}{38} = \frac{7}{10} \qquad\qquad q_1 = V_1 y_1 = \frac{19}{5}\frac{3}{19} = \frac{3}{5}$$

$$p_2 = V_1 x_2 = \frac{19}{5}\frac{3}{38} = \frac{3}{10} \qquad\qquad q_2 = V_2 y_2 = \frac{19}{5}\frac{2}{19} = \frac{2}{5}$$

Since the optimal strategies are the same for A and A_1 we get

$$P^* = [p_1\, p_2] = \begin{bmatrix} \frac{7}{10} & \frac{3}{10} \end{bmatrix} \qquad Q^* = \begin{bmatrix} q_1 \\ q_2 \end{bmatrix} = \begin{bmatrix} \frac{3}{5} \\ \frac{2}{5} \end{bmatrix}$$

and the value of matrix game A is

$$V = V_1 - 4 = \frac{19}{5} - 4 = -\frac{1}{5}$$

Step 5. To verify that the value of the matrix game defined by A is indeed $-\frac{1}{5}$ the student should show that

$$P^* A Q^* = -\tfrac{1}{5} \qquad\qquad\qquad \blacksquare$$

The steps leading to a solution of a 2×2 matrix game using linear programming are summarized as follows:

To find

$$p^* = [p_1\, p_2], \qquad Q^* = \begin{bmatrix} q_1 \\ q_2 \end{bmatrix}, \qquad \text{and} \qquad V$$

for the nonstrictly determined game

$$A = \begin{bmatrix} a & b \\ c & d \end{bmatrix}$$

we proceed as follows:

Step 1. If necessary, add a suitable constant k to each entry of the matrix A to obtain a matrix

$$A_1 = \begin{bmatrix} a_1 & b_1 \\ c_1 & d_1 \end{bmatrix} \quad \begin{matrix} a_1 = a + k & b_1 = b + k \\ c_1 = c + k & d_1 = d + k \end{matrix}$$

in which all entries are positive. The strategies P^* and Q^* for the new matrix game are the same. The value of the new game is V_1 and $V = V_1 - k$.

Step 2. Set up the two linear programming problems:

Minimize $\quad x = x_1 + x_2$

subject to
$$a_1 x_1 + c_1 x_2 \geq 1$$
$$b_1 x_1 + d_1 x_2 \geq 1$$
$$x_1, x_2 \geq 0$$

Maximize $\quad y = y_1 + y_2$

subject to
$$a_1 y_1 + b_1 y_2 \leq 1$$
$$c_1 y_1 + d_1 y_2 \leq 1$$
$$y_1, y_2 \geq 0$$

Step 3. Use the geometric method on the two problems listed in Step 2 to solve for

$$x_1, x_2 \quad \text{and} \quad y_1, y_2.$$

Step 4. Use the solutions in Step 3 to find

$$V_1 = \frac{1}{x_1 + x_2} \quad \text{or} \quad V_1 = \frac{1}{y_1 + y_2}$$

$$P^* = [p_1 p_2] = [V_1 x_1 \ V_1 x_2] \quad Q^* = \begin{bmatrix} q_1 \\ q_2 \end{bmatrix} = \begin{bmatrix} V_1 y_1 \\ V_1 y_2 \end{bmatrix} \quad V = V_1 - k$$

Step 5. Check your solutions by using the formula

$$P^* A Q^* = V$$

SIMPLEX SOLUTION OF AN $m \times n$ MATRIX GAME

The method just illustrated for 2×2 matrix games generalizes to any matrix game. We observed that to find the solution we need to solve a linear programming problem and its dual. In general, for an $m \times m$ matrix game the simplex method is used. The advantage of using this method is that, since the two linear programming problems are mutually dual, the solution of one automatically gives the solution to the other. We list the steps needed to solve a 2×3 matrix game; the generalization to an $m \times n$ matrix requires the same steps.

Simplex Solution of a 2×3 Matrix Game
To find

$$P^* = [p_1 \, p_2], \qquad Q^* = \begin{bmatrix} q_1 \\ q_2 \\ q_3 \end{bmatrix}, \qquad \text{and} \qquad V$$

for the nonstrictly determined game

$$A = \begin{bmatrix} a & b & c \\ d & e & f \end{bmatrix}$$

we proceed as follows:

Step 1. If necessary, add a suitable constant k to each entry of the matrix A to obtain a matrix

$$A_1 = \begin{bmatrix} a_1 & b_1 & c_1 \\ d_1 & e_1 & f_1 \end{bmatrix}$$

in which all entries are positive or zero. The strategies P^* and Q^* for the new matrix game are the same. The value of the new game is V_1 and $V = V_1 - k$.

Step 2. Set up the linear programming problem and its dual

Minimize $\quad x = x_1 + x_2$ Maximize $\quad y = y_1 + y_2 + y_3$

subject to $a_1 x_1 + d_1 x_2 \geq 1$ subject to $a_1 y_1 + b_1 y_2 + c_1 y_3 \leq 1$

$\qquad\qquad b_1 x_1 + e_1 x_2 \geq 1$ $\qquad\qquad d_1 y_1 + e_1 y_2 + f_1 y_3 \leq 1$

$\qquad\qquad c_1 x_1 + f_1 x_2 \geq 1$ $\qquad\qquad\qquad y_1, y_2, y_3 \geq 0$

$\qquad\qquad\qquad x_1, x_2 \geq 0$

Step 3. Use the simplex method to solve the maximum problem. (The solution to the minimization problem is automatically achieved because of the principle of duality.)

Step 4. Use the solution in Step 3 to find

$$V_1 = \frac{1}{x_1 + x_2} \qquad \text{or} \qquad V_1 = \frac{1}{y_1 + y_2 + y_3}$$

$$P^* = [p_1 \, p_2] = [V_1 x_1 \quad V_1 x_2] \qquad Q^* = \begin{bmatrix} q_1 \\ q_2 \\ q_3 \end{bmatrix} = \begin{bmatrix} V_1 y_1 \\ V_1 y_2 \\ V_1 y_3 \end{bmatrix}$$

$$V = V_1 - k$$

Step 5. Check your solution by using the formula

$$P^* A Q^* = V$$

Example 2

Solve this matrix game using the simplex method.

$$A = \begin{bmatrix} 1 & 5 & -1 \\ 3 & -1 & 4 \end{bmatrix}$$

Solution

Step 1. Since there are negative entries in this matrix, we add a suitable constant $k = 3$ to each entry of the given matrix to obtain

$$A_1 = \begin{bmatrix} 4 & 8 & 2 \\ 6 & 2 & 7 \end{bmatrix}$$

Step 2. Apply the simplex method to

$$\text{Maximize} \quad y = y_1 + y_2 + y_3$$

subject to

$$4y_1 + 8y_2 + 2y_2 \le 1$$
$$6y_1 + 2y_2 + 7y_3 \le 1$$

Thus we obtain the following sequence of tableaux

$$\begin{bmatrix} & & & s_1 & s_2 & \\ 4 & \boxed{8} & 2 & 1 & 0 & 1 \\ 6 & 2 & 7 & 0 & 1 & 1 \\ \hline -1 & -1 & -1 & 0 & 0 & 0 \end{bmatrix}$$

$$\begin{bmatrix} \frac{1}{2} & 1 & \frac{1}{4} & \frac{1}{8} & 0 & \frac{1}{8} \\ 5 & 0 & \boxed{\frac{13}{2}} & -\frac{1}{4} & 1 & \frac{3}{4} \\ \hline -\frac{1}{2} & 0 & -\frac{3}{4} & \frac{1}{8} & 0 & \frac{1}{8} \end{bmatrix}$$

$$\begin{bmatrix} \frac{4}{13} & 1 & 0 & \frac{7}{52} & -\frac{1}{26} & \frac{5}{52} \\ \frac{10}{13} & 0 & 1 & -\frac{1}{26} & \frac{2}{13} & \frac{3}{26} \\ \hline \frac{1}{13} & 0 & 0 & \frac{5}{52} & \frac{3}{26} & \frac{11}{52} \end{bmatrix}$$

Step 3. From the final tableau we read the optimal solution:

$$y_1 = 0 \qquad y_2 = \tfrac{5}{52} \qquad y_3 = \tfrac{3}{26} \qquad y = \tfrac{11}{52}$$

and

$$x_1 = \tfrac{5}{52} \qquad x_2 = \tfrac{3}{26}$$

Step 4. Use the solutions in Step 3 to find the value V_1 and the optimal strategies:

$$V_1 = \frac{1}{y} = \tfrac{52}{11}$$

$$P^* = [p_1 \, p_2] = [V_1 x_1 \quad V_1 x_2] = [\tfrac{5}{11} \quad \tfrac{6}{11}]$$

$$Q^* = \begin{bmatrix} q_1 \\ q_2 \\ q_3 \end{bmatrix} = \begin{bmatrix} V_1 y_1 \\ V_1 y_2 \\ V_1 y_3 \end{bmatrix} = \begin{bmatrix} 0 \\ \frac{5}{11} \\ \frac{6}{11} \end{bmatrix}$$

$$V = V_1 - 3 = \tfrac{19}{11}$$

Step 5. Verify your answer by checking

$$P^*AQ^* = \begin{bmatrix} \frac{5}{11} & \frac{6}{11} \end{bmatrix} \begin{bmatrix} 1 & 5 & -1 \\ 3 & -1 & 4 \end{bmatrix} \begin{bmatrix} 0 \\ \frac{5}{11} \\ \frac{6}{11} \end{bmatrix}$$

$$= \tfrac{19}{11} = V \qquad\qquad\qquad ■$$

We conclude this section by showing that for a 2×3 matrix

$$A = \begin{bmatrix} a & b & c \\ d & e & f \end{bmatrix}$$

with a saddle point at a, the saddle-point row and column are optimal strategies for the two players. We first note that because of the saddle point, we have $a \leq b$, $a \leq c$, and $a \geq d$. In the initial tableau

$$\left[\begin{array}{ccc|cc|c} ⓐ & b & c & 1 & 0 & 1 \\ d & e & f & 0 & 1 & 1 \\ \hline -1 & -1 & -1 & 0 & 0 & 0 \end{array} \right]$$

the entry a can be chosen as pivot, yielding a second tableau:

$$\left[\begin{array}{ccc|cc|c} 1 & \dfrac{b}{a} & \dfrac{c}{a} & \dfrac{1}{a} & 0 & \dfrac{1}{a} \\[2ex] 0 & e - \dfrac{d}{a}b & f - \dfrac{d}{a}c & -\dfrac{d}{a} & 1 & 1 - \dfrac{d}{a} \\[2ex] \hline 0 & -1 + \dfrac{b}{a} & -1 + \dfrac{c}{a} & \dfrac{1}{a} & 0 & \dfrac{1}{a} \end{array} \right]$$

This is a final tableau, yielding $x_1 = 1/a$, $x_2 = 0$; $y_1 = 1/a$, $y_2 = 0$, $y_3 = 0$; and $x = y = 1/a$. Hence $p_1 = 1$, $p_2 = 0$; $q_1 = 1$, $q_2 = 0$, $q_3 = 0$; and $V = a$, giving the desired conclusion.

Exercise 10.5

Answers to odd-numbered problems begin on page A-36.

A   In Problems 1–4, solve the given 2×2 matrix game by using the geometric approach to linear programming.

1. $\begin{bmatrix} 2 & 0 \\ -2 & 6 \end{bmatrix}$

2. $\begin{bmatrix} 4 & -1 \\ 1 & 2 \end{bmatrix}$

3. $\begin{bmatrix} -1 & 1 \\ 1 & -1 \end{bmatrix}$

4. $\begin{bmatrix} 6 & -5 \\ -3 & 9 \end{bmatrix}$

In Problems 5–8, use the simplex approach to solve the matrix games.

5. $\begin{bmatrix} 2 & -1 & 0 \\ -1 & 2 & 1 \end{bmatrix}$

6. $\begin{bmatrix} -2 & 2 & 4 \\ 3 & 1 & -1 \end{bmatrix}$

7. $\begin{bmatrix} 3 & -2 \\ -3 & 4 \\ 0 & 2 \end{bmatrix}$

8. $\begin{bmatrix} -4 & 0 \\ -3 & -4 \\ 0 & -5 \end{bmatrix}$

B   In Problems 9 and 10, remove recessive rows and columns, and then solve them by using the geometric approach to linear programming.

9. $\begin{bmatrix} -6 & -4 & 2 \\ 2 & -1 & 2 \\ -3 & 4 & 4 \end{bmatrix}$

10. $\begin{bmatrix} -6 & 2 & 4 \\ 4 & 4 & 2 \\ 1 & 3 & -1 \end{bmatrix}$

APPLICATIONS   11. **Investment Strategy** An investor has a choice of two types of investments, A and B, depending on two possible states of the economy, inflation and recession. The following payoff matrix shows the estimated percentage increases in the value of the investments over the coming year for each possible state of the economy.

$$\begin{array}{c} \\ \text{Invest in } A \\ \text{Invest in } B \end{array} \begin{array}{c} \text{Inflation} \quad \text{Recession} \\ \begin{bmatrix} 12 & 7 \\ -3 & 22 \end{bmatrix} \end{array}$$

(a) Find the investors optimal strategy $P^* = [p_1 \; p_2]$.

(b) Find the value of the game.

(c) What return can the investor expect if she follows the optimal strategy?

12. **Investment Strategy** An investor has a choice of three types of investments, corporate bonds, municipal bonds, and stocks, depending on two possible states of the economy, inflation and recession. The following payoff matrix shows the estimated percentage increases in the value of each investment over the coming year for each possible state of the economy.

$$\begin{array}{c} \\ \text{Corporate bonds} \\ \text{Municipal bonds} \\ \text{Stocks} \end{array} \begin{array}{c} \text{Inflation} \quad \text{Recession} \\ \begin{bmatrix} 5 & -1 \\ -1 & 4 \\ 1 & 3 \end{bmatrix} \end{array}$$

(a) Find the investors optimal strategy $P^* = [p_1 \; p_2]$.

(b) Find the value of the game.

(c) What return can the investor expect if he follows the optimal strategy?

13. **Toy Production** A toy manufacturer is about to set up next year's production schedule for three types of models: economy, regular, and deluxe. The quantity of each model to be produced depends on the state of the economy. Records over the past 10 years indicate that if the economy is slowing down the manufacturer will net 3, 2, and 1 million dollars, respectively, on sales of economy, regular and deluxe models. If the economy is growing the manufacturer will net -2, 0, and 3 million dollars, respectively, on sales of economy, regular, and deluxe models.

(a) Set up a payoff matrix for this problem.

(b) Find the manufacturer's optimal strategy.

(c) Find the value of the game.

(d) How should the manufacturer allocate its resources for each model to maximize return irrespective of how the economy performs in the coming year?

CHAPTER REVIEW

Important Terms and Formula		
two-person game	strictly determined	dominant row
zero-sum game	saddle point	recessive row
payoff	fair game	reduced matrix
game matrix	pure strategy	dominant column
strategy	mixed strategy	recessive column
best strategy	expected payoff	
value	optimal strategy	

$$E = PAQ \qquad E = PAQ = \frac{a_{11} \cdot a_{22} - a_{12} \cdot a_{21}}{a_{11} + a_{22} - a_{12} - a_{21}}$$

True-False Questions

(Answers on page A-36)

T F 1. The value of a strictly-determined game is unique.

T F 2. In a two-person, zero-sum game, whatever is gained (lost) by Player I is lost (gained) by Player II.

T F 3. A reduced matrix can be formed by removing dominant rows.

T F 4. In mixed strategy games, the value of the game depends on the strategy each player uses.

Fill in the Blanks

(Answers on page A-36)

1. Each entry of a matrix game is called a _____ .

2. In a non-strictly determined game, when optimal strategies are used, the expected payoff is called the _____ of the game.

3. If a matrix A contains a column c* whose entries are greater than or equal to the corresponding entries of another column c, then c is said to _____ c*.

Review Exercises

Answers to Odd-Numbered Problems begin on page A-36.

1. Determine which of the following two-person zero-sum games are strictly determined. For those that are, find the value of the game.

(a) $\begin{bmatrix} 5 & 3 \\ 2 & 4 \end{bmatrix}$ (b) $\begin{bmatrix} 29 & 15 \\ 79 & 3 \end{bmatrix}$ (c) $\begin{bmatrix} 50 & 75 \\ 30 & 15 \end{bmatrix}$

(d) $\begin{bmatrix} 7 & 14 \\ 9 & 13 \end{bmatrix}$ (e) $\begin{bmatrix} 0 & 2 & 4 \\ 4 & 6 & 10 \\ 16 & 14 & 12 \end{bmatrix}$

2. Find the expected payoff of the game below for the given strategies:

$$\begin{bmatrix} -1 & 1 \\ 1 & -1 \end{bmatrix}$$

(a) $P = \begin{bmatrix} \frac{1}{3} & \frac{2}{3} \end{bmatrix}$; $Q = \begin{bmatrix} 1 \\ 0 \end{bmatrix}$ (b) $P = \begin{bmatrix} 0 & 1 \end{bmatrix}$; $Q = \begin{bmatrix} \frac{1}{2} \\ \frac{1}{2} \end{bmatrix}$

(c) $P = \begin{bmatrix} \frac{1}{2} & \frac{1}{2} \end{bmatrix}$; $Q = \begin{bmatrix} \frac{1}{2} \\ \frac{1}{2} \end{bmatrix}$

3. Show that if a 2×2 or 2×3 matrix game has a saddle point, then either one row dominates the other or one column dominates another column.

4. Give an example to show that the result of Problem 3 is not true for 3×3 matrix games.

5. Find the optimal strategy for each player in the following games. Assume Player I plays rows and entries denote winnings of Player I. What is the value of each game?

(a) $\begin{bmatrix} 4 & 6 & 3 \\ 1 & 2 & 5 \end{bmatrix}$ (b) $\begin{bmatrix} 1 & 6 \\ 5 & 2 \\ 7 & 4 \end{bmatrix}$

(c) $\begin{bmatrix} 2 & 1 \\ 4 & 0 \\ 3 & 4 \end{bmatrix}$ (d) $\begin{bmatrix} 0 & 3 & 2 \\ 4 & 2 & 3 \end{bmatrix}$

APPLICATIONS   6. **Investment Strategy** An investor has a choice of two investments, A and B. The percentage gain of each investment over the next year depends on whether the economy is "up" or "down." This investment information is displayed in the following payoff matrix.

	Economy up	Economy down
Invest in A	−5	20
Invest in B	18	0

(a) Find the best investment allocation if the investor wishes to maximize the minimum expected gain. That is, find the row player's optimal strategy in the corresponding matrix game.

(b) Find the percentage gain the investor is assured of when using this optimal strategy. That is, find the value of the corresponding matrix game.

7. **Investment Strategy** There are two possible investments, A and B, and two possible states of the economy, "inflation" and "recession." The estimated percentage increases in the value of the investments over the coming year for each possible state of the economy are shown in the following payoff matrix.

	Inflation	Recession
Invest in A	10	5
Invest in B	−5	20

(a) Find the investor's optimal strategy $P = \begin{bmatrix} p_1 & p_2 \end{bmatrix}$.

(b) Find the value E of this game.

(c) What return can the investor expect if the optimal strategy is followed?

8. **Real Estate Development** A real estate developer has bought a large tract of

land in Cook County. He is considering using some of the land for apartments, some for a shopping center, and some for houses. It is not certain whether the Cook County government will build a highway near his property or not. His financial advisor provides an estimate for the percentage profit to be made in each case and these percentages are given in the following table:

Builder	Government	
	Highway	*No highway*
Apartments	25%	5%
Shopping center	20%	15%
Houses	10%	20%

What percentage of the land should he use for each of the three categories assuming that the government of Cook County is an active opponent?

9. **Investment Strategy** You are interested in two investments, A and B. If the economy goes into a state of inflation next year, investment A will go up in value by 10 percent and investment B will go down in value by 5 percent. If the economy is in a state of recession, investment A will go up in value by 5 percent and investment B will go up in value by 20 percent.

(a) Think of this situation as a game in which you are the row player and find your optimal strategy $P = [p_1 \quad p_2]$ and the value E of the game.

(b) Suppose you have a certain amount of money and you invest p_1 of it in investment A and p_2 of it in investment B, where p_1 and p_2 are the fractions in your optimal strategy in part (a). What is your expected percentage profit in this case?

10. **Betting Strategy** The Pistons are going to play the Bulls in a basketball game. If you place your bet in Detroit you can get 3 to 2 odds for a bet on the Bulls and if you place your bet in Chicago you can get 1 to 1 odds for a bet on the Pistons. This information is displayed in the following payoff matrix in which the entries represent the payoffs in dollars for each $1 bet.

$$\begin{array}{cc} & \begin{matrix} \text{Bulls} & \text{Pistons} \\ \text{win} & \text{win} \end{matrix} \\ \begin{matrix} \text{Bet on Bulls} \\ \text{Bet on Pistons} \end{matrix} & \begin{bmatrix} 1.5 & -1 \\ -1 & 1 \end{bmatrix} \end{array}$$

(a) Think of this situation as a game in which you are the row player and find your optimal mixed strategy $P = [p_1 \quad p_2]$ and the value E of the game.

(b) Suppose you have k dollars to bet and you bet p_1 of it on the Bulls and p_2 of it on the Pistons where p_1 and p_2 are the fractions in your optimal strategy in part (a). By at least how much can you expect to come out ahead, no matter who wins the game?

11

MARKOV CHAINS

* This section may be omitted without loss of continuity.

11.1

AN INTRODUCTION TO MARKOV CHAINS

MARKOV CHAINS AND TRANSITION MATRICES
COMPUTING STATE DISTRIBUTIONS

In Chapter 8 we introduced Bernoulli trials. In discussing Bernoulli trials, we made the assumption that the outcome of each experiment is *independent* of the outcome of any previous experiment.

Here we will discuss another type of probabilistic model, called a *Markov chain,* where there is some connection between one trial and the next. Markov chains have been shown to have applications in many areas, among them business, psychology, sociology and biology.

Loosely speaking, a Markov chain or process is one in which what happens next is governed by what happened immediately before. At any stage a Markov experiment is in one of a finite number of states, with the next stage of the experiment consisting of movement to a possibly different state. The probability of moving to a certain state depends only on the state previously occupied and does not vary with time. Here are some situations that may be thought of as Markov chains:

1. There are yearly population shifts between a city and its surrounding suburbs. At any time, a person is either in the city or in the suburbs. So there are two states here with movement between them. As we track the population over time we can ask: What percentage is where?

2. Several detergents compete in the market. Each year there is some shift of customer loyalty from one brand to another. At any point in time the state of a customer would correspond to the brand of detergent he or she uses.

3. A psychology experiment consists of placing a mouse in a maze composed of rooms. The mouse moves from room to room, and we think of this movement as moving from state to state.

Let us elaborate on situation 3 with the idea of introducing some basic notions.

MARKOV CHAINS AND TRANSITION MATRICES

Consider the maze consisting of four connecting rooms shown in Figure 1. The rooms are numbered 1, 2, 3, 4 for convenience and each room contains pulsating lights of a different color. The experiment consists of releasing a mouse in a particular room and observing its behavior.

We assume an observation is made whenever a movement occurs or after a fixed time interval, whichever comes first. Since the movement of the mouse is random in nature we will use probabilistic terms to describe it.

We will refer to the rooms as states. For example, the probability p_{12} that the mouse moves from state 1 to state 2 might be $p_{12} = \frac{1}{2}$, while the probability of moving from state 1 to state 3 might be $p_{13} = \frac{1}{4}$. We are using p_{ij} to represent the probability of moving from state i to state j. So, p_{11} would represent the probability

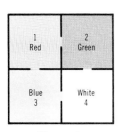

Figure 1

that the mouse remains in room 1 for one observation interval. Here we must have $p_{14} = 0$, since there is no direct passage between rooms 1 and 4.

A convenient way of displaying the probabilities p_{ij} would be to store them in a matrix. The resulting matrix

$$P = \begin{bmatrix} p_{11} & p_{12} & p_{13} & p_{14} \\ p_{21} & p_{22} & p_{23} & p_{24} \\ p_{31} & p_{32} & p_{33} & p_{34} \\ p_{41} & p_{42} & p_{43} & p_{44} \end{bmatrix}$$

is called the *transition matrix* of the experiment.

In our example, since there are four states, P is a 4×4 matrix. Were we to assign values to the remaining probabilities, a possible choice for the transition matrix P might be

$$P = \begin{matrix} & \begin{matrix} 1 & 2 & 3 & 4 \end{matrix} \\ \begin{matrix} 1 \\ 2 \\ 3 \\ 4 \end{matrix} & \begin{bmatrix} \frac{1}{4} & \frac{1}{2} & \frac{1}{4} & 0 \\ \frac{1}{6} & \frac{2}{3} & 0 & \frac{1}{6} \\ \frac{1}{3} & 0 & \frac{1}{3} & \frac{1}{3} \\ 0 & \frac{1}{4} & \frac{1}{2} & \frac{1}{4} \end{bmatrix} \end{matrix}$$

where the rows and columns are indexed by the four states (rooms). Thus P records the probabilities of where the mouse will go next given that we know where it is now. This essential idea of movement from state to state with attached probabilities can be generalized.

Markov Chain **A Markov chain is a sequence of experiments each of which results in one of a finite number of states that we label 1, 2, . . . , m. The probability that a given state is entered depends only on the state previously occupied.**

As before, we let

$$p_{ij} = \text{probability of moving from state } i \text{ to state } j$$

In a Markov chain with m states, the *transition matrix* $P = [p_{ij}]$ is an $m \times m$ matrix:

$$P = \begin{bmatrix} p_{11} & p_{12} & \cdots & p_{1m} \\ \cdot & & & \cdot \\ \cdot & & & \cdot \\ \cdot & & & \cdot \\ p_{m1} & p_{m2} & \cdots & p_{mm} \end{bmatrix}$$

Notice that P is a square matrix with entries that are always nonnegative since they represent probabilities. Also, the sum of the entries in every row is 1 since, as in the mouse and maze example, once the mouse is in a given room, it either stays there or moves to one of the other rooms with probability 1.

COMPUTING STATE DISTRIBUTIONS

The transition matrix contains the information necessary to predict what happens next, given that we know what happened before. It remains to specify what the state of affairs was at the start of the experiment. For example, what room was the mouse placed in at the start? We introduce a special row vector for this purpose.

In a Markov chain with m states, the *initial probability distribution* is a $1 \times m$ row vector $v^{(0)}$ whose ith entry is the probability the experiment was in state i at the start.

For example, if the mouse was equally likely to be placed in any one of the rooms at the start, then we would have $v^{(0)} = [\frac{1}{4} \quad \frac{1}{4} \quad \frac{1}{4} \quad \frac{1}{4}]$. Whereas, if it was decided to always place the mouse initially in room 1, then we would write $v^{(0)} = [1 \quad 0 \quad 0 \quad 0]$.

A *probability vector* is a vector whose entries are nonnegative and sum up to 1.

We conclude that the initial probability distribution $v^{(0)}$ of a Markov chain is a probability row vector.

Example 1

Look at the maze in Figure 1. Suppose we assign the following transition probabilities:

$$\text{From room 1 to} \begin{Bmatrix} 1 & 2 & 3 & 4 \\ \frac{1}{3} & \frac{1}{3} & \frac{1}{3} & 0 \end{Bmatrix}$$

Here, the mouse starts in room 1, and $\frac{1}{3}$ of the time it remains there during the time interval of observation, $\frac{1}{3}$ of the time it enters room 2, and $\frac{1}{3}$ of the time it enters room 3. Since it cannot go to room 4 directly from room 1, the probability assignment is 0. Similarly, the transition probabilities in moving from room 2, room 3, and room 4 may be given as follows:

$$\text{From room 2 to} \begin{Bmatrix} 1 & 2 & 3 & 4 \\ \frac{1}{3} & \frac{1}{3} & 0 & \frac{1}{3} \end{Bmatrix} \qquad \text{From room 3 to} \begin{Bmatrix} 1 & 2 & 3 & 4 \\ \frac{1}{3} & 0 & \frac{1}{3} & \frac{1}{3} \end{Bmatrix}$$

$$\text{From room 4 to} \begin{Bmatrix} 1 & 2 & 3 & 4 \\ 0 & \frac{1}{3} & \frac{1}{3} & \frac{1}{3} \end{Bmatrix}$$

Find the transition matrix P. If the initial placement of the mouse is in room 4, find the initial probability distribution. What are the probabilities of being in each room after two observations?

Solution

The transition matrix P is

$$P = [p_{ij}] = \begin{array}{c} \\ 1 \\ 2 \\ 3 \\ 4 \end{array} \begin{array}{cccc} 1 & 2 & 3 & 4 \\ \begin{bmatrix} \frac{1}{3} & \frac{1}{3} & \frac{1}{3} & 0 \\ \frac{1}{3} & \frac{1}{3} & 0 & \frac{1}{3} \\ \frac{1}{3} & 0 & \frac{1}{3} & \frac{1}{3} \\ 0 & \frac{1}{3} & \frac{1}{3} & \frac{1}{3} \end{bmatrix} \end{array}$$

Next, since the initial placement of the mouse is in room 4, the initial probability distribution is

$$v^{(0)} = [p_1^{(0)} \quad p_2^{(0)} \quad p_3^{(0)} \quad p_4^{(0)}] = [0 \quad 0 \quad 0 \quad 1]$$

To answer the last question, we use a probability tree. See Figure 2. The num-

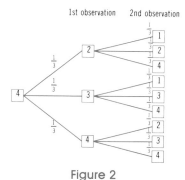

Figure 2

bers in each square refer to the room occupied. From this tree diagram we deduce, for example, that the mouse will be in state 1 after two observations with probability

$$p_{41}^{(2)} = \frac{1}{3} \cdot \frac{1}{3} + \frac{1}{3} \cdot \frac{1}{3} = \frac{2}{9}$$

Similarly,

$$p_{42}^{(2)} = \frac{1}{3} \cdot \frac{1}{3} + \frac{1}{3} \cdot \frac{1}{3} = \frac{2}{9} \tag{1}$$

$$p_{43}^{(2)} = \frac{1}{3} \cdot \frac{1}{3} + \frac{1}{3} \cdot \frac{1}{3} = \frac{2}{9}$$

$$p_{44}^{(2)} = \frac{1}{3} \cdot \frac{1}{3} + \frac{1}{3} \cdot \frac{1}{3} + \frac{1}{3} \cdot \frac{1}{3} = \frac{1}{3}$$

We could record the results of Example 1 by writing $v^{(2)} = [\frac{2}{9} \quad \frac{2}{9} \quad \frac{2}{9} \quad \frac{1}{3}]$. Using this notation, the row vector $v^{(k)}$ would be used to record the probabilities of being in the various states after k trials. The ith entry of $v^{(k)}$ is the probability of being in state i at stage k. For example, $v^{(4)} = [\frac{1}{8} \quad \frac{1}{8} \quad 0 \quad \frac{3}{4}]$ would say that the probability of being in state 2 at stage 4 of the experiment is $\frac{1}{8}$.

Instead of following the procedure of the above example, it is possible to compute $v^{(k)}$ directly from the transition matrix P and the initial probability distribution $v^{(0)}$.

In a Markov chain the probability distribution $v^{(k)}$ after k observations satisfies

$$v^{(k)} = v^{(k-1)}P \tag{2}$$

where P is the transition matrix. For example, this says

$$v^{(1)} = v^{(0)}P$$
$$v^{(2)} = v^{(1)}P$$
$$v^{(3)} = v^{(2)}P, \text{ and so on}$$

Thus the succeeding distribution can always be derived from the previous one by multiplying by the transition matrix.

Example 2
Using the information from Example 1, find the probability distribution after two observations.

Solution
We had $v^{(0)} = [0 \quad 0 \quad 0 \quad 1]$, so by (2)

$$v^{(1)} = v^{(0)}P = [0 \quad 0 \quad 0 \quad 1] \begin{bmatrix} \frac{1}{3} & \frac{1}{3} & \frac{1}{3} & 0 \\ \frac{1}{3} & \frac{1}{3} & 0 & \frac{1}{3} \\ \frac{1}{3} & 0 & \frac{1}{3} & \frac{1}{3} \\ 0 & \frac{1}{3} & \frac{1}{3} & \frac{1}{3} \end{bmatrix} = [0 \quad \tfrac{1}{3} \quad \tfrac{1}{3} \quad \tfrac{1}{3}]$$

Using (2) again,

$$v^{(2)} = v^{(1)}P = [0 \quad \tfrac{1}{3} \quad \tfrac{1}{3} \quad \tfrac{1}{3}] \begin{bmatrix} \frac{1}{3} & \frac{1}{3} & \frac{1}{3} & 0 \\ \frac{1}{3} & \frac{1}{3} & 0 & \frac{1}{3} \\ \frac{1}{3} & 0 & \frac{1}{3} & \frac{1}{3} \\ 0 & \frac{1}{3} & \frac{1}{3} & \frac{1}{3} \end{bmatrix} = [\tfrac{2}{9} \quad \tfrac{2}{9} \quad \tfrac{2}{9} \quad \tfrac{1}{3}]$$

This way of obtaining $v^{(2)}$ agrees with the results found in (1).   ∎

A few computations using (2) lead to another result.

$$v^{(1)} = v^{(0)}P$$
$$v^{(2)} = v^{(1)}P = [v^{(0)}P]P = v^{(0)}P^2$$
$$v^{(3)} = v^{(2)}P = [v^{(0)}P^2]P = v^{(0)}P^3$$
$$v^{(4)} = v^{(3)}P = [v^{(0)}P^3]P = v^{(0)}P^4$$

In each line above we have substituted the result of the preceding line. These calculations can clearly be continued to produce the following observation.

$$v^{(k)} = v^{(0)}P^k \tag{3}$$

where P^k is the kth power of the transition matrix. Thus, the conclusion of Example 2 could also have been arrived at by squaring the transition matrix and computing

$$v^{(2)} = v^{(0)}P^2 = [0 \quad 0 \quad 0 \quad 1] \begin{bmatrix} \frac{1}{3} & \frac{2}{9} & \frac{2}{9} & \frac{2}{9} \\ \frac{2}{9} & \frac{1}{3} & \frac{2}{9} & \frac{2}{9} \\ \frac{2}{9} & \frac{2}{9} & \frac{1}{3} & \frac{2}{9} \\ \frac{2}{9} & \frac{2}{9} & \frac{2}{9} & \frac{1}{3} \end{bmatrix} = [\tfrac{2}{9} \quad \tfrac{2}{9} \quad \tfrac{2}{9} \quad \tfrac{1}{3}]$$

Equations (2) and (3) are two formulas for finding $v^{(k)}$. Equation (2) has the advantage that it lends itself better to a computer program.

Example 3

Population Movement Suppose that the city of Glenwood is experiencing a movement of its population to the suburbs. At present, 85% of the total population lives in the city and 15% lives in the suburbs. But each year 7% of the city people move to the suburbs, whereas only 1% of the suburb people move back to the city. Assuming that the total population (city and suburbs together) remains constant, what percent of the total will remain in the city after 5 years?

Solution

This problem can be expressed as a sequence of experiments in which each experiment measures the proportion of people in the city and the proportion of people in the suburbs.

In the $(n + 1)$st year, these proportions will depend for their value only on the proportions in the nth year and not on the proportions found in earlier years. Thus, we have an experiment that can be represented as a Markov chain.

The initial probability distribution for this system is

$$\begin{array}{cc} \text{City} & \text{Suburbs} \end{array}$$
$$v^{(0)} = [.85 \quad .15]$$

That is, initially, 85% of the people reside in the city and 15% in the suburbs.

The transition matrix P is

$$P = \begin{array}{c} \\ \text{City} \\ \text{Suburbs} \end{array} \begin{bmatrix} \overset{\text{City}}{.93} & \overset{\text{Suburbs}}{.07} \\ .01 & .99 \end{bmatrix}$$

That is, each year 7% of the city people move to the suburbs (so that 93% remain in the city) and 1% of the suburb people move to the city (so that 99% remain in the suburbs).

To find the probability distribution after 5 years, we need to compute $v^{(5)}$:

$$v^{(1)} = v^{(0)}P = [.85 \quad .15]\begin{bmatrix} .93 & .07 \\ .01 & .99 \end{bmatrix} = [.792 \quad .208]$$

$$v^{(2)} = v^{(1)}P = [.792 \quad .208]\begin{bmatrix} .93 & .07 \\ .01 & .99 \end{bmatrix} = [.73864 \quad .26136]$$

$$v^{(3)} = v^{(2)}P = [.73864 \quad .26136]\begin{bmatrix} .93 & .07 \\ .01 & .99 \end{bmatrix} = [.68955 \quad .31045]$$

$$v^{(4)} = v^{(3)}P = [.68955 \quad .31045]\begin{bmatrix} .93 & .07 \\ .01 & .99 \end{bmatrix} = [.64439 \quad .35561]$$

$$v^{(5)} = v^{(4)}P = [.64439 \quad .35561]\begin{bmatrix} .93 & .07 \\ .01 & .99 \end{bmatrix} = [.60284 \quad .39716]$$

Thus, by the properties (2) and (3), the probability distribution after 5 years is

$$v^{(5)} = v^{(0)}P^5 = v^{(4)}P = [.60284 \quad .39716]$$

Thus, after 5 years, 60.28% of the residents live in the city and 39.72% live in the suburbs. ∎

This example leads us to inquire whether the situation in Glenwood ever stabilizes. That is, after a certain number of years is an equilibrium reached? Also, does the equilibrium, if attained, depend on what the initial distribution of the population was or is it independent of the initial state? We deal with these questions in the next section.

Exercise 11.1

Answers to Odd-Numbered Problems begin on page A-38.

A   **1.** Explain why the matrix below cannot be the transition matrix for a Markov chain.

$$\begin{bmatrix} 0 & 1 & 0 \\ \frac{1}{3} & \frac{1}{3} & \frac{1}{3} \\ \frac{1}{2} & -\frac{1}{2} & \frac{1}{2} \end{bmatrix}$$

2. Explain why the matrix below cannot be the transition matrix for a Markov chain.

$$\begin{bmatrix} 1 & \frac{1}{2} & \frac{1}{3} & \frac{1}{4} \\ 0 & 1 & 0 & 0 \\ 0 & \frac{1}{2} & \frac{1}{2} & 0 \\ 1 & 0 & 0 & 0 \end{bmatrix}$$

3. Consider a Markov chain with transition matrix

$$\begin{array}{cc} & \begin{array}{cc} \text{State 1} & \text{State 2} \end{array} \\ \begin{array}{c} \text{State 1} \\ \text{State 2} \end{array} & \begin{bmatrix} \frac{1}{3} & \frac{2}{3} \\ \frac{1}{4} & \frac{3}{4} \end{bmatrix} \end{array}$$

(a) What does the entry $\frac{2}{3}$ in this matrix represent?

(b) Assuming that the system is initially in state 1, find the probability distribution one observation later.

(c) Assuming that the system is initially in state 2, find the probability distribution one observation later.

(d) Draw a probability tree to find the probability distribution two observations later.

4. Consider a Markov chain with transition matrix

$$\begin{array}{cc} & \begin{array}{cc} \text{State 1} & \text{State 2} \end{array} \\ \begin{array}{c} \text{State 1} \\ \text{State 2} \end{array} & \begin{bmatrix} .3 & .7 \\ .4 & .6 \end{bmatrix} \end{array}$$

(a) What does the entry .4 in the matrix represent?

(b) Assuming that the system is initially in state 1, find the probability distribution two observations later.

(c) Assuming that the system is initially in state 2, find the probability distribution two observations later.

5. Consider the transition matrix of Problem 4. If the initial probability distribution is [.25   .75], what is the probability distribution after two observations?

6. Consider a Markov chain with transition matrix

$$P = \begin{bmatrix} .7 & .2 & .1 \\ .6 & .2 & .2 \\ .4 & .1 & .5 \end{bmatrix}$$

If the initial distribution is [.25   .25   .5], what is the probability distribution in the next observation?

B   **7.** Find the values of a, b, and c that will make the following matrix a transition matrix for a Markov chain:

$$\begin{bmatrix} .2 & a & .4 \\ b & .6 & .2 \\ 0 & c & 0 \end{bmatrix}$$

8. In the maze of Figure 1, if the initial probability distribution is $v^{(0)} = [\tfrac{1}{2} \quad 0 \quad \tfrac{1}{2} \quad 0]$, find the probability distribution after two observations.

9. In Example 3, if the initial probability distribution for Glenwood is $v^{(0)} = [.7 \quad .3]$, what is the population distribution after 5 years?

C   **10.** If A is a transition matrix, what about A^2? A^3? What do you conjecture about A^n?

11. Let

$$A = \begin{bmatrix} a_{11} & a_{12} \\ a_{21} & a_{22} \end{bmatrix}$$

be a transition matrix and

$$u = [u_1 \quad u_2]$$

be a probability row vector. Prove that uA is a probability vector.

12. Let

$$P = \begin{bmatrix} p_{11} & p_{12} \\ p_{21} & p_{22} \end{bmatrix}$$

be a transition matrix. Prove that $v^{(k)} = v^{(k-1)}P$.

APPLICATIONS   **13.** Voting Patterns   The voting pattern for a certain group of cities is such that 60% of the Democratic (D) mayors were succeeded by Democrats and 40% by Republicans (R). Also, 30% of the Republican mayors were succeeded by Democrats and 70% by Republicans.

(a)   Explain why the above is a Markov chain.

(b)   Set up the 2×2 matrix P with columns and rows labeled D and R to display these transitions.

(c)   Compute P^2 and P^3.

14. Consumer Loyalty   A new rapid transit system has just been installed. It is anticipated that each week 90% of the commuters who used the rapid transit will continue to do so. Of those who traveled by car, 20% will begin to use the rapid transit instead.

(a)   Explain why the above is a Markov chain.

(b)   Set up the 2×2 matrix P with columns and rows labeled R (rapid transit) and C (car) to display these transitions.

(c)   Compute P^2 and P^3.

15. Market Penetration   A company is promoting a certain product, say Brand X wine. The result of this is that 75% of the people drinking Brand X wine over any given period of 1 month, continue to drink it the next month; of those people drinking other brands of wine in the period of 1 month, 35% change over to the promoted wine the next month. We would like to know what fraction of wine drinkers will drink Brand X after 2 months if 50% drink Brand X wine now.

1	2	3
4	5	6
7	8	9

16. **Maze Experiment** Consider the maze with nine rooms shown in the figure. The system consists of the maze and a mouse. We assume that the following learning pattern exists: If the mouse is in room 1, 2, 3, 4, or 5, it moves with equal probability to any room that the maze permits; if it is in room 8, it moves directly to room 9; if it is in room 6, 7, or 9, it remains in that room.

   (a) Explain why the above experiment is a Markov chain.

   (b) Construct the transition matrix P.

17. **Consumer Loyalty** Suppose that, during the year 1991, 45% of the drivers in a certain metropolitan area had Travelers automobile insurance, 30% had General American insurance, and 25% were insured by some other companies. Suppose also that a year later: *(1)* of those who had been insured by Travelers in 1991, 92% continued to be insured by Travelers, but 8% had switched their insurance to General American; *(2)* of those who had been insured by General American in 1991, 90% continued to be insured by General American, but 4% had switched to Travelers and 6% had switched to some other companies; *(3)* of those who had been insured by some other companies in 1991, 82% continued but 10% had switched to Travelers, and 8% had switched to General American. Use these data to answer the following questions:

   (a) What percentage of drivers in the metropolitan area were insured by Travelers and General American in 1992?

   (b) If these trends continued for one more year, what percentage of the drivers were insured by Travelers and General American in 1993?

18. A professor either walks or drives to a university. He never drives 2 days in a row, but if he walks 1 day, he is just as likely to walk the next day as to drive his car. Show that this forms a Markov chain and give the transition matrix.

11.2

REGULAR MARKOV CHAINS

POWERS OF THE TRANSITION MATRIX
REGULAR CHAINS
FIXED PROBABILITY VECTOR AND EQUILIBRIUM
MODEL: SOCIAL MOBILITY

A fundamental question about Markov chains is: What happens in the long run? Is it the case that the distribution into states tends to stabilize over time? In this section, we investigate the conditions under which a Markov chain produces an *equilibrium*, or *steady-state*, situation. We will also give techniques for finding this equilibrium distribution, when it exists.

POWERS OF THE TRANSITION MATRIX

We begin by examining *powers* of the transition matrix P. Recall that the (i, j)th entry p_{ij} of P is the probability of moving from state i to state j in any one step. Are there corresponding interpretations for the entries in any *power* P^2, P^3, P^4, . . . of the transition matrix?

To motivate the answer we use the data of Problem 15 in Exercise 11.1. There the transition matrix is

$$
P = \begin{array}{c} \\ \text{Brand } X, E_1 \\ \text{Other brands, } E_2 \end{array}
\begin{array}{cc}
\overset{\displaystyle \text{Brand } X}{\overset{\displaystyle E_1}{}} & \overset{\displaystyle \text{Other brands}}{\overset{\displaystyle E_2}{}}
\end{array}
\begin{bmatrix}
.75 & .25 \\
.35 & .65
\end{bmatrix}
$$

A probability tree depicting 3 stages of this experiment is given in Figure 3.

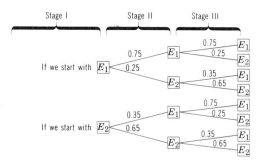

Figure 3

The probability $p_{11}^{(2)}$ of proceeding from E_1 to E_1 in two stages is

$$p_{11}^{(2)} = (.75)(.75) + (.25)(.35) = .65$$

The probability $p_{12}^{(2)}$ from E_1 to E_2 in two stages is

$$p_{12}^{(2)} = (.75)(.25) + (.25)(.65) = .35$$

The probability $p_{21}^{(2)}$ from E_2 to E_1 in two stages is

$$p_{21}^{(2)} = (.35)(.75) + (.65)(.35) = .49$$

The probability $p_{22}^{(2)}$ from E_2 to E_2 in two stages is

$$p_{22}^{(2)} = (.35)(.25) + (.65)(.65) = .51$$

If we square the matrix P, we obtain

$$
P^2 = \begin{bmatrix}
(.75)(.75) + (.25)(.35) & (.75)(.25) + (.25)(.65) \\
(.35)(.75) + (.65)(.35) & (.35)(.25) + (.65)(.65)
\end{bmatrix}
$$

$$
= \begin{bmatrix}
.65 & .35 \\
.49 & .51
\end{bmatrix}
$$

and we notice that

$$
P^2 = \begin{bmatrix}
p_{11}^{(2)} & p_{12}^{(2)} \\
p_{21}^{(2)} & p_{22}^{(2)}
\end{bmatrix}
$$

Thus, in this example, the *square* of the transition matrix gives the probabilities for moving from one state to another state in *two* stages.

This is true in general, and higher powers of the transition matrix carry similar interpretations. More precisely,

> If P is the transition matrix of a Markov chain, then the (i, j)th entry of P^n (nth power of P) gives the probability of passing from state i to state j in n stages.

REGULAR CHAINS

We now turn to the question involving the long-term behavior of a Markov chain and begin by observing the results of an example we have used before.

Example 1

Take the transition matrix P of Problem 15 of Exercise 11.1 and compute some of its powers.

Solution

Some of the powers of the transition matrix P are given below.

$$P = \begin{bmatrix} .7500 & .2500 \\ .3500 & .6500 \end{bmatrix} \qquad P^7 = \begin{bmatrix} .5840 & .4159 \\ .5823 & .4176 \end{bmatrix}$$

$$P^2 = \begin{bmatrix} .6500 & .3500 \\ .4900 & .5100 \end{bmatrix} \qquad P^8 = \begin{bmatrix} .5836 & .4163 \\ .5829 & .4170 \end{bmatrix}$$

$$P^3 = \begin{bmatrix} .6100 & .3900 \\ .5460 & .4540 \end{bmatrix} \qquad P^9 = \begin{bmatrix} .5834 & .4165 \\ .5831 & .4168 \end{bmatrix}$$

$$P^4 = \begin{bmatrix} .5940 & .4060 \\ .5683 & .4316 \end{bmatrix} \qquad P^{10} = \begin{bmatrix} .5833 & .4166 \\ .5832 & .4167 \end{bmatrix}$$

$$P^5 = \begin{bmatrix} .5876 & .4124 \\ .5773 & .4226 \end{bmatrix} \qquad P^{11} = \begin{bmatrix} .5833 & .4166 \\ .5833 & .4166 \end{bmatrix} \leftarrow$$

$$P^6 = \begin{bmatrix} .5850 & .4149 \\ .5809 & .4190 \end{bmatrix} \qquad P^{12} = \begin{bmatrix} .5833 & .4166 \\ .5833 & .4166 \end{bmatrix} \leftarrow$$

No change, so we stop here ∎

We notice the interesting fact that the powers P^n seem to be converging and stabilizing around a fixed matrix. Furthermore, all rows of that matrix are identical. Also, if we let

$$\mathbf{t} = [.5833 \quad .4166]$$

and compute $\mathbf{t}P$ we find that

$$\mathbf{t}P = [.5833 \quad .4166] \begin{bmatrix} .75 & .25 \\ .35 & .65 \end{bmatrix} = [.5833 \quad .4166]$$

That is, $\mathbf{t}P = \mathbf{t}$. So, the row vector $\mathbf{t}$ is fixed by P.

Will this always happen? And what interpretation can be placed on the vector $\mathbf{t}$? Before giving the answers, we need a definition.

Regular Markov Chain   **A Markov chain is said to be *regular* if for some power of its transition matrix P, all of the entries are positive.**

Example 2

The transition matrix

$$P = \begin{bmatrix} \frac{1}{2} & \frac{1}{2} \\ 1 & 0 \end{bmatrix}$$

is regular since the square of P, namely,

$$P^2 = \begin{bmatrix} \frac{3}{4} & \frac{3}{4} \\ \frac{1}{2} & \frac{1}{2} \end{bmatrix}$$

has only positive entries.   ∎

Example 3

The matrix

$$P = \begin{bmatrix} 1 & 0 \\ \frac{3}{4} & \frac{1}{4} \end{bmatrix}$$

is not regular since every power of P will always have $p_{12} = 0$. The matrix

$$\begin{bmatrix} 0 & 1 \\ 1 & 0 \end{bmatrix}$$

is another example of a transition matrix that is not regular.   ∎

It is regular Markov chains that exhibit the behavior found in Example 1. We state the following result.

Let P be the transition matrix of a regular Markov chain. Then,

(a)   The matrices P^n approach a fixed matrix T, as n gets large. We write $P^n \to T$ (as n gets large).

(b)   The rows of T are all identical and equal to a probability row vector $\mathbf{t}$.

(c) $\mathbf{t}$ is the unique probability vector that satisfies $\mathbf{t}P = \mathbf{t}$.

FIXED PROBABILITY VECTOR AND EQUILIBRIUM

Because of (c), $\mathbf{t}$ is called the *fixed probability* vector of the transition matrix P. What is the significance of this vector? The answer is that it tells us what the long run distribution into states will be. More precisely, for a regular Markov chain

---

(d)   If $v^{(0)}$ is any initial distribution, then $v^{(n)} \to \mathbf{t}$ as n gets large.

Thus, $\mathbf{t}$ can be thought of as representing a limit state, since the state distributions

$v^{(n)}$ will approach **t** as time goes on. **t** can also be thought of as an equilibrium state, because once the system is in state **t** it stays there.

For example, the computations in Example 1 show that in the long run 58.33% of wine drinkers will be drinking Brand X wine.

An important point to note is that it is not necessary to compute higher and higher power P^n of the transition matrix to find T and, in the process, **t**. Result (c) allows us to find **t** by solving a system of equations.

Example 4

Find the fixed probability vector for Example 3 in Section 1.

Solution

Let $\mathbf{t} = [t_1 \quad t_2]$ be the desired fixed probability vector. Then $t_1 + t_2 = 1$ and

$$[t_1 \quad t_2] \begin{bmatrix} .93 & .07 \\ .01 & .99 \end{bmatrix} = [t_1 \quad t_2]$$

Or,

$$[.93t_1 + .01t_2 \quad .07t_1 + .99t_2] = [t_1 \quad t_2]$$

Equating corresponding entries yields the following system of two equations in two unknowns

$$.93t_1 + .01t_2 = t_1$$
$$.07t_1 + .99t_2 = t_2$$

Or,

$$-.07t_1 + .01t_2 = 0$$
$$.07t_1 - .01t_2 = 0$$

As it stands this system has infinitely many solutions. However, when we include the equation

$$t_1 + t_2 = 1$$

the resulting system is

$$-.07t_1 + .01t_2 = 0$$
$$.07t_1 - .01t_2 = 0$$
$$t_1 + t_2 = 1$$

which will have a unique solution, namely,

$$t_1 = \frac{1}{8} \qquad t_2 = \frac{7}{8}$$

In obtaining the fixed probability vector $[\frac{1}{8} \quad \frac{7}{8}]$, we learn that in the long run, $\frac{1}{8}$ of the population will live in the city while $\frac{7}{8}$ will be suburbanites. ■

Example 5

Find the fixed probability vector **t** of the transition matrix

$$P = \begin{bmatrix} \frac{1}{2} & 0 & \frac{1}{2} \\ \frac{1}{2} & \frac{1}{2} & 0 \\ \frac{1}{3} & \frac{1}{3} & \frac{1}{3} \end{bmatrix}$$

Solution

Let $\mathbf{t} = [t_1 \quad t_2 \quad t_3]$ be the fixed vector. Then $t_1 + t_2 + t_3 = 1$ and

$$[t_1 \quad t_2 \quad t_3] \begin{bmatrix} \frac{1}{2} & 0 & \frac{1}{2} \\ \frac{1}{2} & \frac{1}{2} & 0 \\ \frac{1}{3} & \frac{1}{3} & \frac{1}{3} \end{bmatrix} = [t_1 \quad t_2 \quad t_3]$$

So,

$$\tfrac{1}{2}t_1 + \tfrac{1}{2}t_2 + \tfrac{1}{3}t_3 = t_1$$
$$\tfrac{1}{2}t_2 + \tfrac{1}{3}t_3 = t_2$$
$$\tfrac{1}{2}t_1 + \tfrac{1}{3}t_3 = t_3$$

or

$$-\tfrac{1}{2}t_1 + \tfrac{1}{2}t_2 + \tfrac{1}{3}t_3 = 0$$
$$-\tfrac{1}{2}t_2 + \tfrac{1}{3}t_3 = 0$$
$$\tfrac{1}{2}t_1 - \tfrac{2}{3}t_3 = 0$$

If we add the condition that $t_1 + t_2 + t_3 = 1$, we obtain

$$-3t_1 + 3t_2 + 2t_3 = 0$$
$$-3t_2 + 2t_3 = 0$$
$$3t_1 - 4t_3 = 0$$
$$t_1 + t_2 + t_3 = 1$$

The solution of this system

$$\mathbf{t} = [\tfrac{4}{9} \quad \tfrac{2}{9} \quad \tfrac{1}{3}]$$

is the fixed vector: ■

A nice feature is that the long run distribution **t** can be found simply by solving a system of equations. A more remarkable feature is that the same equilibrium is reached no matter what the initial distribution was. We sketch the reason for this in the 2×2 case.

Suppose $v^{(0)} = [a \quad b]$ is *any* initial distribution. Let $\mathbf{t} = [t_1 \quad t_2]$. Then, as we indicated,

$$P^n \to T = \begin{bmatrix} t_1 & t_2 \\ t_1 & t_2 \end{bmatrix}$$

Now $v^{(n)} = v^{(0)}P^n$. So, as n gets large,

$$v^{(n)} = v^{(0)}P^n \to v^{(0)}T$$

But

$$v^{(0)}T = [a \quad b]\begin{bmatrix} t_1 & t_2 \\ t_1 & t_2 \end{bmatrix} = [(a+b)t_1 \quad (a+b)t_2]$$

Since $v^{(0)}$ is a probability vector, it follows that $a + b = 1$.

Thus

$$v^{(n)} \rightarrow v^{(0)}T = [t_1 \quad t_2] = \mathbf{t}$$

That is, the same $\mathbf{t}$ is reached regardless of the entries used in $v^{(0)}$.

Example 6

Consumer Loyalty Consider a certain community in a well-defined area with three grocery stores, I, II, and III. Within this community (we assume that the population is fixed) there always exists a shift of customers from one grocery store to another. A study was made on January 1, and it was found that $\frac{1}{4}$ of the population shopped at Store I, $\frac{1}{3}$ at Store II, and $\frac{5}{12}$ at Store III. Each month Store I retains 90% of its customers and loses 10% of them to Store II. Store II retains 5% of its customers and loses 85% of them to Store I and 10% of them to Store III. Store III retains 40% of its customers and loses 50% of them to Store I and 10% to Store II. The transition matrix P is

$$P = \begin{array}{c} \\ \text{I} \\ \text{II} \\ \text{III} \end{array} \begin{array}{ccc} \text{I} & \text{II} & \text{III} \\ \begin{bmatrix} .90 & .10 & 0 \\ .85 & .05 & .10 \\ .50 & .10 & .40 \end{bmatrix} \end{array}$$

We would like to answer the following questions:

(a) What proportion of customers will each store retain by February 1?
(b) By March 1?
(c) Assuming the same pattern continues, what will be the long-run distribution of customers among the three stores?

Solution
(a) To answer the first question, we note that the initial probability distribution is $v^{(0)} = [\frac{1}{4} \quad \frac{1}{3} \quad \frac{5}{12}]$. By February 1, the probability distribution is

$$v^{(1)} = v^{(0)}P = [\frac{1}{4} \quad \frac{1}{3} \quad \frac{5}{12}]\begin{bmatrix} .90 & .10 & .00 \\ .85 & .05 & .10 \\ .50 & .10 & .40 \end{bmatrix} = [.7167 \quad .0833 \quad .2000]$$

(b) To find the probability distribution after 2 months (March 1), we compute $v^{(2)}$:

$$v^{(2)} = v^{(0)}P^2 = [\frac{1}{4} \quad \frac{1}{3} \quad \frac{5}{12}]\begin{bmatrix} .895 & .095 & .010 \\ .857 & .098 & .045 \\ .735 & .095 & .170 \end{bmatrix} = [.8155 \quad .0956 \quad .0882]$$

(c)   To find the long run distribution we determine the fixed probability vector **t** of the regular transition matrix P. Let $\mathbf{t} = [t_1 \quad t_2 \quad t_3]$, where $t_1 + t_2 + t_3 = 1$. Then

$$[t_1 \quad t_2 \quad t_3] \begin{bmatrix} .90 & .10 & .00 \\ .85 & .05 & .10 \\ .50 & .10 & .40 \end{bmatrix} = [t_1 \quad t_2 \quad t_3]$$

$$[t_1 \quad t_2 \quad t_3] = [.8889 \quad .0952 \quad .0159]$$

Thus, in the long run Store I will have about 88.9% of all customers, Store II will have 9.5%, and Store III will have 1.6%.   ∎

Example 7

Spread of Rumor A United States Senator has determined whether to vote Yes or No on an important bill pending in Congress and conveys this decision to an aide. The aide then passes this news on to another individual, who passes it on to a friend, and so on, each time to a new individual. Assume p is the probability that any one person passes on the information opposite to the way he heard it. Then $1 - p$ is the probability a person passes on the information the same way he heard it. With what probability will the nth person receive the information as a Yes vote?

Solution
This can be viewed as a Markov chain model. Although it is not intuitively obvious, we shall find that the answer is essentially independent of p.

To obtain the transition matrix, we observe that two states are possible: A Yes is heard or a No is heard. The transition from a Yes to a No or from a No to a Yes occurs with probability p. The transition from Yes to Yes or from No to No occurs with probability $1 - p$. Thus, the transition matrix P is

$$\begin{array}{cc} & \begin{array}{cc} \text{Yes} & \text{No} \end{array} \\ \begin{array}{c} \text{Yes} \\ \text{No} \end{array} & \begin{bmatrix} 1 - p & p \\ p & 1 - p \end{bmatrix} \end{array}$$

The probability that the nth person will receive the information in one state or the other is given by successive powers of the matrix P, that is, by the matrices P^n. In fact, the answer in this case rapidly approaches $t_1 = \frac{1}{2}, t_2 = \frac{1}{2}$, after any considerable number of people are involved. This can easily be shown by verifying that the fixed probability vector is $[\frac{1}{2} \quad \frac{1}{2}]$. Hence, no matter what the Senator's initial decision is, eventually (after enough information exchange), half the people hear that the Senator is going to vote Yes and half hear that the Senator is going to vote No.   ∎

An interesting interpretation of this result applies to successive roll calls in Congress on the same issue. We let p represent the probability that a member changes his or her mind on an issue from one roll call to the next. Then the above example shows that if enough roll calls are forced, the Congress will eventually be near evenly split on the issue. This could perhaps serve as a model for explaining the parliamentary device of delaying actions by a minority.

MODEL: SOCIAL MOBILITY*

This model is based on an article by S. J. Prais, Department of Applied Economics, Cambridge University, who analyzed the movement among social classes in late 1940s England.

In the example, the following assumptions are made.

1. Class is treated as if it related only to the male side of the family line. This is largely because in these studies social class is measured by the occupation of the father.

2. The influence of one's ancestors in determining one's class is transmitted entirely through one's father so that, if the influence of one's father has been taken into account, then the total influence of one's ancestors is accounted for.

A social transition matrix representing England in 1949 is given in Table 1.

Table 1

The Social Transition Matrix for England, 1949

	1	2	3	4	5	6	7
1. Professional and high administrative	.388	.146	.202	.062	.140	.047	.015
2. Managerial and executive	.107	.267	.227	.120	.206	.053	.020
3. High grade supervisory and non-manual	.035	.101	.188	.191	.357	.067	.061
4. Lower grade supervisory and non-manual	.021	.039	.112	.212	.430	.124	.062
5. Skilled manual and routine non-manual	.009	.024	.075	.123	.473	.171	.125
6. Semiskilled manual	.000	.013	.041	.088	.391	.312	.155
7. Unskilled manual	.000	.008	.036	.083	.364	.235	.274

The element in the ith row and jth column of this matrix, denoted by p_{ij}, gives the proportion of fathers in the ith social class whose sons move into the jth social class. Furthermore, it is supposed that if there is uncertainty in the tracing of a family line through time, then p_{ij} represents the probability of transition by a family from class i into class j in the interval of one generation. For example, p_{42} indicates that .039 of the sons of fathers in class 4 (lower grade supervisory and nonmanual) move into class 2 (managerial and executive). The equilibrium probability vector for the matrix in Table 1 was found by Prais to be

$$[.023 \quad .042 \quad .088 \quad .127 \quad .409 \quad .182 \quad .129]$$

Prais compared the above result with the actual data and obtained the figures listed in Table 2.

---

* S. J. Prais, "Measuring Social Mobility," *Journal of the Royal Statistical Society,* **118** (1955), pp. 55–66.

Table 2
Actual and Equilibrium Distributions of the Social Classes in England

Class	Actual Distribution		Equilibrium Distribution (3)
	Fathers (1)	Sons (2)	
1. Professional	.037	.029	.023
2. Managerial	.043	.046	.042
3. Higher grade nonmanual	.098	.094	.088
4. Lower grade nonmanual	.148	.131	.127
5. Skilled manual	.432	.409	.409
6. Semiskilled manual	.131	.170	.182
7. Unskilled manual	.111	.121	.129

The equilibrium distribution depends only on the structural propensities of the society and not on the distribution of the population among the classes found at any instant. The equilibrium distribution is also independent of the unit of time in which the elements of P are measured. Suppose, for example, that observations were taken showing the relationship between the social status of grandson and grandfather. Every element of the transition matrix would then be different, since it would refer to a transition during a period of two generations instead of one generation. However, the equilibrium distribution corresponding to such a matrix would be unchanged. For, if the matrix relating the status of sons to fathers is P, that relating those of grandsons to grandfathers will be P^2 (provided, of course, that nothing has happened to change the characteristics of the society in the period considered), and when these matrices are raised to the nth power, they obviously tend to the same value as n gets very large.

It can be shown that the average number of generations spent by a family in social class j is $1/(1 - p_{jj})$, where p_{jj} is the jth diagonal element of the social transition matrix. So, Prais' model would predict that a family of unskilled manual laborers would occupy that profession for aproximately $1/(1 - .274) = 1.38$ generations.

Exercise 11.2   *Answers to Odd-Numbered Problems begin on page A-38.*

A   In Problems 1–6 determine which of the given matrices are regular. For those that are, find the fixed probability vector.

1. $\begin{bmatrix} \frac{1}{2} & \frac{1}{2} \\ 1 & 0 \end{bmatrix}$

2. $\begin{bmatrix} \frac{1}{2} & \frac{1}{2} \\ 0 & 1 \end{bmatrix}$

3. $\begin{bmatrix} 0 & 1 \\ \frac{1}{4} & \frac{3}{4} \end{bmatrix}$

4. $\begin{bmatrix} \frac{1}{3} & \frac{2}{3} \\ 1 & 0 \end{bmatrix}$

5. $\begin{bmatrix} 1 & 0 & 0 \\ \frac{1}{4} & \frac{1}{2} & \frac{1}{4} \\ 0 & 1 & 0 \end{bmatrix}$

6. $\begin{bmatrix} \frac{1}{4} & \frac{3}{4} & 0 \\ \frac{1}{2} & 0 & \frac{1}{2} \\ 0 & 1 & 0 \end{bmatrix}$

7. Show that the transition matrix P of Example 7 has a fixed probability vector $[\frac{1}{2} \quad \frac{1}{2}]$.

8. Verify the result we obtained in Example 6, part (c).

APPLICATIONS   9. Consumer Loyalty A grocer stocks his store with three types of detergents, A, B, C. When Brand A is sold out, the probability is .7 that he stocks up with Brand A again.

When he sells out Brand B, the probability is .8 that he will stock up again with Brand B. Finally, when he sells out Brand C, the probability is .6 that he will stock up with Brand C again. When he switches to another detergent, he does so with equal probability for the remaining two brands. Find the transition matrix. In the long run, how does he stock up with detergents?

10. **Consumer Loyalty** A housewife buys three kinds of cereal: A, B, C. She never buys the same cereal in successive weeks. If she buys Cereal A, then the next week she buys Cereal B. However, if she buys either B or C, then the next week she is three times as likely to buy A as the other brand. Find the transition matrix. In the long run, how often does she buy each of the three brands?

11. **Voting Patterns** In England, of the sons of members of the Conservative party, 70% vote Conservative and the rest vote Labor. Of the sons of Laborites, 50% vote Labor, 40% vote Conservative, and 10% vote Socialist. Of the sons of Socialists, 40% vote Socialist, 40% vote Labor, and 20% vote Conservative. What is the probability that the grandson of a Laborite will vote Socialist? What is the membership distribution in the long run?

12. **Educational Trends** Use the data given in the table and assume that the indicated trends continue in order to answer the questions below.

		Maximum Education Children Achieve		
		College	H.S.	E.S.
Highest educational level of parents	College	80%	18%	2%
	High school	40%	50%	10%
	Elementary school	20%	60%	20%

(a) What is the transition matrix?
(b) What is the probability that a grandchild of a college graduate is a college graduate?
(c) What is the probability that the grandchild of a high school graduate only finishes elementary school?
(d) If at present 30%, 40%, and 30% of the population are college, high school, and elementary school graduates, respectively, what will be the distribution of the grandchildren of the present population?
(e) What will the long-run distribution be?

13. **Family Traits** The probabilities that a blonde mother will have a blonde, brunette, or redheaded daughter are .6, .2 and .2, respectively. The probabilities that a brunette mother will have a blonde, brunette, or redheaded daughter are .1, .7, and .2, respectively. And the probabilities that a redheaded mother will have a blonde, brunette, or redheaded daughter are .4, .2, and .4, respectively. What is the probability that a blonde woman is the grandmother of a brunette? If the population of women is now 50% brunettes, 30% blondes, and the rest redheads, what will the distribution be:

(a) After two generations?
(b) In the long run?

11.3

ABSORBING MARKOV CHAINS

ABSORBING STATES
GAMBLER'S RUIN PROBLEM
EXPECTED LENGTH OF STAY
MODEL: THE RISE AND FALL OF STOCK PRICES

We have already seen examples of transition matrices in our presentation of Markov chains in which there are states that are impossible to leave. Such states are called *absorbing*. For instance, in Problem 16 in Exercise 11.1, room 9 is absorbing since once room 9 is reached, the probability of leaving it and passing to some different room is 0. Similarly, rooms 6 and 7 are absorbing states since it is impossible to leave these rooms once they have been reached.

ABSORBING STATES

Absorbing State; Absorbing Chain   **In a Markov chain, if p_{ij} denotes the probability of going from state E_i to state E_j, then E_i is called an *absorbing state* if $p_{ii} = 1$. A Markov chain is said to be an *absorbing chain* if and only if it contains at least one absorbing state and it is possible to go from *any* nonabsorbing state to an absorbing state in one or more stages.**

Thus, an absorbing state will capture the process and will not allow any state to pass from it.

In general, chains described by stochastic matrices can oscillate indefinitely from state to state in such a way that they exhibit no long-term trend. One such example is a Markov process chain having the nonregular matrix

$$\begin{bmatrix} 0 & 1 \\ 1 & 0 \end{bmatrix}$$

as its transition matrix. The idea of introducing absorbing states is to reduce the degree of oscillation since when an absorbing state is reached, the process no longer changes. That is, absorbing chain matrices exhibit a long-term trend. Furthermore, we can determine this trend using a simple computational technique.

When working with an absorbing Markov chain, it is convenient to rearrange the states so that the absorbing states come first and then the nonabsorbing states follow.

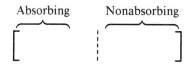

Once this rearrangement takes place, the transition matrix can be subdivided into four submatrices:

Absorbing   Nonabsorbing

$$\begin{bmatrix} I & \vdots & 0 \\ \hdashline S & \vdots & Q \end{bmatrix}$$

Here, I is an identity matrix, 0 denotes a matrix having all 0 entries, and the matrices S and Q are the two submatrices corresponding to the absorbing and nonabsorbing states. For example, the absorbing transition matrix

$$\begin{array}{c} \\ E_1 \\ E_2 \\ E_3 \\ E_4 \\ E_5 \end{array} \begin{array}{ccccc} E_1 & E_2 & E_3 & E_4 & E_5 \\ \begin{bmatrix} 0 & .5 & 0 & 0 & .5 \\ 0 & 0 & .9 & 0 & .1 \\ 0 & 0 & 0 & .7 & .3 \\ 0 & 0 & 0 & 1 & 0 \\ 0 & 0 & 0 & 0 & 1 \end{bmatrix} \end{array}$$

is first rearranged to get

$$\begin{array}{c} \\ E_4 \\ E_5 \\ E_1 \\ E_2 \\ E_3 \end{array} \begin{array}{ccccc} E_4 & E_5 & E_1 & E_2 & E_3 \\ \begin{bmatrix} 1 & 0 & 0 & 0 & 0 \\ 0 & 1 & 0 & 0 & 0 \\ 0 & .5 & 0 & .5 & 0 \\ 0 & .1 & 0 & 0 & .9 \\ .7 & .3 & 0 & 0 & 0 \end{bmatrix} \end{array}$$

Then the partitioned matrix is

$$\begin{bmatrix} 1 & 0 & \vdots & 0 & 0 & 0 \\ 0 & 1 & \vdots & 0 & 0 & 0 \\ \hdashline 0 & .5 & \vdots & 0 & .5 & 0 \\ 0 & .1 & \vdots & 0 & 0 & .9 \\ .7 & .3 & \vdots & 0 & 0 & 0 \end{bmatrix}$$

Here,

$$S = \begin{bmatrix} 0 & .5 \\ 0 & .1 \\ .7 & .3 \end{bmatrix} \qquad Q = \begin{bmatrix} 0 & .5 & 0 \\ 0 & 0 & .9 \\ 0 & 0 & 0 \end{bmatrix}$$

Example 1

Consider the Markov chains with P_1 and P_2 as their transition matrices:

$$P_1 = \begin{array}{c} E_1 \\ E_2 \\ E_3 \\ E_4 \end{array} \begin{array}{cccc} E_1 & E_2 & E_3 & E_4 \\ \begin{bmatrix} .4 & .2 & .4 & 0 \\ 0 & 1 & 0 & 0 \\ .1 & 0 & .5 & .4 \\ .1 & 0 & .3 & .6 \end{bmatrix} \end{array} \qquad P_2 = \begin{array}{c} E_1 \\ E_2 \\ E_3 \end{array} \begin{array}{ccc} E_1 & E_2 & E_3 \\ \begin{bmatrix} 1 & 0 & 0 \\ 0 & \frac{1}{4} & \frac{3}{4} \\ 0 & \frac{1}{3} & \frac{2}{3} \end{bmatrix} \end{array}$$

Test to see whether either or both are absorbing chains.

Solution

The chain having P_1 as its transition matrix is absorbing since the second state is an absorbing state and it is possible to pass from each of the other states to the second. Specifically, it is possible to pass from the first state directly to the second state and from either the third or the fourth state to the first state and then to the second. On the other hand, the matrix P_2 is an example of a nonabsorbing matrix since it is impossible to go from the nonabsorbing state E_2 to the absorbing state E_1. ∎

GAMBLER'S RUIN PROBLEM

Consider the following game involving two players, sometimes called the *gambler's ruin problem*. Player I has $3 and Player II has $2. They flip a fair coin; if it is a head, Player I pays Player II $1, and if it is a tail, Player II pays Player I $1. The total amount of money in the game is, of course, $5. We would like to know how long the game will last, that is, how long it will take for one of the players to go broke or win all the money. (This game can easily be generalized by assuming that Player I has M dollars and Player II has N dollars.)

In this experiment, how much money a player has after any given flip of the coin depends only on how much he had after the previous flip and will not depend (directly) on how much he had in the preceding stages of the game. This experiment can thus be represented by a Markov chain.

For the *gambler's ruin* problem, the game does not have to involve flipping a coin. That is, the probability that Player I wins may not equal the probability that Player II wins. Also, questions can be raised as to what happens if the stakes are doubled, how long the game can be expected to last, and so on.

Suppose the coin being flipped is fair so that a probability of $\frac{1}{2}$ is assigned to each event. The states are the amounts of money each player has at each stage of the game. Each player can increase or decrease the amount of money he has by only $1 at a time.

The transition matrix P is then of the following form:

$$
P = \begin{array}{c@{}c}
 & \begin{array}{cccccc} 0 & 1 & 2 & 3 & 4 & 5 \end{array} \\
\begin{array}{c} 0 \\ 1 \\ 2 \\ 3 \\ 4 \\ 5 \end{array} &
\left[\begin{array}{cccccc}
1 & 0 & 0 & 0 & 0 & 0 \\
\frac{1}{2} & 0 & \frac{1}{2} & 0 & 0 & 0 \\
0 & \frac{1}{2} & 0 & \frac{1}{2} & 0 & 0 \\
0 & 0 & \frac{1}{2} & 0 & \frac{1}{2} & 0 \\
0 & 0 & 0 & \frac{1}{2} & 0 & \frac{1}{2} \\
0 & 0 & 0 & 0 & 0 & 1
\end{array} \right]
\end{array}
$$

Notice that p_{00} is the probability of having $0 given that a player has started with $0. This is a sure event, since, once a player is in state 0, he stays there forever, (he is broke). Similarly p_{55} represents the probability of having $5, given that a player started with $5, which is again a sure event (the player has won all the money).

With regard to this problem, the following questions are of interest:

(a) Given that one gambler is in a nonabsorbing state, what is the expected number of times that he will hold between $1 and $4 inclusive before the termination of the game? That is, on the average, how many times will the process be in nonabsorbing states?

(b) What is the expected length of the process (game)?

(c) What is the probability that an absorbing state is reached (that is, that one gambler will eventually be wiped out)?

To answer the above questions, let's look at the transition matrix P. Rearrange this matrix so that the two absorbing states will appear in the first 2 rows:

$$P = \begin{array}{c} \\ 0 \\ 5 \\ 1 \\ 2 \\ 3 \\ 4 \end{array} \begin{array}{cccccc} 0 & 5 & 1 & 2 & 3 & 4 \\ \left[\begin{array}{cc|cccc} 1 & 0 & 0 & 0 & 0 & 0 \\ 0 & 1 & 0 & 0 & 0 & 0 \\ \hline \frac{1}{2} & 0 & 0 & \frac{1}{2} & 0 & 0 \\ 0 & 0 & \frac{1}{2} & 0 & \frac{1}{2} & 0 \\ 0 & 0 & 0 & \frac{1}{2} & 0 & \frac{1}{2} \\ 0 & \frac{1}{2} & 0 & 0 & \frac{1}{2} & 0 \end{array} \right] \end{array}$$

If we let $I_2, \mathbf{0}, S$, and Q denote the matrices

$$I_2 = \begin{bmatrix} 1 & 0 \\ 0 & 1 \end{bmatrix} \qquad \mathbf{0} = \begin{bmatrix} 0 & 0 & 0 & 0 \\ 0 & 0 & 0 & 0 \end{bmatrix}$$

$$S = \begin{bmatrix} \frac{1}{2} & 0 \\ 0 & 0 \\ 0 & 0 \\ 0 & \frac{1}{2} \end{bmatrix} \qquad Q = \begin{bmatrix} 0 & \frac{1}{2} & 0 & 0 \\ \frac{1}{2} & 0 & \frac{1}{2} & 0 \\ 0 & \frac{1}{2} & 0 & \frac{1}{2} \\ 0 & 0 & \frac{1}{2} & 0 \end{bmatrix}$$

we can write the matrix P in the form

$$P = \left[\begin{array}{c|c} I_2 & \mathbf{0} \\ \hline S & Q \end{array} \right]$$

As we indicated earlier we can do this to any matrix representing an absorbing Markov chain. If r of the states are absorbing, and s of the states are nonabsorbing, then the transition matrix P can be written as

$$P = \left[\begin{array}{c|c} I_r & \mathbf{0} \\ \hline S & Q \end{array} \right]$$

where I_r is the $r \times r$ identity matrix, $\mathbf{0}$ is the zero matrix of dimension $r \times s$, S is of dimension $s \times r$, and Q is of dimension $s \times s$.

To answer the questions raised about the data of the gambler's ruin problem, we need the following result:

For an absorbing Markov chain that has a transition matrix P of the form

$$P = \left[\begin{array}{c|c} I_r & \mathbf{0} \\ \hline S & Q \end{array} \right]$$

where S is of dimensions $s \times r$ and Q is of dimension $s \times s$, define the matrix T as

$$T = [I_s - Q]^{-1} \qquad (1)$$

The entries of T give the expected number of times the process is in each nonabsorbing state, provided the process began in a nonabsorbing state.

The matrix T given in (1) is called the *fundamental matrix* of an absorbing Markov chain.

In the gambler's ruin problem, the fundamental matrix T is

$$T = \left[\begin{bmatrix} 1 & 0 & 0 & 0 \\ 0 & 1 & 0 & 0 \\ 0 & 0 & 1 & 0 \\ 0 & 0 & 0 & 1 \end{bmatrix} - \begin{bmatrix} 0 & \frac{1}{2} & 0 & 0 \\ \frac{1}{2} & 0 & \frac{1}{2} & 0 \\ 0 & \frac{1}{2} & 0 & \frac{1}{2} \\ 0 & 0 & \frac{1}{2} & 0 \end{bmatrix} \right]^{-1} = \begin{array}{c} 1 \\ 2 \\ 3 \\ 4 \end{array} \begin{array}{cccc} 1 & 2 & 3 & 4 \\ \left[\begin{matrix} 1.6 & 1.2 & .8 & .4 \\ 1.2 & 2.4 & 1.6 & .8 \\ .8 & 1.6 & 2.4 & 1.2 \\ .4 & .8 & 1.2 & 1.6 \end{matrix} \right] \end{array}$$

This provides the answers to question (a). The entry .8 in row 3, column 1, indicates that .8 is the expected number of times the player will have $1 if he started with $3. In the fundamental matrix T, the column headings indicate present money, while the row headings indicate money started with.

EXPECTED LENGTH OF STAY

The expected number of steps before absorption for each nonabsorbing state is found by adding the entries in the corresponding row of the fundamental matrix T.

So to answer question (b), we again look at the matrix T. The expected number of games before absorption (when one of the players wins or loses all the money) can be found by adding the entries in each row of T. Thus, if a player starts with $3, the expected number of games before absorption is

$$.8 + 1.6 + 2.4 + 1.2 = 6.0$$

If a player starts with $1, the expected number of games before absorption is

$$1.6 + 1.2 + .8 + .4 = 4.0$$

Our last result deals with the probabilities of being absorbed.

The (i, j)th entry in the matrix product $T \cdot S$ gives the probability that, starting in nonabsorbing state i, we reach the absorbing state j.

Thus, to answer question (c), we find the product of the matrices T and S:

$$T \cdot S = \begin{bmatrix} 1.6 & 1.2 & .8 & .4 \\ 1.2 & 2.4 & 1.6 & .8 \\ .8 & 1.6 & 2.4 & 1.2 \\ .4 & .8 & 1.2 & 1.6 \end{bmatrix} \begin{bmatrix} \frac{1}{2} & 0 \\ 0 & 0 \\ 0 & 0 \\ 0 & \frac{1}{2} \end{bmatrix} = \begin{array}{c} 1 \\ 2 \\ 3 \\ 4 \end{array} \begin{array}{cc} 0 & 5 \\ \left[\begin{matrix} .8 & .2 \\ .6 & .4 \\ .4 & .6 \\ .2 & .8 \end{matrix} \right] \end{array}$$

The entry in row 3, column 2, indicates the probability is .6 that the player starting with $3 will win all the money. The entry in row 2, column 1, indicates the probability is .6 that a player starting with $2 will lose all his money.

The above techniques are applicable in general to any transition matrix P of an absorbing Markov chain.

In the gambler's ruin problem, we assumed Player I started with $3 and Player II with $2. Furthermore, we assumed the probability of Player I winning $1 was .5. Table 3 gives probabilities for ruin and expected length for other kinds of betting situations for which the bet is 1 unit.

Table 3

Probability That Player I Wins	Amount of Units Player I Starts with	Amount of Units Player II Starts with	Probability That Player I Goes Broke	Expected Length of Game	Expected Gain of Player I
.50	9	1	.1	9	0
.50	90	10	.1	900	0
.50	900	100	.1	90,000	0
.50	8000	2000	.2	16,000,000	0
.45	9	1	.210	11	−1.1
.45	90	10	.866	765.6	−76.6
.45	99	1	.182	171.8	−17.2
.40	90	10	.983	441.3	−88.3
.40	99	1	.333	161.7	−32.3

Suppose, for example, that Player I starts with $90 and Player II with $10, with Player I having a probability of .45 of winning (the game being unfavorable to Player I). If at each trial, the stake is $1, Table 3 shows that the probability is .866 that Player I is ruined. If the same game is played with Player I having $9 to start and Player II having $1 to start, the probability that Player I is ruined drops to .210.

*MODEL: THE RISE AND FALL OF STOCK PRICES

This example from M. Dryden*, involves using Markov chains to analyze the movement of stock prices. The stocks (shares) were those whose daily progress was recorded by the London *Financial Times.* At any point in time a given stock was classified as being in one of three states: increase (I), decrease (D), or no change (N), depending on whether its price had risen, fallen, or remained the same compared to the previous trading day. The intent of the model was to study the long range movement of stocks through these states. Data reporting the past history of stock prices over a period of 1097 trading days were used to statistically fit a transition matrix to the model. The transition matrix used was

$$P = \begin{array}{c} \\ I \\ D \\ N \end{array} \begin{array}{ccc} I & D & N \\ \left[\begin{array}{ccc} .586 & .073 & .340 \\ .070 & .639 & .292 \\ .079 & .064 & .857 \end{array}\right] \end{array}$$

Thus, a share that increased one day would have probability .586 of also increasing the next day and a probability of .340 of registering no change the next day. Note that the tendency to remain in the same state is strongest for the no-change state, since p_{33} is the largest of the diagonal elements.

* Adapted from Myles M. Dryden, "Share Price Movements: A Markovian Approach," *Journal of Finance,* **24** (March 1969), pp. 49–60.

The fixed probability vector **t** for the transition matrix can be computed to be

$$\mathbf{t} = [.156 \quad .154 \quad .687]$$

Thus, in the long run, a stock will have increased its price 15.6% of the time, decreased its price 15.4% of the time, and remained unchanged 68.7% of the time.

How long will a stock increase before its price begins to fall? Or, in general, how long will a stock be in a given state before it moves to another one? These questions are highly reminiscent of those encountered while studying absorbing chains, yet here the transition matrix P is not absorbing. Thus the techniques used earlier do not apply. The way out is to engage in some wishful thinking.

Suppose we wanted to know on the average how long a stock would spend in states I or N before arriving in state D. We could then assume state D is an absorbing one and replace the current probability p_{22} with 1 and make all other elements in the second row of P zero. The new matrix P' we have created is

$$P' = \begin{array}{c} \\ I \\ D \\ N \end{array}\begin{array}{ccc} I & D & N \\ \left[\begin{array}{ccc} .586 & .073 & .340 \\ 0 & 1 & 0 \\ .079 & .064 & .857 \end{array}\right] \end{array}$$

which is the matrix of an absorbing chain.

Rearranging and partitioning this matrix, we obtain

$$\begin{array}{c} \\ D \\ I \\ N \end{array}\begin{array}{ccc} D & I & N \\ \left[\begin{array}{c|cc} 1 & 0 & 0 \\ \hline .073 & .586 & .340 \\ .064 & .079 & .857 \end{array}\right] \end{array}$$

so that

$$Q = \begin{array}{c} \\ I \\ N \end{array}\begin{array}{cc} I & N \\ \left[\begin{array}{cc} .586 & .340 \\ .079 & .857 \end{array}\right] \end{array}$$

and the fundamental matrix is

$$(I - Q)^{-1} = \left[\begin{array}{cc} .414 & -.340 \\ -.079 & .143 \end{array}\right]^{-1} = \begin{array}{c} \\ I \\ N \end{array}\begin{array}{cc} I & N \\ \left[\begin{array}{cc} 4.42 & 10.51 \\ 2.44 & 12.80 \end{array}\right] \end{array}$$

The entries of $(I - Q)^{-1}$ are interpreted as follows. The (i, j)th entry gives the average time spent in state j having started in state i before reaching the absorbing state D for the first time.

So, a stock currently increasing would spend an average of 4.42 days increasing and 10.51 days not changing before declining. Hence, a total of 14.93 days on the average would elapse before an increasing stock first began to decrease. Likewise, if a stock is currently exhibiting no change, the second row of the above matrix shows that an average of $2.44 + 12.80 = 15.24$ days would go by before it began to decrease.

Suppose a stock is increasing and then either levels off or declines. How long will it take before it starts rising again? In general, given we are in state i and leave, what is the average time that elapses before we return again to state i? The answer, which we state without proof, can be succinctly given:

> The average amount of time elapsed between visits to state i (called *mean recurrence time*) is given by the reciprocal of the ith component of the fixed probability vector **t**.

Here,

$$\mathbf{t} = [.156 \quad .154 \quad .687]$$

So we can then compute

State	Mean Recurrence Time (Days)
I	6.41
D	6.49
N	1.46

In this model, then, days on which a share's price fails to change follow each other fairly closely.

Answers to Odd-Numbered Problems begin on page A-38.

Exercise 11.3

In Problems 1–6 state which of the given matrices are absorbing Markov chains.

A 1. $\begin{bmatrix} 0 & 1 \\ \frac{1}{4} & \frac{3}{4} \end{bmatrix}$ 2. $\begin{bmatrix} 1 & 0 \\ \frac{1}{3} & \frac{2}{3} \end{bmatrix}$ 3. $\begin{bmatrix} 1 & 0 & 0 \\ \frac{1}{8} & \frac{5}{8} & \frac{2}{8} \\ 0 & 0 & 1 \end{bmatrix}$

4. $\begin{bmatrix} 0 & 0 & 1 \\ 1 & 0 & 0 \\ 0 & 1 & 0 \end{bmatrix}$ 5. $\begin{bmatrix} 0 & 1 & 0 & 0 \\ 1 & 0 & 0 & 0 \\ 0 & 0 & 1 & 0 \\ \frac{1}{4} & 0 & \frac{3}{4} & 0 \end{bmatrix}$ 6. $\begin{bmatrix} \frac{1}{3} & \frac{1}{3} & 0 & \frac{1}{3} \\ 0 & \frac{1}{4} & \frac{1}{4} & \frac{1}{2} \\ 0 & 0 & 1 & 0 \\ 0 & \frac{1}{2} & 0 & \frac{1}{2} \end{bmatrix}$

B 7. Find the fundamental matrix T of the absorbing Markov chain in Problem 3. Also, find S and $T \cdot S$.

8. Follow the directions of Problem 7 for the matrix in Problem 6.

9. Suppose that for a certain absorbing Markov chain the fundamental matrix T is found to be

$$\begin{array}{c c} & \begin{array}{ccc} \$1 & \$2 & \$3 \end{array} \\ \begin{array}{c} \$1 \\ \$2 \\ \$3 \end{array} & \begin{bmatrix} 1.5 & .5 & .8 \\ 1.2 & 2.3 & .6 \\ .3 & 1.8 & 2.1 \end{bmatrix} \end{array}$$

(a) What is the expected number of times a person will have $3 given that he started with $1? With $2?

(b) If a player starts with $3, how many games can he expect to play before absorption?

10. For the data in Problem 9, suppose that we are given the following matrix S:

$$S = \begin{bmatrix} \frac{1}{2} & 0 \\ 0 & 0 \\ 0 & \frac{1}{2} \end{bmatrix}$$

where column 1 stands for state $0 and column 2 for state $4. What is the probability that the player will be absorbed in state $0 if he started with $3?

APPLICATIONS

11. **Gambler's Ruin Problem** A person repeatedly bets $1 each day. If he wins, he wins $1 (he receives his bet of $1 plus winnings of $1). He stops playing when he goes broke, or when he accumulates $3. His probability of winning is .4 and of losing is .6. What is the probability of eventually accumulating $3 if he starts with $1? With $2?

12. **Admissions Analysis** The following data were obtained from the admissions office of a two-year junior college. Of the first-year class (F), 75% became sophomores (S) the next year and 25% dropped out (D). Of those who were sophomores during a particular year, 90% graduated (G) by the following year and 10% dropped out.

(a) Set up a Markov chain with states D, F, G, and S which describes the situation.
(b) How many are absorbing?
(c) Determine the matrix T.
(d) Determine the probability that an entering first-year student will eventually graduate.

13. **Gambler's Ruin Problem** Marsha wants to purchase a $4000 used car, but only has $1000 available. Not wishing to finance the purchase, she makes a series of wagers in which the winnings equal whatever is bet. The probability of winning is .4 and the probability of losing is .6. In a daring strategy Marsha decides to bet all her money or at least enough to obtain $4000 until she loses everything or has $4000. That is, if she has $1000, she bets $1000; if she has $2000, she bets $2000.

(a) What is the expected number of wagers placed before the game ends?
(b) What is the probability that Marsha is wiped out?
(c) What is the probability that Marsha wins the amount needed to purchase the car?

14. Answer the questions in Problem 13 if the probability of winning is .5.

15. Answer the questions in Problem 13 if the probability of winning is .6.

16. **Targeting** Three armored cars, A, B, and C, are engaged in a three-way battle. Armored car A has probability $\frac{1}{3}$ of destroying its target, B has probability $\frac{1}{2}$ of destroying its target, and C has probability $\frac{1}{6}$ of destroying its target. The armored cars fire at the same time and each fires at the strongest opponent not yet destroyed. Using as states the surviving cars at any round, set up a Markov chain and answer the following questions:

(a) How many states are in this chain?
(b) How many are absorbing?
(c) Find the expected number of rounds fired.
(d) Find the probability that A survives.

17. **Stock Price Model** In the model, a stock currently showing no change would, on the average, stay how long in that state?

18. Stock Price Model By making state I an absorbing state compute the average number of days elapsed before a stock currently decreasing would start to increase again.

*11.4

AN APPLICATION TO GENETICS

Most of this section is based on the work of Gregor Mendel. The Mendelian theory of genetics states that many traits of an offspring are determined by the genes of the parents. Each parent has a pair of genes and the basic assumption of Mendel's theory is that the offspring inherits one gene from each parent in a random, independent way.

In the most simple cases genes are of two types: *dominant,* denoted by A, and *recessive,* denoted by a. There are four possible pairings of the two types of genes: AA, Aa, aA, and aa. However, genetically, the two genotypes Aa and aA are the same. An individual having the genotype AA is called *dominant* or *homozygous;* an individual with the genotype Aa is called *hybrid* or *heterozygous;* and one with the genotype aa is called *recessive.*

Let's consider some of the possibilities. If both parents are dominant (homozygous), their offspring must be dominant (homozygous); if both parents are recessive, their offspring are recessive; and if one parent is dominant (homozygous) and one recessive, their offspring are hybrid (heterozygous).

If one parent is dominant (AA) and the other is hybrid (Aa), the offspring must get a dominant gene A from the dominant parent and either a dominant gene A or a recessive gene a from the hybrid parent. In this case, the probability is $\frac{1}{2}$ that the offspring will be dominant (AA) and the probability is $\frac{1}{2}$ that the offspring will be hybrid (Aa).

Similarly, if one parent is recessive (aa) and the other is hybrid (Aa), the probability is $\frac{1}{2}$ that the offspring will be recessive (aa) and is $\frac{1}{2}$ that the offspring will be hybrid (Aa).

If both parents are hybrid (heterozygous), the offspring have equal probability of getting a dominant gene or recessive gene from each parent. Thus, the probability that the offspring will be dominant (homozygous) is $\frac{1}{4}$; recessive, $\frac{1}{4}$; and hybrid (heterozygous), $\frac{1}{2}$. See Figure 4.

---

* This section may be omitted without loss of continuity.
GREGOR JOHANN MENDEL (1822–1884), an Austrian monk, discovered the first laws of heredity and thereby laid the foundation for the science of genetics. For many years, Mendel taught science in the technical high school at Brunn, Austria, without a teacher's license. The reason—he failed the *biology* portion of the license examination!

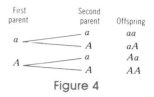

Figure 4

Example 1

Suppose we start with one parent whose genotype is unknown and another whose genotype is known, say hybrid (heterozygous). Their offspring is mated with a person whose genotype is hybrid (heterozygous). This mating procedure is continued. In the long run, what is the genotype of the offspring?

Solution

Since the genotype of an offspring depends solely on the genotype of the parents, such a mating process is an example of a Markov chain. Label the possible states in the process by D (dominant), H (hybrid), and R (recessive). The transition matrix P is

$$P = \begin{array}{c} \\ D \\ H \\ R \end{array} \begin{array}{ccc} D & H & R \\ \begin{bmatrix} \frac{1}{2} & \frac{1}{2} & 0 \\ \frac{1}{4} & \frac{1}{2} & \frac{1}{4} \\ 0 & \frac{1}{2} & \frac{1}{2} \end{bmatrix} \end{array} \tag{1}$$

The entries of P are obtained as follows: The first row $[\frac{1}{2} \quad \frac{1}{2} \quad 0]$ of P gives the probabilities that the offspring will be D, H, R, respectively, when the unknown parent is dominant (AA); the second row $[\frac{1}{4} \quad \frac{1}{2} \quad \frac{1}{4}]$ of P gives the probabilities that the offspring will be D, H, R, respectively, when the unknown parent is hybrid (Aa); the third row $[0 \quad \frac{1}{2} \quad \frac{1}{2}]$ of P gives probabilities that the offspring will be D, H, R, respectively when the unknown parent is recessive (aa).

Now, P is regular since the entries of P^2 are all positive:

$$P^2 = \begin{bmatrix} \frac{3}{8} & \frac{1}{2} & \frac{1}{8} \\ \frac{1}{4} & \frac{1}{2} & \frac{1}{4} \\ \frac{1}{8} & \frac{1}{2} & \frac{3}{8} \end{bmatrix}$$

The fixed probability vector of P is found to be

$$\mathbf{t} = [\tfrac{1}{4} \quad \tfrac{1}{2} \quad \tfrac{1}{4}] \tag{2}$$

Thus, in the long run, no matter what the genotype of the unknown parent, the probability that the genotype of the offspring will be dominant (homozygous) is $\frac{1}{4}$; hybrid (heterozygous), $\frac{1}{2}$; and recessive, $\frac{1}{4}$. ■

Example 2

Brother – Sister Mating Problem In the so-called *brother – sister mating model,* two individuals are mated, and, from among their direct descendants, two individuals of opposite sex are selected at random. These are mated, and the process continues indefinitely. With three possible genotypes, AA, Aa, aa, for each parent, we have to distinguish six combinations of offspring as follows:

$$E_1: \quad AA \times AA \qquad E_2: \quad AA \times Aa \qquad E_3: \quad Aa \times Aa$$
$$E_4: \quad Aa \times aa \qquad E_5: \quad aa \times aa \qquad E_6: \quad AA \times aa$$

where, for example, E_4: $Aa \times aa$ indicates the mating of a hybrid (Aa) with a recessive (aa). The transition matrix for this experiment is

$$
\begin{array}{c c}
 & \begin{array}{c c c c c c} E_1 & E_2 & E_3 & E_4 & E_5 & E_6 \end{array} \\
\begin{array}{c} E_1 \\ E_2 \\ E_3 \\ E_4 \\ E_5 \\ E_6 \end{array} &
\left[\begin{array}{c c c c c c}
1 & 0 & 0 & 0 & 0 & 0 \\
\frac{1}{4} & \frac{1}{2} & \frac{1}{4} & 0 & 0 & 0 \\
\frac{1}{16} & \frac{1}{4} & \frac{1}{4} & \frac{1}{4} & \frac{1}{16} & \frac{1}{8} \\
0 & 0 & \frac{1}{4} & \frac{1}{2} & \frac{1}{4} & 0 \\
0 & 0 & 0 & 0 & 1 & 0 \\
0 & 0 & 1 & 0 & 0 & 0
\end{array}\right]
\end{array}
\qquad (3)
$$

We obtain the entries in (3) using the following reasoning: States E_1 and E_5 both have 1 on the diagonal and 0 for all other elements in the same row, since crossing two dominants (homozygous) always yields a dominant (homozygous); likewise, crossing two recessives always yields a recessive. When the process is in one of the other states, say row E_3, we have

$$E_3: \quad Aa \times Aa \qquad P(AA) = \tfrac{1}{4} \qquad P(aA) = \tfrac{1}{2} \qquad P(aa) = \tfrac{1}{4}$$

$$
\begin{array}{l l l}
E_1: & AA \times AA & \tfrac{1}{4} \times \tfrac{1}{4} = \tfrac{1}{16} \\
E_2: & AA \times Aa & 2 \times \tfrac{1}{4} \times \tfrac{1}{2} = \tfrac{1}{4} \\
E_3: & Aa \times Aa & \tfrac{1}{2} \times \tfrac{1}{2} = \tfrac{1}{4} \\
E_4: & Aa \times aa & 2 \times \tfrac{1}{2} \times \tfrac{1}{4} = \tfrac{1}{4} \\
E_5: & aa \times aa & \tfrac{1}{4} \times \tfrac{1}{4} = \tfrac{1}{16} \\
E_6: & AA \times aa & 2 \times \tfrac{1}{4} \times \tfrac{1}{4} = \tfrac{1}{8}
\end{array}
$$

The process described above is an example of an absorbing Markov chain. The states E_1 and E_5 are absorbing states.

Now, let us perform the same calculations for the transition matrix in (3) as we did for the gambler's ruin problem in Section 11.3. The transition matrix P is rewritten as follows:

$$
P =
\begin{array}{c c}
 & \begin{array}{c c c c c c} E_1 & E_5 & E_2 & E_3 & E_4 & E_6 \end{array} \\
\begin{array}{c} E_1 \\ E_5 \\ E_2 \\ E_3 \\ E_4 \\ E_6 \end{array} &
\left[\begin{array}{c c | c c c c}
1 & 0 & 0 & 0 & 0 & 0 \\
0 & 1 & 0 & 0 & 0 & 0 \\ \hline
\frac{1}{4} & 0 & \frac{1}{2} & \frac{1}{4} & 0 & 0 \\
\frac{1}{16} & \frac{1}{16} & \frac{1}{4} & \frac{1}{4} & \frac{1}{4} & \frac{1}{8} \\
0 & \frac{1}{4} & 0 & \frac{1}{4} & \frac{1}{2} & 0 \\
0 & 0 & 0 & 1 & 0 & 0
\end{array}\right]
\end{array}
= \begin{bmatrix} I_2 & \mathbf{0} \\ S & Q \end{bmatrix}
$$

where the matrices I_2, $\mathbf{0}$, S, and Q are

$$I_2 = \begin{bmatrix} 1 & 0 \\ 0 & 1 \end{bmatrix} \qquad \mathbf{0} = \begin{bmatrix} 0 & 0 & 0 & 0 \\ 0 & 0 & 0 & 0 \end{bmatrix}$$

$$S = \begin{bmatrix} \frac{1}{4} & 0 \\ \frac{1}{16} & \frac{1}{16} \\ 0 & \frac{1}{4} \\ 0 & 0 \end{bmatrix} \qquad Q = \begin{bmatrix} \frac{1}{2} & \frac{1}{4} & 0 & 0 \\ \frac{1}{4} & \frac{1}{4} & \frac{1}{4} & \frac{1}{8} \\ 0 & \frac{1}{4} & \frac{1}{2} & 0 \\ 0 & 1 & 0 & 0 \end{bmatrix}$$

The fundamental matrix T is

$$T = [I_4 - Q]^{-1} = \begin{array}{c} \\ E_2 \\ E_3 \\ E_4 \\ E_6 \end{array} \begin{array}{cccc} E_2 & E_3 & E_4 & E_6 \\ \begin{bmatrix} \frac{8}{3} & \frac{4}{3} & \frac{2}{3} & \frac{1}{6} \\ \frac{4}{3} & \frac{8}{3} & \frac{4}{3} & \frac{1}{3} \\ \frac{2}{3} & \frac{4}{3} & \frac{8}{3} & \frac{1}{6} \\ \frac{4}{3} & \frac{8}{3} & \frac{4}{3} & \frac{4}{3} \end{bmatrix} \end{array}$$

The product of the fundamental matrix and S is

$$T \cdot S = \begin{bmatrix} \frac{8}{3} & \frac{4}{3} & \frac{2}{3} & \frac{1}{6} \\ \frac{4}{3} & \frac{8}{3} & \frac{4}{3} & \frac{1}{3} \\ \frac{2}{3} & \frac{4}{3} & \frac{8}{3} & \frac{1}{6} \\ \frac{4}{3} & \frac{8}{3} & \frac{4}{3} & \frac{4}{3} \end{bmatrix} \begin{bmatrix} \frac{1}{4} & 0 \\ \frac{1}{16} & \frac{1}{16} \\ 0 & \frac{1}{4} \\ 0 & 0 \end{bmatrix} = \begin{array}{c} E_2 \\ E_3 \\ E_4 \\ E_6 \end{array} \begin{array}{cc} E_1 & E_5 \\ \begin{bmatrix} \frac{3}{4} & \frac{1}{4} \\ \frac{1}{2} & \frac{1}{2} \\ \frac{1}{4} & \frac{3}{4} \\ \frac{1}{2} & \frac{1}{2} \end{bmatrix} \end{array}$$

Genetically, the matrix $T \cdot S$ can be interpreted to mean that after a large number of inbred matings, a person is either in state E_1 or state E_5. That is, only pure genotypes remain, while the mixed genotype (hybrid heterozygous) will disappear. Notice also that if one starts in the state E_4: $Aa \times aa$, which has 3 recessive genes and 1 dominant gene, the probability for ending up in the state E_1: $AA \times AA$ is $\frac{1}{4}$, which is the ratio of dominant genes to total genes.

From the fundamental matrix, we can find the expected number of generations needed to pass from a nonabsorbing state to either absorbing state. Thus, the expected number of generations needed to pass from E_3 to either E_1 or E_5 is

$$\frac{4}{3} + \frac{8}{3} + \frac{4}{3} + \frac{1}{3} = \frac{17}{3} = 5\frac{2}{3}$$ ∎

Finally, consider a genetic experiment in which a large population is randomly mated. We assume that males and females have the same proportion of each genotype and that male and female offspring are equally likely to occur. It would seem a logical conclusion of Mendel's law that after a large number of matings, the recessive genotype must disappear. However, the mere fact that recessive genotypes continue to exist implies that this is not the case. This seeming discrepancy in the theory was resolved early in the twentieth century, by the famous mathematician G. H. Hardy who proved that the proportion of genotypes in a population stabilizes after one generation.*

G. H. HARDY (1877–1947), an English mathematician, is credited with the discovery in 1908 of the Hardy–Weinberg law in genetics. He published many brilliant articles in number theory as well.

* G. H. Hardy, "Mendelian Proportions in a Mixed Population," *Science,* N.S. **28** (1908), pp. 49–50.

Exercise 11.4

Answers to Odd-Numbered Problems begin on page A-39.

1. In Example 1, prove that the fixed probability vector **t** for the transition matrix P of (1) is given by (2).

2. In Example 1, suppose the known genotype is dominant (homozygous) and each offspring is mated with a person having a dominant (homozygous) genotype.

   (a) Find the transition matrix P.
   (b) Find the fixed probability vector. Interpret the answer.
   (c) Find the fundamental matrix T.
   (d) What is the expected number of generations needed to pass from each nonabsorbing stage?

3. Answer the same questions given in Problem 2 if the known genotype is recessive.

CHAPTER REVIEW

Important Terms and Formulas

Markov chain	transition probability	equilibrium distribution
initial state	probability vector	absorbing state
initial probability distribution	regular Markov chain	absorbing Markov chain
transition matrix	fixed probability vector	fundamental matrix
	equilibrium state	

$$v^{(k)} = v^{(k-1)}P$$
$$v^{(k)} = v^{(0)}P^k$$

$$P = \left[\begin{array}{c|c} I_r & 0 \\ \hline S & Q \end{array}\right]$$

$$T = [I_s - Q]^{-1}$$

True-False Questions

(Answers are on page A-39)

T   F   1. The matrix $\begin{bmatrix} 1 & 0 \\ -1 & 1 \end{bmatrix}$ is a transition matrix for a Markov chain.

T   F   2. The transition matrix $\begin{bmatrix} \frac{1}{2} & \frac{1}{2} \\ 0 & 1 \end{bmatrix}$ is regular.

T   F   3. The matrix $\begin{bmatrix} 0 & 1 \\ \frac{1}{3} & \frac{2}{3} \end{bmatrix}$ is an absorbing Markov chain.

T   F   4. If P is a transition matrix of a Markov chain, then P^2 gives the probability of moving from one state to another state in two stages.

Fill in the Blanks

(Answers are on page A-39)

1. In a Markov chain with m states, the initial probability distribution is a _____ row vector.

2. A probability vector is a vector whose entries are _____ and sum up to _____.

3. In a Markov chain with transition matrix P the probability distribution $v^{(k)}$ after k observations satisfies _____.

4. A Markov chain is said to be regular if for some power of its transition matrix P, all entries are _____.

Review Exercises   *Answers to Odd-Numbered Problems begin on page A-39.*

A   1. Find the fixed probability vector of:

   (a) $\begin{bmatrix} \frac{1}{4} & \frac{3}{4} \\ \frac{1}{2} & \frac{1}{2} \end{bmatrix}$

   (b) $\begin{bmatrix} \frac{1}{3} & \frac{2}{3} \\ \frac{2}{3} & \frac{1}{3} \end{bmatrix}$

B   2. Define and explain in words the meaning of a *regular* transition matrix. Give an example of such a matrix and of a matrix that is not regular.

APPLICATIONS   3. **Customer Loyalty** Three beer distributors, A, B, and C, each presently holds $\frac{1}{3}$ of the beer market. Each wants to increase its share of the market, and to do so, each introduces a new brand. During the next year, it is learned that:

   (a) A keeps 50% of its business and loses 20% to B and 30% to C.
   (b) B keeps 40% of its business and loses 40% to A and 20% to C.
   (c) C keeps 25% of its business and loses 50% to A and 25% to B.

   Assuming this trend continues, after 2 years what share of the market does each distributor have? In the long run, what is each distributor's share?

4. If the current share of the market for each beer distributor A, B, and C in Problem 3 is A: 25%, B: 25%, C: 50%, and the market trend is the same, answer the same questions.

5. **Marketing** A representative of a book publishing company has to cover three universities, U_1, U_2, and U_3. She never sells at the same university in successive months. If she sells at University U_1, then the next month she sells at U_2. However, if she sells at either U_2 or U_3, then the next month she is three times as likely to sell at U_1 as at the other university. In the long run, how often does she sell at each of the universities?

6. **Family Traits** Assume that the probability of a fat father having a fat son is .7 and that of a skinny father having a skinny son is .4. What is the probability of a fat father being the great grandfather of a fat great grandson? In the long run, what will be the distribution? Does it depend on the initial physical state of the fathers?

7. **Gambler's Ruin Problem** Suppose a man has $2, which he is going to bet $1 at a time until he either loses all his money or he has $5. Assume he wins with a probability of .45 and he loses with a probability of .55. Construct the transition probability for this game and answer the questions as stated on page 544 in Section 11.3. *Hint:* The fundamental matrix T of the transition matrix P is

$$T = \begin{bmatrix} 1.584282 & 1.062331 & .635281 & .285876 \\ 1.298405 & 2.360736 & 1.411736 & .635281 \\ .94899 & 1.725454 & 2.360736 & 1.062331 \\ .52194 & .949000 & 1.298405 & 1.584281 \end{bmatrix}$$

PART THREE

DISCRETE MATHEMATICS

12

LOGIC AND LOGIC CIRCUITS

12.1

PROPOSITIONS

In this chapter, we survey many of the fundamental concepts found in the area of mathematics called *logic.* There are several reasons for studying logic. Two of the more important ones are *(1)* to gain proficiency in correct mathematical reasoning and *(2)* to apply the tool of logic to practical situations such as the design of *logic circuits* used in computers and other electronic devices.

In mathematics, the words "not" and "or" and the phrases "if . . . , then . . . ," "if and only if," and so on are used extensively. A knowledge of the exact meaning of these words is necessary before we can make precise the laws of inference and deduction that are constantly used in mathematics. The study of logic will enable you to gain a basic understanding of what constitutes a mathematical argument. This will eliminate common errors made in mathematical, as well as nonmathematical, arguments.

We hope this chapter will give you some indication of the usefulness of logic in uncovering ambiguities and nonsequiturs. Furthermore, we hope this chapter offers evidence in favor of Church's remark that "the value of logic . . . is not that it supports a particular system, but that the process of logical organization of any system (empiricist or otherwise) serves to test its internal consistency, to verify its logical adequacy to its declared purpose, and to isolate and clarify the assumptions on which it rests." *

In the last two sections of this chapter, we apply the concepts of logic to a problem in the application of conflicting rules and to the design of electronic circuits.

PROPOSITIONS

English and similar languages are composed of various words and phrases with distinct functions that have a bearing on the meaning of the sentences in which they occur.

English sentences may be classified as *declarative, interrogative, exclamatory, or imperative.* In the study of logic, we assume that we are able to recognize a declarative sentence (or *statement*) and form an opinion as to whether it is true or false.

Proposition **A *proposition* is a declarative sentence that can be meaningfully classified as either true or false.**

---

* Church, *Introduction to Mathematical Logic,* Vol. 1, Princeton University Press, Princeton, N.J., 1956, p. 55.

Example 1
The price of an IBM-AT personal computer was $1800 on June 16, 1987.
 This is a proposition, although few of us can say whether it is true or false. ■

Example 2
The earth is round.
 This sentence records a possible fact about reality and is a proposition. Some people would classify this proposition as true and others as false, depending on what the word *round* means to them. (Does it mean simply "curved," or "perfectly spherical?") Only in an ideal situation can we unequivocally state to which truth category a proposition belongs. ■

Example 3
What is the exchange rate from United States dollars to German marks?
 This is not a proposition — it is an interrogative sentence. ■

Example 4
The prices of most stocks on the New York Stock Exchange rose during the period 1929–1931.
 This is a proposition. It happens to be false. ■

In an ideal situation, a proposition could be easily and decisively classified as true or false. However, as Examples 1 and 2 illustrate, it is often difficult to classify propositions in this way, because of unclear meanings, ambiguous situations, differences of opinion, etc. In the mathematical treatment of logic, we avoid these difficulties by assuming "for the sake of argument" either the truth or the falsity of certain propositions to draw conclusions about other propositions, using *symbolic logic.*

COMPOUND PROPOSITIONS
Consider the proposition "Jones is handsome and Smith is selfish." This sentence is obtained by joining the two propositions "Jones is handsome," "Smith is selfish" by the word "and."

Compound Proposition; Connectives **A *compound proposition* is a proposition formed by connecting two or more propositions or by negating a single proposition. The words and phrases (or symbols) used to form compound propositions are called *connectives.***

Some of the connectives used in English are: *or; either . . . or; and; but; if . . . , then; not.*

Conjunction **Let *p* and *q* denote propositions. The compound proposition *p and q* is called the *conjunction of p and q* and is denoted symbolically by**

$$p \wedge q$$

We define $p \wedge q$ to be true when both p and q are true and to be false otherwise.

Example 5

Consider the two statements

> p: Washington, D.C., is the capital of the United States.
>
> q: Hawaii is the fiftieth state of the United States.

The conjunction of p and q is

> $p \wedge q$:   Washington, D.C., is the capital of the United States, and Hawaii is the fiftieth state of the United States.

Since both p and q are true statements, we conclude that the compound statement $p \wedge q$ is true.   ■

Example 6

Consider the two statements

> p: Washington, D.C., is the capital of the United States.
>
> q: Vermont is the largest of the fifty states.

The compound proposition $p \wedge q$ is

> $p \wedge q$:   Washington, D.C., is the capital of the United States, and Vermont is the largest of the fifty states.

Since the statement q is false, the compound statement $p \wedge q$ is false—even though p is true.   ■

Inclusive Disjunction   **Let p and q be any propositions. The compound proposition p or q is called the *inclusive disjunction of p and q* and is denoted symbolically by**

$$p \vee q$$

We define $p \vee q$ to be true if *at least one* of the propositions p, q is true. That is, $p \vee q$ is true if both p and q are true, if p is true and q is false, or if p is false and q is true. It is false only if both p and q are false.

Example 7

Consider the propositions

> p: XYZ Company is the largest producer of nails in the world.
>
> q: Mines. Ltd., has three uranium mines in Nevada.

The compound proposition $p \vee q$ is

> $p \vee q$:   XYZ Company is the largest producer of nails in the world or Mines. Ltd. has three uranium mines in Nevada.   ■

EXCLUSIVE DISJUNCTION

The English word "or" can be used in two different ways—as an inclusive ("and/or") or exclusive ("either/or") disjunction. The correct meaning is usually inferred from the context in which the word is used. However, when it is important to be precise (as it often is in mathematics, business, science, etc.) we must carefully distinguish between the two meanings of "or."

Exclusive Disjunction **Let p and q be any proposition. The *exclusive disjunction of p and q*, read as *either p or q but not both* is denoted symbolically by**

$$p \veebar q$$

We define $p \veebar q$ to be true if exactly one of the propositions p, q is true. That is, $p \veebar q$ is true if p is true and q is false, or if p is false and q is true. It is false if both p and q are false or if both p and q are true.

Example 8

Consider the propositions

p: XYZ Company earned $3.20 per share in 1980.

q: XYZ Company paid a dividend of $1.20 per share in 1980.

The inclusive disjunction of p and q is

$p \vee q$: XYZ Company earned $3.20 per share in 1980 or XYZ Company paid a dividend of $1.20 per share in 1980, or both.

The exclusive disjunction of p and q is

$p \veebar q$: XYZ Company earned $3.20 per share in 1980 or XYZ Company paid a dividend of $1.20 per share in 1980, but *not* both.   ■

Example 9

Consider the compound proposition

p: This weekend I will date Caryl or Mary.

The use of the connective "or" is not clear here. If the "or" means inclusive disjunction, then at least one and possibly both girls will be dated. If the "or" means exclusive disjunction, then only one girl will be dated.   ■

NEGATION; QUANTIFIERS

Negation **If p is any proposition, the *negation of p*, denoted by**

$$\sim p$$

and read as *not p*, is a proposition that is false when p is true and true when p is false.

The negation of p is sometimes called the *denial* of p. The symbol $\sim$ is called the *negation operator*.

The definition of $\sim p$ assumes that p and $\sim p$ cannot both be true. In classical logic, this assumption is known as the *law of contradiction.*

Example 10

Consider the proposition

p: One share of XYZ stock is worth less than \$85.

The negation of p is

$\sim p$:   One share of XYZ stock is worth at least \$85.

The sentence "A share of XYZ stock is worth more than \$85" is *not* a correct statement of the negation of p.   ■

A *quantifier* is a word or phrase telling how many (Latin *quantus*). English quantifiers include "all," "none," "some," and "not all." The quantifiers "all," "every," and "each" are interchangeable. The quantifiers "some," "there exist(s)," and "at least one" are also interchangeable.

Example 11

The following propositions all have the same meaning:

p: All people are intelligent.

q: Every person is intelligent.

r: Each person is intelligent.

s: Any person is intelligent.   ■

Example 12

The negation of the proposition

p: All students are intelligent.

is

$\sim p$:   Some students are not intelligent.

$\sim p$:   There exists a student who is not intelligent.

$\sim p$:   At least one student is not intelligent.

The negation of

q: No student is intelligent.

is

$\sim q$:   At least one student is intelligent.

Note that "No student is intelligent" is *not* the negation of p; "All students are intelligent" is *not* the negation of q.   ■

Exercise 12.1   *Answers to Odd-Numbered Problems begin on page A-40.*

In Problems 1–8 determine which are propositions.

1. The cost of shell egg futures was up on June 18, 1980.
2. The gross national product exceeded one billion dollars in 1935.
3. What a portfolio!
4. Why did you buy XYZ Company stock?
5. The earnings of XYZ Company doubled last year.
6. Where is the new mine of Mines, Ltd?
7. Jones is guilty of murder in the first degree.
8. What a hit!

In Problems 9–16 negate each proposition.

9. A fox is an animal.
10. The outlook for bonds is not good.
11. I am buying stocks and bonds.
12. Mike is selling his apartment building and his business.
13. No one wants to buy my house.
14. Everyone has at least one television set.
15. Some people have no car.
16. Jones is permitted not to see that all votes are not counted.

In Problems 17–24 let p denote "John is an economics major" and let q denote "John is a sociology minor." State each proposition as a simple sentence.

17. $p \lor q$ 18. $p \underline{\lor} q$ 19. $p \land q$
20. $\sim p$ 21. $\sim p \lor \sim q$ 22. $\sim(\sim q)$
23. $\sim p \lor q$ 24. $\sim p \land q$

12.2

TRUTH TABLES

TRUTH VALUES AND TRUTH TABLES
LOGICAL EQUIVALENCE; THE LAWS OF LOGIC

TRUTH VALUES AND TRUTH TABLES

The *truth value* of a proposition is either *true* (denoted by T) or *false* (denoted by F). A *truth table* is a table that shows the truth value of a compound proposition for all possible cases.

For example, consider the conjunction of any two propositions p and q. Recall that $p \wedge q$ is false if either p is false or q is false, or if both p and q are false. There are four possible cases.

1. p is true and q is true.
2. p is true and q is false.
3. p is false and q is true.
4. p is false and q is false.

These four cases are listed in the first two columns of Table 1, which is the truth table for $p \wedge q$. For convenience, the cases for p and q will always be listed in this order.

The truth values of $\sim p$ are given in Table 2.

Using the previous definitions of inclusive disjunction and exclusive disjunction, we obtain truth tables for $p \vee q$ and $p \veebar q$. See Table 3.

Table 1

	p	q	$p \wedge q$
Case 1	T	T	T
Case 2	T	F	F
Case 3	F	T	F
Case 4	F	F	F

Table 2

p	$\sim p$
T	F
F	T

Table 3

p	q	$p \vee q$	$p \veebar q$
T	T	T	F
T	F	T	T
F	T	T	T
F	F	F	F

Besides using the connectives $\wedge$, $\vee$, $\veebar$, $\sim$ one at a time to form compound propositions, we can use them together to form more complex statements.

For example, Table 4 is the truth table for $(p \vee q) \veebar (\sim p)$. The parentheses are used to indicate that $\vee$ and $\sim$ are applied before $\veebar$.

Observe that the first columns of the table are for the component propositions $p, q, \ldots$ and that there are enough rows in the table to allow for all possible combinations of T and F. (For two components p and q, 4 rows are necessary; for three components p, q, and r, 8 rows would be necessary, and so on; for n statements, 2^n rows are needed.)

Table 4 has five columns, and each column corresponds to a stage in construct-

Table 4

p	q	$p \vee q$	$\sim p$	$(p \vee q) \veebar (\sim p)$
T	T	T	F	T
T	F	T	F	T
F	T	T	T	F
F	F	F	T	T

ing the compound proposition, beginning with the simple components p, q, . . . At each stage, one or more components constructed in earlier stages are combined, until we obtain the final result in the last column.

We first outline the truth table for the given compound proposition:

p	q	$p \vee q$	$\sim p$	$(p \vee q) \underline{\vee} (\sim p)$
T	T			
T	F			
F	T			
F	F			

Each stage of the proposition is written on the top row, to the right of its intermediate stages; there is a column underneath each component or connective. Truth values are then entered in the truth table, one step at a time:

Stage 1

p	q	$p \vee q$	$\sim p$	$(p \vee q) \underline{\vee} (\sim p)$
T	T	T		
T	F	T		
F	T	T		
F	F	F		

Stage 2

p	q	$p \vee q$	$\sim p$	$(p \vee q) \underline{\vee} (\sim p)$
T	T	T	F	
T	F	T	F	
F	T	T	T	
F	F	F	T	

Stage 3

p	q	$p \vee q$	$\sim p$	$(p \vee q) \underline{\vee} (\sim p)$
T	T	T	F	T
T	F	T	F	T
F	T	T	T	F
F	F	F	T	T

The truth table of the compound proposition then consists of the original columns under p and q and the fifth column entered into the table in the last stage.

Example 1
Construct the truth table for $(p \vee \sim q) \wedge p$. The component parts of this proposition are p, $\sim q$, and $p \vee \sim q$.

p	q	$\sim q$	$p \vee \sim q$	$(p \vee \sim q) \wedge p$
T	T	F	T	T
T	F	T	T	T
F	T	F	F	F
F	F	T	T	F

The parentheses are used to indicate that the $\vee$ is applied before $\wedge$ and the $\sim$ applies only to q. Notice that the truth table is the same as for p alone. ∎

Example 2

Construct the truth table for $\sim(p \vee q) \vee (\sim p \wedge \sim q)$.

p	q	$\sim p$	$\sim q$	$p \vee q$	$\sim(p \vee q)$	$(\sim p \wedge \sim q)$	$\sim(p \vee q) \vee (\sim p \wedge \sim q)$
T	T	F	F	T	F	F	F
T	F	F	T	T	F	F	F
F	T	T	F	T	F	F	F
F	F	T	T	F	T	T	T

The parentheses indicate that the first $\sim$ symbol negates $p \vee q$ [not p alone and not $(p \vee q) \vee (\sim p \wedge \sim q)$]. Notice that the statements $\sim(p \vee q)$, $\sim p \wedge \sim q$, and $\sim(p \vee q) \vee (\sim p \wedge \sim q)$ all have the same truth table. ∎

The next example is of a truth table involving three components, p, q, and r.

Example 3

Construct the truth table for $p \wedge (q \vee r)$.

p	q	r	$q \vee r$	$p \wedge (q \vee r)$
T	T	T	T	T
T	T	F	T	T
T	F	T	T	T
T	F	F	F	F
F	T	T	T	F
F	T	F	T	F
F	F	T	T	F
F	F	F	F	F

∎

LOGICAL EQUIVALENCE; THE LAWS OF LOGIC

Very often, two propositions stated in different ways have the same meaning. For example, in law, "Jones agreed and is obligated to paint Smith's house" has the same meaning as "It was agreed and contracted by Jones that Jones would paint the house belonging to Smith and Jones is therefore required to paint the aforesaid house."

Logically Equivalent **If two propositions a and b have the same truth values in every possible case, the propositions are called *logically equivalent*. This relationship is denoted by $a \equiv b$.**

Example 4

Show that $\sim(p \wedge q)$ is logically equivalent to $\sim p \vee \sim q$.

Solution

Construct the truth table as shown below.

1	2	3	4	5	6	7
p	q	$p \wedge q$	$\sim(p \wedge q)$	$\sim p$	$\sim q$	$\sim p \vee \sim q$
T	T	T	F	F	F	F
T	F	F	T	F	T	T
F	T	F	T	T	F	T
F	F	F	T	T	T	T

Since the entries in columns 4 and 7 of the truth table are the same, the two propositions are logically equivalent. ■

Example 5

Show that $\sim p \wedge \sim q$ is logically equivalent to $\sim(p \vee q)$.

Solution

Construct the truth table as shown. The entries under $\sim p \wedge \sim q$ and $\sim(p \vee q)$ are the same, so the two propositions are logically equivalent.

p	q	$p \vee q$	$\sim p$	$\sim q$	$\sim p \wedge \sim q$	$\sim(p \vee q)$
T	T	T	F	F	F	F
T	F	T	F	T	F	F
F	T	T	T	F	F	F
F	F	F	T	T	T	T

■

The following laws are especially useful in the study of logic circuits. They can be proved using truth tables.

> **Idempotent Laws** For any proposition p,
> $$p \wedge p \equiv p \qquad p \vee p \equiv p$$

Here, $p \wedge p \equiv p$ since $p \wedge p$ is true when p is true and false when p is false. A similar argument shows that $p \vee p \equiv p$.

> **Commutative Laws** For any two propositions p and q,
> $$p \wedge q \equiv q \wedge p \qquad p \vee q \equiv q \vee p$$

The commutative laws state that, if two or more propositions are combined using the connective *and,* changing the order in which the components are connected does not change the meaning of the compound proposition. The same is true of the connective *or.*

For example, if we consider the two statements

p: Mrs. Jones is attractive.

q: Mr. Jones is intelligent.

we see that the compound propositions

$p \wedge q$:   Mrs. Jones is attractive and Mr. Jones is intelligent.

$q \wedge p$:   Mr. Jones is intelligent and Mrs. Jones is attractive.

have the same meaning.

> **Associative Laws** For any three propositions p, q, r,
> $$(p \wedge q) \wedge r \equiv p \wedge (q \wedge r) \qquad (p \vee q) \vee r \equiv p \vee (q \vee r)$$

Because of the associative laws, it is possible to omit the parentheses when using the same connective more than once. For instance, we can write

$$p \wedge q \wedge r \qquad \text{for} \qquad (p \wedge q) \wedge r$$

and

$$p \wedge q \wedge r \wedge s \qquad \text{for} \qquad [(p \wedge q) \wedge r] \wedge s$$

and similarly for $\vee$. Note, however, that the parentheses cannot be omitted when using $\wedge$ and $\vee$ together; see Problems 34 and 35.

> **Distributive Laws** For any three propositions p, q, r,
> $$p \vee (q \wedge r) \equiv (p \vee q) \wedge (p \vee r)$$
> $$p \wedge (r \vee q) \equiv (p \wedge r) \vee (p \wedge q)$$

Here, $p \vee (q \wedge r) \equiv (p \vee q) \wedge (p \vee r)$ means that *p or (q and r)* has the same meaning as *(p or q) and (p or r)*. Also, $p \wedge (r \vee q) \equiv (p \wedge r) \vee (p \wedge q)$ means that the compound proposition *p and (r or q)* has the same meaning as *(p and r) or (p and q)*.

Example 6

Consider the three propositions

p: Betsy will do her homework.

q: Betsy will wash her car.

r: Betsy will read a book.

The first distributive law, namely,

$$p \vee (q \wedge r) \equiv (p \vee q) \wedge (p \vee r)$$

says that these two statements are logically equivalent:

1. Betsy will do her homework, or she will wash her car and read a book.
2. Betsy will do her homework or wash her car, and Betsy will do her homework or read a book. ∎

> **De Morgan's Laws** For any two propositions p and q,
>
> $$\sim (p \vee q) \equiv \sim p \wedge \sim q \qquad \sim (p \wedge q) \equiv \sim p \vee \sim q$$

That is, the compound proposition *p or q* is false only when *p* and *q* are both false. Similarly, *p and q* is false when either *p* or *q* (or both) is false. De Morgan's laws are proved using the truth tables in Examples 4 and 5.

Example 7

Negate the compound statements:

(a) The first child is a girl and the second child is a boy.

(b) Tonight I will study or I will go bowling.

Solution

(a) Let p and q represent the components:

p: The first child is a girl.

q: The second child is a boy.

AUGUSTUS DE MORGAN (1806–1871), British mathematician and logician, was born in Madura, India, the son of a British army officer. He was graduated from Trinity College in Cambridge, England in 1827, but was denied a teaching position there for refusing to subscribe to religious tests. He was, however, appointed to a mathematics professorship at the newly opened University of London. He is best known for his work *Formal Logic,* which appeared in 1847. He also wrote papers on the foundations of algebra, philosophy of mathematical methods, and probability, as well as several successful elementary textbooks.

To negate the statement $p \wedge q$ is to find $\sim(p \wedge q)$. By De Morgan's law,

$$\sim(p \wedge q) \equiv \sim p \vee \sim q$$

The negation can be stated as

> The first child is not a girl or the second child is not a boy.

(b)   As above, we represent p and q as

> p: I will study.
>
> q: I will go bowling.

We want to negate the statement $p \vee q$. By De Morgan's law,

$$\sim(p \vee q) \equiv \sim p \wedge \sim q$$

The negation can be stated as

> Tonight I will not study and I will not go bowling.   ■

Absorption Laws For any two propositions p and q,

$$p \vee (p \wedge q) \equiv p \qquad p \wedge (p \vee q) \equiv p$$

Here, $p \vee (p \wedge q) \equiv p$ means that p *or* (p *and* q) is logically equivalent to p. Similarly, $p \wedge (p \vee q) \equiv p$ means that p *and* (p *or* q) is logically equivalent to p.

Example 8
Use truth tables to prove the distributive law, $p \vee (q \wedge r) \equiv (p \vee q) \wedge (p \vee r)$.

Solution
Since the entries in the last two columns of the truth table below are the same, the two propositions are logically equivalent.

p	q	r	$q \wedge r$	$p \vee q$	$p \vee r$	$p \vee (q \wedge r)$	$(p \vee q) \wedge (p \vee r)$
T	T	T	T	T	T	T	T
T	T	F	F	T	T	T	T
T	F	T	F	T	T	T	T
T	F	F	F	T	T	T	T
F	T	T	T	T	T	T	T
F	T	F	F	T	F	F	F
F	F	T	F	F	T	F	F
F	F	F	F	F	F	F	F

Example 9

Use truth tables to prove the idempotent law, $p \wedge p \equiv p$.

Solution

The truth table is

p	$p \wedge p$
T	T
F	F

■

Exercise 12.2

Answers to Odd-Numbered Problems begin on page A-40.

A   In Problems 1–16 construct a truth table for each compound proposition.

1. $p \vee \sim q$ 2. $\sim p \vee \sim q$

3. $\sim p \wedge \sim q$ 4. $\sim p \wedge q$

5. $\sim(\sim p \wedge q)$ 6. $(p \vee \sim q) \wedge \sim p$

7. $\sim(\sim p \vee \sim q)$ 8. $(p \vee \sim q) \wedge (q \wedge \sim p)$

9. $(p \vee \sim q) \wedge p$ 10. $p \wedge (q \vee \sim q)$

11. $(p \veebar q) \wedge (p \wedge \sim q)$ 12. $(p \wedge \sim q) \vee (q \wedge \sim p)$

13. $(p \wedge q) \vee (\sim p \wedge \sim q)$ 14. $(p \wedge q) \vee (p \wedge r)$

15. $(p \wedge \sim q) \veebar r$ 16. $(\sim p \vee q) \wedge \sim r$

In Problems 17–22 construct a truth table for each law.

17. Idempotent laws   18. Commutative laws

19. Associative laws   20. Distributive laws

21. Absorption laws   22. De Morgan's laws

In Problems 23–26 show that the given propositions are logically equivalent.

23. $p \wedge (\sim q \vee q)$ and p 24. $p \vee (q \wedge \sim q)$ and p

25. $\sim(\sim p)$ and p 26. $p \wedge q$ and $q \wedge p$

In Problems 27–30 construct a truth table for each proposition.

27. $p \wedge (q \wedge \sim p)$

28. $(p \wedge q) \vee p$

29. $[(p \wedge q) \vee (\sim p \wedge \sim q)] \wedge p$

30. $(\sim p \wedge \sim q \wedge r) \vee (p \wedge q \wedge r)$

B   In Problems 31–33 use the propositions

         p: Smith is an exconvict.

         q: Smith is rehabilitated.

to give examples, using English sentences, of each law.

31. Idempotent laws   **32.** Commutative laws

33. De Morgan's laws

34. Use the distributive and commutative laws to show that

$$(p \vee q) \wedge r \equiv (p \wedge r) \vee (q \wedge r)$$

35. Use the distributive and commutative laws to show that

$$(p \wedge q) \vee r \equiv (p \vee r) \wedge (q \vee r)$$

36. The statement

> The actor is intelligent or handsome and talented.

could be interpreted to mean either

> *a:*   The actor is intelligent, or he is handsome and talented;

or

> *b:*   The actor is intelligent or handsome, and he is talented.

Describe a case in which *a* is true and *b* is false.

37. The statement

> Michael will sell his car and buy a bicycle or rent a truck.

could be interpreted to mean either

> *a:*   Michael will sell his car, and he will buy a bicycle or rent a truck;

or

> *b:*   Michael will sell his car and buy a bicycle, or he will rent a truck.

Find a case in which *b* is true and *a* is false.

In Problems 38–40 use De Morgan's laws to negate each proposition.

38. Mike can hit the ball well and he can pitch strikes.

39. Katy is a good volleyball player and is not conceited.

40. The baby is crying or talking all the time.

12.3

IMPLICATIONS; THE BICONDITIONAL CONNECTIVE; TAUTOLOGIES

THE CONDITIONAL CONNECTIVE
CONVERSE, CONTRAPOSITIVE, AND INVERSE
THE BICONDITIONAL CONNECTIVE
TAUTOLOGIES
SUBSTITUTION

THE CONDITIONAL CONNECTIVE

Consider the following compound proposition: 'If I get an A in Math, then I will continue to study." The above sentence states a condition under which I will continue to study. Another example of such a proposition is: "If we get an offer of $100,000, we will sell the house."

Such propositions occur frequently in mathematics, and an understanding of their nature is extremely important.

Implication; Conditional Connective **If p and q are any two propositions, then we call the proposition**

If p, then q

an *implication* or *conditional statement* and the connective *if . . . , then* the *conditional connective.*

We denote the conditional connective symbolically by $\Rightarrow$ and the implication *If p, then q* by $p \Rightarrow q$. In the implication $p \Rightarrow q$, p is called the *hypothesis* and q is called the *conclusion*. The implication $p \Rightarrow q$ can also be read as follows:

1. p implies q.
2. p is sufficient for q.
3. p only if q.
4. q is necessary for p.
5. q, if p.

To arrive at a truth table for implication, we consider the following situation. Suppose we make the statement

If XYZ common stock reaches $90 per share, it will be sold.

When is the statement true and when is it false? Clearly, if XYZ stock reaches $90 per share and it is not sold, the implication is false. It is also clear that if XYZ stock reaches $90 per share and it is sold, the implication is true. In other words, if the hypothesis and conclusion are both true, the implication is true; if the hypothesis is true and the conclusion false, the implication is false.

But what happens when the hypothesis is false? We say that the implication is true (in this case, the stock could be sold anyway or it might not be sold. In either case, we would not say that a person making the above statement was a liar. That is, we would not accuse the person of making a false statement. For this reason we say that an implication with a false hypothesis is not false and, therefore, is true.) The truth table for implication is given in Table 5.

Table 5

p	q	$p \Rightarrow q$
T	T	T
T	F	F
F	T	T
F	F	T

As in the previous section, we express a compound proposition symbolically by replacing each component statement and connective by an appropriate symbol.

For example, denoting "I study" and "I will pass" by a and b, respectively, the proposition "If I study, then I will pass" is written as $a \Rightarrow b$. This proposition can also be read as "A sufficient condition for passing is to study."

To understand "implication" better, look at an implication as a conditional promise. If the promise is broken, the implication is false; otherwise, it is true. For this reason, the only circumstance under which the implication $p \Rightarrow q$ is false is when p is true and q is false.

Example 1

Consider the implication

If you are a thief, then you will go to jail.

If you are a thief and you do go to jail, the implication is true. If you are a thief and you do not go to jail, the promise is broken; thus, the implication is false. If you are not a thief, we have no way of telling what would happen if you were a thief. Since the premise is not tested, it is not broken; thus, the implication is true. ■

The word *then* in an implication merely serves to separate the conclusion from the hypothesis — it can be, and often is, omitted.

The implication $p \Rightarrow q$ can be expressed in the symbols defined previously. In fact, Table 6 shows that $p \Rightarrow q$ is logically equivalent to the compound proposition $\sim p \vee q$.

Table 6

p	q	$\sim p$	$\sim p \vee q$	$p \Rightarrow q$
T	T	F	T	T
T	F	F	F	F
F	T	T	T	T
F	F	T	T	T

CONVERSE, CONTRAPOSITIVE, AND INVERSE

Suppose we start with the implication $p \Rightarrow q$ and then interchange the roles of p and q, obtaining the implication, "If q, then p."

Converse **The implication *If q, then p* is called the *converse* of the implication *If p, then q*. That is, $q \Rightarrow p$ is the converse of $p \Rightarrow q$.**

The truth tables for the implication $p \Rightarrow q$ and its converse $q \Rightarrow p$ are compared in Table 7.

Table 7

p	q	$p \Rightarrow q$	$q \Rightarrow p$
T	T	T	T
T	F	F	T
F	T	T	F
F	F	T	T

Notice that $p \Rightarrow q$ and $q \Rightarrow p$ are not equivalent. As an illustration, consider the following example.

Example 2
Consider the statements

p: You are a thief.

q: You will go to jail.

The implication $p \Rightarrow q$ states that

If you are a thief, you will go to jail.

The converse of this implication, namely, $q \Rightarrow p$, states that

If you go to jail, you are a thief.

To say that thieves go to jail is not the same as saying that everyone who goes to jail is a thief.   ■

This example illustrates that the truth of an implication does not imply the truth of its converse. Many common fallacies in thinking arise from confusing an implication with its converse.

Contrapositive **The implication *If not q, then not p*, written as $\sim q \Rightarrow \sim p$, is called the *contrapositive* of the implication $p \Rightarrow q$.**

Example 3
Consider the statements

p: You are a thief.

q: You will go to jail.

The implication $p \Rightarrow q$ states "If you are a thief, then you will go to jail." The contrapositive, namely $\sim q \Rightarrow \sim p$, is "If you do not go to jail, then you are not a thief."   ■

Inverse **The implication *If not p, then not q*, written as $\sim p \Rightarrow \sim q$, is called the *inverse* of the implication *If p, then q*.**

Example 4
Consider the statements

p: You are a thief.

q: You will go to jail.

The implication $p \Rightarrow q$ states "If you are a thief, then you will go to jail." The inverse of $p \Rightarrow q$, namely $\sim p \Rightarrow \sim q$, is "If you are not a thief, then you will not go to jail."   ■

Example 5

The truth tables for $p \Rightarrow q$, $q \Rightarrow p$, $\sim p \Rightarrow \sim q$, and $\sim q \Rightarrow \sim p$ are given in Table 8.

Table 8

State-ments		Implication	Converse			Inverse	Contrapositive
p	q	$p \Rightarrow q$	$q \Rightarrow p$	$\sim p$	$\sim q$	$\sim p \Rightarrow \sim q$	$\sim q \Rightarrow \sim p$
T	T	T	T	F	F	T	T
T	F	F	T	F	T	T	F
F	T	T	F	T	F	F	T
F	F	T	T	T	T	T	T

Notice that the entries under Implication and Contrapositive are the same; also, the entries under Converse and Inverse are the same. We conclude that

$$p \Rightarrow q \equiv \sim q \Rightarrow \sim p \qquad q \Rightarrow p \equiv \sim p \Rightarrow \sim q$$

Thus, we have shown that an implication and its contrapositive are logically equivalent. Also, the converse and inverse of an implication are logically equivalent. ∎

THE BICONDITIONAL CONNECTIVE

The compound statement "p if and only if q" is another way of stating the conjunction of two implications: "if p, then q and if q, then p." For example, consider

p: Jill is happy.

q: Jack is attentive.

If we say "Jill is happy if, and only if, Jack is attentive," we mean that if Jill is happy, then Jack is attentive, and if Jack is attentive, then Jill is happy. In symbols, we could write this as

$$(p \Rightarrow q) \wedge (q \Rightarrow p)$$

Biconditional Connective **The connective *if and only if* is called the *biconditional connective* and is denoted by the symbol ⟺. The compound proposition**

$$p \Leftrightarrow q \qquad (p \text{ if and only if } q)$$

is equivalent to

$$(p \Rightarrow q) \wedge (q \Rightarrow p)$$

The statement $p \Leftrightarrow q$ may also be read as "**p is necessary and sufficient for q,**" or as "**p implies q and q implies p.**" (The abbreviation "iff" for "if and only if" is also sometimes used.)

We can restate the definition of the biconditional connective by its truth table (Table 9).

Table 9

p	q	$p \Rightarrow q$	$q \Rightarrow p$	$(p \Rightarrow q) \wedge (q \Rightarrow p)$	$p \Leftrightarrow q$
T	T	T	T	T	T
T	F	F	T	F	F
F	T	T	F	F	F
F	F	T	T	T	T

Example 6

Let p and q denote the statements

p: Joe is happy.

q: Joe is not studying

Then "A necessary condition for Joe to be happy is that Joe is not studying" means "If Joe is happy, he is not studying." Moreover, "A sufficient condition for Joe to be happy is that Joe is not studying" means "If Joe is not studying, he is happy."

Thus, if both implications $(p \Rightarrow q, q \Rightarrow p)$ are true statements, then a necessary and sufficient condition for Joe to be happy is that Joe is not studying.

In other words, Joe is happy if and only if he is not studying. ∎

TAUTOLOGIES

In Section 12.2 we saw how compound propositions can be obtained from simple propositions by using connectives. By using symbols of grouping (parentheses and brackets), we can form more complicated propositions. We denote by $P(p, q, \ldots)$ a compound proposition, where $p, q, \ldots$ are the components of the compound proposition. Some examples of such propositions are

$$\sim(p \wedge q) \qquad p \vee \sim q \qquad (p \wedge \sim q) \qquad (\sim p \wedge q) \sim p \vee \sim (\sim q \wedge \sim r)$$

Ordinarily, when we write a compound proposition, we cannot be certain of the truth of that proposition unless we know the truth or falsity of the component propositions. But a compound proposition that is true regardless of the truth values of its components can be constructed. Propositions of this kind are called *tautologies*.

Tautology **A *tautology* is a compound proposition $P(p, q, \ldots)$ that is true in every possible case.**

Examples of tautologies are

$$p \vee \sim p \qquad p \Rightarrow (p \vee q)$$

Tables 10 and 11 show the truth tables of these tautologies.

Table 10

p	$\sim p$	$p \vee \sim p$
T	F	T
F	T	T

Table 11

p	q	$p \vee q$	$p \Rightarrow (p \vee q)$
T	T	T	T
T	F	T	T
F	T	T	T
F	F	F	T

Many other examples of tautologies are provided by the logical equivalences in Section 12.2. For instance, to say that

$$\sim(p \vee q) \equiv \sim p \wedge \sim q$$

is to say that the biconditional proposition

$$\sim(p \vee q) \Leftrightarrow \sim p \wedge \sim q$$

is a tautology.

SUBSTITUTION

Let $P(p, q, \ldots)$ be a compound proposition with p as one of its components. If we replace p with another proposition h, then we obtain a new compound proposition $P(h, q, \ldots)$. It should be clear that $P(p, q, \ldots)$ has the same truth value as $P(h, q, \ldots)$ whenever h has the same truth value as p. This leads to the following principle:

The Law of Substitution **Suppose p and h are propositions and $h \equiv p$. If h is substituted for p in the compound proposition $P(p, q, \ldots)$, a logically equivalent proposition is obtained:**

$$P(p, q, \ldots) \equiv P(h, q, \ldots)$$

The law of substitution is often used to obtain new versions of existing tautologies. It is particularly useful in mathematics for constructing tautologies of a certain kind, known as *valid arguments* or *proofs*. This is the topic of Section 12.4.

Exercise 12.3

Answers to Odd-Numbered Problems begin on page A-42.

A   In Problems 1–10 write the converse, contrapositive, and inverse of each statement.

1. $\sim p \Rightarrow q$ 2. $\sim p \Rightarrow \sim q$ 3. $\sim q \Rightarrow \sim p$ 4. $p \Rightarrow \sim q$
5. If it is raining, the grass is wet.
6. It is raining if it is cloudy.
7. If it is not raining, it is not cloudy.
8. If it is not cloudy, then it is not raining.

9. Rain is sufficient for it to be cloudy.

10. Rain is necessary for it to be cloudy.

11. Give a verbal sentence that describes

    (a) $p \Rightarrow q$ (b) $q \Rightarrow p$ (c) $\sim p \Rightarrow q$

    using the components

    p: Jack studies psychology.

    q: Mary studies sociology.

12. Show that

    $$p \Rightarrow q \equiv \sim q \Rightarrow \sim p$$

    using the fact that $p \Rightarrow q \equiv \sim p \vee q$.

13. Show that

    $$p \Rightarrow (q \vee r) \equiv (p \wedge \sim q) \Rightarrow r$$

    (a) using a truth table; (b) using De Morgan's laws and the fact that $p \Rightarrow q \equiv \sim p \vee q$.

14. Show that

    $$(p \wedge q) \Rightarrow r \equiv (p \wedge \sim r) \Rightarrow \sim q$$

    using the same two methods as in Problem 13.

In Problems 15–24 construct a truth table for each statement.

15. $\sim p \vee (p \wedge q)$ 16. $\sim p \wedge (p \vee q)$

17. $p \vee (\sim p \wedge q)$ 18. $(p \vee q) \wedge \sim q$

19. $\sim p \Rightarrow q$ 20. $(p \vee q) \Rightarrow p$

21. $\sim p \vee p$ 22. $p \wedge \sim p$

23. $p \wedge (p \Rightarrow q)$ 24. $p \vee (p \Rightarrow q)$

B   In Problems 25–28 prove that the following only have truth value T.

25. $p \wedge (q \wedge r) \Leftrightarrow (p \wedge q) \wedge r$ 26. $p \vee (q \vee r) \Leftrightarrow (p \vee q) \vee r$

27. $p \wedge (p \vee q) \Leftrightarrow p$ 28. $p \vee (p \wedge q) \Leftrightarrow p$

In Problems 29–34 let p be "The examination is hard" and q be "The grades are low." Write each symbolically.

29. If the examination is hard, the grades are low.

30. The examination is not hard nor are the grades low.

31. The grades are not low, and the examination is not hard.

32. The examination is not hard, and the grades are low.

33. The grades are low only if the examination is hard.

34. The grades are low if the examination is hard.

12.4

ARGUMENTS

VALID ARGUMENTS

In an argument or a proof it is important to realize that we are not concerned with the *truth* of the conclusion, but rather with whether the conclusion does or does not follow from the premises. If the conclusion follows from the premises, we say that our reasoning is *valid;* if it does not, we say that our reasoning is *invalid.* For example, from the two

Premises: (1) All college students are intelligent

(2) All freshmen are college students,

follows the

Conclusion: All freshmen are intelligent

Now the last statement certainly is not regarded generally as true, but the reasoning leading to it is valid. *If both of the premises had been true, the conclusion also would have been true.*

Argument   **An *argument* or *proof* consists of a set of propositions $p_1, p_2, \ldots, p_n$, called the *premises or hypotheses,* and a proposition q, called the *conclusion.* An argument is *valid* if and only if the conclusion is true whenever the premises are all true. An argument that is *not* valid is called a *fallacy* or an *invalid* argument.**

DIRECT AND INDIRECT PROOF

We will limit our discussion to two types of argument: direct proof and indirect proof.

In a *direct proof* we go through a chain of propositions, beginning with the hypotheses and leading to the desired conclusion.

Example 1
Suppose it is true that

Either John obeyed the law or John was punished but not both.

and

John was not punished.

Prove that

John obeyed the law.

Solution

Direct proof: Let p and q denote the propositions

p: John obeyed the law.

q: John was punished.

We can write the premises as

$$p \veebar q \quad \text{and} \quad \sim q$$

Since $\sim q$ is true then, by the law of contradiction (see page 564), q is false. Since $p \veebar q$ is true, either p is true or q is true. Thus, p must be true, so John obeyed the law. ∎

More examples of direct proof are given later, but before we look at them, let's discuss two laws of logic that are useful in direct proofs.

> **Law of Detachment** If the implication $p \Rightarrow q$ is true, and if p is true, then q must be true.

See Table 5, Section 12.3, for an illustration of this law.

> **Law of Syllogism** Let p, q, r be three propositions. If
>
> $$p \Rightarrow q \quad \text{and} \quad q \Rightarrow r$$
>
> are both true, then $p \Rightarrow r$ is true.

Table 12 illustrates this law.

Table 12

p	q	r	$p \Rightarrow q$	$q \Rightarrow r$	$p \Rightarrow r$	$(p \Rightarrow q) \wedge (q \Rightarrow r)$	$(p \Rightarrow q) \wedge (q \Rightarrow r) \Rightarrow (p \Rightarrow r)$
T	T	T	T	T	T	T	T
T	T	F	T	F	F	F	T
T	F	T	F	T	T	F	T
T	F	F	F	T	F	F	T
F	T	T	T	T	T	T	T
F	T	F	T	F	T	F	T
F	F	T	T	T	T	T	T
F	F	F	T	T	T	T	T

Example 2
Suppose it is true that

It is snowing.

and

If it is warm, then it is not snowing.

and

If it is not warm, then I cannot go swimming.

Prove that

I cannot go swimming.

Solution
Direct proof: Let p, q, r represent the statements

p: It is snowing.

q: It is warm.

r: I can go swimming.

Our premises are the propositions

$$p \qquad q \Rightarrow {\sim}p \qquad {\sim}q \Rightarrow {\sim}r$$

We want to prove that

$${\sim}r$$

is true. Since $q \Rightarrow {\sim}p$ is true, its contrapositive

$$p \Rightarrow {\sim}q$$

is also true. Using the law of syllogism, since $p \Rightarrow {\sim}q$ and ${\sim}q \Rightarrow {\sim}r$, we see that

$$p \Rightarrow {\sim}r$$

But we know p is true. By the law of detachment, ${\sim}r$ is true. That is, I cannot go swimming. ∎

Example 3
Suppose it is true that

If Dan comes, so will Bill.

and

If Sandy will not come, then Bill will not come.

Show that

If Dan comes, then Sandy will come.

Solution

Let p, q, r denote the propositions

> p: Dan comes.
> q: Bill will come.
> r: Sandy will come.

The premises are

$$p \Rightarrow q \quad \text{and} \quad \sim r \Rightarrow \sim q$$

If we assume $\sim r \Rightarrow \sim q$ is true, then the contrapositive $q \Rightarrow r$ is true also. Thus

$$p \Rightarrow q \quad \text{and} \quad q \Rightarrow r$$

By the law of syllogism it is true that

$$p \Rightarrow r$$

That is, if Dan comes, then Sandy will come. ∎

Example 4

Suppose it is true that

> If I enjoy studying, then I will study.

and

> I will do my homework or I will not study.

and

> I will not do my homework.

Show that

> I do not enjoy studying.

Solution

Let p, q, r denote the three propositions

> p: I enjoy studying.
> q: I will study.
> r: I will do my homework.

Then we know that

$$\sim r \quad r \vee \sim q \quad p \Rightarrow q$$

are true. We want to prove that $\sim p$ is true. Since $\sim r$ is true, then r is false. Also, either r or $\sim q$ is true. Thus, $\sim q$ is true. Since $p \Rightarrow q$, we have

$$\sim q \Rightarrow \sim p$$

is true. Hence, $\sim p$ must be true. ∎

In the examples of direct proof that we have just seen, one proceeds by the laws of logic from the premises to the conclusion.

In any valid proof, one shows that an implication $p \Rightarrow q$ is a tautology. To do this by the method of *indirect proof,* we tentatively suppose that q is false (equivalently, that $\sim q$ is true) and show that we are thereby led to a *contradiction*—that is, a logically impossible situation. This contradiction can be resolved only by abandoning the supposition that the conclusion is false. Therefore, it must be true.

An indirect proof is also called a *proof by contradiction.* The concept is illustrated in the next two examples.

Example 5
Prove the result of Example 1 using an indirect proof.

Solution
To prove that the conclusion p is true, we suppose instead that p is false, hoping to be able to show that this leads to a contradiction.

Thus we assume that

$$\sim p \qquad \sim q \qquad p \veebar q$$

all are true. Either p is true or q is true. But both p and q are false. This is impossible. Thus, we have reached a contradiction, which means that p must be true.   ∎

Example 6
Suppose that

> If I am lazy, I do not study.

and

> I study or I enjoy myself.

and

> I do not enjoy myself.

Prove that

> I am not lazy.

Solution
Let p, q, r be the statements

p: I am lazy.
q: I study.
r: I enjoy myself.

Assume that

$$\sim r \qquad p \Rightarrow \sim q \qquad q \vee r$$

are true.

We want to show that $\sim p$, the conclusion, is true. In an indirect proof, we assume that $\sim p$ is false; that is, that p is true. If p and $p \Rightarrow \sim q$ are true, then $\sim q$ is true (by the law of detachment). That is, q is false. One of the premises is $\sim r$; thus, r is also false. But $q \vee r$ is true, which is impossible if q and r are both false. This contradiction tells us that we have incorrectly assumed that $\sim p$ is false. Thus, $\sim p$ is true.   ■

MODEL: LIFE INSURANCE*

The following problem was first studied by Edmund C. Berkeley in 1936 and was solved by the use of principles of logic. His employer, the Prudential Life Insurance Company, had the procedure that when any policyholder requested a change in the schedule of premium payments, one of two rules was involved as company policy. The question was raised as to whether these two rules were logically equivalent. That is, did both rules give rise to the same payment arrangements or did the use of one rule over the other give different payment arrangements?

Berkeley was of the opinion that the two rules, in some instances, gave different directions to the policyholder. An example of a typical clause found in one of the rules was: If a policyholder was making premium payments several times a year with one of the payments falling due on the policy anniversary and if he requested the schedule be changed to one annual payment on the anniversary date, and if his payments were in full up to a date not the anniversary date and if he made this request more than 2 months after the issue date and if his request also came within 2 months after the policy anniversary date, then a certain action was to be taken!

Berkeley replaced this complicated part of one of the rules by the compound proposition

$$p \wedge q \wedge r \wedge s \wedge t \Rightarrow A$$

where p, q, r, s, t are the five statements and A is the action called for. By doing the same with all parts of both rules and by using the laws of logic, Berkeley was able to demonstrate an inconsistency of application of one rule over the other. In fact, there turned out to be four situations in which contradictory actions occurred.

As a result of Berkeley's effort, Prudential replaced the two cumbersome rules by a simple one.

Since this incident, similar situations involving a maze of if's, and's, but's, and implications, particularly in the areas of legal contracts between corporations, have been checked for accuracy and consistency (to eliminate loopholes, etc.) by the use of logic.

Exercise 12.4   *Answers to Odd-Numbered Problems begin on page A-43.*

A   Prove the statements in Problems 1–4 first by using a direct proof and then by using an indirect proof.

   1. When it rains, John does not go to school. John is going to school. Show that it is not raining.

* John E. Pfeiffer, "Symbolic Logic," *Scientific American* (December 1960).

2. If I do not go to work, I will go fishing. I will not go fishing. Show that I will go to work.

3. If Smith is elected president, Kuntz will be elected secretary. If Kuntz is elected secretary, then Brown will not be elected treasurer. Smith is elected president. Show that Brown is not elected treasurer.

4. Either Katy is a good girl or Mike is a good boy. If Danny cries, then Katy is not a good girl. Mike is not a good boy. Does Danny cry?

B In Problems 5–7 determine whether the arguments are valid.

5. Hypotheses: When students study, they receive good grades.
   These students do not study.
   Conclusion: These students do not receive good grades.

6. Hypotheses: If Danny is affluent, he is either a snob or a hypocrite but not both.
   Danny is a snob and is not a hypocrite.
   Conclusion: Danny is not affluent.

7. Hypotheses: If Tami studies, she will not fail this course.
   If she does not play with her dolls too often, she will study.
   Tami failed the course.
   Conclusion: She played with her dolls too often.

8. Hypotheses: If John invests his money wisely, he will be rich.
   If John is rich, he will buy an expensive car.
   John bought an expensive car.
   Conclusion: John invested his money wisely.

12.5

LOGIC CIRCUITS

CIRCUITS AND GATES
REDESIGNING CIRCUITS
NAND, NOR, AND XOR

CIRCUITS AND GATES

A *logic circuit* is a type of electrical circuit widely used in computers and other electronic devices (such as calculators, digital watches, and compact disc players). The simplest logic circuits, called *gates,* have the following properties.

1. Current flows to the circuit through one or two connectors called *input lines,* and from the circuit through a connector called the *output line.*

2. The current in any of the input or output lines may have either of two possible voltage levels. The higher level is denoted by 1 and the lower level by 0. (Level 1 is sometimes referred to as *On* or *True;* Level 0 may be referred to as *Off* or *False.*)

3. The voltage level of the output line depends on the voltage level(s) of the

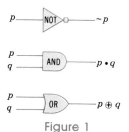

Figure 1

input line(s), according to rules of logic similar to the principles discussed in Section 12.3.

The three most basic types of gates are the *inverter* (or *NOT gate*), the *AND gate,* and the *OR gate.*

The standard symbols for inverters, AND gates, and OR gates appear in Figure 1. In each case, the input lines appear at the left of the diagram and the output line at the right. The label on each line denotes the voltage level or *truth value* of that line; the symbols $\sim p$, pq (or $p \wedge q$), and $p \oplus q$ (or $p \vee q$) are defined by the output tables, in Tables 13, 14, and 15. Note that $\sim p = 1 - p$ and

$$p \oplus q = \begin{cases} p & \text{if} & p \geq q \\ q & \text{if} & q > p \end{cases} = \max\{p, q\}$$

Table 13

p	$\sim p$
1	0
0	1

$(\sim p = 1 - p)$

Table 14

p	q	pq
1	1	1
1	0	0
0	1	0
0	0	0

Table 15

p	q	$p \oplus q$
1	1	1
1	0	1
0	1	1
0	0	0

Computer designers and electrical engineers frequently use one of the symbols $\tilde{p}$, $\bar{p}$, or p' in place of $\sim p$.

More complex logic circuits are constructed by connecting two or more gates.

Example 1

Construct a logic circuit whose inputs are p, q, r and whose output is $(pq) \oplus \sim r$. When is this output equal to 1?

Solution

Apparently, we need to connect an AND gate, an inverter, and an OR gate. Figure 2 shows how.

(Just as we write $ab + c$ for $(ab) + c$ in ordinary algebra, we write $pq \oplus r$ for $(pq) \oplus r$ and $pq \oplus \sim r$ for $(pq) \oplus (\sim r)$ in circuit algebra. Note that $p(q \oplus r) \neq pq \oplus r$.)

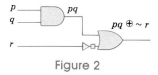

Figure 2

Table 16

p	q	r	pq	$\sim r$	$pq \oplus \sim r$
1	1	1	1	0	1
1	1	0	1	1	1
1	0	1	0	0	0
1	0	0	0	1	1
0	1	1	0	0	0
0	1	0	0	1	1
0	0	1	0	0	0
0	0	0	0	1	1

Table 16 is the output table for $pq \oplus \sim r$. This table confirms what we would expect: The output is 1 when $p = q = 1$ (regardless of whether r is 0 or 1) and when $r = 0$ (regardless of the values of p and q). ∎

REDESIGNING CIRCUITS

Since the output tables for circuits have the same form as the truth tables for propositions (with 1 and 0 corresponding to T and F, pq corresponding to $p \wedge q$, and $p \oplus q$ corresponding to $p \vee q$), the principles of logic formulated in Section 12.3 apply. We can translate these principles as follows:

---

1. Idempotent laws

$$pp = p \qquad\qquad p \oplus p = p$$

2. Associative laws

$$(pq)r = p(qr) \qquad\qquad (p \oplus q) \oplus r = p \oplus (q \oplus r)$$

3. Commutative laws

$$pq = qp \qquad\qquad p \oplus q = q \oplus p$$

4. Distributive laws

$$p \oplus qr = (p \oplus q)(p \oplus r) \qquad p(q \oplus r) = pq \oplus pr$$

5. De Morgan's laws

$$\sim(p \oplus q) = (\sim p)(\sim q) \qquad \sim(pq) = \sim p \oplus \sim q$$

6. Absorption laws

$$p \oplus pq = p \qquad\qquad p(p \oplus q) = p$$

These principles are often useful in simplifying the design of circuits. If two circuits perform the same function and one has fewer gates and connectors than the other, the simpler circuit is preferred because it is less expensive to manufacture or purchase, and also because it is less likely to fail due to overheating or physical stress. Very often it will also be more efficient in terms of power consumption and speed of operation.

Example 2

Redesign the circuit of Example 1 so that $r = p$; simplify the circuit.

Solution

We can easily modify the circuit as in Figure 3. The solid dot is used in circuit diagrams to show where a connector branches.

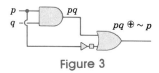

Figure 3

Using the distributive laws, we can simplify:

$$pq \oplus \sim p = (p \oplus \sim p)(q \oplus \sim p) = 1 \cdot (q \oplus \sim p) = q \oplus \sim p$$

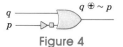

Figure 4

(Recall that $\sim p = 1$ if $p = 0$, and vice versa.) The circuit can therefore be replaced with the one in Figure 4. ■

Example 3
Find the output of the circuit in Figure 5 and simplify its design.

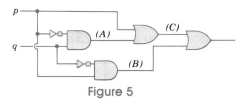

Figure 5

Solution
The output is $B \oplus C$, where

$$A = (\sim p)q, \qquad B = p(\sim q), \qquad C = p \oplus A$$

Thus

$$C = p \oplus (\sim p)q = (p \oplus \sim p)(p \oplus q) = 1 \cdot (p \oplus q) = p \oplus q$$

and the output is

$$B \oplus C = p(\sim q) \oplus (p \oplus q)$$

This expression can be simplified to

$$p(\sim q) \oplus (p \oplus q) = [p(\sim q) \oplus p] \oplus q = p \oplus q$$

Thus the circuit can be replaced by a single OR gate. ■

NAND, NOR, AND XOR
The standard symbols for NAND, NOR, and XOR gates, and their outputs, are shown in Figure 6. (The names are abbreviations for NOT-AND, NOT-OR, and Exclusive-OR, respectively.) Table 17 is the output table for XOR.

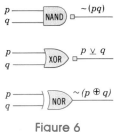

Figure 6

Table 17

p	q	$p \vee q$
1	1	0
1	0	1
0	1	1
0	0	0

Exercise 12.5   *Answers to Odd-Numbered problems begin on page A-44.*

A   In Problems 1–4 determine when the output of each circuit is 1; use a truth table if necessary.

1.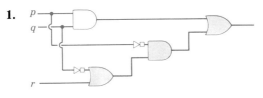

2.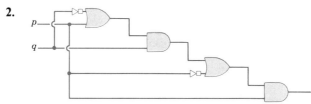

3.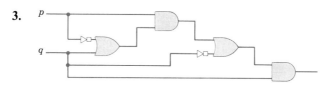

4.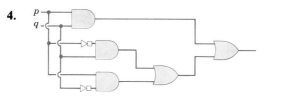

B   In Problems 5–8 construct a circuit corresponding to each expression.

5. $(\sim p \oplus \sim q)(p \oplus q)$ **6.** $(p \oplus \sim q)(\sim p)$

7. $\sim(p \oplus q)(\sim p)$ **8.** $(\sim p \oplus q)[(\sim p)(\sim q)]$

9. Design simpler circuits having the same outputs as the circuits in Problems 1, 3, 5, and 7.

10. Design simpler circuits having the same outputs as the circuits in Problems 2, 4, 6, and 8.

11. Design a logic circuit that can be turned On or Off from either of two switches. (That is, if the inputs are p and q, the output can be changed from 1 to 0 or from 0 to 1 by changing the level of either p or q, but not both.)

12. Design a circuit, consisting of at least two gates, whose output is always 1.

13. Design a circuit, consisting of at least two gates, whose output is always 0.

14. Using an OR gate, an AND gate, and an Inverter, design a circuit that will work as an XOR gate. [*Hint:* What combination of $\cdot$, $\oplus$, and $\sim$ has the same truth table as $\underline{\vee}$?]

15. Design a circuit that will work as an OR gate, using

(a) two inverters and a NAND gate;

(b) three NAND gates.

[*Hint:* (a) Use De Morgan's laws. (b) How can a NAND gate be substituted for an Inverter?]

16. Design a circuit that will work as an AND gate using

(a)   two inverters and a NOR gate;

(b)   three NOR gates

C   17. Show that a circuit whose output is

$$pq \oplus pr \oplus q(\sim r)$$

can be replaced by one whose output is

$$pr \oplus q(\sim r)$$

[*Hint:* $pq = pq(r \oplus \sim r) = pqr \oplus pq(\sim r)$]

18. Show that

(a) $(p \oplus q)(p \oplus r)(q \oplus \sim r) = (p \oplus r)(q \oplus \sim r)$

(b) $(p \oplus r)(q \oplus \sim r) = p(\sim r) \oplus qr$

CHAPTER REVIEW

Important Terms and Formulas		
proposition	De Morgan's laws	law of substitution
connective	absorption laws	argument
conjunction ($\wedge$)	implication ($\Rightarrow$)	fallacy
inclusive disjunction ($\vee$)	conditional connective ($\Rightarrow$)	direct proof
exclusive disjunction ($\underline{\vee}$)	hypothesis (premise)	law of detachment
negation ($\sim$)	conclusion	law of syllogisms
proof by contradiction	sufficient condition	indirect proof
quantifier	necessary condition	logic circuit
truth value	converse	inverter
logically equivalent ($\equiv$)	contrapositive	AND gate
idempotent laws	inverse	OR gate
commutative laws	biconditional connective ($\Leftrightarrow$)	NAND gate
associative laws		NOR gate
distributive laws	tautology	XOR gate

$$p \wedge q \equiv q \wedge p$$
$$p \vee q \equiv q \vee p$$
$$(p \wedge q) \wedge r \equiv p \wedge (q \wedge r)$$
$$(p \vee q) \vee r \equiv p \vee (q \vee r)$$

$$p \vee (q \wedge r) \equiv (p \vee q) \wedge (p \vee r)$$
$$p \wedge (r \vee q) \equiv (p \wedge r) \vee (p \wedge q)$$
$$\sim (p \wedge q) \equiv \sim p \vee \sim q$$
$$\sim (p \vee q) \equiv \sim p \wedge \sim q$$

True-False Questions	*(Answers are on page A-45)*

T   F   1. The negation of the statement "Some salesmen are intelligent" is "All salesmen are intelligent."

T   F   2. The statement $\sim(p \wedge q)$ is logically equivalent to $\sim p \vee \sim q$.
T   F   3. The statement $\sim(p \vee q)$ is logically equivalent to $\sim p \wedge \sim q$.
T   F   4. The statement $\sim p \Rightarrow q$ is logically equivalent to $q \Rightarrow p$.
T   F   5. The statement $p \Leftrightarrow q$ is logically equivalent to $(p \Rightarrow q) \wedge (\sim q \Rightarrow \sim p)$.

Fill in the Blanks   *(Answers are on page A-45)*

1. The compound proposition *p or q* is called the inclusive disjunction of *p* and *q* and is denoted by ――――――.
2. The negation of *p* is denoted by ――――――.
3. If two propositions have the same truth values in every possible case, the propositions are said to be ――――――.
4. The two parts of an argument are the ―――――― and the ――――――.
5. The output of a logic circuit is either ―――――― or ――――――.

Review Exercises   *Answers to Odd-Numbered Problems begin on page A-45.*

In Problems 1–4 circle each correct answer; some questions may have more than one correct answer.

A   **1.** Which of the following negate the statement below?

p: All people are rich.

(a) Some people are rich.
(b) Some people are poor.
(c) Some people are not rich.
(d) No person is rich.

2. Which of the following negate the statement below?

p: It is either hot or humid.

(a) It is neither hot nor humid.
(b) Either it is not hot or it is not humid.
(c) It is not hot and it is not humid.
(d) It is hot, but not humid.

3. Which of the following statements are logically equivalent to the statement below?

$$(\sim p \vee q) \wedge r$$

(a) $(p \Rightarrow q) \wedge r$ (b) $\sim p \vee (q \wedge r)$
(c) $(\sim p \Rightarrow q) \wedge r$ (d) $(\sim p \vee r) \wedge (q \wedge r)$

4. Which of the following statements are logically equivalent to the statement below?

$$p \wedge \sim q$$

(a) $\sim q \wedge p$ (b) $p \vee q$
(c) $\sim p \vee q$ (d) $q \Rightarrow \sim p$

In Problems 5–8 negate each proposition.

5. Some people are rich.

6. All people are rich.

7. Danny is not tall and Mary is short.

8. Neither Mike nor Katy is big.

In Problems 9–12 construct a truth table for each compound proposition.

9. $(p \land q) \lor \sim p$

10. $(p \lor q) \land p$

11. $\sim p \lor (p \lor \sim q)$

12. $\sim p \Rightarrow (p \lor q)$

In Problems 13–15 let p stand for "I will pass the course" and let q stand for "I will do homework regularly." Put each statement into symbolic form.

13. I will pass the course, if I do homework regularly.

14. Passing this course is a sufficient condition for me to do homework regularly.

15. I will pass this course if and only if I do homework regularly.

16. Write the converse, contrapositive, and inverse of the statement

If it is not sunny, it is cold.

17. Prove the following by the use of direct proof: If I do not paint the house, I will go bowling. I will not go bowling. Show that I will paint the house.

18. Using reasons from logic, give a valid argument to answer the question, "Is Katy a good girl?" Given that:

    (a)   Mike is a bad boy or Danny is crying.

    (b)   If Katy is a good girl, then Mike is not a bad boy.

    (c)   Danny is not crying.

19. Determine whether the following are logically equivalent:

$$\sim p \lor q \qquad p \Rightarrow q$$

20. Determine whether the two statements below are logically equivalent:

$$(p \Rightarrow q) \land (\sim q \lor p) \qquad p \Leftrightarrow q$$

21. Show that the output of an XOR gate is equal to

$$(p \oplus q)[\sim(pq)]$$

Use this fact to construct an XOR gate by connecting an OR gate, an AND gate, and a NAND gate.

22. Construct an XOR gate by connecting two NOR gates and an AND gate.

13

RELATIONS, FUNCTIONS, AND INDUCTION

13.1

RELATIONS

DEFINITION OF RELATION
APPLICATIONS TO COMPUTER SCIENCE
PROPERTIES OF RELATIONS

In this section we introduce relations, a concept fundamental in mathematics and its applications. The word relation is a common term used in mathematics to indicate a relationship between two objects. Relationships occur everywhere. Two people belonging to the same family may be related as father-and-son, mother-and-son, brother-and-sister, husband-and-wife, and so forth. In mathematics, two numbers may be related by being equal, or by one being greater than the other. Two sets may be related as one being a subset of the other, etc. These are only a few examples of relations. Later in this section, and in the exercises, we present more examples. First we start with a formal definition.

DEFINITION OF RELATION

Relation **Let A and B be two nonempty sets. A *relation R* from A to B is a set R of ordered pairs (a,b) where $a \in A$ and $b \in B$. For every such ordered pair we write aRb, read "a is related to b by R." If the relation R is from the set A to itself we say R is a *relation on A.** *

Example 1
Let $A = \{1,2,7\}$ and $B = \{2,5\}$. List the elements of each relation R defined below.

(a) $a \in A$ is related to $b \in B$, that is, aRb if, and only if, $a < b$.
(b) $a \in A$ is related to $b \in B$, that is, aRb if, and only if, a and b are both odd numbers.
(c) $a \in A$ is related to $b \in B$, that is, aRb if, and only if, $(a + b)$ is an even number.

Solution
(a) Since $1 \in A$ is less than $2 \in B$, then $1R2$. Similarly $1R5$ and $2R5$. Therefore

$$R = \{(1,2), (1,5), (2,5)\}$$

Note that $7 \in A$ is not related to $5 \in B$, since $7 \not< 5$.
(b) Since $1 \in A$ and $5 \in B$ are both odd, then $1R5$. Similarly $7R5$. Therefore

$$R = \{(1,5), (7,5)\}$$

Note that $2 \in A$ is not related to $5 \in B$ since 2 is not odd.

---

* R is sometimes called a *binary relation R* from A to B since the elements of the set R are ordered *pairs*.

(c)   Here, $1 \in A$ is related to $5 \in B$ since $1 + 5 = 6$ is even. Therefore, $1R5$. Similarly, $2R2$ and $7R5$. Therefore,

$$R = \{(1,5), (2,2), (7,5)\}$$

Note that $1 + 2 = 3$ is not even; thus $1 \in A$ is not related to $2 \in B$.   ■

Example 2

Let A be the set of all integers. We define a relation R on A as follows: aRb if, and only if, a and b have the same remainder when divided by 2.* Is $5R3$? Is $-1R1$? Is $7R4$?

Solution

Since both $5 \div 2$ and $3 \div 2$ give 1 as a remainder, then $5R3$. Since $-1 = -1 \cdot 2 + 1$ and $1 = 0 \cdot 2 + 1$, we conclude that both $-1 \div 2$ and $1 \div 2$ give 1 as a remainder; therefore, $-1R1$. Since $7 \div 2$ gives 1 as a remainder and $4 \div 2$ gives 0 as a remainder, 7 is not related to 4.   ■

APPLICATIONS TO COMPUTER SCIENCE

In computer science the binary digits 0 and 1 are called *bits*. An ordered collection of consecutive bits is called a *binary word*. Since computers store information in the form of binary words, they are important in the field of computer science. Of special importance are binary words of length 7, called *bytes,*† because a byte is the smallest addressable memory area in the computer, that is, the smallest unit that can be located in memory.

The sets S and S^* defined below are referred to in the exercises and in later sections of this chapter.

Let $S = \{0,1\}$. We use 0 and 1 to form binary words. Some binary words are 0, 00, 01, 11, 10, 000, 001, 010, 100, 110, 1101, 10011, and 11000011.

Now let S^* denote the set of all binary words. S^*, clearly, contains an infinite number of elements and hence makes its listing impractical. Next, for every word a in S^* let

$$\text{length } (a) = \text{ the number of bits in } a$$

For example, if $a = 101011$, then length $(a) = 6$. With this in mind, we can define a relation R on S^* by considering the number of bits in each element in S^* as follows: For a and b in S^*, aRb if, and only if, length $(a) = $ length (b). Thus, for instance, $(000) R (111)$ since length $(000) = $ length $(111) = 3$, and $(1001) R (0110)$ since both 1001 and 0110 have length equal to 4.

Example 3

Let the set S^* and the relation R be the ones defined above. Let 01 be in S^*. List all elements in S^* that are related to 01 under R.

---

* This relation is called the "congruence modulo 2" relation and may be denoted by $a \equiv b$ (mod 2).

† On some computers, such as the IBM computers, a byte is 8 bits long.

Solution

01 is of length 2. The binary words in S^* related to 01 under R are also of length 2. The list of all such words is 00, 01, 10, 11. ∎

PROPERTIES OF RELATIONS

A relation R on a set A often satisfies certain properties. Three of these properties, reflexivity, symmetry, and transitivity, are defined as follows:

1. R is *reflexive* if, for each element $a \in A$, we have aRa.
2. R is *symmetric* if, whenever aRb, we have bRa.
3. R is *transitive* if, whenever aRb and bRc, we have aRc.

Example 4

Let R be a relation defined on the set $A = \{a,b,c\}$. For each definition of R given below determine whether R is reflexive, symmetric, or transitive.

(a) $R = \{(a,a), (a,b), (b,b), (c,c), (b,c)\}$.
(b) $R = \{(c,a), (a,c), (c,c), (c,b), (b,c)\}$.
(c) $R = \{(a,b), (b,a), (a,c), (c,a), (b,c), (c,b), (a,a), (b,b), (c,c)\}$.

Solution

(a) Since (a,a), (b,b), and $(c,c) \in R$, then R is reflexive. Because $(a,b) \in R$, but $(b,a) \notin R$, then R is not symmetric. And because (a,b) and $(b,c) \in R$, but $(a,c) \notin R$, then R is not transitive.
(b) This relation is not reflexive because $(a,a) \notin R$. It is not transitive because (a,c) and (c,a) are in R, but (a,a) is not. However, it is symmetric. (Why?).
(c) R is reflexive, symmetric, and transitive. (Why?). ∎

Example 5

Let $A = \{1,2,3\}$ and let R be a relation defined on A. Suppose further that the ordered pairs $(1,1)$, $(1,2)$, and $(2,3)$ belong to R. What ordered pairs, besides the ones given above, must belong to R in order to make R

(a) reflexive
(b) symmetric
(c) transitive

Solution

(a) R will be reflexive if every element in A is related to itself. The ordered pairs $(2,2)$ and $(3,3)$ must also belong to R to make it reflexive.
(b) R will be symmetric if whenever (a,b) is in R, so is (b,a). Obviously, $(1,1)$ satisfies this condition. Now since $(1,2) \in R$ then, for R to be symmetric, $(2,1)$ must also belong to R. Similarly, since $(2,3) \in R$ then, for R to be symmetric, $(3,2)$ must also belong to R.

(c) R will be transitive if whenever (a,b) and (b,c) are in R, so is (a,c). Now $(1,2)$ and $(2,3) \in R$ implies that, for R to be transitive, $(1,3)$ must also belong to R.

■

Exercise 13.1

Answers to Odd-Numbered Problems begin on page A-46.

A   1.   Let $A = \{x,y,z\}$, $B = \{1,3,5\}$. A relation R from A to B is defined as

$$R = \{(x,5), (y,1), (z,3)\}$$

Answer true or false:

$xR5,$ $yR3,$ $zR3,$ $zR1,$ $yR1,$ $xR1,$ $xR3,$ $yR5,$ $zR5$

2.   Let $A = \{1,2,3\}$, $B = \{a,b,c,d\}$. A relation R from A to B is defined as

$$R = \{(1,a), (1,c), (2,b), (3,a), (3,c)\}$$

Answer true or false:

$1Ra,$ $1Rb,$ $1Rc,$ $2Ra,$ $2Rb,$ $2Rc,$ $3Ra,$ $3Rb,$ $3Rc$

3.   Let $A = \{2,3,5\}$, $B = \{2,4,6,10\}$. A relation R from A to B is given as follows:

$2R2,$ $2R4,$ $2R6,$ $2R10,$ $3R6,$ $5R10$

Write R as a set of ordered pairs.

4.   Let $C = \{\text{Apple, TRS, IBM, Atari}\}$, $M = \{1000, 360, 370, 800, \text{II}\}$. Let R, the relation from C to M, be given as follows:

$(\text{Apple}) R (\text{II}), (\text{TRS}) R (1000), (\text{IBM}) R (370), (\text{Atari}) R (800)$

Write R as a set of ordered pairs.

5.   Let $D = \{1,2,3,4,5\}$ and $C = \{1,4,9,16,25,36,49,64,81\}$. The relation R from D to C is defined as follows:

$$xRy \text{ means } y = x^2$$

Replace the "?" by the appropriate value:

$1R?,$ $2R?,$ $3R?,$ $4R?,$ $5R?,$ $?R4,$ $?R25$

6.   Let $F = \{-2, -10, -1, 0, 1, 2\}$. The relation R on F is defined as follows:

$$aRb \text{ means } a \geq b$$

Answer true or false:

$-2R-2,$ $-2R-1,$ $-1R-2,$ $-1R0,$ $0R0,$ $1R2,$ $2R1$

7.   The *Cartesian product* of two nonempty sets A and B, denoted by $A \times B$, is the set consisting of all ordered pairs (a,b) with $a \in A$ and $b \in B$. That is,

$$A \times B = \{(a,b) \mid a \in A, b \in B\}$$

Let $A = \{1,2,5\}$ and $B = \{2,4,7\}$.

(a)   Find the Cartesian product $A \times B$.

(b)   Define the relation "less than," call it R, from the set A to the set B. Write R as a set of ordered pairs.

(c)   What can you say about the set R and $A \times B$?

8. Repeat Parts b and c of Problem 7, but now let R be the relation "greater than or equal to."

9. Let $S = \{a,b\}$ and $S^2 =$ set of all words on S of length 2.

   (a) List all elements of S^2.
   (b) The relation L on S^2 is defined by: vLw means that the first letter in v is the same as the first letter in w where v and w are in S^2. Write L as a set of ordered pairs.

10. Repeat Problem 9b for L defined as follows: vLw means first letter of v is not the same as first letter of w.

In Problems 11–18 write the set of ordered pairs in the relation R.

11. Let $A = \{1,2,3\}$, $B = \{2,4,6,8\}$. Define R to be the relation from the set A to the set B given by: aRb means a is a factor of b.

12. Let $A = \{1,2,3,4,5,6\}$. Let R be a relation on A given by: aRb means $a \equiv b \pmod 2$ (see footnote page 598).

13. Let $A = \{1,2,3,4,5,6,7,8,9,10\}$; R is a relation on A given by: aRb means a and b are either both even or both odd.

14. Let $A = \{1,2,3,4,5\}$; R is a relation on A given by: aRb means ab is even.

15. Let $A = \{0,1,2,3\}$ and R on A be defined as follows: aRb means $a \le b$.

16. Let $A = \{1,2,3,4,5,9\}$ and R on A be defined as follows: aRb means $b = a + 2$.

17. Let $D = \{0,1,2,3,4,5,9,15,16\}$ and R on D be defined as follows: xRy means $x = \sqrt{y}$.

18. Let $I = \{-4,-2,-1,0,1,2,4\}$ and R on I be defined as: xRy means $x = y^2$.

B   For a relation R from the set A to the set B, the *inverse relation* of R, denoted R^{-1}, is a relation from B to A defined as follows:

$$bR^{-1}a \text{ means } aRb$$

For Problems 19–22 find R^{-1}.

19. $R = \{(1,1), (1,2), (2,5), (3,7)\}$.
20. $R = \{(a,a), (a,b), (b,a), (b,c)\}$.
21. $R = \{(*,42), (H,72), (/,47), (x,88)\}$.
22. R is the relation "$\le$" on integers.

In Problems 23–30 determine whether the given relation R on the set A is reflexive, symmetric, or transitive.

23. Let R on a set A of real numbers be defined as: aRb means $a = b$.

24. Let R on the set A of real numbers be defined as: aRb means $a < b$.

25. Let R be the relation of congruence modulo 2 on the set A of integers (see Example 2).

26. Let A be the set of integers and R a relation on A be defined as: aRb means a divides b (or a is a factor of b).

27. Let A be the set of all lines in the plane, and R be defined by: l_1Rl_2 means l_1 is a line parallel to l_2.

28. R is defined as follows: if an arrow connects a point to another, then the 2 points are related. (See the figure.) That is,

$$R = \{(a,a), (a,b), (b,b), (b,a)\}$$

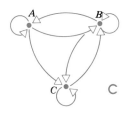

29. *R* is defined as follows: if an arrow connects a point to another, then the 2 points are related. (See the figure.) That is,

$$R = \{(A,A),\ (A,B),\ (B,A),\ (B,B),\ (B,C),\ (C,B),\ (C,C),\ (A,C)\}$$

30. Let $A = \{1,2,3\}$ and $R = \{(2,2)\}$.

31. Consider the set $A = \{t,u,v,w\}$ and the relation *R* on *A* defined as $R = \phi$. (That is, *R* is the empty relation. In other words, no relation exists between any pair of elements of *A*.) Show that *A* is symmetric and transitive but not reflexive.

13.2

FUNCTIONS

| DEFINITION OF A FUNCTION
| SPECIAL TYPES OF FUNCTIONS

In this section we discuss one of the most important concepts in mathematics and its applications: the concept of a function. As we will see, a function is a special type of a relation.

DEFINITION OF A FUNCTION

Function Let *A* and *B* be two nonempty sets. A *function f* from *A* into *B* is a relation from *A* to *B* that relates every element in *A* to only one element in *B*. The set *A* is called the *domain* of the function. For each element *a* in *A*, the corresponding element *b* in *B*, related to *a* by *f*, is called the *image* of *a*. The set of all images of the elements of the domain is called the *range* of the function.

Since there may be elements in *B* that are the image of no *a* in *A*, it follows that the range of a function is a subset of *B*.

Functions are often denoted by letters such as *f*, *F*, *g*, and *G*. If *f* is a function from *A* into *B*, then for each element *a* in *A* the corresponding image in the set *B* is denoted by the symbol $f(a)$ and is read "*f* of *a*." (Note that $f(a)$ does not mean *f* times *a*). We can think of a function *f* as associating elements in *A* to elements in *B* and visualize *f* by means of a simple diagram as we do in Figure 1.

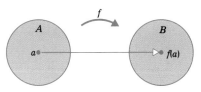

Figure 1

Example 1

Let $A = \{1,3,5\}$ and $B = \{2,4\}$. Define the function f from A into B, as shown in Figure 2. Find $f(1), f(3), f(5)$, and the range of f.

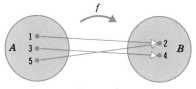

Figure 2

Solution

Following the arrows, we have

$$f(1) = 2 \qquad f(3) = 4 \qquad f(5) = 2 \qquad \text{Range of } f = \{2,4\}$$ ■

Example 2

Let $A = \{1,2,3\}$ and $B = \{a,b,c,d,e\}$. Define the function f from A into B, as shown in Figure 3. Find $f(1), f(2), f(3)$, and the range of f.

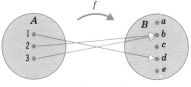

Figure 3

Solution

Following the arrows, we find

$$f(1) = d \qquad f(2) = b \qquad f(3) = b \qquad \text{Range of } f = \{b,d\}$$

Because b and d are the only elements in B that are assigned to elements in A, we observe that the range of f is a proper subset of B. ■

Example 3

The diagrams in Figure 4 show a correspondence by which elements of $A = \{1,2,3\}$ are associated to elements in $B = \{w,x,y,z\}$. Which of these diagrams define a function from A into B?

Solution

In Figure 4a not all elements in A are associated with elements in B. For example, the element 3 in A has no image in B. This is indicated by the absence of an arrow going from 3 in A to an element in B. Thus 4a does not define a function from A into B. In Figure 4b every element in A is associated with a unique element in B. Thus Figure 4b defines a function. In Figure 4c the uniqueness of images is

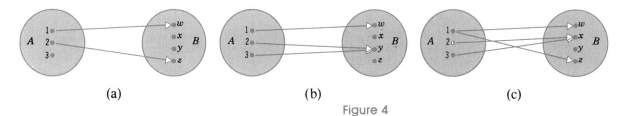

(a)   (b)   (c)

Figure 4

violated. For example, the element $1 \in A$ has two images. This is shown by the two arrows emanating from $1 \in A$ to w and to $z \in B$. Thus Figure 4c does not define a function. ∎

When the sets A and B are infinite, a complete arrow diagram of a function f cannot be drawn. In such cases functions are sometimes defined by a formula, and the images of f can then be found from the formula defining f.

Example 4

Let f be a function from the set of real numbers to itself defined by the formula

$$f(x) = 2x - 1$$

Find $f(0), f(0.5), f(-1), f(1), f(-10), f(100)$.

Solution

$f(0)$ is obtained from substituting 0 for x in $2x - 1$. That is,

$$f(0) = 2(0) - 1 = -1$$

Similarly,

$$f(0.5) = 2(0.5) - 1 = 0$$
$$f(-1) = 2(-1) - 1 = -3$$
$$f(1) = 2(1) - 1 = 1$$
$$f(-10) = 2(-10) - 1 = -21$$
$$f(100) = 2(100) - 1 = 199$$

∎

Example 5

Hamming Distance Function In a branch of computer science, called *coding theory*, a function called the *Hamming distance function* is of importance. Refer to Section 8.4 Chapter 8. This function gives a measure of the difference between two binary words that have the same length. We define the Hamming distance function H as follows: Let (u,v) be a pair of binary words of the same length. We compare u and v position by position and define

$H(u,v) = $ the number of positions in which u and v have different bits

For example,

$$H(0101, 1010) = 4$$

because 0101 and 1010 differ in all four positions. On the other hand,

$$H(1110, 1010) = 1$$

because 1110 and 1010 differ only in one position, namely, the second position.

■

SPECIAL TYPES OF FUNCTIONS

In Example 1 we notice that the two elements 1 and 5 in the domain A are assigned the same element 2 in the range B. And in Example 2 we notice that the range of f is not the entire set B. This shows that functions can be of different types. In the remainder of this section we introduce functions that have special properties. These functions are important in mathematics and its applications.

In the next four definitions we consider f to be a function from A into B.

One-to-One Function **A function f from A into B is *one-to-one*, or *injective*, if for all elements a, b in A such that $a \neq b$ we have $f(a) \neq f(b)$. That is, if no two elements of A are assigned to the same element in B, or, equivalently, if each element of the range corresponds to exactly one element of the domain, then f is one-to-one.**

Example 6
Let $A = \{1,2,3\}$, $B = \{a,b,c,d\}$, and let $f(1) = a, f(2) = d, f(3) = c$. Is f one-to-one?

Solution
Yes, because the different elements 1, 2, 3, in A are assigned to the different elements a, d, c, respectively, in B.   ■

Example 7
Figure 5 defines a function f from the set $A = \{a,b\}$ into the set $B = \{x,y\}$. Is f one-to-one?

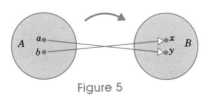

Figure 5

Solution
Yes, because the two arrows show that a is assigned to y and that b, which is different from a, is assigned to x, which is different from y.   ■

Example 8
Let $f(x) = x^2$, x any real number. Show that f is not one-to-one.

Solution
We need only show that two distinct numbers are assigned to the same number under f. Choose the distinct values $x_1 = 1$ and $x_2 = -1$. Now

$$f(x_1) = f(1) = (1)^2 = 1$$
$$f(x_2) = f(-1) = (-1)^2 = 1$$

Thus, we see that $f(1) = f(-1) = 1$. Therefore, f is not one-to-one. ∎

Onto **A function f from A into B is *onto*, or *surjective* if every element of B is the image of some element in A, that is, if B = range of f.**

Example 9
The diagrams in Figure 6 define the functions f and g from the set $A = \{1,2,3,4\}$ into the set $B = \{a,b,c,d\}$. Which of these functions is onto?

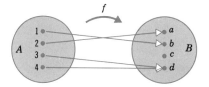 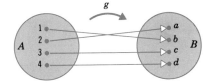

Figure 6

Solution
Under f, $c \in B$ is not the image of any element in A. Thus, f is not onto. Under g, every element in B is an image. Therefore, g is onto. ∎

Example 10
Let $f(x) = x^2$, x any real number. Show that f is not onto.

Solution
Since we cannot find a real number whose square is negative, then the set of negative real numbers is not in the range. Thus f is not onto. ∎

If a function f is both one-to-one and onto we say that f is *bijective*.

If a function f is bijective, then there will be another function g that will "undo" what f did. In a way, g will be the opposite of f as the following definition shows:

The function g from B to A is called the *inverse* of f if $g(b) = a$, where $f(a) = b$. g is usually denoted by f^{-1}, read "f inverse."

Note that if f is one-to-one and onto then the inverse g of f exists. Also, the range of f is the domain of g and the range of g is the domain of f.

Example 11

Let $A = \{1,2,3\}$, $B = \{a,b,c\}$. Define f from A into B as

$$f(1) = a, \qquad f(2) = b, \qquad f(3) = c$$

Since f is both one-to-one and onto, it is bijective. Therefore, f^{-1} from B into A exists and is defined as

$$f^{-1}(a) = 1, \qquad f^{-1}(b) = 2, \qquad f^{-1}(c) = 3$$ ■

Exercise 13.2

Answers to Odd-Numbered Problems begin on page A-46.

A   1. Let $A = \{a,b,c\}$ and $B = \{1,2,3,4\}$. Let f be a function from A into B, as shown in the figure below. Find $f(a), f(b), f(c)$. Also find the range of f.

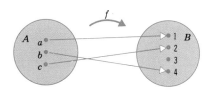

2. Let $A = \{0,1,2\}$ and $B = \{1,10,100\}$. Let f be a function from A into B, as shown in the figure below. Find $f(0), f(1), f(2)$. Also find the range of f.

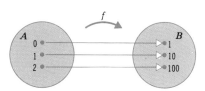

3. Let f be a function from the set of real numbers into itself defined by

$$f(x) = 5 - x$$

Find $f(0), f(5), f(-5), f(10), f(-10)$.

4. Let f be a function defined by

$$f(x) = \frac{-1}{x - 1}$$

where x is any real number and $x \neq 1$. Find $f(-2), f(-1), f(0), f(2)$.

In Problems 5–8, f assigns elements of $A = \{x,y,z\}$ into elements of $B = \{1,2,3,4\}$. State whether f defines a function. If it doesn't explain why.

5. $f(x) = 3, f(x) = 1, f(z) = 4$.

6. $f(x) = 1, f(y) = 2, f(z) = 3.$

7. $f(x) = 4, f(y) = 2, f(z) = 4.$

8. $f(y) = 2, f(y) = 3, f(z) = 1.$

In Problems 9–12, $A = \{0,1,2\}$ and $B = \{1,2,3,4\}$. Use the correspondence pictured in the accompanying figures to determine whether f is a function.

9.

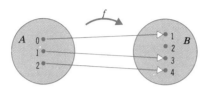

10.

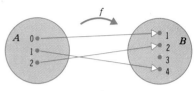

11.

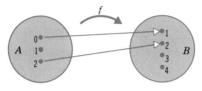

12.

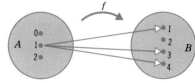

B   **13.** Let l be a relation from the set of all binary words into the set $M = \{0,1,2,3, \ldots\}$, defined by

$$l(w) = \text{length of } w$$

(see Section 13.1). Is l a function?

14. Let f be a relation that sends binary words into the set $\{0,1\}$ defined by:

$$f(w) = \begin{cases} 1, \text{ if } w \text{ contains an odd number of 0s} \\ 0, \text{ if } w \text{ contains an odd number of 1s} \end{cases}$$

Is f a function?

15. Let f be a relation that sends binary words into the set $M = \{0,1,2,3, \ldots\}$ defined by

$$f(w) = \text{number of 0s in } w$$

Is f a function?

16. Let H be the Hamming distance function. (See Example 5.) Find $H(1000, 0001)$, $H(1011, 1011)$, $H(10, 01)$, $H(101, 010)$, $H(1010101, 1010111)$.

In Problems 17–20, f is a function from A into B pictured in the accompanying figures. State whether f is one-to-one, onto, or bijective.

17.

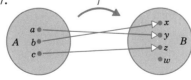

18.

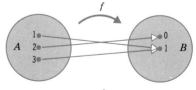

19.

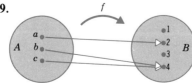

20.

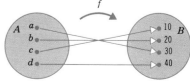

C   **21.** Show that the Hamming distance function on the set of all pairs of binary words of length n is not one-to-one.

22. Show that the Hamming distance function on the set of all pairs of binary words of length n is not onto. [*Hint:* Can you find binary words u and v of length n such that $H(u,v) \geq n$?]

23. Let f be a function from the set of integers into the set of integers given by the formula $f(n) = 2n$. Is f onto?

24. Let f be the function from the set A into the set B given in the figure below. (a) Find $f(1), f(2), f(3)$. (b) Show that f is bijective.

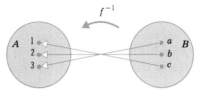

25. Refer to the figure of Problem 24. Find $f^{-1}(a), f^{-1}(b), f^{-1}(c)$.

26. Let f be a bijective function defined by

$$f(1) = 1, \qquad f(2) = 4, \qquad f(3) = 9, \qquad f(4) = 16$$

Find

$$f^{-1}(1), \qquad f^{-1}(4), \qquad f^{-1}(9), \qquad f^{-1}(16).$$

Let f be a function from A into B and let g be a function from B into C. The *composition of f and g,* denoted by $g \circ f$, read "g of f," results in a new function from A into C and is given by

$$(g \circ f)(a) = g(f(a))$$

where a is in A. That is, the function $g \circ f$ associates a in A to $f(a)$ in B first and then associates the element $f(a)$ in B to the element $g(f(a))$ in C. See the figure below.

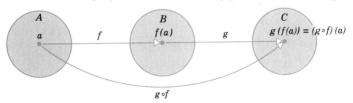

27. Use the figure below to find

$$(g \circ f)(1), \qquad (g \circ f)(2), \qquad (g \circ f)(3)$$

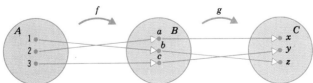

28. If f is a bijective function from A into B, then f has an inverse f^{-1} from B into A. This means that $f^{-1} \circ f$ is a function from A into A. Such a function has the property

$$(f^{-1} \circ f)(a) = a \qquad \text{and} \qquad (f \circ f^{-1})(a) = a$$

We call this the *identity function.* Let $f(x) = x + 1$ be a function from the set of real numbers to the set of real numbers. If $f^{-1}(x) = x - 1$, show that $(f^{-1} \circ f)(x) = x$ and $(f \circ f^{-1})(x) = x$.

13.3

SEQUENCES

In this section we present a special type of function — one whose domain is the set of numbers $\{0,1,2,3, \ldots\}$. These functions are called *sequences*. Sequences are useful in both mathematics and computer science.

Let S be a set. A *sequence* is a function s from the set $\{0,1,2,3, \ldots\}$ into S. If S is the set of real numbers, then s is called a *sequence of reals*.

For every number n in the set $\{0,1,2,3, \ldots\}$, $s(n)$ will be the value assigned to n by the function s. We call $s(n)$ the nth term of the sequence and denote it by s_n [instead of $s(n)$]. We will also use the symbol (s_n) to denote the sequence itself.

Example 1
Let the sequence (s_n) be given by

$$s_n = n^2$$

List the first four terms of (s_n).

Solution

$$s_0 = 0^2 = 0$$
$$s_1 = 1^2 = 1$$
$$s_2 = 2^2 = 4$$
$$s_3 = 3^2 = 9$$

Example 2
Let $s_n = (-1)^n$. Write out the first four terms of (s_n).

Solution

$$s_0 = (-1)^0 = 1$$
$$s_1 = (-1)^1 = -1$$
$$s_2 = (-1)^2 = 1$$
$$s_3 = (-1)^3 = -1$$

Note that the range of s is $\{-1,1\}$.

Example 3

Let (s_n) be a sequence given by $s_n = a$. List s_0, s_1, s_2, s_3, etc.

Solution

$s_0 = a$, $s_1 = a$, $s_2 = a$, $s_3 = a$, etc.
Such a sequence is called a *constant sequence*.

Example 4

For $(s_n) = \left(\dfrac{1}{n}\right)$, $n \neq 0$, write out the first four terms of s_n.

Solution

$$s_1 = \frac{1}{1} = 1, \qquad s_2 = \frac{1}{2}, \qquad s_3 = \frac{1}{3}, \qquad s_4 = \frac{1}{4}$$

Sometimes values of a sequence may be matrices as the following example shows.

Example 5

Let the matrix-values sequence (M_n) be given by

$$M_n = \begin{bmatrix} 1+n & -n \\ 0 & (-1)^n \end{bmatrix}$$

Write out the terms M_0, M_1, and M_2.

Solution

$$M_0 = \begin{bmatrix} 1+0 & 0 \\ 0 & (-1)^0 \end{bmatrix} = \begin{bmatrix} 1 & 0 \\ 0 & 1 \end{bmatrix}$$

$$M_1 = \begin{bmatrix} 1+1 & -1 \\ 0 & (-1)^1 \end{bmatrix} = \begin{bmatrix} 2 & -1 \\ 0 & -1 \end{bmatrix}$$

$$M_2 = \begin{bmatrix} 1+2 & -2 \\ 0 & (-1)^2 \end{bmatrix} = \begin{bmatrix} 3 & -2 \\ 0 & 1 \end{bmatrix}$$

Exercise 13.3

Answers to Odd-Numbered Problems begin on page A-46.

A

1. Suppose a sequence (s_n) is defined by $s_n = (-1)^n/n$, $n \neq 0$. Write out s_1, s_2, s_3, s_4, s_{100}.
2. Answer the question in Problem 1 for $s_n = (-1)^{n+1}/n$.
3. Consider a sequence (b_n) given by $b_n = n/(n+1)$
   (a) List b_0, b_1, b_2, b_3, b_4, b_5
   (b) Find $b_{n+1} - b_n$, for $n = 0, 1, 2$
4. Answer the question in Problem 3 for $b_n = -n/(n+1)$.

5. Find the first 8 terms of the sequence $a_k = 2^k$.

6. Answer the question in Problem 5 for $a_k = 2^{-k}$.

7. List the first 7 terms of the sequence $s_n = n!$.

B   8. Let $M_n = \begin{bmatrix} 1 & n^2 \\ n & 1 \end{bmatrix}$. Find M_0, M_1, M_2, and M_3.

9. Let $M_n = \begin{bmatrix} 1 & 1-n & 0 \\ 1-n & n & n-1 \\ 0 & n-1 & n+1 \end{bmatrix}$. Find M_0, M_1, M_2, and M_3.

10. Let $M_n = \begin{bmatrix} 0 & 1 \\ 1 & 0 \end{bmatrix}^n$ where $\begin{bmatrix} 0 & 1 \\ 1 & 0 \end{bmatrix}^n$ stands for $\begin{bmatrix} 0 & 1 \\ 1 & 0 \end{bmatrix}$ multiplied by itself n times. Find M_1, M_2, M_3, and M_{10}.

C   In Problems 11–16 write a formula for the nth term.

11. $1, -1, 1, -1, \ldots$

12. $2, 0, 2, 0, 2, 0, \ldots$

13. $1, 3, 5, 7, 9, \ldots$

14. $0, 5, 25, 125, \ldots$

15. $1, \dfrac{1}{2}, \dfrac{1}{3}, \dfrac{1}{4}, \dfrac{1}{5}, \ldots$

16. $1, \dfrac{1}{3}, \dfrac{1}{9}, \dfrac{1}{27}, \ldots$

17. **Towers of Hanoi*** The number of moves required in solving the puzzle of the Towers of Hanoi for n disks is given by

$$(s_n) = 2^n - 1$$

Find $s_0, s_1, s_2, s_3, s_4, s_5, s_6, s_7$, and s_8.

18. Suppose we have a set of straight lines where none of the lines are parallel and no three of the lines go through the same point. The number of distinct regions that such a set of n lines divides the plane into is given by the nth term of the sequence

$$(s_n) = \binom{n+1}{2} + 1, \ n \geq 1$$

Find s_1, s_2, s_3, and s_4.

19. The Fibonacci sequence† is given by

$$(s_n) = \frac{1}{\sqrt{5}} \left[\left(\frac{1+\sqrt{5}}{2} \right)^{n+1} - \left(\frac{1-\sqrt{5}}{2} \right)^{n+1} \right]$$

Find $s_0, s_1, s_2, s_3, s_4, s_5, s_6, s_7$, and s_8.

13.4

MATHEMATICAL INDUCTION

Mathematical induction is used to establish that mathematical statements involving positive integers are true for *all* positive integers. For example, we can use mathematical induction to prove that the following statement:

* See Example 3, Section 13.5.

† See Example 2, Section 13.5.

$$1 + 2 + 3 + \cdots + n = \frac{n(n+1)}{2} \tag{1}$$

holds for all positive integers n.

Before describing the method of mathematical induction, let us try to realize its power. To do this, we use Equation (1) and substitute the various possible values of $n = 1, 2, 3, \ldots$ to construct the following table:

Value of n	Left-Hand Side of Equation (1)	Right-Hand Side of Equation (1)	Formula in Equation (1)
1	1	$\frac{1(1+1)}{2} = 1$	Holds
2	$1 + 2 = 3$	$\frac{2(2+1)}{2} = 3$	Holds
3	$1 + 2 + 3 = 6$	$\frac{3(3+1)}{2} = 6$	Holds
4	$1 + 2 + 3 + 4 = 10$	$\frac{4(4+1)}{2} = 10$	Holds
etc.			

We make two observations about this table:

1. It is impossible to substitute all possible values of n because there is an infinite number of them (as many as the number of positive integers). This means that our table can never be complete, and so Equation (1) can never be proven by substituting all the various values of n.
2. Although the pattern of the right-hand side column suggests that $n(n+1)/2$ is valid, we can never be sure that Equation (1) does not fail for some untried value of n.

By mathematical induction, however, we will prove that Equation (1) holds for all positive integers n. We will show this in Example 2.

In the remainder of this section we use $S(n)$ to denote the statement that we wish to prove is true.

Example 1

Let $S(n)$ be the statement

$$1 + 2 + 3 + \cdots + n = \frac{n(n+1)}{2}$$

(a) Write $S(1)$. Is $S(1)$ true?
(b) Write $S(2)$. Is $S(2)$ true?
(c) Write $S(k)$. (k is a positive integer.)
(d) Write $S(k+1)$.

Solution

It is clear that $S(n)$ is the statement about the sum of the first n positive integers.

(a) $S(1)$ is the statement about the sum of the first positive integer. By Formula (1), $S(1)$ is the statement

$$1 = \frac{1(1+1)}{2}$$

which is true.

(b) $S(2)$ is the statement about the sum of the first two positive integers. By Formula (1), $S(2)$ is the statement

$$1 + 2 = \frac{2(2+1)}{2}$$

which is also true.

(c) $S(k)$ is the statement about the sum of the first k positive integers. By Formula (1), $S(k)$ is the statement

$$1 + 2 + 3 + \cdots + k = \frac{k(k+1)}{2}$$

(d) $S(k+1)$ is the statement about the sum of the first $(k+1)$ positive integers. This is obtained by substituting $k + 1$ for n in Equation (1), to get

$$1 + 2 + 3 + \cdots + k + (k+1) = \frac{(k+1)[(k+1)+1]}{2} \qquad (2)$$

∎

We are now ready to state the principle of mathematical induction.

Principle of Mathematical Induction   Let $S(n)$ be a statement that involves positive integers $n = 1, 2, 3, \ldots$. If we can show that the following two conditions are satisfied:

> **Condition I** The statement $S(1)$ is true
>
> **Condition II** Let k be any positive integer. If assuming $S(k)$ is true implies that $S(k+1)$ is also true,

then $S(n)$ must be true for *all* positive integers.

Notice that from Condition I $S(1)$ is true. By Condition II $S(2)$ is true. Again using Condition II $S(3)$ is true and so on, for all natural numbers.

Example 2

Show that

$$1 + 2 + 3 + \cdots + n = \frac{n(n+1)}{2} \qquad (1)$$

is true for all positive integers n.

Solution

We need to show first that $S(1)$ is true. Because $1 = 1(1 + 1)/2$, Condition I is satisfied. To show that Condition II is true, we assume that Formula (1) holds for some positive integer k. That is, we assume $S(k)$ is true for some positive integer k and, based on this assumption, we must show that $S(k + 1)$ is true. We look at the sum of the first $k + 1$ positive integers:

$$1 + 2 + 3 + \cdots + k + (k + 1) = [1 + 2 + 3 + \cdots + k] + (k + 1)$$

$$\underset{\uparrow}{=} \frac{k(k + 1)}{2} + (k + 1)$$

by our assumption that $S(k)$ is true

$$= \frac{k(k + 1) + 2(k + 1)}{2}$$

$$\underset{\uparrow}{=} \frac{(k + 1)[k + 2]}{2}$$

Take $(k + 1)$ as a common factor

$$= \frac{(k + 1)[(k + 1) + 1]}{2}$$

which is $S(k + 1)$. (See Formula 2.) Thus, Condition II is also satisfied. And by the principle of mathematical induction, Equation (1) is true for all positive integers n.

■

Example 3

Let $S(n)$ be the following statement describing the sum of the first n positive odd integers:

$$1 + 3 + 5 + \cdots + (2n - 1) = n^2 \tag{3}$$

(a)   Write $S(1)$ and show it is true.
(b)   Write $S(2)$ and show it is true.
(c)   Write $S(k)$.
(d)   Write $S(k + 1)$.
(e)   Show, using mathematical induction, that $S(n)$ is true for all positive integers n.

Solution

(a) $S(1)$ describes the sum of the first positive odd integer; Equation (3) gives $1 = 1^2$. This is true; therefore, $S(1)$ is true.
(b) $S(2)$ describes the sum of the first two positive odd integers; Equation (3) gives $1 + 3 = 2^2$. Because both sides are equal to 4, $S(2)$ is true.

(c) $S(k)$ describes the sum of the first k positive odd integers. Equation (3) gives
$1 + 3 + 5 + \cdots + (2k - 1) = k^2$.

(d) $S(k + 1)$ describes the sum of the first $k + 1$ positive odd integers. Equation (3) gives

$$1 + 3 + 5 + \cdots + (2k - 1) + (2k + 1) = (k + 1)^2$$

(e) Since, from (a) above $S(1)$ is true, Condition I is satisfied. To show that Condition II is satisfied we assume that $S(k)$ is true [see (c) above] and we show that this implies $S(k + 1)$ [see (d) above] is true. So we consider the sum of the first $(k + 1)$ positive odd integers:

$$1 + 3 + 5 + \cdots + (2k - 1) + (2k + 1)$$
$$= \underbrace{[1 + 3 + 5 + \cdots + (2k - 1)]}_{\substack{= k^2 \text{ since we} \\ \text{assumed } S(k) \text{ holds}}} + (2k + 1)$$
$$= k^2 + 2k + 1$$
$$= (k + 1)^2$$

Thus, $S(k + 1)$ holds and so Condition II is satisfied, and Formula (3) is true for all positive integers. ■

Exercise 13.4   *Answers to Odd-Numbered Problems begin on page A-47.*

A   **1.** Let $S(n)$ be the statement

$$n < 2^n, \qquad n \text{ is a positive integer}$$

(a) Write $S(1)$. Is $S(1)$ true?
(b) Write $S(2)$. Is $S(2)$ true?
(c) Write $S(k)$.
(d) Write $S(k + 1)$.

2. Let $S(n)$ be the statement

$$1 \cdot 2 + 2 \cdot 3 + 3 \cdot 4 + \cdots + n \cdot (n + 1) = \frac{n(n + 1)(n + 2)}{3}$$

(a) Write $S(1)$. Is $S(1)$ true?
(b) Write $S(2)$. Is $S(2)$ true?
(c) Write $S(k)$.
(d) Write $S(k + 1)$.

3. Let $S(n)$ be the statement

$$1^2 + 2^2 + 3^2 + \cdots + n^2 = \frac{n(n + 1)(2n + 1)}{6}$$

(a)   Write $S(1)$. Is $S(1)$ true?
(b)   Write $S(5)$. Is $S(5)$ true?
(c)   Write $S(k)$.
(d)   Write $S(k+1)$.

4. Let $S(n)$ be the statement

$$\frac{1}{1\cdot 2}+\frac{1}{2\cdot 3}+\frac{1}{3\cdot 4}+\cdots+\frac{1}{n(n+1)}=\frac{n}{n+1}$$

(a)   Write $S(1)$. Is $S(1)$ true?
(b)   Write $S(3)$. Is $S(3)$ true?
(c)   Write $S(k)$.
(d)   Write $S(k+1)$.

5. Let $S(n)$ be the statement

$$1+2+2^2+\cdots+2^n=2^{n+1}-1$$

(a)   Write $S(1)$. Is $S(1)$ true?
(b)   Write $S(5)$. Is $S(5)$ true?
(c)   Write $S(k)$.
(d)   Write $S(k+1)$.

6. Let $S(n)$ be the statement

$$-1+1-1+1-\cdots+(-1)^n=\frac{(-1)^n-1}{2}$$

(a)   Write $S(1)$. Is $S(1)$ true?
(b)   Write $S(2)$. Is $S(2)$ true?
(c)   Write $S(100)$. Is $S(100)$ true?
(d)   Write $S(k)$.
(e)   Write $S(k+1)$.

B   In Problems 7–12 prove by mathematical induction that the statement is true for all positive integers n.

7. $1\cdot 2+2\cdot 3+3\cdot 4+\cdots+n\cdot(n+1)=\dfrac{n(n+1)(n+2)}{3}$

8. $1^2+2^2+3^2+\cdots+n^2=\dfrac{n(n+1)(2n+1)}{6}$

9. $\dfrac{1}{1\cdot 2}+\dfrac{1}{2\cdot 3}+\dfrac{1}{3\cdot 4}+\cdots+\dfrac{1}{n\cdot(n+1)}=\dfrac{n}{n+1}$

10. $2+4+6+\cdots+2n=n\cdot(n+1)$

11. $2+5+8+\cdots+(3n-1)=\dfrac{n(3n+1)}{2}$

12. $n<2^n$

13.5

RECURRENCE RELATIONS

Thus far we have defined a sequence by giving a general formula for its nth term or by writing a few of its terms. Often this approach is not possible. An alternative is to write the sequence by finding a relationship among its terms. Such a relationship is called a *recurrence relation*. Sequences defined this way normally arise when each term of the sequence depends upon the previous terms of the sequence. Defining a sequence recursively requires giving both the recurrence relation and specifying the first few terms of the sequence. Specifying the first few terms of the sequence is called setting *initial conditions*. An example will help clarify these concepts.

Example 1

Compute the first six terms of the sequence defined by the following recurrence relation and initial conditions:

(a)   Recurrence relation: $s_n = 2s_{n-1} + 1$,   for all integers $n \geq 1$
   Initial condition: $s_0 = 1$

(b)   Recurrence relation: $s_n = s_{n-1} + s_{n-2}$,   for all integers $n \geq 2$
   Initial conditions: $s_0 = 1$, $s_1 = 2$

(c)   Recurrence relation: $s_n = s_{n-1}^2 + ns_{n-2} + s_{n-3}$,   for all integers $n \geq 3$
   Initial conditions: $s_0 = 1$, $s_1 = 2$, $s_2 = -1$

Solution

(a)   Here the initial condition is given as $s_0 = 1$. Therefore we need to compute s_1, s_2, s_3, s_4, and s_5. Substituting $n = 1$ in the recurrence relation gives

$$s_1 = 2s_{1-1} + 1 = 2s_0 + 1 = 2(1) + 1 = 3$$
$$\uparrow$$
$$s_0 = 1$$

Similarly,

$$s_2 = 2s_{2-1} + 1 = 2s_1 + 1 = 2(3) + 1 = 7$$
$$\uparrow$$
$$s_1 = 3$$
$$s_3 = 2s_{3-1} + 1 = 2s_2 + 1 = 2(7) + 1 = 15$$
$$\uparrow$$
$$s_2 = 7$$
$$s_4 = 2s_{4-1} + 1 = 2s_3 + 1 = 2(15) + 1 = 31$$
$$\uparrow$$
$$s_3 = 15$$

and

$$s_5 = 2s_{5-1} + 1 = 2s_4 + 1 = 2(31) + 1 = 63$$
$$\uparrow$$
$$s_4 = 31$$

Thus the first six terms of this sequence are 1, 3, 7, 15, 31, 63.

(b) Here the initial conditions are given as $s_0 = 1$ and $s_1 = 2$; therefore, we need to compute s_2, s_3, s_4, and s_5. Substituting $n = 2$ in the recurrence relation gives

$$s_2 = s_{2-1} + s_{2-2} = s_1 + s_0 = 2 + 1 = 3$$
$$\uparrow$$
$$s_0 = 1 \text{ and } s_1 = 2$$

Similarly,

$$s_3 = s_{3-1} + s_{3-2} = s_2 + s_1 = 3 + 2 = 5$$
$$\uparrow$$
$$s_1 = 2 \text{ and } s_2 = 3$$
$$s_4 = s_{4-1} + s_{4-2} = s_3 + s_2 = 5 + 3 = 8$$
$$\uparrow$$
$$s_2 = 3 \text{ and } s_3 = 5$$

and

$$s_5 = s_{5-1} + s_{5-2} = s_4 + s_3 = 8 + 5 = 13$$
$$\uparrow$$
$$s_3 = 5 \text{ and } s_4 = 8$$

Thus the first six terms of this sequence are 1, 2, 3, 5, 8, 13.

(c) Here the initial conditions are given as $s_0 = 1$, $s_1 = 2$, and $s_2 = -1$; therefore, we need to compute s_3, s_4, and s_5. Substituting $n = 3$ in the recurrence relation gives

$$s_3 = s_{3-1}^2 + 3s_{3-2} + s_{3-3} = s_2^2 + 3s_1 + s_0 = (-1)^2 + 3(2) + 1 = 1 + 6 + 1 = 8$$
$$\uparrow$$
$$s_0 = 1, \ s_1 = 2, \text{ and } s_2 = -1$$

Similarly,

$$s_4 = s_{4-1}^2 + 4s_{4-2} + s_{4-3} = s_3^2 + 4s_2 + s_1 = 8^2 + 4(-1) + 2$$
$$\uparrow$$
$$s_1 = 2, \ s_2 = -1, \text{ and } s_3 = 8$$
$$= 64 - 4 + 2 = 62,$$

and

$$s_5 = s_{5-1}^2 + 5s_{5-2} + s_{5-3} = s_4^2 + 5s_3 + s_2 = (62)^2 + 5(8) - 1$$
$$\uparrow$$
$$s_2 = -1, \ s_3 = 8, \text{ and } s_4 = 62$$
$$= 3844 + 40 - 1 = 3883$$

Thus the first six terms of this sequence are 1, 2, -1, 8, 62, 3883. ∎

These examples lead us to the following definition.

A *recurrence relation* for the sequence $s_0, s_1, s_2, s_3, \ldots$ is a formula that relates each term s_n to previous terms of the sequence. The *initial condition* for a recurrence relation is a set of values of the first few terms of the sequence.

The next two examples are two well-known sequences that are defined recursively.

Example 2

In 1202 A.D., a European mathematician named Leonardo Fibonacci posed the following problem:

Assume we start with a single pair of newborn rabbits and that each pair does not reproduce during its first month of life, but after its first month, it produces one new pair (a male and a female) at the end of every month. If we further assume that no rabbits die, how many pairs of rabbits will there be at the end of n months?

Solution

Let r_n denote the number of rabbit pairs at the end of the nth month. Since we started with one pair, then we have $r_0 = 1$. And, since this pair is not fertile during its first month, then at the end of the first month we will still have (this) one pair, that is, $r_1 = 1$. At the end of the second month this pair will produce another pair and so there will be two pairs, that is, $r_2 = r_0 + r_1 = 1 + 1 = 2$. At the end of the third month the one-month-old pair does not produce but the old pair does, the one we started with; thus there will be one newborn pair in addition to the previous two pairs, that is,

$$r_3 = r_2 + r_1 = 2 + 1 = 3$$

At the end of the fourth month there will be two pairs that are reproductive (as many as there was at the end of the second month, which is 2), which will produce two pairs, and the one-month-old pair that cannot produce. Thus,

$$r_4 = r_3 + r_2 = 3 + 2 = 5$$

Continuing in this manner we find that at the end of the nth month there will be r_{n-2} new pairs born plus r_{n-1} pairs (those include the one month olds). That is,

Recurrence relation $r_n = r_{n-1} + r_{n-2}$

Initial condition $r_0 = 1, \quad r_1 = 1$ ■

Fibonacci numbers occur in many applications. For example, the number of binary words that do not contain the pattern 00, follows a Fibonacci sequence.

Example 3

The Towers of Hanoi   This is a puzzle about three poles and n disks of increasing sizes called the Towers of Hanoi puzzle. Initially, all of the disks (all have holes at their centers) are placed on the first pole, as shown in Figure 7*a*. We want to transfer all the disks from the first pole to the third, so that they appear as in Figure 7*b*. If we can move only one disk at a time and if we are not allowed to place a larger disk on top of a smaller one, what is the minimum number of moves required to transfer a tower of *n* disks from the first pole to the third pole?

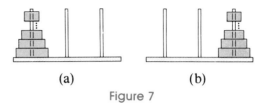

(a)   (b)

Figure 7

Solution
We begin by letting m_n be the minimum number of moves required to move the *n* disks from the first pole to the third. Obviously, $m_0 = 0$. $m_1 = 1$ because we can move one disk from the first pole to the third in one move. Notice that this is the minimum number of moves required to transfer one disk from the first pole to the third. To find m_2 we proceed as follows: First, we move the smaller disk from the first pole to the second leaving the larger one on the first pole (Figures 8*a*, 8*b*). Second, we move the larger disk from the first pole to the third (Figure 8*c*). Third, and last, we move the smaller disk from the second pole to the third (Figure 8*d*).

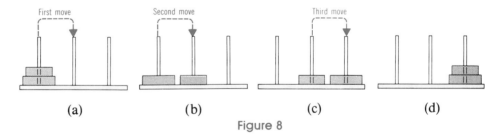

(a)   (b)   (c)   (d)

Figure 8

Thus a minimum of three moves is needed to transfer two disks from the first pole to the third. This is the minimum number of moves because if we follow a different set of moves to transfer the two disks from the first pole to the third without violating the rules of the puzzle, we end up with more than three moves (Try it!). Thus m_2 requires the following: m_1 moves from the first pole to the second (moves for transferring the smaller disk from the first pole to the second), one move of the larger disk from the first pole to the third, and another m_1 moves from the second pole to the third (moves for transferring the smaller disk from the second pole to the third). Thus

$$m_2 = 2m_1 + 1 = 2 \cdot 1 + 1 = 3$$

For a tower of three disks the minimum number of moves m_3 required to transfer the three disks from the first pole to the third is computed as follows: m_2 moves are required to transfer the top two disks from the first pole to the second (Figures 9*a*, 9*b*), one move of the large disk from the first pole to the third (Figure 9*c*), and

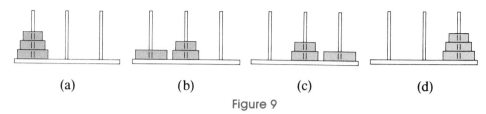

(a)   (b)   (c)   (d)

Figure 9

another m_2 moves of the smaller two disks from the second pole to the third (Figure 9d). Thus

$$m_3 = 2m_2 + 1 = 2 \cdot 3 + 1 = 7$$

Continuing our reasoning* in this way, it can be shown that

| Recurrence relation | $m_n = 2m_{n-1} + 1$ | for all integers $n \geq 2$ |
| Initial condition | $m_0 = 0,$ $m_1 = 1$ | |

More examples on recursively defined sequences are found in the exercises.

Exercise 13.5

Answers to Odd-Numbered Problems begin on page A-48.

In Problems 1–14 a recurrence relation and an initial condition are given that define a sequence (s_n). Find the first six terms of (s_n).

A   **1.** $s_n = 2s_{n-1}$ for all integers $n \geq 1$
   $s_0 = 1$

   2. $s_n = s_{n-1} + n$ for all integers $n \geq 1$
   $s_0 = 1$

   3. $s_n = ns_{n-1} - 1$ for all integers $n \geq 1$
   $s_0 = 1$

   4. $s_n = s_{n-1} + 2^n$ for all integers $n \geq 1$
   $s_0 = -1$

   5. $s_n = (n + 1)s_{n-1}$ for all integers $n \geq 1$
   $s_0 = 2$

   6. $s_n = s_{n-1} + n^2$ for all integers $n \geq 1$
   $s_0 = 1$

   7. $s_n = s_{n-2} + s_{n-1} + \binom{n}{2}$ for all integers $n \geq 2$
   $s_0 = 1, s_1 = 1$

   8. $s_n = 2s_{n-2} + s_{n-1}^2$ for all integers $n \geq 2$
   $s_0 = 1, s_1 = -1$

   9. $s_n = s_{n-1}s_{n-2}$ for all integers $n \geq 2$
   $s_0 = 1, s_1 = 2$

---

* A proof of this requires mathematical induction.

10. $s_n = (s_{n-1} + s_{n-2})^{1/2}$ for all integers $n \geq 2$
    $s_0 = 3, s_1 = 6$

11. $s_n = s_{n-1} + s_{n-2} + s_{n-3}$ for all integers $n \geq 3$
    $s_0 = 1, s_1 = 2, s_2 = 2$

12. $s_n = s_{n-1} + s_{n-2} - 2s_{n-3}$ for all integers $n \geq 3$
    $s_0 = 2, s_1 = 1, s_2 = 1$

13. $s_n = ns_{n-1} + s_{n-2} - s_{n-3}$ for all integers $n \geq 3$
    $s_0 = 1, s_1 = -1, s_2 = -1$

14. $s_n = (-1)^n s_{n-1} + ns_{n-2} + n^2 s_{n-3}$ for all integers $n \geq 3$
    $s_0 = 1, s_1 = 1, s_2 = 1$

B 15. Use the recurrence relation and initial condition for the Fibonacci number sequence $r_0, r_1, r_2, r_3, \ldots$ defined in Example 2 to compute $r_{10}, r_{11}, r_{12}, r_{13}$, and r_{14}.

16. Use the recurrence relation and initial condition for the Towers of Hanoi sequence $m_0, m_1, m_2, m_3, \ldots$ defined in Example 3 to compute m_6, m_7, m_8, and m_9.

C 17. (a) Write all binary words of lengths 0, 1, 2, 3, and 4 that do not contain the bit pattern 00.
       (b) For all integers $n \geq 2$, let $s_n =$ the number of binary words of length n that do not contain the pattern 00. Find a recurrence relation relating s_n to s_{n-1} and s_{n-2}.

18. Repeat Problem 17 for the binary words that do not contain the bit pattern 11.

19. Suppose $1000 is invested in an account paying 10% interest compounded annually. Assume that no withdrawals are made, and let A_n be the amount in the account after n years.
    (a) Compute A_0, A_1, A_2, A_3, and A_4.
    (b) Find a recurrence relation relating A_n to A_{n-1} for all integers $n \geq 1$.

20. Repeat Problem 19 for $2000 with annually compound interest of 5.5%.

21. Consider a set of n lines in a plane no two of which are parallel and no three of which intersect at the same point. Let p_n denote the number of distinct regions formed by these lines. See the figure.

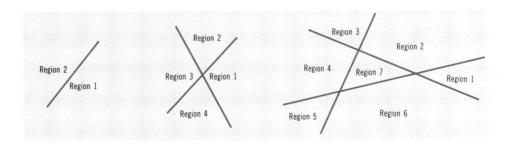

    (a) Find p_1, p_2, p_3, and p_4.
    (b) Find a recurrence relation relating p_n to p_{n-1} for all integers $n \geq 2$.

CHAPTER REVIEW

<table>
<tr><td rowspan="2">Important Terms</td><td>relation</td><td>domain</td><td>composition of functions</td></tr>
<tr><td>bit</td><td>image</td><td>identity function</td></tr>
<tr><td></td><td>binary word</td><td>range</td><td>inverse function</td></tr>
<tr><td></td><td>byte</td><td>one-to-one</td><td>sequence</td></tr>
<tr><td></td><td>reflexive</td><td>injective</td><td>mathematical induction</td></tr>
<tr><td></td><td>symmetric</td><td>onto</td><td>recurrence relation</td></tr>
<tr><td></td><td>transitive</td><td>surjective</td><td>initial condition</td></tr>
<tr><td></td><td>function</td><td>bijective</td><td></td></tr>
</table>

True-False Questions

(Answers are on page A-48)

T F   1. For all relations R from A to B and all $a \in A$, $b \in B$ if aRb, then bRa.

T F   2. The range and the domain of all functions are the same.

T F   3. A bijective function is both one-to-one and onto.

T F   4. One way to define a sequence is to give a formula for its nth term.

T F   5. Condition II of the principle of mathematical induction states that if the statement $S(k)$ is true for some positive integer k, then so is $S(k + 1)$.

T F   6. Every recurrence relation must have some initial condition.

Fill in the Blanks

(Answers are on page A-48)

1. A relation R from A to B is the set R of _____ _____ where $a \in A$ $b \in B$.

2. A _____ from the set X into the set Y is a relation that associates with each element of X exactly one element of Y.

3. A one-to-one and onto function is called _____.

4. A sequence is a function whose domain is the set of _____ _____.

5. _____ _____ is used to prove that a statement $S(n)$, which involves natural numbers, is true for all n.

6. A formula that relates each term of a sequence to previous terms is called a _____ _____. Specifying the first few terms is giving the _____ _____.

Review Exercises

Answers to Odd-Numbered Problems begin on page A-49.

A   **1.** Let $A = \{1,2,3,8,9,27\}$ and R be a relation on A defined by aRb if, and only if, a is the cube of b. Write R as a set of ordered pairs.

   2. State the inverse relation R^{-1} of the relation R of Problem 1. Write R^{-1} as a set of ordered pairs.

3. Which of the following relations defines a function from $A = \{0,1,2\}$ to $B = \{0,1,2,3\}$?

   (a) $f(0) = 0, f(1) = 3, f(2) = 1$.
   (b) $g(0) = 1, g(1) = 1, g(2) = 1$.
   (c) $h(0) = 1, h(1) = 2$.
   (d) $F(0) = 3, F(1) = 2, F(0) = 1, F(2) = 3$.

B 4. Suppose $S(n)$ is the statement $1 + 3 + 3^2 + 3^3 + \cdots + 3^{n-1} = \frac{1}{2}(3^n - 1)$.
   Find $S(1), S(2), S(3)$. Use mathematical induction to prove that $S(n)$ is true for all n.

5. Find the first 10 terms of each sequence

   (a) $a_k = (-1)^k + k$
   (b) $a_k = (-1)^k \cdot k$

6. Find the first 5 terms of the sequence defined by the following recurrence relation and initial conditions:

$$s_n = (n - 2)s_{n-2} + s_{n-1}, \qquad n \geq 2$$
$$s_0 = 1, \qquad s_1 = 0$$

7. Find the first 7 terms of the sequence defined by the following recurrence relation and initial conditions:

$$s_n = s_{n-3} \cdot (s_{n-2} + s_{n-3}), \qquad n \geq 3$$
$$s_0 = 2, \qquad s_1 = 2, \qquad s_2 = 1$$

8. Suppose that you start with a single pair of newborn rabbits and that each pair reproduces a new pair (a male and a female) every two months except the first two months of its life. If no rabbits die, how many rabbits will there be at the end of the

   (a) Fourth month?
   (b) Sixth month?
   (c) Tenth month?

14

GRAPHS AND TREES

14.1

GRAPHS

DEFINITION OF GRAPH
FORMULA FOR DEGREES AND EDGES

Many situations that occur in computer science, operations research, physical sciences, and economics are of a combinatorial nature and can be analyzed by using techniques found in a relatively new area of mathematics called *graph theory*. In this chapter, we present a brief introduction and explain some of the more elementary results of graph theory and their applications.

Before giving a formal definition of a graph, we point out that the term *graph* has two quite different meanings. One definition of a graph is the one we studied in Chapter 1 and used when we graphed straight lines and linear inequalities.

DEFINITION OF GRAPH

In this chapter a *graph* will mean a set of points (called *vertices*) with one or more curves or lines (called *edges*) connecting a point to itself or a pair of points. For a given pair of vertices, our concern is not what the edge looks like, but rather whether the two vertices have an edge joining them or not. For example, the two diagrams in Figure 1 represent the same graph: each has 4 vertices v_1, v_2, v_3, and v_4 and 8 edges $e_1, e_2, e_3, e_4, e_5, e_6, e_7$, and e_8. Note that the edges e_5 and e_6 intersect in the graph in Figure 1*a*, but their intersection is not a vertex. Edges e_2 and e_7 are two different edges connecting the same pair of vertices (v_2 and v_3). Such edges are called *parallel edges*. (Note that this is a departure from the usual definition of parallel lines.) Edge e_8 is called a *loop* because it connects a vertex, v_3, to itself.

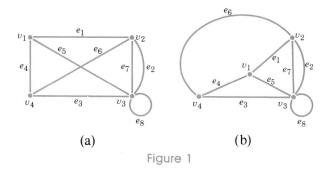

(a)   (b)

Figure 1

Graph A *graph G consists of the following two finite sets: (1) a nonempty set V of vertices and (2) a set E of edges, where each edge either connects two vertices or connects a vertex to itself.*

A vertex that is not connected to any other vertex or to itself is called an *isolated vertex.*

Example 1

For the graph given in Figure 2

(a) Write the vertex set.
(b) Write the edge set.
(c) Find all isolated vertices.
(d) Find all loops.
(e) Find all parallel edges.

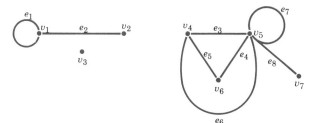

Figure 2

Solution

(a) The vertex set is

$$V = \{v_1, v_2, v_3, v_4, v_5, v_6, v_7\}$$

(b) The edge set is

$$E = \{e_1, e_2, e_3, e_4, e_5, e_6, e_7, e_8\}$$

(c) v_3 is the only vertex that is isolated.
(d) e_1 and e_7 are the only loops.
(e) e_3 and e_6 are parallel edges since they both connect v_4 and v_5. ■

A graph with no parallel edges and no loops is called a *simple graph*.

Example 2

Draw all simple graphs with 3 vertices.

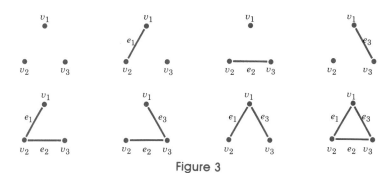

Figure 3

Solution

In Figure 3, we list all possible graphs having 3 vertices that are free of parallel edges and loops.   ■

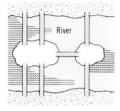

Figure 4

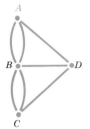

Figure 5

Example 3

Königsberg Bridge This is one of the oldest problems involving graphs. The town of Königsberg (now Kaliningrad in the Soviet Union) is crossed by the Pregel river with two islands in the river. The islands are connected to each other by one bridge. The larger island is connected to each bank by two bridges. The smaller island is connected to each bank by one bridge. Thus, there is a total of seven bridges. See Figure 4. The townspeople wondered whether one can begin at one of the banks or islands, walk across each bridge exactly once, and return to the starting point. This problem is equivalent to the following: Let each mass of land be represented by a vertex and each bridge by an edge to obtain a graph as in Figure 5 where A and C are the river banks, B is the larger island, D is the smaller island, and the 7 edges represent the 7 bridges. Can one start at either one of the vertices A, B, C, or D, traverse each edge in Figure 5 exactly once and return to the starting vertex? A Swiss mathematician named Leonhard Euler (1707–1783) studied this problem and in 1736 was able to answer it. We give his answer in Section 14.3.   ■

Example 4

Computer, telephone, and transportation route networks (to name a few) can be represented by graphs. Inspecting or analyzing their graphs usually determines connecting edges for optimization purposes. Figure 6 shows a communication network.

Figure 6   ■

Example 5

Pascal Data Types Relationships can be depicted by graphs. For example, the hierarchy of data types in the Pascal language is given in the graph in Figure 7.   ■

FORMULA FOR DEGREES AND EDGES

Examples 3, 4, and 5 show the usefulness of graphs. In the remainder of this section we discuss some definitions and results associated with a graph.

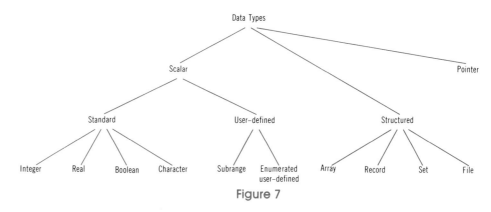

Figure 7

Let G be a graph and v a vertex of G. The _degree of v_, denoted deg(v), is the number of edges emanating from v. An edge that is a loop is counted twice.

Note that a vertex v is isolated if and only if $\deg(v) = 0$.

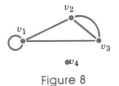

Figure 8

Example 6
Find the degree of each vertex in the graph G shown in Figure 8.

Solution
There are two edges connecting the vertex v_1 to the vertices v_2 and v_3, respectively, and there is a loop at the vertex v_1; thus $\deg(v_1) = 4$. Similarly, $\deg(v_2) = 3$; $\deg(v_3) = 3$; and $\deg(v_4) = 0$, since v_4 is not connected to any of the other vertices. ∎

It turns out that there is a relation between the number of edges in a graph and the sum of the degrees of the vertices in the graph. If we add all the degrees, every edge will be counted twice. Thus we have:

Let G be a graph with vertices $v_1, v_2, \ldots, v_n$. Then the sum of the degrees of all the vertices of G equals twice the number of edges in G. That is,

$$\deg(v_1) + \deg(v_2) + \cdots + \deg(v_n) = 2 \cdot \text{(the number of edges in } G) \quad (1)$$

Note that this theorem says that the sum of the degrees of the vertices is even, and hence the number of odd degree vertices must be even.

Example 7
Can we draw a graph G with 3 vertices v_1, v_2, and v_3 where

(a) $\deg(v_1) = 1$, $\deg(v_2) = 2$, and $\deg(v_3) = 2$.
(b) $\deg(v_1) = 2$, $\deg(v_2) = 1$, and $\deg(v_3) = 1$.
(c) $\deg(v_1) = 0$, $\deg(v_2) = 0$, and $\deg(v_3) = 4$.

Justify your answers, and if yes, draw G.

Solution

(a) No, since $\deg(v_1) + \deg(v_2) + \deg(v_3) = 1 + 2 + 2 = 5$, which is an odd number. Then, by the above theorem, such a graph does not exist.

(b) Yes, $\deg(v_1) + \deg(v_2) + \deg(v_3) = 4$, which is an even number. Then, G can be either one of the two graphs shown in Figure 9. In each case

$$\deg(v_1) = 2, \qquad \deg(v_2) = 1, \qquad \text{and } \deg(v_3) = 1$$

Figure 9

(c) Yes, $\deg(v_1) + \deg(v_2) + \deg(v_3) = 4$, which is an even number. Then G is the graph shown in Figure 10. Here $\deg(v_1) = 0$, $\deg(v_2) = 0$, and $\deg(v_3) = 4$. (Because of the two loops at v_3.)

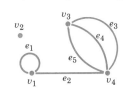

Figure 10

Exercise 14.1

Answers to Odd-Numbered Problems begin on page A-49.

In Problems 1 and 2 write the vertices and edges of each graph and find all loops, all isolated vertices, and all parallel edges.

A   **1.**

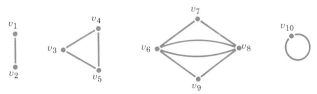

2.

3. Draw all simple graphs that have two vertices.

4. Draw all simple graphs that have four vertices and six edges.

5. Find the degree of each vertex of the graphs in Problems 1 and 2.

6. Can we draw a graph with vertices $v_1, v_2, v_3, v_4, v_5, v_6, v_7$, and v_8 of degrees 2, 2, 3, 4, 5, 5, 6, and 8, respectively? Justify your answer.

7. Let G be a graph with vertices v_1, v_2, v_3, v_4, v_5, and v_6 of degrees 1, 2, 3, 3, 4, and 5, respectively. How many edges does G have? Justify your answer.

8. Can we draw a graph with vertices v_1, v_2, v_3, and v_4 of degrees:

(a)   1, 2, 3, and 3, respectively?

(b)   1, 1, 1, and 4, respectively?

(c)   1, 2, 3, and 4, respectively?

Justify your answers

B   9.   Can we draw a simple graph with vertices v_1, v_2, v_3, and v_4 of degrees 1, 2, 3, and 4, respectively? Justify your answer.

10.   Can we draw a simple graph with vertices v_1, v_2, v_3, v_4, and v_5 of degrees:

(a)   2, 3, 3, 3, and 5, respectively?

(b)   1, 1, 1, 2, and 3, respectively?

Justify your answers.

11.   Give an example of a simple graph (a) having no vertices of even degree (b) having no vertices of odd degree.

In Problems 12 and 13, use the following definition:

The *complete graph of order n,* denoted by K_n, is the graph that has *n* vertices and every vertex is connected to every other vertex by exactly one edge.

C   12.   Draw K_1, K_2, K_3, K_4, and K_5.

13.   How many edges does K_6 have?

In Problems 14 and 15 use the following definition:

A graph *G* is called *subgraph* of a graph *H* if every vertex in *G* is also a vertex in *H* and every edge in *G* is also an edge in *H*.

14.   Find five subgraphs of

15.   Find three subgraphs that contain all vertices, of

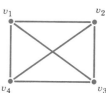

In Problems 16–18, use the following definition:

Let *G* be a simple graph. The *complement* of *G* is the simple graph $\overline{G}$ with the same vertices as *G*, but contains only those edges that are missing in *G*.

16.   Find the complement of

17. Find the complement of

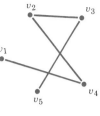

18. Find the complement of

14.2

PATHS AND CONNECTEDNESS

In this section we introduce some definitions and give some results regarding a special type of graph, the *connected graph*. Intuitively speaking, a graph is connected if it is possible to go from any vertex to any other vertex by traversing the edges of the graph. Many problems in graph theory involve the question of whether or not a graph is connected.

Going from a vertex to another in a graph amounts to following a sequence of neighboring edges. For example, in the graph in Figure 11 we can go from the vertex v_1 to the vertex v_3 by traversing the edges e_1 and e_2. This we write as an alternating sequence of adjacent vertices and edges: $v_1 e_1 v_2 e_2 v_3$. Or we can take a longer route as follows $v_1 e_7 v_4 e_4 v_5 e_5 v_5 e_6 v_1 e_7 v_4 e_3 v_3$. Obviously, there are several other routes that can also be followed to go from v_1 to v_3.

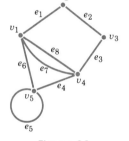

Figure 11

Let u and v be two vertices in a graph G. A *path* from u to v is an alternating sequence of adjacent vertices and edges of G. This sequence begins at u and ends in v.

If u and v are the same vertex (that is, $u = v$), then the path with no edges is *trivial* and is denoted by u or v.

A *simple path* from u to v is a path with no repeated vertices.

A *circuit* (or *cycle*) is a path that begins and ends at the same vertex and has no repeated edges.

A *simple circuit* is a path with no repeated edges and no repeated vertices other than the first and the last.

Example 1
Consult the graph in Figure 12 to determine which of the following sequences are paths, simple paths, circuits, and simple circuits.

(a) $v_1 e_1 v_2 e_6 v_4 e_3 v_3 e_2 v_2$ (b) $v_1 e_1 v_2 e_2 v_3 e_3 v_4 e_4 v_5$

(c) $v_1 e_8 v_4 e_3 v_3 e_7 v_1 e_8 v_4$ (d) $v_5 e_5 v_1 e_8 v_4 e_3 v_3 e_2 v_2 e_6 v_4 e_4 v_5$

(e) $v_2 e_2 v_3 e_3 v_4 e_4 v_5 e_5 v_1 e_1 v_2$

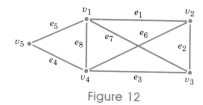

Figure 12

Solution

(a) This is a path from v_1 to v_2. Vertex v_2 repeats; therefore it is not a simple path. The first and the last vertices are different; therefore it is not a circuit.

(b) This is a path from v_1 to v_5. No vertex repeats; therefore this is a simple path. (Note that a simple path is never a circuit.)

(c) Here, edge e_8 repeats. Therefore this is just a path from v_1 to v_4.

(d) This path starts and ends at v_5. It has no repeated edges. Vertex v_4 repeats; therefore, it is a circuit.

(e) This path starts and ends at v_2. It has no repeated edges and no repeated vertices (other than v_2); therefore it is a simple circuit. ■

We are now ready to define a connected graph.

Connected Graph Let G be a graph. Then G is said to be a *connected graph* if for every pair of distinct vertices u and v ($u \neq v$) in G, there is a path between u and v.

Example 2

Let G_1, G_2, and G_3 be the graphs given in Figure 13. Determine which of these graphs is connected.

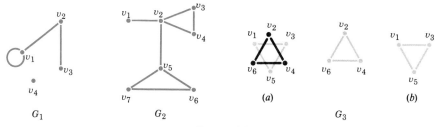

Figure 13

Solution

In G_1 there is no path from the vertex v_4 to any other vertex. Thus G_1 is not connected. In G_2 there is a path between every distinct pair of vertices. Thus G_2 is connected. For G_3, (a) and (b) in Figure 13 are the same graph. (b) shows that G_3 consists of two separate parts. That is, there is no path between, say, v_1 and v_2. Thus G_3 is not connected. You should not be misled by the graph of G_3 given in (a) which seems to show that G_3 is connected. (Recall that two edges may intersect but not necessarily at a vertex.) ■

We end this section by stating without proof two results, that may help in determining the connectedness of a given graph.

> **Theorem I** Let G be a connected graph with n vertices. Then G must have at least $n - 1$ edges.

> **Theorem II** Let G be a simple graph with n vertices (that is, G has no loops and no parallel edges). If G has more than $\binom{n-1}{2}$ edges, then it must be connected.

Exercise 14.2   *Answers to Odd-Numbered Problems begin on page A-49.*

In Problems 1–4 a graph is given. Identify the specified paths as paths, not simple; simple paths; circuits; or simple circuits.

A   **1.** (a) $v_3 e_3 v_4 e_4 v_5 e_6 v_2 e_7 v_4$
      (b) $v_1 e_5 v_5 e_6 v_2 e_1 v_1 e_5 v_5$
      (c) $v_1 e_1 v_2 e_6 v_5 e_4 v_4 e_3 v_3$
      (d) $v_3 e_2 v_2 e_1 v_1 e_5 v_5 e_6 v_2 e_7 v_4 e_3 v_3$
      (e) $v_5 e_6 v_2 e_2 v_3 e_3 v_4 e_4 v_5$

2. (a) $v_3 e_4 v_3$
      (b) $v_1 e_1 v_2 e_2 v_3$
      (c) $v_2 e_1 v_1 e_3 v_3 e_4 v_3 e_2 v_2$

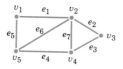

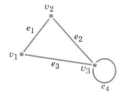

3. (a) $v_1 e_1 v_2 e_2 v_1$
      (b) $v_1 e_1 v_2 e_5 v_3 e_4 v_1$
      (c) $v_1 e_3 v_1$
      (d) $v_1 e_1 v_2 e_6 v_2 e_2 v_1 e_3 v_1$

4. (a) $v_1 e_1 v_2 e_2 v_3 e_3 v_4 e_8 v_1$
      (b) $v_1 e_{10} v_6 e_6 v_2 e_2 v_3 e_7 v_1 e_1 v_2 e_6 v_6$

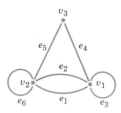

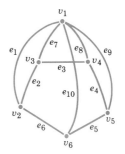

5. For the graph find:

(a) Four different simple paths.

(b) Four different circuits that are not simple.

(c) Four different simple circuits.

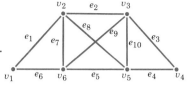

6. Find six simple paths from v_1 to v_4 in the graph.

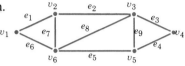

7. For the graph in Problem 6, find a circuit that is not simple, contains all vertices, and begins at v_1.

8. For the graph in Problem 6, how many simple circuits that contain all vertices and begin at v_1 are there?

9. Draw a simple circuit consisting of

(a) Only one edge

(b) Only two edges

10. Draw a graph that is connected, but removing one edge makes it disconnected.

B 11. Let G be a connected graph and let e be an edge. If removing e from G results in a disconnected graph, then e is called a *bridge*. Find all bridges for each of the following graphs.

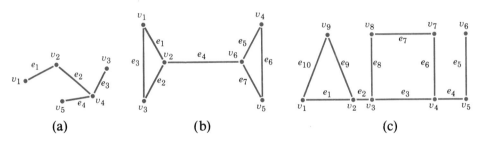

(a)   (b)   (c)

12. Can there be a connected graph G with $n-1$ edges and n vertices where the deletion of one edge disconnects G? [*Hint:* Use Theorem (I).]

(Refer to Problem 14 of Section 14.1). A subgraph H of G is a *connected component* of G if and only if

1. H is connected.

2. G does not have a connected subgraph that contains H as its subgraph.

13. Find a connected component of the following graph.

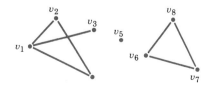

14. Repeat Problem 13 for the following graphs.

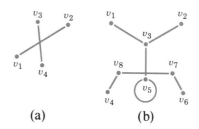

(a)   (b)

15. Suppose G is a simple graph with

    (a) 6 vertices and 11 edges. Can G be disconnected? Why?

    (b) 6 vertices and 10 edges. Can G be disconnected? Why?

C **16.** Use induction to prove Theorem I.

14.3

EULERIAN AND HAMILTONIAN CIRCUITS

EULERIAN CIRCUIT
CRITERION FOR AN EULERIAN CIRCUIT
FLEURY'S ALGORITHM
HAMILTONIAN CIRCUITS

In this section we present results that will help us solve the Königsberg bridge problem and problems similar to it. As noted in Section 14.1, Euler studied this problem and in 1736 published its solution. We begin with the following definition of an *Eulerian circuit.*

EULERIAN CIRCUIT

Let G be a connected graph. A circuit in G that contains every edge in G is called an *Eulerian circuit.*

Thus, an Eulerian circuit is a path that begins and ends at the same vertex, traverses every vertex at least once, and every edge exactly once.

Example 1
Consider the graphs G_1 to G_6 given in Figure 14.

    G_1 contains no Eulerian circuit because there is no edge connecting vertex v_4 to the rest of the graph; that is, because G_1 is not connected. G_2 contains no Eulerian

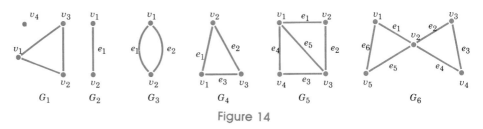

Figure 14

circuit because any circuit will use e_1 twice. The circuit $v_1 e_1 v_2 e_2 v_1$ in G_3 is Eulerian. In G_4 the circuit $v_1 e_1 v_2 e_2 v_3 e_3 v_1$ is Eulerian. G_5 contains no Eulerian circuit (try it!). $v_1 e_1 v_2 e_2 v_3 e_3 v_4 e_4 v_2 e_5 v_5 e_6 v_1$ is an Eulerian circuit contained in G_6. ∎

We make the following observations regarding Example 1:

1. G_1 suggests that a disconnected graph cannot contain an Eulerian circuit.
2. G_2 and G_5, both of which contain no Eulerian circuit, have vertices of odd degree. Both vertices in G_2 are of degree 1. Vertices v_1 and v_3 in G_5 are of degree 3.
3. G_3 and G_6, which contain an Eulerian circuit, have all vertices of even degree. In G_3, v_1, and v_2 are of degree 2. In G_6, v_1, v_3, v_4, and v_5 are all of degree 2, and v_2 is of degree 4.

These observations are not accidental. Indeed, there is a criterion for determining when a graph has an Eulerian circuit. We state this criterion without proving it. Later we will present an algorithm for finding an Eulerian circuit for those graphs that possess one.

CRITERION FOR AN EULERIAN CIRCUIT

> **Theorem I** Let G be a graph. G contains an Eulerian circuit if, and only if
>
> 1. G is connected,
> 2. Every vertex in G is of even degree.

Example 2
Show that the graphs in Figure 15 contain no Eulerian circuit.

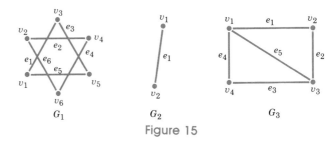

Figure 15

Solution

G_1 is not connected. Therefore by Theorem I it cannot contain an Eulerian circuit. G_2 is connected, but vertices v_1 and v_2 are of degree 1, which is odd. Therefore, by Theorem I, G_2 cannot contain an Eulerian circuit.

G_3 is also connected but vertices v_1 and v_3 are each of degree 3, which is odd. Therefore, by Theorem I, G_3 cannot contain an Eulerian circuit. ∎

We are ready, at this point, to give the solution to the Königsberg bridge problem outlined in Section 14.1, Example 3.

Example 3

In Section 14.1, Example 3, it was pointed out that the town of Königsberg can be represented by a graph, such as graph G in Figure 16, where each vertex represents a mass of land and each edge represents a bridge. The problem can then be stated as follows: Is there a route that starts at a vertex, traverses every edge, exactly once, and returns to that vertex? In other words, does graph G contain an Eulerian circuit?

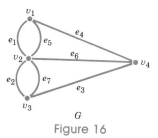

G

Figure 16

Solution

In the graph G in Figure 16, every vertex is of odd degree. Vertices v_1, v_4, and v_3 are of degree 3, and vertex v_2 is of degree 5. Therefore, by Theorem I, G does not contain an Eulerian circuit. That is, it is not possible to start at one of the banks or islands, cross every bridge exactly once, and return to the starting place. ∎

Our discussion continues with an algorithm that gives a way to construct Eulerian circuits when they exist.

FLEURY'S ALGORITHM

If G is a graph that contains an Eulerian circuit, then this circuit can be obtained as follows.

1. Select any vertex in G as an initial vertex. From this initial vertex traverse edges in G, removing all edges that are traversed and all isolated vertices that are generated in the process.
2. If edge e is a bridge connecting vertices u and v (see Problem 11, Section 14.2), then e can be selected only if the removal of e causes either u or v or both to become isolated.

We illustrate this procedure by applying it to a graph that contains an Eulerian circuit.

Example 4

Use Fleury's algorithm to find an Eulerian circuit of the graph G given in Figure 17.

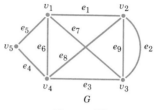

Figure 17

Solution

G is connected and every vertex in G is of even degree. Therefore, by Theorem I, G contains an Eulerian circuit. To obtain this circuit we start at v_1 (we could start at any other vertex) and traverse edge, e_1, e_7, e_6, or e_5. We choose edge e_1 and, thus, remove it. Figure 18a shows edge e_1 removed. (We denote removed edges by dashed lines.) At v_2, there are three untraversed edges (e_2, e_8, and e_9; these are the solid lines at v_2 in Figure 18a. We can choose to traverse any one of them. We choose e_8 and thus remove it. Figure 18b shows e_8 removed. Next, we traverse e_3 and remove it. Figure 18c shows e_3 removed. Note that we could equally choose e_4

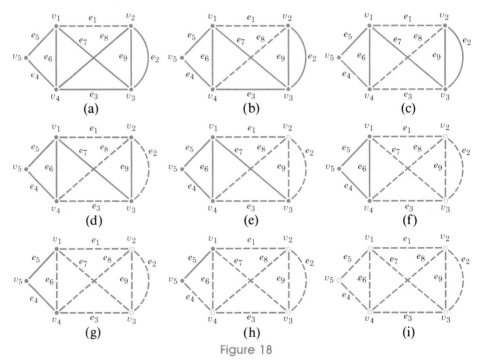

Figure 18

or e_6. At this point we are at v_3 with edges e_2, e_9, and e_7 untraversed. Inspecting the remaining graph (Figure 18c), that is, the graph without the dashed (removed) edges, we find that e_7 is a bridge, which means that its removal disconnects the remaining graph. So, we do not choose e_7 (unless we have no other choice). This means that we can only choose either edge e_9 or e_2. We select e_2, traverse, and remove it (Figure 18d). We are back at v_2. The only untraversed edge at this vertex is e_9. We traverse e_9 and remove it. This leaves v_2 isolated from the rest of the graph. In Figure 18e we draw v_2 as a blank circle to denote that it is a removed vertex. We proceed with our edge traversal in the following order: e_7, e_6, e_4, and e_5 as diagrammed in Figure 18f to 18i. Thus the circuit that we obtain is

$$v_1 e_1 v_2 e_8 v_4 e_3 v_3 e_2 v_2 e_9 v_3 e_7 v_1 e_6 v_4 e_4 v_5 e_5 v_1$$

Remember that this is not the only Eulerian circuit in G. ∎

HAMILTONIAN CIRCUITS

A related problem to that of finding an Eulerian circuit, in which all edges of a graph appear exactly once, is that of finding a circuit in which all vertices of a graph appear exactly once. Such a circuit is called a *Hamiltonian circuit* honoring the Irish mathematician Sir William R. Hamilton (1805–1864) who introduced this problem in 1859 (see Problem 10). Note that a Hamiltonian circuit includes every vertex of a graph exactly once (except that the first and last are the same) and may or may not include every edge.

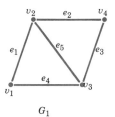

G_1

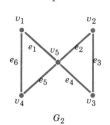

G_2

Figure 19

Example 5

Which of the graphs given in Figure 19 has a Hamiltonian circuit. Give the circuits for the graphs that contain them.

Solution

In G_1, we can start at v_1 and traverse edge e_1 to v_2, e_2 to v_4, e_3 to v_3, and then e_4 to v_1. Thus, G_1 has a Hamiltonian circuit given by $v_1 e_1 v_2 e_2 v_4 e_3 v_3 e_4 v_1$. Notice how all vertices appear in this circuit but not all edges. Edge e_5 is not used in this circuit.

For G_2 we argue as follows: If we start at v_1, v_2, v_3, or v_4 and visit every vertex, vertex v_5 will be visited twice. And if we start at v_5 itself, then in going from the vertices v_1 or v_4 to v_3 or v_2, respectively, or vice-versa, v_5 will be passed again (since we began at v_5). To complete the circuit we must come back to v_5, thus traversing it three times. All this means that G_2 has no Hamiltonian circuit. ∎

While there is a criterion for determining whether or not a graph contains an Eulerian circuit (Theorem I), a similar criterion does not exist for Hamiltonian circuits. There is, however, a procedure that can be used to determine that some graphs contain Hamiltonian circuits. We state this procedure without proof.

> **Theorem II** Let G be a simple connected graph with n vertices where $n \geq 3$. If the sum of the degrees of each pair of its nonadjacent vertices is greater than or equal to n, then G has a Hamiltonian circuit.

Note that Theorem II does *not* say that if a simple graph G with $n(\geq 3)$ vertices has a Hamiltonian circuit, then the sum of the degrees of each pair of its nonadjacent vertices *must* be greater than or equal to n. This is clear from the following example.

Example 6

For the graph in Figure 20, $n = 5$ and the circuit $v_1e_1v_2e_2v_3e_3v_4e_4v_5e_5v_1$ is Hamiltonian. Yet the sum of the degrees of any two nonadjacent vertices is 4, which is not greater than or equal to 5. For instance, $\deg(v_1) + \deg(v_3) = 4$, $\deg(v_1) + \deg(v_4) = 4$, and so on. ∎

Thus, Theorem II is a sufficient, but not a necessary, condition for a simple graph to possess a Hamiltonian circuit.

Example 7

In the graph G_1 in Figure 19, the sum of the degrees of each pair of nonadjacent vertices is 4, $[\deg(v_1) + \deg(v_4)]$, which is equal to the number of vertices. Thus G_1 has a Hamiltonian circuit (see Example 5). ∎

Figure 20

Exercise 14.3

Answers to Odd-Numbered Problems begin on page A-50.

A   **1.** Determine which of the following graphs contain a Eulerian circuit. If it does, then find an Eulerian circuit for the graph.

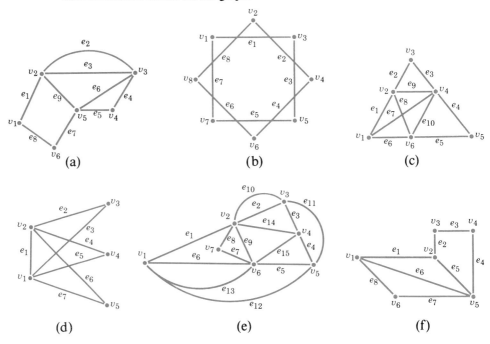

2. Does the complete graph K_4 (see Problem 12 in Exercise 14.1) contain an Eulerian circuit?

3. Does the complete graph K_5 contain an Eulerian circuit?

4. Does the complete graph K_n contain an Eulerian circuit? Explain.

B 5. A city consists of two land masses, situated on both banks of a river and three islands connected to each other and to the banks as shown in the figure. Is there a way to start at any point and make a round trip through all land masses crossing each bridge exactly once? If so, how can this be done?

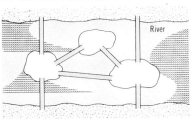

6. Give an example of a graph for which every vertex has an even degree but which does not contain an Eulerian circuit.

7. The following is a floor plan of a certain house. (*A* denotes the outside of the house.) Is it possible to start outside the house, or in any one of the rooms, take a tour through the house passing through each doorway, exactly once, and return to the starting place. [*Hint:* Represent the floor plan of the house with a graph where the outside and each room is represented by a vertex and each doorway by an edge.]

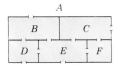

8. Repeat Problem 7 for the following floor plan.

9. Find a Hamiltonian circuit for each of the following graphs.

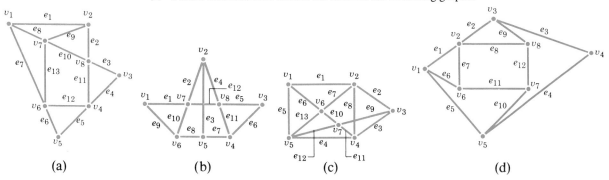

(a)   (b)   (c)   (d)

10. **Hamilton's Puzzle** In 1859 Sir William Hamilton presented the following problem: Suppose each vertex in the graph represents a big international city and that each

edge represents a transportation route. Can a salesperson make a round trip where he or she visits every city exactly once?

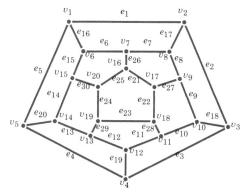

11. Show that the following graphs do not have a Hamiltonian circuit.

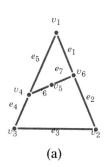

(a)

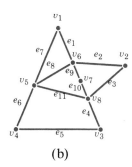

(b)

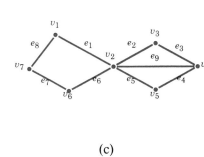

(c)

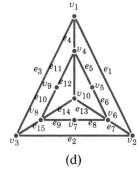

(d)

12. Determine whether or not the graph contains a Hamiltonian circuit. Find such a circuit for the graph that has one.

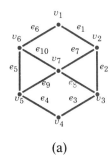

(a)

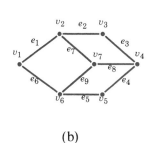

(b)

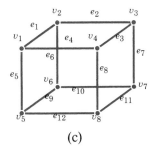

(c)

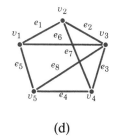

(d)

C **13.** Give an example of a graph that contains both Eulerian and Hamiltonian circuits.

14. Give an example of a graph that contains an Eulerian circuit but does not contain a Hamiltonian circuit.

15. Give an example of a graph that contains a Hamiltonian circuit but does not contain an Eulerian circuit.

16. Which of the following complete graphs has a Hamiltonian circuit? (Refer to Problem 12, Exercise 14.1)

$$K_2, K_3, K_n \quad \text{where } n > 3$$

14.4

TREES

ROOTED TREE
APPLICATIONS IN COMPUTER SCIENCE

Another class of graphs, called *trees,* are graphs that have no circuits (thus they cannot have parallel edges or loops). Trees arise frequently in computer science. In this section we present a brief exposure to the topic of trees.

Tree **Let T be a graph. T is called a *tree* if, and only if,**

(a) T is connected.
(b) T contains no circuits (except trivial ones).

T is called a *trivial tree* if it consists of a single vertex.

If after relabeling all the edges and vertices of two trees, they correspond, we say the trees are *identical.*

Example 1
Draw all distinct trees that have

(a) one vertex
(b) two vertices
(c) three vertices
(d) four vertices

Solution
(a) A tree consisting of one vertex is the trivial tree T_1 shown in Figure 21a.
(b) Figure 21b shows the tree T_2 that has two vertices.
(c) Figure 21c shows the tree T_3 with three vertices.
(d) Here, the tree T_4 may look like either one of the trees drawn in Figure 21d.

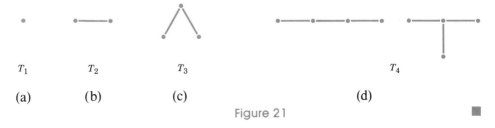

T_1 $\qquad$ T_2 $\qquad$ T_3 $\qquad\qquad\qquad\qquad$ T_4

(a) $\qquad$ (b) $\qquad$ (c) $\qquad\qquad\qquad\qquad$ (d)

Figure 21

Example 2
State why each graph in Figure 22 does not represent a tree.

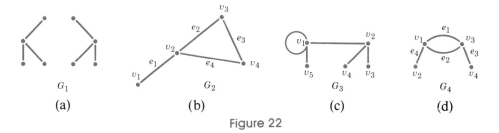

Figure 22

Solution

In Figure 22 the graph G_1 in (a) is not connected, so it is not a tree. The graph G_2 in (b) has the nontrivial circuit $v_2 e_2 v_3 e_3 v_4 e_4 v_2$. Therefore, it is not a tree. The graph G_3 in (c) contains a loop at v_1. Thus, it is not a tree. The graph G_4 in (d) has two edges e_1 and e_2 that are parallel. Hence, G_4 is not a tree. ∎

The next theorem, which we present without a proof, is a useful way to characterize trees. In essence, it says that a tree is a connected graph with the fewest number of edges, meaning that if one edge is removed, then the graph becomes disconnected.

> **Theorem I** Let G be a connected graph that has n vertices. G is a tree if, and only if, it has exactly $n - 1$ edges.

ROOTED TREE

A special type of tree, called a *rooted tree,* is useful in computer science. We next give the definition of this tree and some terms related to it. Let T be a tree and let u, v, and w be vertices in T.

1. If v is distinguished from the other vertices in T, then T is called a *rooted* tree and v is called a *root.*
2. If u is adjacent to v (v need not necessarily be a root) but farther from the root than v, then u is called the *child* of v. If u and w are the only children of v with u located to the left of v and w to the right of v, then u and w are called, respectively, the *left* and the *right children* of v.
3. If T is rooted and if every vertex in T has left and right children, either a left or a right child, or no children, then T is called a *binary tree.*
4. If the vertex u has no children, then u is called a *leaf* (or a *terminal* vertex). If u, which is not a root, has either one or two children, then u is called an *internal vertex.*
5. The *descendants* of the vertex u is the set consisting of all the children of u together with the descendants of those children.

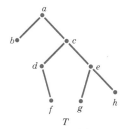

Figure 23

Example 3

Consider the rooted tree T in Figure 23.

(a) What is the root of T?

(b)   Is *T* a binary tree? If so, find the left and the right children of every vertex.

(c)   Find the leaves and the internal vertices of *T*.

(d)   Find the descendants of the vertices *a* and *c*.

Solution

(a)   Vertex *a* is distinguished as the only vertex located at the top of the tree. Therefore, *a* is the root.

(b)   Yes, every vertex has two children, one child, or no children. The following table indicates the children of each vertex.

Vertex	Left Child	Right Child
a	*b*	*c*
b	None	None
c	*d*	*e*
d	None	*f*
e	*g*	*h*
f	None	None
g	None	None
h	None	None

(c)   The leaves are those vertices that have no children. These are *b, f, g*, and *h*. The internal vertices are those that have either one child or two children. These are *c, d*, and *e*.

(d)   The descendants of *a* are *b, c, d, e, f, g, h*. The descendants of *c* are *d, e, f, g, h*.   ■

Let *T* be a binary tree, and let *u* and *v* be two vertices in *T*. The *subtree rooted at u* is the tree consisting of the root *u*, all its descendants, and all the edges connecting them. If *u* is the left child of *v*, then the subtree rooted at *u* is called the *left subtree of v rooted at u*, and if *u* is the right child of *v*, then the subtree rooted at *u* is called the *right subtree of v rooted at u*. If *u* is a leaf, then the subtree rooted at *u* is called a *trivial subtree.*

Example 4

Consider the binary tree in Figure 24.

(a)   Find the left and right subtrees of *v*.

(b)   Find the subtree rooted at *r*.

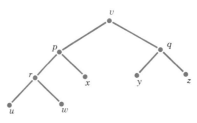

Figure 24

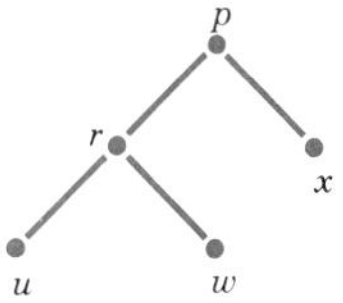

Figure 25

Figure 26

Figure 27

Solution
(a)   The left subtree of v is given in Figure 25.
        The right subtree of v is given in Figure 26.
(b)   The subtree rooted at r is given in Figure 27.   ■

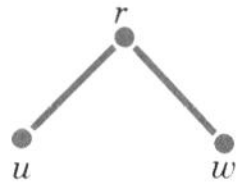

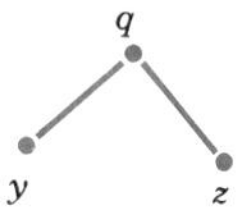

APPLICATIONS IN COMPUTER SCIENCE

Algebraic Expressions   Binary trees are used to represent algebraic expressions. The vertices of the tree are labeled with the numbers, variables, or operations that make up the expression. The leaves of the tree can only be labeled with numbers or variables. Operations such as addition, subtraction, multiplication, division, or exponentiation can only be assigned to internal vertices. The operation at each vertex operates on its left and right subtrees from left to right. The next two examples illustrate such uses of binary trees.

Example 5
Use a binary tree to represent the expression

(a) $x * y$ ("*" means multiplication)
(b) $(x + y)/z$ ("/" means division)
(c) $[(x - y) ** 2]/(x + y)$ ("**" means exponentiation)

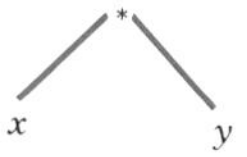

Figure 28

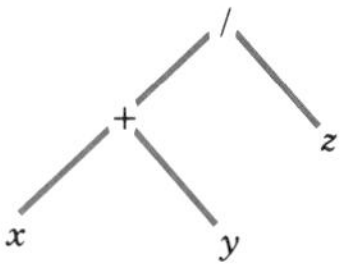

Figure 29

Solution
(a)   In this expression the first term is x, the second term is y, and the operation is *. Therefore, our tree, shown in Figure 28, must be rooted at * and must have two subtrees, one for each term.
(b)   Here, we first add x and y and then divide by z. This means our tree must have / as its root and two subtrees: a left subtree rooted at + that adds x to y and a right subtree rooted at z. The tree is shown in Figure 29.
(c)   To get the first term, which is $(x - y) ** 2$, we first subtract y from x, then square. To get the second term, which is $x + y$, we add x to y. Finally, we divide these two terms. This means our tree must be rooted at / and must have two subtrees as shown in Figure 30.   ■

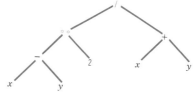

Figure 30

The Binary Search Tree Suppose we have a set of numbers. We will call them *keys.* We are interested in two of the many operations that can be performed on this set:

1. Ordering (or sorting) the set.
2. Searching the ordered set to locate a certain key and, in the event of not finding the key in the set, adding it at the right position so that the ordering of the set is maintained.

In the next example we describe a method that uses trees to perform the operations outlined above. This method involves storing the list in a tree where each element is stored at a vertex.

Example 6
(a)   Use a binary tree to store the elements of the following list of numbers in an increasing order: 7, 10, 21, 3, 24, 23.
(b)   Use the tree constructed in (a) to search for the number 19 in the list. If the number is not found in the list, update the list by adding the number to it.

Solution
(a)   We begin by selecting any number from the list to be the root of our binary tree. Say we select 10 to be this root. We draw the left and the right children of 10 as shown in Figure 31a.

Then, we pick another number from the list, say, 3. Now, 3 is less than 10,

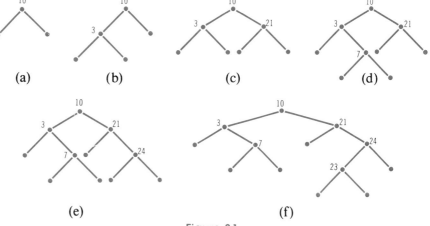

Figure 31

so we label the left child of 10 with 3 and then draw the left and the right children of 3, as Figure 31*b* shows.

Next, we pick another number from the list, say, 21. To position 21 on the tree, we start at the root 10 and compare it to 21. Since 21 is greater than 10, we move down to the right child of 10. This child is unlabeled, so we label it with 21 and then draw the left and the right children of this vertex. See Figure 31*c*.

We continue in this fashion. We choose our next element from the list, say, 7. We again start at the root 10 and compare it to 7. Since 7 is less than 10 we move down to the left child of 10, which is the vertex labeled 3. We compare 7 to 3. Since 7 is greater than 3, then we move further down to the right child of 3. This child is unlabeled, so we label it with 7 and draw the left and the right children of 7. See Figure 31*d*.

Say our next choice from the list is the number 24. Because 24 is greater than 10, we move down to the right child 21 of 10. Because 24 is greater than 21, we move further down to the right child of 21. This child is unlabeled. Therefore, we label it 21, and then draw the left and the right children of 24. See Figure 31*e*.

Finally, the last number left in the list is 23. We start at the root 10. 23 is greater than 10, so we move down to the right child of 10, which is labeled 21. Since 23 is greater than 21, then we move further down to the right child of 21, which is labeled 24. Because 23 is less than 24, we move even further down to the left child of 24, which is unlabeled. So, we label it with 23 and then draw the right and the left children of 23. See Figure 31*f*. Figure 31*f* is called a *binary search tree.*

(b) To search for the number 19 in the order list, we begin at the root 10. Because 19 is greater than 10, we move down to the right child of 10, which is labeled 21. Because 19 is less than 21, we move down to the left child of 21, which is unlabeled. See Figure 31*f*. Since we have reached an unlabeled vertex, we conclude that the number 19 is not found in the list, and we label this vertex 19 and draw the left and the right children of 19. See Figure 32. This adds the key 19 to the list while maintaining the increasing order of the numbers in the list.

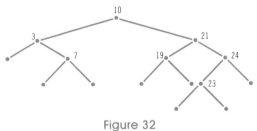

Figure 32

Exercise 14.4   *Answers to Odd-Numbered Problems begin on page A-50.*

A   **1.** Draw three distinct trees that have five vertices.

   2. Draw three distinct binary trees that have five vertices.

3. Draw three distinct binary trees that have seven vertices.
4. Draw three distinct binary trees that have eight vertices.

In Problems 5 – 12 either draw a graph with the given properties or explain why such a graph cannot exist.

5. A tree that has 7 vertices and 7 edges.
6. A connected graph that has 7 vertices and 7 edges.
7. A tree that has 6 vertices with the sum of the degrees of the vertices being 10.
8. A tree that has (a) 6 vertices and 8 edges, (b) 6 edges and 8 vertices.
9. A tree with all vertices of degree 2.
10. A tree that has 10 vertices of degree 3, 3, 3, 3, 1, 1, 1, 1, 1, 1.
11. A connected graph that has 5 edges and 6 vertices and that is not a tree. (Theorem I!)
12. A tree that has 7 vertices.
13. Does a tree with 120 vertices and 119 edges exist? Explain.
14. Does a connected graph with 3 vertices and 1 edge exist? Explain.
15. Draw three distinct rooted trees that have 4 vertices.
16. Draw four distinct rooted trees that have 5 vertices.

In Problems 17 and 18, find all leaves (or terminal vertices) and internal vertices for the given graph.

17.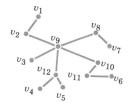

18.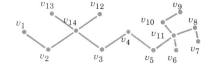

19. In a tree, can a leaf (or terminal vertex) be of degree 2? Explain.
20. In a tree, can an internal vertex be of degree 1? Explain.
21. Give an example of a nontrivial tree that has no internal vertices.

In Problems 22 and 23, given a tree T, find:

(a) All descendants of the vertices x and y.
(b) The left and the right subtrees of the vertices x and y.
(c) The subtree rooted at x.

22.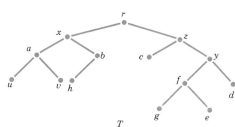

23.

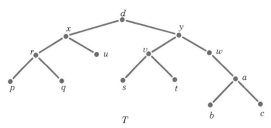

T

24. Give an example of a binary tree that has neither a left nor a right subtree.

B In Problems 25–28, a binary tree is given. Find the algebraic expression represented by the tree.

25.

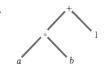

26.

27.

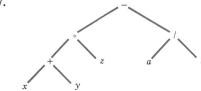

28.

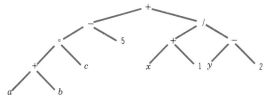

In Problems 29 and 30, an algebraic expression is given. Use a binary tree to represent the expression.

29. $[(a + b)/c] + d$

30. $\{1/[(a + b) ** 2]\} - [(a + 1) * c] + b$

In Problems 31 and 32, a list is given. For this list:

(a) Construct a binary search tree that stores the set.

(b) Use the tree constructed in (a) to search for the given key. If the key is not found in the set, then add it to the set at the right location.

31. Bob, Mary, Peter, Frank, Michael, Sue, Angela. (The order must be alphabetic). Key is Paul.

32. 25, 99, 112, 56, 72, 65. (The order must be increasing.) Key is 132.

C Use the definition below in Problems 33 and 34.

(Spanning Tree) **Let G be a connected graph. If a subgraph T of G is a tree that contains all vertices of G, then T is called a *spanning tree* for G.**

33. (a) Is the tree (i)
a spanning tree for the graph (ii)?
(b) Is the tree (i)
a spanning tree for the graph (ii)?

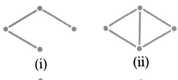

(i)   (ii)

34. Find two spanning trees for the graph below.

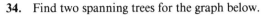

(i)   (ii)

14.5

DIRECTED GRAPHS

DEFINITION OF A DIRECTED GRAPH
DEFINITION OF A DIRECTED PATH
ORIENTABLE GRAPHS

So far our discussion in this chapter has been about different types of graphs that are undirected, meaning that all edges have no directions; thus crossing an edge is possible in both directions, back and forth. In this section we discuss a new type of graph called a *directed graph*. In a directed graph every edge is assigned a certain direction, so that crossing an edge is possible only in the direction assigned. Directed graphs are used in many situations. For example, one-way streets are represented by directed graphs where the vertices are the intersections and the edges are the streets. Flowcharts can be viewed as directed graphs where the vertices represent the instructions, and the edges represent the flow of control.

DEFINITION OF A DIRECTED GRAPH

Directed Graph **Let D be a graph. If every edge in D has a direction, then D is called a *directed graph* or *digraph* and its edges are called *arcs*. The vertex where an arc starts is called the *initial point* and the vertex where the arc ends is called the *terminal point*. When the directions of the edges in D are disregarded, the resulting graph obtained is called the *underlying graph* of D.**

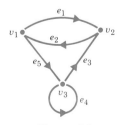

Figure 33

Example 1
Consider the digraph in Figure 33.

(a) Give the initial and terminal points of each arc.
(b) Draw the underlying graph.

Solution

(a) The following table gives all the arcs with their initial and terminal points.

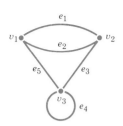

Figure 34

Arc	Initial Point	Terminal Point
e_1	v_1	v_2
e_2	v_2	v_1
e_3	v_3	v_2
e_4	v_3	v_3
e_5	v_1	v_3

(b)   Figure 34 shows the underlying graph.   ■

Example 2
Draw the digraph with the following specifications:

Arc	Initial Point	Terminal Point
e_1	v_1	v_2
e_2	v_3	v_2
e_3	v_3	v_4
e_4	v_4	v_1
e_5	v_2	v_4
e_6	v_4	v_2
e_7	v_3	v_3

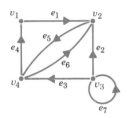

Figure 35

Solution
Figure 35 shows the graph.   ■

We mentioned that each vertex in a digraph has edges coming into it and (or) going out of it. This prompts us to define the following terms.

Let v be a vertex in the digraph D. The *indegree of v*, denoted by indeg(v), is the number of arcs in D whose terminal point is v. The *outdegree of v*, denoted by outdeg(v), is the number of arcs in D whose initial point is v.

Example 3
In the digraph in Figure 36, find the indegree and the outdegree of each vertex.

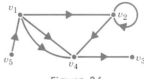

Figure 36

Solution
The answer is given in the following table:

Vertex	Indegree	Outdegree
v_1	1	3
v_2	2	2
v_3	1	0
v_4	3	1
v_5	0	1

Note that the loop at v_2 is counted as an arc coming out and going into v_2. ∎

DEFINITION OF A DIRECTED PATH

A *directed path* in a digraph D is a sequence of vertices and edges such that the terminal point of one arc is the initial point of the next. If, in D, a directed path exists from the vertex u to the vertex v, then v is said to be *reachable* from u.

Example 4
Consider the graph in Figure 37.

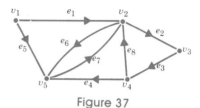

Figure 37

(a) Find three distinct directed paths from v_1 to v_5.
(b) Is v_1 reachable from v_4? Is v_1 reachable from v_5?

Solution
(a) One directed path from v_1 to v_5 is $v_1 e_1 v_2 e_6 v_5$. A second directed path is $v_1 e_1 v_2 e_2 v_3 e_3 v_4 e_4 v_5$. A third directed path is $v_1 e_1 v_2 e_2 v_3 e_3 v_4 e_8 v_2 e_6 v_5$.
(b) Since the indegree of v_1, is 0, then v_1 is not reachable from any other vertex. ∎

   Digraphs where every vertex is reachable from any other vertex are important. We discuss them next.

Let D be a digraph. If every vertex in D is reachable from any other vertex in D, then D is called *strongly connected*. And if the underlying graph of D is connected, then D is said to be *weakly connected*.

Example 5
Which of the two digraphs in Figure 38 is strongly connected and which one is weakly connected?

Figure 38

Solution

In D_1, v_3 is not reachable from v_1 or v_2, and the underlying graph is connected. Therefore, D_1 is weakly connected. In D_2, every vertex is reachable from the others. Therefore, D_2 is strongly connected.   ■

ORIENTABLE GRAPHS

Suppose a graph represents a network of streets in a certain town where the vertices represent the intersections of the streets and the edges represent the streets connecting these intersections. An important question is whether or not the graph can be directed so that the streets can be made into one-way streets and still be able to get from any intersection to any other. That is, we want to determine whether or not the resulting digraph is strongly connected. We refer to this process of directing the edges of a graph as *orientation.*

Let G be a graph. If every edge in G can be given a direction such that the result is a strongly connected digraph, then G is said to be *orientable*. An edge of a connected graph is called a *bridge* if its removal results in a disconnected graph.

The following theorem, stated without proof, is a criterion that characterizes an orientable graph.

> **Theorem I** A graph G is orientable if and only if it is connected and has no bridges.

Next, we present a procedure that produces an orientation for an orientable graph G. We remind you that a spanning tree of a graph G is a subgraph of G which is a tree that contains all the vertices of G. (See Problem 33, Section 14.4.)

Our procedure consists of the following steps.

Step 1. This step is referred to as a *depth-first search.* It involves the generation of a spanning tree T of the graph G as follows: We start at a vertex and search through the edges and vertices of G for a path. Then we extend this path as long as possible to a spanning tree. Because such a tree must include all vertices of G, if, during our search, a vertex is reached beyond which no new vertices can be reached, we back up along the path obtained, as far as necessary, to a vertex where branching to a new vertex is possible. We continue this process until a spanning tree of G is formed. During the search every vertex and edge encountered are numbered. Such numbering is called a *depth-first numbering* of G. Example 6 below explains further the process described in this step.

Step 2. We assign directions to the edges of T formed in Step 1 in the following way: Each edge of the tree T is directed from lower to higher depth-first number vertices.

Step 3. We direct all other vertices (not included in the tree T) from higher to lower depth-first number vertices.

We clarify this procedure by considering the following example.

Example 6

Determine whether or not the graph drawn in Figure 39 is orientable. If it is, then use the above procedure to orient the edges so that the resulting graph is strongly connected.

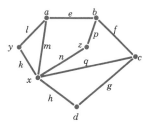

Figure 39

Solution

Since the graph is connected and has no bridges, then it is orientable. To orient it we proceed as follows

Step 1. Our aim is to generate a spanning tree. So, we begin by selecting any vertex in the graph, we pick vertex a and label it v_1. Then we traverse edge e to vertex b labeling them e_1 and v_2, respectively. We continue in this fashion, avoiding all edges that will produce a circuit, to obtain the path $v_1 e_1 v_2 e_2 v_3 e_3 v_4 e_4 v_5$. See Figure 40$a$. In this figure we denote the untraversed edges by dashed lines and the unvisited vertices by a blank circle. In this path, we stop at vertex d (labeled v_5) because we cannot go any further to reach a new vertex. Notice that if we continue the path from v_5 to the already visited vertex v_3 we would form the circuit whose vertices are v_3, v_4, and v_5 and thus no longer have a tree. Remember that trees are circuit-free connected graphs. Now, because no new vertices can be reached from v_5, we back up to vertex v_4 from which we can branch to either vertex y or vertex z. We choose to traverse edge k to vertex y and label them e_5 and v_6, respectively. See Figure 40b.

At v_6 no new vertices can be branched. So, here again we back up to v_4 from which we reach vertex z. We traverse edge n to this vertex and label them e_6 and v_7, respectively. See Figure 40c.

This completes the generation of our spanning tree because the tree with vertices v_1, v_2, v_3, v_4, v_5, v_6, and v_7 and edges e_1, e_2, e_3, e_4, e_5, and e_6, which we just formed, is a subgraph of the given graph, connected, and contains all vertices of the given graph. Thus both our depth-first search and depth-first numbering are complete.

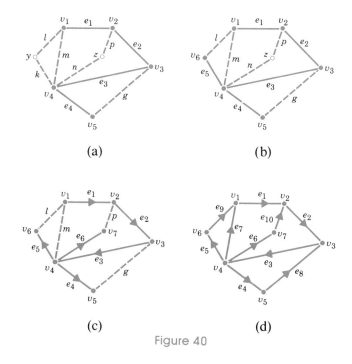

(a)   (b)

(c)   (d)

Figure 40

Step 2. We direct the edges of the generated spanning tree from lower to higher numbered vertices as shown in Figure 40c. Note that no direction is assigned to the dashed lines at this point.

Step 3. We, finally, direct the edges that are not part of our spanning tree (formed in Step 1), that is, we direct the dashed lines in Figure 40c from higher to lower numbered vertices. See Figure 40d.

The digraph thus produced is strongly connected because every vertex is reachable from any other vertex. ■

Exercise 14.5   *Answers to Odd-Numbered Problems begin on page A-52.*

A   In Problems 1 and 2 a digraph is given.

(a)   Find the initial and terminal points of each arc.

(b)   Find the indegrees and the outdegrees of all vertices.

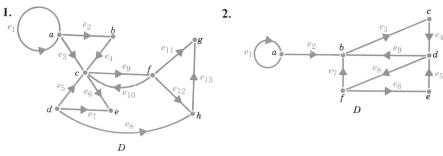

1.   2.

In Problems 3 and 4 draw the digraph with the given specifications.

3.

Arc	Initial Point	Terminal Point
e_1	v_1	v_2
e_2	v_2	v_2
e_3	v_2	v_3
e_4	v_1	v_3
e_5	v_1	v_4
e_6	v_4	v_1

4.

Arc	Initial Point	Terminal Point
e_1	v_2	v_1
e_2	v_3	v_2
e_3	v_3	v_4
e_4	v_4	v_1
e_5	v_1	v_3
e_6	v_3	v_1
e_7	v_4	v_2

In Problems 5 and 6 draw the digraph with the given specifications. (Note that the sum of the indegrees, or outdegrees, of all vertices is equal to the number of the arcs in the digraph.)

5.

Vertex	Indegree	Outdegree
v_1	0	1
v_2	2	2
v_3	2	0
v_4	2	1
v_5	1	3
v_6	1	1

6.

Vertex	Indegree	Outdegree
v_1	1	2
v_2	2	0
v_3	2	2
v_4	0	1

7. Draw a digraph with three vertices where each vertex has indegree 2.
8. Draw a digraph with three vertices where each vertex has outdegree 2.

In Problems 9 and 10 a digraph D is given.

(a) Is v_5 reachable from v_1?
(b) Is v_1 reachable from v_5?

Justify your answers.

9.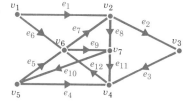

10.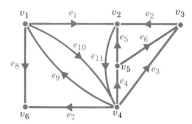

11. Find three distinct directed paths from v_1 to v_6 in the digraph in Problem 9.
12. Is the digraph in Problem 9 strongly connected? Justify your answer.
13. Is the digraph in Problem 10 strongly connected? Justify your answer.

B   14. A digraph D is called a *tournament* if it is loop-free and has exactly one arc between any two vertices. Thus, the vertices may represent contestants competing in a match, and an arc from vertex v may mean u won over v. Given the following tournament:

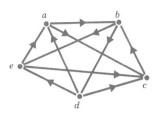

(a) What contestant won the greatest number of games?

(b) Find a path that does not start with the contestant in (a) and that contains all contestants.

15. Let $A = \{2,3,4,9,36\}$ and let R define the relation on A as follows: xRy if and only if x divides y. Draw a digraph D that represents R where the vertices of D are the elements of A and an arc from u to v in D means uRv.

16. Repeat Problem 15 for $A = \{1,2,5,8,9\}$ and R defined on A as follows: aRb if and only if a is less than b.

In Problems 17–20 determine whether or not the given graph is orientable, and if it is, then use the procedure given in this section to orient the edges so that the resulting graph is strongly connected.

17.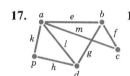

18.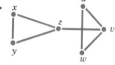

19.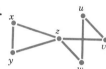

20.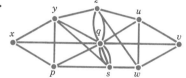

CHAPTER REVIEW

Important Terms and Formulas

vertex
edge
graph
parallel edges
loop
isolated vertex
simple graph
degree
complete graph
subgraph
complement
connected
path
trivial
simple path
circuit

simple circuit
connected graph
Eulerian circuit
Fleury's algorithm
Hamiltonian circuit
tree
trivial tree
rooted tree
child
binary tree
leaf
internal vertex
descendants
subtree
binary search tree
directed graph

arc
initial point
terminal point
underlying graph
indegree
outdegree
directed path
reachable
strongly connected
weakly connected
orientable
bridge
depth-first search
depth-first numbering
tournament

$$\deg(v_1) + \deg(v_2) + \cdots + \deg(v_n) = 2 \text{ (the number of edges in G)}$$

True-False Questions   *(Answers on page A-53)*

T   F   1. The sum of the degrees of the vertices of a graph may be any number.

T   F   2. If a graph G has n vertices and $n - 1$ edges, then G is necessarily connected.

T   F   3. If the degree of every vertex in a graph G is even, then G must contain an Eulerian circuit.

T   F   4. A Hamiltonian circuit need not include every edge.

T   F   5. No tree can contain a circuit.

T   F   6. In a strongly connected digraph D any vertex in D is reachable from any other vertex in D.

Fill in the Blanks   *(Answers on page A-53)*

1. A graph consists of a set V of ——————— and a set E of ———————.

2. G is a simple graph if G has no ——————— and no ——————— edges.

3. A circuit that contains every edge is said to be ———————.

4. Trees are graphs that are ———————-free.

5. In a digraph, every edge has a ———————.

6. A graph G is orientable if its edges can be ——————— in such a way that the resulting digraph is ——————— ———————.

Review Exercises   *Answers to Odd-Numbered Problems begin on page A-53.*

A   **1.** Can we draw a simple graph with vertices v_1, v_2, v_3, and v_4 of degrees 2, 2, 3, and 3, respectively? Justify your answer.

2. Draw the complete graph K_8.

3. Consult the figure.
Find

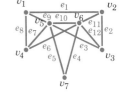

(a)   a simple path between v_1 and v_3 containing v_7.

(b)   a circuit starting at v_1 and containing v_7 and v_3.

(c)   a simple circuit starting at v_1 and containing v_7.

4. How many simple circuits, that contain all vertices, are there in K_4?

B   **5.** The Figure shows the outline of a city built on both banks of the river and on the three islands connected to the banks and to each other by the bridges shown. Can a tourist make a round-trip through all land masses crossing each bridge once? Justify your answer.

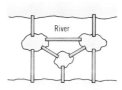

6. Does the graph in the Figure contain a Hamiltonian circuit? Justify your answer.

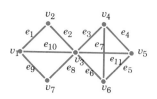

7. Use Fleury's algorithm to find an Eulerian circuit for the graph shown in the Figure.

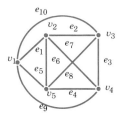

8. (a) Use a binary tree to store alphabetically the elements of the following list: one, two, three, four, five, six.

   (b) Use the tree in (a) to search for the word ten. If this word is not in the list, add it to the list at the right position without losing the alphabetical order.

9. Use a binary tree to represent the following algebraic expression

$$[(a - b)/d] + (d - 1)/a$$

10. Draw a rooted tree that has 3 internal vertices and 2 leaves.

11. Use the graph in the Figure to find two spanning trees.

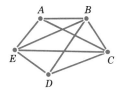

12. Draw a digraph with 5 vertices where each vertex has outdegree 2.

13. Is the digraph in the Figure strongly connected? Why or why not?

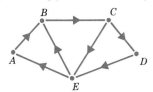

14. Determine whether or not the graph shown in the Figure is orientable. If it is, use the procedure of Section 14.5 to orient the edges in such a way that the resulting graph is strongly connected.

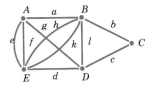

ARITHMETIC AND GEOMETRIC SEQUENCES

ARITHMETIC SEQUENCES

An arithmetic sequence is a sequence of numbers in which the sum of any term and a constant produces the next term. The constant d is called the *common difference*. For example, the sequence

$$3, 7, 11, 15, \ldots$$

is an arithmetic sequence with common difference 4.

Observe that an arithmetic sequence is completely determined if the first term and the common difference are known. In fact, if

$$a_1, a_2, a_3, \ldots, a_n, \ldots$$

is an arithmetic sequence with a_1 the first term and d the common difference, then the terms of this sequence can be written

$$a_1 = a_1$$
$$a_2 = a_1 + d$$
$$a_3 = a_2 + d = (a_1 + d) + d = a_1 + 2d$$
$$a_4 = a_3 + d = (a_1 + 2d) + d = a_1 + 3d$$

and in general the nth term of the arithmetic sequence is

$$a_n = a_1 + (n - 1)d \tag{1}$$

Example 1

Find the 11th term of the arithmetic sequence whose first term is 6 and whose common difference is 7.

Solution

Here $a_1 = 6$ and $n = 11$, and $d = 7$, so that by (1)

$$a_n = a_1 + (n - 1)d$$
$$a_{11} = 6 + (11 - 1)7$$
$$a_{11} = 76$$

■

If S_n represents the sum of the first n terms of an arithmetic sequence with a_1 as the first term and d the common difference, then

$$S_n = a_1 + a_2 + a_3 + \cdots + a_n$$

or

$$S_n = a_1 + (a_1 + d) + (a_1 + 2d) + \cdots + [a_1 + (n-1)d] \qquad (2)$$

The terms of an arithmetic sequence can also be generated by starting with a_n and subtracting d to obtain preceding terms. Cast in this form

$$S_n = a_n + (a_n - d) + (a_n - 2d) + \cdots + [a_n - (n-1)d] \qquad (3)$$

Adding corresponding terms of (2) and (3) yields

$$2S_n = (a_1 + a_n) + (a_1 + a_n) + \cdots + (a_1 + a_n) = n(a_1 + a_n)$$

Since

$$a_n = a_1 + (n-1)d$$

We may write

$$S_n = \frac{n}{2}(a_1 + [a_1 + (n-1)d])$$

$$= \frac{n}{2}[2a_1 + (n-1)d]$$

Given an arithmetic sequence with first term a_1, nth term a_n, and common difference d, then the sum of the first n terms is

$$S_n = \frac{n}{2}[2a_1 + (n-1)d] \qquad (4)$$

Example 2

Find the sum of the first 10 terms of the arithmetic sequence with $a_1 = 5$ and $d = 8$.

Solution

We use (4) with $n = 10$

$$S_{10} = \frac{10}{2}[2(5) + (10-1)8]$$

$$= 410 \qquad \blacksquare$$

Example 3

Fresh Starts had sales of $800,000 in its first year of operations. If the sales increased by $50,000 per year thereafter; find (a) Fresh Starts' sales in the sixth year, and (b) its total sales over the first six years of operation.

Solution

(a) We have here an arithmetic sequence with $a_1 = 800,000$, $d = 50,000$, and $n = 6$, so that

$$a_6 = 800,000 + (6 - 1)50,000 = 1,050,000$$

or

$$\$1,050,000$$

(b) To find the total sales over the first 6 years we need to find the sum of the first six terms of the arithmetic sequence

$$S_6 = \frac{6}{2}[2(800,000) + (6 - 1)50,000]$$

$$= \$5,550,000$$

or

$$= \$5,550,000 \qquad \blacksquare$$

GEOMETRIC SEQUENCES

A geometric sequence is a sequence in which the product of any term and a nonzero constant produces the next term. The constant term is called the common ratio and is customarily denoted by r.

For instance,

$$3, 6, 12, 24, 48, \ldots,$$

is a geometric sequence with common ratio 2.

A geometric sequence is completely determined by the first term and r, the common ratio. The terms of the geometric sequence can be written

$$a_1 = a_1$$
$$a_2 = a_1 r$$
$$a_3 = a_2 r = (a_1 r)r = a_1 r^2$$
$$a_4 = a_3 r = (a_1 r^2)r = a_1 r^3$$

and in general the nth term of the geometric sequence is

$$a_n = a_1 r^{n-1} \qquad (5)$$

Example 4

Find the eighth term of the geometric sequence whose first term is 3 and whose common ratio is 2.

Solution

$$a_8 = 3 \cdot 2^{8-1}$$
$$= 384$$

The eighth term is 384. $\qquad \blacksquare$

Example 5

Find the common ratio for the geometric sequence

$$4, 6, 9, 13.5, \ldots$$

Solution

The common ratio is obtained by dividing any term by the immediately preceding term. Then

$$r = \frac{6}{4} = \frac{9}{6} = \frac{13.5}{9} = \frac{3}{2}$$ ∎

We proceed as we did with arithmetic sequences to derive a formula for the sum S_n of the first n terms of a geometric sequence,

$$S_n = a_1 + a_2 + a_3 + \cdots + a_n$$

Using (5) the formula becomes

$$S_n = a_1 + a_1 r + a_1 r^2 + \cdots + a_1 r^{n-1} \qquad (6)$$

Multiplying each side of this equation by r

$$r S_n = a_1 r + a_1 r^2 + a_1 r^3 + \cdots a_1 r^n \qquad (7)$$

subtracting (7) from (6) gives

$$S_n - r S_n = a_1 + a_1 r + a_1 r^2 + \cdots + a_1 r^{n-1}$$
$$\qquad\qquad - a_1 r - a_1 r^2 - a_1 r^3 - \cdots - a_1 r^{n-1} - a_1 r^n$$
$$S_n(1 - r) = a_1 - a_1 r^n$$

Given a geometric sequence with first term a_1, nth term a_n, and common ratio $r \neq 1$, then the sum of the first n terms is

$$S_n = \frac{a_1(1 - r^n)}{1 - r}$$

Example 6

Find the sum of the first five terms of the geometric sequence whose first term is 3 and whose common ratio is 4.

Solution

The first five terms are 3, 12, 48, 192, and 768.
In the formula for the sum we set $a_1 = 3$, $r = 4$, and $n = 5$:

$$S_5 = \frac{3(1 - 4^5)}{1 - 4}$$
$$= \frac{3(1 - 1024)}{-3}$$
$$= 1023$$ ∎

Example 7

After doing \$100,000 business volume her first year, a dress shop owner plotted a 12% increase each year over the previous year. What was her business volume during the fifth year? What was her total business volume for the 5 years?

Solution

The 5-year sequence of business volumes is geometric: the volume in any year is 12% more than the previous year's volume. It amounts to 112% of the previous years' volume. During the fifth year,

$$a_5 = a_1 r^{5-1}$$
$$= 100,000 \, (1.12)^4$$
$$= 100,000 \, (1.57)$$
$$= 157,000.$$

That volume is approximately \$157,000.
    Over the 5-year period,

$$S_5 = \frac{a_1(1 - r^5)}{1 - r}$$
$$= \frac{100,000(1 - 1.12^5)}{1 - 1.12}$$
$$= \frac{-76,200}{-0.12}$$
$$= 635,000$$

The 5-year business volume is about \$635,000. ■

Exercise

Answers to odd numbered problems begin on page A-53.

ARITHMETIC
SEQUENCES

In Problems 1–8 write the first five terms of the arithmetic sequence, given the first term and common difference.

A

1. $a_1 = 1, d = 3$
2. $a_1 = 3, d = 2$
3. $a_1 = 5, d = -2$
4. $a_1 = 7, d = -4$
5. $a_1 = 72, d = 16$
6. $a_1 = 45, d = 24$
7. $a_1 = 2, d = -7$
8. $a_1 = 4, d = -11$

In Problems 9–16 find the specified term of each arithmetic sequence.

9. $a_1 = 5, d = 3; a_{12}$
10. $a_1 = 7, d = 4; a_{18}$
11. $a_1 = 6, d = -7; a_{15}$
12. $a_1 = 46, d = -9; a_{11}$
13. $3, 11, 19, 27, \ldots ; a_{21}$
14. $50, 58, 66, 74, \ldots ; a_{16}$
15. $63, 56, 49, 42, \ldots ; a_8$
16. $2, -4, -10, -16, \ldots ; a_{10}$

B
17. Find the first five terms of the arithmetic sequence whose fourth and ninth terms are 5 and 29, respectively.

18. Find the first five terms of the arithmetic sequence whose third term and eleventh term are 8.5 and 28.5, respectively.

19. What is the sum of the first 250 even positive integers?

20. What is the sum of the first 250 odd positive integers?

C **21.** Show that the sum of the first n odd positive integers is n^2.

22. Show that the sum of the first n even positive integers is $n^2 + n$.

APPLICATIONS **27.** If beginning pay at a job is $200 a week and the employer gives a $5 raise every 3 months, what is the weekly pay after 3 years?

28. In 1985, tuition at Great Lakes University was $6000 per academic year. If the tuition charge increases $800 each year, what will the tuition be in 1993?

29. The Split and Spare Bowling Lanes offered $5250 in prize money for the top five finishers in a tournament. If any prize is to be $225 more than the next lower prize, how much is each prize?

30. A total of $4000 in prize money is to be distributed to the four top winners in an essay contest in such a way that consecutive prizes differ by $300. How much is each prize?

GEOMETRIC
SEQUENCES

In Problems 1–6, for those of the following sequence that are geometric, find the common ratio r.

A **1.** 2, 4, 8, 16, 32 . . .

2. 4, 12, 36, 108, . . . ,

3. $\frac{1}{3}, \frac{2}{3}, \frac{3}{3}, \frac{4}{3}, \frac{5}{3}, \ldots$

4. 1, -3, 7, -11, . . .

5. 1, $\frac{-1}{2}, \frac{1}{4}, \frac{-1}{8}, \frac{1}{16}$ . . .

6. -1, 1, 3, 5, 7, . . . ,

In Problems 7–18 write the first five terms of each geometric sequence, given the first term and the common ratio.

7. $a_1 = 7, r = 2$

8. $a_1 = 8, r = 3$

9. $a_1 = 15, r = \frac{1}{5}$

10. $a_1 = 24, r = \frac{1}{6}$

11. $a_1 = -\frac{3}{49}, r = 7$

12. $a_1 = -\frac{7}{125}, r = 5$

13. $a_1 = 4, r = -3$

14. $a_1 = -5, r = -2$

15. $a_1 = 6.3, r = -1$

16. $a_1 = \sqrt{3}, r = 1$

17. $a_1 = 1, r = 0.12$

18. $a_1 = 1, r = -1.2$

In Problems 19–26 find the specified term of each geometric sequence.

19. $a_1 = 5, r = 4; a_7$

20. $a_1 = 13, r = 5; a_8$

21. $a_1 = \frac{5}{3}, r = -\frac{2}{3}; a_9$

22. $a_1 = \frac{6}{35}, r = -\frac{7}{3}; a_8$

23. $\frac{2}{3}, \frac{4}{3}, \frac{8}{3}, \frac{16}{3}, \ldots; a_{10}$

24. $\frac{1}{5}, \frac{1}{15}, \frac{1}{45}, \frac{1}{135}, \ldots; a_{12}$

25. $-16, 24, -36, 54, \ldots a_8$

26. $625, -250, 100, -40, \ldots; a_9$

In Problems 27–34 find S_7 for each geometric sequence.

27. $1, -\frac{1}{2}, \frac{1}{4}, -\frac{1}{8}, \ldots$

28. $1, \frac{1}{2}, \frac{1}{4}, \frac{1}{8}, \ldots$

29. $\frac{2}{3}, 2, 6, 18, \ldots$

30. $-\frac{5}{7}, \frac{5}{6}, -\frac{35}{36}, \frac{245}{216}, \ldots$

31. $a_1 = 15, r = \frac{1}{2}$

32. $a_1 = 12, r = \frac{2}{3}$

33. $a_1 = -16, r = \frac{3}{2}$

34. $a_1 = 20, r = \frac{5}{4}$

APPLICATIONS

35. An $8000 loan is to be repaid in five installments. If the first installment is $5000 and each succeeding installment is half the preceding installment, what is the total amount paid?

36. By continually introducing new products, the Composite Computer Company has reduced the price on its basic financial transaction handling configuration, called The Factor, 20% each year from the preceding year's price. If the 1985 price was $8000, what is the 1990 price?

37. The number of transactions per second that a Composite Computer can regularly handle has increased by 15% each year from the previous year's capacity. How does the 1995 capacity compare with the 1990 capacity?

38. In a psychology experiment, each subject is able to pass along about 70% of the information received. Approximately what percent of the information given to a first subject is passed along by the fifth subject?

TABLES

Table 1

Mathematics of Finance

$i = 0.0025\ (\tfrac{1}{4}\%)$							
n	$(1+i)^n$	$s_{\overline{n}|i}$	$a_{\overline{n}|i}$	n	$(1+i)^n$	$s_{\overline{n}|i}$	$a_{\overline{n}|i}$
1	1.002 500	1.000 000	0.997 506	51	1.135 804	54.321 654	47.826 604
2	1.005 006	2.002 500	1.992 525	52	1.138 644	55.457 459	48.704 842
3	1.007 519	3.007 506	2.985 062	53	1.141 490	56.596 102	49.580 890
4	1.010 038	4.015 025	3.975 124	54	1.144 344	57.737 593	50.454 753
5	1.012 563	5.025 063	4.962 718	55	1.147 205	58.881 936	51.326 437
6	1.015 094	6.037 625	5.947 848	56	1.150 073	60.029 141	52.195 947
7	1.017 632	7.052 719	6.930 522	57	1.152 948	61.179 214	53.063 288
8	1.020 176	8.070 351	7.910 745	58	1.155 830	62.332 162	53.928 467
9	1.022 726	9.090 527	8.888 524	59	1.158 720	63.487 993	54.791 489
10	1.025 283	10.113 253	9.863 864	60	1.161 617	64.646 713	55.652 358
11	1.027 846	11.138 536	10.836 772	61	1.164 521	65.808 329	56.511 080
12	1.030 416	12.166 383	11.807 254	62	1.167 432	66.972 850	57.367 661
13	1.032 992	13.196 799	12.775 316	63	1.170 351	68.140 282	58.222 106
14	1.035 574	14.229 791	13.740 963	64	1.173 277	69.310 633	59.074 420
15	1.038 163	15.265 365	14.704 203	65	1.176 210	70.483 910	59.924 608
16	1.040 759	16.303 529	15.665 040	66	1.179 150	71.660 119	60.772 676
17	1.043 361	17.344 287	16.623 481	67	1.182 098	72.839 270	61.618 630
18	1.045 969	18.387 648	17.579 533	68	1.185 053	74.021 368	62.462 474
19	1.048 584	19.433 617	18.533 200	69	1.188 016	75.206 421	63.304 213
20	1.051 205	20.482 201	19.484 488	70	1.190 986	76.394 437	64.143 853
21	1.053 834	21.533 407	20.433 405	71	1.193 964	77.585 423	64.981 400
22	1.056 468	22.587 240	21.379 955	72	1.196 948	78.779 387	65.816 858
23	1.059 109	23.643 708	22.324 145	73	1.199 941	79.976 335	66.650 232
24	1.061 757	24.702 818	23.265 980	74	1.202 941	81.176 276	67.481 528
25	1.064 411	25.764 575	24.205 466	75	1.205 948	82.379 217	68.310 751
26	1.067 072	26.828 986	25.142 609	76	1.208 963	83.585 165	69.137 907
27	1.069 740	27.896 059	26.077 416	77	1.211 985	84.794 128	69.962 999
28	1.072 414	28.965 799	27.009 891	78	1.215 015	86.006 113	70.786 034
29	1.075 096	30.038 213	27.940 041	79	1.218 053	87.221 129	71.607 017
30	1.077 783	31.113 309	28.867 871	80	1.221 098	88.439 181	72.425 952
31	1.080 478	32.191 092	29.793 388	81	1.224 151	89.660 279	73.242 845
32	1.083 179	33.271 570	30.716 596	82	1.227 211	90.884 430	74.057 700
33	1.085 887	34.354 749	31.637 503	83	1.230 279	92.111 641	74.870 524
34	1.088 602	35.440 636	32.556 112	84	1.233 355	93.341 920	75.681 321
35	1.091 323	36.529 237	33.472 431	85	1.236 438	94.575 275	76.490 095
36	1.094 051	37.620 560	34.386 465	86	1.239 529	95.811 713	77.296 853
37	1.096 787	38.714 612	35.298 220	87	1.242 628	97.051 242	78.101 599
38	1.099 528	39.811 398	36.207 700	88	1.245 735	98.293 871	78.904 339
39	1.102 277	40.910 927	37.114 913	89	1.248 849	99.539 605	79.705 076
40	1.105 033	42.013 204	38.019 863	90	1.251 971	100.788 454	80.503 816
41	1.107 796	43.118 237	38.922 557	91	1.255 101	102.040 425	81.300 565
42	1.110 565	44.226 033	39.822 999	92	1.258 239	103.295 526	82.095 327
43	1.113 341	45.336 598	40.721 196	93	1.261 384	104.553 765	82.888 106
44	1.116 125	46.449 939	41.617 154	94	1.264 538	105.815 150	83.678 909
45	1.118 915	47.566 064	42.510 876	95	1.267 699	107.079 688	84.467 740
46	1.121 712	48.684 979	43.402 370	96	1.270 868	108.347 387	85.254 603
47	1.124 517	49.806 692	44.291 641	97	1.274 046	109.618 255	86.039 504
48	1.127 328	50.931 208	45.178 695	98	1.277 231	110.892 301	86.822 448
49	1.130 146	52.058 536	46.063 536	99	1.280 424	112.169 532	87.603 440
50	1.132 972	53.188 683	46.946 170	100	1.283 625	113.449 956	88.382 483

Table 1

Continued

			$i = 0.005\ (\tfrac{1}{2}\%)$				
n	$(1+i)^n$	$s_{\overline{n}|i}$	$a_{\overline{n}|i}$	n	$(1+i)^n$	$s_{\overline{n}|i}$	$a_{\overline{n}|i}$
1	1.005 000	1.000 000	0.995 025	51	1.289 642	57.928 389	44.918 195
2	1.010 025	2.005 000	1.985 099	52	1.296 090	59.218 031	45.689 747
3	1.015 075	3.015 025	2.970 248	53	1.302 571	60.514 121	46.457 459
4	1.020 151	4.030 100	3.950 496	54	1.309 083	61.816 692	47.221 353
5	1.025 251	5.050 250	4.925 866	55	1.315 629	63.125 775	47.981 445
6	1.030 378	6.075 502	5.896 384	56	1.322 207	64.441 404	48.737 757
7	1.035 529	7.105 879	6.862 074	57	1.328 818	65.763 611	49.480 305
8	1.040 707	8.141 409	7.822 959	58	1.335 462	67.092 429	50.239 110
9	1.045 911	9.182 116	8.779 064	59	1.342 139	68.427 891	50.984 189
10	1.051 140	10.228 026	9.730 412	60	1.348 850	69.770 031	51.725 561
11	1.056 396	11.279 167	10.677 027	61	1.355 594	71.118 881	52.463 245
12	1.061 678	12.335 562	11.618 932	62	1.362 372	72.474 475	53.197 258
13	1.066 986	13.397 240	12.556 151	63	1.369 184	73.836 847	53.927 620
14	1.072 321	14.464 226	13.488 708	64	1.376 030	75.206 032	54.654 348
15	1.077 683	15.536 548	14.416 625	65	1.382 910	76.582 062	55.377 461
16	1.083 071	16.614 230	15.339 925	66	1.389 825	77.964 972	56.096 976
17	1.088 487	17.697 301	16.258 632	67	1.396 774	79.354 797	56.812 912
18	1.093 929	18.785 788	17.172 768	68	1.403 758	80.751 571	57.525 285
19	1.099 399	19.879 717	18.082 356	69	1.410 777	82.155 329	58.234 115
20	1.104 896	20.979 115	18.987 419	70	1.417 831	83.566 105	58.939 418
21	1.110 420	22.084 011	19.887 979	71	1.424 920	84.983 936	59.641 212
22	1.115 972	23.194 431	20.784 059	72	1.432 044	86.408 856	60.339 514
23	1.121 552	24.310 403	21.675 681	73	1.439 204	87.840 900	61.034 342
24	1.127 160	25.431 955	22.562 866	74	1.446 401	89.280 104	61.725 714
25	1.132 796	26.559 115	23.445 638	75	1.453 633	90.726 505	62.413 645
26	1.138 460	27.691 911	24.324 018	76	1.460 901	92.180 138	63.098 155
27	1.144 152	28.830 370	25.198 028	77	1.468 205	93.641 038	63.779 258
28	1.149 873	29.974 522	26.067 689	78	1.475 546	95.109 243	64.456 974
29	1.155 622	31.124 395	26.933 024	79	1.482 924	96.584 790	65.131 317
30	1.161 400	32.280 017	27.794 054	80	1.490 339	98.067 714	65.802 305
31	1.167 207	33.441 417	28.650 800	81	1.497 790	99.558 052	66.469 956
32	1.173 043	34.608 624	29.503 284	82	1.505 279	101.055 842	67.134 284
33	1.178 908	35.781 667	30.351 526	83	1.512 806	102.561 122	67.795 308
34	1.184 803	36.960 575	31.195 548	84	1.520 370	104.073 927	68.453 042
35	1.190 727	38.145 378	32.035 371	85	1.527 971	105.594 297	69.107 505
36	1.196 681	39.336 105	32.871 016	86	1.535 611	107.122 268	69.758 711
37	1.202 664	40.532 785	33.702 504	87	1.543 289	108.657 880	70.406 678
38	1.208 677	41.735 449	34.529 854	88	1.551 006	110.201 169	71.051 421
39	1.214 721	42.944 127	35.353 089	89	1.558 761	111.752 175	71.692 956
40	1.220 794	44.158 847	36.172 228	90	1.566 555	113.310 936	72.331 300
41	1.226 898	45.379 642	36.987 291	91	1.574 387	114.877 490	72.966 467
42	1.233 033	46.606 540	37.798 300	92	1.582 259	116.451 878	73.598 475
43	1.239 198	47.839 572	38.605 274	93	1.590 171	118.034 137	74.227 338
44	1.245 394	49.078 770	39.408 232	94	1.598 121	119.624 308	74.853 073
45	1.251 621	50.324 164	40.207 196	95	1.606 112	121.222 430	75.475 694
46	1.257 879	51.575 785	41.002 185	96	1.614 143	122.828 542	76.095 218
47	1.264 168	52.833 664	41.793 219	97	1.622 213	124.442 684	76.711 660
48	1.270 489	54.097 832	42.580 318	98	1.630 324	126.064 898	77.325 035
49	1.276 842	55.368 321	43.363 500	99	1.638 476	127.695 222	77.935 358
50	1.283 226	56.645 163	44.142 786	100	1.646 668	129.333 698	78.542 645

Table 1

Continued

			$i = 0.0075$ ($\frac{3}{4}$%)				
n	$(1+i)^n$	$s_{\overline{n}\mid i}$	$a_{\overline{n}\mid i}$	n	$(1+i)^n$	$s_{\overline{n}\mid i}$	$a_{\overline{n}\mid i}$
1	1.007 500	1.000 000	0.992 556	51	1.463 854	61.847 214	42.249 575
2	1.015 056	2.007 500	1.977 723	52	1.474 833	63.311 068	42.927 618
3	1.022 669	3.022 556	2.955 556	53	1.485 894	64.785 901	43.600 614
4	1.030 339	4.045 225	3.926 110	54	1.497 038	66.271 796	44.268 599
5	1.038 067	5.075 565	4.889 440	55	1.508 266	67.768 834	44.931 612
6	1.045 852	6.113 631	5.845 598	56	1.519 578	69.277 100	45.589 689
7	1.053 696	7.159 484	6.794 638	57	1.530 975	70.796 679	46.242 868
8	1.061 599	8.213 180	7.736 613	58	1.542 457	72.327 659	46.891 184
9	1.069 561	9.274 779	8.671 576	59	1.554 026	73.870 111	47.534 674
10	1.077 583	10.344 339	9.599 580	60	1.565 681	75.424 137	48.173 374
11	1.085 664	11.421 922	10.520 675	61	1.577 424	76.989 818	48.807 319
12	1.093 807	12.507 586	11.434 913	62	1.589 254	78.567 242	49.436 545
13	1.102 010	13.601 393	12.342 345	63	1.601 174	80.156 496	50.061 086
14	1.110 276	14.703 404	13.243 022	64	1.613 183	81.757 670	50.680 979
15	1.118 603	15.813 679	14.136 995	65	1.625 281	83.370 852	51.296 257
16	1.126 992	16.932 282	15.024 313	66	1.637 471	84.996 134	51.906 955
17	1.135 445	18.059 274	15.905 025	67	1.649 752	86.633 605	52.513 107
18	1.143 960	19.194 718	16.779 181	68	1.662 125	88.283 356	53.114 746
19	1.152 540	20.338 679	17.646 830	69	1.674 591	89.945 482	53.711 907
20	1.161 184	21.491 219	18.508 020	70	1.687 151	91.620 073	54.304 622
21	1.169 893	22.652 403	19.362 799	71	1.699 804	93.307 223	54.892 925
22	1.178 667	23.822 296	20.211 215	72	1.712 553	95.007 028	55.476 849
23	1.187 507	25.000 963	21.053 315	73	1.725 397	96.719 580	56.056 426
24	1.196 414	26.188 471	21.889 146	74	1.738 337	98.444 977	56.631 688
25	1.205 387	27.384 884	22.718 755	75	1.751 375	100.183 314	57.202 668
26	1.214 427	28.590 271	23.542 189	76	1.764 510	101.934 689	57.769 397
27	1.223 535	29.804 698	24.359 493	77	1.777 744	103.699 199	58.331 908
28	1.232 712	31.028 233	25.170 713	78	1.791 077	105.476 943	58.890 231
29	1.241 957	32.260 945	25.975 893	79	1.804 510	107.268 021	59.444 398
30	1.251 272	33.502 902	26.775 080	80	1.818 044	109.072 531	59.994 440
31	1.260 656	34.754 174	27.568 318	81	1.831 679	110.890 575	60.540 387
32	1.270 111	36.014 830	28.355 650	82	1.845 417	112.722 254	61.082 270
33	1.279 637	37.284 941	29.137 122	83	1.859 258	114.567 671	61.620 119
34	1.289 234	38.564 578	29.912 776	84	1.873 202	116.426 928	62.153 965
35	1.298 904	39.853 813	30.682 656	85	1.887 251	118.300 130	62.683 836
36	1.308 645	41.152 716	31.446 805	86	1.901 405	120.187 381	63.209 763
37	1.318 460	42.461 361	32.205 266	87	1.915 666	122.088 787	63.731 774
38	1.328 349	43.779 822	32.958 080	88	1.930 033	124.004 453	64.249 900
39	1.338 311	45.108 170	33.705 290	89	1.944 509	125.934 486	64.764 169
40	1.348 349	46.446 482	34.446 938	90	1.959 092	127.878 995	65.274 609
41	1.358 461	47.794 830	35.183 065	91	1.973 786	129.838 087	65.781 250
42	1.368 650	49.153 291	35.913 713	92	1.988 589	131.811 873	66.284 119
43	1.378 915	50.521 941	36.638 921	93	2.003 503	133.800 462	66.783 245
44	1.389 256	51.900 856	37.358 730	94	2.018 530	135.803 965	67.278 655
45	1.399 676	53.290 112	38.073 181	95	2.033 669	137.822 495	67.770 377
46	1.410 173	54.689 788	38.782 314	96	2.048 921	139.856 164	68.258 439
47	1.420 750	56.099 961	39.486 168	97	2.064 288	141.905 085	68.742 867
48	1.431 405	57.520 711	40.184 782	98	2.079 770	143.969 373	69.223 689
49	1.442 141	58.952 116	40.878 195	99	2.095 369	146.049 143	69.700 932
50	1.452 957	60.394 257	41.566 447	100	2.111 084	148.144 512	70.174 623

Table 1

Continued

		$i = 0.01$ (1%)					
n	$(1+i)^n$	$s_{\overline{n}|i}$	$a_{\overline{n}|i}$	n	$(1+i)^n$	$s_{\overline{n}|i}$	$a_{\overline{n}|i}$
1	1.010 000	1.000 000	0.990 099	51	1.661 078	66.107 814	39.798 136
2	1.020 100	2.010 000	1.970 395	52	1.677 689	67.768 892	40.394 194
3	1.030 301	3.030 100	2.940 985	53	1.694 466	69.446 581	40.984 351
4	1.040 604	4.060 401	3.901 966	54	1.711 410	71.141 047	41.568 664
5	1.051 010	5.101 005	4.853 431	55	1.728 525	72.852 457	42.147 192
6	1.061 520	6.152 015	5.795 476	56	1.745 810	74.580 982	42.719 992
7	1.072 135	7.213 535	6.728 195	57	1.763 268	76.326 792	43.287 121
8	1.082 857	8.285 671	7.651 678	58	1.780 901	78.090 060	43.848 635
9	1.093 685	9.368 527	8.566 018	59	1.798 710	79.870 960	44.404 589
10	1.104 622	10.462 213	9.471 305	60	1.816 697	81.669 670	44.955 038
11	1.115 668	11.566 835	10.367 628	61	1.834 864	83.486 367	45.500 038
12	1.126 825	12.682 503	11.255 077	62	1.853 212	85.321 230	46.039 642
13	1.138 093	13.809 328	12.113 740	63	1.871 744	87.174 443	46.573 903
14	1.149 474	14.947 421	13.003 703	64	1.890 462	89.046 187	47.102 874
15	1.160 969	16.096 896	13.865 053	65	1.909 366	90.936 649	47.626 608
16	1.172 579	17.257 864	14.717 874	66	1.928 460	92.846 015	48.145 156
17	1.184 304	18.430 443	15.562 251	67	1.197 745	94.774 475	48.658 570
18	1.196 147	19.614 748	16.398 269	68	1.967 222	96.722 220	49.166 901
19	1.208 109	20.810 895	17.226 008	69	1.986 894	98.689 442	49.670 199
20	1.220 190	22.019 004	18.045 553	70	2.006 763	100.676 337	50.168 514
21	1.232 392	23.239 194	18.856 983	71	2.026 831	102.683 100	50.661 895
22	1.244 716	24.471 586	19.660 379	72	2.047 099	104.709 931	51.150 391
23	1.257 163	25.716 302	20.455 821	73	2.067 570	106.757 031	51.634 051
24	1.269 735	26.973 465	21.243 387	74	2.088 246	108.824 601	52.112 922
25	1.282 432	28.243 200	22.023 156	75	2.109 128	110.912 847	52.587 051
26	1.295 256	29.525 632	22.795 204	76	2.130 220	113.021 975	53.056 486
27	1.308 209	30.820 888	23.559 608	77	2.151 522	115.152 195	53.521 274
28	1.321 291	32.129 097	24.316 443	78	2.173 037	117.303 717	53.981 459
29	1.334 504	33.450 388	25.065 785	79	2.194 768	119.476 754	54.437 088
30	1.347 849	34.784 892	25.807 708	80	2.216 715	121.671 522	54.888 206
31	1.361 327	36.132 740	26.542 285	81	2.238 882	123.888 237	55.334 858
32	1.374 941	37.494 068	27.269 589	82	2.261 271	126.127 119	55.777 087
33	1.388 690	38.869 009	27.989 693	83	2.283 884	128.388 391	56.214 937
34	1.402 577	40.257 699	28.702 666	84	2.306 723	130.672 274	56.648 453
35	1.416 603	41.660 276	29.408 580	85	2.329 790	132.978 997	57.077 676
36	1.430 769	43.076 878	30.107 505	86	2.353 088	135.308 787	57.502 650
37	1.445 076	44.507 647	30.799 510	87	2.376 619	137.661 875	57.923 415
38	1.459 527	45.952 724	31.484 663	88	2.400 385	140.038 494	58.340 015
39	1.474 123	47.412 251	32.163 033	89	2.424 389	142.438 879	58.752 490
40	1.488 864	48.886 373	32.834 686	90	2.448 633	144.863 267	59.160 881
41	1.503 752	50.375 237	33.499 689	91	2.473 119	147.311 900	59.565 229
42	1.518 790	51.878 989	34.158 108	92	2.497 850	149.785 019	59.965 573
43	1.533 978	53.397 779	34.810 008	93	2.522 829	152.282 869	60.361 954
44	1.549 318	54.931 757	35.455 454	94	2.548 057	154.805 698	60.754 410
45	1.564 811	56.481 075	36.094 508	95	2.573 538	157.353 755	61.142 980
46	1.580 459	58.045 885	36.727 236	96	2.599 273	159.927 293	61.527 703
47	1.596 263	59.626 344	37.353 699	97	2.625 266	162.526 565	61.908 617
48	1.612 226	61.222 608	37.973 959	98	2.651 518	165.151 831	62.285 759
49	1.628 348	62.834 834	38.588 079	99	2.678 033	167.803 349	62.659 168
50	1.644 632	64.463 182	39.196 118	100	2.704 814	170.481 383	63.028 879

Table 1

Continued

			$i = 0.0125 (1\frac{1}{4}\%)$				
n	$(1+i)^n$	$s_{\overline{n}|i}$	$a_{\overline{n}|i}$	n	$(1+i)^n$	$s_{\overline{n}|i}$	$a_{\overline{n}|i}$
1	1.012 500	1.000 000	0.987 654	51	1.884 285	70.742 812	37.543 581
2	1.025 156	2.012 500	1.963 115	52	1.907 839	72.627 097	38.067 734
3	1.037 971	3.037 656	2.926 534	53	1.931 687	74.534 936	38.585 417
4	1.050 945	4.075 627	3.878 058	54	1.955 833	76.466 623	39.096 708
5	1.064 082	5.126 572	4.817 835	55	1.980 281	78.422 456	39.601 687
6	1.077 383	6.190 654	5.746 010	56	2.005 034	80.402 737	40.100 431
7	1.090 850	7.268 038	6.662 726	57	2.030 097	82.407 771	40.593 019
8	1.104 486	8.358 888	7.568 124	58	2.055 473	84.437 868	41.079 524
9	1.118 292	9.463 374	8.462 345	59	2.081 167	86.493 341	41.560 024
10	1.132 271	10.581 666	9.345 526	60	2.107 181	88.574 508	42.034 592
11	1.146 424	11.713 937	10.217 803	61	2.133 521	90.681 689	42.503 300
12	1.160 755	12.860 361	11.079 312	62	2.160 190	92.815 210	42.966 223
13	1.175 264	14.021 116	11.930 185	63	2.187 193	94.975 400	43.423 430
14	1.189 955	15.196 380	12.770 553	64	2.214 532	97.162 593	43.874 992
15	1.204 829	16.386 335	13.600 546	65	2.242 214	99.377 125	44.320 980
16	1.219 890	17.591 164	14.420 292	66	2.270 242	101.619 339	44.761 462
17	1.235 138	18.811 053	15.229 918	67	2.298 620	103.889 581	45.196 506
18	1.250 577	20.046 192	16.029 549	68	2.327 353	106.188 201	45.626 178
19	1.266 210	21.296 769	16.819 308	69	2.356 444	108.515 553	46.050 547
20	1.282 037	22.562 979	17.599 316	70	2.385 900	110.871 998	46.469 676
21	1.298 063	23.845 016	18.369 695	71	2.415 724	113.257 898	46.883 630
22	1.314 288	25.143 078	19.130 563	72	2.445 920	115.673 621	47.292 474
23	1.330 717	26.457 367	19.882 037	73	2.476 494	118.119 542	47.696 271
24	1.347 351	27.788 084	20.624 235	74	2.507 450	120.596 036	48.095 082
25	1.364 193	29.135 435	21.357 269	75	2.538 794	123.103 486	48.488 970
26	1.381 245	30.499 628	22.081 253	76	2.570 528	125.642 280	48.877 995
27	1.398 511	31.880 873	22.796 299	77	2.602 660	128.212 809	49.262 218
28	1.415 992	33.279 384	23.502 518	78	2.635 193	130.815 469	49.641 696
29	1.433 692	34.695 377	24.200 018	79	2.668 133	133.450 662	50.016 490
30	1.451 613	36.129 069	24.888 906	80	2.701 485	136.118 795	50.386 657
31	1.469 759	37.580 682	25.569 290	81	2.735 254	138.820 280	50.752 254
32	1.488 131	39.050 441	26.241 274	82	2.769 444	141.555 534	51.113 337
33	1.506 732	40.538 571	26.904 962	83	2.804 062	144.324 978	51.469 963
34	1.525 566	42.045 303	27.560 456	84	2.839 113	147.129 040	51.822 185
35	1.544 636	43.570 870	28.207 858	85	2.874 602	149.968 153	52.170 060
36	1.563 944	45.115 506	28.847 267	86	2.910 534	152.842 755	52.513 639
37	1.583 493	46.679 449	29.478 783	87	2.946 916	155.753 289	52.852 977
38	1.603 287	48.292 642	30.102 501	88	2.983 753	158.700 206	53.188 125
39	1.623 328	49.886 229	30.718 520	89	3.021 049	161.683 958	53.519 136
40	1.643 619	51.489 557	31.326 933	90	3.058 813	164.705 008	53.846 060
41	1.664 165	53.133 177	31.927 835	91	3.097 048	167.763 820	54.168 948
42	1.684 967	54.797 341	32.521 319	92	3.135 761	170.860 868	54.487 850
43	1.706 029	56.482 308	33.107 475	93	3.174 958	173.996 629	54.802 815
44	1.727 354	58.188 337	33.686 395	94	3.214 645	177.171 587	55.113 892
45	1.748 946	59.915 691	34.258 168	95	3.254 828	180.386 232	55.421 127
46	1.770 808	61.664 637	34.822 882	96	3.295 513	183.641 059	55.724 570
47	1.792 943	63.435 445	35.380 624	97	3.336 707	186.936 573	56.024 267
48	1.815 355	65.228 388	35.931 481	98	3.378 416	190.273 280	56.320 264
49	1.838 047	67.043 743	36.475 537	99	3.420 646	193.651 696	56.612 606
50	1.861 022	68.881 790	37.012 876	100	3.463 404	197.072 342	56.901 339

Table 1

Continued

$i = 0.015$ ($1\frac{1}{2}\%$)

n	$(1+i)^n$	$s_{\overline{n}|i}$	$a_{\overline{n}|i}$	n	$(1+i)^n$	$s_{\overline{n}|i}$	$a_{\overline{n}|i}$
1	1.015 000	1.000 000	0.985 222	51	2.136 821	75.788 070	35.467 673
2	1.030 225	2.015 000	1.955 883	52	2.168 873	77.924 892	35.928 742
3	1.045 678	3.045 225	2.912 200	53	2.201 406	80.093 765	36.382 997
4	1.061 364	4.090 903	3.854 385	54	2.234 428	82.295 171	36.830 539
5	1.077 284	5.152 267	4.782 645	55	2.267 944	84.529 599	37.271 467
6	1.093 443	6.229 551	5.697 187	56	2.301 963	86.797 543	37.705 879
7	1.109 845	7.322 994	6.598 214	57	2.336 493	89.099 506	38.133 871
8	1.126 493	8.432 839	7.485 925	58	2.371 540	91.435 999	38.555 538
9	1.143 390	9.559 332	8.360 517	59	2.407 113	93.807 539	38.970 973
10	1.160 541	10.702 722	9.222 185	60	2.443 220	96.214 652	39.380 269
11	1.177 949	11.863 262	10.071 118	61	2.479 868	98.657 871	39.783 516
12	1.195 618	13.041 211	10.907 505	62	2.517 066	101.137 740	40.180 804
13	1.213 552	14.236 830	11.731 532	63	2.554 822	103.654 806	40.572 221
14	1.231 756	15.450 382	12.543 382	64	2.593 144	106.209 628	40.957 853
15	1.250 232	16.682 138	13.343 233	65	2.632 042	108.802 772	41.337 786
16	1.268 986	17.932 370	14.131 264	66	2.671 522	111.434 814	41.712 105
17	1.288 020	19.201 355	14.907 649	67	2.711 595	114.106 336	42.080 891
18	1.307 341	20.489 376	15.672 561	68	2.752 269	116.817 931	42.444 228
19	1.326 951	21.796 716	16.426 168	69	2.793 553	119.570 200	42.802 195
20	1.346 855	23.123 667	17.168 639	70	2.835 456	122.363 753	43.154 872
21	1.367 058	24.470 522	17.900 137	71	2.877 988	125.199 209	43.502 337
22	1.387 564	25.837 580	18.620 824	72	2.921 158	128.077 197	43.844 667
23	1.408 377	27.225 144	19.330 861	73	2.964 975	130.998 355	44.181 938
24	1.429 503	28.633 521	20.030 405	74	3.009 450	133.963 331	44.514 224
25	1.450 945	30.063 024	20.719 611	75	3.054 592	136.972 781	44.841 600
26	1.472 710	31.513 969	21.398 632	76	3.100 411	140.027 372	45.164 138
27	1.494 800	32.986 678	22.067 617	77	3.146 917	143.127 783	45.481 910
28	1.517 222	34.481 479	22.726 717	78	3.194 120	146.274 700	45.794 985
29	1.539 981	35.998 701	23.376 076	79	3.242 032	149.468 820	46.103 433
30	1.563 080	37.538 681	24.015 838	80	3.290 663	152.710 852	46.407 323
31	1.586 526	39.101 762	24.646 146	81	3.340 023	156.001 515	46.706 723
32	1.610 324	40.688 288	25.267 139	82	3.390 123	159.341 536	47.001 697
33	1.634 479	42.298 612	25.878 954	83	3.440 975	162.731 661	47.292 313
34	1.658 996	43.933 092	26.481 728	84	3.492 590	166.172 636	47.578 633
35	1.683 881	45.592 088	27.075 595	85	3.544 978	169.665 226	47.860 722
36	1.709 140	47.275 969	27.660 684	86	3.598 153	173.210 204	48.138 643
37	1.734 777	48.985 109	28.237 127	87	3.652 125	176.808 357	48.412 456
38	1.760 798	50.719 885	28.805 052	88	3.706 907	180.460 482	48.682 222
39	1.787 210	52.480 684	29.364 583	89	3.762 511	184.167 390	48.948 002
40	1.814 018	54.267 894	29.915 845	90	3.818 949	187.929 900	49.209 855
41	1.841 229	56.081 912	30.458 961	91	3.876 233	191.748 849	49.467 837
42	1.868 847	57.923 141	30.994 050	92	3.934 376	195.625 082	49.722 007
43	1.896 880	59.791 988	31.521 232	93	3.993 392	199.559 458	49.972 421
44	1.925 333	61.688 868	32.040 622	94	4.053 293	203.552 850	50.219 134
45	1.954 213	63.614 201	32.552 337	95	4.114 092	207.606 142	50.462 201
46	1.983 526	65.568 414	33.056 490	96	4.175 804	211.720 235	50.701 675
47	2.013 279	67.551 940	33.553 192	97	4.238 441	215.896 038	50.937 611
48	2.043 478	69.565 219	34.042 554	98	4.302 017	220.134 479	51.170 060
49	2.074 130	71.608 698	32.524 683	99	4.366 547	224.436 496	51.399 074
50	2.105 242	73.682 828	34.999 688	100	4.432 046	228.803 043	51.624 704

Table 1

Continued

			$i = 0.0175$ (1¾%)				
n	$(1+i)^n$	$s_{\overline{n}|i}$	$a_{\overline{n}|i}$	n	$(1+i)^n$	$s_{\overline{n}|i}$	$a_{\overline{n}|i}$
1	1.017 500	1.000 000	0.982 801	51	2.422 453	81.283 014	33.554 014
2	1.035 306	2.017 500	1.948 699	52	2.464 846	83.705 466	33.959 719
3	1.053 424	3.052 806	2.897 984	53	2.507 980	86.170 312	34.358 446
4	1.071 859	4.106 230	3.830 943	54	2.551 870	88.678 292	34.750 316
5	1.090 617	5.178 089	4.747 855	55	2.596 528	91.230 163	35.135 446
6	1.109 702	6.268 706	5.648 998	56	2.641 967	93.826 690	35.513 951
7	1.129 122	7.378 408	6.534 641	57	2.688 202	96.468 658	35.885 947
8	1.148 882	8.507 530	7.405 053	58	2.735 245	99.156 859	36.251 545
9	1.168 987	9.656 412	8.260 494	59	2.783 112	101.892 104	36.610 855
10	1.189 444	10.825 399	9.101 223	60	2.831 816	104.675 216	36.963 986
11	1.210 260	12.014 844	9.927 492	61	2.881 373	107.507 032	37.311 042
12	1.231 439	13.225 104	10.739 550	62	2.931 797	110.388 405	37.652 130
13	1.252 990	14.456 543	11.537 641	63	2.983 104	113.320 202	37.987 351
14	1.274 917	15.709 533	12.322 006	64	3.034 308	116.303 306	38.316 807
15	1.297 228	16.984 449	13.092 880	65	3.088 426	119.338 614	38.640 597
16	1.319 929	18.281 677	13.850 497	66	3.142 473	122.427 039	38.958 817
17	1.343 028	19.601 607	14.595 083	67	3.197 466	125.569 513	39.271 565
18	1.366 531	20.944 635	15.326 863	68	3.253 422	128.766 979	39.578 934
19	1.390 445	22.311 166	16.046 057	69	3.310 357	132.020 401	39.881 016
20	1.414 778	23.701 611	16.752 881	70	3.368 288	135.330 758	40.177 903
21	1.439 537	25.116 389	17.447 549	71	3.427 233	138.699 047	40.469 683
22	1.464 729	26.555 926	18.130 269	72	3.487 210	142.126 280	40.756 445
23	1.490 361	28.020 655	18.801 248	73	3.548 236	145.613 490	41.038 276
24	1.516 443	29.511 016	19.460 686	74	3.610 330	149.161 726	41.315 259
25	1.542 981	31.027 459	20.108 782	75	3.673 511	152.772 056	41.587 478
26	1.569 983	32.570 440	20.745 732	76	3.737 797	156.445 567	41.855 015
27	1.597 457	34.140 422	21.371 726	77	3.803 209	160.183 364	42.117 951
28	1.625 413	35.737 880	21.986 955	78	3.869 765	163.986 573	42.376 364
29	1.653 858	37.363 293	22.591 602	79	3.937 486	167.856 338	42.630 334
30	1.682 800	39.017 150	23.185 849	80	4.006 392	171.793 824	42.879 935
31	1.712 249	40.699 950	23.769 876	81	4.076 504	175.800 216	43.125 243
32	1.742 213	42.412 200	24.343 859	82	4.147 843	179.876 720	43.366 332
33	1.772 702	44.154 413	24.907 970	83	4.220 430	184.024 563	43.603 275
34	1.803 725	45.927 115	25.462 378	84	4.294 287	188.244 992	43.836 142
35	1.835 290	47.730 840	26.007 251	85	4.369 437	192.539 280	44.065 005
36	1.867 407	49.566 129	26.542 753	86	4.445 903	196.908 717	44.289 931
37	1.900 087	51.433 537	27.069 045	87	4.523 706	201.354 620	44.510 989
38	1.933 338	53.333 624	27.586 285	88	4.602 871	205.878 326	44.728 244
39	1.967 172	55.266 962	28.094 629	89	4.683 421	210.481 196	44.941 764
40	2.001 597	57.234 134	28.594 230	90	4.765 381	215.164 617	45.151 610
41	2.036 625	59.235 731	29.085 238	91	4.848 775	219.929 998	44.357 848
42	2.072 266	61.272 357	29.567 801	92	4.933 629	224.778 773	45.560 539
43	2.108 531	63.344 623	30.042 065	93	5.019 967	229.712 401	45.759 743
44	2.145 430	65.453 154	30.508 172	94	5.107 816	234.732 368	45.955 521
45	2.182 975	67.598 584	30.966 263	95	5.197 203	239.840 185	46.147 933
46	2.221 177	69.781 559	31.416 474	96	5.288 154	245.037 388	46.337 035
47	2.260 048	72.002 736	31.858 943	97	5.380 697	250.325 542	46.522 884
48	2.299 599	74.262 784	32.293 801	98	5.474 859	255.706 239	46.705 537
49	2.339 842	76.562 383	32.721 181	99	5.570 669	261.181 099	46.885 049
50	2.380 789	78.902 225	33.141 209	100	5.668 156	266.751 768	47.061 473

Table 1

Continued

$i = 0.02\ (2\%)$							
n	$(1+i)^n$	$s_{\overline{n}|i}$	$a_{\overline{n}|i}$	n	$(1+i)^n$	$s_{\overline{n}|i}$	$a_{\overline{n}|i}$
1	1.020 000	1.000 000	0.980 392	51	2.745 420	87.270 989	31.787 849
2	1.040 400	2.020 000	1.941 561	52	2.800 328	90.016 409	32.144 950
3	1.061 208	3.060 400	2.883 883	53	2.856 335	92.816 737	32.495 049
4	1.082 432	4.121 608	3.807 729	54	2.913 461	95.673 072	32.838 283
5	1.104 081	5.204 040	4.713 460	55	2.971 731	98.586 534	33.174 788
6	1.126 162	6.308 121	5.601 431	56	3.031 165	101.558 264	33.504 694
7	1.148 686	7.434 283	6.471 991	57	3.091 789	104.589 430	33.828 131
8	1.171 659	8.582 969	7.325 481	58	3.153 624	107.681 218	34.145 226
9	1.195 093	9.754 628	8.162 237	59	3.216 697	110.834 843	34.456 104
10	1.218 994	10.949 721	8.982 585	60	3.281 031	114.051 539	34.760 887
11	1.243 374	12.168 715	9.786 848	61	3.346 651	117.332 570	35.059 693
12	1.268 242	13.412 090	10.575 341	62	3.413 584	120.679 222	35.352 640
13	1.293 607	14.680 331	11.348 374	63	3.481 856	124.092 806	35.639 843
14	1.319 479	15.973 938	12.106 249	64	3.551 493	127.574 662	35.921 415
15	1.345 868	17.293 417	12.849 264	65	3.622 523	131.126 155	36.197 466
16	1.372 786	18.639 285	13.577 709	66	3.694 974	134.748 679	36.468 103
17	1.400 241	20.012 071	14.291 872	67	3.768 873	138.443 652	36.733 435
18	1.428 246	21.412 312	14.992 031	68	3.844 251	142.212 525	36.993 564
19	1.456 811	22.840 559	15.678 462	69	3.921 136	146.056 776	37.248 592
20	1.485 947	24.297 370	16.351 433	70	3.999 558	149.977 911	37.498 619
21	1.515 666	25.783 317	17.011 209	71	4.079 549	153.977 469	37.743 744
22	1.545 980	27.298 984	17.658 048	72	4.161 140	158.057 019	37.984 063
23	1.576 899	28.844 963	18.292 204	73	4.244 363	162.218 159	38.219 670
24	1.608 437	30.421 862	18.913 926	74	4.329 250	166.462 522	38.450 657
25	1.640 606	32.030 300	19.523 456	75	4.415 835	170.791 773	38.677 114
26	1.673 418	33.670 906	20.121 036	76	4.504 152	175.207 608	38.899 132
27	1.706 886	35.344 324	20.706 898	77	4.594 235	179.711 760	39.116 796
28	1.741 024	37.051 210	21.281 272	78	4.686 120	184.305 996	39.330 192
29	1.775 845	38.792 235	21.844 385	79	4.779 842	188.992 115	39.539 404
30	1.811 362	40.568 079	22.396 456	80	4.875 439	193.771 958	39.744 514
31	1.847 589	42.379 441	22.937 702	81	4.972 948	198.647 397	39.945 602
32	1.884 541	44.227 030	23.468 335	82	5.072 407	203.620 345	40.142 747
33	1.922 231	46.111 570	23.988 564	83	5.173 855	208.692 752	40.336 026
34	1.960 676	48.033 802	24.498 592	84	5.277 332	213.866 607	40.525 516
35	1.999 890	49.994 478	24.998 619	85	5.382 879	219.143 939	40.711 290
36	2.039 887	51.994 367	25.488 842	86	5.490 536	224.526 818	40.893 422
37	2.080 685	54.034 255	25.969 453	87	5.600 347	230.017 354	41.071 982
38	2.122 299	56.114 940	26.440 641	88	5.712 354	235.617 701	41.247 041
39	2.164 745	58.237 238	26.902 589	89	5.826 601	241.330 055	41.418 668
40	2.208 040	60.401 983	27.355 479	90	5.943 133	247.156 656	41.586 929
41	2.252 200	62.610 023	27.799 489	91	6.061 996	253.099 789	41.751 891
42	2.297 244	64.862 223	28.234 794	92	6.183 236	259.161 785	41.913 619
43	2.343 189	67.159 468	28.661 562	93	6.306 900	265.345 021	42.072 175
44	2.390 053	69.502 657	29.079 963	94	6.433 038	271.651 921	42.227 623
45	2.437 854	71.892 710	29.490 159	95	6.561 699	278.084 960	42.380 023
46	2.486 611	74.330 564	29.892 314	96	6.692 933	284.646 659	42.529 434
47	2.536 344	76.817 176	30.286 582	97	6.826 792	291.339 592	42.675 916
48	2.587 070	79.353 519	30.673 120	98	6.963 328	298.166 384	42.819 525
49	2.638 812	81.940 590	31.052 078	99	7.102 594	305.129 712	42.960 319
50	2.691 588	84.579 401	31.423 606	100	7.244 646	312.232 306	43.098 352

Table 1

Continued

		$i = 0.0225$ $(2\frac{1}{4}\%)$					
n	$(1 + i)^n$	$s_{\overline{n}|i}$	$a_{\overline{n}|i}$	n	$(1 + i)^n$	$s_{\overline{n}|i}$	$a_{\overline{n}|i}$
1	1.022 500	1.000 000	0.997 995	51	3.110 492	93.799 664	30.155 889
2	1.045 506	2.022 500	1.934 470	52	3.180 479	96.910 157	30.470 307
3	1.069 030	3.068 006	2.869 897	53	3.252 039	100.090 635	30.777 806
4	1.093 083	4.137 036	3.784 740	54	3.325 210	103.342 674	31.078 539
5	1.117 678	5.230 120	4.679 453	55	3.400 027	106.667 885	31.372 654
6	1.142 825	6.347 797	5.554 477	56	3.476 528	110.067 912	31.660 298
7	1.168 539	7.490 623	6.410 246	57	3.554 750	113.544 440	31.941 611
8	1.194 831	8.659 162	7.247 185	58	3.634 732	117.099 190	32.216 735
9	1.221 715	9.853 993	8.065 706	59	3.716 513	120.733 922	32.485 804
10	1.249 203	11.075 708	8.866 216	60	3.800 135	124.450 435	32.748 953
11	1.277 311	12.324 911	9.649 111	61	3.885 638	128.250 570	33.006 311
12	1.306 050	13.602 222	10.414 779	62	3.973 065	132.136 208	33.258 006
13	1.335 436	14.908 272	11.163 598	63	4.062 459	136.109 272	33.504 162
14	1.365 483	16.243 708	11.895 939	64	4.153 864	140.171 731	33.744 902
15	1.396 207	17.609 191	12.612 166	65	4.247 326	144.325 595	33.980 344
16	1.427 621	19.005 398	13.312 631	66	4.342 891	148.572 920	34.210 605
17	1.459 743	20.433 020	13.997 683	67	4.440 606	152.915 811	34.435 800
18	1.492 587	21.892 763	14.667 661	68	4.540 519	157.356 417	34.656 039
19	1.526 170	23.385 350	15.322 896	69	4.642 681	161.896 937	34.871 432
20	1.560 509	24.911 520	15.963 712	70	4.747 141	166.539 618	35.082 085
21	1.595 621	26.472 029	16.590 428	71	4.853 952	171.286 759	35.288 103
22	1.631 522	28.067 650	17.203 352	72	4.963 166	176.140 711	35.489 587
23	1.668 231	29.699 172	17.802 790	73	5.074 837	181.103 877	35.686 638
24	1.705 767	31.367 403	18.389 036	74	5.189 021	186.178 714	35.879 352
25	1.744 146	33.073 170	18.962 383	75	5.305 774	191.367 735	36.067 826
26	1.783 390	34.817 316	19.523 113	76	5.425 154	196.673 509	36.252 153
27	1.823 516	36.600 706	20.071 504	77	5.547 220	202.098 663	36.432 423
28	1.864 545	38.424 222	20.607 828	78	5.672 032	207.645 883	36.608 727
29	1.906 497	40.288 767	21.132 350	79	5.799 653	213.317 916	36.781 151
30	1.949 393	42.195 264	21.645 330	80	5.930 145	219.117 569	36.949 781
31	1.993 255	44.144 657	22.147 022	81	6.063 574	225.047 714	37.114 700
32	2.038 103	46.137 912	22.637 674	82	6.200 004	231.111 288	37.275 990
33	2.083 960	48.176 015	23.117 530	83	6.339 504	237.311 292	37.433 731
34	2.130 849	50.259 976	23.586 826	84	6.482 143	243.650 796	37.588 001
35	2.178 794	52.390 825	24.045 796	85	6.627 991	250.132 939	37.738 877
36	2.227 816	54.569 619	24.494 666	86	6.777 121	256.760 930	37.886 432
37	2.277 942	56.797 435	24.933 658	87	6.929 606	263.538 051	38.030 740
38	2.329 196	59.075 377	25.362 991	88	7.085 522	270.467 657	38.171 873
39	2.381 603	61.404 573	25.782 876	89	7.244 947	277.553 179	38.309 900
40	2.435 189	63.786 176	26.193 522	90	7.407 958	284.798 126	38.444 890
41	2.489 981	66.221 365	26.595 132	91	7.574 637	292.206 083	38.576 910
42	2.546 005	68.711 346	26.987 904	92	7.745 066	299.780 720	38.706 024
43	2.603 290	71.257 351	27.372 033	93	7.919 330	307.525 786	38.832 298
44	2.661 864	73.860 642	27.747 710	94	8.097 515	315.445 117	38.955 792
45	2.721 756	76.522 506	28.115 120	95	8.279 709	323.542 632	39.076 569
46	2.782 996	79.244 262	28.474 444	96	8.466 003	331.822 341	39.194 689
47	2.845 613	82.027 258	28.825 863	97	8.656 488	340.288 344	39.310 209
48	2.909 640	84.872 872	29.169 548	98	8.851 259	348.944 831	39.423 187
49	2.975 107	87.782 511	29.505 670	99	9.050 412	357.796 090	39.533 680
50	3.042 046	90.757 618	29.834 396	100	9.254 046	366.846 502	39.641 741

Table 1

Continued

	$i = 0.025\ (2\tfrac{1}{2}\%)$

n	$(1 + i)^n$	$s_{\overline{n}\mid i}$	$a_{\overline{n}\mid i}$	n	$(1 + i)^n$	$s_{\overline{n}\mid i}$	$a_{\overline{n}\mid i}$
1	1.025 000	1.000 000	0.975 610	51	3.523 036	100.921 458	28.646 158
2	1.050 625	2.025 000	1.927 424	52	3.611 112	104.444 494	28.923 081
3	1.076 891	3.075 625	2.856 024	53	3.701 390	108.055 606	29.193 249
4	1.103 813	4.152 516	3.761 974	54	3.793 925	111.756 996	29.456 829
5	1.131 408	5.256 329	4.645 828	55	3.888 773	115.550 921	29.713 979
6	1.159 693	6.387 737	5.508 125	56	3.985 992	119.439 694	29.964 858
7	1.188 686	7.547 430	6.349 391	57	4.085 642	123.425 687	30.209 617
8	1.218 403	8.736 116	7.170 137	58	4.187 783	127.511 329	30.448 407
9	1.248 863	9.954 519	7.970 866	59	4.292 478	131.699 112	30.681 373
10	1.280 085	11.203 382	8.752 064	60	4.399 790	135.991 590	30.908 656
11	1.312 087	12.483 466	9.514 209	61	4.509 784	140.391 380	31.130 397
12	1.344 889	13.795 553	10.257 765	62	4.622 529	144.901 164	31.346 728
13	1.378 511	15.140 442	10.983 185	63	4.738 092	149.523 693	31.557 784
14	1.412 974	16.518 953	11.690 912	64	4.856 545	154.261 786	31.763 691
15	1.448 298	17.931 927	12.381 378	65	4.977 958	159.118 330	31.964 577
16	1.484 506	19.380 225	13.055 003	66	5.102 407	164.096 289	32.160 563
17	1.521 618	20.864 730	13.712 198	67	5.229 967	169.198 696	32.351 769
18	1.559 659	22.386 349	14.353 364	68	5.360 717	174.428 663	32.538 311
19	1.598 650	23.946 007	14.978 891	69	5.494 734	179.789 380	32.720 303
20	1.638 616	25.544 658	15.589 162	70	5.632 103	185.284 114	32.897 857
21	1.679 582	27.183 274	16.184 549	71	5.772 905	190.916 217	33.071 080
22	1.721 571	28.862 856	16.765 413	72	5.917 228	196.689 122	33.240 078
23	1.764 611	30.584 427	17.332 110	73	6.065 159	202.606 351	33.404 954
24	1.808 726	32.349 038	17.884 986	74	6.216 788	208.671 509	33.565 809
25	1.853 944	34.157 764	18.424 376	75	6.372 207	214.888 297	33.722 740
26	1.900 293	36.011 708	18.950 611	76	6.531 513	221.260 504	33.875 844
27	1.947 800	37.912 001	19.464 011	77	6.694 800	227.792 017	34.025 214
28	1.996 495	39.859 801	19.964 889	78	6.862 170	234.486 818	34.170 940
29	2.046 407	41.856 296	20.453 550	79	7.033 725	241.348 988	34.313 113
30	2.097 568	43.902 703	20.930 293	80	7.209 568	248.382 713	34.451 817
31	2.150 007	46.000 271	21.395 407	81	7.389 807	255.592 280	34.587 139
32	2.203 757	48.150 278	21.849 178	82	7.574 552	262.982 087	34.719 160
33	2.258 851	50.354 034	22.291 881	83	7.763 916	270.556 640	34.847 961
34	2.315 322	52.612 885	22.723 786	84	7.948 014	278.320 556	34.973 620
35	2.373 205	54.928 207	23.145 157	85	8.156 964	286.278 569	35.096 215
36	2.432 535	57.301 413	23.556 251	86	8.360 888	294.435 534	35.215 819
37	2.493 349	59.733 948	23.957 318	87	8.569 911	302.796 422	35.332 507
38	2.555 682	62.227 297	24.348 603	88	8.784 158	311.366 333	35.446 348
39	2.619 574	64.782 979	24.730 344	89	9.003 762	320.150 491	35.557 413
40	2.685 064	67.402 554	25.102 775	90	9.228 856	329.154 253	35.665 768
41	2.752 190	70.087 617	25.466 122	91	9.459 578	338.383 110	35.771 481
42	2.820 995	72.839 808	25.820 607	92	9.696 067	347.842 687	35.874 616
43	2.891 520	75.660 803	26.166 446	93	9.938 469	357.538 755	35.975 235
44	2.963 808	78.552 323	26.503 849	94	10.186 931	367.477 223	36.073 400
45	3.037 903	81.516 131	26.833 024	95	10.441 604	377.664 154	36.169 171
46	3.113 851	84.554 034	27.154 170	96	10.702 644	388.105 758	36.262 606
47	3.191 697	87.667 885	27.467 483	97	10.970 210	398.808 402	36.353 762
48	3.271 490	90.859 582	27.773 154	98	11.244 465	409.778 612	36.442 694
49	3.353 277	94.131 072	28.071 369	99	11.525 577	421.023 077	36.529 458
50	3.437 109	97.484 349	28.362 312	100	11.813 716	432.548 654	36.614 105

Table 1

Continued

		$i = 0.03$ (3%)					
n	$(1+i)^n$	$s_{\overline{n}|i}$	$a_{\overline{n}|i}$	n	$(1+i)^n$	$s_{\overline{n}|i}$	$a_{\overline{n}|i}$
1	1.030 000	1.000 000	0.970 874	51	4.515 423	117.180 773	25.951 227
2	1.060 900	2.030 000	1.913 470	52	4.650 886	121.696 197	26.166 240
3	1.092 727	3.090 900	2.828 611	53	4.790 412	126.347 082	26.374 990
4	1.125 509	4.183 627	3.717 098	54	4.934 125	131.137 495	26.577 660
5	1.159 274	5.309 136	4.579 707	55	5.082 149	136.071 620	26.774 428
6	1.194 052	6.468 410	5.417 191	56	5.234 613	141.153 768	26.965 464
7	1.229 874	7.662 462	6.230 283	57	5.391 651	146.388 381	27.150 936
8	1.266 770	8.892 336	7.019 692	58	5.553 401	151.780 033	27.331 005
9	1.304 773	10.159 106	7.786 109	59	5.720 003	157.333 434	27.505 831
10	1.343 916	11.463 879	8.530 203	60	5.891 603	163.053 437	27.675 564
11	1.384 234	12.807 796	9.252 624	61	6.068 351	168.945 040	27.840 353
12	1.425 761	14.192 030	9.954 004	62	6.250 402	175.013 391	28.000 343
13	1.468 534	15.617 790	10.634 955	63	6.437 914	181.263 793	28.155 673
14	1.512 590	17.086 324	11.296 073	64	6.631 051	187.701 707	28.306 478
15	1.557 967	18.598 914	11.937 935	65	6.829 983	194.332 758	28.452 892
16	1.604 706	20.156 881	12.561 102	66	7.034 882	201.162 741	28.595 040
17	1.652 848	21.761 588	13.166 118	67	7.245 929	208.197 623	28.733 049
18	1.702 433	23.414 435	13.753 513	68	7.463 307	215.443 551	28.867 038
19	1.753 506	25.116 868	14.323 799	69	7.687 206	222.906 858	28.997 124
20	1.806 111	26.870 374	14.877 475	70	7.917 822	230.594 064	29.123 421
21	1.860 295	28.676 486	15.415 024	71	8.155 357	238.511 886	29.246 040
22	1.916 103	30.536 780	15.936 917	72	8.400 017	246.667 242	29.365 088
23	1.973 587	32.452 884	16.443 608	73	8.652 016	255.067 259	29.480 668
24	2.032 794	34.426 470	16.935 542	74	8.911 578	263.719 277	29.592 881
25	2.093 778	36.459 264	17.413 148	75	9.178 926	272.630 856	29.701 826
26	2.156 591	38.553 042	17.876 842	76	9.454 293	281.809 781	29.807 598
27	2.221 289	40.709 634	18.327 031	77	9.737 922	291.264 075	29.910 290
28	2.287 928	42.930 923	18.764 108	78	10.030 060	301.001 997	30.009 990
29	2.356 566	45.218 850	19.188 455	79	10.330 962	311.032 057	30.106 786
30	2.427 262	47.575 416	19.600 441	80	10.640 891	321.363 019	30.200 763
31	2.500 080	50.002 678	20.000 428	81	10.960 117	332.003 909	30.292 003
32	2.575 083	52.502 759	20.388 766	82	11.288 921	342.964 026	30.380 586
33	2.652 335	55.077 841	20.765 792	83	11.627 588	354.252 947	30.466 588
34	2.731 905	57.730 177	21.131 837	84	11.976 416	365.880 536	30.550 086
35	2.813 862	60.462 082	21.487 220	85	12.335 709	377.856 952	30.631 151
36	2.898 278	63.275 944	21.832 252	86	12.705 780	390.192 660	30.709 855
37	2.985 227	66.174 223	22.167 235	87	13.086 953	402.898 440	30.786 267
38	3.074 783	69.159 449	22.492 462	88	13.479 562	415.985 393	30.860 454
39	3.167 027	72.234 233	22.808 215	89	13.883 949	429.464 955	30.932 479
40	3.262 038	75.401 260	23.114 772	90	14.300 467	443.348 904	31.002 407
41	3.359 899	78.663 298	23.412 400	91	14.729 481	457.649 371	31.070 298
42	3.460 696	82.023 196	23.701 359	92	15.171 366	472.378 852	31.136 212
43	3.564 517	85.483 892	23.981 902	93	15.626 507	487.550 217	31.200 206
44	3.671 452	89.048 409	24.254 274	94	16.095 302	503.176 724	31.262 336
45	3.781 596	92.719 861	24.518 713	95	16.578 161	519.272 026	31.322 656
46	3.895 044	96.501 457	24.775 449	96	17.075 506	535.850 186	31.381 219
47	4.011 895	100.396 501	25.024 708	97	17.587 771	552.925 692	31.438 077
48	4.132 252	104.408 396	25.266 707	98	18.115 404	570.513 463	31.493 279
49	4.256 219	108.540 648	25.501 657	99	18.658 866	588.628 867	31.546 872
50	4.383 906	112.796 867	25.729 764	100	19.218 632	607.287 733	31.598 905

Table 1

Continued

	$i = 0.035 \ (3\frac{1}{2}\%)$						
n	$(1+i)^n$	$s_{\overline{n}|i}$	$a_{\overline{n}|i}$	n	$(1+i)^n$	$s_{\overline{n}|i}$	$a_{\overline{n}|i}$
1	1.035 000	1.000 000	0.966 184	51	5.780 399	136.582 837	23.628 616
2	1.071 225	2.035 000	1.899 694	52	5.982 713	142.363 236	23.795 765
3	1.108 718	3.106 225	2.801 637	53	6.192 108	148.345 950	23.957 260
4	1.147 523	4.214 943	3.673 079	54	6.408 832	154.538 058	24.113 295
5	1.187 686	5.362 466	4.515 052	55	6.633 141	160.946 890	24.264 053
6	1.229 255	6.550 152	5.328 553	56	6.865 301	167.580 031	24.409 713
7	1.272 279	7.779 408	6.114 544	57	7.105 587	174.445 332	24.550 448
8	1.316 809	9.051 687	6.873 956	58	7.354 282	181.550 919	24.686 423
9	1.362 897	10.368 496	7.607 687	59	7.611 682	188.905 201	24.817 800
10	1.410 599	11.731 393	8.316 605	60	7.878 091	196.516 883	24.944 734
11	1.459 970	13.141 992	9.001 551	61	8.153 824	204.394 974	25.067 376
12	1.511 069	14.601 962	9.663 334	62	8.439 208	212.548 798	25.185 870
13	1.563 956	16.113 030	10.302 738	63	8.734 580	220.988 006	25.300 358
14	1.618 695	17.676 986	10.920 520	64	9.040 291	229.722 586	25.410 974
15	1.675 349	19.295 681	11.517 411	65	9.356 701	238.762 876	25.517 849
16	1.733 986	20.971 030	12.094 117	66	9.684 185	248.119 577	25.621 110
17	1.794 676	22.705 016	12.651 321	67	10.023 132	257.803 762	25.720 880
18	1.857 489	24.499 691	13.189 682	68	10.373 941	267.826 894	25.817 275
19	1.922 501	26.357 180	13.709 837	69	10.737 029	278.200 835	25.910 411
20	1.989 789	28.279 682	14.212 403	70	11.112 825	288.937 865	26.000 397
21	2.059 431	30.269 471	14.697 974	71	11.501 774	300.050 690	26.087 340
22	2.131 512	32.328 902	15.167 125	72	11.904 336	311.552 464	26.171 343
23	2.206 114	34.460 414	15.620 410	73	12.320 988	323.456 800	26.252 505
24	2.283 328	36.666 528	16.058 368	74	12.752 223	335.777 788	26.330 923
25	2.363 245	38.949 857	16.481 515	75	13.198 550	348.530 011	26.406 689
26	2.445 959	41.313 102	16.890 352	76	13.660 500	361.728 561	26.479 892
27	2.531 567	43.759 060	17.285 365	77	14.138 617	375.389 061	26.550 621
28	2.620 172	46.290 627	17.667 019	78	14.633 469	389.527 678	26.618 957
29	2.711 878	48.910 799	18.035 767	79	15.145 640	404.161 147	26.684 983
30	2.806 794	51.622 677	18.392 045	80	15.675 738	419.306 787	26.748 776
31	2.905 031	54.429 471	18.736 276	81	16.224 388	434.982 524	26.810 411
32	3.006 708	57.334 502	19.068 865	82	16.792 242	451.206 913	26.869 963
33	3.111 942	60.341 210	19.390 208	83	17.379 970	467.999 155	26.927 500
34	3.220 860	63.453 152	19.700 684	84	17.988 269	485.379 125	26.983 092
35	3.333 590	66.674 013	20.000 661	85	18.617 859	503.367 394	27.036 804
36	3.450 266	70.007 603	20.290 494	86	19.269 484	521.985 253	27.088 699
37	3.571 025	73.457 869	20.570 525	87	19.943 916	541.254 737	27.138 840
38	3.696 011	77.028 895	20.841 087	88	20.641 953	561.198 653	27.187 285
39	3.825 372	80.724 906	21.102 500	89	21.364 421	581.840 606	27.234 092
40	3.959 260	84.550 278	21.355 072	90	22.112 176	603.205 027	27.279 316
41	4.097 834	88.509 537	21.599 104	91	22.886 102	625.317 203	27.323 010
42	4.241 258	92.607 371	21.834 883	92	23.687 116	648.203 305	27.365 227
43	4.389 702	96.848 629	22.062 689	93	24.516 165	671.890 421	27.406 017
44	4.543 342	101.238 331	22.282 791	94	25.374 230	696.406 585	27.445 427
45	4.702 359	105.781 673	22.495 450	95	26.262 329	721.780 816	27.483 504
46	4.866 941	110.484 031	22.700 918	96	27.181 510	748.043 145	27.520 294
47	5.037 284	115.350 973	22.899 438	97	28.132 863	775.224 655	27.555 839
48	5.213 589	120.388 257	23.091 244	98	29.117 513	803.357 517	27.590 183
49	5.396 065	125.601 846	23.276 564	99	30.136 626	832.475 031	27.623 366
50	5.584 927	130.997 910	23.455 618	100	31.191 408	862.611 657	27.653 425

Table 1

Continued

		$i = 0.04$ (4%)					
n	$(1 + i)^n$	$s_{\overline{n}|i}$	$a_{\overline{n}|i}$	n	$(1 + i)^n$	$s_{\overline{n}|i}$	$a_{\overline{n}|i}$
1	1.040 000	1.000 000	0.961 538	51	7.390 951	159.773 767	21.617 485
2	1.081 600	2.040 000	1.886 095	52	7.686 589	167.164 718	21.747 582
3	1.124 864	3.121 600	2.775 091	53	7.994 052	174.851 306	21.872 675
4	1.169 859	4.246 464	3.629 895	54	8.313 814	182.845 359	21.992 957
5	1.216 653	5.416 323	4.451 822	55	8.646 367	191.159 173	22.108 612
6	1.265 319	6.632 975	5.242 137	56	8.992 222	109.805 540	22.219 819
7	1.315 932	7.898 294	6.002 055	57	9.351 910	208.797 762	22.326 749
8	1.368 569	9.214 226	6.732 745	58	9.725 987	218.149 672	22.429 567
9	1.423 312	10.582 795	7.435 332	59	10.115 026	227.875 659	22.528 430
10	1.480 244	12.006 107	8.110 896	60	10.519 627	237.990 685	22.623 490
11	1.539 454	13.486 351	8.760 477	61	10.940 413	248.510 312	22.714 894
12	1.601 032	15.025 805	9.385 074	62	11.378 029	259.450 725	22.802 783
13	1.665 074	16.626 838	9.985 648	63	11.833 150	270.828 754	22.887 291
14	1.731 676	18.291 911	10.563 123	64	12.306 476	282.661 904	22.968 549
15	1.800 944	20.023 588	11.118 387	65	12.798 735	294.968 380	23.046 682
16	1.872 981	21.824 531	11.632 296	66	13.310 685	307.767 116	23.121 810
17	1.947 900	23.697 512	12.165 669	67	13.843 112	321.077 800	23.194 048
18	2.025 817	25.645 413	12.659 297	68	14.396 836	334.920 912	23.263 507
19	2.106 849	27.671 229	13.133 939	69	14.972 710	349.317 749	23.330 296
20	2.191 123	29.778 079	13.590 326	70	15.571 618	364.290 459	23.394 515
21	2.278 768	31.969 202	14.029 160	71	16.194 483	379.862 077	23.456 264
22	2.369 919	34.247 970	14.451 115	72	16.842 262	396.056 560	23.515 639
23	2.464 716	36.617 889	14.856 842	73	17.515 953	412.898 823	23.572 730
24	2.563 304	39.082 604	15.246 963	74	18.216 591	430.414 776	23.627 625
25	2.665 836	41.645 908	15.622 080	75	18.945 255	448.631 367	23.680 408
26	2.772 470	44.311 745	15.982 769	76	19.703 065	467.576 621	23.731 162
27	2.883 369	47.084 214	16.329 586	77	20.491 187	487.279 686	23.779 963
28	2.998 703	49.967 583	16.663 063	78	21.310 835	507.770 873	23.826 688
29	3.118 651	52.966 286	16.983 715	79	22.163 268	529.081 708	23.872 008
30	3.243 398	56.084 938	17.292 033	80	23.049 799	551.244 977	23.915 392
31	3.373 133	59.328 335	17.588 494	81	23.971 791	574.294 776	23.957 108
32	3.508 059	62.701 469	17.873 552	82	24.930 663	598.266 567	23.997 219
33	3.648 381	66.209 527	18.147 646	83	25.927 889	623.197 230	24.035 787
34	3.794 316	69.857 909	18.411 198	84	26.965 005	649.125 119	24.072 872
35	3.946 089	73.652 225	18.664 613	85	28.043 605	676.090 123	24.108 531
36	4.103 933	77.598 314	18.908 282	86	29.165 349	704.133 728	24.142 818
37	4.268 090	81.702 246	19.142 579	87	30.331 963	733.299 078	24.175 787
38	4.438 813	85.970 336	19.367 864	88	31.545 242	763.631 041	24.207 487
39	4.616 366	90.409 150	19.584 485	89	32.807 051	795.176 282	24.237 969
40	4.801 021	95.025 516	19.792 774	90	34.119 333	827.983 334	24.267 276
41	4.993 061	99.826 536	19.993 052	91	35.484 107	862.102 667	24.295 459
42	5.192 784	104.819 598	20.185 627	92	36.903 471	897.586 774	24.322 557
43	5.400 495	110.012 382	20.370 795	93	38.379 610	934.490 244	24.348 612
44	5.616 515	115.412 877	20.548 841	94	39.914 794	972.869 854	24.373 666
45	5.841 176	121.029 392	20.720 040	95	41.511 386	1012.784 648	24.397 756
46	6.074 823	126.870 568	20.884 654	96	43.171 841	1054.296 034	24.420 919
47	6.317 816	132.945 390	21.042 936	97	44.898 715	1097.467 876	24.443 191
48	6.570 528	139.263 206	21.195 131	98	46.694 664	1142.366 591	24.464 607
49	6.833 349	145.833 734	21.341 472	99	48.562 450	1189.061 254	24.485 199
50	7.106 683	152.667 084	21.482 185	100	50.504 948	1237.623 705	24.504 999

Table 1

Continued

			$i = 0.045$ $(4\frac{1}{2}\%)$				
n	$(1 + i)^n$	$s_{\overline{n}|i}$	$a_{\overline{n}|i}$	n	$(1 + i)^n$	$s_{\overline{n}|i}$	$a_{\overline{n}|i}$
1	1.045 000	1.000 000	0.956 938	51	9.439 105	187.535 665	19.867 950
2	1.092 025	2.045 000	1.872 668	52	9.863 865	106.974 770	19.969 330
3	1.141 166	3.137 025	2.748 964	53	10.307 739	206.838 634	20.066 345
4	1.192 519	4.278 191	3.587 526	54	10.771 587	217.146 373	20.159 181
5	1.246 182	5.470 710	4.389 977	55	11.256 308	227.917 959	20.248 021
6	1.302 260	6.716 892	5.157 872	56	11.762 842	239.174 268	20.333 034
7	1.360 862	8.019 152	5.892 701	57	12.292 170	250.937 110	20.414 387
8	1.422 101	9.380 014	6.595 886	58	12.845 318	263.229 280	20.492 236
9	1.486 095	10.802 114	7.268 790	59	13.423 357	276.074 597	20.566 733
10	1.552 969	12.288 209	7.912 718	60	14.027 408	289.497 954	20.638 022
11	1.622 853	13.841 179	8.528 917	61	14.658 641	303.525 362	20.706 241
12·	1.695 881	15.464 032	9.118 581	62	15.318 280	318.184 031	20.771 523
13	1.772 196	17.159 913	9.682 852	63	16.007 603	333.502 283	20.833 993
14	1.851 945	18.932 109	10.222 825	64	16.727 945	349.509 868	20.893 773
15	1.935 282	20.784 054	10.739 546	65	17.480 702	366.237 831	20.950 979
16	2.022 370	22.719 337	11.234 015	66	18.267 334	383.718 533	21.005 722
17	2.113 377	24.741 707	11.707 191	67	19.089 364	401.985 867	21.058 107
18	2.208 479	26.855 084	12.159 992	68	19.948 385	421.075 231	21.108 236
19	2.307 860	29.063 562	12.593 294	69	20.846 063	441.023 617	21.156 207
20	2.411 714	31.371 423	13.007 936	70	21.784 136	461.869 680	21.202 112
21	2.520 241	33.783 137	13.404 724	71	22.764 422	483.653 815	21.246 040
22	2.633 652	36.303 378	13.784 425	72	23.788 821	506.418 237	21.288 077
23	2.752 166	38.937 030	14.147 775	73	24.859 318	530.207 057	21.328 303
24	2.876 014	41.689 196	14.495 478	74	25.977 987	555.066 375	21.366 797
25	3.005 434	44.565 210	14.828 209	75	27.146 996	581.044 362	21.403 634
26	3.140 679	47.570 645	15.146 611	76	28.368 611	608.191 358	21.438 884
27	3.282 010	50.711 324	15.451 303	77	29.645 199	636.559 969	21.472 616
28	3.429 700	53.993 333	15.742 874	78	30.979 233	666.205 168	21.504 896
29	3.584 036	57.423 033	16.021 889	79	32.373 298	697.184 401	21.535 785
30	3.745 318	61.007 070	16.288 889	80	33.830 096	729.557 699	21.565 345
31	3.913 857	64.752 388	16.544 391	81	35.352 451	763.387 795	21.593 632
32	4.089 981	68.666 245	16.788 891	82	36.943 311	798.740 246	21.620 700
33	4.274 030	72.756 226	17.022 862	83	38.605 760	835.683 557	21.646 603
34	4.466 362	77.030 256	17.246 758	84	40.343 019	874.289 317	21.671 390
35	4.667 348	81.496 618	17.461 012	85	42.158 455	914.632 336	21.695 110
36	4.877 378	86.163 966	17.666 041	86	44.055 586	956.790 791	21.717 809
37	5.096 860	91.041 344	17.862 240	87	46.038 087	1000.846 377	21.739 530
38	5.326 219	96.138 205	18.049 990	88	48.109 801	1046.884 464	21.760 316
39	5.565 899	101.464 424	18.229 656	89	50.274 742	1094.994 265	21.780 207
40	5.816 365	107.030 323	18.401 584	90	52.537 105	1145.269 007	21.799 241
41	6.078 101	112.846 688	18.566 109	91	54.901 275	1197.806 112	21.817 455
42	6.351 615	118.924 789	18.723 550	92	57.371 832	1252.707 387	21.834 885
43	6.637 438	125.276 404	18.874 210	93	59.953 565	1310.079 219	21.851 565
44	6.936 123	131.913 842	19.018 383	94	62.651 475	1370.032 784	21.867 526
45	7.248 248	138.849 965	19.156 347	95	65.470 792	1432.684 259	21.882 800
46	7.574 420	146.098 214	19.288 371	96	68.416 977	1498.155 051	21.897 417
47	7.915 268	153.672 633	19.414 709	97	71.495 741	1566.572 028	21.911 403
48	8.271 456	161.587 902	19.535 607	98	74.713 050	1638.067 770	21.924 788
49	8.643 671	169.859 357	19.651 298	99	78.075 137	1712.780 819	21.937 596
50	9.032 636	178.503 028	19.762 008	100	81.588 518	1790.855 956	21.949 853

Table 1

Continued

	$i = 0.05\ (5\%)$				$i = 0.06\ (6\%)$		
n	$(1+i)^n$	$s_{\overline{n}|i}$	$a_{\overline{n}|i}$	n	$(1+i)^n$	$s_{\overline{n}|i}$	$a_{\overline{n}|i}$
1	1.050 000	1.000 000	0.952 381	1	1.060 000	1.000 000	0.943 396
2	1.102 500	2.050 000	1.859 410	2	1.123 600	2.060 000	1.833 393
3	1.157 625	3.152 500	2.723 248	3	1.191 016	3.183 600	2.673 012
4	1.215 506	4.310 125	3.545 951	4	1.262 477	4.374 616	3.465 106
5	1.276 282	5.525 631	4.329 477	5	1.338 226	5.637 093	4.212 364
6	1.340 096	6.801 913	5.075 692	6	1.418 519	6.975 319	4.917 324
7	1.407 100	8.142 008	5.786 373	7	1.503 630	8.393 838	5.582 381
8	1.477 455	9.549 109	6.463 213	8	1.593 848	9.897 468	6.209 794
9	1.551 328	11.026 564	7.107 822	9	1.689 479	11.491 316	6.801 692
10	1.628 895	12.577 893	7.721 735	10	1.790 848	13.180 795	7.360 087
11	1.710 339	14.206 787	8.306 414	11	1.898 299	14.971 643	7.886 875
12	1.795 856	15.917 127	8.863 252	12	2.012 196	16.869 941	8.383 844
13	1.885 649	17.712 983	9.393 573	13	2.132 928	18.882 138	8.852 683
14	1.979 932	19.598 632	9.898 641	14	2.260 904	21.015 066	9.294 984
15	2.078 928	21.578 564	10.379 658	15	2.396 558	23.275 970	9.712 249
16	2.182 875	23.657 492	10.837 770	16	2.540 352	25.672 528	10.105 895
17	2.292 018	25.040 366	11.274 066	17	2.692 773	28.212 880	10.477 260
18	2.406 619	28.132 385	11.689 587	18	2.854 339	30.905 653	10.827 603
19	2.526 950	30.539 004	12.085 321	19	3.025 600	33.759 992	11.158 116
20	2.653 298	33.065 954	12.462 210	20	3.207 135	36.785 591	11.469 921
21	2.785 963	35.719 252	12.821 153	21	3.399 564	39.992 727	11.764 077
22	2.925 261	38.505 214	13.163 003	22	3.603 537	43.392 290	12.041 582
23	3.071 524	41.430 475	13.488 574	23	3.819 750	46.995 828	12.303 379
24	3.225 100	44.501 999	13.798 642	24	4.048 935	50.815 577	12.550 358
25	3.386 355	47.727 099	14.093 945	25	4.291 871	54.864 512	12.783 356
26	3.555 673	51.113 454	14.375 185	26	4.549 383	59.156 383	13.003 166
27	3.733 456	54.669 126	14.643 034	27	4.822 346	63.705 766	13.210 534
28	3.920 129	58.402 583	14.898 127	28	5.111 687	68.528 112	13.406 164
29	4.116 136	62.322 712	15.141 074	29	5.418 388	73.639 798	13.590 721
30	4.321 942	66.438 848	15.372 451	30	5.743 491	79.058 186	13.764 831
31	4.538 039	70.760 790	15.592 810	31	6.088 101	84.801 677	13.929 086
32	4.764 941	75.298 829	15.802 677	32	6.453 387	90.889 778	14.084 043
33	5.003 189	80.063 771	16.002 549	33	6.840 590	97.343 165	14.230 230
34	5.253 348	85.066 959	16.192 904	34	7.251 025	104.183 755	14.368 141
35	5.516 015	90.320 307	16.374 194	35	7.686 087	111.434 780	14.498 246
36	5.791 816	95.836 323	16.546 852	36	8.147 252	119.120 867	14.620 987
37	6.081 407	101.628 139	16.711 287	37	8.636 087	127.268 119	14.736 780
38	6.385 477	107.709 546	16.867 893	38	9.154 252	135.904 206	14.846 019
39	6.704 751	114.095 023	17.017 041	39	9.703 507	145.058 458	14.949 075
40	7.039 989	120.799 774	17.159 086	40	10.285 718	154.761 966	15.046 297
41	7.391 988	127.839 763	17.294 368	41	10.902 861	165.047 684	15.138 016
42	7.761 588	135.231 751	17.423 208	42	11.557 033	175.950 545	15.224 543
43	8.149 667	142.993 339	17.545 912	43	12.250 455	187.507 577	15.306 173
44	8.557 150	151.143 006	17.662 773	44	12.985 482	199.758 032	15.383 182
45	8.985 008	159.700 156	17.774 070	45	13.764 611	212.743 514	15.455 832
46	9.434 258	168.685 164	17.880 066	46	14.590 487	226.508 125	15.524 370
47	9.905 971	178.119 422	17.981 016	47	15.465 917	241.098 612	15.589 028
48	10.401 270	188.025 393	18.077 158	48	16.393 872	256.564 529	15.650 027
49	10.921 333	198.426 663	18.168 722	49	17.377 504	272.958 401	15.707 572
50	11.467 400	209.347 996	18.255 925	50	18.420 154	290.335 905	15.761 861

Table 1

Continued

	$i = 0.07 (7\%)$				$i = 0.08 (8\%)$		
n	$(1 + i)^n$	$s_{\overline{n}|i}$	$a_{\overline{n}|i}$	n	$(1 + i)^n$	$s_{\overline{n}|i}$	$a_{\overline{n}|i}$
1	1.070 000	1.000 000	0.934 579	1	1.080 000	1.000 000	0.925 925
2	1.144 900	2.070 000	1.808 018	2	1.166 400	2.080 000	1.783 265
3	1.225 043	3.214 900	2.624 316	3	1.259 712	3.246 400	2.577 097
4	1.310 796	4.439 943	3.387 211	4	1.360 489	4.506 112	3.312 127
5	1.402 552	5.750 739	4.100 197	5	1.469 328	5.866 601	3.992 710
6	1.500 730	7.153 291	4.766 540	6	1.586 874	7.335 929	4.622 880
7	1.605 781	8.654 021	5.389 289	7	1.713 824	8.922 803	5.206 370
8	1.718 186	10.259 803	5.971 299	8	1.850 930	10.636 628	5.746 639
9	1.838 459	11.977 989	6.515 232	9	1.999 005	12.487 558	6.246 888
10	1.967 151	13.816 448	7.023 582	10	2.158 925	14.486 562	6.710 081
11	2.104 852	15.783 599	7.498 674	11	2.331 639	16.645 487	7.138 964
12	2.252 192	17.888 451	7.942 686	12	2.518 170	18.977 126	7.536 078
13	2.409 845	20.140 643	8.357 651	13	2.719 624	21.495 297	7.903 776
14	2.578 534	22.550 488	8.745 468	14	2.937 194	24.214 920	8.244 237
15	2.759 032	25.129 022	9.107 914	15	3.172 169	27.152 114	8.559 479
16	2.952 164	27.888 054	9.446 649	16	3.425 943	30.324 283	8.851 369
17	3.158 815	30.840 217	9.763 223	17	3.700 018	33.750 226	9.121 638
18	3.379 932	33.999 033	10.059 087	18	3.996 019	37.450 244	9.371 887
19	3.616 528	37.378 965	10.335 595	19	4.315 701	41.446 263	9.603 599
20	3.869 684	40.995 492	10.594 014	20	4.660 957	45.761 964	9.818 147
21	4.140 562	44.865 177	10.835 527	21	5.033 834	50.422 921	10.016 803
22	4.430 402	49.005 739	11.061 240	22	5.436 540	55.456 755	10.200 744
23	4.740 530	53.436 141	11.272 187	23	5.871 464	60.893 296	10.371 059
24	5.072 367	58.176 671	11.469 334	24	6.341 181	66.764 759	10.528 758
25	5.427 433	63.249 038	11.653 583	25	6.848 475	72.105 940	10.674 776
26	5.807 353	68.676 470	11.825 779	26	7.396 353	79.954 415	10.809 978
27	6.213 868	74.483 823	11.986 709	27	7.988 061	87.350 768	10.935 165
28	6.648 838	80.697 691	12.137 111	28	8.627 106	95.338 830	11.051 078
29	7.114 257	87.346 529	12.277 674	29	9.317 275	103.965 936	11.158 406
30	7.612 255	94.460 786	12.409 041	30	10.062 657	113.283 211	11.257 783
31	8.145 113	102.073 041	12.531 814	31	10.867 669	123.345 868	11.349 799
32	8.715 271	110.218 154	12.646 555	32	11.737 083	134.213 537	11.434 999
33	9.325 340	118.933 425	12.753 790	33	12.676 050	145.950 620	11.513 888
34	9.978 114	128.258 765	12.854 009	34	13.690 134	158.626 670	11.586 934
35	10.676 581	138.236 878	12.947 672	35	14.785 344	172.316 804	11.654 568
36	11.423 942	148.913 460	13.035 208	36	15.968 172	187.102 148	11.717 193
37	12.223 618	160.337 402	13.117 017	37	17.245 626	203.070 320	11.775 179
38	13.079 271	172.561 020	13.193 473	38	18.625 276	220.315 945	11.828 869
39	13.994 820	185.640 292	13.264 928	39	20.115 298	238.941 221	11.878 582
40	14.974 458	199.635 112	13.331 709	40	21.724 522	259.056 519	11.924 613
41	16.022 670	214.609 570	13.394 120	41	23.462 483	280.781 040	11.967 235
42	17.144 257	230.632 240	13.452 449	42	25.339 482	304.243 523	12.006 699
43	18.344 355	247.776 496	13.506 962	43	27.366 640	329.583 005	12.043 240
44	19.628 460	266.120 851	13.557 908	44	29.555 972	356.949 646	12.077 074
45	21.002 452	285.749 311	13.605 522	45	31.920 449	386.505 617	12.108 402
46	22.472 623	306.751 763	13.650 020	46	34.474 085	418.426 067	12.137 409
47	24.045 707	329.224 386	13.691 608	47	37.232 012	452.900 152	12.164 267
48	25.728 907	353.270 093	13.730 474	48	40.210 573	490.132 164	12.189 136
49	27.529 930	378.999 000	13.766 799	49	43.427 419	530.342 737	12.212 163
50	29.457 025	406.528 929	13.800 746	50	46.901 613	573.770 156	12.233 485

Table 2

Normal Curve Table
Z = Z-score

An entry in the table is the area under the curve between $Z = 0$ and a positive value of Z. Areas for negative values of Z are obtained by symmetry.

Area corresponding to Z

Z	0.00	0.01	0.02	0.03	0.04	0.05	0.06	0.07	0.08	0.09
0.0	0.0000	0.0040	0.0080	0.0120	0.0160	0.0199	0.0239	0.0279	0.0319	0.0359
0.1	0.0398	0.0438	0.0478	0.0517	0.0557	0.0596	0.0636	0.0675	0.0714	0.0753
0.2	0.0793	0.0832	0.0871	0.0910	0.0948	0.0987	0.1026	0.1064	0.1103	0.1141
0.3	0.1179	0.1217	0.1255	0.1293	0.1331	0.1368	0.1406	0.1433	0.1480	0.1517
0.4	0.1554	0.1591	0.1628	0.1664	0.1700	0.1736	0.1772	0.1808	0.1844	0.1879
0.5	0.1915	0.1950	0.1985	0.2019	0.2054	0.2088	0.2123	0.2157	0.2190	0.2224
0.6	0.2257	0.2291	0.2324	0.2357	0.2389	0.2422	0.2454	0.2486	0.2517	0.2549
0.7	0.2580	0.2611	0.2642	0.2673	0.2703	0.2734	0.2764	0.2794	0.2823	0.2852
0.8	0.2881	0.2910	0.2939	0.2967	0.2995	0.3023	0.3051	0.3078	0.3106	0.3133
0.9	0.3159	0.3186	0.3212	0.3238	0.3264	0.3289	0.3315	0.3340	0.3365	0.3389
1.0	0.3413	0.3438	0.3461	0.3485	0.3508	0.3531	0.3554	0.3577	0.3599	0.3621
1.1	0.3642	0.3665	0.3686	0.3708	0.3729	0.3749	0.3770	0.3790	0.3810	0.3830
1.2	0.3849	0.3869	0.3888	0.3907	0.3925	0.3944	0.3962	0.3980	0.3997	0.4015
1.3	0.4032	0.4049	0.4066	0.4082	0.4099	0.4115	0.4131	0.4147	0.4162	0.4177
1.4	0.4192	0.4207	0.4222	0.4236	0.4251	0.4265	0.4279	0.4292	0.4306	0.4319
1.5	0.4332	0.4345	0.4357	0.4370	0.4382	0.4394	0.4406	0.4418	0.4429	0.4441
1.6	0.4452	0.4463	0.4474	0.4484	0.4495	0.4505	0.4515	0.4525	0.4535	0.4545
1.7	0.4554	0.4564	0.4573	0.4582	0.4591	0.4599	0.4608	0.4616	0.4625	0.4633
1.8	0.4641	0.4649	0.4656	0.4664	0.4671	0.4678	0.4686	0.4693	0.4699	0.4706
1.9	0.4713	0.4719	0.4726	0.4732	0.4738	0.4744	0.4750	0.4756	0.4761	0.4767
2.0	0.4772	0.4778	0.4783	0.4788	0.4793	0.4798	0.4803	0.4808	0.4812	0.4817
2.1	0.4821	0.4826	0.4830	0.4834	0.4838	0.4842	0.4846	0.4850	0.4854	0.4857
2.2	0.4861	0.4864	0.4868	0.4871	0.4875	0.4878	0.4881	0.4884	0.4887	0.4890
2.3	0.4893	0.4896	0.4898	0.4901	0.4904	0.4906	0.4909	0.4911	0.4913	0.4916
2.4	0.4918	0.4920	0.4922	0.4925	0.4927	0.4929	0.4931	0.4932	0.4934	0.4936
2.5	0.4938	0.4940	0.4941	0.4943	0.4945	0.4946	0.4948	0.4949	0.4951	0.4952
2.6	0.4953	0.4955	0.4956	0.4957	0.4959	0.4960	0.4961	0.4962	0.4963	0.4964
2.7	0.4965	0.4966	0.4967	0.4968	0.4969	0.4970	0.4971	0.4972	0.4973	0.4974
2.8	0.4974	0.4975	0.4976	0.4977	0.4977	0.4978	0.4979	0.4979	0.4980	0.4981
2.9	0.4981	0.4982	0.4982	0.4983	0.4984	0.4984	0.4985	0.4985	0.4986	0.4986
3.0	0.4987	0.4987	0.4987	0.4988	0.4988	0.4989	0.4989	0.4989	0.4990	0.4990

ANSWERS TO ODD-NUMBERED PROBLEMS

CHAPTER 1

Exercise 1.1 (page 14)

1. 0.5   **3.** 1.625   **5.** 1.333 . . .   **7.** 0.1666 . . .   **9.** 45%   **11.** 112%   **13.** 6%   **15.** 0.25%
17. 0.42   **19.** 0.002   **21.** 0.00001   **23.** 0.734   **25.** $\frac{1}{20}$ **27.** $\frac{3}{4}$ **29.** $\frac{26}{27}$ **31.** $\frac{2}{3}$ **33.** 150   **35.** 18
37. 105   **39.** $12.08   **41.** $x = 1$ **43.** $x = 6$ **45.** $x = -1$ **47.** $x = -4$ **49.** $x = 3$ **51.** $x = 1$
53. $x = 4$ or $x = -3$ **55.** $x = 3$ or $x = 2$ **57.** $x = 4$ or $x = -4$ **59.** $x = 9$ **61.** $x = \frac{1}{8}$
63. $x = -9$ **65.** $x = 2$ **67.** $x = 2$ **69.** $x = 5$ **71.** $\frac{1}{3} > 0.33$ **73.** $3 = \sqrt{9}$ **75.** $x \leq -1$

77. $x \geq -1$ **79.** $x \geq 1$ **81.** $x \leq -4$ **83.** $x < 3$ 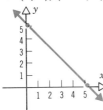   **85.** $x \leq 5$

87. $x < \frac{19}{6}$ **89.** $x < -\frac{8}{3}$ **91.** 700   **93.** 67.8%

Exercise 1.2 (page 26)

1. $A = (4, 2)$; $B = (6, 2)$; $C = (5, 3)$; $D = (-2, 1)$; $E = (-2, -3)$; $F = (3, -2)$; $G = (6, -2)$; $H = (5,0)$
3. 4   **5.** 1

7. $y = x - 3$

x	0	3	2	-2	4	-4
y	-3	0	-1	-5	1	-7

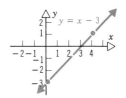

9. $2x - y = 6$

x	0	3	2	-2	4	-4
y	-6	0	-2	-10	2	-14

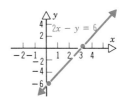

11. $6x - 4y = 1$ **13.** $6x - 4y = 5$ **15.** $2x - y = -7$ **17.** $x = -2$ **19.** $15x + 30y = -2$
21. $(0, 3), (3, 0)$ **23.** $(0, 6), (-30, 0)$ **25.** $(0, -\frac{10}{3}), (\frac{5}{3}, 0)$ **27.** $(0, -8), (2, 0)$ **29.** $(4, 0)$
31. $(0, 5), (5, 0)$ **33.** $(0, 4), (12, 0)$

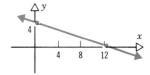

35. $(0, -8), (-6, 0)$

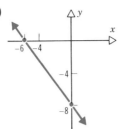

37. $(0, 5), (-\frac{5}{2}, 0)$

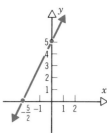

39. $(0, 8), (-40, 0)$

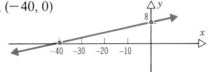

41. $(0, 9), (\frac{3}{2}, 0)$

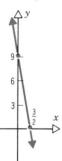

43. $(4, 0)$

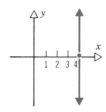

45.

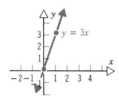

$y = 3x$

47.

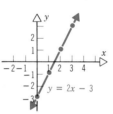

$y = 2x - 3$

49.

$y = 0$
$x - \text{axis}$

51.

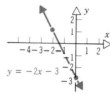

$y = -2x - 3$

53.

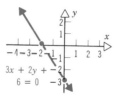

$3x + 2y + 6 = 0$

55.

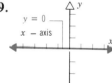

$y = -2x - 4$
$y = -2x - 3$

The lines are parallel.

57. 1 **59.** 2 **61.** $\frac{37}{14}$ **63.** $2x - y + 7 = 0$

65. $2x + 3y + 1 = 0$ **67.** $x - 2y + 5 = 0$

69. $3x + y - 3 = 0$ **71.** $x - 2y - 2 = 0$

73. $x - 1 = 0$ **75.** Slope $= \frac{3}{2}$, y-intercept $= -3$

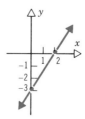

77. Slope $= -\frac{1}{2}$, y-intercept $= 2$

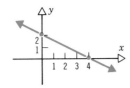

79. Slope undefined, no y-intercept

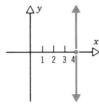

81. $x = 2y$ **83.** $x + y = 2$

85. $F° = \frac{9}{5}C° + 32$ or $C° = \frac{5}{9}(F° - 32)$;   **87.** $C = \$138.40$ **89.** $w = 5h - 190$
$\frac{5}{9}(70 - 32) = \frac{190}{9} = 21.111 \ldots$

91. (a) Oil $= -8t + 16,240$ (in thousands of barrels)   **93.** (a) $y = \frac{4}{3}t - 1546$
 (b) 2030   (b) 54%

95. (a) Sat $= -7t + 14,487$
 (b) 529

Exercise 1.3 (page 34)

1. $m_1 = m_2 = -1$ **3.** $m_1 = m_2 = \frac{2}{3}$ **5.** $(x, y) = (3, 2)$ **7.** $(x, y) = (9, \frac{13}{2})$ **9.** $(x, y) = (-\frac{1}{9}, \frac{7}{3})$
11. $(x, y) = (1, 2)$ **13.** $(x, y) = (0, 2)$ **15.** $(x, y) = (1, 3)$ **17.** $(x, y) = (-2, 0)$ **19.** No solution
21. Infinitely many solutions   **23.** One solution   **25.** No solution
27. $m_1 = \frac{1}{3}, m_2 = -3, m_1 m_2 = -1$ **29.** $m_1 = -\frac{1}{2}, m_2 = 2, m_1 m_2 = -1$
31. $m_1 = -\frac{1}{4}, m_2 = 4, m_1 m_2 = -1$ **33.** $y = 2x$ **35.** $3x = y$ **37.** $x + 2y = -5$
39. $15x + 30y = 19$ **41.** $t = 4$ **43.** $5x - 19y = 85$ **45.** $y = -3$
47. $m_1 = \dfrac{b}{a}, m_2 = -\dfrac{a}{b}, m_1 m_2 = -1$

Exercise 1.4 (page 42)

1. (a) $1000 + 180t$ (b) 1090   (c) 1180   (d) 1360   **3.** $x = 30$

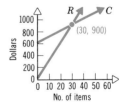

5. $x = 500$

7. $x = 1200$ items

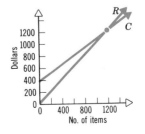

9. (a) \$80,000   (b) \$95,000   (c) \$105,000   (d) \$120,000   **11.** $p = \$1$ **13.** $p = \$10$

15. $p = 1$

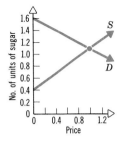

17. Costs equal for 8000 units. For 6000 units manufacturer should purchase components; at a cost of \$48,000.

19.

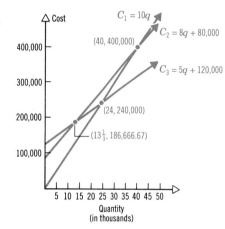

Chapter Review

True–False Questions (page 44)
1. F   **2.** T   **3.** T   **4.** F   **5.** F   **6.** T   **7.** F   **8.** T   **9.** F   **10.** F

Fill in the Blanks (page 45)
1. x-coordinate, y-coordinate   **2.** Repeating, terminates   **3.** Undefined, zero   **4.** Negative
5. Parallel   **6.** Identical   **7.** Perpendicular   **8.** Intersecting   **9.** A straight line   **10.** $t = aw + b$

Review Exercises (page 45)
1. $x = -7$ **3.** $x = -\frac{11}{3}$ **5.** $x = -11$ **7.** $x \le 3$ **9.** $x \ge -1$ **11.**

13.

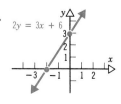

15. $x + 2y - 5 = 0$ **17.** $3x + 2y = 0$

19. $2x - y + 2 = 0$

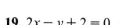

21. $x - y + 2 = 0$ **23.** Slope $= -\frac{9}{2}$
    y-intercept $= 9$

25. Slope $= -2$
    y-intercept $= \frac{9}{2}$

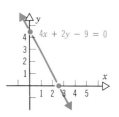

27. No solution   **29.** One solution   **31.** Infinitely many solutions

33. \$46,000 in bonds; \$44,000 in bank   **35. (a)**
    (b)

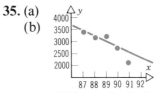

(c) $y - 2800 = -200(x - 90)$
(d) 2400

(Using only the last two digits of the date)

Mathematical Questions (page 47)
1. b **2.** d **3.** d **4.** b **5.** b **6.** b **7.** d **8.** c **9.** c **10.** c **11.** b **12.** b

CHAPTER 2

Exercise 2.1 (page 61)
1. $x = 16$ **3.** $x = 1$ **5.** $x = 3$ **7.** $x = 2$ **9.** $x = -4$ **11.** $x = 5$ **13.** $x = 7$ **15.** $x = 0$
    $y = 6$ $y = 1$ $y = -1$ $y = -4$ $y = 5$ $y = 2$ $y = \frac{3}{2}$ $y = 0$
17. $x = \frac{281}{95}$ **19.** $x = -\frac{427}{53}$ **21.** $x = 23$ **23.** $x = 2$ **25.** $x = 5$ **27.** No solution
    $y = -\frac{59}{19}$ $y = \frac{90}{53}$ $y = 7$ $y = -1$ $y = 3$
29. No solution   **31.** $x = 2$ **33.** $x = \frac{499}{182}$ **35.** $x = -2$
        $y = 1$ $y = -\frac{79}{26}$ $y = -7$
37. Infinitely many solutions (identical systems)   **39.** $x = 4$ **41.** $x = t$ **43.** $x = 8$
        $y = -\frac{1}{3}$ $y = -4t - 3$ $y = -2$
45. $x = -2$ **47.** $x = y = \frac{1}{2}$ **49.** No solution   **51.** $a = -8$ **53.** $x = 2$
    $y = 10$ $z = \frac{1}{4}$ $b = 7$ $y = -1$
            $c = \frac{1}{2}$ $z = 0$
55. $x = -1$ **57.** $x = 1$ **59.** $x = 1$
    $y = 6$ $y = 2$ $y = 1$
    $z = -2$ $z = -2$ $z = 1$
61. 20 caramels, 30 creams; increase the number of caramels to increase profits
63. \$31,250 in bonds; \$18,750 in savings certificates   **65.** 8 nickels
67. 30 cc of 15% acid; 70 cc of 5% acid   **69.** 3260 adults
71. First investment $=$ \$37,500; second investment $=$ \$12,500
73. 100 acres of corn, 900 acres of soya   **75.** 40 sets of \$20 sets, 160 sets of \$15 sets
77. 5 units of Food I   **79.** 4 assorted cartons   **81.** $x = 3$
    5 units of Food II   8 mixed cartons $y = 2$
    $\frac{2}{3}$ units of Food III   5 single cartons $z = 4$

Exercise 2.2 (page 74)

1. $\begin{bmatrix} 2 & -3 & | & 5 \\ 1 & -1 & | & 3 \end{bmatrix}$ **3.** $\begin{bmatrix} 2 & 1 & | & -6 \\ 1 & 1 & | & -1 \end{bmatrix}$ **5.** $\begin{bmatrix} 2 & -1 & -1 & | & 0 \\ 1 & -1 & -1 & | & 1 \\ 3 & -1 & 0 & | & 2 \end{bmatrix}$ **7.** $\begin{bmatrix} 2 & -3 & 1 & | & 7 \\ 1 & 1 & -1 & | & 1 \\ 2 & 2 & -3 & | & -4 \end{bmatrix}$

9. $\begin{bmatrix} 4 & -1 & 2 & -1 & | & 4 \\ 1 & 1 & 0 & 0 & | & -6 \\ 0 & 2 & -1 & 1 & | & 5 \end{bmatrix}$ **11.** $3x - y = 4$ **13.** $5y + 2z = 3$
$\qquad\qquad\qquad\qquad\qquad\qquad\qquad\quad 2x + 5y = 2 \qquad\qquad y + 3z = 1$
$\qquad\qquad\qquad\qquad\qquad\qquad\qquad\qquad\qquad\qquad\qquad\qquad x + 2y + 4z = 6$

15. $R_1 = r_2$ Interchange rows **17.** $R_1 = 4r_1$ **19.** $\begin{bmatrix} 1 & 0 & 2 \\ 0 & 1 & 4 \\ 3 & 6 & 9 \end{bmatrix}$ **21.** $\begin{bmatrix} 0 & 6 & 3 \\ 0 & 1 & 4 \\ 1 & 0 & 2 \end{bmatrix}$
$\quad\;\; R_2 = r_1$

23. $\begin{bmatrix} 3 & 6 & 9 \\ 1 & 3 & 7 \\ 1 & 0 & 2 \end{bmatrix}$ **25.** $x = 2$ **27.** $x = 2$ **29.** $x = 2$ **31.** $x = 2$ **33.** $x = \frac{1}{2}$ **35.** $x = \frac{3}{2}$
$\qquad\qquad\qquad\qquad\quad y = 4 \qquad y = 1 \qquad y = 1 \qquad y = -3 \qquad y = \frac{1}{3} \qquad y = \frac{2}{3}$

37. $x = 2$ **39.** $x = \frac{2}{3}$ **41.** $x = 1$ **43.** $x = -1$ **45.** $x = 2$ **47.** $x = \frac{1}{2}$ **49.** $x = \frac{1}{3}$
$\quad\;\; y = 3 \qquad\; y = \frac{1}{3} \qquad\; y = 4 \qquad y = 1 \qquad\; y = -1 \qquad y = \frac{1}{4} \qquad y = \frac{2}{3}$
$\qquad\qquad\qquad\qquad\qquad z = 0 \qquad\; z = 2 \qquad\; z = 1 \qquad\; z = \frac{3}{4} \qquad z = 1$

51. $x = 1, y = 2$ **53.** No solution **55.** Infinitely many solutions $(2x - 3y = 6)$ **57.** No solution
$\quad\;\; z = 0, w = 1$

59. Unique solution $(x = 1, y = 1)$ **61.** No solution $(3 \neq 0)$

63. Infinitely many solutions $(x = 0, y = -6 + z)$

65. Industry I $= 200$ **67.** $1000 invested at 6% **69.** $x = 20, y = 12, z = 6$
$\qquad$ Industry II $= 100 \qquad\quad$ $2200 invested at 7%
$\qquad$ Industry III $= 300 \qquad\quad$ $1800 invested at 8%

Exercise 2.3 (page 86)

1. Not in reduced row-echelon form   **3.** Not in reduced row-echelon form

5. Not in reduced row-echelon form   **7.** In reduced row-echelon form

9. In reduced row-echelon form   **11.** Infinitely many solutions   **13.** One solution

15. Infinitely many solutions   **17.** Infinitely many solutions   **19.** Infinitely many solutions

21. $x = 2, y = 1$ **23.** $x = 1, y = -3$ **25.** $x_1 = 3, x_2 = 2$ **27.** $x_1 = \frac{13}{4}, x_2 = -\frac{1}{2}$

29. $x_1 = 3, x_2 = 2, x_3 = -4$ **31.** $x_1 = -17, x_2 = 24, x_3 = 33, x_4 = 14$

33. There are infinitely many solutions, and we can solve for $x_1, x_2,$ and x_3 in terms of x_4: $x_1 = \frac{4}{3} + \frac{1}{3}x_4$;
$\quad\; x_2 = \frac{14}{15} + \frac{11}{15}x_4; x_3 = -\frac{16}{15} - \frac{4}{15}x_4$

35. The system is inconsistent.   **37.** The system is inconsistent.

39.

No. of Liters 10% Solution	No. of Liters 30% Solution	No. of Liters 50% Solution
55	15	30
50	25	25
45	35	20
40	45	15
35	55	10
30	65	5

41. 2 packages of first type **43.** 4 special pair packs **45.** 2000 of first species
     10 packages of second type   1 darkening pack $6000 - 2$ (# of third species)
     4 packages of third type   2 economy packs $0 \leq$ # of third species arbitrary ≤ 3000

Exercise 2.4 (page 97)

1. 2×2 **3.** 2×3 **5.** 2×1 **7.** 1×1
9. False. Two equal matrices must have the same dimensions. **11.** True **13.** True **15.** True

17. $x = 4, z = 3$ **19.** $x = 5, y = 1$ **21.** $\begin{bmatrix} 5 & 4 \\ 2 & 9 \end{bmatrix}$ **23.** $\begin{bmatrix} 26 & 12 \\ 4 & 4 \\ 14 & 0 \end{bmatrix}$ **25.** $\begin{bmatrix} 3 & -5 & -10 \\ -9 & 6 & -13 \\ 9 & 9 & 11 \\ 7 & -23 & 4 \end{bmatrix}$

27. $\begin{bmatrix} 3 & -5 & 4 \\ 5 & 3 & 3 \end{bmatrix}$ **29.** $\begin{bmatrix} 13 & -6 & -7 \\ -6 & 1 & -7 \end{bmatrix}$ **31.** $\begin{bmatrix} 9 & -5 & -6 \\ 1 & 1 & -3 \end{bmatrix}$ **33.** $\begin{bmatrix} -2 & -17 & 32 \\ 28 & 14 & 23 \end{bmatrix}$

35. $\begin{bmatrix} 5 & -2 & 3 \\ -12 & 1 & -5 \end{bmatrix}$ **37.** $x = 4, y = -11, z = 6$

39. (a) $\begin{bmatrix} \frac{5}{2} \\ -1 \\ 4 \end{bmatrix}$ (b) $\begin{bmatrix} \frac{3}{2} \\ -1 \\ 2 \end{bmatrix}$ (c) $\begin{bmatrix} \frac{5}{4} \\ -\frac{1}{2} \\ 2 \end{bmatrix}$ (d) $\begin{bmatrix} \frac{11}{2} \\ 6 \\ 4 \end{bmatrix}$ (e) $\begin{bmatrix} \frac{1}{2} \\ -2 \\ -1 \end{bmatrix}$ (f) $\begin{bmatrix} \frac{9}{8} \\ \frac{3}{2} \\ \frac{1}{2} \end{bmatrix}$

41.

	$\frac{1}{2}''$	$1''$	$2''$
Steel	25	45	35
Aluminum	13	20	23

or

	Steel	Aluminum
$\frac{1}{2}''$	25	13
$1''$	45	20
$2''$	35	23

43.

	<$15,000	>$15,000
Dem	351	203
Rep	271	215
Ind	73	55

or

	Dem	Rep	Ind
<$15,000	351	271	73
>$15,000	203	215	55

Exercise 2.5 (page 105)

1. Dimension $BA = 3 \times 4$ **3.** AB is not defined. **5.** $(BA)C$ is not defined.
7. $BA + A$ is defined. Dimension is 3×4 **9.** $DC + B$ is defined. Dimension is 3×3

11. $\begin{bmatrix} -1 & 10 & -8 \\ -4 & 16 & -8 \end{bmatrix}$ **13.** $\begin{bmatrix} 11 & 5 \\ 13 & -9 \end{bmatrix}$ **15.** $\begin{bmatrix} 6 & 10 \\ 8 & 2 \\ -4 & 5 \end{bmatrix}$ **17.** $\begin{bmatrix} -6 & 42 & -9 \\ 1 & 20 & 6 \\ -7 & 4 & -18 \end{bmatrix}$

19. $\begin{bmatrix} 3 & -1 \\ 4 & 2 \end{bmatrix}$ **21.** $\begin{bmatrix} 8 & 4 & 22 \\ 4 & 32 & 16 \end{bmatrix}$ **23.** $\begin{bmatrix} -14 & 7 \\ -20 & -6 \end{bmatrix}$ **25.** $\begin{bmatrix} 10 & 30 & 37 \\ 15 & 16 & 50 \\ -6 & 20 & -8 \end{bmatrix}$

27. $AB = \begin{bmatrix} 4 & -2 \\ 6 & 4 \end{bmatrix}$ $BA = \begin{bmatrix} 7 & -3 \\ 7 & 1 \end{bmatrix}$ **29.** $A = \begin{bmatrix} 2 & 1 \\ -\frac{1}{2} & -\frac{1}{2} \end{bmatrix}$

31. $\begin{bmatrix} 1 & 2 & 5 \\ 2 & 4 & 10 \\ -1 & -2 & -5 \end{bmatrix} \begin{bmatrix} 1 & 2 & 5 \\ 2 & 4 & 10 \\ -1 & -2 & -5 \end{bmatrix} = \begin{bmatrix} 0 & 0 & 0 \\ 0 & 0 & 0 \\ 0 & 0 & 0 \end{bmatrix}$

33. The possibilities are: $a = 0$, $b = 0$; $a = -1$, $b = 0$; $a = -\frac{1}{2}$, $b = \frac{1}{2}$; $a = -\frac{1}{2}$, $b = \frac{1}{2}$.   **35.** $[\frac{1}{3}, \quad \frac{2}{3}]$

37. $\begin{bmatrix} 1 & -3 \\ 9 & -2 \end{bmatrix}$ **39.** $\begin{bmatrix} 8 & 0 \\ 76 & 27 \end{bmatrix}$ **41.** $\begin{bmatrix} -56 & -27 \\ -4 & -18 \end{bmatrix}$ **43.** A is a 3×3 matrix

45. (a) PQ represents the matrix of raw materials needed to fill order: $PQ = [138 \ 189 \ 97 \ 399]$

(b) QC represents the matrix of costs to produce each product: $QC = \begin{bmatrix} 311 \\ 653 \\ 614 \end{bmatrix}$

(c) PQC represents the total cost to produce the order: $PQC = 13{,}083$

Exercise 2.6 (page 116)

1. $\begin{bmatrix} 1 & 2 \\ 2 & 3 \end{bmatrix}\begin{bmatrix} -3 & 2 \\ 2 & -1 \end{bmatrix} = \begin{bmatrix} 1 & 0 \\ 0 & 1 \end{bmatrix} = I_2$ **3.** $\begin{bmatrix} -1 & -2 \\ 3 & 4 \end{bmatrix}\begin{bmatrix} 2 & 1 \\ -\frac{3}{2} & -\frac{1}{2} \end{bmatrix} = \begin{bmatrix} 1 & 0 \\ 0 & 1 \end{bmatrix} = I_2$

5. $\begin{bmatrix} 1 & 2 & 3 \\ 2 & 3 & 4 \\ 1 & 2 & 1 \end{bmatrix}\begin{bmatrix} -\frac{5}{2} & 2 & -\frac{1}{2} \\ 1 & -1 & 1 \\ \frac{1}{2} & 0 & -\frac{1}{2} \end{bmatrix} = \begin{bmatrix} 1 & 0 & 0 \\ 0 & 1 & 0 \\ 0 & 0 & 1 \end{bmatrix} = I_3$ **7.** $\begin{bmatrix} 3 & -5 \\ -1 & 2 \end{bmatrix}$ **9.** $\begin{bmatrix} 4 & -1 \\ 3 & -1 \end{bmatrix}$

11. $\begin{bmatrix} 1.5 & -0.5 \\ -2 & 1 \end{bmatrix}$ **13.** $\begin{bmatrix} 0 & 0 & 1 \\ 0 & 1 & 0 \\ 1 & 0 & 0 \end{bmatrix}$ **15.** $\begin{bmatrix} \frac{4}{9} & \frac{1}{9} & \frac{1}{9} \\ \frac{4}{3} & -\frac{2}{3} & \frac{1}{3} \\ \frac{7}{9} & -\frac{5}{9} & \frac{4}{9} \end{bmatrix}$ **17.** $\begin{bmatrix} 1 & -1 & 2 \\ -1 & 2 & -3 \\ -1 & 1 & -1 \end{bmatrix}$

19. $\begin{bmatrix} 2 & -1 & -1 & -2 \\ -1 & 1 & 1 & 2 \\ -2 & 1 & 2 & 3 \\ -1 & 1 & 1 & 1 \end{bmatrix}$ **21.** $\begin{bmatrix} 4 & 6 & | & 1 & 0 \\ 2 & 3 & | & 0 & 1 \end{bmatrix} \rightarrow \begin{bmatrix} 4 & 6 & | & 1 & 0 \\ \boxed{0 \quad 0} & | & -\frac{1}{2} & 0 \end{bmatrix}$

23. $\begin{bmatrix} -8 & 4 & | & 1 & 0 \\ -4 & 2 & | & 0 & 1 \end{bmatrix} \rightarrow \begin{bmatrix} -8 & 4 & | & 1 & 0 \\ \boxed{0 \quad 0} & | & -\frac{1}{2} & 0 \end{bmatrix}$ **25.** $\begin{bmatrix} 1 & 1 & 1 & | & 1 & 0 & 0 \\ 3 & -4 & 2 & | & 0 & 1 & 0 \\ 0 & -0 & -0 & | & 0 & 0 & 1 \end{bmatrix}$ **27.** $\begin{bmatrix} 2 & -1 \\ -1 & 1 \end{bmatrix}$

29. $\begin{bmatrix} \frac{1}{3} & \frac{1}{3} \\ 0 & \frac{1}{2} \end{bmatrix}$ **31.** No inverse   **33.** $x = 2$, $y = 4$ **35.** $x = 2$, $y = 1$ **37.** $x = \frac{1}{2}$, $y = \frac{1}{3}$

39. $x = 2$, $y = 3$ **41.** $x = 1$, $y = 1$ **43.** $x = 2$, $y = 1$ **45.** $x = 1$, $y = 4$, $z = 0$

47. $x = 2$, $y = -1$, $z = 1$ **49.** $x = \frac{1}{3}$, $y = \frac{2}{3}$, $z = 1$ **51.** $x_1 = 2$, $x_2 = 1$, $x_3 = -1$

53. $x_1 = 2$, $x_2 = -1$, $x_3 = 3$ **55.** $x = 7$, $y = -1$, $z = -4$ **57.** $x_1 = 3$, $x_2 = -2$, $x_3 = 1$, $x_4 = 1$

59. $x_1 = 2$, $x_2 = 1$, $x_3 = -1$, $x_4 = 0$

Exercise 2.7A (page 124)

1. $A = C = \$10{,}000$, $B = \frac{3}{4}C = \$7500$ **3.** $A = \frac{6}{15}C$, $B = \frac{11}{15}C$, where $C = \$10{,}000$ **5.** $\begin{bmatrix} 203 \\ 166.977 \\ 137.847 \end{bmatrix}$

7. $x_1 = 0.8x_4 = 8000$, $x_2 = 0.72x_4 = 7200$, $x_3 = 0.48x_4 = 4800$, $x_4 = \$10{,}000$ **9.** $x = \begin{bmatrix} 160 \\ 75.38 \end{bmatrix}$

Exercise 2.7B (page 128)

1. (a) Using I: 41 23 70 45 41 23 62 41 64

36 19 11 59 39 7 4 94 60

Using II: 13 69 −21 20 98 −35 1 123 18
              8 44 −13 1 32 0 1 141 24

(b) Using I: 85 50 71 43 90 54 89 61 43
            24 59 32 45 24 69 41 67 43

Using II: 20 140 −27 15 153 −12 15 138 −16
              5 62 −5 18 132 −21 13 139 −7

(c) Using I: 64 36 49 31 75 47 65 37 72 43 75
            47 57 35 77 46 95 57 24 13 39 22

Using II: 20 93 −35 13 145 −7 19 141 −23 14 146 −9
              9 120 −2 15 159 −11 9 89 −6 5 171 16

3. You flunk

Exercise 2.7C (page 133)

1. $\begin{bmatrix} 4 & 3 \\ 1 & 1 \\ 2 & 0 \end{bmatrix}$ **3.** $\begin{bmatrix} 1 & 0 & 1 \\ 11 & 12 & 4 \end{bmatrix}$ **5.** [8   6   3]   **7.** (a) $y = \frac{54}{35}x + \frac{27}{5}$ (b) 17.74   **9.** $y = \frac{334}{223}x + \frac{8058}{223}$

11. (2) is symmetric; yes

Chapter Review

True–False Questions (page 135)

1. T   **2.** F   **3.** F   **4.** T   **5.** T   **6.** F   **7.** F   **8.** T   **9.** F

Fill in the Blanks (page 135)

1. 3×2 **2.** One, infinite   **3.** Rows, columns   **4.** Inverse   **5.** Columns, rows   **6.** Row operations
7. Zero matrix   **8.** Identity matrix   **9.** 3×3 **10.** 5×5

Review Exercises (page 136)

1. $\begin{bmatrix} -1 & 3 & 16 \\ 3 & 15 & 8 \\ 5 & 10 & 29 \end{bmatrix}$ **3.** $\begin{bmatrix} -3 & 9 & 48 \\ 9 & 45 & 24 \\ 15 & 30 & 87 \end{bmatrix}$ **5.** $\begin{bmatrix} -9 & -9 & -6 \\ -3 & 3 & -6 \\ -3 & -6 & 39 \end{bmatrix}$ **7.** $\begin{bmatrix} -20 & 0 & 70 \\ 10 & 80 & 30 \\ 20 & 40 & 210 \end{bmatrix}$

9. $\begin{bmatrix} -\frac{7}{2} & -\frac{3}{2} & \frac{25}{2} \\ 3 & \frac{9}{2} & \frac{11}{2} \\ -\frac{37}{2} & -10 & 19 \end{bmatrix}$ **11.** $\begin{bmatrix} 19 & 36 & 38 \\ 26 & 77 & 73 \\ 73 & 160 & 206 \end{bmatrix}$ **13.** $\begin{bmatrix} 16 & 32 & 27 \\ 16 & 10 & 19 \\ -104 & -80 & -113 \end{bmatrix}$ **15.** $\begin{bmatrix} \frac{1}{3} & 0 \\ \frac{2}{3} & 1 \end{bmatrix}$

17. $\begin{bmatrix} 1 & -3 & 2 \\ -3 & 3 & -1 \\ 2 & -1 & 0 \end{bmatrix}$ **19.** $\begin{bmatrix} \frac{1}{16} & \frac{1}{32} & \frac{1}{4} \\ \frac{3}{16} & \frac{3}{32} & -\frac{1}{4} \\ -\frac{3}{16} & \frac{13}{32} & \frac{1}{4} \end{bmatrix}$ **21.** $\left[\begin{array}{ccc|ccc} 1 & 2 & -3 & 1 & 0 & 0 \\ 0 & -2 & 14 & -4 & 1 & 0 \\ 0 & 0 & 0 & 0 & 0 & 1 \end{array}\right]$

The matrix has no inverse.

23. $\left[\begin{array}{ccc|c} 1 & 1 & -1 & 2 \\ 0 & 1 & -1 & 1 \\ 0 & 0 & 0 & \frac{1}{2} \end{array}\right]$ **25.** $x_1 = \frac{23}{16}, x_2 = -\frac{73}{32}, x_3 = -\frac{9}{4}$ **27.** $x_1 = 29, x_2 = 8, x_3 = -24$

There is no solution.

29. $x_1 = \frac{10}{7} + \frac{3}{7}x_3, x_2 = -\frac{9}{7} + \frac{5}{7}x_3$; $x_1 = \frac{10}{7}, x_2 = -\frac{9}{7}, x_3 = 0$; $x_1 = \frac{13}{7}, x_2 = -\frac{4}{7}, x_3 = 1$; $x_1 = \frac{16}{7}, x_2 = \frac{1}{7}, x_3 = 2$

31. $x_1 = 1 - \frac{3}{5}x_3,\ x_2 = 2 + \frac{4}{5}x_3;\ x_1 = 1,\ x_2 = 2,\ x_3 = 0;\ x_1 = \frac{2}{5},\ x_2 = \frac{14}{5},\ x_3 = 1;\ x_1 = -\frac{1}{5},\ x_2 = \frac{18}{5},\ x_3 = 2$

33. $\begin{bmatrix} 1 & 0 & | & 1 \\ 0 & 1 & | & \frac{1}{2} \\ \underline{0} & \underline{0} & | & \frac{7}{2} \end{bmatrix}$ There is no solution.   **35.** $y = -z,\ x = w$ **37.** A correct answer.

CHAPTER 3

Exercise 3.2 (page 148)

1. $x \geq 0$

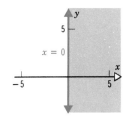

3. $x \geq 0,\ y \geq 0$

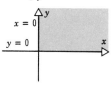

5. $2x - 3y \leq -6$

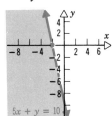

7. $5x + y \leq -10$

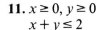

9. $x \geq 5$

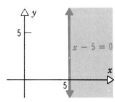

11. $x \geq 0,\ y \geq 0$ Bounded,
$x + y \leq 2$ Vertices: $(0, 0)$, $(0, 2)$, $(2,0)$

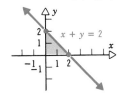

13. $x \geq 0,\ y \geq 0$ Bounded,
$x + y \geq 2$ Vertices: $(0, 2)$,
$2x + 3y \leq 6$ $(2, 0)$, $(3, 0)$

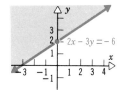

15. $x \geq 0,\ y \geq 0$ Bounded,
$2 \leq x + y,\ x + y \leq 8$ Vertices: $(0, 2)$, $(0, 8)$,
$2x + y \leq 10$ $(2, 6)$, $(5, 0)$, $(2, 0)$

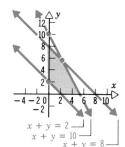
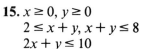

17. $x \geq 0$, $y \geq 0$
$x + y \geq 2$, $2x + 3y \leq 12$
$3x + y < 12$
Bounded, Vertices:
$(0, 2)$, $(0, 4)$, $(\frac{22}{7}, \frac{12}{7})$, $(4, 0)$, $(2, 0)$

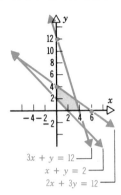

$3x + y = 12$
$x + y = 2$
$2x + 3y = 12$

19. $x \geq 0$, $y \geq 0$
$1 \leq x + 2y$, $x + 2y \leq 10$
Bounded, Vertices: $(0, 1)$,
$(0, 5)$, $(10, 0)$, $(1, 0)$

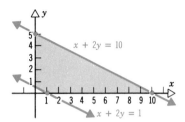

$x + 2y = 10$
$x + 2y = 1$

23. $x \geq 0$, $y \geq 0$ Bounded,
$8x + y \leq 20$ Vertices: $(0, 0)$ $(0, \frac{8}{5})$
$6x - 5y \geq -8$ $(2, 4)$, $(\frac{5}{2}, 0)$

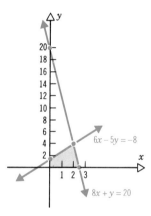

$6x - 5y = -8$
$8x + y = 20$

21. $x \geq 0$, $y \geq 0$ Unbounded,
$x + 2y \geq 5$ Vertices: $(0, 16)$,
$5x + y \geq 16$ $(3, 1)$, $(5, 0)$

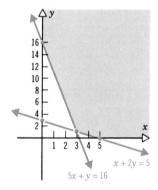

$x + 2y = 5$
$5x + y = 16$

27. $x \geq 0$, $y \geq 0$ Unbounded
$x + 2y \geq 12$ Vertices: $(2, 6)$
$x + y \geq 8$ $(4, 4)$, $(10, 1)$
$x \geq 2$
$y \geq 1$

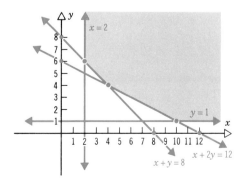

$x = 2$
$y = 1$
$x + 2y = 12$
$x + y = 8$

25. $x \geq 0$, $y \geq 0$ Bounded,
$3x + 5y \leq 38$ Vertices: $(0, 6)$
$x - y \leq -6$ $(0, \frac{38}{5})$, $(1, 7)$

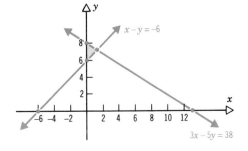

$x - y = -6$
$3x - 5y = 38$

29. Letting x = # low-grade packages
$\qquad y$ = # high-grade packages
$\qquad x \geq 0, y \geq 0$
$\qquad 4x + 8y \leq 1600 \Longleftrightarrow x + 2y \leq 400$
$\qquad 12x + 8y \leq 1920 \Longleftrightarrow 3x + 2y \leq 480$

Bounded,
Vertices: (0, 0), (40, 180)
(0, 200), (160, 0)

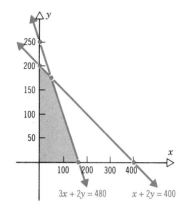

31. Let x = no. of dumpsters
$\qquad y$ = no. of tankers
$\qquad x \geq 0, y \geq 0$
$\qquad 3x + 2y \leq 80$
$\qquad 4x + 3y \leq 120$

Vertices: (0, 0), (0, 40), $(\frac{80}{3}, 0)$

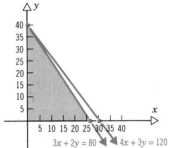

33. Let x = no. of units of first grain
$\qquad y$ = no. of units of second grain
$\qquad x \geq 0, y \geq 0$
$\qquad x + 2y \geq 5$
$\qquad 5x + y \geq 16$

Vertices: (0, 16), (3, 1), (5, 0)

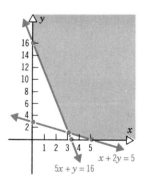

35. Let x = no. of ounces of food A
$\qquad y$ = no. of ounces of food B
$\qquad x \geq 0, y \geq 0$
$\qquad 5x + 4y \geq 85$
$\qquad 3x + 3y \geq 70$
$\qquad 2x + 3y \geq 50$

Vertices: $(0, \frac{70}{3})$, $(20, \frac{10}{3})$, (25, 0)

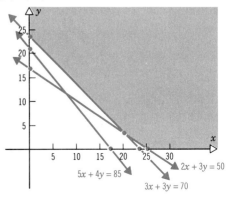

Exercise 3.3 (page 163)

1. The maximum is 38 at (7, 8).   **3.** The maximum is 71 at (7, 8).
The minimum is 10 at (2, 2).   The minimum is 16 at (8, 1).
5. The maximum is 55 at (7, 8).
The minimum is 14 at points on the line segment joining (2, 2) and (8, 1).
7. The minimum is 14 at (4, 2).   **9.** The minimum is 12 at (0, 6).
11. The minimum is 20 at (4, 2).   **13.** The maximum is 14 at (0, 2).
15. The maximum is 15 at (3, 0).   **17.** The maximum is 56 at (0, 8).
19. $z = 5x + 7y$ has no maximum under the conditions.   **21.** The minimum is 4 at (2, 0).
23. The minimum is 4 at (2, 0).   **25.** The minimum is $\frac{3}{2}$ at (0, $\frac{1}{2}$).
27. The minimum is $\frac{20}{3}$ at ($\frac{10}{3}$, $\frac{10}{3}$).   **29.** The minimum is 20 at (0, 10); the maximum is 50 at (10, 0).
31. The minimum is $\frac{70}{3}$ at ($\frac{10}{3}$, $\frac{10}{3}$); the maximum is 40 at (0, 10).   **33.** The maximum is 26 at (6, 10).
35. The minimum is 38 at (6, 4).   **37.** The maximum is $\frac{191}{2}$ at ($\frac{7}{2}$, $\frac{13}{2}$).
39. 90 packages of the low-grade mixture and 105 packages of the high-grade mixture.
41. The maximum income is $2550 with $7500 invested in first security and $22,500 invested in second security.
43. 15 units of the first product and 25 units of the second product maximizes profit at $2100.
45. Two weeks for the first repairman and three weeks for the second repairman minimizes costs at $1150.
47. The maximum legal profit is $1450, obtained with a production of 700 units of A and 400 units of B.
49. He should add 5 ounces of Supplement I and 1 ounce of Supplement II to each 100 ounces of feed.
51. The 500 pounds of picnic patties and $83\frac{1}{3}$ pounds of hamburger should be made.
53. The price of the high-grade carpet should be between $520 and $620 per roll.

Chapter Review

True – False Questions (page 167)
1. T   **2.** F   **3.** T   **4.** T   **5.** T   **6.** F

Fill in the Blanks (page 168)
1. Half plane   **2.** Objective function   **3.** Feasible   **4.** Bounded   **5.** Vertex

Review Exercises (page 168)
1. $x - 3y < 0$ **3.** $5x + y \geq 10$ **5.** $x \geq 0, y \geq 0, 3x + 2y \leq 12, x + y \geq 1$
Bounded; Vertices (0, 1),
(0, 6), (4, 0), (1, 0)

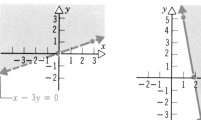

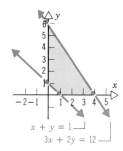

7. $x \geq 0$, $y \geq 0$, $x + 2y \geq 4$, $3x + y \geq 6$
Unbounded; Vertices $(0, 2)$, $(0, 6)$,
$(\frac{8}{5}, \frac{6}{5})$, $(4, 0)$, $(2, 0)$

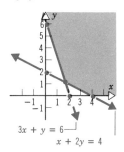

$3x + y = 6$
$x + 2y = 4$

9. $x \geq 0$, $y \geq 0$, $3x + 2y \geq 6$, $3x + 2y \leq 12$, $x + 2y \leq 8$
Bounded; Vertices $(0, 3)$,
$(2, 3)$, $(4, 0)$, $(2, 0)$

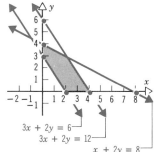

$3x + 2y = 6$
$3x + 2y = 12$
$x + 2y = 8$

Problems 11 – 18 use the following graph:
$x \geq 0$, $y \geq 0$, $x + 2y \leq 40$, $x + y \geq 10$
The vertices are $(0, 10)$, $(0, 20)$, $(\frac{40}{3}, \frac{40}{3})$, $(20, 0)$, $(10, 0)$.

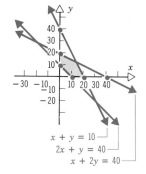

$x + y = 10$
$2x + y = 40$
$x + 2y = 40$

11. Maximum is $\frac{80}{3}$ at $(\frac{40}{3}, \frac{40}{3})$.   **13.** Minimum is 20 at $(0, 10)$.
15. Maximum is 40 at any point on the line segment between $(\frac{40}{3}, \frac{40}{3})$ and $(20, 0)$.
17. Maximum is 20 at $(10, 0)$.
19. Maximum is 235 at $(5, 8)$; minimum is 60 at any point on the line segment between $(0, 3)$ and $(4, 0)$.
21. Maximum is 195 at $(5, 6)$; minimum is 150 at $(2, 6)$.   **23.** The maximum is 1950 at $(250, 566\frac{2}{3})$.
25. The maximum profit is \$160 and occurs for a production of 20 dozen cans of food A and 10 dozen cans of food B.
27. She should buy 7.5 pounds of A and 11.25 pounds of B.
29. Maximum profit of \$1760 with 8 downhill skis and 24 cross-country skis.

Mathematical Questions (page 170)
 1. b **2.** a **3.** c **4.** c **5.** d **6.** c

CHAPTER 4

Exercise 4.1 (page 182)
 1. Standard form   **3.** Not in standard form   **5.** Not in standard form   **7.** Not in standard form
 9. Standard form   **11.** Cannot be modified so as to be in standard form.
13. Cannot be modified so as to be in standard form.   **15.** Can be modified to:

Maximize $P = 2x_1 + x_2 + 3x_3$
Subject to the constraints
$x_1 - x_2 - x_3 \leq 6$ $\quad x_1 \geq 0$, $x_2 \geq 0$
$-2x_1 + 3x_2 \leq 12$ $\quad x_3 \geq 0$

17.

	x_1	x_2	x_3	s_1	s_2	s_3	P	
	5	2	1	1	0	0	0	20
	6	1	4	0	1	0	0	24
	1	1	4	0	0	1	0	16
	-2	-1	-3	0	0	0	1	0

19.

	x_1	x_2	s_1	s_2	s_3	P	
	2.2	-1.8	1	0	0	0	5
	0.8	1.2	0	1	0	0	2.5
	1	1	0	0	1	0	0.1
	-3	-5	0	0	0	1	0

21.

	x_1	x_2	x_3	s_1	s_2	P	
	1	1	1	1	0	0	50
	3	2	1	0	1	0	10
	-2	-3	-1	0	0	1	0

23.

	x_1	x_2	x_3	s_1	s_2	s_3	P	
	3	1	4	1	0	0	0	5
	1	1	0	0	1	0	0	5
	2	-1	1	0	0	1	0	6
	-3	-4	-2	0	0	0	1	0

Exercise 4.2 (page 197)

1. $s_1 = 300 - x_1 - 2x_2$
$s_2 = 480 - 3x_1 - 2x_2$
$P = x_1 + x_2$

$$\begin{bmatrix} \frac{1}{2} & 1 & \frac{1}{2} & 0 & 0 & 150 \\ 2 & 0 & -1 & 1 & 0 & 180 \\ \hline 0 & 0 & 1 & 0 & 1 & 300 \end{bmatrix}$$

$x_2 = 150 - \frac{1}{2}x_1 - \frac{1}{2}s_1$
$s_2 = 180 - 2x_1 + s_1$
$P = 300 - s_1$

3. $s_1 = 24 - x_1 - 2x_2 - 4x_3$
$s_2 = 32 - 2x_1 + x_2 - x_3$
$s_3 = 18 - 3x_1 - 2x_2 - 4x_3$
$P = x_1 + 2x_2 + 3x_3$

$$\begin{bmatrix} -2 & 0 & 0 & 1 & 0 & -1 & 0 & 6 \\ \frac{7}{2} & 0 & 3 & 0 & 1 & \frac{1}{2} & 0 & 41 \\ \frac{3}{2} & 1 & 2 & 0 & 0 & \frac{1}{2} & 0 & 9 \\ \hline 2 & 0 & 1 & 0 & 0 & 1 & 1 & 18 \end{bmatrix}$$

$s_1 = 6 + 2x_1 + s_3$
$s_2 = 41 - \frac{7}{2}x_1 - 3x_3 - \frac{1}{2}s_3$
$x_2 = 9 - \frac{3}{2}x_1 - 2x_3 - \frac{1}{2}s_3$
$P = 18 - 2x_1 - x_3 - s_3$

5. $s_1 = 20 + 3x_1 - x_3$
$s_2 = 24 - 2x_1 - x_4$
$s_3 = 28 + 3x_2 - x_3$
$s_4 = 24 + 3x_2 - x_4$
$P = x_1 + 2x_2 + 3x_3 + 4x_4$

$$\begin{bmatrix} 0 & 0 & 1 & \frac{3}{2} & 1 & \frac{3}{2} & 0 & 0 & 0 & 56 \\ 1 & 0 & 0 & \frac{1}{2} & 0 & \frac{1}{2} & 0 & 0 & 0 & 12 \\ 0 & -3 & 1 & 0 & 0 & 0 & 1 & 0 & 0 & 28 \\ 0 & -3 & 0 & 1 & 0 & 0 & 0 & 1 & 0 & 24 \\ \hline 0 & -2 & -3 & -\frac{7}{2} & 0 & \frac{1}{2} & 0 & 0 & 1 & 12 \end{bmatrix}$$

$s_1 = 56 - x_3 - \frac{3}{2}x_4 - s_2$
$x_1 = 12 - \frac{1}{2}x_4 - \frac{1}{2}s_2$
$s_3 = 28 - 3x_2 - x_3$
$s_4 = 24 + 3x_2 - x_4$
$P = 12 + 2x_2 + 3x_3 + \frac{7}{2}x_4 - \frac{1}{2}s_2$

7. (b) Requires additional pivoting; the pivot element is in row 1, column 1.

9. (a) Final tableau; $P = \frac{256}{7}$, $x_1 = \frac{32}{7}$, $x_2 = 0$.

11. (c) No solution; all entries in the pivot column are negative.

13. $P = \frac{204}{7}$, $x_1 = \frac{24}{7}$, $x_2 = \frac{12}{7}$ **15.** $P = 8$, $x_1 = \frac{2}{3}$, $x_2 = \frac{2}{3}$ **17.** $P = 6$, $x_1 = 2$, $x_2 = 0$, $x_3 = 0$

19. No solution, since all the ratios for column two are negative.

21. $P = 30$, $x_1 = 0$, $x_2 = 0$, $x_3 = 10$ **23.** $P = 42$, $x_1 = 1$, $x_2 = 10$, $x_3 = 0$, $x_4 = 0$

25. $P = 40$, $x_1 = 20$, $x_2 = 0$, $x_3 = 0$ **27.** $P = 800$, $x_1 = 80$, $x_2 = x_3 = 0$

29. $P = 50$, $x_1 = 0$, $x_2 = 15$, $x_3 = 5$, $x_4 = 0$ **31.** $P = 190$, $A = 0$, $B = 40$, $C = 75$

33. $P = \$275,000$ for 200,000 gallons of regular, 0 gallons of premium, and 25,000 gallons of super-premium.

35. $P = \$7500$ with $\$45,000$ in stocks, $\$15,000$ in corporate bonds, and $\$30,000$ in municipal bonds.

37. $P = \$14,400$, $A = 180$, $B = 20$, $C = 0$

39. $P = \$1500$, 300 of Jean I, 0 of Jean II, and 100 of Jean III.

41. $x_1 =$ number of cans of Can I; $x_2 =$ number of cans of Can II; $x_3 =$ number of cans of Can III; $P =$ maximum revenue. $P = 800$, $x_1 = 0$, $x_2 = 100$, $x_3 = 20$.

43. $\$12,000$ is the maximum profit when there are 1200 television console cabinets, 0 stereo system cabinets, and 0 radio cabinets.

45. $\$30,000$ is the maximum profit when 0 TVs are made in Chicago, 375 TVs are made in New York, and 0 TVs are made in Denver.

Exercise 4.3 (page 212)

1. Standard form   **3.** Not in standard form   **5.** Standard form

7. Maximize $P = 2y_1 + 6y_2$ subject to $y_1 + 2y_2 \le 2$, $y_1 + 3y_2 \le 3$, $y_1 \ge 0$.

9. Maximize $P = 5y_1 + 4y_2$ subject to $y_1 + 2y_2 \le 3$, $y_1 + y_2 \le 1$, $y_1 \le 1$, $y_1 \ge 0$, $y_2 \ge 0$.

11. $C = 6$, $x_1 = 0$, $x_2 = 2$ **13.** $C = 12$, $x_1 = 0$, $x_2 = 4$ **15.** $C = \frac{21}{5}$, $x_1 = \frac{8}{5}$, $x_2 = 0$, $x_3 = \frac{13}{5}$

17. $C = 5$, $x_1 = 1$, $x_2 = 1$, $x_3 = 0$, $x_4 = 0$

19. The minimum cost is $\$7.50$, when there are 0 units of Food I, 7.5 units of Food II, and 0 units of Food III.

21. The minimum cost is $\$0.22$, when 2 P pills and 4 Q pills are taken as supplements.

23. The minimum cost is $\$290$, when $A = 20$ units, $B = 30$ units, and $C = 150$ units.

25. The minimum cost is $\$70.30$ when 4 Lunch no. 1 and 5 Lunch no. 3 are ordered.

Exercise 4.4 (page 224)

1. $P = 20$, $x_1 = 5$, $x_2 = 0$ **3.** $P = 100$, $x_1 = 0$, $x_2 = 10$ **5.** $P = 30$, $x_1 = 10$, $x_2 = 0$

7. $C = 275$, $x_1 = 25$, $x_2 = 100$ **9.** $C = 12$, $x_1 = 4$, $x_2 = 6$ **11.** $C = 30$, $x_1 = 3$, $x_2 = 9$

13. $P = 44$, $x_1 = 4$, $x_2 = 8$ **15.** $C = 14$, $x_1 = 4$, $x_2 = 2$, $x_3 = 10$

17. $P = 27$, $x_1 = 9$, $x_2 = 0$, $x_3 = 0$ **19.** $C = \frac{20}{3}$, $x_1 = 0$, $x_2 = 0$, $x_3 = \frac{20}{3}$ **21.** $P = 7$, $x_1 = 1$, $x_2 = 2$

23. $C = 48$, $x_1 = 8$, $x_2 = 2$ **25.** $P = 20$, $x_1 = 0$, $x_2 = 0$, $x_3 = 1$

27. $C = \frac{74}{9}$, $x_1 = \frac{19}{54}$, $x_2 = \frac{7}{3}$, $x_3 = \frac{29}{27}$, $x_4 = \frac{25}{18}$

29. The minimum cost is $\$150,000$ when 100 units are shipped from M1 to A1, 300 units are shipped from M1 to A2, 400 units are shipped from M2 to A1, and 0 units are shipped from M2 to A2.

31. 55 units $W_1 \to R_1$, 75 units $W_2 \to R_2$; Cost $= \$965$

33. The minimum cost is $\$120$ for both $A = 0$ with $B = 15$, and for $A = 2$ with $B = 14$.

Chapter Review

True–False Questions (page 228)

1. T   **2.** F   **3.** T   **4.** F   **5.** T

Fill in the Blanks (page 228)

1. Slack variables   **2.** Column   **3.** $\ge$ **4.** Dual   **5.** Von Neuman duality principle

Review Exercises (page 229)

1. (a) Further pivoting needed.   (b) Final tableau.   (c) Further pivoting needed.

3. $C = 7$, $x_1 = 3$, $x_2 = 1$ **5.** $P = 22$, 500, $x = 0$, $y = 100$, $z = 50$

7. $P = 352$, $x_1 = 0$, $x_2 = \frac{6}{5}$, $x_3 = \frac{28}{5}$ **9.** $C = 350$, $x_1 = 0$, $x_2 = 50$, $x_3 = 50$

11. $P = 150$, $x_1 = 0$ $x_2 = 50$ **13.** $P = 12{,}250$, $x_1 = 0$, $x_2 = 5$, $x_3 = 25$

15. The maximum profit is $\approx \$5714.29$ for cultivation of 0 acres of corn, 0 acres of wheat, and $\frac{1000}{7} \approx 143$ acres of soybeans.

  1. c **2.** d **3.** c **4.** a **5.** b **6.** a **7.** c **8.** d

CHAPTER 5

Exercise 5.1 (page 239)
  1. $10   **3.** 10%   **5.** $110   **7.** 10%   **9.** $25   **11.** $45   **13.** $150   **15.** 10%   **17.** $33\frac{1}{3}$%   **19.** $13\frac{1}{3}$%
21. $P = \dfrac{I}{rt}$ **23.** $P = \dfrac{A}{1 + rt}$ **25.** $A = \dfrac{P}{1 - rt}$ **27.** $1160   **29.** $1680   **31.** $A = \$1263.16$
33. $A = \$1428.57$ **35.** Simple discount rate of 9% is better.   **37.** $471.70
39. (a) $850   (b) $3400   **41.** $166.67 goes for principal and $100 toward interest.
43. Since the interest charged on the simple loan is $34, while the interest charged on the discounted loan is $35.41, she should take the simple loan rate.
45. $30,000   **47.** 8.4%

Exercise 5.2 (page 250)
  1. 1.104   **3.** 1.442   **5.** 1.077   **7.** 1%   **9.** 3.5%   **11.** 12%   **13.** 8%   **15.** $1348.18   **17.** $545
19. $854.36   **21.** $95.14   **23.** $456.97   **25.** (a) $1295.03   (b) $1308.65   (c) $1302.26
27. (a) $1266.77   (b) $1425.76   (c) $1604.71   **29.** (a) $3947.05   (b) $3115.83
31. (a) 8.16%   (b) 12.68%   **33.** $33\frac{1}{3}$% simple interest   **35.** Between 11 and 12 years.   **37.** 6.82%
39. $6\frac{1}{4}$% compounded annually is better.   **41.** 9% compounded monthly is better   **43.** $109,395.56
45. (a) $300   (b) $323.06   **47.** $604.51   **49.** $29,137.83   **51.** $42,640.10   **53.** 19,918
55. Yes (7.25%)   **57.** $18,508.09   **59.** $10,810.76   **61.** 9.1%

Exercise 5.3 (page 260)
  1. $10,949.72   **3.** $2977.81   **5.** $32.81   **7.** $33.58   **9.** $n = 12$ **11.** $1593.74   **13.** $5073.00
15. $8356.36   **17.** $62,822.56   **19.** $9126.56   **21.** $524.04   **23.** $22,192.08   **25.** $201,575
27. $16,555.75

Exercise 5.4 (page 268)
  1. $4463.24   **3.** $4629.03   **5.** $264.36   **7.** $262.86   **9.** 36   **11.** $15,723.40   **13.** $856.60
15. $85,795.43   **17.** $39,272.59   **19.** $470.73
21. 5 years (20 years remaining): $18,699.94 10 years (15 years remaining): $24,493.23
23. $P = 2008.18$ **25.** $A = \$25,906$
27. The principal of either loan is $95,000 ($= \$120,000 - \$25,000$).

	Monthly Payment
8%, 240 months	($95,000)(0.008364) = $794.58
9%, 300 months	($95,000)(0.008392) = $797.24

The 9% loan requires the larger monthly payment. Since the payments are larger and there are more of them, the total amount of the payments, and therefore the total interest to be paid, is also larger.

	Equity After 10 Years
8%, 120 months remaining	$120,000 - ($794.58)(82.42148) = $54,509.54
9%, 180 months remaining	$120,000 - ($797.24)(98.5934) = $41,397.40

After 10 years, the equity from the 8%, 20-year loan is larger.
29. $335.46   **31.** $31,432.23   **33.** $332.79   **35.** $474.01; $4752.43
37. For a 30-year mortgage, $R = $966.37, $I = $235,892.50
For a 15-year mortgage, $R = $1189.89, $I = $102,180.41

Chapter Review

True–False Questions (page 271)
1. T   **2.** F   **3.** T   **4.** T   **5.** T

Fill in the Blanks (page 271)
1. Present value, future value   **2.** 10.3%   **3.** 2.5%   **4.** Present value   **5.** Amortization schedule

Review Exercises (page 271)
1. $1240   **3.** 1   **5.** $1444.89   **7.** $4160.03   **9.** $35,198.63   **11.** $n = 6$ **13.** $I = $436
15. $A = $125.12   **17.** (a) $I = $1080   (b) $A = $1044.55
19. The compound interest loan cost less $P = $71.36.   **21.** $P = $378.12
23. $545.22 monthy payments; total interest $= $103,566; equity after 5 years ($n = $ 240 months remaining): $= $23,502.
25. $1049, $21,575.52   **27.** $P = $119,432   **29.** $P = $108,004   **31.** $10,078.44   **33.** $37.98
35. $141.22   **37.** 9.38%   **39.** $1156.60   **41.** $2087.09   **43.** $330.74

Mathematical Questions (page 274)
1. b **2.** c **3.** b **4.** b **5.** d **6.** a **7.** c

CHAPTER 6

Exercise 6.1 (page 290)
1. None of these   **3.** None of these   **5.** $\subset, \subseteq$ **7.** $\subset, \subseteq$ **9.** $\subset, \subseteq$
11. $A \subseteq C$. This is called the *transitive law; $A \subset C$; $A \subset C$*
13. $\{a, b, c\}, \{a, b\}, \{a, c\}, \{b, c\}, \{a\}, \{b\}, \{c\}, \varnothing$; all except $\{a, b, c\}$ are proper subsets
15. $2^5 = 32; 2^6 = 64, 2^7 = 128$ **17.** $\{3\}$ **19.** $\{1, 2, 3, 5, 7\}$ **21.** $\{3, 5\}$ **23.** $\{1, 2, 3, 4, 5, 6, 7\}$
25. (a) $\{0, 1, 2, 3, 5, 7, 8\}$ (b) $\{5\}$ (c) $\{5\}$ (d) $\{0, 1, 2, 3, 4, 6, 7, 8, 9\}$ (e) $\{4, 6, 9\}$
(f) $\{0, 1, 5, 7\}$ (g) $\varnothing$ (h) $\{5\}$
27. (a) $\{b, c, d, e, f, g\}$ (b) $\{c\}$ (c) $\{a, h, i, j, \ldots, z\}$ (d) $\{a, b, d, e, f, \ldots, z\}$
29. (a)   (b)   (c)   (d)

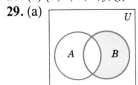   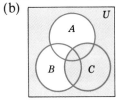   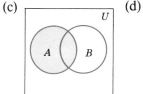

(e) 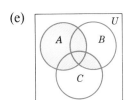   (f)   (g) 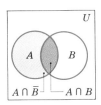   (h)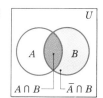

31. $A \cap E = \{x|x$ is a customer of IBM and is a member of the Board of Directors of IBM$\}$
33. $A \cup D = \{x|x$ is a customer of IBM or is a stockholder of IBM$\}$
35. $M \cap S = \{$All male college students who smoke$\}$
37. $\overline{M} \cap \overline{S} = \{$All female college students who do not smoke$)$ **39.** 4   **41.** 10   **43.** 4   **45.** 0
47. 4   **49.** 4   **51.** $c[A \cap (B \cap C)] = c(\{8\}) = 1$ **53.** $c(A \cup B) = 5$ **55.** 2   **57.** 10   **59.** 452
61. 36   **63.** 46   **65.** 24   **67.** 3   **69.** 63
71. 109 = Number of male seniors who are not on the dean's list
    97 = Number of female seniors who are not on the dean's list
   369 = Number of female students who are not seniors and not on the dean's list
    24 = Number of female seniors on the dean's list
    73 = Number of female students on the dean's list who are not seniors
    89 = Number of male students on the dean's list who are not seniors
   347 = Number of male students who are not seniors and not on the dean's list
     0 = Number of male seniors on the dean's list
73. (a) 42   (b) 9   (c) 5   (d) 2   (e) 44   (f) 31
75. Total for English, according to the diagram, is at least 303, whereas the staff member indicated that
    the total taking English was 281.   **77.** 68

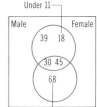

Exercise 6.2 (page 298)
  1. 8   **3.** 24   **5.** 864   **7.** 36   **9.** 1320   **11.** 1200   **13.** 720, 40320   **15.** 360, 1296   **17.** 16
19. 5040   **21.** $2^{35} \approx 3.436 \times 10^{10}$ **23.** (a) 6,760,000   (b) 3,407,040   (c) 3,276,000   **25.** 120
27. 16   **29.** $3^8 = 6561$ **31.** (a) 720   (b) 1000   (c) 810

Exercise 6.3 (page 305)
  1. 60   **3.** 120   **5.** 90   **7.** 9   **9.** 28   **11.** 42   **13.** 8   **15.** 1   **17.** 56   **19.** 1
21. (a) 720   (b) 120   (c) 24   **23.** 5040   **25.** 15120   **27.** 19,958,400   **29.** 3,368,253,000
31. 32,760   **33.** 90

Exercise 6.4 (page 310)
  1. 15   **3.** 21   **5.** 5   **7.** 28   **9.** 56   **11.** 2380
13. If the 3 officers are of equal rank, the answer is $C(25, 3) = 2300$. If there are 3 distinct offices (e.g.,
    president, vice-president, secretary), the answer is $P(25, 3) = 13,800$.
15. 15   **17.** 60

19. 26,046,720 (Order is not important if we assume that specific positions will be assigned to the 7 linemen after the team is formed.)

21. 10   **23.** 1,192,052,400   **25.** 75,287,520   **27.** 35   **29.** 4   **31.** 1,217,566,350   **33.** 60

35. 10,626   **37.** $\dfrac{50!}{35!}$ **39.** 5040

Exercise 6.5 (page 317)

1. (a) 1024   (b) 210   (c) 56   (d) 968   **3.** (a) 120   (b) 63   (c) 35   (d) 1   **5.** 70   **7.** 1260

9. 4,989,600   **11.** 2100   **13.** (a) 280   (b) 280   (c) 640   **15.** $\dfrac{30!}{(6!)^5} \simeq 1.37 \times 10^{18}$ **17.** 93

19. 826   **21.** 128; 128

Exercise 6.6 (page 325)

1. $(x + y)^5 = x^5 + 5x^4y + 10x^3y^2 + 10x^2y^3 + 5xy^4 + y^5$

3. $(x + 3y)^3 = x^3 + 3x^2(3y) + 3x(3y)^2 + (3y)^3 = x^3 + 9x^2y + 27xy^2 + 27y^3$

5. $(2x - y)^4 = (2x)^4 + 4(2x)^3(-y) + 6(2x)^2(-y)^2 + 4(2x)(-y)^3 + (-y)^4$
$\qquad\qquad = 16x^4 - 32x^3y + 24x^2y^2 - 8xy^3 + y^4$

7. 10   **9.** 405   **11.** 32   **13.** 1023   **15.** 512

17. $\binom{10}{7} = \binom{9}{7} + \binom{9}{6} = \binom{8}{7} + \binom{8}{6} + \binom{9}{6} = \binom{7}{7} + \binom{7}{6} + \binom{8}{6} + \binom{9}{6} = \binom{6}{6} + \binom{7}{6} + \binom{8}{6} + \binom{9}{6}$

19. $\binom{12}{6}$ by Example 5 Sec. 6.6

Chapter Review

True–False Questions (page 326)

1. T   **2.** T   **3.** F   **4.** F   **5.** T   **6.** T   **7.** T   **8.** F

Fill in the Blanks (page 326)

1. Disjoint   **2.** Permutation   **3.** Combination   **4.** Pascal   **5.** Binomial coefficients
6. Binomial theorem

Review Exercises (page 327)

1. None of these   **3.** None of these   **5.** None of these   **7.** ϵ **9.** $\subset$ **11.** None of these
13. $\subset, \subseteq$ **15.** None of these   **17.** (a) {3, 6, 8, 9}   (b) {6}   (c) {2, 3, 6, 7}   **19.** 3
21. (a) 45   (b) 33   (c) 50   **23.** 120   **25.** 10   **27.** 6   **29.** 72   **31.** Maximum: 12; 6
33. 218,400   **35.** (a) 525   (b) 1715   **37.** 12,441,600   **39.** (a) 4845   (b) 5700   (c) 7805
41. 924 paths from A to B **43.** 240   **45.** 30   **47.** 24

CHAPTER 7

Exercise 7.2 (page 343)

1. (a) {H, T}   (b) {0, 1, 2}   (c) {M, D}   **3.** {$HHH, HHT, HTH, HTT, THH, THT, TTH, TTT$}
5. {$HH1, HH2, HH3, HH4, HH5, HH6, HT1, HT2, HT3, HT4, HT5, HT6, TH1, TH2, TH3, TH4,$
$TH5, TH6, TT1, TT2, TT3, TT4, TT5, TT6$}

7. {*RA, RB, RC, GA, GB, GC*}   **9.** {*RR, RG, GR, GG*}

11. {*AA*1, *AA*2, *AA*3, *AA*4, *AB*1, *AB*2, *AB*3, *AB*4, *AC*1, *AC*2, *AC*3, *AC*4, *BA*1, *BA*2, *BA*3, *BA*4, *BB*1, *BB*2, *BB*3, *BB*4, *BC*1, *BC*2, *BC*3, *BC*4, *CA*1, *CA*2, *CA*3, *CA*4, *CB*1, *CB*2, *CB*3, *CB*4, *CC*1, *CC*2, *CC*3, *CC*4}

13. {*RA*1, *RA*2, *RA*3, *RA*4, *RB*1, *RB*2, *RB*3, *RB*4, *RC*1, *RC*2, *RC*3, *RC*4, *GA*1, *GA*2, *GA*3, *GA*4, *GB*1, *GB*2, *GB*3, *GB*4, *GC*1, *GC*2, *GC*3, *GC*4}

15. 16   **17.** 216   **19.** 1326   **21.** 1, 2, 3, 6, and 7   **23.** 2   **25.** $\frac{1}{4}$ **27.** $\frac{1}{12}$

29. {*HHHH, HHHT, HHTH, HHTT, HTHH, HTHT, HTTH, HTTT, THHH, THHT, THTH, THTT, TTHH, TTHT, TTTH, TTTT*}; assign $\frac{1}{16}$ to each simple event.

31. {*HTTT, TTTT*}

33. {*HHHT, HHTH, HHTT, HTHH, HTHT, HTTH, THHH, THHT, THTH, TTHH*}

35. $P(A) = \frac{1}{18}$ **37.** $\frac{1}{9}$ **39.** $\frac{1}{6}$ **41.** Probability of heads $\frac{3}{4}$, probability of tails $\frac{1}{4}$ **43.** $\frac{1}{2}$ **45.** $\frac{5}{6}$ **47.** $\frac{1}{2}$

49. (i) $P(E) = \frac{3}{4}$,   (ii) $P(E) = \frac{5}{9}$;   (i) $P(F) = \frac{1}{2}$,   (ii) $P(F) = \frac{4}{9}$;   (i) $P(G) = \frac{1}{4}$,   (ii) $P(G) = \frac{4}{9}$

51. {*RRR, RRL, RLR, RLL, LRR, LRL, LLR, LLL*}. Probability of each simple event is $\frac{1}{8}$. (We assume that the rat does not know left from right.)

   (a) $P(E) = \frac{3}{8}$ (b) $P(F) = \frac{1}{8}$ (c) $P(G) = \frac{1}{2}$ (d) $P(H) = \frac{1}{2}$

53. (a) The number on the red die is three times the number on the green die.

   (b) The number on the red die is 1 larger than the number on the green die.

   (c) The number on the red die is smaller than or equal to the number on the green die.

   (d) The sum of the numbers on the two dice is 8.

   (e) The number on the green die is the square of the number on the red die.

   (f) The numbers on the two dice are the same.

55. $P(C_1) = \frac{1}{4}$, $P(C_2) = \frac{1}{2}$, $P(C_3) = \frac{1}{4}$, $P(C_1 \cup C_2) = \frac{3}{4}$, $P(C_1 \cup C_3) = \frac{1}{2}$

Exercise 7.3 (page 353)

1. .8   **3.** .5   **5.** .35

7. The events "Sum is 2" and "Sum is 12" are mutually exclusive. The probability of obtaining a 2 or a 12 is $\frac{1}{36} + \frac{1}{36} = \frac{1}{18}$.

9. 0.35   **11.** 0.2   **13.** (a) 0.7   (b) 0.4   (c) 0.2   (d) 0.3   **15.** (a) 0.68   (b) 0.58   (c) 0.32

17. (a) 0.57   (b) 0.95   (c) 0.83   (d) 0.38   (e) 0.29   (f) 0.05   (g) 0.78   (h) 0.71   **19.** $\frac{3}{4}$ **21.** $\frac{5}{12}$

23. $\frac{1}{2}$ **25.** The odds for E are 7 to 3; the odds against E are 3 to 7.

27. The odds for F are 4 to 1; the odds against F are 1 to 4.   **29.** $\frac{100}{54}$ **31.** 1 to 5; 1 to 17; 2 to 7

33. $\frac{11}{15}$. The odds are 11 to 4.

35. $P(A \cup B \cup C) = P(A \cup B) + P(C) - P[(A \cup B) \cap C]$

$= P(A) + P(B) - P(A \cap B) + P(C) - P[(A \cap C) \cup (B \cap C)]$

$= P(A) + P(B) + P(C) - P(A \cap B) - (P(A \cap C) + P(B \cap C) - P[(A \cap C) \cap (B \cap C)])$

$= P(A) + P(B) + P(C) - P(A \cap B) - (P(A \cap C) - P(B \cap C) + P(A \cap B \cap C)$

Exercise 7.4 (page 361)

1. $\frac{1}{52}$ **3.** $\frac{1}{4}$ **5.** $\frac{3}{13}$ **7.** $\frac{5}{13}$ **9.** $\frac{12}{13}$ **11.** $\frac{3}{23}$ **13.** $\frac{7}{23}$ **15.** $\frac{8}{23}$ **17.** $\frac{11}{23}$ **19.** $\frac{1}{6}$ **21.** $\frac{65}{150} \approx 0.433$

23. 0.236   **25.** 0.0298   **27.** Probability is greater than 0.99999.

29. Probability that all 5 are defective: $\approx .0000028$. Probability that at least 2 are defective: $\approx .103$.

31. (a) $\dfrac{1287}{2,598,960}$ (b) ≈ 0.0107 (c) ≈ 0.2743

33. (a) ≈ 0.0000015 (b) ≈ 0.0000123 (c) ≈ 0.00024 (d) ≈ 0.0014 (e) 0.0076

Exercise 7.5 (page 364)

1. 0.2051 **3.** 0.6907 **5.** $\frac{1}{19}$ **7.** $\frac{1}{5!} = \frac{1}{120}$ **9.** $\frac{1}{2}$

Chapter Review

True – False Questions (page 365)

1. T   **2.** F   **3.** T   **4.** T   **5.** F

Fill in the Blanks (page 366)

1. $\{c, d\}$ **2.** 2   **3.** 1, 0   **4.** For   **5.** Equally likely   **6.** Mutually exclusive   **7.** 0.4

Review Exercises (page 366)

1. $S = \{BB, BG, GB, GG\}$ **3.** (a) $\frac{10}{91}$ (b) $\frac{45}{91}$ (c) $\frac{55}{91}$

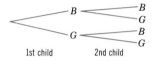

1st child   2nd child

5. (a) 0.21 (b) 0.23 (c) 0.1825 **7.** 0.25 **9.** (a) $\frac{1}{2}$ (b) $\frac{11}{24}$ (c) $\frac{13}{24}$ **11.** $\frac{2}{3}$ **13.** $\frac{7}{13}$

Mathematical Questions (page 368)

1. e **2.** c **3.** c **4.** c **5.** b

CHAPTER 8

Exercise 8.1 (page 376)

1. 0.69 **3.** 0.9 **5.** 0 **7.** 0.40 **9.** 0.24 **11.** 0.10 **13.** 0.08 **15.** $\frac{5}{12}$ **17.** $\frac{1}{3}$ **19.** $\frac{1}{2}; \frac{1}{4}$ **21.** $\frac{1}{2}$
23. 0.1748 **25.** (a) $\frac{1}{2}$ (b) $\frac{2}{3}$ **27.** (a) $\frac{1}{26}$ (b) $\frac{1}{2}$ (c) $\frac{1}{13}$
29. (a) $\frac{5}{11}$ (b) $\frac{6}{23}$ (c) $\frac{5}{11}$ (d) $\frac{5}{12}$ (c) $\frac{14}{15}$ (f) $\frac{1}{4}$
31. (a) 0.718 (b) 0.330 (c) 0.071 (d) 0.194 (e) 0.236 (f) 0.357 **33.** $P(2 \text{ girl}|\text{first girl}) = \frac{1}{2}$
35. $\frac{1}{16}$; yes $\frac{1}{8}$ **37.** $\frac{32}{75}$ **39.** $\frac{25}{204}$ **41.** $\frac{1}{13}$ **43.** $P(F) \cdot P(E|F) = P(E \cap F) = P(E) \cdot P(F|E)$
45. Since $F = (E \cap F) \cup (\overline{E} \cap F)$ and $(E \cap F) \cap (\overline{E} \cap F) = \varnothing$,

$$\frac{P(E \cap F)}{P(F)} + \frac{P(\overline{E} \cap F)}{P(F)} = \frac{P(E \cap F) + P(\overline{E} \cap F)}{P(F)} = \frac{P(F)}{P(F)} = 1$$

47. Given $P(E|F) = P(E)$ and $P(E) \neq 0$. From Problem 43 we know that:
$$P(F) \cdot P(E|F) = P(E) \cdot P(F|E), \text{ then substituting for } P(E|F).$$
$$P(F) \cdot P(E) = P(E) \cdot P(F|E)$$
$$P(F) = P(F|E)$$

49. 0.866 **51.** 0.67 **53.** 0.5

Exercise 8.2 (page 386)

1. 0.15 **3.** 0.125 **5.** No   **7.** (a) 0.3 (b) 0.5 (c) 0.15 (d) 0.65 **9.** $\frac{4}{147}$ **11.** 0.5, no
13. No   **15.** Yes   **17.** (a) $\frac{1}{4}$ (b) $\frac{1}{4}$ (c) $\frac{1}{16}$ **19.** No
21. $P(A) = \frac{1}{4} + \frac{1}{4} = \frac{1}{2}$ A and B are independent; $P(A \cap B) = P(2) = \frac{1}{4} = P(A)P(B)$;
$\ P(B) = \frac{1}{4} + \frac{1}{4} = \frac{1}{2}$ A and C are independent; $P(A \cap C) = P(1) = \frac{1}{4} = P(A)P(C)$;
$\ P(C) = \frac{1}{4} + \frac{1}{4} = \frac{1}{2}$ B and C are independent; $P(B \cap C) = P(3) = \frac{1}{4} = P(B)P(C)$.

23. Consider Problem 21.

(a) A and B are not mutually exclusive, but they are independent.

(b) Let $D = \{4\}$. Then: $P(D) = \frac{1}{12}$; $A \cap D = \emptyset$. Hence, they are mutually exclusive, but since $P(A \cap D) = P(\emptyset) = 0 \neq \frac{1}{12} \cdot \frac{1}{2}$, they are not independent.

(c) Let $E = \{1, 4\}$. Then: $A \cap E = \{1\}$; $P(E) = \frac{1}{4} + \frac{1}{12} = \frac{1}{3}$; $P(A \cap E) = P(1) = \frac{1}{4} \neq \frac{1}{2} \cdot \frac{1}{3}$. Thus, A and E are not independent and are not mutually exclusive.

25. $P(F) = 0$.

$E \cap F \subset F$; hence, $P(E \cap F) \leq P(F) = 0$

Thus, $P(E \cap F) = 0$. But

$P(E)P(F) = P(E) \cdot 0 = 0$.

Hence, E and F are independent.

27. If $E \cap F = \emptyset$, then $P(E \cap F) = P(\emptyset) = 0$. But, by independence, $P(E \cap F) = P(E)P(F)$. Hence, $P(E)P(F) = 0$, which implies that $P(E) = 0$ or $P(F) = 0$—a contradiction. Thus, we have $E \cap F \neq \emptyset$.

29. (a) Let $E = $ at least one ace, $F = $ no ace.

$P(E) = 1 - P(F) = 1 - 0.4823 = 0.5177$

(b) Let $G = $ at least one pair of aces, $H = $ no pair of aces.

$P(G) = 1 - P(H) = 1 - 0.509 = 0.491$

31. (a) $\frac{9}{16}$ (b) $\frac{1}{16}$ (c) $\frac{3}{8}$ **33.** (a) $\frac{27}{64}$ (b) $\frac{9}{64}$ **35.** (a) 0.4096 (b) 0.1536 (c) 0.9728

37. (a) $\frac{25}{81}$ (b) $\frac{40}{81}$ **39.** (a) 0.1231 (b) 0.0097 (c) no (d) no (e) no (f) no

41. Let $V = $ votes for her

(a) $P(V) \cdot P(V) = \frac{4}{5} \cdot \frac{4}{5} = \frac{16}{25}$

(b) $P(\overline{V}) \cdot P(\overline{V}) = \frac{1}{5} \cdot \frac{1}{5} = \frac{1}{25}$

(c) $P(V) \cdot P(\overline{V}) + P(\overline{V}) \cdot P(V) = \frac{4}{5} \cdot \frac{1}{5} + \frac{1}{5} \cdot \frac{4}{5} = \frac{8}{25}$

Exercise 8.3 (page 398)

1. $\frac{12}{31}$ **3.** $\frac{7}{31}$ **5.** $\frac{12}{31}$ **7.** .017 **9.** .018 **11.** $P(A_1|E) = \frac{3}{17} \approx .176$ $P(A_2|E) = \frac{14}{17} \approx .824$

13. $P(A_1|E) = \frac{5}{18} \approx .278$ $P(A_2|E) = \frac{9}{18} = .500$ $P(A_3|E) = \frac{4}{18} \approx 0.222$

15. $P(A_1|E) = \frac{8}{29} \approx .276$ $P(A_2|E) = \frac{20}{29} \approx .690$ $P(A_3|E) = \frac{1}{29} \approx 0.034$

17. $P(A_2|E) = 0$; $P(A_3|E) = .065$; $P(A_4|E) = 0$; $P(A_5|E) = 0.065$

19. $P(U_1|E) = \frac{1}{3} \approx 0.333$ $P(U_{II}|E) = 0.20$ $P(U_{III}|E) = 0.467$ **21.** $P(E|F) = \dfrac{P(E \cap F)}{P(F)} = \dfrac{P(F)}{P(F)} = 1$

23. 0.80 **25.** $P(M|C) = 0.945$ **27.** $P(D|V) = 0.385$, $P(R|V) = 0.39$, $P(I|V) = 0.225$

29. 0.39, 0.21, 0.41 **31.** 0.466, 0.343 **33.** 0.217

Exercise 8.4 (page 409)

1. 0.0250 **3.** 0.0811 **5.** 0.0916 **7.** 0.2968 **9.** 0.22 . . . **11.** ≈ 0.5787 **13.** ≈ 0.3292

15. 0.0368 **17.** 0.2362 **19.** 0.0273 **21.** 0.03125 **23.** 0.3634 **25.** 0.1098 **27.** 0.1608

29. 0.3125

31. (a)

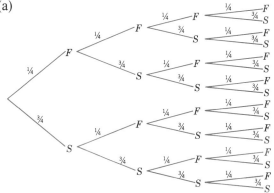

(b) .2109375 **33.** $341/12^5$

(c) .2109375

35. Recall that $\binom{n}{k} = \binom{n}{n-k}$ thus,

$$b(n, n-k; 1-p) = \binom{n}{n-k}(1-p)^{n-k}[1-(1-p)]^{n-(n-k)}$$

$$= \binom{n}{k}(1-p)^{n-k}p^k = b(n, k; p)$$

37. 0.0007 **39.** (a) 0.2793 (b) 0.0515 (c) 0.3366 (d) 0.9942 **41.** 0.1225 **43.** 0.6242
45. 0.6481 **47.** (a) 0.1094 (b) 0.4660 **49.** 0.9647

Exercise 8.5 (page 414)

1. $P(X=0) = \frac{1}{4}$, $P(X=1) = \frac{1}{2}$, $P(X=2) = \frac{1}{4}$
3. $P(X=0) = \frac{1}{8}$, $P(X=1) = \frac{3}{8}$, $P(X=2) = \frac{3}{8}$, $P(X=3) = \frac{1}{8}$
5. $P(X=0) = 0.216$, $P(X=1) = 0.432$, $P(X=2) = 0.288$, $P(X=3) = 0.064$

Exercise 8.6 (page 422)

1. 1.2 **3.** 44,560 **5.** She should pay $0.80 for a fair game.
7. He should pay $1.67 for a fair game. **9.** By 35¢ **11.** (a) $0.75 (b) No (c) Lose $2
13. It is not a fair bet. **15.** The outcomes are   **17.** $7

$e_1 =$ Heart not an ace, $p_1 = \frac{12}{52} = \frac{3}{13}$, $m_1 = 0.25$
$e_2 =$ Ace not ace of hearts, $p_2 = \frac{3}{52}$, $m_2 = 0.35$
$e_3 =$ Ace of hearts, $p_3 = \frac{1}{52}$, $m_3 = 0.75$
$e_4 =$ Neither an ace nor a heart, $p_4 = \frac{36}{52} = \frac{9}{13}$, $m_4 = -.15$
No, because her expected value is -0.012.

19. $E = np = 2000(\frac{1}{6}) = 333.3$ **21.** 10 **23.** 2.73 tosses **25.** 1.25
27. $E_1 = m_1 p_1 + m_2 p_2 + \cdots + m_n p_n$. Multiplying each value by k we get:

$$E_2 = km_1 p_1 + km_2 p_2 + \cdots + km_n p_n = k[m_1 p_1 + m_2 p_2 + \cdots + m_n p_n] = k \cdot E_1$$

Thus, we can see that the expected value of the new experiment, E_2, is k times the original expected value E_1. Add to each outcome used in figuring E_1 the constant k. The new expected value is:

$$E_3 = (m_1 + k)p_1 + (m_2 + k)p_2 + \cdots + (m_n + k)p_n = m_1 p_1 + kp_1 + m_2 p_2 + kp_2 + \cdots$$
$$+ m_n p_n + kp_n = m_1 p_1 + m_2 p_2 + \cdots + m_n p_n + kp_1 + kp_2 + \cdots + kp_n$$
$$= E_1 + k(p_1 + p_2 + \cdots + p_n)$$

Since $p_1 + p_2 + \cdots + p_n = 1$, we get $E_3 = E_1 + k$. Thus, we see that the expected value of the new experiment, E_3, is the expected value of the original experiment, E_1, plus k.
29. The second site has a higher expected profit.

Exercise 8.7 (page 428)

1. The expected number of customers is 9. The optimal number of cars is 9. The expected daily profit is 48.40.
3.

Group Size	$p^n - \frac{1}{n}$
2	$(.95)^2 - .5 = 0.4025$
3	$(.95)^3 - .333 = 0.524$
4	$(.95)^4 - .25 = 0.565$
5	$(.95)^5 - .2 = 0.574$
6	$(.95)^6 - .167 = 0.568$

5. (a) $E(x) = 75,000 - 75,000(.05)^x - 500x.$ (b) Two divers should be used.

Chapter Review

True–False Questions (page 430)
1. F   **2.** F   **3.** F   **4.** T   **5.** F   **6.** T

Fill in the Blanks (page 430)
1. Independent   **2.** Conditional probability
3. (a) The intersection of any two of the subsets is empty.
   (b) Each subset is nonempty.
   (c) $A_1 \cup A_2 \cup \cdots \cup A_n = S$
4. Bayes' formula   **5.** (c) Independent   (d) Same   **6.** A number   **7.** Expected value

Review Exercises (page 431)
1. $P(E \cap F) = 0.075$, $P(F) = 0.125$ **3.** (a) 0   (b) 0   (c) 0   (d) 0.75
5. (a) 0.075   (b) 0.5   (c) 0.375   (d) 0.225   (e) 0.225   **7.** (a) 0.7   (b) 0.9   (c) 0.63   (d) 0.97
9. (a) $\frac{1}{3}$ (b) $\frac{9}{38}$ (c) 0.56   **11.** $b(5, 0; .2) = 0.3277$; $b(5, 1; .2) = 0.4096$; $b(5, 2; .2) = 0.2048$;
$b(5, 3; .2) = 0.0512$; $b(5,4; .2) = 0.0064$; $b(5, 5; .2) = 0.0003$
13. (a) $\frac{1}{4096}$ (b) 0.3871   (c) 3871 to 6129   **15.** 29.52
17. (a) Expected value for five tickets is $0.90.   **19.** $\frac{76}{7}$ **21.** Game is not fair, since $E = -\frac{1}{37}$
   (b) She paid $0.35 extra.
23. 1.5   **25.** (a) $\frac{138}{400}$ (b) $\frac{168}{400}$ (c) $\frac{104}{400}$ (d) $\frac{160}{400}$ (e) $\frac{147}{400}$ (f) $\frac{138}{400} \cdot \frac{85}{400} = \frac{1173}{16,000} \neq \frac{4}{400}$
27. (a) 0.0250   (b) 0.0625   (c) 0.4   **29.** 0.163   **31.** 0.325

Mathematical Questions (page 435)
1. *b*   **2.** *d*   **3.** *d*   **4.** *b*   **5.** *a*   **6.** *a*   **7.** *a*   **8.** *d*   **9.** *b*   **10.** *b*   **11.** *c*   **12.** *d*   **13.** *d*

CHAPTER 9

Exercise 9.1 (page 441)
1. A poll should be taken either door-to-door or by means of the telephone.
3. A poll should be taken door-to-door in which people are asked to fill out a questionnaire.
5. The data should be gathered from all different kinds of banks.
7. (a) Asking a group of children if they like candy to determine what percentage of people like candy.
   (b) Asking a group of people over 65 their opinion toward Medicare to determine the opinion of
      people in general about Medicare.
9. By taking a poll downtown, you would question mostly people who are either shopping or working
   downtown. For instance, you would question few students.

Exercise 9.2 (page 448)
1. (a) 250   (b) 249   (c) 275   (d) 50   (e) 33   (f) The 5th
   (g) 752   (h) Histogram   (i) Frequency polygon

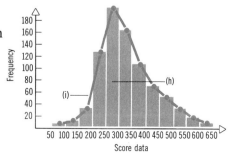

3. (a) Line chart

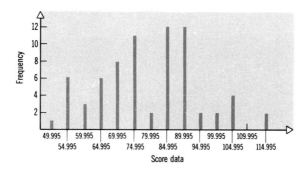

(b) Histogram   (c) Frequency polygon   (d) Cumulative frequency

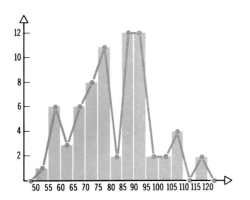

Class Interval	Tally	*f*	*cf*
114.995–119.995	‖	2	71
109.995–114.995		0	69
104.995–109.995	‖‖	4	69
99.995–104.995	‖	2	65
94.995–99.995	‖	2	63
89.995–94.995	⋕ ⋕ ‖	12	61
84.995–89.995	⋕ ⋕ ‖	12	49
79.995–84.995	‖	2	37
74.995–79.995	⋕ ⋕ ‖	11	35
69.995–74.995	⋕ ‖‖	8	24
64.995–69.995	⋕ ‖	6	16
59.995–64.995	‖‖	3	10
54.995–59.995	⋕ ‖	6	7
49.995–54.995	‖	1	1

(e) Cumulative frequency distribution

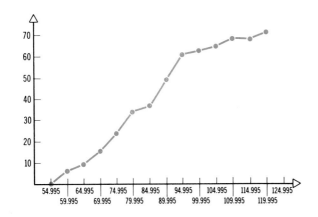

5. (a) Range $= 296 - 78 = 218$

Score	Tally	f	Score	Tally	f	Score	Tally	f
296	|	1	175	|	1	137	||	2
289	|	1	172	|	1	136	||	2
256	|	1	171	||	2	134	|	1
245	||	2	169	|||	3	132	||	2
240	|	1	166	||	2	131	||||	4
232	|	1	165	||	2	130	|	1
230	|	1	162	|	1	129	|||	3
224	||	2	161	||	2	128	||	2
222	|	1	158	|	1	127	|||	3
218	||	2	157	|	1	126	|	1
212	|	1	156	||	2	123	||	2
211	|	1	155	|	1	122	|	1
207	|	1	154	||	2	119	||	2
204	|	1	153	|||	3	116	|||	3
202	|	1	152	|	1	115	|	1
198	||	2	149	|	1	113	|	1
194	|	1	148	|	1	112	|	1
192	|	1	146	||	2	111	|	1
190	|||	3	145	||	2	110	|	1
188	|	1	144	||	2	108	|	1
185	|||	3	142	|	1	105	|	1
184	|	1	141	|	1	100	|	1
178	|	1	140	|	1	95	|	1
176	||	2	138	|	1	91	|	1
						78	|	1

(b) Line chart

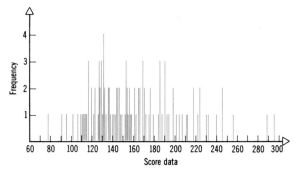

(c) Histogram   (d) Frequency polygon

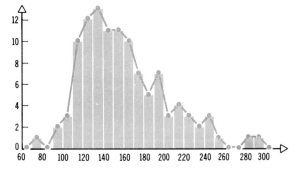

(e) Cumulative (less than) frequency

Class Interval	Tally	f	cf	Class Interval	Tally	f	cf
290.5–300.5	I	1	110	170.5–180.5	HHT II	7	80
280.5–290.5	I	1	109	160.5–170.5	HHT HHT	10	73
270.5–280.5		0	108	150.5–160.5	HHT HHT I	11	63
260.5–270.5		0	108	140.5–150.5	HHT HHT	10	52
250.5–260.5	I	1	108	130.5–140.5	HHT HHT III	13	42
240.5–250.5	II	2	107	120.5–130.5	HHT HHT IIII	14	29
230.5–240.5	II	2	105	110.5–120.5	HHT IIII	9	15
220.5–230.5	IIII	4	103	100.5–110.5	III	3	6
210.5–220.5	IIII	4	99	90.5–100.5	II	2	3
200.5–210.5	III	3	95			0	1
190.5–200.5	IIII	4	92	70.5–80.5	I	1	1
180.5–190.5	HHT III	8	88				

(f)

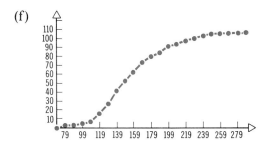

(g) Cumulative (more than) frequency

Class Interval	Tally	f	cf	Class Interval	Tally	f	cf
70.5–80.5	I	1	110	180.5–190.5	HHT III	8	30
80.5–90.5		0	109	190.5–200.5	IIII	4	22
90.5–100.5	II	2	109	200.5–210.5	III	3	18
100.5–110.5	III	3	107	210.5–220.5	IIII	4	15
110.5–120.5	HHT IIII	9	104	220.5–230.5	IIII	4	11
120.5–130.5	HHT HHT IIII	14	95	230.5–240.5	II	2	7
130.5–140.5	HHT HHT III	13	81	240.5–250.5	II	2	5
140.5–150.5	HHT HHT	10	68	250.5–260.5	I	1	3
150.5–160.5	HHT HHT I	11	58	260.5–270.5		0	2
160.5–170.5	HHT HHT	10	47	270.5–280.5		0	2
170.5–180.5	HHT II	7	37	280.5–290.5	I	1	2
				290.5–300.5	I	1	1

(h)

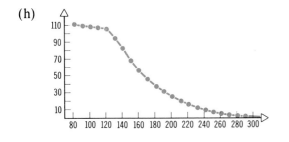

7.

Class Interval	Tally	f	cf
1600–3500	|	1	50
1400–1599	||	2	49
1300–1399	||	2	47
1200–1299	||	2	45
1100–1199	||||	4	43
1000–1099	||||	4	39
900–999	||| ||	7	35
800–899	||| ||| |||	13	28
700–799	||| |||		11	15
400–699	||||	4	4

Exercise 9.3 (page 452)

1.

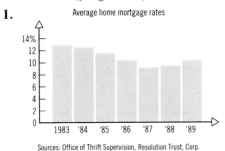

3.

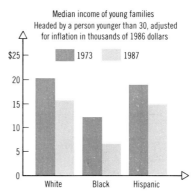

5.

7.

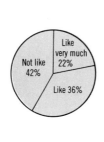

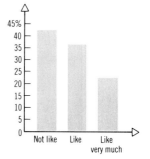

Exercise 9.4 (page 459)

1. Mean = 31.25; median = 30.5; no mode   **3.** Mean = 70.4; median = 70; mode = 55
5. Mean = 76.2; median = 75; no mode   **7.** Mean = $\bar{X}$ = 7.8%; median = 7.5%   **9.** 206.49
11. 39,300; median = 36,000. The median describes the situation more realistically, since it is closer to the salary of most of the faculty members in the sample.

13. Mean = $1626.10; median = $1605.20
15. 1970 mean = $53,225; 1977 mean = $59,650; 1983 mean = $74,300
17. For Table 5, $C_{75} \approx 91.77$, $C_{40} \approx 77.0$. For Table 6, $C_{75} \approx 93.03$, $C_{40} \approx 76.52$

19. Let 325 be the assumed mean; $n = 753$, $\Sigma fx' = -13$, $\overline{X} = 325 + \left(\dfrac{-13}{753}\right)(50) \approx 324.1$.

Exercise 9.5 (page 466)
1. (b)   **3.** ≈ 6.53 **5.** ≈ 5.196 **7.** ≈ 12.47 **9.** ≈ 66.68 **11.** $\overline{X} = 31.9$; $\sigma = 7.80$
13. Since the mean of stream II is 2038 versus a mean of 2000 for stream I, stream II is preferred.
15. 91.77   **17.** (a) 75%   (b) 64%   (c) $88\frac{8}{9}\%$ (d) 25%   (e) $11\frac{1}{9}\%$ **19.** ($24.37, $78.13)

Exercise 9.6 (page 475)
1. $\overline{X} = 8$; $\sigma = 1$ **3.** $\overline{X} = 18$; $\sigma = 1$

5. $Z = \dfrac{7 - 13.1}{9.3} \approx -0.6559$, $Z = \dfrac{9 - 13.1}{9.3} \approx -0.4409$, $Z = \dfrac{13 - 13.1}{9.3} \approx -0.0108$

$Z = \dfrac{15 - 13.1}{9.3} \approx 0.2043$, $Z = \dfrac{29 - 13.1}{9.3} \approx 1.7097$, $Z = \dfrac{37 - 13.1}{9.3} \approx 2.5699$

$Z = \dfrac{41 - 13.1}{9.3} = 3.0000$

7. 0.3085   **9.** 0.0688
11. (a) 0.3133   (b) 0.3642   (c) 0.4938   (d) 0.4987   (e) 0.2734   (f) 0.4896   (g) 0.2881
(h) 0.4988

13. $Z = \dfrac{x - \overline{X}}{\sigma}$; $\overline{X} = 64$, $\sigma = 2$.   (a) 68.26% of the women are between 62 and 66 inches tall;

$(0.6826)(2000) = 1365.2 \approx 1365$ women   (b) ≈ 1909 women   (c) ≈ 1995 women
15. (a) 1.04%   (b) According to the table, Z must be close to 1.04 if the proportional area is 35%
(≈ 0.3508). $(1.04)\sigma = (1.04)(5.2) \approx 5.4$ pounds. Thus, we expect 70% of the students to be within 5.4
pounds of the mean, or between 134.6 and 135.4 pounds.
17. $b(15, 0; .3) \approx 0.0047$
$b(15, 1; .3) \approx 0.0305$
$b(15, 2; .3) \approx 0.0916$
$b(15, 3; .3) \approx 0.1700$
$b(15, 4; .3) \approx 0.2186$
$b(15, 5; .3) \approx 0.2061$
$b(15, 6; .3) \approx 0.1472$
$b(15, 7; .3) \approx 0.0811$
$b(15, 8; .3) \approx 0.0348$
$b(15, 9; .3) \approx 0.0116$
$b(15, 10; .3) \approx 0.0030$
$b(15, 11; .3) \approx 0.0006$
$b(15, 12; .3) \approx 0.0001$
$b(15, 13; .3) \approx 0.0000$
$b(15, 14; .3) \approx 0.0000$
$b(15, 15; .3) \approx 0.0000$

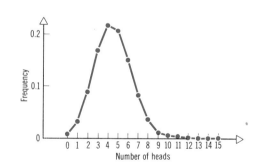

19. 0.7698   **21.** 0.5   **23.** 0.0287   **25.** Kathleen has the highest relative standing.
27. The number of shoes he should expect to replace out of 1000 is 239.

Chapter Review

True–False Questions (page 478)
  1. F   **2.** T   **3.** F   **4.** T   **5.** T

Fill in the Blanks (page 478)
  1. Mean, median, mode   **2.** Standard deviation   **3.** Bell shape   **4.** Z score
  5. Random outcome lies between $\overline{X} + k$ and $\overline{X} - k$.

Review Exercises (page 478)
  1. (a)

Score	Tally	Frequency	Score	Tally	Frequency
100	||	2	66	||	2
99	|	1	63	||	2
95	|	1	60	|	1
92	|	1	55	|	1
90	|	1	52	||	2
89	|	1	48	|	1
87	||	2	44	|	1
85	||	2	42	|	1
83	|	1	33	|	1
82	|	1	30	|	1
80	|||	3	26	|	1
78	||	2	21	|	1
77	|	1	20	|	1
75	|	1	19	|	1
74	|	1	17	|	1
73	||	2	14	||	2
72	||	2	12	|	1
70	|	1	10	|	1
69	|	1	8	|	1

Range $= 100 - 8 = 92$

  (b) Line chart

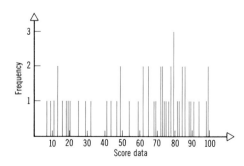

(c) Histogram
(d) Frequency
     polygon

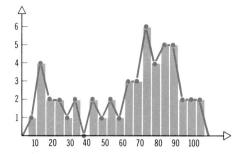

(e) Cumulative (more than) frequency

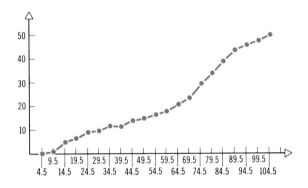

Class Interval	Tally	f	cf	Class Interval	Tally	f	cf
99.5–104.5	||	2	2	49.5–54.5	||	2	35
94.5–99.5	||	2	4	44.5–49.5	|	1	36
89.5–94.5	||	2	6	39.5–44.5	||	2	38
84.5–89.5	||||	5	11	34.5–39.5		0	38
79.5–84.5	ⅼ卅	5	16	29.5–34.5	||	2	40
74.5–79.5	||||	4	20	24.5–29.5	|	1	41
69.5–74.5	卅 |	6	26	19.5–24.5	||	2	43
64.5–69.5	|||	3	29	14.5–19.5	||	2	45
59.5–64.5	|||	3	32	9.5–14.5	||||	4	49
54.5–59.5	|	1	33	4.5–9.5	|	1	50

(f) Cumulative (less than) frequency

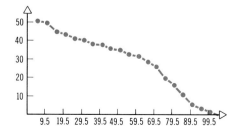

Class Interval	f	cf	Class Interval	f	cf
99.5–104.5	2	50	49.5–54.5	2	17
94.5–99.5	2	48	44.5–49.5	1	15
89.5–94.5	2	46	39.5–44.5	2	14
84.5–89.5	5	44	34.5–39.5	0	12
79.5–84.5	5	39	29.5–34.5	2	12
74.5–79.5	4	34	24.5–29.5	1	10
69.5–74.5	6	30	19.5–24.5	2	9
64.5–69.5	3	24	14.5–19.5	2	7
59.5–64.5	3	21	9.5–14.5	4	5
54.5–59.5	1	18	4.5–9.5	1	1

3.

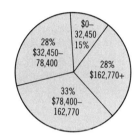

28% $32,450–78,400
15% $0–32,450
28% $162,770+
33% $78,400–162,770

5. The mean is a poor measure in (b) because it gives too much importance to the extreme value 195.

7. $A = \{20, 15, 5, 0\}; B = \{12, 11, 10, 9, 8\}$: Both sets have a mean of 10. The standard deviation for the first set is $\sigma_1 = \sqrt{250/5} \approx 7.07$. The standard deviation for the second set is $\sigma_2 = \sqrt{10/5} \approx 1.41$.

9. ≈ 2.77

11. (a) $(0.6827)(600) = 409.62$ (b) $[(0.4987) - (0.3413)](600) = 94.44$ (c) $(0.4972)(600) = 298.32$

13. (a) $(0.4970) - (0.4115) = 0.0855$ (b) $(0.4599) - (0.3849) = 0.0750$ **15.** At least 0.75

Mathematical Questions (page 480)

1. e **2.** b **3.** c

CHAPTER 10

Exercise 10.1 (page 485)

1. Let the entries denote Tami's winnings in cents.

Laura chooses columns and Tami chooses rows: Tami
$$\begin{array}{c} \\ I \\ II \end{array} \begin{bmatrix} -10 & 10 \\ 10 & -10 \end{bmatrix}$$
(Laura: I, II)

3. The entries denote Tami's winnings in cents. Tami
$$\begin{array}{c} 1 \\ 4 \\ 7 \end{array} \begin{bmatrix} -20 & 50 & -80 \\ 50 & -80 & 110 \\ -80 & 110 & -140 \end{bmatrix}$$
(Laura: 1, 4, 7)

5. Strictly determined; value is -1.   **7.** Strictly determined; value is 2.

9. Not strictly determined.   **11.** Strictly determined; value is 2.   **13.** Not strictly determined.

15. $0 \leq a \leq 3$ (There is no saddle point in row 1, unless $a \leq 3$; in row 2, unless $a \leq -9$; in row 3, unless $a \leq -5$; in column 1, unless $a \geq 0$; in column 2, unless $a \geq 8$; in column 3, unless $a \geq 5$. Thus, there is no saddle point unless $0 \leq a \leq 3$. But if $0 \leq a \leq 3$, then there is a saddle point in row 1, column 1.)

17. $a \leq 0 \leq b$ or $b \leq 0 \leq a$ (The matrix $\begin{bmatrix} a & 0 \\ 0 & b \end{bmatrix}$ is strictly determined if and only if there is a saddle point; a is a saddle point if and only if $a = 0$; b is a saddle point if and only if $b = 0$. The 0 in row 1, column 2 is a saddle point if and only if $0 \leq a$ and $0 \geq b$. The 0 in row 2, column 1 is a saddle point if and only if $0 \leq b$ and $0 \geq a$.)

Exercise 10.2 (page 489)

1. $E = 1.42$ **3.** $E = \frac{9}{4}$ **5.** $E = \frac{19}{8}$ **7.** $E = \frac{17}{9}$ **9.** $E = \frac{1}{3}$.

11. The nonstrictly determined games are those without saddle points. If $a_{11} = a_{12}$ or $a_{21} = a_{22}$, then the game is strictly determined. (See Problem 16 in Exercise 10.1)

(a) If $a_{11} > a_{12}$, then $a_{12} < a_{22}$ to prevent a_{12} from being a saddle point. This means that $a_{21} < a_{22}$ to prevent a_{22} from being a saddle point. Also, $a_{11} > a_{21}$ to prevent a_{21} from being a saddle point.

(b) If $a_{11} < a_{12}$, then $a_{21} > a_{11}$ to prevent a_{11} from being a saddle point. This means that $a_{22} < a_{21}$ to prevent a_{21} from being a saddle point. Also, $a_{12} > a_{22}$ to prevent a_{22} from being a saddle point.

Exercise 10.3 (page 496)

1. The optimal strategy for Player I is $[0.75 \quad 0.25]$.
The optimal strategy for Player II is $[0.25 \quad 0.75]$.
The value of the game is

$$E = 1.75$$

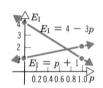

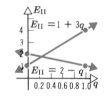

3. The optimum strategy for Player I is $[\frac{1}{6} \quad \frac{5}{6}]$.
The optimum strategy for Player II is $[\frac{1}{3} \quad \frac{2}{3}]$.
The value of the game is

$$E = PAQ = [\frac{1}{6} \quad \frac{5}{6}] \begin{bmatrix} -3 & 2 \\ 1 & 0 \end{bmatrix} \begin{bmatrix} \frac{1}{3} \\ \frac{2}{3} \end{bmatrix} = \frac{1}{3}$$

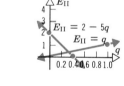

5. The optimum strategy for Player I is $[\frac{5}{8} \quad \frac{3}{8}]$.
The optimum strategy for Player II is $[\frac{5}{8} \quad \frac{3}{8}]$.
The value of the game is

$$E = PAQ = [\frac{5}{8} \quad \frac{3}{8}] \begin{bmatrix} 2 & -1 \\ -1 & 4 \end{bmatrix} \begin{bmatrix} \frac{5}{8} \\ \frac{3}{8} \end{bmatrix} = \frac{7}{8}$$

7. $p_1 = \frac{3}{8}$; $p_2 = \frac{5}{8}$;
$q_1 = \frac{1}{2}$; $q_2 = \frac{1}{2}$;
$V = 1.5$

The game favors the Democrat.

9. If $a_{11} + a_{22} - a_{12} - a_{21} = 0$, then the game is strictly determined. Otherwise, from Problem 11 in Exercise 10.2, we must have either:

(a) $a_{11} - a_{12} > 0$ and $a_{22} - a_{21} > 0$; hence, $a_{11} + a_{22} - a_{12} - a_{21} > 0$, or

(b) $a_{11} - a_{12} < 0$ and $a_{22} - a_{21} < 0$; hence, $a_{11} + a_{22} - a_{12} - a_{21} < 0$.

11.   Opponent

		Deserted	Busy
Spy	Deserted	-100	30
	Busy	10	-2

$p_1 = \frac{6}{71};\ \ p_2 = \frac{65}{71};\ \ q_1 = \frac{16}{71};\ \ q_2 = \frac{55}{71}$

The value is $\frac{50}{71}$.

Exercise 10.4 (page 507)

1. Row 2 dominates row 3; the reduced matrix is $\begin{bmatrix} 8 & 3 & 8 \\ 6 & 5 & 4 \end{bmatrix}$.

Column 2 dominates column 1; the reduced matrix is $\begin{bmatrix} 3 & 8 \\ 5 & 4 \end{bmatrix}$. $\begin{array}{ll} p_1 = \frac{1}{6} & q_1 = \frac{2}{3} \\ p_2 = \frac{5}{6} & q_2 = \frac{1}{3} \\ V = \frac{14}{3} \end{array}$

3. Column 3 dominates columns 1 and 4; the reduced matrix is $\begin{bmatrix} 1 & 0 \\ -2 & 1 \end{bmatrix}$. $\begin{array}{ll} p_1 = \frac{3}{4} & q_1 = \frac{1}{4} \\ p_2 = \frac{1}{4} & q_2 = \frac{3}{4} \\ V = \frac{1}{4} \end{array}$

5. Column 3 dominates column 1; column 2 dominates column 4; the reduced matrix is $\begin{bmatrix} -4 & 2 \\ 6 & -5 \end{bmatrix}$.

$\begin{array}{ll} p_1 = \frac{11}{17} & q_1 = \frac{7}{17} \\ p_2 = \frac{6}{17} & q_2 = \frac{10}{17} \\ V = -\frac{8}{17} \end{array}$

7. Row 1 dominates row 3; the reduced matrix is $\begin{bmatrix} 4 & -5 & 5 \\ -6 & 3 & 3 \end{bmatrix}$.

Column 2 dominates column 3; the reduced matrix is $\begin{bmatrix} 4 & -5 \\ -6 & 3 \end{bmatrix}$. $\begin{array}{ll} p_1 = \frac{1}{2} & q_1 = \frac{4}{9} \\ p_2 = \frac{1}{2} & q_2 = \frac{5}{9} \\ V = -1 \end{array}$

9. Row 3 dominates rows 2 and 4; the reduced matrix is $\begin{bmatrix} 1 & 3 & 0 \\ 0 & 4 & 1 \end{bmatrix}$.

Column 1 dominates column 2; the reduced matrix is $\begin{bmatrix} 1 & 0 \\ 0 & 1 \end{bmatrix}$. $\begin{array}{ll} p_1 = \frac{1}{2} & q_1 = \frac{1}{2} \\ p_2 = \frac{1}{2} & q_2 = \frac{1}{2} \\ V = \frac{1}{2} \end{array}$

11. Row 2 dominates row 3; the reduced matrix is $\begin{bmatrix} 4 & 3 & -1 \\ 1 & 1 & 4 \end{bmatrix}$.

Column 2 dominates column 1; the reduced matrix is $\begin{bmatrix} 3 & -1 \\ 1 & 4 \end{bmatrix}$. $\begin{array}{ll} p_1 = \frac{3}{7} & q_1 = \frac{5}{7} \\ p_2 = \frac{4}{7} & q_2 = \frac{2}{7} \\ V = \frac{13}{7} \end{array}$

13. Let the thief be in area A with probability q_1. He is then in area B with probability $(1 - q_1)$. The detectives have six choices. The expected values for the probabilities for the detectives to find and arrest a thief for each of the six choices are:

① $E_I = -0.24q_1 + 0.75$
② $E_I = 0.28q_1 + 0.36$
③ $E_I = -0.72q_1 + 0.91$
④ $E_I = -0.02q_1 + 0.60$
⑤ $E_I = -0.48q_1 + 0.85$
⑥ $E_I = -0.20q_1 + 0.76$

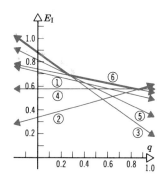

The intersection of lines 2 and 6 gives the optimum strategy of the thief.

$q_1 = \frac{5}{6}$ $q_2 = 1 - q_1 = \frac{1}{6}$

Eliminating rows 1, 3, 4, and 5, we get:

$p_1 = \frac{5}{12}$ $p_2 = \frac{7}{12}$ $V = 0.5933 \ldots$

Exercise 10.5 (page 517)

1. $P^* = [\frac{4}{5} \ \frac{1}{5}], Q^* = \begin{bmatrix} \frac{3}{5} \\ \frac{2}{5} \end{bmatrix}, V = \frac{6}{5}$ **3.** $P^* = [\frac{1}{2} \ \frac{1}{2}], Q^* = \begin{bmatrix} \frac{1}{2} \\ \frac{1}{2} \end{bmatrix}, V = 0$ (Game is fair.)

5. $P^* = [\frac{1}{2} \ \frac{1}{2}], Q^* = \begin{bmatrix} \frac{1}{2} \\ \frac{1}{2} \\ 0 \end{bmatrix}, V = \frac{1}{2}$ **7.** $P^* = [\frac{2}{7} \ 0 \ \frac{5}{7}], Q^* = \begin{bmatrix} \frac{4}{7} \\ \frac{3}{7} \end{bmatrix}, V = \frac{6}{7}$

9. $P^* = [\frac{7}{10} \ \frac{3}{10}], Q^* = \begin{bmatrix} \frac{1}{2} \\ \frac{1}{2} \end{bmatrix}, V = \frac{1}{2}$ **11.** (a) $[\frac{5}{6} \ \frac{1}{6}]$ (b) $\frac{19}{2}$ (c) 9.5%

13. (a)

	Growing economy	Slowing economy
Economy model	-2	3
Regular model	0	2
Deluxe model	3	1

(b) $P^* = [\frac{2}{7} \ 0 \ \frac{5}{7}]$ (c) $\frac{11}{7}$
(d) The manufacturer should utilize $\frac{5}{7}$ of its resources for the deluxe model, $\frac{2}{7}$ of its resources for the economy model, and produce no regular models.

Chapter Review

True–False Questions (page 519)
1. T   **2.** T   **3.** T   **4.** F

Fill in the Blanks (page 519)
1. Payoff   **2.** Value   **3.** Strictly determined, nonstrictly determined   **4.** Dominant

Review Exercises (page 519)
1. (a) Not strictly determined.   (b) Strictly determined; value is 15.
(c) Strictly determined; value is 50.   (d) Strictly determined; value is 9.
(e) Strictly determined; value is 12.

3. Let's first examine the 2×2 matrix $\begin{bmatrix} a & b \\ c & d \end{bmatrix}$.

Let a be the saddle point. We know that $a \leq b$ and $a \geq c$. If $d \leq c$, then $b \geq d$ and row 1 dominates row 2. If $d \geq c$, then column 1 dominates column 2. Similar reasoning would have shown the desired result if we had chosen a saddle point other than a.

Now consider the matrix $\begin{bmatrix} a & b & c \\ d & e & f \end{bmatrix}$.

Let a be the saddle point. We know that $a \leq b$, $a \leq c$, and $a \geq d$. If $d \leq e$, then column 1 dominates column 2. If $d \leq f$, then column 1 dominates column 3. If $d \geq e$ and $d \geq f$, then row 1 dominates row 2. Similar reasoning would have shown the desired result if we had chosen a saddle point other than a.

5. (a) Column 1 dominates column 2; the reduced matrix is $\begin{bmatrix} 4 & 3 \\ 1 & 5 \end{bmatrix}$.
$\quad p_1 = \frac{4}{5} \quad q_1 = \frac{2}{5}$
$\quad p_2 = \frac{1}{5} \quad q_2 = \frac{3}{5}$
$\quad V = \frac{17}{5}$

(b) Row 3 dominates row 2; the reduced matrix is $\begin{bmatrix} 1 & 6 \\ 7 & 4 \end{bmatrix}$.
$\quad p_1 = \frac{3}{8} \quad q_1 = \frac{1}{4}$
$\quad p_2 = \frac{5}{8} \quad q_2 = \frac{3}{4}$
$\quad V = \frac{19}{4}$

(c) Row 3 dominates row 1; the reduced matrix is $\begin{bmatrix} 4 & 0 \\ 3 & 4 \end{bmatrix}$.
$\quad p_1 = \frac{1}{5} \quad q_1 = \frac{4}{5}$
$\quad p_2 = \frac{4}{5} \quad q_2 = \frac{1}{5}$
$\quad V = \frac{16}{5}$

(d) Let Player I play row 1 with probability p. He then plays row 2 with probability $(1 - p)$. Player II has 3 choices. The expected earnings for Player I for each of the 3 choices are:

① $E_1 = 0p + 4(1 - p) = 4 - 4p$
② $E_1 = 3p + 2(1 - p) = 2 + p$
③ $E_1 = 2p + 3(1 - p) = 3 - p$

By examining the graphs of these equations, we see that the intersection of lines 1 and 2 gives the optimum strategy for Player I: $p = \frac{2}{5}$, $1 - p = \frac{3}{5}$.

Eliminating column 3, we get $\begin{bmatrix} 0 & 3 \\ 4 & 2 \end{bmatrix}$.

$q_1 = \dfrac{2 - 3}{2 - 4 - 3} = \frac{1}{5} \qquad q_2 = \frac{4}{5}$

Player I should select row 1 with probability $\frac{2}{5}$ and row 2 with probability $\frac{3}{5}$. Player II should select column 1 with probability $\frac{1}{5}$, column 2 with probability $\frac{4}{5}$, and never select column 3.

$V = \frac{12}{5}$

7. (a) $P^* = [\frac{5}{6}, \frac{1}{6}]$
(b) $V = 7.5$
(c) 7.5% increase

9. (a) $P^* = [\frac{13}{18}, \frac{5}{18}]$, (b) $V = \frac{35}{6}$
(c) 5.8%

CHAPTER 11

Exercise 11.1 (page 529)

1. The sum of the entries in row 3 is not equal to 1; there is a negative entry in row 3, column 2.

3. (a) The probability of a change from state 1 to state 2 is $\frac{2}{3}$.   (b) $[\frac{1}{3} \quad \frac{2}{3}]$ (c) $[\frac{1}{4} \quad \frac{3}{4}]$

   (d) $A^{(0)} = [0 \quad 1]$

   $p_{21}^{(2)} = \frac{1}{4} \cdot \frac{1}{3} + \frac{3}{4} \cdot \frac{1}{4} = \frac{13}{48}$

   $p_{22}^{(2)} = \frac{1}{4} \cdot \frac{2}{3} + \frac{3}{4} \cdot \frac{3}{4} = \frac{35}{48}$

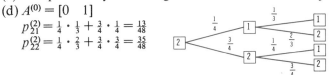

5. [.3625 .6375]   **7.** $a = .4$ $b = .2$ $c = 1$ **9.** [.5040 .4960]

11. $uA = [u_1 a_{11} + u_2 a_{21} \quad u_1 a_{12} + u_2 a_{22}]$

   $(uA)_1 + (uA)_2 = (u_1 a_{11} + u_2 a_{21}) + (u_1 a_{12} + u_2 a_{22})$

   $= u_1(a_{11} + a_{12}) + u_2(a_{21} + a_{22})$

   $= u_1 + u_2 = 1$

13. (a) The probability that a Democratic candidate is elected depends only on whether the previous mayor was a Democrat or a Republican (and similarly for a Republican candidate).

   (b) $\begin{array}{c} \\ D \\ R \end{array} \begin{array}{cc} D & R \\ \begin{bmatrix} .6 & .4 \\ .3 & .7 \end{bmatrix} \end{array}$ (c) $\begin{bmatrix} .48 & .52 \\ .39 & .61 \end{bmatrix}$; $\begin{bmatrix} .444 & .556 \\ .417 & .583 \end{bmatrix}$

15. 57% will drink brand X after 2 months.   **17.** $\begin{array}{c} \\ T \\ G \\ \text{Other} \end{array} \begin{array}{ccc} T & G & \text{Other} \\ \begin{bmatrix} .92 & .08 & 0 \\ .04 & .90 & .06 \\ .10 & .08 & .82 \end{bmatrix} \end{array} = P;$ $[.45 \quad .30 \quad .25] = A^{(0)}$

   (a) 77.7%   (b) 79.758%

Exercise 11.2 (page 540)

1. Regular: $[\frac{2}{3} \quad \frac{1}{3}]$ is the fixed vector.   **3.** Regular. $[t_1 \quad t_2] = [\frac{1}{5} \quad \frac{4}{5}]$ **5.** Not regular.

7. If $[t_1 \quad t_2] \begin{bmatrix} 1-p & p \\ p & 1-p \end{bmatrix} = [t_1 \quad t_2]$, then: $\left. \begin{array}{c} (1-p)t_1 + pt_2 = t_1 \\ pt_1 + (1-p)t_2 = t_2 \end{array} \right\}$ $t_1 = t_2$, $t_1 + t_2 = 1$, $t_1 = \frac{1}{2}$, $t_2 =$

9. $P = \begin{array}{c} \\ A \\ B \\ C \end{array} \begin{array}{ccc} A & B & C \\ \begin{bmatrix} .7 & .15 & .15 \\ .1 & .8 & .1 \\ .2 & .2 & .6 \end{bmatrix} \end{array}$ He stocks Brand A 30.77% of the time, Brand B 46.15% of the time, Brand C 23.08% of the time.

11. The probability that the grandson of a Laborite will vote Socialist is .09.

   The fixed probability vector is $[t_1 \quad t_2 \quad t_3] = [.5532 \quad .3830 \quad .0638]$.

13. 30%   (a) [.377 .375 .248]   (b) $[t_1 \quad t_2 \quad t_3] = [.35 \quad .40 \quad .25]$

Exercise 11.3 (page 549)

1. Nonabsorbing (no absorbing states)   **3.** Absorbing

5. Nonabsorbing (absorbing state 3 is not accessible to states 1 and 2).

7. $\begin{array}{c} \\ 1 \\ 3 \\ 2 \end{array}\begin{array}{c} 1 \quad 3 \quad 2 \\ \begin{bmatrix} 1 & 0 & 0 \\ 0 & 1 & 0 \\ \frac{1}{8} & \frac{2}{8} & \frac{5}{8} \end{bmatrix} \end{array}$ $S = [\frac{1}{8} \quad \frac{2}{8}]; \quad T \cdot S = [\frac{1}{3} \quad \frac{2}{3}]$ **9.** (a) $T_{13} = .8; \quad T_{23} = .6$ (b) 4.2

11. With a stake of \$1, the probability is $\frac{4}{19}$. With \$2, the probability is $\frac{10}{19}$.

13. (a) Expected number of wagers is 1.4.
   (b) The probability that she is wiped out is .84.
   (c) The probability that she wins is .16.

15. (a) Expected number of wagers is 1.6.
   (b) $T \cdot S = \begin{bmatrix} .64 & .36 \\ .4 & .6 \end{bmatrix}$ The probability that she is wiped out is .64.
   (c) The probability that she wins is .36.

17. 12.80

Exercise 11.4 (page 555)

1. $[\frac{1}{4} \quad \frac{1}{2} \quad \frac{1}{4}] \begin{bmatrix} \frac{1}{2} & \frac{1}{2} & 0 \\ \frac{1}{4} & \frac{1}{2} & \frac{1}{4} \\ 0 & \frac{1}{2} & \frac{1}{2} \end{bmatrix} = [\frac{1}{4} \quad \frac{1}{2} \quad \frac{1}{4}]$

3. (a) $P = \begin{array}{c} D \\ H \\ R \end{array}\begin{array}{c} D \quad H \quad R \\ \begin{bmatrix} 0 & 1 & 0 \\ 0 & \frac{1}{2} & \frac{1}{2} \\ 0 & 0 & 1 \end{bmatrix} \end{array}$ (b) P is not regular, but a fixed probability vector does exist. It is [0 0 1]. This indicates that in the long run the unknown genotype will be R.

(c) $\begin{array}{c} R \\ H \\ D \end{array}\begin{array}{c} R \quad H \quad D \\ \begin{bmatrix} 1 & 0 & 0 \\ \frac{1}{2} & \frac{1}{2} & 0 \\ 0 & 1 & 0 \end{bmatrix} \end{array}$ $T = \begin{array}{c} H \\ D \end{array}\begin{array}{c} H \quad D \\ \begin{bmatrix} 2 & 0 \\ 2 & 1 \end{bmatrix} \end{array}$ (d) If the unknown is D to start, three stages are required. If the unknown is H to start, two stages are required.

Chapter Review

True–False Questions (page 555)
   1. F   **2.** F   **3.** F   **4.** T

Fill in the Blanks (page 555)
   1. Probability   **2.** $m \times m$ **3.** Nonnegative   **4.** $v^{(k)} = v^{(0)}P^k$ **5.** Positive

Review Exercises (page 556)
   1. (a) $[\frac{2}{5} \quad \frac{3}{5}]$ is the fixed vector.   (b) $[\frac{1}{2} \quad \frac{1}{2}]$ is the fixed vector.
   3. After 2 years [.4717 .2692 .2592]. In the long run, A's share is $\frac{80}{169}$, B's share is $\frac{45}{169}$, and C's share is $\frac{44}{169}$.
   5. In the long run, she sells $\frac{11}{31}$ of the time at U_1, $\frac{16}{31}$ of the time at U_2, and $\frac{4}{31}$ of the time at U_3.

7. Transition matrix:

	0	1	2	3	4	5
0	1	0	0	0	0	0
1	.55	0	.45	0	0	0
2	0	.55	0	.45	0	0
3	0	0	.55	0	.45	0
4	0	0	0	.55	0	.45
5	0	0	0	0	0	1

or

	0	5	1	2	3	4
0	1	0	0	0	0	0
5	0	1	0	0	0	0
1	.55	0	0	.45	0	0
2	0	0	.55	0	.45	0
3	0	0	0	.55	0	.45
4	0	.45	0	0	.55	0

The expected length of the game is 5.706158. The probability that he is wiped out is .7142275.

CHAPTER 12

Exercise 12.1 (page 565)

1. Proposition   **3.** Not a proposition   **5.** Proposition   **7.** Proposition
9. A fox is not an animal.   **11.** I am not buying stocks and bonds.
13. Someone wants to buy my house.   **15.** Every person has a car.
17. John is an economics major or a sociology major.
19. John is an economics major and a sociology major.
21. John is not an economics major or he is not a sociology major.
23. John is not an economics major or he is a sociology major.

Exercise 12.2 (page 573)

1.

p	q	$\sim q$	$p \vee \sim q$
T	T	F	T
T	F	T	T
F	T	F	F
F	F	T	T

3.

p	q	$\sim p$	$\sim q$	$\sim p \wedge \sim q$
T	T	F	F	F
T	F	F	T	F
F	T	T	F	F
F	F	T	T	T

5.

p	q	$\sim p$	$\sim p \wedge q$	$\sim(\sim p \wedge q)$
T	T	F	F	T
T	F	F	F	T
F	T	T	T	F
F	F	T	F	T

7.

p	q	$\sim p$	$\sim q$	$\sim p \vee \sim q$	$\sim(\sim p \vee \sim q)$
T	T	F	F	F	T
T	F	F	T	T	F
F	T	T	F	T	F
F	F	T	T	T	F

9.

p	q	$\sim q$	$p \vee \sim q$	$(p \vee \sim q) \wedge p$
T	T	F	T	T
T	F	T	T	T
F	T	F	F	F
F	F	T	T	F

11.

p	q	$\sim q$	$p \underline{\vee} q$	$p \wedge \sim q$	$(p \underline{\vee} q) \wedge (p \wedge \sim q)$
T	T	F	F	F	F
T	F	T	T	T	T
F	T	F	T	F	F
F	F	T	F	F	F

13.

p	q	$\sim p$	$\sim q$	$p \wedge q$	$\sim p \wedge \sim q$	$(p \wedge q) \vee (\sim p \wedge \sim q)$
T	T	F	F	T	F	T
T	F	F	T	F	F	F
F	T	T	F	F	F	F
F	F	T	T	F	T	T

15.

p	q	r	~q	p∧~q	(p∧~q)∨r
T	T	T	F	F	T
T	T	F	F	F	F
T	F	T	T	T	F
T	F	F	T	T	T
F	T	T	F	F	T
F	T	F	F	F	F
F	F	T	T	F	T
F	F	F	T	F	F

17.

p	p∧p	p∨p
T	T	T
F	F	F

Since each column is the same, $p \equiv p \wedge p \equiv p \vee p$

19.

p	q	r	p∧q	q∧r	(p∧q)∧r	p∧(q∧r)
T	T	T	T	T	T	T
T	T	F	T	F	F	F
T	F	T	F	F	F	F
T	F	F	F	F	F	F
F	T	T	F	T	F	F
F	T	F	F	F	F	F
F	F	T	F	F	F	F
F	F	F	F	F	F	F

The last two columns are the same, so $(p \wedge q) \wedge r \equiv p \wedge (q \wedge r)$.

p	q	r	p∨q	q∨r	(p∨q)∨r	p∨(q∨r)
T	T	T	T	T	T	T
T	T	F	T	T	T	T
T	F	T	T	T	T	T
T	F	F	T	F	T	T
F	T	T	T	T	T	T
F	T	F	T	T	T	T
F	F	T	F	T	T	T
F	F	F	F	F	F	F

The last two columns are the same, so $(p \vee q) \vee r \equiv p \vee (q \vee r)$.

21.

① p	② q	③ p∨q	④ p∧q	⑤ p∧(p∨q)	⑥ p∨(p∧q)
T	T	T	T	T	T
T	F	T	F	T	T
F	T	T	F	F	F
F	F	F	F	F	F

Since columns 1 and 5 are the same, $p \equiv p \wedge (p \vee q)$.
Since columns 1 and 6 are the same, $p \equiv p \vee (p \wedge q)$.

23.

① p	② q	③ ~q	④ ~q∨q	⑤ p∧(~q∨q)
T	T	F	T	T
T	F	T	T	T
F	T	F	T	F
F	F	T	T	F

Since columns 1 and 5 are the same, $p \equiv p \wedge (\sim q \vee q)$.

25.

① p	② ~p	③ ~(~p)
T	F	T
F	T	F

Since columns 1 and 3 are the same, $p \equiv \sim(\sim p)$.

27.

p	q	$\sim p$	$q \wedge (\sim p)$	$p \wedge (q \wedge \sim p)$
T	T	F	F	F
T	F	F	F	F
F	T	T	T	F
F	F	T	F	F

29.

p	q	$\sim p$	$\sim q$	$p \wedge q$	$\sim p \wedge \sim q$	$(p \wedge q) \vee (\sim p \wedge \sim q)$	$[(p \wedge q) \vee (\sim p \wedge \sim q)] \wedge p$
T	T	F	F	T	F	T	T
T	F	F	T	F	F	F	F
F	T	T	F	F	F	F	F
F	F	T	T	F	T	T	F

31. Smith is an exconvict and he is an exconvict. $\equiv$ Smith is an exconvict or he is an exconvict. $\equiv$ Smith is an exconvict.

33. "It is not true that Smith is an exconvict or rehabilitated" means the same as the statement "Smith is not an exconvict and he is not rehabilitated." "It is not true that Smith is an exconvict and he is rehabilitated" means the same as "Smith is not an exconvict or he is not rehabilitated."

35. $(p \wedge q) \vee r \equiv r \vee (p \wedge q) \equiv (r \vee p) \wedge (r \vee q) \equiv (p \vee r) \wedge (r \vee q) \equiv (p \vee r) \wedge (q \vee r)$

37. Use a truth table to show that b is true and a is false if Michael rents a truck and does not sell his car.

39. Katy is not a good volleyball player or she is conceited.

Exercise 12.3 (page 580)

1. $\sim p \Rightarrow q$; Converse: $q \Rightarrow \sim p$; Contrapositive: $\sim q \Rightarrow p$; Inverse: $p \Rightarrow \sim q$.

3. $\sim q \Rightarrow \sim p$; Converse: $\sim p \Rightarrow \sim q$; Contrapositive: $p \Rightarrow q$; Inverse: $q \Rightarrow p$.

5. If it is raining, the grass is wet.
Converse: If the grass is wet, it is raining.
Contrapositive: If the grass is not wet, it is not raining.
Inverse: If it is not raining, the grass is not wet.

7. If it is not raining, it is not cloudy,"
Converse: If it is not cloudy, it is not raining.
Contrapositive: If it is cloudy, it is raining.
Inverse: If it is raining, it is cloudy.

9. If it is raining, it is cloudy.
Converse: If it is cloudy, it is raining.
Contrapositive: If it is not cloudy, it is not raining.
Inverse: If it is not raining, it is not cloudy.

11. (a)   If Jack studies psychology, then Mary studies sociology.
(b)   If Mary studies sociology, then Jack studies psychology.
(c)   If Jack does not study psychology, then Mary studies sociology.

13. (a)

p	q	r	$q \vee r$	$p \Rightarrow (q \vee r)$	$p \wedge \sim q$	$(p \wedge \sim q) \Rightarrow r$
T	T	T	T	T	F	T
T	T	F	T	T	F	T
T	F	T	T	T	T	T
T	F	F	F	F	T	F
F	T	T	T	T	F	T
F	T	F	T	T	F	T
F	F	T	T	T	F	T
F	F	F	F	T	F	T

(b) $p \Rightarrow (q \vee r) \equiv \sim p \vee (q \vee r)$;
$(p \wedge \sim q) \Rightarrow r \equiv \sim (p \wedge \sim q) \vee r$
$\equiv (\sim p \vee \sim(\sim q)) \vee r$
$\equiv (\sim p \vee q) \vee r$
$\equiv \sim p \vee (q \vee r)$

15.

p	q	$\sim p$	$p \wedge q$	$\sim p \vee (p \wedge q)$
T	T	F	T	T
T	F	F	F	F
F	T	T	F	T
F	F	T	F	T

17.

p	q	$\sim p$	$\sim p \wedge q$	$p \vee (\sim p \wedge q)$
T	T	F	F	T
T	F	F	F	T
F	T	T	T	T
F	F	T	F	F

19.

p	q	$\sim p$	$\sim p \Rightarrow q$
T	T	F	T
T	F	F	T
F	T	T	T
F	F	T	F

21.

p	$\sim p$	$\sim p \vee p$
T	F	T
F	T	T

23.

p	q	$p \Rightarrow q$	$p \wedge (p \Rightarrow q)$
T	T	T	T
T	F	F	F
F	T	T	F
F	F	T	F

25.

p	q	r	$q \wedge r$	$p \wedge (q \wedge r)$	$p \wedge q$	$(p \wedge q) \wedge r$	$p \wedge (q \wedge r) \Leftrightarrow (p \wedge q) \wedge r$
T	T	T	T	T	T	T	T
T	T	F	F	F	T	F	T
T	F	T	F	F	F	F	T
T	F	F	F	F	F	F	T
F	T	T	T	F	F	F	T
F	T	F	F	F	F	F	T
F	F	T	F	F	F	F	T
F	F	F	F	F	F	F	T

27.

p	q	$p \vee q$	$p \wedge (p \vee q)$	$p \wedge (p \vee q) \Leftrightarrow p$
T	T	T	T	T
T	F	T	T	T
F	T	T	F	T
F	F	F	F	T

29. $p \Rightarrow q$ **31.** $\sim p \wedge \sim q$ **33.** $q \Rightarrow p$

Exercise 12.4 (page 587)

1. Let p and q be the statements, p: It is raining, q: John is going to school. Assume that $p \Rightarrow \sim q$ and q are true statements.

Prove: $\sim p$ is true.
Direct: $p \Rightarrow \sim q$ is true.
  Also, its contrapositive $q \Rightarrow \sim p$ is true and q is true.
  Thus, $\sim p$ is true by the law of detachment.
Indirect: Assume $\sim p$ is false.
  Then p is true; $p \Rightarrow \sim q$ is true.
  Thus, $\sim q$ is true by the law of detachment.
  But q is true, and we have a contradiction.
  The assumption is false and $\sim p$ is true.

3. Let p, q, and r be the statements, p: Smith is elected president; q: Kuntz is elected secretary; r: Brown is elected treasurer. Assume that $p \Rightarrow q$, $q \Rightarrow \sim r$, and p are true statements.

Prove: $\sim r$ is true.
Direct: $p \Rightarrow q$ and $q \Rightarrow \sim r$ are true.
 So $p \Rightarrow \sim r$ is true by the law of syllogism, and p is true.
 Thus, $\sim r$ is true by the law of detachment.

Indirect: Assume $\sim r$ is false.
 Then r is true; $p \Rightarrow q$ is true; $q \Rightarrow \sim r$ is true.
 So, $p \Rightarrow \sim r$ is true by the law of syllogism.
 $r \Rightarrow \sim p$, its contrapositive, is true.
 Thus, $\sim p$ is true by the law of detachment.
 But p is true, and we have a contradiction.
 The assumption is false and $\sim r$ is true.

5. Not valid   **7.** Valid

Exercise 12.5 (page 592)

1. The output $pq[\sim p(\sim q \oplus r)]$ is 1 when
  1. $p = q = 1$
  2. $p = 0$ and $q = 0$
  3. $p = 0$ and $r = 1$

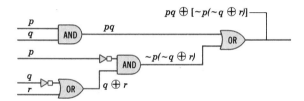

3. The output is $(\sim q \oplus [p(\sim p \oplus q)])q = (\sim q)q \oplus p(\sim p \oplus q)q = p(\sim p \oplus q)q = [p(\sim p) \oplus pq]q = pqq = pq$, which is 1 if and only if p and q are both 1.

5. 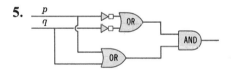   **7.**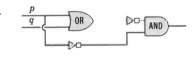

9. (For Problem 3):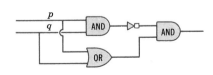

(For Problem 5):

p	q	$\sim p \oplus \sim q$	$p \oplus q$	$(\sim p \oplus \sim q)(p \oplus q)$
1	1	0	1	0
1	0	1	1	1
0	1	1	1	1
0	0	1	0	0

$(\sim p \oplus \sim q)(p \oplus q) = [\sim(pq)](p \oplus q)$

(For Problem 7): $\sim(p \oplus q) \sim p = \sim[(p \oplus q) \oplus p] = \sim[p \oplus q]$

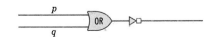

11. The truth table for this circuit is either   Thus two possible circuits are:

p	q		or	p	q	
1	1	1		1	1	0
1	0	0		1	0	1
0	1	0		0	1	1
0	0	1		0	0	0

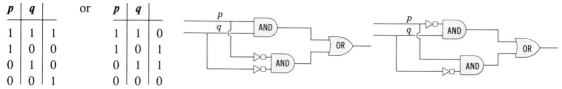

13.

$p(\sim p) = 0$

15. $p \vee q \equiv \sim(\sim p \sim q)$ (a)   (b)

17. $pq \oplus pr \oplus q(\sim r) = pqr \oplus pq(\sim r) \oplus pr \oplus q(\sim r) = pr(q \oplus 1) \oplus (p \oplus 1)[q(\sim r)] = pr(1) \oplus 1[q(\sim r)]$
$= pqr \oplus pr \oplus pq(\sim r) \oplus q(\sim r) = pr \oplus q(\sim r)$

Chapter Review

True – False Questions (page 593)
1. F   **2.** F   **3.** T   **4.** F   **5.** T

Fill in the Blanks (page 594)
1. $p \vee q$ **2.** $\sim p$ **3.** Logically equivalent   **4.** Hypothesis, conclusion   **5.** 0, 1

Review Exercises (page 594)
1. (c)   **3.** (a)   **5.** Nobody is rich.   **7.** Danny is tall or Mary is not short.

9.

p	q	$\sim p$	$p \wedge q$	$(p \wedge q) \vee \sim p$
T	T	F	T	T
T	F	F	F	F
F	T	T	F	T
F	F	T	F	T

11.

p	q	$\sim p$	$\sim q$	$p \vee \sim q$	$\sim p \vee (p \vee \sim q)$
T	T	F	F	T	T
T	F	F	T	T	T
F	T	T	F	F	T
F	F	T	T	T	T

13. $q \Rightarrow p$ **15.** $p \Leftrightarrow q$

17. Let p be the statement "I paint the house" and let q be the statement "I go bowling." Assume $\sim p \Rightarrow q$ and $\sim q$ are true.
Prove: p is true.
Since $\sim p \Rightarrow q$ is true, its contrapositive $\sim q \Rightarrow p$ is true. We have $\sim q$ is true and hence, by the law of detachment, p is true.

19.

p	q	$\sim p$	$\sim p \vee q$	$p \Rightarrow q$	$\sim p \vee q \equiv p \Rightarrow q$
T	T	F	T	T	
T	F	F	F	F	
F	T	T	T	T	
F	F	T	T	T	

21.

p	q	$(p \oplus q)[\sim(pq)]$	$p \veebar q$
1	1	0	0
1	0	1	1
0	1	1	1
0	0	0	0

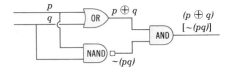

CHAPTER 13

Exercise 13.1 (page 600)

1. True, false, true, false, true, false, false, false, false.

3. {(2, 2), (2, 4), (2, 6), (2, 10), (3, 6), (5, 10)}. **5.** 1, 4, 9, 16, 25, 2, 5.

7. (a) {(1, 2), (1, 4), (1, 7), (2, 2), (2, 4), (2, 7), (5, 2), (5, 4), (5, 7)}.

(b) {(1, 2), (1, 4), (1, 7), (2, 4), (2, 7), (5, 7)}.

(c) R is a subset of $A \times B$.

9. (a) {aa, ab, ba, bb} (b) {(aa, ab), (ab, aa), (ba, bb), (bb, ba)}.

11. {(1, 2), (1, 4), (1, 6), (1, 8), (2, 2), (2, 4), (2, 6), (2, 8), (3, 6)}.

13. {(1, 1), (2, 2), (3, 3), (4, 4), (5, 5), (6, 6), (7, 7), (8, 8), (9, 9), (10, 10), (1, 3), (3, 1), (1, 5), (5, 1), (1, 7), (7, 1), (1, 9), (9, 1), (2, 4), (4, 2), (2, 6), (6, 2), (2, 8), (8, 2), (2, 10), (10, 2), (3, 5), (5, 3), (3, 7), (7, 3), (3, 9), (9, 3), (4, 6), (6, 4), (4, 8), (8, 4), (4, 10), (10, 4), (5, 7), (7, 5), (5, 9), (9, 5), (6, 8), (8, 6), (6, 10), (10, 6), (7, 9), (9, 7), (8, 10), (10, 8)}.

15. {(0, 0), (0, 1), (0, 2), (0, 3), (1, 1), (1, 2), (1, 3), (2, 2), (2, 3), (3, 3)}.

17. {(0, 0), (1, 1), (2, 4), (3, 9), (4, 16)}. **19.** {(1, 1), (2, 1), (5, 2), (7, 3)}.

21. {(42, '*'), (72, 'H'), (47, '/'), (88, 'x')}. **23.** Reflexive, symmetric, transitive.

25. Reflexive, symmetric, transitive. **27.** Reflexive, symmetric, transitive.

29. Reflexive. Not symmetric because $(A, C) \in R$ but $(C, A) \notin R$. Not transitive because $(C, B) \in R$ and (B, A), but $(C, A) \notin R$.

31. Since we cannot find a and b in A such that aRb; then R is symmetric. Similarly, since we cannot find a, b, and c in A such that aRb and bRc, then R is transitive. And since $(t, t) \notin R$, etc., then R is not reflexive.

Exercise 13.2 (page 607)

1. 1, 4, 2. Range = {1, 2, 4}. **3.** 5, 0, 10, -5, 15.

5. Does not define a function, because f assigns two different values (1 and 3) to x.

7. Yes, f defines a function. **9.** Yes, f is a function. **11.** No, f is not a function. Domain $\neq A$.

13. Yes. **15.** Yes **17.** One-to-one not onto, therefore, not bijective.

19. Neither one-to-one nor onto, therefore not bijective.

21. For example H(10, 10) = 0 = H(01, 01). Thus, H is not one-to-one.

23. No. Odd integers in the range are not associated with any integers in the domain. **25.** 3, 2, 1.

27. $(g \circ f)(1) = g[f(1)] = g(b) = z$.

$(g \circ f)(2) = g[f(2)] = g(a) = x$.

$(g \circ f)(3) = g[f(3)] = g(c) = y$.

Exercise 13.3 (page 611)

1. $-1, \frac{1}{2}, -\frac{1}{3}, \frac{1}{4}, \frac{1}{100}$.

3. (a) $0, \frac{1}{2}, \frac{2}{3}, \frac{3}{4}, \frac{4}{5}, \frac{5}{6}$.

(b) $n = 0,\ b_1 - b_0 = \frac{1}{2} - 0 = \frac{1}{2}.$
       $n = 1,\ b_2 - b_1 = \frac{2}{3} - \frac{1}{2} = \frac{1}{6}.$
       $n = 2,\ b_3 - b_2 = \frac{3}{4} - \frac{2}{3} = \frac{1}{12}$

5. 1, 2, 4, 8, 16, 32, 64, 128.   **7.** 1, 1, 2, 6, 24, 120, 720.

9. $M_0 = \begin{bmatrix} 1 & 1 & 0 \\ 1 & 0 & -1 \\ 0 & -1 & 1 \end{bmatrix}$ $M_1 = \begin{bmatrix} 1 & 0 & 0 \\ 0 & 1 & 0 \\ 0 & 0 & 2 \end{bmatrix}$ $M_2 = \begin{bmatrix} 1 & -1 & 0 \\ -1 & 2 & 1 \\ 0 & 1 & 3 \end{bmatrix}$ $M_3 = \begin{bmatrix} 1 & -2 & 0 \\ -2 & 3 & 2 \\ 0 & 2 & 4 \end{bmatrix}$

11. $(-1)^n,\ n = 0, 1, 2, \ldots$ **13.** $2n + 1,\ n = 0, 1, 2, \ldots$ **15.** $\dfrac{1}{n+1},\ n = 0, 1, 2, \ldots$

17. 0, 1, 3, 7, 15, 31, 63, 127.   **19.** 1, 1, 2, 3, 5, 8, 13, 21, 34.

Exercise 13.4 (page 616)

1. (a) $1 < 2$. Yes.   (b) $2 < 2^2$. Yes.   (c) $k < 2^k$.   (d) $(k + 1) < 2^{k+1}$.

3. (a) $1^2 = \dfrac{1(1 + 1)(2 + 1)}{6}$. Yes.

(b) $1^2 + 2^2 + 3^2 + 4^2 + 5^2 = \dfrac{5(5 + 1)(10 + 1)}{6}$. Yes.

(c) $1^2 + 2^2 + 3^2 + \cdots + k^2 = \dfrac{k(k + 1)(2k + 1)}{6}$

(d) $1^2 + 2^2 + 3^2 \cdots + k^2 + (k + 1)^2 = \dfrac{(k + 1)[(k + 1) + 1][2(k + 1) + 1]}{6} = \dfrac{(k + 1)(k + 2)(2k + 3)}{6}$

5. (a) $1 + 2 = 2^2 - 1$. Yes.
(b) $1 + 2 + 2^2 + 2^3 + 2^4 + 2^5 = 2^{5+1} - 1$. Yes.
(c) $1 + 2 + 2^2 + \cdots + 2^k = 2^{k+1} - 1$
(d) $1 + 2 + 2^2 + \cdots + 2^k + 2^{k+1} = 2^{(k+1)+1} - 1$

7. Since $1 \cdot 2 = \dfrac{1(1 + 1)(1 + 2)}{3}$ then $s(1)$ is true and condition 1 is true. Assume that $s(k)$ is true. Show

$s(k + 1)$ is true. $s(k + 1)$ states:

$$1 \cdot 2 + 2 \cdot 3 + 3 \cdot 4 + \cdots + k \cdot (k + 1) + (k + 1)[(k + 1) + 1] = \dfrac{(k + 1)[(k + 1) + 1][(k + 1) + 2]}{3} \quad (1)$$

Using our assumption that $s(k)$ is true the left hand side of

$$(1) = \dfrac{k(k + 1)(k + 2)}{3} + (k + 1)[(k + 1) + 1] = \dfrac{k(k + 1)(k + 2) + 3(k + 1)(k + 2)}{3}$$

$$= \dfrac{(k + 1)(k + 2)(k + 3)}{3} = \dfrac{(k + 1)[(k + 1) + 1][(k + 1) + 2]}{3}$$

which shows that $s(k + 1)$ is true and thus condition II is true.

9. Since $\dfrac{1}{1 \cdot 2} = \dfrac{1}{1 + 1}$, then $s(1)$ is true and thus condition I is satisfied. Assume $s(k)$ is true. Show that

$s(k + 1)$ is also true. $s(k + 1)$ states:

(1) $\dfrac{1}{1\cdot 2}+\dfrac{1}{2\cdot 3}+\dfrac{1}{3\cdot 4}+\cdots+\dfrac{1}{(k+1)[(k+1)+1]}=\dfrac{k+1}{(k+1)+1}$

Since by assumption $s(k)$ is true then the left hand side of (1) gives

$$\dfrac{k}{k+1}+\dfrac{1}{(k+1)[(k+1)+1]}=\dfrac{k[(k+1)+1]+1}{(k+1)[(k+1)+1]}=\dfrac{k(k+2)+1}{(k+1)[(k+1)+1]}$$

$$=\dfrac{k^2+2k+1}{(k+1)[(k+1)+1]}=\dfrac{(k+1)(k+1)}{(k+1)[(k+1)+1]}$$

$$=\dfrac{k+1}{[(k+1)+1]}$$

which is the right hand side of (1). Thus $s(k+1)$ is true and so condition II is satisfied.

11. Since $2=1(3+1)/2$, then $s(1)$ is true and thus condition 1 is satisfied. Assume $s(k)$ is true show $s(k+1)$ is also true. $s(k+1)$ states:

(1) $2+5+8+\cdots+[3(k+1)-1]=\dfrac{(k+1)[3(k+1)+1]}{2}$

By our assumption that $s(k)$ is true the left-hand side of (1) is then

$$\dfrac{k(3k+1)}{2}+3(k+1)-1=\dfrac{k(3k+1)+6(k+1)-2}{2}=\dfrac{3k^2+7k+4}{2}=\dfrac{(k+1)(3k+4)}{2}$$

$$=\dfrac{(k+1)(3k+3+1)}{2}=\dfrac{(k+1)[3(k+1)+1]}{2}$$

which is the right-hand side of (1). Thus $s(k+1)$ is true and so condition II is satisfied.

Exercise 13.5 (page 622)

1. 1, 2, 4, 8, 16, 32   **3.** 1, 0, -1, -4, -17, -86 **5.** 2, 4, 12, 48, 240, 1440   **7.** 1, 1, 3, 7, 16, 33
9. 1, 2, 2, 4, 8, 32   **11.** 1, 2, 2, 5, 9, 16   **13.** 1, -1, -1, -5, -20, -104
15. 89, 144, 233, 377, 610
17. (a) The word that contains no bits is of length 0 that does not contain the bit pattern 00.
  The words 0 and 1 are of length 1 that do not contain the pattern 00.
  The words 01, 10, 11 are of length 2 that do not contain the pattern 00.
  The words 010, 101, 011, 110, 111 are of length 3 that do not contain the bit pattern 00.
  The words 1010, 1101, 1110, 1111, 0101, 1011, 0111, 0110 are of length 4 that do not contain the bit pattern 00.
  (b) Recurrence relation $s_n=s_{n-1}+s_{n-2}$ with $s_0=1$ $s_1=2$ as initial conditions.
19. (a) $1000, $1100, $1210, $1331, $1464.1.   **21.** (a) 2, 4, 7, 11.
  (b) $A_n=A_{n-1}+0.1A_{n-1}$ (recurrence relations)   (b) $p_n=p_{n-1}+n$ (recurrence relation)
  $A_0=1,000$ (initial condition). $p_1=2$ (initial condition)

Chapter Review

True–False Questions (page 624)
1. F   **2.** F   **3.** T   **4.** T   **5.** F   **6.** T

Fill in the Blanks (page 624)

1. Ordered pairs (a, b)   **2.** Function   **3.** Bijective   **4.** Natural numbers
5. Mathematical induction   **6.** Recurrence relation, initial condition

Review Exercises (page 624)

1. {(1, 1), (8, 2), (27, 3)}   **3.** (a) and (b)
5. (a) 1, 0, 3, 2, 5, 4, 7, 6, 9, 8   (b) 0, -1, 2, -3, 4, -5, 6, -7, 8, -9
7. 2, 2, 1, 4, 12, 60, 960

CHAPTER 14

Exercise 14.1 (page 631)

1. Vertices: v_1, v_2, v_3, v_4. Edges: e_1, e_2, e_3, e_4, e_5. Loop: e_1. Isolated vertex: v_2. Parallel edges: e_3, e_4, e_5.
3. v_1 v_2; v_1 v_2

5.

Vertex	Degree
v_1	3
v_2	0
v_3	3
v_4	4

Vertex	Degree
v_1	1
v_2	1
v_3	2
v_4	2
v_5	2
v_6	4
v_7	2
v_8	4
v_9	2
v_{10}	2

7. 9, as the following figure shows:

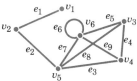

9. No, for $\deg(v_4) = 4$ there must be a loop.   **11.** (a) v_1 v_2 (b)   **13.** 15

15.   **17.**

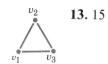

Exercise 14.2 (page 635)

1. (a) Path, not simple   (b) Path, not simple   (c) Simple path   (d) Circuit   (e) Simple circuit
3. (a) Simple circuit   (b) Simple circuit   (c) Simple circuit   (d) Path, not simple

5. (a) $v_1 e_1 v_2$
$v_5 e_5 v_6 e_9 v_3$
$v_1 e_6 v_6 e_5 v_5 e_4 v_4$
$v_2 e_2 v_3 e_3 v_4 e_4 v_5$

(b) $v_1 e_1 v_2 e_7 v_6 e_5 v_5 e_{10} v_3 e_9 v_6 e_6 v_1$
$v_1 e_1 v_2 e_8 v_5 e_5 v_6 e_9 v_3 e_2 v_2 e_7 v_6 e_6 v_1$
$v_1 e_1 v_2 e_7 v_6 e_9 v_3 e_{10} v_5 e_5 v_6 e_6 v_1$
$v_2 e_8 v_5 e_{10} v_3 e_3 v_4 e_4 v_5 e_5 v_6 e_9 v_3 e_2 v_2$

(c) $v_1 e_1 v_2 e_7 v_6 e_6 v_1$
$v_3 e_{10} v_5 e_4 v_4 e_3 v_3$
$v_2 e_8 v_5 e_{10} v_3 e_9 v_6 e_7 v_2$
$v_6 e_5 v_5 e_4 v_4 e_3 v_3 e_2 v_2 e_1 v_1 e_6 v_6$

7. $v_1 e_1 v_2 e_2 v_3 e_9 v_5 e_4 v_4 e_3 v_3 e_8 v_6 e_6 v_1$ **9. (a)**   **(b)**

11. (a) e_1, e_2, e_3, e_4.   **(b)** e_4.   **(c)** e_2, e_4, e_5.   **13.**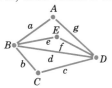
15. (a) No, because $11 > \binom{5}{2} = 10$ (Theorem II).
(b) Yes, because $10 \not> \binom{5}{2} = 10$ (Theorem II).

Exercise 14.3 (page 642)

1. (a) Every vertex is of even degree. By Theorem 1 there is a Eulerian circuit.
$v_1 e_1 v_2 e_2 v_3 e_4 v_4 e_5 v_5 e_6 v_3 e_3 v_2 e_9 v_5 e_7 v_6 e_8 v_1$ is an Eulerian circuit.
(b) The graph is not connected, therefore, no Eulerian circuit.
(c) Degree of $v_1 = 3$ is not even, therefore, no Eulerian circuit.
(d) Every vertex is of even degree. Therefore the graph contains a Eulerian circuit. The circuit
$v_1 e_1 v_2 e_2 v_3 e_3 v_1 e_5 v_4 e_4 v_2 e_6 v_5 e_7 v_1$ is Eulerian.
(e) Every vertex is of even degree. Therefore the graph contains an Eulerian circuit. The circuit
$v_1 e_1 v_2 e_{10} v_3 e_{11} v_5 e_{12} v_1 e_{13} v_6 e_5 v_5 e_4 v_4 e_3 v_3 e_2 v_2 e_{14} v_4 e_{15} v_6 e_9 v_2 e_8 v_7 e_7 v_6 e_6 v_1$ is Eulerian.
(f) Degree of $v_1 = 3$ is not even. Therefore, the graph does not contain an Eulerian circuit.
3. Yes, each vertex is of even degree.
5.

Yes. Let A and C in the figure be the land masses; B, E, and D be the three islands; and a, b, c, d, e, f, and g be the seven bridges. Note that each vertex has even degree. Therefore, the graph contains a Eulerian circuit. The following circuit is the desired round-trip:

$$AaBbCcDdBeEfDgA.$$

7. No, the corresponding graph has vertex B (vertex representing room B) of degree, 3 which is not even.
9. (a) $v_1 e_1 v_2 e_9 v_7 e_{10} v_8 e_3 v_3 e_4 v_4 e_5 v_5 e_6 v_6 e_7 v_1$
(b) $v_1 e_1 v_7 e_2 v_2 e_4 v_8 e_5 v_3 e_6 v_4 e_7 v_5 e_8 v_6 e_9 v_1$
(c) $v_1 e_1 v_2 e_2 v_3 e_3 v_4 e_{11} v_7 e_{10} v_6 e_{13} v_5 e_5 v_1$
(d) $v_4 e_3 v_3 e_9 v_8 e_{12} v_7 e_{11} v_6 e_7 v_2 e_1 v_1 e_5 v_5 e_4 v_4$

11. Use Theorem 2,
(a) $\deg(v_1) + \deg(v_5) = 4 < 6$, number of vertices.
(b) $\deg(v_2) + \deg(v_7) = 4 < 8$, number of vertices.
(c) $\deg(v_1) + \deg(v_6) = 4 < 7$, number of vertices.
(d) $\deg(v_5) + \deg(v_{10}) = 4 < 10$, number of vertices.

13.   **15.**

Exercise 14.4 (page 650)

1.

3.

5. By Theorem 1, such a tree cannot exist.
7. Cannot exist. A graph with the given specifications must contain a circuit.
9. Cannot exist. Such a graph must contain a circuit.
11. Cannot exist. Such a graph *must* be a tree.   **13.** Yes. Use Theorem 1
15.   **17.** Leaves; $v_1, v_3, v_4, v_5, v_6, v_7$.
Internal vertices: $v_2, v_9, v_8, v_{10}, v_{11}, v_{12}$.

19. No. By definition a leaf is connected to the rest of a tree by only one edge.

21. 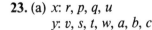   **23.** (a) x: r, p, q, u
y: v, s, t, w, a, b, c
    (b)

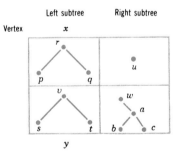

    (c)

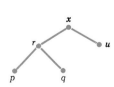

25. $a * b + 1$ **27.** $(x + y) * z - a/b$

29.

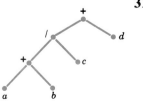

31.

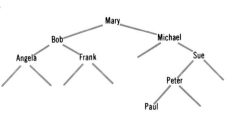

33. (a) Yes.   (b) Yes.

Exercise 14.5 (page 658)

1. (a)

Arc	Initial Point	Terminal Point
e_1	a	a
e_2	a	b
e_3	a	c
e_4	b	c
e_5	d	c
e_6	c	e
e_7	d	e
e_8	d	h
e_9	c	f
e_{10}	f	c
e_{11}	f	g
e_{12}	f	h
e_{13}	h	g

(b)

Vertex	Indegree	Outdegree
a	1	3
b	1	1
c	4	2
d	0	3
e	2	0
f	1	3
g	2	0
h	2	1

3. 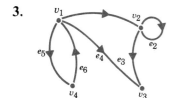   **5.** 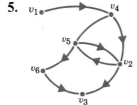   **7.**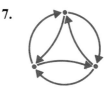

9. (a) Yes. $v_1 e_1 v_2 e_8 v_7 e_{10} v_5$ is a directed path from v_1 to v_5.
(b) No. No arcs go into v_1, that is, $\text{indeg}(v_1) = 0$.

11. $v_1 e_1 v_2 e_8 v_7 e_{10} v_5 e_5 v_6$; $v_1 e_6 v_6$; $v_1 e_1 v_2 e_8 v_7 e_{11} v_4 e_{12} v_6$

13. No. Since outdegree $(v_6) = 0$, then no other vertex is reachable from v_6.   **15.**

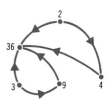

17. Graph is connected and contains no bridges. Therefore it is orientable. The digraph in the figure is the desired strongly connected digraph.

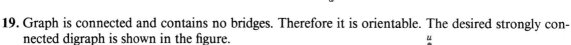

19. Graph is connected and contains no bridges. Therefore it is orientable. The desired strongly connected digraph is shown in the figure.

Chapter Review

True–False Questions (page 661)
1. F   **2.** F   **3.** T   **4.** T   **5.** T   **6.** T

Fill in the Blanks (page 661)
1. Vertices, edges   **2.** Loop, parallel   **3.** Eulerian   **4.** Circuit   **5.** Direction
6. Directed, strongly connected.

Review Exercises (page 661)
1. No. Such a simple graph must at most have 6 edges.   **3.** (a) $v_1 e_9 v_5 e_5 v_7 e_4 v_6 e_{12} v_3$
(b) $v_1 e_9 v_5 e_5 v_7 e_4 v_6 e_{12} v_3 e_3 v_5 e_9 v_1$
(c) $v_1 e_9 v_5 e_5 v_7 e_4 v_6 e_{11} v_2 e_1 v_1$

5. No. The graph in the figure represents the city where the vertices A and B are the two banks, C, D, and E are the three islands and the edges are the bridges. Since $\deg(B) = 3$ which is odd, then the graph does not contain a Eulerian circuit.

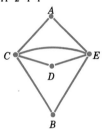

7. $v_1 e_1 v_2 e_2 v_3 e_8 v_5 e_6 v_2 e_7 v_4 e_9 v_1 e_{10} v_3 e_3 v_4 e_4 v_5 e_5 v_1$ is an Eulerian circuit.   **9.**

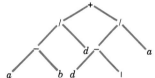

11.

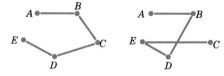

13. Yes. Every vertex is reachable from the others.

APPENDIX ANSWERS

Arithmetic Sequences (page 667)
1. 1, 4, 7, 10, 13   **3.** 5, 3, 1, -1, -3 **5.** 72, 88, 104, 120, 136   **7.** 2, -5, -12, -19, -26
9. 38   **11.** -92 **13.** 163   **15.** 14   **17.** $-\frac{47}{5}$, $-\frac{23}{5}$, $\frac{1}{5}$, 5, $\frac{49}{5}$ **19.** 62,750

21. $S_n = \dfrac{n}{2}[2 \cdot 1 + 2(n-1)]$ **23.** 255   **25.** 1500, 1275, 1050, 825, 600

$$= \frac{n}{2}(2 + 2n - 2)$$

$$= \frac{n}{2}(2n) = n^2$$

Geometric Sequences (page 668)

1. 2   **3.** Arithmetic progression   **5.** $-\frac{1}{2}$ **7.** 7, 14, 28, 56, 112   **9.** 15, 3, $\frac{3}{5}$, $\frac{3}{25}$, $\frac{3}{125}$
11. $-\frac{3}{49}$, $-\frac{3}{7}$, -3, -21, -147 **13.** 4, -12, 36, -108, 324   **15.** 6.3, -6.3, 6.3, -6.3, 6.3
17. 1, 0.12, 0.0144, 0.0017, 0.0002   **19.** 20,480   **21.** $\frac{2}{234,375}$ **23.** $\frac{1024}{3}$ **25.** $\frac{2187}{8}$ **27.** $\frac{43}{64}$ **29.** $\frac{2186}{3}$
31. $\frac{1905}{64}$ **33.** -8716 **35.** \$9687.50   **37.** 175% of 1990 capacity

INDEX